Hypo-Analytic Structures: Local Theory

Princeton Mathematical Series

1. The Classical Groups *by Hermann Weyl*
3. An Introduction to Differential Geometry *by Luther Pfahler Eisenhart*
4. Dimension Theory *by W. Hurewicz and H. Wallman*
6. The Laplace Transform *by D. V. Widder*
7. Integration *by Edward J. McShane*
8. Theory of Lie Groups: I *by C. Chevalley*
9. Mathematical Methods of Statistics *by Harald Cramér*
10. Several Complex Variables *by S. Bochner and W. T. Martin*
11. Introduction to Topology *by S. Lefschetz*
12. Algebraic Geometry and Topology *Edited by R. H. Fox, D. C. Spencer, and A. W. Tucker*
14. The Topology of Fibre Bundles *by Norman Steenrod*
15. Foundations of Algebraic Topology *by Samuel Eilenberg and Norman Steenrod*
16. Functionals of Finite Riemann Surfaces *by Menahem Schiffer and Donald C. Spencer*
17. Introduction to Mathematical Logic, Vol. I *by Alonzo Church*
19. Homological Algebra *by H. Cartan and S. Eilenberg*
20. The Convolution Transform *by I. I. Hirschman and D. V. Widder*
21. Geometric Integration Theory *by H. Whitney*
22. Qualitative Theory of Differential Equations *by V. V. Nemytskii and V. V. Stepanov*
23. Topological Analysis *by Gordon T. Whyburn* (revised 1964)
24. Analytic Functions *by Ahlfors, Behnke and Grauert, Bers, et al.*
25. Continuous Geometry *by John von Neumann*
26. Riemann Surfaces *by L. Ahlfors and L. Sario*
27. Differential and Combinatorial Topology *Edited by S. S. Cairns*
28. Convex Analysis *by R. T. Rockafellar*
29. Global Analysis *Edited by D. C. Spencer and S. Iyanaga*
30. Singular Integrals and Differentiability Properties of Functions *by E. M. Stein*
31. Problems in Analysis *Edited by R. C. Gunning*
32. Introduction to Fourier Analysis on Euclidean Spaces *by E. M. Stein and G. Weiss*
33. Étale Cohomology *by J. S. Milne*
34. Pseudodifferential Operators *by Michael E. Taylor*
36. Representation Theory of Semisimple Groups: An Overview Based on Examples *by Anthony W. Knapp*
37. Foundations of Algebraic Analysis *by Masaki Kashiwara, Takahiro Kawai, and Tatsuo Kimura. Translated by Goro Kato*
38. Spin Geometry *by H. Blaine Lawson, Jr., and Marie-Louise Michelsohn*
39. Topology of 4-Manifolds *by Michael H. Freedman and Frank Quinn*
40. Hypo-Analytic Structures: Local Theory *by François Treves*

François Treves

HYPO-ANALYTIC STRUCTURES

LOCAL THEORY

PRINCETON UNIVERSITY PRESS
PRINCETON, NEW JERSEY

Published by Princeton University Press, 41 William Street,
Princeton, New Jersey 08540
In the United Kingdom: Princeton University Press, Oxford

Library of Congress Cataloging-in-Publication Data

Treves, François, 1930–
Hypo-analytic structures : local theory / François Treves.
p. cm.—(Princeton mathematical series)
Includes bibliographical references and index.
ISBN 0-691-08744-X
1. Differential equations, Partial. 2. Manifolds (Mathematics)
3. Vector fields. I. Title. II. Series.
QA377.T682 1991 515′.353—dc20 91-18210

This book has been composed in Linotron Times Roman

Princeton University Press books are printed on acid-free paper, and meet the guidelines for permanence and durability of the Committee on Production Guidelines for Book Longevity of the Council on Library Resources

Printed in the United States of America

10 9 8 7 6 5 4 3 2 1

Designed by Laury A. Egan

To Louis Nirenberg and to Laurent Schwartz

IN ADMIRATION AND FRIENDSHIP

Contents

V

Involutive Structures with Boundary

VI

Local Integrability and Local Solvability in Elliptic Structures

VII

Examples of Nonintegrability and of Nonsolvability

VIII

Necessary Conditions for the Vanishing of the Cohomology. Local Solvability of a Single Vector Field

Preface

The purpose of this book is to give an organized presentation of a number of results, some classical, some relatively recent, about overdetermined systems of linear PDE defined by complex, smooth vector fields in a real, smooth manifold $\mathcal{M}$. Actually, the last chapter of the present volume extends part of the results to systems of *nonlinear* first-order differential equations.

The presentation is arranged in three echelons, of increasing depth: at level 1 we set down the minimal requirements on the structures to be studied and we bring to the fore the "satellite" concepts that flow naturally from those requirements (chap. I). Our minimal demands are two. The first is a kind of stability condition, in that the system of vector fields must have a constant rank—i.e., they must be sections of a vector subbundle $\mathcal{V}$ (called the *tangent structure bundle*) of the complex(ified) tangent bundle $\mathbb{C}T\mathcal{M}$. The second is a requirement of involution, or formal integrability, as it is often called: the commutation bracket of two smooth sections of $\mathcal{V}$ must also be a section of $\mathcal{V}$. There is a dual viewpoint we adopt frequently: in it the focus is on the orthogonal of $\mathcal{V}$, which is a vector subbundle T' of the complex cotangent bundle $\mathbb{C}T^*\mathcal{M}$ (T' is called the *cotangent structure bundle*). The involutive nature of the structure translates into the fact that T' is *closed* in the old sense of E. Cartan: the differential of any smooth section of T' belongs to the ideal generated by T' in the exterior algebra $\Lambda\mathbb{C}T^*\mathcal{M}$. The datum of the pair of vector bundles $\mathcal{V}$, $T' = \mathcal{V}^{\perp}$, is what we call a *formally integrable* or, perhaps more often, an *involutive structure* on the manifold $\mathcal{M}$.

Notions of formal integrability naturally lead to the question of true integrability, at least in the small, and thus to the *locally integrable structures*, defined as those involutive structures in which the cotangent structure bundle T' is locally generated by *exact* differentials. The *local* picture is a kind of diptych: on one side stands a local basis of $\mathcal{V}$, n smooth, complex vector fields $L_1,\ldots,L_n$, linearly independent over the complex field and which can be assumed to commute; on the other side we have a complete set of *first integrals*, that is to say, $m = \dim \mathcal{M} - n$ smooth, complex-valued functions $Z_1,\ldots,Z_m$ which are solutions of the homogeneous equations

$$L_j h = 0, j = 1,\ldots,n, \tag{1}$$

and whose differentials are $\mathbb{C}$-linearly independent.

In a locally integrable structure one can do much more than in a formally integrable one. This is shown in chapter II: one can approximate (locally) every solution of the homogeneous equations (1) by polynomials, with complex (constant) coefficients, $\mathcal{P}(Z)$ [we write systematically $Z = (Z_1,\ldots,Z_m)$]; one can represent (locally) every distribution solution as a sum of appropriate derivatives of $\mathcal{C}^r$ solutions (for any $r < +\infty$). And this kind of result can be generalized to the differential forms that are closed in the differential complex associated with the given structure.

In chapter III the third level is reached. That there is a third level reflects the fact that, in studying the equations (1), or the associated inhomogeneous equations, in a locally integrable structure, the set of first integrals one uses remains essentially unchanged. This is obvious in the approximation and representation formulas of chapter II. Once it is accepted that the first integrals be kept fixed (up to biholomorphic substitutions), novel questions emerge. A consequence of the approximation by polynomials $\mathcal{P}(Z)$ is that all continuous solutions are locally constant on the preimages of points in complex space $\mathbb{C}^m$ under the map Z; these preimages, if viewed as germs of sets, are invariants of the locally integrable structure (we refer to them as the *fibres* of the structure at the point about which the analysis takes place). It follows that any classical solution h is locally the pullback of a continuous function $\tilde{h}$ on the image of Z. The fact that h is a solution of the equations (1) is reflected in the fact that $\tilde{h}$ is a solution of the "tangential" Cauchy-Riemann equations on the image of Z. This observation leads to a range of exciting questions, without parallels in the theory of higher-order differential equations: Is the pushforward $\tilde{h}$ the restriction of a holomorphic function in a full neighborhood (in complex space) of the image of Z? Is $\tilde{h}$ the boundary value of a holomorphic function in an open set whose boundary contains the image of Z? Are there structures in which any of these properties is true of every solution? I am sure that the reader senses the potential in these questions. The above properties are unchanged under local biholomorphisms of the target space, $\mathbb{C}^m$, of the map Z. Whence the concept of a *hypo-analytic structure* on the manifold $\mathcal{M}$: an atlas of hypo-analytic charts (U,Z) whose domains U cover $\mathcal{M}$, and such that the mappings $Z : \mathrm{U} \to \mathbb{C}^m$ (always of class $\mathcal{C}^\infty$ in this book) agree on overlaps up to biholomorphisms. The two best-known examples of hypo-analytic structures are the analytic structures and the tangential Cauchy-Riemann (CR) structure that a generic submanifold of $\mathbb{C}^m$ inherits from the ambient complex structure. A locally integrable structure might underlie different hypo-analytic structures (these will vary with inequivalent choices of the first integrals). But there are certain locally integrable structures that underlie a unique hypo-analytic structure: in such a structure every solution is (locally) a holomorphic function of the first integrals. We call them *hypocomplex*.

Chapter IV is almost entirely devoted to the construction of *normal forms* of the first integrals. The statement and the proof we have chosen to give seem fairly natural generalizations to hypo-analytic structures (in formal, or in convergent, power series) of those in Bloom and Graham [1], valid for the tangential CR structures of the generic submanifolds of complex space.

Chapter V does what must be done: it extends some of the preceding concepts to a manifold with boundary. At this juncture an unexpected dichotomy arises, between those structures in which the boundary is noncharacteristic and those in which it is totally characteristic. The definition of a distribution solution is profoundly different in each of these two situations.

The first five chapters make up the "foundational" portion of the exposition. Much of it consists of definitions and of lists of elementary properties of the entities that have been defined. This is not the kind of stuff that makes for entertaining reading; but it seems to be an unavoidable part of mathematical theory building. As the exposition progresses and the structures become richer, from merely involutive to locally integrable to hypo-analytic, the wealth of "theorems" gradually increases and so does the sophistication of the arguments.

The next two chapters, VI and VII, are devoted to significant classes of involutive structures. The importance of some of these classes, such as the complex structures, goes immeasurably far, beyond the scope of this book. Here the standpoint is mainly that of local integrability and of local solvability. The latter refers to the possibility of locally solving the inhomogeneous equations

$$\mathrm{L}_j u = f_j, j = 1,\ldots,n, \tag{2}$$

for any choice of $\mathscr{C}^\infty$ right-hand sides f_j provided they satisfy the obvious compatibility conditions

$$\mathrm{L}_j f_k = \mathrm{L}_k f_j, j, k = 1,\ldots,n. \tag{3}$$

Elliptic structures, which comprise complex structures, are locally integrable: this is the content of the Newlander-Nirenberg theorem. This can be rephrased by saying that, in a manifold $\mathcal{M}$ equipped with an elliptic structure, there exist local coordinates $x_1,\ldots,x_\nu$, $y_1,\ldots,y_\nu$, $t_1,\ldots,t_{n-\nu}$ such that a set of first integrals is provided by the functions $z_j = x_j + \sqrt{-1}y_j$; and the partial derivatives $\partial/\partial\bar{z}_j$, $\partial/\partial t_k$ $(1 \le j \le \nu,\ 1 \le k \le n-\nu)$ can be selected as vector fields L_j. In other words, the manifold $\mathcal{M}$ is locally isomorphic to $\mathbb{C}^\nu \times \mathbb{R}^{n-\nu}$; the differential complex associated with the elliptic structure is locally isomorphic to the sum $d_t + \bar{\partial}_z$ of the De Rham complex in the variables t and of the Dolbeault complex in the variables z. From there it is easy enough to construct a *homotopy operator* in the complex, based on the Bochner-Matinelli (and Koppelman-Leray) formulas in z-space and the standard radial (or conic) homotopy formula for d_t. The Newlander-Nirenberg theorem has far-reaching consequences, some of which are encountered right away at the end of Chapter VI and some only at the end of the book, in Chapter X, when we deal with nonlinear equations. It should also be underscored that not only is the space $\mathbb{C}^m \times \mathbb{R}^n$, equipped with its standard elliptic structure, the local model for all elliptic structures; but furthermore, every hypo-analytic manifold can be locally embedded as a kind of generic submanifold of $\mathbb{C}^m \times \mathbb{R}^n$ whose hypo-analytic structure is obtained by restricting the complex coordinates z_j.

Not every involutive structure is locally integrable, and local solvability does not hold in every locally integrable structure. Chapter VII presents the basic ''counterexamples'' to local integrability and to local solvability. The Lewy equation, and its two-dimensional analogue, the Mizohata equation, are the most celebrated examples of nonsolvable equations. Nirenberg's perturbations of the Lewy and Mizohata equations were the first examples of non-integrability (or of *homogeneous nonsolvability*, as one might prefer to say). Chapter VIII again takes up the question of nonsolvability but from a different perspective, and with the purpose of finding necessary conditions in order that local solvability be valid (when either the tangent structure bundle $\mathcal{V}$ or the cotangent structure bundle T′ are line bundles, the necessary conditions are also sufficient). This approach is customary since the pioneering work Hörmander [1]: by an argument from Functional Analysis one derives an a priori estimate from the hypothesis of local solvability, or of local exactness if one deals with higher levels in the differential complex. Once this is done one endeavors to show that the a priori estimate cannot hold unless a certain condition is satisfied. In some instances the condition bears on the first integrals; in others, like condition $(\mathcal{P})$ for a single vector field, it bears on the *symbol* of the vector fields.

Perhaps the most attractive aspect of hypo-analyticity is its accessibility to the methods of Harmonic Analysis—so far through a variant of the Fourier integral, called the Fourier-Brós-Iagolnitzer, or FBI, transform. One possible definition, a simple one, tightly fitting the applications to hypo-analyticity, and the basic properties of this transform are given in chapter IX. A few rudimentary applications are described: how it serves to characterize hypo-analyticity, as well as $\mathcal{C}^\infty$ regularity, and to establish the propagation of hypo-analyticity by ''elliptic'' submanifolds. The FBI integral is a powerful tool in the analysis of *microlocal* singularities, as will be shown in volume 2 of the book. In the present volume we keep strictly to the local point of view.

As I indicated at the beginning of this preface, chapter X is concerned with nonlinear complex equations, specifically, with involutive systems of such equations. Our approach is by linearization (via the Hamiltonian lift) and exploitation of the results for linear equations gathered in chapter II. Under the hypothesis that the Hamiltonian structure is locally integrable, one can establish the uniqueness in the Cauchy problem, and the approximation of solutions of class $\mathcal{C}^2$ in the fully nonlinear case, and of class $\mathcal{C}^1$ in the quasilinear case.

Even within the confines of the local viewpoint adopted in volume 1 there are still important items that are missing from the presentation. To name only two, of great interest: the reader will not find anything relating to the ''Kohn Laplacian'' associated with the system of vector fields $L_1,\ldots,L_n$; and not a word is said about the relation between general involutive structures and the

Lie algebras of nilpotent Lie groups. On both counts the author had nothing new to say; and it appeared that an adequate treatment would require a large amount of space. On the first topic we refer the reader to the extant texts (*e.g.*, chapters 22 and 23 of Hörmander [4]).

Still on the subject of the associated Laplacian it should be added that the underlying "philosophy" of the present book is that of a *direct* approach to the system of equations (1) or (2). Such a philosophy is buttressed by the feeling that, given the level of generality at which much of the analysis is carried out, the direct approach has a better chance of leading somewhere; and also, that any result which can be obtained by recourse to the Laplacian (without being a result strictly *about the Laplacian*, like those relating to the $\bar{\partial}$-Neumann problem) must be obtainable directly. I hope this view will be vindicated in volume 2, when the microlocal methods, applied to equations (1) and (2), will be shown to yield the regularity results one expects, and more.

In any event, the theory described in this book is very much "open-ended." Not only are there a large number of loose ends, of internal questions that remain to be settled, but the possible directions of outward exploration are dizzying: toward systems of higher-order equations (of principal type); toward more general overdetermined systems, to name only the two most visible.

Of the theorems in this book for which the author can claim some credit, most were the product of a collaboration, lasting almost a decade, with M. S. Baouendi. Other resulted from joint work with C. H. Chang, P. Cordaro, the late Ch. Goulaouic, N. Hanges, H. Jacobowitz, L. Nirenberg, and L. P. Rothschild. The conception of the whole theory was born of this collaborative research and nourished by its progress.

I am especially indebted to Paulo Cordaro for pointing out errors and suggesting improvements, particularly in chapters V and VIII. I also want to thank Shif Berhanu and Tejinder Neelon for alerting me to embarrassing mistakes in chapters IV and VIII. Needless to say, the errors that remain in the text are entirely the responsibility of the author.

Hypo-Analytic Structures: Local Theory

I

Formally and Locally Integrable Structures. Basic Definitions

A look at the contents of this first chapter should make perfectly clear what its title suggests: this is the chapter in which the fundamental concepts of the theory of involutive and locally integrable structures are laid out. Due to the high density of definitions some of the sections (e.g., I.3, I.4) are for reference rather than for reading. Here we highlight the main concepts introduced in chapter I, in the hope of helping readers select those in which they might want to delve further.

First, of course, the notion of *involutive* (or *formally integrable*) *structure* on a $\mathscr{C}^\infty$ manifold $\mathcal{M}$: it is the datum of a vector subbundle $\mathcal{V}$ of the complexified tangent bundle $\mathbb{C}T\mathcal{M}$ satisfying the Frobenius condition, $[\mathcal{V},\mathcal{V}] \subset \mathcal{V}$. The structure is *locally integrable* if the subbundle T' of $\mathbb{C}T^*\mathcal{M}$ orthogonal to $\mathcal{V}$ for the natural duality between tangent and cotangent vectors is locally spanned by exact differentials (section I.1) The *characteristic set* of the structure is defined in section I.2, in accordance with the traditional definition in PDE theory: it is the subset T° of the real tangent bundle $T^*\mathcal{M}$ consisting of the common zeros of the symbols of the sections of $\mathcal{V}$ (thus $T^\circ = T' \cap T^*\mathcal{M}$). In particular, the structure is called *elliptic* when $T^\circ = 0$, the zero section of the tangent bundle.

An involutive structure on the manifold $\mathcal{M}$ assigns a distinctive role to certain submanifolds of $\mathcal{M}$. Consider for instance the *noncharacteristic* submanifolds, whose conormal bundles intersect the characteristic set T° solely along the zero section. Any distribution solution of the equations $Lh = 0$, as L ranges over the $\mathscr{C}^\infty$ sections of $\mathcal{V}$, is a $\mathscr{C}^\infty$ function transversally to a noncharacteristic submanifold Σ, with values in the space of distributions tangential to Σ (Proposition I.4.3); the trace of h on Σ is well defined. An especially important role is played, throughout the theory, by the *maximally real* submanifolds of $\mathcal{M}$, i.e., the submanifolds whose complex tangent space intersects $\mathcal{V}$ only at 0 and whose dimension is maximum (i.e., it is equal to the complex codimension of the fibres of $\mathcal{V}$). When the structure of $\mathcal{M}$ is *complex*, which means that $\mathbb{C}T\mathcal{M} = \mathcal{V}\oplus\overline{\mathcal{V}}$ ($\overline{\mathcal{V}}$ is the complex conjugate of $\mathcal{V}$), "maximally real" is short for "totally real of maximum dimension" (i.e., of real dimension equal to the complex dimension of $\mathcal{M}$). In a locally integrable structure the maximally real submanifolds are the minimal uniqueness sets for the "initial value" (or Cauchy) problem (Corollary II.3.6).

Next, the *Levi form* (section I.8). It associates to a characteristic point $(x_0,w_0) \in T^\circ$ (with $x_0 \in \mathcal{M}$, $w_0 \neq 0$) the hermitian quadratic form on the fibre $\mathcal{V}_{x_0}$, $\mathfrak{Q}_{(x_0,w_0)}(\mathbf{v}) = (2\imath)^{-1}\langle w_0,[\mathrm{L},\overline{\mathrm{L}}]\,|_{x_0}\rangle$, where L is any smooth section of $\mathcal{V}$ in some neighborhood of x_0 whose value at x_0 is equal to $\mathbf{v} \in \mathcal{V}_{x_0}$. The role of the Levi form in the study of complex manifolds and in that of tangential Cauchy-Riemann (i.e., CR) structures hardly needs recalling.

The next item on our list has not had a visible role in the theory of CR structures, at least not until recently: we refer to the structure's *orbits* in an open subset Ω of $\mathcal{M}$ (section I.11). The real parts $\mathcal{R}e\mathrm{L}$ of the $\mathcal{C}^\infty$ sections of $\mathcal{V}$ over Ω, $\mathrm{L} \in \mathcal{C}^\infty(\Omega;\mathcal{V})$, generate a Lie algebra $\mathfrak{g}(\mathcal{V})$ for the commutation bracket. The orbit Σ_p through an arbitrary point $p \in \mathcal{M}$ is the result of the composite action, on p, of the flow of all the vector fields belonging to $\mathfrak{g}(\mathcal{V})$ (Definition I.11.1); Σ_p is an immersed, connected submanifold of $\mathcal{M}$ whose tangent space contains (but is not necessarily equal to) the freezing of $\mathfrak{g}(\mathcal{V})$. Keep in mind that the dimensions of the orbits are not all equal. The significance of the orbits in a locally integrable structure becomes apparent in chapter II: if an orbit of $\mathcal{V}$ in Ω intersects the support of a *solution h* in Ω (of all the homogeneous equations $\mathrm{L}h = 0$, $\mathrm{L} \in \mathcal{C}^\infty(\Omega;\mathcal{V})$) it is perforce contained in supp h (Theorem II.3.3); given any orbit Σ of $\mathcal{V}$ in Ω through some point p, there is a distribution solution in an open neighborhood $\mathrm{U} \subset \Omega$ of p whose support is contained in $\Sigma \cap \mathrm{U}$ (Theorem II.3.4).

The notion of a *real* involutive structure is classical, and obvious: the tangent structure bundle $\mathcal{V}$ is spanned (over $\mathbb{C}$) by its intersection with the real tangent bundle $T\mathcal{M}$ (i.e., $\mathcal{V}$ admits local bases consisting of real vector fields). The Frobenius theorem (Theorem I.10.1) asserts that any real involutive structure is locally integrable. The orbits of a real structure are the integral manifolds of $\mathcal{V}$. Every continuous solution is constant on each leaf of this foliation. The leaves all have the same real dimension, n, equal to the rank (i.e., the complex fibre dimension) of $\mathcal{V}$.

The Frobenius theorem is also valid for any *analytic* involutive structure: in this case both the manifold $\mathcal{M}$ and the tangent structure bundle $\mathcal{V}$ are assumed to be real-analytic (but $\mathcal{V}$ is a *complex* vector bundle). Every involutive structure of class $\mathcal{C}^\omega$ is locally integrable (Theorem I.10.5). This follows from the natural holomorphic analogue, when $\mathcal{M}$ is a complex-analytic manifold and $\mathcal{V}$ a holomorphic vector subbundle of $T^{1,0}\mathcal{M}$. The integral manifolds of $\mathcal{V}$ foliate $\mathcal{M}$; they are holomorphic submanifolds of $\mathcal{M}$, all with the same complex dimension, equal to the rank of $\mathcal{V}$. Locally $\mathcal{V}$ always admits "first integrals"

$Z_1,\ldots,Z_m$, which are holomorphic functions (Theorem I.10.3). Every holomorphic solution is constant on every leaf of $\mathcal{V}$. One can study locally an analytic involutive structure by embedding it in its complexification; the integral manifolds of the holomorphic extension of the structure intersect the (real) manifold $\mathcal{M}$ along analytic sets that are not necessarily smooth, yet are important: all analytic solutions are locally constant on those intersections. In chapter II it is even shown that all continuous solutions are locally constant on any such set (called a *fibre* of the structure—a concept shown in chapter II to be valid for any locally integrable structure, not merely for the real-analytic ones). It should also be mentioned that the orbits in an analytic involutive structure are the so-called *Nagano leaves*; they are integral manifolds of $\mathfrak{g}(\mathcal{V})$ (still, their dimensions are not all the same).

No doubt the reader is familiar with the *differential complex* associated with the $\mathcal{C}^\infty$ structure of $\mathcal{M}$, or with its complex structure when $\mathcal{M}$ is a complex manifold. In the former case $\mathcal{V} = \mathbb{C}T\mathcal{M}$; the differential complex is the *De Rham complex*, in which the differential operator is the *exterior derivative d*. In the latter case $\mathcal{V} = T^{0,1}\mathcal{M}$; one is led to the study of the *Dolbeault* (or *Cauchy-Riemann*) *complex*; the differential operator is the Cauchy-Riemann operator $\bar\partial$. It is not difficult to define the differential complex attached to a general involutive structure. From our viewpoint, close to PDE theory, it is an object of primary interest: indeed, we propose to lay the foundations for the solvability theory of the corresponding differential equations, that is to say, the cohomology theory of the differential complex, and for the study of their solutions when they exist. The basic definition is coordinate free, and global, as it should be (see section I.6). But it is also important to have at one's disposal concrete, and perforce (at the present level of generality) local, representations. Such are found in sections I.5 and I.6. Local representations are also useful in dealing with the characteristic set, the Levi form, etc. As shown in sections I.7 and I.9 they are substantially simpler in locally integrable structures.

I.1 Involutive Systems of Linear PDE Defined by Complex Vector Fields. Formally and Locally Integrable Structures

Let Ω be an open set in a $\mathcal{C}^\infty$ manifold $\mathcal{M}$ and consider a finite family of vector fields $L_1,\ldots,L_n$, with smooth (i.e., $\mathcal{C}^\infty$) complex coefficients in Ω. Write $N = \dim \mathcal{M}$ and suppose that there are local coordinates in Ω, $x_1,\ldots,x_N$. Then

$$L_j = \sum_{k=1}^{N} c_{jk}(x)\partial/\partial x_k,\ c_{jk} \in \mathcal{C}^\infty(\Omega).$$

Our purpose is to study the *homogeneous equations* defined by these vector fields

$$L_j u = 0, j = 1,\ldots,n, \tag{I.1.1}$$

as well as the *inhomogeneous equations*

$$L_j v = f_j, j = 1,\ldots,n. \tag{I.1.2}$$

The goal is to establish the most basic properties of their solutions: existence, uniqueness, regularity, approximation, and so on. For some of these properties (uniqueness, for instance) one might have to submit the solutions to additional requirements, such as boundary conditions.

Before embarking on such a research program one may ask what is known about such systems of equations in PDE theory. The answer to this question is "in general, not much." In some particular cases, a lot is known; it suffices, at this stage, to mention the Cauchy-Riemann equations. Even a cursory look at the problems, and at what is classically known, makes it obvious that some kind of restrictions must be imposed on the class of systems we propose to study. For instance, a lot is known about the *local solvability* of a single equation (I.1.2), i.e., when $n = 1$, *provided the vector field does not vanish anywhere*. Very little is known otherwise. It is therefore advisable to impose the natural generalization of this condition when n is any integer ≥ 1:

the vector fields L_j *are linearly independent at every point.* (I.1.3)

Of course Condition (I.1.3) demands $n \leq N = \dim \mathcal{M}$.

Another requirement originates from the theory of *real* vector fields. Look first at the homogeneous equations (I.1.1): it is clear that any solution will also be a solution of the equations

$$[L_j,L_k]\, u = L_j L_k u - L_k L_j u = 0,$$
$$[L_j,[L_k,L_\ell]]u = 0, \text{ etc.}$$

In other words, the relevant vector fields are the members of the Lie algebra $\mathcal{L}$ (for the commutation bracket) generated by the L_j; and rather than dealing with the L_j we should deal with a basis of $\mathcal{L}$ (regarded as a linear space). But it may well happen that such a basis will contain vector fields that are not linearly independent at every point (example: $L_1 = \partial/\partial x_1$, $L_2 = \partial/\partial x_2 + x_1 x_3 \partial/\partial x_3$), and we would then be faced with the difficulties arising from this fact. We must therefore assume that there is a basis of the vector space $\mathcal{L}$ that consists of vector fields that are linearly independent at every point. But then we may as well take this basis to be the system $L_1,\ldots,L_n$ we start with. This means that to (I.1.3) we adjoin the *Frobenius condition*:

for every pair j, $k = 1,\ldots,n$, *and at every point,* (I.1.4)
the commutation bracket $[L_j,L_k]$ *belongs to* $\mathrm{Span}(L_1,\ldots,L_n)$.

In accordance with classical terminology we shall say that the system $\{L_1,\ldots,L_n\}$ is *involutive*, or that it is a *system in involution*. Condition (I.1.4) has important implications bearing on the inhomogeneous equations (I.1.2). First, it is clear that these equations demand that the right-hand sides satisfy the *compatibility conditions* (I.1.6) below. Suppose that the Frobenius condition (I.1.4) translates into the relations

$$[L_j, L_k] = \sum_{\ell=1}^{n} c_{jk}^{\ell} L_\ell, \; 1 \le j, k \le n, \tag{I.1.5}$$

with $c_{jk}^{\ell} \in \mathscr{C}^\infty(\Omega)$. Then, applying L_k to both sides of (I.1.2), exchanging j and k, and taking (I.1.5) into account yields:

$$L_j f_k - L_k f_j = \sum_{\ell=1}^{n} c_{jk}^{\ell} f_\ell, \; 1 \le j, k \le n. \tag{I.1.6}$$

Furthermore, to the equations (I.1.2) we may adjoin the equations

$$L_j u_k - L_k u_j = f_{jk}, \; 1 \le j, k \le n; \tag{I.1.7}$$

$$L_j u_{k\ell} - L_k u_{j\ell} + L_\ell u_{jk} = f_{jk\ell}, \; 1 \le j, k, \ell \le n; \tag{I.1.8}$$

etc. The equations (I.1.7) stand to (I.1.2) as the equations $curl\, \vec{u} = F$ stand to the equations $grad\ u = \vec{f}$. The right-hand sides must of course satisfy compatibility conditions deriving from (I.1.5). Thus our system of vector fields $\{L_1,\ldots,L_n\}$ defines a differential operator $L^{(q)}$ acting on multiplets $(u_J)_{|J|=q}$ of functions (or distributions) in Ω, and transforming them into multiplets $(f_K)_{|K|=q+1}$. We are using the multi-index notation, which will be standard throughout the book: J is a strictly increasing sequence of integers $1 \le j_1 < \cdots < j_q \le n$; $|J|$ denotes its length, presently equal to q. It is a straightforward consequence of the Frobenius condition (I.1.4) that

$$L^{(q+1)} \circ L^{(q)} = 0. \tag{I.1.9}$$

In other words, the sequence of differential operators $L^{(q)}$ ($q = 0, 1,\ldots$) forms a *differential complex*. This will be formalized later by using differential forms. The present situation generalizes that of the gradient, as alluded to above; switching to differential forms leads to the dual notion, the *exterior derivative*, which gives rise to the *De Rham complex*. Likewise the Cauchy-Riemann vector fields give rise to the *Cauchy-Riemann complex* (sometimes called the *Dolbeault* complex).

It is clear that the kind of properties of solutions of the system of Equations (I.1.1) or (I.1.2) we propose to study will not depend on the choice of local coordinates x_i. This is why we have taken the open set Ω in a manifold, not in Euclidean space $\mathbb{R}^N$—notwithstanding the fact that most of the considerations and results in this book will be local. There is also another kind of invariance that must be kept in mind when dealing with equations (I.1.1) or (I.1.2): invariance under linear substitutions of the vector fields L_j by vector fields

$$\mathrm{L}_j^{\#} = \sum_{k=1}^{n} \gamma_{jk}\mathrm{L}_k,$$

provided the functions $\gamma_{jk} \in \mathscr{C}^\infty(\Omega)$ are the entries of a nonsingular matrix. This means that what matters here is not the system of vector fields L_j but rather the vector bundle (over Ω) they generate, that is to say, the vector bundle whose *fibre* at an arbitrary point $p \in \Omega$ is the linear span (over $\mathbb{C}$) of the complex vectors $\mathrm{L}_j|_p$ tangent to $\mathcal{M}$ at p. Clearly this is a vector subbundle, over Ω, of the *complexified* (or complex, as we shall more often say) tangent vector bundle of $\mathcal{M}$. The dimension (equal to n) of the $\mathbb{C}$-linear span of the vector fields L_j is often called the *rank* of the vector bundle; we shall also refer to it as its *fibre dimension*.

Actually it is convenient to assume that the vector bundle in question, from now on denoted by $\mathcal{V}$, is defined over the whole of $\mathcal{M}$ and not just over Ω. Over Ω there are smooth sections of $\mathcal{V}$, namely $\mathrm{L}_1,\ldots,\mathrm{L}_n$, which span the fibre of $\mathcal{V}$ at each point—what we shall call a *smooth basis* of $\mathcal{V}$ over Ω. (In general there will not exist any smooth basis of $\mathcal{V}$ over the whole of $\mathcal{M}$.) In this language what was said earlier can be rephrased by saying that any other smooth basis of $\mathcal{V}$ over Ω could be substituted for $\{\mathrm{L}_1,\ldots,\mathrm{L}_n\}$ in the equations (I.1.1) or (I.1.2). A different basis may be better suited to the analysis of certain aspects of these equations.

In summary, we see that our basic object of study is a vector subbundle $\mathcal{V}$ of $\mathbb{C}T\mathcal{M}$ endowed with the Frobenius property, which we abbreviate as follows:

$$[\mathcal{V},\mathcal{V}] \subset \mathcal{V}, \tag{I.1.10}$$

meaning that the commutation bracket of two smooth sections of $\mathcal{V}$ over some open subset of $\mathcal{M}$ is again a section of $\mathcal{V}$ over the same subset.

There is a dual description of the preceding structure. The duality referred to here is the one between tangent vectors (the values of vector fields at a given point) and cotangent vectors (the values of differential forms at the same point). The orthogonal, in the sense of this duality, of the vector subbundle $\mathcal{V} \subset \mathbb{C}T\mathcal{M}$ is a vector subbundle of the complex cotangent bundle $\mathbb{C}T^*\mathcal{M}$ of $\mathcal{M}$ which we shall always call T' (we also write $\mathrm{T}' = \mathcal{V}^\perp$, $\mathcal{V} = \mathrm{T}'^\perp$). The fibre of T' at an arbitrary point of $\mathcal{M}$ is the orthogonal of the fibre of $\mathcal{V}$ in the cotangent space to $\mathcal{M}$ at that point. The fibre dimension of T' is equal to $m = N - n$.

The following extends to $\mathbb{C}T^*\mathcal{M}$ the definition introduced in E. Cartan [1] for vector subbundles of the *real* cotangent bundle $T^*\mathcal{M}$:

DEFINITION I.1.1. *We say that the vector subbundle* T' *of* $\mathbb{C}T^*\mathcal{M}$ *is* closed *if, given any smooth section* φ *of* T' *over an open subset* Ω *of* $\mathcal{M}$, *each point of* Ω *has an open neighborhood* $U \subset \Omega$ *in which there are m smooth sections of* T',

$\varpi_1,\dots,\varpi_m$, *and an equal number of smooth differential forms*, $\psi_1,\dots,\psi_m$, *such that, in the set* U,

$$d\varphi = \psi_1\wedge\varpi_1 + \cdots + \psi_m\wedge\varpi_m. \tag{I.1.11}$$

It is clear that, in the above definition, we may assume the ϖ_i to be linearly independent, i.e., we may take $\{\varpi_1,\dots,\varpi_m\}$ to be any smooth basis of T′ over U—if such a basis exists (of course such bases do exist as soon as U is sufficiently small). When the ϖ_i are linearly independent (I.1.11) is equivalent to

$$\varpi_1\wedge\cdots\wedge\varpi_m\wedge d\varphi \equiv 0. \tag{I.1.12}$$

The reason for introducing Definition I.1.1 lies in the following:

Proposition I.1.1. *For the vector subbundle $\mathcal{V}$ of $\mathbb{C}T\mathcal{M}$ to have the Frobenius property* (I.1.10) *it is necessary and sufficient that the vector subbundle* T′ *of* $\mathbb{C}T^*\mathcal{M}$ *be closed.*

Proof. Let φ be a complex differential form, $\mathbf{v}_j$ ($j = 1, 2$) two complex vector fields in Ω, all smooth. Denote by $\langle\ ,\ \rangle$ the duality bracket between p-vectors and p-covectors (here $p = 1, 2$). We make use of the following identity:

$$\langle d\varphi,\mathbf{v}_1\wedge\mathbf{v}_2\rangle = \mathbf{v}_1\langle\varphi,\mathbf{v}_2\rangle - \mathbf{v}_2\langle\varphi,\mathbf{v}_1\rangle - \langle\varphi,[\mathbf{v}_1,\mathbf{v}_2]\rangle \tag{I.1.13}$$

valid in Ω.

Suppose that φ is a $\mathcal{C}^\infty$ section of T′ while $\mathbf{v}_1$ and $\mathbf{v}_2$ are $\mathcal{C}^\infty$ sections of $\mathcal{V}$. The equation (I.1.13) reduces to

$$\langle d\varphi,\mathbf{v}_1\wedge\mathbf{v}_2\rangle = -\langle\varphi,[\mathbf{v}_1,\mathbf{v}_2]\rangle. \tag{I.1.14}$$

To say that (I.1.10) holds is to say that the right-hand side in (I.1.14) vanishes for all possible choices of φ, $\mathbf{v}_1$, $\mathbf{v}_2$. To say that T′ is closed is to say that the left-hand side in (I.1.14) does. ∎

In the sequel we shall refer to either $\mathcal{V}$ or T′ as the *structure bundle*. To differentiate between them we shall often refer to $\mathcal{V}$ as the *tangent structure bundle* and to T′ as the *cotangent structure bundle*. The datum of $\mathcal{V}$ and T′ is often called a *formally integrable* structure on $\mathcal{M}$; we shall also use the name "*involutive structure.*"

Definition I.1.2. *We shall say that the structure defined by* T′ *on* $\mathcal{M}$ *is* locally integrable *if every point of* $\mathcal{M}$ *has an open neighborhood in which* T′ *is spanned by exact forms.*

In Definition I.1.2 "exact forms" may be replaced by "closed forms" since every closed form is locally exact, by the Poincaré lemma.

Let $du_1,\dots,du_r$ be smooth exact forms in an open neighborhood U of $\mathfrak{p} \in \mathcal{M}$

that span T′ over U. Out of the set of cotangent vectors $du_j|_p$ $(1 \leq j \leq r)$ we can select a basis of $T'|_p$, $du_{j_1}|_p,\ldots,du_{j_m}|_p$. The one-forms $du_{j_1},\ldots,du_{j_m}$ remain linearly independent in a neighborhood $V \subset U$ of p, and since $m =$ rank of T′, they span T′ over V. They form what we shall refer to as a *smooth basis* of T′ over V.

We shall eventually see that not every formally integrable structure is locally integrable.

DEFINITION I.1.3. *Let the manifold $\mathcal{M}$ be equipped with a formally integrable structure in which the structure bundles are $\mathcal{V}$ and* T′. *By a* distribution solution *in an open set $\Omega \subset \mathcal{M}$ we shall mean a distribution u in Ω such that* $Lu = 0$ *in Ω for all $\mathcal{C}^\infty$ sections* L *of $\mathcal{V}$ in Ω.*

If a distribution solution in Ω is a $\mathcal{C}^r$ function (with $0 \leq r \leq +\infty$ or $r = \omega$), we shall simply say that it is a $\mathcal{C}^r$ solution. Likewise we shall speak of H^s solutions, L^p solutions, etc. A $\mathcal{C}^1$ solution will often be called a *classical solution*. In this case its differential is a continuous section of T′ over Ω. Actually one can define the distribution sections of a complex vector bundle (see Schwartz [1], p. 339). The differential of any distribution solution in Ω is a distribution section of T′ over Ω.

The following statement is an immediate consequence of the chain rule:

PROPOSITION I.1.2. *Let ν be an integer ≥ 1, $h_1,\ldots,h_\nu$ be $\mathcal{C}^1$ solutions in an open subset Ω of $\mathcal{M}$, F a holomorphic function in an open subset of $\mathbb{C}^\nu$ containing the image of Ω under the map*

$$p \rightarrow h(p) = (h_1(p),\ldots,h_\nu(p)).$$

Then $F \circ h$ is a $\mathcal{C}^1$ solution in Ω.

In particular the classical solutions in Ω form an algebra. If one of them, h, does not vanish anywhere in Ω, $1/h$ is a $\mathcal{C}^1$ solution.

What has been said in the present section in the "$\mathcal{C}^\infty$ category" can be repeated within the "analytic category." In this book "analytic" will always stand for "real-analytic." When meaning "complex-analytic" we shall always say "holomorphic." Often, instead of "analytic" we shall also say "of class $\mathcal{C}^\omega$" or simply $\mathcal{C}^\omega$. If $\mathcal{M}$ is an analytic manifold, the structure bundles $\mathcal{V}$ and T′ can be taken to be analytic, which means that, locally, they are spanned by analytic vector fields and differential forms respectively. We say then that $\mathcal{M}$ is equipped with an *analytic formally integrable* (or *involutive*) *structure*. If locally T′ is spanned by the differentials of analytic (complex-valued) functions, we say that $\mathcal{M}$ is equipped with an *analytic locally integrable structure*.

I.2. The Characteristic Set. Partial Classification of Formally Integrable Structures

In dealing with the structure bundles $\mathcal{V}$ and T′ a few elementary facts about real/complex linear algebra will be of help. Let E be a complex linear subspace of $\mathbb{C}^N$, $\overline{\mathrm{E}}$ its complex conjugate. There is a linear subspace (over $\mathbb{C}$) of E, E_1, such that

$$\mathrm{E} = (\mathrm{E}\cap\overline{\mathrm{E}})\oplus\mathrm{E}_1 \tag{I.2.1}$$

where $\oplus$ stands for the direct sum. The vectors in $\mathrm{E}^\circ = \mathrm{E}\cap\mathbb{R}^N$ are the fixed points of the complex conjugate map in E. Thus $\mathrm{E}^\circ \subset \mathrm{E}\cap\overline{\mathrm{E}}$ and

$$\mathrm{E}_1\cap\overline{\mathrm{E}}_1 = \mathrm{E}_1\cap\mathbb{R}^N = \{0\}. \tag{I.2.2}$$

The real-part map $z \to \mathscr{R}e\, z = (z + \bar{z})/2$ is a retraction of $\mathrm{E}\cap\overline{\mathrm{E}}$ onto E°, and the null space of this map in $\mathrm{E}\cap\overline{\mathrm{E}}$ is equal to $\imath\mathrm{E}^\circ$. As a consequence $\mathrm{E}\cap\overline{\mathrm{E}} = \mathrm{E}^\circ\oplus(\imath\mathrm{E}^\circ)$ or, equivalently,

$$\mathrm{E}\cap\overline{\mathrm{E}} = \mathrm{E}^\circ\otimes_{\mathbb{R}}\mathbb{C}. \tag{I.2.3}$$

In particular we see that $\dim_{\mathbb{R}} \mathrm{E}^\circ = \dim_{\mathbb{C}} (\mathrm{E}\cap\overline{\mathrm{E}})$.

$\mathscr{R}e$ maps E_1 onto a 2ν-dimensional subspace of $\mathbb{R}^N$ that intersects E° only at 0. Thus $\mathscr{R}e$ maps E onto a linear subspace of $\mathbb{R}^N$ whose (real) dimension is equal to $2\nu + d$, and which we denote by $\mathscr{R}e\mathrm{E}$; $\mathscr{R}e\mathrm{E}$ spans $\mathrm{E} + \overline{\mathrm{E}}$ over the complex numbers. We recall that

$$\dim_{\mathbb{C}} \mathrm{E} + \dim_{\mathbb{C}} \overline{\mathrm{E}} = \dim_{\mathbb{C}} (\mathrm{E} + \overline{\mathrm{E}}) + \dim_{\mathbb{C}} (\mathrm{E}\cap\overline{\mathrm{E}}). \tag{I.2.4}$$

Let $\{\mathbf{e}_1,\ldots,\mathbf{e}_\nu\}$ be a linear basis of E_1; then $\{\bar{\mathbf{e}}_1,\ldots,\bar{\mathbf{e}}_\nu\}$ is a linear basis of $\overline{\mathrm{E}}_1$ and moreover, since $\mathrm{E}_1\cap\overline{\mathrm{E}}_1 = \{0\}$, the 2ν vectors $\mathbf{e}_j$, $\bar{\mathbf{e}}_k$ are $\mathbb{C}$-linearly independent (this is the same as saying that the real vectors $\mathscr{R}e\mathbf{e}_j$, $\mathscr{I}m\mathbf{e}_k$ are linearly independent). We select any basis $\mathbf{e}_{\nu+1},\ldots,\mathbf{e}_{\nu+d}$ of E° to obtain a linear basis of E consisting of vectors $\mathbf{e}_1,\ldots,\mathbf{e}_\nu,\mathbf{e}_{\nu+1},\ldots,\mathbf{e}_{\nu+d}$ (thus $\dim_{\mathbb{C}} \mathrm{E} = \nu + d$) such that

$$\begin{gathered}\mathbf{e}_1,\ldots,\mathbf{e}_\nu,\bar{\mathbf{e}}_1,\ldots,\bar{\mathbf{e}}_\nu,\mathbf{e}_{\nu+1},\ldots,\mathbf{e}_{\nu+d} \textit{ are } \mathbb{C}\textit{-linearly independent;}\\ \mathbf{e}_{\nu+1},\ldots,\mathbf{e}_{\nu+d} \textit{ are real.}\end{gathered} \tag{I.2.5}$$

We shall apply these considerations to the fibres of T′. If x is an arbitrary point of $\mathcal{M}$ and ζ a complex cotangent vector to $\mathcal{M}$ at x, we write $\zeta = \xi + \imath\eta$ with ξ, η real cotangent vectors to $\mathcal{M}$ at x; then $\bar{\zeta} = \xi - \imath\eta$. The complex conjugate map $(x,\zeta) \to (x,\bar{\zeta})$ induces a vector bundle isomorphism of T′ onto $\overline{\mathrm{T}}'$. The intersections $\mathrm{T}'\cap\overline{\mathrm{T}}'$, $\mathrm{T}^\circ = \mathrm{T}'\cap\mathrm{T}^*\mathcal{M}$ are *not*, in general, vector bundles: the dimensions of their fibres might vary from point to point. We have

$$\mathrm{T}'\cap\overline{\mathrm{T}}' = \mathrm{T}^\circ\otimes_{\mathbb{R}}\mathbb{C}. \tag{I.2.6}$$

(All pairing of vector bundles over one and the same manifold, such as tensor products and direct sums, will be carried out *fibrewise*. In (I.2.6) the factor $\mathbb{C}$ at the right stands for the constant bundle $\mathcal{M} \times \mathbb{C}$.)

The set T° plays a very important role in PDE theory. Notice that if L is any section of $\mathcal{V}$ in some open set Ω and $(x,\xi) \in \mathrm{T}^\circ|_\Omega$ then $\langle \xi, \mathrm{L}|_x \rangle = 0$ (as always, the duality bracket between tangent and cotangent vectors is denoted by $\langle , \rangle$). This prompts us to introduce the following

DEFINITION I.2.1. *Let* L *be a vector field in an open subset* Ω *of* $\mathcal{M}$. *We shall call* symbol *of* L *and denote by* $\sigma(\mathrm{L})$ *the function* $(x,\xi) \to \langle \xi, \mathrm{L}|_x \rangle$ *in* $\mathrm{T}^*\mathcal{M}|_\Omega$.

Definition I.2.1 differs from the standard definition of the symbol of a differential operator by a factor of $\imath = \sqrt{-1}$. Suppose that there is a system of local coordinates $x_1,\ldots,x_N$ in Ω; call $\xi_1,\ldots,\xi_N$ the dual coordinates in the cotangent spaces to $\mathcal{M}$ at points of Ω. If the expression of the vector field L in these coordinates is

$$\mathrm{L} = \sum_{k=1}^{N} c_k(x)\partial/\partial x_k,$$

the expression of its symbol is

$$\sigma(\mathrm{L})(x,\xi) = \sum_{k=1}^{N} c_k(x)\xi_k. \tag{I.2.7}$$

We see that T° *is the set of common zeros of all the symbols* $\sigma(\mathrm{L})$ *of all sections* L *of* $\mathcal{V}$ (say, over $\mathcal{M}$). Although it is customary to exclude the zero section from the characteristic set, we shall not do so in this book, for reasons of convenience.

DEFINITION I.2.2. *By the* characteristic set *of the formally integrable structure defined by* $\mathcal{V}$ (or T′) *on* $\mathcal{M}$ *we shall mean the subset* T° *of* $\mathrm{T}^*\mathcal{M}$.

At a given point $x_0 \in \mathcal{M}$ the fibre T'_{x_0} has a linear basis of the kind (I.2.5). It means that, in some open neighborhood of x_0, T′ is spanned by $\nu + d = m$ smooth differential one-forms $\varphi_1,\ldots,\varphi_\nu$, $\varpi_1,\ldots,\varpi_d$, such that

$$\varphi_1\wedge\cdots\wedge\varphi_\nu\wedge\overline{\varphi}_1\wedge\cdots\wedge\overline{\varphi}_\nu\wedge\varpi_1\wedge\cdots\wedge\varpi_d \neq 0; \tag{I.2.8}$$

$$\varpi_1,\ldots,\varpi_d \textit{ are real at the point } x_0. \tag{I.2.9}$$

The covectors $\varpi_j|_{x_0}$, $j = 1,\ldots,d$, form a basis of T° at x_0. This may not be true at nearby points as the ϖ_j may not be real there.

We could have also applied the linear algebra of the beginning to the vector bundle $\mathcal{V}$. In this case the fibre at a point $x \in \mathcal{M}$ of $\mathbb{C}T\mathcal{M}$ plays the role of $\mathbb{C}^N$, that of $T\mathcal{M}$ plays the role of $\mathbb{R}^N$ and $E = \mathcal{V}_x$. This allows us to define the subset $\mathcal{V}^0$ of $T\mathcal{M}$; we have

$$\mathcal{V}\cap\overline{\mathcal{V}} = \mathcal{V}^0\otimes_{\mathbb{R}}\mathbb{C}. \qquad \text{(I.2.10)}$$

An arbitrary point x_0 of $\mathcal{M}$ has an open neighborhood in which there is a smooth basis $L_1,\ldots,L_n$ of $\mathcal{V}$ with the following properties:

$$L_1,\ldots,L_\nu,\overline{L}_1,\ldots,\overline{L}_\nu,L_{\nu+1},\ldots,L_n \textit{ are linearly independent;} \qquad \text{(I.2.11)}$$

$$L_{\nu+1},\ldots,L_n \textit{ are real at } x_0. \qquad \text{(I.2.12)}$$

In connection with the decomposition of T' we note that $T'\cap\overline{T}'$ is the orthogonal of $\mathcal{V} + \overline{\mathcal{V}}$, and that $T' + \overline{T}'$ is the orthogonal of $\mathcal{V}\cap\overline{\mathcal{V}}$ (here + stands for the fibrewise vector sum, not necessarily direct, of the vector bundles).

The complex conjugate vector bundles $\overline{T}'$ and $\overline{\mathcal{V}}$, and the characteristic set $T°$, allow us to define four important classes of formally integrable structures:

Definition I.2.3. *The formally integrable structure defined on* $\mathcal{M}$ *by the vector bundles* T' *and* $\mathcal{V}$ *will be called*

real *if* $T' = \overline{T}'$ (i.e., $\mathcal{V} = \overline{\mathcal{V}}$);

complex *if* $\mathbb{C}T^*\mathcal{M} = T'\oplus\overline{T}'$ (i.e., $\mathbb{C}T\mathcal{M} = \mathcal{V}\oplus\overline{\mathcal{V}}$);

elliptic *if* $T'\cap\overline{T}' = 0$ (i.e., $\mathbb{C}T\mathcal{M} = \mathcal{V} + \overline{\mathcal{V}}$);

Cauchy-Riemann (abbreviated to CR) *if* $\mathbb{C}T^*\mathcal{M} = T' + \overline{T}'$ (i.e., $\mathcal{V}\cap\overline{\mathcal{V}} = 0$).

When the structure is *real* we have

$$T' = T°\otimes_{\mathbb{R}}\mathbb{C},\ \mathcal{V} = \mathcal{V}^0\otimes_{\mathbb{R}}\mathbb{C}, \qquad \text{(I.2.13)}$$

which means that in the neighborhood of every point of $\mathcal{M}$ there is a smooth basis of $\mathcal{V}$ consisting of real vector fields, and one of T' consisting of real one-forms.

When the structure is *elliptic*, we have $T° = 0$; the characteristic set of the structure reduces to 0, wherein lies the justification for the name "elliptic." Note that in any elliptic structure we have

$$m \le n. \qquad \text{(I.2.14)}$$

Locally T' is spanned by smooth one-forms $\varphi_1,\ldots,\varphi_m$ such that

$$\varphi_1\wedge\cdots\wedge\varphi_m\wedge\overline{\varphi}_1\wedge\cdots\wedge\overline{\varphi}_m \neq 0. \qquad \text{(I.2.15)}$$

If the structure on $\mathcal{M}$ defined by $\mathcal{V}$ and T' is CR, we must have

$$n \le m. \qquad \text{(I.2.16)}$$

Locally, $\mathcal{V}$ is spanned by smooth vector fields $L_1,\ldots,L_n$ such that $L_1,\ldots,L_n,\overline{L}_1,\ldots,\overline{L}_n$ are linearly independent.

When the structure defined by $\mathcal{V}$ or T′ is *complex*, we must have

$$m = n = \frac{1}{2}\dim \mathcal{M}. \tag{I.2.17}$$

In our terminology a complex structure is both elliptic and CR. But many authors exclude the complex structures from the class of CR structures. It should also be mentioned that what we have just defined as a complex structure is often referred to as an *almost-complex* structure, a reference often followed by the remark that every almost-complex structure is a complex structure, by which it is meant that the structure in question is locally integrable (Definition I.1.2). That this is so is stated in the Newlander-Nirenberg theorem (see section V.5).

We briefly describe some examples of formally integrable structures in the classes we have just introduced.

EXAMPLE I.2.1. In Euclidean space $\mathbb{R}^N$ consider the vector fields $\partial/\partial x_1,\ldots,$ $\partial/\partial x_n$. They span a vector subbundle $\mathcal{V}$ of $\mathbb{C}T\mathbb{R}^N$ whose orthogonal T′ in $\mathbb{C}T^*\mathbb{R}^N$ is spanned by the differentials $dx_{n+1},\ldots,dx_N$. This defines a real formally integrable structure on $\mathbb{R}^N$. We do not exclude the case $n = 0$, i.e., $T' = \mathbb{C}T^*\mathbb{R}^N$, nor the case $n = N$, i.e., when $T' = 0$ and the structure is elliptic. ■

EXAMPLE I.2.2. The archetype of all complex structures is the one defined on $\mathbb{R}^{2n}$ by the *Cauchy-Riemann vector fields*

$$\partial/\partial \bar{z}_j = \tfrac{1}{2}(\partial/\partial x_j + \imath\partial/\partial y_j),\ j = 1,\ldots,n \tag{I.2.18}$$

(in the present context the real coordinates in $\mathbb{R}^{2n}$ are denoted by $x_1,\ldots,x_n,y_1,\ldots,y_n$ and one writes $\bar{z}_j = x_j + \imath y_j$). The tangent structure bundle, spanned by $\partial/\partial\bar{z}_1,\ldots,\partial/\partial\bar{z}_n$, is commonly denoted by $T^{0,1}$; its complex conjugate is denoted by $T^{1,0}$. The orthogonal of $T^{0,1}$ is $T'^{1,0}$, the span of the differentials $dz_1,\ldots,dz_n$. ■

EXAMPLE I.2.3. Denote by $(x_1,\ldots,x_n,y_1,\ldots,y_n,s_1,\ldots,s_d)$ the real coordinates in $\mathbb{R}^{2n+d}$ and call $\mathcal{V}$ the span (over $\mathbb{C}$) of the vector fields $\partial/\partial\bar{z}_j$ $(1 \le j \le n)$ defined as in (I.2.18) and $\partial/\partial s_k$ $(1 \le k \le d)$. The orthogonal of $\mathcal{V}$ in $\mathbb{C}T^*\mathbb{R}^{2n+d}$, T′, is spanned by the differentials $dz_1,\ldots,dz_n$. The structure defined by $\mathcal{V}$ on $\mathbb{R}^{2n+d}$ is elliptic. ■

EXAMPLE I.2.4. Take $\mathbb{R}^{2n+d} = \mathbb{C}^n \times \mathbb{R}^d$ as the base manifold, as in Example I.2.3, but now define $\mathcal{V}$ as the span of the Cauchy-Riemann vector fields $\partial/\partial\bar{z}_j$

alone ($j = 1,\dots,n$). Then $T' = \mathcal{V}^{\perp}$ is spanned by the differentials $dz_1,\dots,dz_n,ds_1,\dots,ds_d$. This defines a CR structure on $\mathbb{C}^n \times \mathbb{R}^d$ to which we shall often refer as the *flat CR structure*. ■

EXAMPLE I.2.5. Let Σ denote a smooth hypersurface in $\mathbb{R}^{2n}$. Identify $\mathbb{R}^{2n}$ to $\mathbb{C}^n$ as in Example I.2.2, by means of the complex coordinates $z_1,\dots,z_n$. Let then T' be the vector subbundle of $\mathbb{C}T^*\Sigma$ spanned by the pullbacks to Σ of the differentials $dz_1,\dots,dz_n$. These pullbacks are linearly independent at all points (and thus the fibre dimension of T' is equal to n). Indeed, since $dz_1,\dots,dz_n$, $d\bar{z}_1,\dots,d\bar{z}_n$ span $\mathbb{C}T^*\mathbb{R}^{2n}$ their pullbacks to Σ span the whole complex cotangent bundle of Σ. If the rank of the pullbacks of $dz_1,\dots,dz_n$ were $< n$, the same would be true of that of their complex conjugates and the rank of $dz_1,\dots,dz_n$, $d\bar{z}_1,\dots,d\bar{z}_n$ would be $\leq 2(n - 1)$. This shows that T' defines a CR structure on Σ. The tangent structure bundle $\mathcal{V}$ on Σ is spanned by the vector fields that are tangent to Σ and that are linear combinations of the Cauchy-Riemann vector fields $\partial/\partial\bar{z}_j$ ($1 \leq j \leq n$). They are called the *tangential Cauchy-Riemann vector fields* on Σ. This is why the name CR is given to the whole class of structures in which $\mathcal{V}\cap\overline{\mathcal{V}} = 0$. ■

REMARK I.2.1. The only formally integrable structure on a manifold $\mathcal{M}$ that is both real and elliptic is the structure defined by $T' = 0$, $\mathcal{V} = \mathbb{C}T\mathcal{M}$. The only structure that is both real and CR is the one defined by $T' = \mathbb{C}T^*\mathcal{M}$, $\mathcal{V} = 0$. ■

The structures in Definition I.2.3 have an important property in common:

PROPOSITION I.2.1. *Suppose that the formally integrable structure defined on* $\mathcal{M}$ *by* $\mathcal{V}$ *and* T' *is any one of the four structures in Definition* I.2.3. *Then* $T'\cap\overline{T}'$, $T' + \overline{T}'$, $\mathcal{V}\cap\overline{\mathcal{V}}$ *and* $\mathcal{V} + \overline{\mathcal{V}}$ *are complex vector bundles, and the characteristic set* T° *is a real vector bundle.*

PROOF. The statements are trivial when the structure is real or when it is complex. We avail ourselves of (I.2.4): in the CR case, it shows that the fibre dimension of $T'\cap\overline{T}'$ is constant, since this is true of that of $T' + \overline{T}'$, everywhere equal to $\dim \mathcal{M}$; in the elliptic case, the fibre dimension of $T' + \overline{T}'$ is everywhere equal to $2m$. ■

In Example I.2.4 the characteristic set T° is the bundle spanned by $ds_1,\dots,$ ds_d. In Example I.2.5 T° is a real *line bundle*. Indeed, (I.2.4) applied to the fibre of T' at an arbitrary point z_0 of the hypersurface Σ yields $\dim_{\mathbb{C}}(T'_{z_0}\cap \overline{T}'_{z_0}) = 2n - \dim \Sigma = 1$, and it suffices to apply (I.2.6).

It is easy to describe a formally integrable structure that does *not* belong to any one of the four classes of Definition I.2.3, yet for which the conclusion in Proposition I.2.1 is valid:

EXAMPLE I.2.6. Denote by $x_1,\dots,x_n$, $y_1,\dots,y_n$, $t_1,\dots,t_\ell$, $s_1,\dots,s_d$ the coordinates in $\mathbb{R}^{2n+\ell+d}$ and call $\mathcal{V}$ the span over $\mathbb{C}$ of the vector fields $\partial/\partial\bar{z}_j$ $(1 \leq j \leq n)$, $\partial/\partial t_k$ $(1 \leq k \leq \ell)$. If none of the numbers n, d, ℓ is equal to zero, the structure defined by $\mathcal{V}$ is neither real nor elliptic nor CR (and a fortiori not complex). The characteristic set is spanned by the differentials $\mathrm{d}s_j$ $(j = 1,\dots,d)$; it is a real vector bundle. ∎

It is also easy to construct a structure for which the conclusion in Proposition I.2.1 is *not* valid:

EXAMPLE I.2.7. In $\mathbb{R}^2$ denote by s, t the coordinates and consider the *Mizohata vector field* $\mathrm{L} = \partial/\partial t - \imath t\partial/\partial s$. It spans a line bundle $\mathcal{V} \subset \mathbb{C}\mathrm{T}\mathbb{R}^2$; its orthogonal, T', is spanned by the differential $\mathrm{d}Z$ of the function $Z = s + \imath t^2/2$. In the region $t \neq 0$ of $\mathbb{R}^2$ $\mathcal{V}$ and T' induce a complex structure; but along the s-axis the characteristic set is equal to the span of $\mathrm{d}s$. ∎

Traditionally, a manifold $\mathcal{M}$ equipped with a complex structure is called a *complex manifold*. A manifold $\mathcal{M}$ equipped with a CR structure is called a *CR manifold*. The real dimension of a complex manifold $\mathcal{M}$ is even; the number $\frac{1}{2}\dim \mathcal{M}$ is denoted by $\dim_{\mathbb{C}} \mathcal{M}$ and called the *complex dimension* of $\mathcal{M}$.

Let Ω be an open subset of the manifold $\mathcal{M}$. If $\mathcal{M}$ is a complex manifold any distribution solution in Ω is a holomorphic function in Ω. If $\mathcal{M}$ is a CR manifold a distribution solution is called a *CR distribution* in Ω, or a *CR function* if the distribution is a function.

I.3. Strongly Noncharacteristic, Totally Real and Maximally Real Submanifolds

Let $\mathcal{M}$ be a $\mathcal{C}^\infty$ manifold, equipped with the involutive structure defined by the vector bundles $\mathcal{V}$,T'. Certain kinds of submanifolds of $\mathcal{M}$ will play special roles in the analysis on the manifold $\mathcal{M}$. Before describing them we must make sure that our terminology about maps and submanifolds is clear and precise.

By a manifold we shall always mean a $\mathcal{C}^\infty$ manifold unless we specify otherwise (for example, when we intend to deal with a real-analytic manifold). In rare instances we shall deal with $\mathcal{C}^r$ manifolds with $r < +\infty$. It ought to be understood once and for all that any manifold we deal with is *countable at infinity* (hence *paracompact* and *Hausdorff*).

Consider two $\mathcal{C}^\infty$ manifolds $\mathcal{X}$, $\mathcal{Y}$ and a $\mathcal{C}^\infty$ mapping $f\colon \mathcal{X} \to \mathcal{Y}$. We denote by $\mathrm{D}f(x) : \mathbb{C}T_x\mathcal{X} \to \mathbb{C}T_{f(x)}\mathcal{Y}$ the *differential of the map f* at the point x, by ${}^t\mathrm{D}f(x)$: $\mathbb{C}T^*_{f(x)}\mathcal{Y} \to \mathbb{C}T^*_x\mathcal{X}$ its *transpose*. We denote by f_* the associated map $\mathbb{C}T\mathcal{X} \to \mathbb{C}T\mathcal{Y}$, $(x,\mathbf{v}) \to (f(x),\mathrm{D}f(x)\mathbf{v})$. If S is any subset of $\mathbb{C}T^*\mathcal{Y}$, we define *the pullback under the map f* as the set

$$f^*\mathrm{S} = \{ (x,\xi) \in \mathbb{C}T^*\mathcal{X};\ \exists\ (y,\eta) \in \mathrm{S} \textit{ such that } y = f(x),\ \xi = {}^t\mathrm{D}f(x)\eta \}. \tag{I.3.1}$$

We recall that the map f is called an *immersion* if $\mathrm{D}f(x)$ is injective for all $x \in \mathcal{X}$. The map f is called a *diffeomorphism* of $\mathcal{X}$ onto $\mathcal{Y}$ if there is a $\mathcal{C}^\infty$ map g: $\mathcal{Y} \to \mathcal{X}$ such that $g \circ f = \mathrm{Id}_{\mathcal{X}}$, the identity map of $\mathcal{X}$, and $f \circ g = \mathrm{Id}_{\mathcal{Y}}$. A diffeomorphism is the same thing as an immersion that is also a homeomorphism.

We say that a subset $\mathcal{X}$ of $\mathcal{M}$ is an *embedded submanifold* (or simply a *submanifold* when there is no risk of misunderstanding) if an arbitrary point x_0 of $\mathcal{X}$ has an open neighborhood U in $\mathcal{M}$ in which there are ν real-valued $\mathcal{C}^\infty$ functions $f_1,\ldots,f_\nu$, *with* ν *independent of* x_0, such that the following is true:

$$\mathrm{d}f_1\wedge\cdots\wedge\mathrm{d}f_\nu \neq 0 \textit{ at every point of } \mathrm{U}; \tag{I.3.2}$$

$$\mathcal{X}\cap\mathrm{U} = \{ x \in \mathrm{U}; f_1(x) = \cdots = f_\nu(x) = 0 \}. \tag{I.3.3}$$

Of course ν is equal to codim $\mathcal{X}$, the *codimension* of $\mathcal{X}$; and dim $\mathcal{M} - \nu$ is equal to the (real) *dimension* of $\mathcal{X}$, dim $\mathcal{X}$.

Example: The subset of $\mathbb{R}^2$ consisting of the points (x_1,x_2) such that $x_1 > 0$, $x_2 - \sin(1/x_1) = 0$, is an embedded submanifold of $\mathbb{R}^2$. ■

The submanifold $\mathcal{X}$ will be called a *closed submanifold* of $\mathcal{M}$ if the subset $\mathcal{X}$ is closed in $\mathcal{M}$. This is equivalent to saying that each point of $\mathcal{M}$ (and not just of $\mathcal{X}$) has an open neighborhood in which there are ν functions with the properties (I.3.2) and (I.3.3).

An *immersed submanifold* of $\mathcal{M}$ is a pair $(\mathcal{X},f)$ consisting of a $\mathcal{C}^\infty$ manifold $\mathcal{X}$ and an immersion $f : \mathcal{X} \to \mathcal{M}$. Often (especially when the map f is injective) the name of immersed submanifold is given to the image $f(\mathcal{X}) \subset \mathcal{M}$. The number dim $\mathcal{M}$ − dim $\mathcal{X}$ is called the *codimension* of $\mathcal{X}$ and is denoted by codim $\mathcal{X}$.

Example: The *cloverleaf* curve is an immersed submanifold of $\mathbb{R}^2$: it is the image of the circle (in which the angular variable is θ) under the map $\theta \to (\cos^3\theta - \cos\theta \sin^2\theta, \cos^2\theta \sin\theta - \sin^3\theta)$. ■

Let $(\mathcal{X},f)$ be an immersed submanifold of $\mathcal{M}$. It follows from the implicit function theorem that *every point* $x \in \mathcal{X}$ *has an open neighborhood* U *such that* $f|_U$ *is a diffeomorphism of* U *onto an embedded submanifold of* $\mathcal{X}$.

An immersion $f : \mathcal{X} \to \mathcal{M}$ is called an *embedding* if there is an embedded submanifold $\mathcal{Y}$ of $\mathcal{M}$ such that f is a diffeomorphism of $\mathcal{X}$ onto $\mathcal{Y}$ (equipped with the manifold structure inherited from $\mathcal{M}$).

Even when the immersion f is injective it need not be a homeomorphism and the image $f(\mathscr{X})$ might not be an embedded submanifold. (We shall sometimes refer to the image of an injective immersion as *an immersed submanifold without self-intersection.*)

Example: It is easy to construct an injective immersion $f: \mathbb{R}^1 \to \mathbb{R}^2$, which coincides with the map $t \to (1/t, \sin t)$ for $t > 2\pi$ and with $t \to (0,t)$ for $t < 1$. In the neighborhood of any point in the segment $(0,t)$, $|t| \leq 1$, the image of f cannot be defined by the vanishing of any function whose differential is $\neq 0$. ■

For submanifolds,

$$\textit{closed} \Rightarrow \textit{embedded} \Rightarrow \textit{immersed}.$$

The first entailment is trivial. The second one is derived by providing the embedded submanifold $\mathscr{X}$ with the manifold structure induced by the ambient manifold $\mathcal{M}$ and then regarding it as the image of the natural embedding of $\mathscr{X}$ into $\mathcal{M}$.

For $x \in \mathscr{X}$ denote by $\mathrm{Ker}_{\mathbb{R}}\, {}^t\mathrm{D}f(x)$ the null space of ${}^t\mathrm{D}f(x)$, i.e., the orthogonal of $\mathrm{D}f(x)(\mathrm{T}_x\mathscr{X})$, in the real cotangent bundle $\mathrm{T}^*_{f(x)}\mathcal{M}$, and by $\mathrm{Ker}_{\mathbb{C}}\, {}^t\mathrm{D}f(x)$ the null space of ${}^t\mathrm{D}f(x)$ *in* $\mathbb{C}\mathrm{T}^*_{f(x)}\mathcal{M}$. The real (resp., complex) dimension of $\mathrm{Ker}_{\mathbb{R}}\, {}^t\mathrm{D}f(x)$ (resp., $\mathrm{Ker}_{\mathbb{C}}\, {}^t\mathrm{D}f(x)$) is equal to codim $\mathscr{X}$. As x ranges over $\mathscr{X}$ the vector spaces $\mathrm{Ker}_{\mathbb{R}}\, {}^t\mathrm{D}f(x)$ (resp., $\mathrm{Ker}_{\mathbb{C}}\, {}^t\mathrm{D}f(x)$) make up a real (resp., complex) vector bundle over $\mathscr{X}$, called the *conormal bundle* (resp., the *complex conormal bundle*) *of* $\mathscr{X}$ in $\mathcal{M}$; we shall denote it by $\mathrm{N}^*_f\mathscr{X}$ (resp., $\mathbb{C}\mathrm{N}^*_f\mathscr{X}$) or simply by $\mathrm{N}^*\mathscr{X}$ (resp., $\mathbb{C}\mathrm{N}^*\mathscr{X}$) when there is no risk of confusion. Of course $\mathrm{N}^*\mathscr{X}$ may, and shall, be regarded as a subbundle of $\mathbb{C}\mathrm{N}^*\mathscr{X}$.

When $\mathscr{X}$ is an embedded submanifold of $\mathcal{M}$ we may apply what precedes, taking f to be the natural embedding of $\mathscr{X}$ into $\mathcal{M}$. Then, for any $x \in \mathscr{X}$, $\mathbb{C}\mathrm{T}_x\mathscr{X}$ is regarded as a vector subspace of $\mathbb{C}\mathrm{T}_x\mathcal{M}$ and $\mathbb{C}\mathrm{N}^*_x\mathscr{X}$ is identified to the orthogonal of this subspace in $\mathbb{C}\mathrm{T}^*_x\mathcal{M}$ (and the same is true with $\mathbb{R}$ substituted for $\mathbb{C}$).

In what follows, whenever we refer to the formally integrable structure of the manifold $\mathcal{M}$, we mean the structure defined by the vector bundles $\mathcal{V}$ and T'.

Let f be a $\mathscr{C}^\infty$ mapping of a manifold $\mathscr{X}$ into $\mathcal{M}$. We call $\mathcal{V}_f$ the orthogonal of $f^*\mathrm{T}'$ for the duality between tangent and cotangent vectors on $\mathscr{X}$. We have

$$\forall\, x \in \mathscr{X},\ \mathrm{D}f(x)(\mathcal{V}_f)_x = \mathcal{V}_{f(x)} \cap \mathrm{D}f(x)(\mathrm{T}_x\mathscr{X}). \tag{I.3.4}$$

PROPOSITION I.3.1. *Let f be a $\mathscr{C}^\infty$ map of the manifold $\mathscr{X}$ into $\mathcal{M}$. The following conditions are equivalent:*

(i) $f^*\mathrm{T}'$ and $\mathcal{V}_f$ are complex vector bundles over $\mathscr{X}$;
(ii) when x varies in $\mathscr{X}$, the dimension of $\mathrm{T}'_{f(x)} \cap \mathrm{Ker}_{\mathbb{C}}\, {}^t\mathrm{D}f(x)$ is constant.

DEFINITION I.3.1. *We say that the* $\mathcal{C}^\infty$ *map* $f: \mathcal{X} \to \mathcal{M}$ *is* compatible *with the formally integrable structure defined by* $\mathcal{V}$ *and* T' *on* $\mathcal{M}$ *if conditions (i) and (ii) in Proposition* I.3.1 *are satisfied.*

We say that an immersed submanifold $(\mathcal{X},f)$ *of* $\mathcal{M}$ *is compatible with the formally integrable structure of* $\mathcal{M}$ *if this is true of the immersion* f.

When we identify the immersed submanifold to its image under f and use the notation $\mathcal{X}$ to mean this image, we write $T'_{\mathcal{X}}$ for the pullback f^*T' and $\mathcal{V}_{\mathcal{X}}$ for the orthogonal $\mathcal{V}_f$. When $\mathcal{X}$ is an embedded submanifold of $\mathcal{M}$, one should be careful to distinguish between $\mathcal{V}_{\mathcal{X}}$ and $\mathcal{V}|_{\mathcal{X}}$. According to (I.3.4), $\mathcal{V}_{\mathcal{X}}$ is the set of pairs $(x,\mathbf{v}) \in \mathcal{V}$ such that $\mathbf{v}$ is tangent to $\mathcal{X}$ at x.

EXAMPLE I.3.1. Let $\mathcal{M}$ be a complex manifold (see end of section I.2). Any immersed submanifold (Σ,f) of $\mathcal{M}$ whose codimension is equal to *one* is compatible with the complex structure of $\mathcal{M}$ (cf. Example I.2.5).

Indeed, if $z \in \Sigma$ the subspace $Df(z)\mathbb{C}T_z\Sigma$ of $\mathbb{C}T_{f(z)}\mathcal{M} = \mathcal{V}_{f(z)}\oplus\overline{\mathcal{V}}_{f(z)}$ is a hyperplane whose intersection with $\mathcal{V}_{f(z)}$ cannot be equal to $\mathcal{V}_{f(z)}$: if it were, its intersection with $\overline{\mathcal{V}}_{f(z)}$ would be equal to $\overline{\mathcal{V}}_{f(z)}$ and we would have $\dim \Sigma = \dim\mathcal{M}$, contrary to the hypothesis. The dimension of this intersection must therefore be equal to $\dim \mathcal{V}_{f(z)} - 1$. ∎

PROPOSITION I.3.2. *If the immersed submanifold* $\mathcal{X}$ *of* $\mathcal{M}$ *is compatible with the formally integrable structure of* $\mathcal{M}$, *the vector bundles* $\mathcal{V}_{\mathcal{X}}$ *and* $T'_{\mathcal{X}}$ *define a formally integrable structure on* $\mathcal{X}$.

PROOF. Since the statement is local in $\mathcal{X}$, we may as well assume that $\mathcal{X}$ is an embedded submanifold of $\mathcal{M}$. Let x_0 be an arbitrary point of $\mathcal{X}$ and let λ_1, λ_2 be two sections of $\mathcal{V}_{\mathcal{X}}$ in some open neighborhood U_0 of x_0 in $\mathcal{X}$. Select an open neighborhood U of x_0 in which there exists a smooth basis $L_1,\ldots,L_n$ of $\mathcal{V}$; contract U_0 so that $U_0 \subset U$. There are $\mathcal{C}^\infty$ functions a_{1k}, a_{2k} $(k = 1,\ldots,n)$ in U_0 such that, in this set,

$$\lambda_1 = \sum_{k=1}^{n} a_{1k}L_k,\ \lambda_2 = \sum_{k=1}^{n} a_{2k}L_k. \tag{I.3.5}$$

For each k, extend a_{1k} and a_{2k} arbitrarily as $\mathcal{C}^\infty$ functions in U; (I.3.5) defines two $\mathcal{C}^\infty$ sections of $\mathcal{V}$ in U. We know then that $[\lambda_1,\lambda_2]$ is a section of $\mathcal{V}$ in U; but in U_0 it is tangent to $\mathcal{X}$ and therefore it is a section of $\mathcal{V}_{\mathcal{X}}$. ∎

DEFINITION I.3.2. *Let* $(\mathcal{X},f)$ *be an immersed submanifold of* $\mathcal{M}$ *compatible with the involutive structure of* $\mathcal{M}$. *We shall refer to the involutive structure on* $\mathcal{X}$ *defined by* $\mathcal{V}_{\mathcal{X}}$ *and* $T'_{\mathcal{X}}$ *as the* pullback to $\mathcal{X}$ *of the involutive structure of* $\mathcal{M}$ *defined by* $\mathcal{V}$ *and* T'. *When* $\mathcal{X}$ *is an embedded submanifold of* $\mathcal{M}$ *we shall also call it the involutive structure* induced by, *or* inherited from, *that of* $\mathcal{M}$.

EXAMPLE I.3.2. Let $\mathcal{M}$ be equipped with a real (resp., CR) structure (Definition I.2.3). Any submanifold $\mathcal{X}$ of $\mathcal{M}$ compatible with the involutive structure of $\mathcal{M}$ inherits from it a real (resp., CR) structure. ■

PROPOSITION I.3.3. *Let $(\mathcal{X}, f)$ be an immersed submanifold of $\mathcal{M}$. For each $x \in \mathcal{X}$ the following two properties are equivalent:*

$$ {}^t\mathrm{D}f(x) \textit{ induces an injection of } \mathrm{T}'_{f(x)} \textit{ into } \mathbb{C}\mathrm{T}^*_x\mathcal{X}; \tag{I.3.6}$$

$$\textit{at } f(x),\ \mathbb{C}\mathrm{T}\mathcal{M} = \mathcal{V} + f_*\mathbb{C}\mathrm{T}\mathcal{X}. \tag{I.3.7}$$

The following two properties are equivalent:

$$ {}^t\mathrm{D}f(x) \textit{ induces a surjection of } \mathrm{T}'_{f(x)} \textit{ onto } \mathbb{C}\mathrm{T}^*_x\mathcal{X}; \tag{I.3.8}$$

$$\textit{at } f(x),\ \mathcal{V} \cap f_*\mathbb{C}\mathrm{T}\mathcal{X} = 0. \tag{I.3.9}$$

The following two properties are equivalent:

$$ {}^t\mathrm{D}f(x) \textit{ induces a bijection of } \mathrm{T}'_{f(x)} \textit{ onto } \mathbb{C}\mathrm{T}^*_x\mathcal{X}; \tag{I.3.10}$$

$$\textit{at } f(x)\ \mathbb{C}\mathrm{T}\mathcal{M} = \mathcal{V} \oplus f_*\mathbb{C}\mathrm{T}\mathcal{X}. \tag{I.3.11}$$

Evident.

DEFINITION I.3.3. *We shall say that the immersed submanifold $(\mathcal{X}, f)$ is* strongly noncharacteristic *if* (I.3.6) *and* (I.3.7) *hold at every point x of $\mathcal{X}$;* totally real *if* (I.3.8) *and* (I.3.9) *hold at every point $x \in \mathcal{X}$;* maximally real *if* (I.3.10) *and* (I.3.11) *hold at every point $x \in \mathcal{X}$.*

As before let m denote the fibre dimension (over $\mathbb{C}$) of T' and n that of $\mathcal{V}$ (thus $m + n = \dim \mathcal{M}$). We note that

if $\mathcal{X}$ is strongly noncharacteristic, then $m \leq \dim \mathcal{X}$ and $\operatorname{codim} \mathcal{X} \leq n$;
if $\mathcal{X}$ is totally real, then $\dim \mathcal{X} \leq m$ and $n \leq \operatorname{codim} \mathcal{X}$;
if $\mathcal{X}$ is maximally real, then $\dim \mathcal{X} = m$ and $\operatorname{codim} \mathcal{X} = n$.

Clearly,

maximally real = totally real and strongly noncharacteristic.

PROPOSITION I.3.4. *Let $\mathcal{X}$ be an embedded submanifold of $\mathcal{M}$. Then, for each $x \in \mathcal{X}$, Conditions* (I.3.6) *and* (I.3.7) *are equivalent to the following one:*

$$\mathrm{T}'_x \cap \mathbb{C}\mathrm{N}^*_x\mathcal{X} = 0. \tag{I.3.12}$$

Conditions (I.3.8) *and* (I.3.9) *are equivalent to*

$$\mathbb{C}\mathrm{T}^*_x\mathcal{M} = \mathbb{C}\mathrm{N}^*_x\mathcal{X} + \mathrm{T}'_x. \tag{I.3.13}$$

Conditions (I.3.10) *and* (I.3.11) *are equivalent to*

$$\mathbb{C}T^*_x\mathcal{M} = \mathbb{C}N^*_x\mathcal{X} \oplus T'_x. \tag{I.3.14}$$

Evident.

If a submanifold is a union of strongly noncharacteristic submanifolds, it is strongly noncharacteristic. If a submanifold is contained in a totally real submanifold, it is totally real.

PROPOSITION I.3.5. *If* $(\mathcal{X},f)$ *is a totally real immersed submanifold of* $\mathcal{M}$, *then the pullback to* $\mathcal{X}$ *of the formally integrable structure of* $\mathcal{M}$ *under the map* f *is a real structure.*

PROOF. If $(\mathcal{X},f)$ is totally real, then (I.3.8) holds at each point $x \in \mathcal{X}$, which is to say $f^*T' = \mathbb{C}T^*\mathcal{X}$, whence the assertion, by Definition I.2.3. ∎

The converse of Proposition I.3.5 is not true:

EXAMPLE I.3.3. In the plane $\mathbb{R}^2$ (where the coordinates are called x,y) consider the vector bundle $\mathcal{V}$ spanned by $L = \partial/\partial x - \imath y\partial/\partial y$. The orthogonal T' of $\mathcal{V}$ is spanned by the differential dZ with $Z = ye^{\imath x}$. Take the submanifold $\mathcal{X}$ to be the x-axis. Along $\mathcal{X}$, T' is spanned by dy which also spans $\mathbb{C}N^*\mathcal{X}$. This contradicts Condition (I.3.13). ∎

Strongly Noncharacteristic, Totally Real, and Maximally Real Submanifolds of a Complex Manifold

In this subsection $\mathcal{M}$ will be a complex manifold. We shall now denote by $T^{0,1}$ the tangent structure bundle and by $T'^{1,0}$ the cotangent one. Define the map $\mathcal{R}e$: $\mathbb{C}T\mathcal{M} \to T\mathcal{M}$ in the obvious manner: if $\mathbf{v}$ is a complex tangent vector to $\mathcal{M}$ at a point z, $\mathcal{R}e(z,\mathbf{v}) = (z,\mathcal{R}e\mathbf{v})$. The map $\mathcal{R}e$ induces an isomorphism of $T^{0,1}$ onto $T\mathcal{M}$. Indeed, this map is injective, for if $\mathcal{R}e\mathbf{v} = 0$, then $\mathbf{v} = -\bar{\mathbf{v}}$ and $\mathbf{v} \in T^{0,1} \cap \overline{T}^{0,1} = 0$ (usually $\overline{T}^{0,1}$ is denoted by $T^{1,0}$). On the other hand, the fibre dimensions of $T\mathcal{M}$ and $T^{0,1}$ are the same. This allows us to define the following $\mathbb{R}$-linear map on $T_z\mathcal{M}$:

$$J_z(\mathcal{R}e\mathbf{v}) = \mathcal{R}e(\imath\mathbf{v}),\ \mathbf{v} \in T^{0,1}_z. \tag{I.3.15}$$

We have $J_z^2 = -\,\mathrm{Id}$. As z ranges over $\mathcal{M}$ the maps J_z make up a vector bundle automorphism of $T\mathcal{M}$. It can be interpreted as a tensor of type (1,1) on $\mathcal{M}$, often called the *Nijenhuis tensor*. The map J enables us to write explicitly the inverse of the map $\mathcal{R}e$: $T^{0,1}_z \to T_z\mathcal{M}$. It is given by

$$\mathbf{u} \to \mathbf{v} = \mathbf{u} - \imath(J_z\mathbf{u}). \tag{I.3.16}$$

Conversely, suppose that we are given a smooth tensor of type (1,1) in $\mathcal{M}$ whose associated bundle map $J : T\mathcal{M} \to T\mathcal{M}$ is an automorphism such that $J_z^2 = -\,\mathrm{Id}$ for each $z \in \mathcal{M}$. By means of (I.3.16) the map J defines an $\mathbb{R}$-linear injection $T\mathcal{M} \to \mathbb{C}T\mathcal{M}$, whose image is a complex vector subbundle $T^{0,1}$ of $\mathbb{C}T\mathcal{M}$. Unless the Nijenhuis tensor satisfies a certain condition this vector subbundle will *not* be involutive. It will not define what we call, in this book, a complex structure. It defines what is sometimes, and rightly so, called an *almost-complex structure*. (In connection with this it should be recalled that any $\mathbb{R}$-linear map J on a real vector space E, satisfying $J^2 = -\,\mathrm{Id}$, is called a complex structure on E.)

We return to our situation, in which J is associated with a complex structure on $\mathcal{M}$. We now consider a submanifold $\mathcal{X}$ of $\mathcal{M}$.

PROPOSITION I.3.6. *The restriction of the bundle map* $\mathcal{R}e$ *is a bijection of* $T^{0,1}\cap\mathbb{C}T\mathcal{X}$ *onto* $T\mathcal{X}\cap JT\mathcal{X}$.

PROOF. We look at the fibres at a point $z \in \mathcal{X}$. If $\mathbf{u} \in T_z\mathcal{X}\cap J_zT_z\mathcal{X}$ set $\mathbf{u}' = -J_z\mathbf{u}$. Then $\mathbf{v} = \mathbf{u} + \imath\mathbf{u}'$ is the image of $\mathbf{u}$ under the map (I.3.16) and thus $\mathbf{v} \in T_z^{0,1}\cap\mathbb{C}T\mathcal{X}$. Conversely, when the latter is true we have $-\imath\mathbf{v} \in T_z^{0,1}\cap\mathbb{C}T\mathcal{X}$, and both $\mathbf{u} = \mathcal{R}e\,\mathbf{v}$ and $\mathbf{u}' = \mathcal{R}e(-\imath\mathbf{v})$ are tangent to $\mathcal{X}$ at z and $\mathbf{u} = J_z\mathbf{u}'$ belongs to $T_z\mathcal{X}\cap J_zT_z\mathcal{X}$. ∎

COROLLARY I.3.1. *The following two conditions are equivalent:*

$$T\mathcal{X}\cap JT\mathcal{X} = 0; \tag{I.3.17}$$

$$\mathbb{C}T\mathcal{X}\cap T^{0,1}|_{\mathcal{X}} = 0. \tag{I.3.18}$$

COROLLARY I.3.2. *In order that the submanifold* $\mathcal{X}$ *be totally real* (Definition I.3.3) *it is necessary and sufficient that the conditions* (I.3.17) *and* (I.3.18) *hold.*

PROOF. (I.3.18) is equivalent to the validity of (I.3.9) for all $x \in \mathcal{X}$ when we take f to be the natural embedding of $\mathcal{X}$ into $\mathcal{M}$. ∎

Corollary I.3.2 shows that the meaning of "totally real" according to Definition I.3.3 agrees with its customary meaning in complex variable theory. Note that all totally real submanifolds of $\mathcal{M}$ have real dimension $\leq n = \dim_{\mathbb{C}} \mathcal{M}$. When they have dimension equal to n they are *maximally real* (see Definition I.3.3 and following remark).

We come now to the *strongly noncharacteristic* submanifolds of $\mathcal{M}$ (Definition I.3.3). The direct sum decomposition $\mathbb{C}T^*\mathcal{M} = T'^{1,0}\oplus T'^{0,1}$ ($T'^{1,0}$ is the cotangent structure bundle, $T'^{0,1}$ its complex conjugate) yields the decomposition $d = \partial + \bar\partial$ of the exterior derivative: if f is a $\mathcal{C}^\infty$ function, ∂f is the $T'^{1,0}$-component of df, $\bar\partial f$ its $T'^{0,1}$-component.

PROPOSITION I.3.7. *Let $\mathcal{X}$ be a submanifold of $\mathcal{M}$. Assume there are ν real-valued $\mathcal{C}^\infty$ functions $f_1,\dots,f_\nu$ in some open set* U *satisfying* (I.3.2) *and* (I.3.3). *The following two properties are equivalent:*

$$\mathcal{X}\cap U \textit{ is a strongly noncharacteristic submanifold of } \mathcal{M}; \quad \text{(I.3.19)}$$

$$\partial f_1\wedge\cdots\wedge\partial f_\nu \neq 0 \textit{ at every point of } U. \quad \text{(I.3.20)}$$

PROOF. Since the f_j are real-valued, (I.3.20) is equivalent to the property that $\bar\partial f_1\wedge\cdots\wedge\bar\partial f_\nu \neq 0$ in U. The latter means that, in U, $df_1,\dots,df_\nu$ *are* $\mathbb{C}$*-linearly independent modulo* $T'^{1,0}$. Since the df_j span $\mathbb{C}N^*\mathcal{X}$ over U, this simply means that (I.3.12) holds. ■

In the theory of several complex variables, Condition (I.3.20) characterizes what are called the *generic submanifolds*.

If $\mathcal{X}$ is generic, then $\operatorname{codim} \mathcal{X} \leq \dim_{\mathbb{C}} \mathcal{M} \leq \dim \mathcal{X}$.

I.4. Noncharacteristic and Totally Characteristic Submanifolds

We continue to look at the submanifolds of the manifold $\mathcal{M}$ equipped with the involutive structure defined by the bundles $\mathcal{V}$ and T′.

DEFINITION I.4.1. *We say that an immersed submanifold* $(\mathcal{X},f)$ *of* $\mathcal{M}$ *is* non-characteristic *at a point* $x \in \mathcal{X}$ *if the map* $(f(x),\xi) \to (x, {}^t\mathrm{D}f(x)\xi)$ *is an injection of the characteristic set* $T^\circ_{f(x)}$ *into* $T^*_x\mathcal{X}$.

If $(\mathcal{X},f)$ is noncharacteristic at $x \in \mathcal{X}$, then $\dim T^\circ_{f(x)} \leq \dim \mathcal{X}$, and $(\mathcal{X},f)$ is noncharacteristic at all points in a neighborhood of x. We say that $\mathcal{X}$ is *non-characteristic* if it is noncharacteristic at every one of its points. It is possible for $\mathcal{X}$ to be noncharacteristic without being compatible with the formally integrable structure of $\mathcal{M}$ (Definition I.3.1).

PROPOSITION I.4.1. *Let $\mathcal{X}$ be an embedded submanifold of $\mathcal{M}$. The following three conditions are equivalent:*

$$\mathcal{X} \textit{ is noncharacteristic;} \quad \text{(I.4.1)}$$

$$N^*\mathcal{X} \cap (T^\circ|_{\mathcal{X}}) = 0; \quad \text{(I.4.2)}$$

$$T\mathcal{M}|_{\mathcal{X}} = T\mathcal{X} + \mathcal{R}e\mathcal{V}|_{\mathcal{X}}. \quad \text{(I.4.3)}$$

PROOF. In Def. I.4.1 we take f to be the natural embedding of $\mathcal{X}$ into $\mathcal{M}$. Condition (I.4.2) is equivalent to the injectivity of the map $T_x^0 \ni \xi \to \xi|_{T_x\mathcal{X}}$ for all $x \in \mathcal{X}$, hence to (I.4.1). On the other hand $T\mathcal{X}$ is the orthogonal of $N^*\mathcal{X}$ in $T\mathcal{M}|_{\mathcal{X}}$ and T° is that of $\mathcal{R}e\mathcal{V}$ in $T^*\mathcal{M}$, which shows that (I.4.3) is the dual version of (I.4.2). ■

PROPOSITION I.4.2. *If an immersed submanifold* $(\mathcal{X},f)$ *is strongly noncharacteristic* (Definition I.3.3), *then it is noncharacteristic. The converse is true when* codim $\mathcal{X} = 1$.

PROOF. That strongly noncharacteristic ⇒ noncharacteristic is an immediate consequence of Definitions I.3.3 and I.4.1.

Assume codim $\mathcal{X} = 1$. It means that, for each $x \in \mathcal{X}$, the orthogonal of $Df(x)\mathbb{C}T_x\mathcal{X}$, i.e., $\mathrm{Ker}_{\mathbb{C}}\, {}^t Df(x)$, is one-dimensional, and thus is spanned by one of its nonzero *real* elements, ξ. If ${}^tDf(x)$ does not induce an injection of $T'_{f(x)}$ into $\mathbb{C}T_x^*\mathcal{X}$, then $T'_{f(x)}$ must contain $\mathrm{Ker}_{\mathbb{C}}\, {}^tDf(x)$, hence must contain ξ; thus $\xi \in T^\circ_{f(x)}$. This means that the restriction of ${}^tDf(x)$ to $T^\circ_{f(x)}$ is not injective. ■

EXAMPLE I.4.1. Take $\mathcal{M} = \mathbb{C}$ equipped with its standard complex structure. Every submanifold of $\mathcal{M}$ is noncharacteristic. But no zero-dimensional submanifold of $\mathcal{M}$ is strongly noncharacteristic. ■

We shall say that a local chart in $\mathcal{M}$, $(U,x_1,\dots,x_N)$, is *adapted to* a submanifold $\mathcal{X}$ of $\mathcal{M}$ *at a point* $\mathfrak{p}$ if $\mathfrak{p} \in \mathcal{X}\cap U$, if all the coordinates x_j vanish at $\mathfrak{p}$ and if

$$U\cap\mathcal{X} = \{x \in U; x_{\nu+1} = \cdots = x_N = 0\} \tag{I.4.4}$$

(thus dim $\mathcal{X} = \nu$). We shall then systematically write

$$x' = (x_1,\dots,x_\nu),\ x'' = (x_{\nu+1},\dots,x_N).$$

PROPOSITION I.4.3. *Let* $\mathcal{X}$ *be a noncharacteristic submanifold of* $\mathcal{M}$, $(U,x_1,\dots,x_N)$ *any local chart in* $\mathcal{M}$ *adapted to* $\mathcal{X}$ *at one of its points,* $\mathfrak{p}$, h *any distribution solution in* U. *Then there is an open neighborhood* U′ *of the origin in* x'*-space and one,* U″, *in* x''*-space, such that* $U' \times U'' \subset U$ *and that*

$$\text{h is a $\mathcal{C}^\infty$ function of x'' in U'' valued in the space of distributions of x' in U'.} \tag{I.4.5}$$

The meaning of (I.4.5) is that, whatever the test-function $\chi \in \mathcal{C}_c^\infty(U')$, $x'' \to \int h(x',x'')\chi(x')dx'$ is a $\mathcal{C}^\infty$ function in U″.

PROOF. We may assume that $\mathcal{V}$ is spanned over U by smooth vector fields L_j

($j = 1,\ldots,n$) and that U is a product set $U' \times U''$ with U' and U'' as in (I.4.5). Form the second-order differential operator

$$\Delta_L = \sum_{j=1}^{n} \bar{L}_j L_j.$$

We have

$$\Delta_L = \mathcal{Q}(x,D'') + \mathcal{R}(x,D), \tag{I.4.6}$$

where $\mathcal{Q}(x,D'')$ is a second-order operator that only involves partial differentiations with respect to the variables x'' and $\mathcal{R}(x,D)$ is an operator that does not contain any second-order differentiation $\partial^2/\partial x_j \partial x_k$ with both $j, k > \nu$. Consider a point (x,ξ) in the conormal bundle of $\mathcal{X}$ above $\mathcal{X} \cap U$; we have $x'' = 0$, $\xi' = 0$. Since the principal symbol of Δ_L is given by

$$\sigma(\Delta_L) = \sum_{j=1}^{n} |\sigma(L_j)|^2, \tag{I.4.7}$$

and since by hypothesis (x,ξ) cannot belong to the characteristic set—i.e., to the set of common zeros of the symbols $\sigma(L_j)$—we must have $\sigma(\mathcal{Q})(x,\xi'')|_{x''=0} \neq 0$. This means that, possibly after contracting U'' about the origin, the second-order differential operator $\mathcal{Q}(x,D'')$ acting in the x'' variables (and depending smoothly on the parameter $x' \in U'$) is elliptic in U''. Since $\Delta_L h = 0$ we have

$$\mathcal{Q}(x,D'')h = -\mathcal{R}(x,D)h \tag{I.4.8}$$

in U. We use Equation (I.4.8) to "trade" differentiability with respect to x' for differentiability with respect to x''. More precisely, one can introduce the local Sobolev spaces $H^{p',p''}_{loc}(U)$ of order p' with respect to x' and of order p'' with respect to x''. Possibly after contracting U further we may assume that $h \in H^{p',p''}_{loc}(U)$ for some $p', p'' \in \mathbb{Z}$. This implies that $\mathcal{R}(x,D)h$ belongs to the space $H^{p'-2,p''-1}_{loc}(U)$, and so does $\mathcal{Q}(x,D'')h$ by virtue of (I.4.8); it follows from standard elliptic theory that $h \in H^{p'-2,p''+1}_{loc}(U)$. Repeating this process proves the assertion. ■

COROLLARY I.4.1. *Let $\mathcal{X}$ be a noncharacteristic submanifold of $\mathcal{M}$, h a distribution solution in an open subset Ω of $\mathcal{M}$ containing $\mathcal{X}$. Then the* trace *of h on $\mathcal{X}$ is a well-defined distribution in $\mathcal{X}$.*

PROOF. Let $(U,x_1,\ldots,x_N)$ be a local chart in $\mathcal{M}$ adapted to $\mathcal{X}$ at an arbitrary point $p \in \mathcal{X}$ and let U' and U'' be as in (I.4.5). Define the trace of h on $U \cap \mathcal{X}$ as the distribution

$$\mathcal{C}^\infty_c(U') \ni \chi \to \langle h,\chi\rangle|_{x''=0} = \int h(x',0)\chi(x')dx'. \tag{I.4.9}$$

Let $y_1,\ldots,y_N$ be a different coordinate system in U, also adapted to $\mathcal{X}$ at p. Write $x' = F(y)$, $x'' = G(y)$. After further contracting U about p we may assume that the following holds:

the Jacobian matrix $\partial F/\partial y'$ is nonsingular at every point of U; (I.4.10)

$G(y) = M(y)y''$, *with M a nonsingular $(n-\nu)\times(n-\nu)$ matrix depending smoothly on y.* (I.4.11)

Then

$$\int h(x',0)\chi(x')dx' =$$
$$\int h(F(y',0),0)\chi(F(y',0))|\det(\partial F/\partial y')(y',0)|dy' = \langle h,\chi\rangle|_{y''=0},$$

in accordance with the rules for changes of variables in distribution theory. This shows that, in U′, the definition of the trace of h is independent of the choice of coordinates. As a consequence, the definitions agree on overlaps of neighborhoods of the kind U′ and they give rise to a distribution globally defined in $\mathcal{X}$. ■

COROLLARY I.4.2. *Let $\mathcal{X}$ and h be as in Corollary* I.4.1. *If supp $h \subset \mathcal{X}$, then $h \equiv 0$.*

PROOF. Same notation as in the proof of Corollary I.4.1. Then, given any $\chi \in \mathcal{C}_c^\infty(\mathrm{U}')$, the function $x'' \to \int h(x)\chi(x')dx'$ cannot have its support equal to $\{0\}$ unless it vanishes identically, which implies at once that $h \equiv 0$ in $\mathrm{U}' \times \mathrm{U}''$. By varying the central point p we conclude that $h \equiv 0$ in a full neighborhood of $\mathcal{X}$. ■

REMARK I.4.1. Proposition I.4.3 is a particular case of a more general result, whose proof is essentially the same as that of Proposition I.4.3: in the notation of that proof let h be a distribution in U such that, for each $j = 1,\ldots,n$, $\mathrm{L}_j h$ is a $\mathcal{C}^\infty$ function of x'' in U″ valued in the space of distributions of x' in U′. Then h has the same property, i.e., (I.4.5) holds. ■

DEFINITION I.4.2. *We say that an immersed submanifold $(\mathcal{X},f)$ of $\mathcal{M}$ is* totally characteristic *if, for every $x \in \mathcal{X}$,*

$$\mathcal{R}_e\mathcal{V}_{f(x)} \subset \mathrm{D}f(x)\mathrm{T}_x\mathcal{X}. \tag{I.4.12}$$

EXAMPLE I.4.2. Take $\mathcal{M} = \mathbb{R}^2$ with coordinates x and t, and let $\mathcal{V}$ be spanned by the vector field $\mathrm{L} = \partial/\partial t - \imath x\partial/\partial x$. The t-axis is totally characteristic. ■

PROPOSITION I.4.4. *In order that the immersed submanifold $(\mathcal{X},f)$ be totally* *characteristic it is necessary and sufficient that the following condition hold, for every* $x \in \mathcal{X}$:

$$\mathcal{V}_{f(x)} \subset Df(x)\mathbb{C}T_x\mathcal{X}. \tag{I.4.13}$$

If $\mathcal{X}$ *is an embedded submanifold of* $\mathcal{M}$, *in order that it be totally characteristic it is necessary and sufficient that the following two equivalent conditions hold:*

$$N^*\mathcal{X} \subset T^\circ|_{\mathcal{X}}; \tag{I.4.14}$$

$$\mathbb{C}N^*\mathcal{X} \subset T'|_{\mathcal{X}}. \tag{I.4.15}$$

PROOF. (I.4.13) ⇒ (I.4.12) trivially. Conversely, if we complexify (I.4.12), we get

$$(\mathcal{V} + \overline{\mathcal{V}})_{f(x)} \subset Df(x)\mathbb{C}T_x\mathcal{X},$$

which implies (I.4.13). In the embedded case, it suffices to note that (I.4.12) and (I.4.14) on the one hand, (I.4.13) and (I.4.15) on the other, are equivalent by duality. ■

I.5. Local Representations

As before, let the manifold $\mathcal{M}$ be equipped with the formally integrable structure defined by the bundles T′, $\mathcal{V}$. Throughout the present section we reason near or at a fixed point, to which we refer as the *origin* and which we denote by 0. All local coordinates defined in some open neighborhood of 0 will vanish at 0.

Let us select, in an open neighborhood U of 0, a smooth basis of T′, $\varphi_1,\ldots,\varphi_\nu$, $\varpi_1,\ldots,\varpi_d$ with Properties (I.2.8) and (I.2.9) (in which we take $x_0 = 0$). Let us select 2ν real-valued $\mathcal{C}^\infty$ functions in U, $x_1,\ldots,x_\nu$, $y_1,\ldots,y_\nu$, such that

$$\textit{at the origin},\ dx_i = \mathcal{R}e\,\varphi_i,\ dy_i = \mathcal{I}m\,\varphi_i\ (1 \leq i \leq \nu). \tag{I.5.1}$$

Below we always write $z_j = x_j + \imath y_j$, $\bar{z}_j = x_j - \imath y_j$. From (I.2.8) we derive that, possibly after contracting U about 0, we will have

$$\varphi_1\wedge\cdots\wedge\varphi_\nu\wedge\varpi_1\wedge\cdots\wedge\varpi_d\wedge d\bar{z}_1\cdots\wedge d\bar{z}_\nu \neq 0 \textit{ in } U. \tag{I.5.2}$$

After further contracting U we may also select $n-\nu$ real-valued $\mathcal{C}^\infty$ functions in U, which we denote by $t_1,\ldots,t_{n-\nu}$, and which are such that

$$\varphi_1,\ldots,\varphi_\nu,\ \varpi_1,\ldots,\varpi_d,\ d\bar{z}_1,\ldots,d\bar{z}_\nu,\ dt_1,\ldots,dt_{n-\nu}\ \textit{form a basis of}\ \mathbb{C}T^*\mathcal{M}\ \textit{over}\ U. \tag{I.5.3}$$

In the open set U the formally integrable structure is real when $\nu = 0$ and the forms ϖ_k are real; it is a complex structure if $\nu = n$, $d = 0$; it is elliptic if $d = 0$; it is a CR structure if $n = \nu$ (Definition I.2.3).

Next we introduce the basis of $\mathbb{C}T\mathcal{M}$ over U that is dual to the basis of $\mathbb{C}T^*\mathcal{M}$ in (I.5.3). We shall call L_j ($j = 1,\ldots,\nu$) the vector fields such that

$$\langle \varphi_i, L_j \rangle = \langle \varpi_k, L_j \rangle = 0, \, i, j = 1,\ldots,\nu, \, k = 1,\ldots,d; \tag{I.5.4}$$

$$L_j\bar{z}_i = \delta_{ij}, \, L_j t_\ell = 0, \, i, j = 1,\ldots,\nu, \, \ell = 1,\ldots,n-\nu, \tag{I.5.5}$$

where δ_{ij} is the Kronecker index. For $j = \nu+1,\ldots,n$, L_j will be the vector field that satisfies (I.5.4) but that, instead of satisfying (I.5.5), satisfies

$$L_j\bar{z}_i = 0, \, L_j t_\ell = \delta_{(j-\nu)\ell}, \, i = 1,\ldots,\nu, \, \ell = 1,\ldots,n-\nu. \tag{I.5.6}$$

In view of (I.5.4), the vector fields $L_1,\ldots,L_n$ make up a smooth basis of $\mathcal{V}$ over U.

We define m vector fields M_i by the relations below. If $i = 1,\ldots,\nu$, we require

$$\langle \varphi_h, M_i \rangle = \delta_{hi}, \, \langle \varpi_k, M_i \rangle = 0, \, h = 1,\ldots,\nu, \, k = 1,\ldots,d; \tag{I.5.7}$$

$$M_i\bar{z}_h = M_i t_\ell = 0, \, h = 1,\ldots,\nu, \, \ell = 1,\ldots,n-\nu. \tag{I.5.8}$$

If $i = \nu+1,\ldots,m$, we still require that (I.5.8) hold, but (I.5.7) must be replaced by

$$\langle \varphi_h, M_i \rangle = 0, \, \langle \varpi_k, M_i \rangle = \delta_{(i-\nu)k}, \, h = 1,\ldots,\nu, \, k = 1,\ldots,d. \tag{I.5.9}$$

It follows from (I.2.9), (I.5.2), and from the "orthonormality relations" (I.5.4) to (I.5.9) that

$$L_1,\ldots,L_n, \bar{L}_1,\ldots,\bar{L}_\nu \textit{ are linearly independent at the origin;} \tag{I.5.10}$$

$$L_{\nu+1},\ldots,L_n, M_{\nu+1},\ldots,M_m \textit{ are real at the origin.} \tag{I.5.11}$$

Let us assume that the differential forms $\varpi_1,\ldots,\varpi_d$ are real throughout U. In passing note that this is equivalent to saying that *over* U, *the characteristic set* $T°$ *is a vector bundle of rank d*. In this case $M_j = \bar{L}_j$ for each $j = 1,\ldots,\nu$. Indeed, complex conjugation transforms the relations (I.5.4)–(I.5.5) into the relations (I.5.7)–(I.5.8) with $\bar{L}$ in the place of M.

Returning to the general case we may state:

PROPOSTION I.5.1. *We have*

$$[L_j, L_k] = 0, \, \forall \, j, k = 1,\ldots,n. \tag{I.5.12}$$

PROOF. By (I.5.3) and (I.5.4) we have

$$[L_j, L_k]\bar{z}_i = [L_j, L_k]\, t_\ell = 0, \, i = 1,\ldots,\nu, \, \ell = 1,\ldots,n-\nu. \tag{I.5.13}$$

Since $[L_j, L_k]$ is a section of $\mathcal{V}$, it is orthogonal to the forms φ_i and ϖ_k and therefore, by (I.5.13), it must vanish identically. ■

Next we introduce d real functions s_k, defined and $\mathcal{C}^\infty$ in U, such that,

$$\textit{at the origin, } \varpi_k = ds_k \; (k = 1,\dots,d). \tag{I.5.14}$$

A consequence of (I.5.14) is that $ds_1,\dots,ds_d$ span the characteristic set at the origin, T_0°. After once again contracting U about 0 we can assert that

$$x_1,\dots,x_\nu,y_1,\dots,y_\nu,s_1,\dots,s_d,t_1,\dots,t_{n-\nu} \textit{ form a coordinate system in } U. \tag{I.5.15}$$

Now, according to (I.5.8) we have

$$[M_h, M_i]\bar{z}_j = [M_h, M_i]t_\ell = 0,$$
$$h, i = 1,\dots,m, j = 1,\dots,\nu, \ell = 1,\dots,n-\nu. \tag{I.5.16}$$

It follows from this that the vector fields $\tilde{M}_j = \partial/\partial z_j$ $(j = 1,\dots,\nu)$ and $\tilde{M}_{\nu+k} = \partial/\partial s_k$ $(k = 1,\dots,d)$ span the same vector subbundle of $\mathbb{C}T\mathcal{M}$ over U as the vector fields M_i. We may then redefine the forms φ_j and ϖ_k so as to satisfy (I.5.7) and (I.5.9) with $\tilde{M}_i$ in the place of M_i. This will have no effect on the conditions (I.5.2). Thus, the orthonormality relations (I.5.4) to (I.5.9) can be preserved while also achieving that

$$[M_h, M_i] = 0, h, i = 1,\dots,m. \tag{I.5.17}$$

We have, however, the following result.

PROPOSITION I.5.2. *The following properties are equivalent:*

There is a $\mathcal{C}^\infty$ *basis of* $\mathbb{C}T\mathcal{M}$ *over an open neighborhood* U *of* 0, $\{L_1,\dots,L_n, M_1,\dots, M_m\}$, *consisting of pairwise commuting vector fields, and such that* $\{L_1,\dots,L_n\}$ *is a basis of* $\mathcal{V}$ *over* U; (I.5.18)

T′ *is spanned by* $\mathcal{C}^\infty$ *closed forms over an open neighborhood* U *of* 0. (I.5.19)

One can rephrase (I.5.19) by saying that the involutive structure of $\mathcal{M}$ *is locally integrable in the neighborhood* U (Definition I.1.2).

PROOF. Let $\{L_1,\dots,L_n, M_1,\dots,M_m\}$ be a smooth basis of $\mathbb{C}T\mathcal{M}$ over U such that $\{L_1,\dots,L_n\}$ is a basis of $\mathcal{V}$ over U. Let $\varpi_1,\dots,\varpi_m$, $\gamma_1,\dots,\gamma_n$ be the (smooth) differential forms in U defined by the conditions:

$$\langle \varpi_i, L_j\rangle = 0, \langle \varpi_i, M_h\rangle = \delta_{ih},$$
$$\langle \gamma_j, L_k\rangle = \delta_{jk}, \langle \gamma_j, M_h\rangle = 0,$$
$$i, h = 1,\dots,m, j, k = 1,\dots,n. \tag{I.5.20}$$

Clearly $\{\varpi_1,\ldots,\varpi_m\}$ is a basis of T′ over U. If we apply (I.1.13) with φ any one of the forms ϖ_i or γ_j, $\mathbf{v}_1$, $\mathbf{v}_2$ any one of the vector fields L_j or M_i, we get (I.1.14).

If all these vector fields commute pairwise, we conclude that $d\varphi$ is orthogonal to all the exterior products $L_j \wedge L_k$, $L_j \wedge M_h$, $M_h \wedge M_i$. The latter make up a basis of the second exterior power of $\mathbb{C}T\mathcal{M}$ over U, hence all the forms ϖ_i and γ_j must be closed. Conversely, if they are closed, (I.1.14) shows that all the brackets of the vector fields, L_j, M_i are orthogonal to all the forms ϖ_i and γ_j, hence must vanish identically in U. ∎

We return to the general case. We may write

$$\varphi_j = \sum_{j'=1}^{\nu} (A_{jj'}dz_{j'} + B_{jj'}d\bar{z}_{j'}) + \sum_{k'=1}^{d} C_{jk'}ds_{k'} + \sum_{\ell=1}^{n-\nu} C'_{j\ell}dt_\ell,$$

$$\varpi_k = \sum_{j'=1}^{\nu} (E_{kj'}dz_{j'} + F_{kj'}d\bar{z}_{j'}) + \sum_{k'=1}^{d} G_{kk'}ds_{k'} + \sum_{\ell=1}^{n-\nu} G'_{k\ell}dt_\ell.$$

Then (I.5.1) and (I.5.14) show that, *at the origin*,

$$A_{jj'} = \delta_{jj'},\ G_{kk'} = \delta_{kk'},$$
$$B_{jj'} = C_{jk} = C'_{j\ell} = E_{kj'} = F_{kj'} = G'_{k\ell} = 0.$$

As a consequence, we may effect a linear substitution of the forms φ_j and ϖ_k so as to achieve

$$\varphi_j = dz_j + \sum_{j'=1}^{\nu} b_{jj'}d\bar{z}_{j'} + \sum_{\ell=1}^{n-\nu} c_{j\ell}dt_\ell,\ j = 1,\ldots,\nu; \tag{I.5.21}$$

$$\varpi_k = ds_k + \sum_{j'=1}^{\nu} b_{(\nu+k)j'}d\bar{z}_{j'} + \sum_{\ell=1}^{n-\nu} c_{(\nu+k)\ell}dt_\ell,\ k = 1,\ldots,d, \tag{I.5.22}$$

with

$$b_{ij}|_0 = c_{i\ell}|_0 = 0,\ i = 1,\ldots,m,\ j = 1,\ldots,\nu,\ \ell = 1,\ldots,n-\nu. \tag{I.5.23}$$

When combined with the orthonormality relations (I.5.4) to (I.5.6) the expressions (I.5.21), (I.5.22) of the basic forms φ_j, ϖ_k yield at once expressions of the vector fields L_j:

$$L_j = \partial/\partial\bar{z}_j - \sum_{j'=1}^{\nu} b_{j'j}\partial/\partial z_{j'} - \sum_{k'=1}^{d} b_{(\nu+k')j}\partial/\partial s_{k'},\ j = 1,\ldots,\nu; \tag{I.5.24}$$

$$L_{\nu+\ell} = \partial/\partial t_\ell - \sum_{j'=1}^{\nu} c_{j'\ell}\partial/\partial z_{j'} - \sum_{k'=1}^{d} c_{(\nu+k')\ell}\partial/\partial s_{k'},\ \ell = 1,\ldots,n-\nu. \tag{I.5.25}$$

By the relations (I.5.7) to (I.5.9), and by inspection of (I.5.21) and (I.5.22), we see that $M_j = \partial/\partial z_j$ for $j = 1,\ldots,\nu$, and $M_{\nu+k} = \partial/\partial s_k$ for $k = 1,\ldots,d$.

In view of (I.5.3) we see that the pullbacks to any submanifold $t = const.$ of U of the forms $\varphi_1,\ldots,\varphi_\nu$, $\varpi_1,\ldots,\varpi_d$, $d\bar{z}_1,\ldots,d\bar{z}_\nu$ span the whole cotangent bundle to that submanifold. It is also obvious, say by (I.5.21), that each $d\bar{z}_j$ belongs to the span of the pullbacks (to $t = const.$) of $\varphi_1,\ldots,\varphi_\nu$, $\bar{\varphi}_1,\ldots,\bar{\varphi}_\nu$. We conclude:

PROPOSITION I.5.3. *The involutive structure of $\mathcal{M}$ induces on each submanifold $t = const.$ of* U *a CR structure.*

Thus we see that the manifold $\mathcal{M}$ is locally fibred by CR submanifolds. Let us take a closer look at the structure induced on a submanifold $t = const.$ Regard

$$\bar{\gamma}_{kj} = -b_{(\nu+k)j} + \sum_{j'=1}^{\nu} \gamma_{kj'} b_{j'j},\ j = 1,\ldots,\nu,\ k = 1,\ldots,d, \qquad \text{(I.5.26)}$$

as an $\mathbb{R}$-linear system of equations in the unknowns $\mathcal{R}e\,\gamma_{kj}$ and $\mathcal{I}m\,\gamma_{kj}$. It has a unique solution provided the entries $b_{j'j}$ are sufficiently small. We may assume this to be the case thanks to (I.5.23). Notice that

$$\gamma_{kj}|_0 = 0. \qquad \text{(I.5.27)}$$

If then we set

$$\theta_k = \varpi_k - \sum_{j=1}^{\nu} \gamma_{kj} dz_j,$$

we check readily that

the pullbacks to any submanifold $t = const.$ of the forms θ_k $(k = 1,\ldots,d)$ are real. (I.5.28)

In other words, throughout U, the differential forms θ_k are real modulo $dt_1,\ldots,dt_{n-\nu}$. Among other things this shows that the jumps, over U, in the fibre dimension of the characteristic set T° over U originate with the variables t_ℓ.

Let us now take U to be a product set

$$\mathrm{U} = \mathcal{A}\times\mathcal{B}\times\mathcal{C} \qquad \text{(I.5.29)}$$

where $\mathcal{A}$, $\mathcal{B}$, and $\mathcal{C}$ are open neighborhoods of the origin in z- space $\mathbb{C}^\nu$, in s-space $\mathbb{R}^d$ and in t-space $\mathbb{R}^{n-\nu}$ respectively. Let $\mathcal{X}_0$ denote a maximally real submanifold of $\mathcal{A}$; this means that $\mathcal{X}_0$ is a totally real submanifold of $\mathcal{A}$ of real dimension ν. Then the pullbacks to $\mathcal{X}_0 \cong \mathcal{X}_0\times\{s_0\}\times\{t_0\}$ of the forms $\varphi_1,\ldots,\varphi_\nu$ span the whole complex cotangent bundle of $\mathcal{X}_0$. It follows at once from this that the pullbacks of the forms $\varphi_1,\ldots,\varphi_\nu$, $\varpi_1,\ldots,\varpi_d$ to

$$X_{t_0} = \mathcal{X}_0\times\mathcal{B}\times\{t_0\} \qquad \text{(I.5.30)}$$

form a basis of $\mathbb{C}T^*X_{t_0}$. We reach the following conclusion:

PROPOSITION I.5.4. *Whatever* $t_0 \in \mathscr{C}$ *the submanifold* X_{t_0} *of* U *is maximally real.*

In particular, for any $(\alpha,\beta) \in \mathbb{R}^2\backslash\{0\}$, the submanifolds

$$\{ (z,s,t) \in U;\ \alpha x+\beta y = 0,\ t = t_0 \} \tag{I.5.31}$$

are maximally real. It is clear, that, unless $n = \nu = 0$, there pass through 0 (which, we recall, is an arbitrary point of $\mathcal{M}$) an infinity of maximally real submanifolds: it is obvious when $\nu > 0$. When $\nu < n$ we can obviously modify our choice of the coordinates t_ℓ.

I.6. The Associated Differential Complex

As before let T′ denote the cotangent structure bundle of the formally integrable structure on $\mathcal{M}$. We recall that the dimension of $\mathcal{M}$ is equal to $N = m + n$. Denote by $\Lambda^k\mathbb{C}T^*\mathcal{M}$ the k-th exterior power of $\mathbb{C}T^*\mathcal{M}$ ($k = 0,1,\ldots$). For any pair of integers $p,\ q \geq 0$ we define the vector subbundle of $\Lambda^{p+q}\mathbb{C}T^*\mathcal{M}$,

$$T'^{p,q},$$

as consisting of those elements (x,ϖ) ($x \in \mathcal{M}$) whose fibre component ϖ is a sum of exterior products $\zeta_1\wedge\cdots\wedge\zeta_{p+q}$ in which at least p of the factors $\zeta_j \in \mathbb{C}T^*_x\mathcal{M}$ belong to T'_x. In other words, in the complex exterior algebra

$$\Lambda\mathbb{C}T^*\mathcal{M} = \bigoplus_{k=0}^{N} \Lambda^k\mathbb{C}T^*\mathcal{M},$$

$T'^{p,q}$ is the homogeneous part of degree $p + q$ in the *ideal* generated by the p-th exterior power of T′, Λ^pT'.

We have:

$$T'^{p+1,q-1} \subset T'^{p,q}, \tag{I.6.1}$$

$$T'^{p,q} = 0 \textit{ if either } p > m \textit{ or } p+q > m+n, \tag{I.6.2}$$

$$T'^{p+1,q-1} = T'^{p,q} \textit{ if } q > n, \tag{I.6.3}$$

$$T'^{1,0} = T',\ T'^{p,0} = \Lambda^pT', \tag{I.6.4}$$

$$T'^{0,q} = \Lambda^q\mathbb{C}T^*\mathcal{M}. \tag{I.6.5}$$

The inclusion (I.6.1) allows us to define the quotient vector bundle

$$\Lambda^{p,q} = T'^{p,q}/T'^{p+1,q-1}, \tag{I.6.6}$$

it being understood that $T'^{p,q} = 0$ if either p or q are < 0. Of course, $\Lambda^{0,0} \cong \mathcal{M} \times \mathbb{C}$; and

$$\Lambda^{p,0} \cong T'^{p,0}. \tag{I.6.7}$$

We derive from (I.6.2) and (I.6.3):

$$\Lambda^{m,q} = T'^{m,q}, \tag{I.6.8}$$

$$\Lambda^{p,q} = 0 \textit{ if either } p > m \textit{ or } q > n. \tag{I.6.9}$$

Let $(x,\dot{\varpi})$ be an arbitrary element of $\Lambda^{p,q}$ and let

$$\varpi = \sum_{|I|=p} \sum_{|J|=q} \zeta_{i_1} \wedge \cdots \wedge \zeta_{i_p} \wedge \tau_{j_1} \wedge \cdots \wedge \tau_{j_q}$$

be a representative of $\dot{\varpi}$ in $T_x'^{p,q}$. We use the *multi-index notation*, also systematically used in the sequel: I stands for the multi-index $\{i_1,\ldots,i_p\}$; $1 \leq i_1 < \cdots < i_p \leq m + n$; $|I|$ stands for the *length* of I, here equal to p. We take here $\zeta_i \in T_x'$ for all $i \in I$. Let $\tau \to \tau^{\#}$ denote the quotient map $\mathbb{C}T^*\mathcal{M} \to \mathbb{C}T^*\mathcal{M}/T'$. Note that $\mathbb{C}T^*\mathcal{M}/T' \cong \mathcal{V}^*$, the *dual* vector bundle of $\mathcal{V}$. To $\dot{\varpi}$ we may assign the element

$$\varpi^{\#} = \sum_{|I|=p} \sum_{|J|=q} (\zeta_{i_1} \wedge \cdots \wedge \zeta_{i_p}) \otimes (\tau^{\#}_{j_1} \wedge \cdots \wedge \tau^{\#}_{j_q})$$

of $\Lambda^p T_x' \otimes \Lambda^q(\mathbb{C}T^*\mathcal{M}/T')$. If we add to any τ_j ($j \in J$) an element of T_x' it has no effect either on $\dot{\varpi}$ or on $\varpi^{\#}$, which shows that the map $\dot{\varpi} \to \varpi^{\#}$ is "canonical." In other words, we have a natural isomorphism

$$\Lambda^{p,q} \cong \Lambda^p T' \otimes \Lambda^q(\mathbb{C}T^*\mathcal{M}/T'). \tag{I.6.10}$$

We derive from this

$$\Lambda^{0,q} \cong \Lambda^q(\mathbb{C}T^*\mathcal{M}/T'), \tag{I.6.11}$$

$$\Lambda^{p,q} \cong \Lambda^{p,0} \otimes \Lambda^{0,q}. \tag{I.6.12}$$

In particular, apply (I.6.12) with $p = m$ and note that $\Lambda^{m,0} \cong \Lambda^m T'$ is a line bundle. It is often called the *canonical bundle* of the involutive structure of $\mathcal{M}$.

PROPOSITION I.6.1. *Let Ω be an open subset of $\mathcal{M}$ in which there is a $\mathcal{C}^\infty$ section σ of the canonical bundle $\Lambda^m T'$ which does not vanish anywhere in Ω. Then, whatever the integer q, $0 \leq q \leq n$, the map*

$$(x,\tau) \to (x,\sigma(x) \wedge \tau) \tag{I.6.13}$$

is an isomorphism of the vector bundle $\Lambda^{0,q}|_\Omega$ onto the vector bundle $\Lambda^{m,q}|_\Omega$.

PROOF. Indeed, any element ϖ of $\Lambda_x^{m,q}$ ($x \in \Omega$) can be written in a unique manner as $\varpi = \sigma(x) \wedge \tau$ with $\tau \in \Lambda_x^{0,q}$. ∎

REMARK I.6.1. The isomorphism (I.6.13) is not "canonical"; it depends on the choice of the nonvanishing section of $\Lambda^{m,0}$. ∎

We are now going to make use of the fact that the cotangent structure bundle T′ is closed (Definition I.1.1). If φ is a smooth section of T′ over an open set $\Omega \subset \mathcal{M}$, its exterior derivative $d\varphi$ is a section of $T'^{1,1}$ over Ω, a property we shall abbreviate by writing

$$dT' \subset T'^{1,1}. \tag{I.6.14}$$

It follows at once from this that if σ is a smooth section of $T'^{p,q}$ over Ω then $d\sigma$ is a section of $T'^{p,q+1}$, i.e.,

$$dT'^{p,q} \subset T'^{p,q+1}. \tag{I.6.15}$$

In the sequel we shall systematically use the following notation: let **E** be a vector bundle over $\mathcal{M}$, Ω an open subset of $\mathcal{M}$; then $\mathcal{C}^r(\Omega;\mathbf{E})$ shall denote the space of $\mathcal{C}^r$ sections of **E** over Ω. This will be valid for any integer $r < +\infty$, as well as for $r = +\infty$. When $\mathcal{M}$ is a real-analytic manifold and **E** is a real-analytic vector bundle over $\mathcal{M}$ the notation will also be used with $r = \omega$: $\mathcal{C}^\omega(\Omega;\mathbf{E})$ will stand for the space of real-analytic sections of **E** over Ω. Also, $\mathcal{C}_c^r(\Omega;\mathbf{E})$ shall denote the subspace of $\mathcal{C}^r(\Omega;\mathbf{E})$ consisting of those sections whose support is compact and contained in Ω ($0 \leq r \leq +\infty$). When **E** is the trivial bundle $\mathbf{E} = \mathcal{M} \times \mathbb{C}$, we write $\mathcal{C}^r(\Omega)$ and $\mathcal{C}_c^r(\Omega)$ for those spaces.

We denote by $\mathcal{D}'(\Omega;\mathbf{E})$ the space of distribution sections of **E**. For the definition of a distribution section of a vector bundle, see Schwartz [1], p. 340. We shall content ourselves with the following remark: suppose the vector bundle **E** admits a $\mathcal{C}^\infty$ basis over Ω, $\{\mathbf{s}_1,\ldots,\mathbf{s}_\rho\}$ (ρ = fibre dimension of **E**); then every distribution section u of **E** over Ω has a unique representation $u = \sum_{j=1}^{\rho} u_j \mathbf{s}_j$ where the u_j are scalar distributions in Ω. By $\mathcal{E}'(\Omega;\mathbf{E})$ we shall mean the space of distribution sections of **E** whose support is compact and contained in Ω.

In accordance with the notation just described, we should call $\mathcal{C}^r(\Omega;\Lambda^p T^*\mathcal{M})$ the space of $\mathcal{C}^r$ *differential forms* of degree p in Ω. We shall however shorten the notation a bit and write $\mathcal{C}^r(\Omega;\Lambda^p)$ instead. Likewise, the space of *p-currents* in Ω shall be denoted by $\mathcal{D}'(\Omega;\Lambda^p)$ (see Schwartz [1], Chap. 9). We also write $\mathcal{C}_c^r(\Omega;\Lambda^p)$, $\mathcal{E}'(\Omega;\Lambda^p)$, and so on.

According to (I.6.15), for each p, $0 \leq p \leq m$, the exterior derivative defines a sequence of linear operators $\mathcal{C}^\infty(\Omega;T'^{p,q}) \to \mathcal{C}^\infty(\Omega;T'^{p,q+1})$ and, by going to the quotients, a sequence of linear operators:

$$d'^{p,q} : \mathscr{C}^\infty(\Omega;\Lambda^{p,q}) \to \mathscr{C}^\infty(\Omega;\Lambda^{p,q+1}) \ (q = 0,1,\ldots). \qquad \text{(I.6.16)}$$

Of course,

$$d'^{p,q+1} \circ d'^{p,q} = 0; \qquad \text{(I.6.17)}$$

moreover, $d'^{p,q}$ *decreases the support of sections*. In other words, for p fixed, the sequence of operators (I.6.16) makes up a *differential complex*.

DEFINITION I.6.1. *We shall refer to the sequence of differential operators* (I.6.16) *as the* differential complex *of degree p over* Ω, *associated with the involutive structure of* $\mathcal{M}$.

We may also deal with the differential complex

$$d'^{p,q} : \mathscr{D}'(\Omega;\Lambda^{p,q}) \to \mathscr{D}'(\Omega;\Lambda^{p,q+1}),\ q = 0,1,\ldots, \qquad \text{(I.6.18)}$$

as well as with the subcomplex

$$d'^{p,q} : \mathscr{C}_c^\infty(\Omega;\Lambda^{p,q}) \to \mathscr{C}_c^\infty(\Omega;\Lambda^{p,q+1}),\ q = 0,1,\ldots, \qquad \text{(I.6.19)}$$

and the one with $\mathscr{E}'$ in the place of $\mathscr{C}_c^\infty$. Often, when there is no risk of confusion, we shall write d' in the place of $d'^{p,q}$.

We shall refer to any element f of *Ker* $d'^{p,q}$ as a *cocycle* of bidegree (p,q); we shall also say that f is d'-*closed*. If $f \in$ *Im* $d'^{p,q-1}$ we shall say that f is a *coboundary* or that f is d'-*exact*.

The cohomology spaces of the above differential complexes are defined in standard fashion: for the complex (I.6.16) we shall use the notation

$$H'^{p,0}(\Omega) = \mathit{Ker}\, d'^{p}, \qquad \text{(I.6.20)}$$

$$H'^{p,q}(\Omega) = \mathit{Ker}\, d'^{p,q}/\mathit{Im}\, d'^{p,q} \ \mathit{if}\ q > 0. \qquad \text{(I.6.21)}$$

Under special circumstances we might be forced to distinguish between the cohomology spaces of the differential complex (I.6.16), to which we may then refer as the $\mathscr{C}^\infty$, or smooth, associated complex, and those of the complex (I.6.18), the distribution associated complex, for instance by writing respectively $\mathscr{C}^\infty H'^{p,q}$ and $\mathscr{D}'H'^{p,q}$ instead of simply $H'^{p,q}$.

REMARK I.6.2. We would like to call the attention of the reader to the case $p = m$. By (I.6.8) we know that

$$\mathscr{C}^\infty(\Omega;\Lambda^{m,q}) \cong \mathscr{C}^\infty(\Omega;T'^{m,q})$$

can be regarded as a space of true differential forms—as opposed to $\mathscr{C}^\infty(\Omega;\Lambda^{p,q})$ with $p < m$, which is a space of *equivalence classes* of differential forms. The operator $d'^{m,q}$ can be equated to the exterior derivative, acting on sections of $T'^{m,q}$.

Let $\psi_1,\ldots,\psi_m$ be a smooth basis of T' over an open neighborhood U of some point of $\mathcal{M}$ consisting of d-closed forms. Then $\psi_1\wedge\cdots\wedge\psi_m$ spans the canonical line bundle $T'^{m,0} \cong \Lambda^m T'$ over U and we have, in U,

$$d(\psi_1\wedge\cdots\wedge\psi_m) = 0. \tag{I.6.22}$$

Let us then identify $\Lambda^{m,q}$ to $\Lambda^m T'\otimes\Lambda^{0,q}$ over U via the map (I.6.13) in which we take $\sigma = \psi_1\wedge\cdots\wedge\psi_m$. Let $x \to \tau(x)$ be a smooth section of $\Lambda^{0,q}$ over U. By (I.6.22) we have

$$d'^{m,q}(\psi_1\wedge\cdots\wedge\psi_m\wedge\tau) = \psi_1\wedge\cdots\wedge\psi_m\wedge d'^{0,q}\tau. \tag{I.6.23}$$

In other words, if $\sigma = \psi_1\wedge\cdots\wedge\psi_m$ the map (I.6.13) induces an isomorphism of the associated differential complexes of degree 0 and of degree m over the neighborhood U. ■

REMARK I.6.3. All that precedes has a *sheaf-theoretic* analogue. We let Ω range over the family of all open subsets of $\mathcal{M}$ and assign to each set Ω the linear space $\mathscr{C}^\infty(\Omega;\Lambda^{p,q})$. To any pair of open sets $\Omega \subset \Omega'$ we assign the restriction mapping $\mathscr{C}^\infty(\Omega';\Lambda^{p,q}) \to \mathscr{C}^\infty(\Omega;\Lambda^{p,q})$. This defines a presheaf. The associated sheaf is the *sheaf of germs* of $\mathscr{C}^\infty$ sections of $\Lambda^{p,q}$, which we shall denote by $\mathcal{C}^\infty(\mathcal{M};\Lambda^{p,q})$. Its stalk at a point $x_0 \in \mathcal{M}$ is the space of germs of $\mathscr{C}^\infty$ sections of $\Lambda^{p,q}$ at x_0. Those germs are equivalence classes of sections, each defined in some neighborhood of x_0, that are equal in some subneighborhood. The operator (I.6.16) defines a linear operator $\mathcal{C}^\infty(\mathcal{M};\Lambda^{p,q}) \to \mathcal{C}^\infty(\mathcal{M};\Lambda^{p,q+1})$ which we shall also denote by d′, or by $d'^{p,q}$ when there is a risk of confusion, and which can also be viewed as a differential operator. The corresponding cohomology space will be denoted by $\mathcal{H}^{p,q}(\mathcal{M})$ or by $\mathcal{C}^\infty\mathcal{H}^{p,q}(\mathcal{M})$.

The analogue is valid for the distribution complex (I.6.18). Indeed, the restriction of currents from Ω' to $\Omega \subset \Omega'$ is well defined, and we may talk of the germs of currents at a point, and define the spaces $\mathcal{D}'(\mathcal{M};\Lambda^{p,q})$, $\mathcal{D}'\mathcal{H}^{p,q}(\mathcal{M})$. ■

EXAMPLE I.6.1. Suppose $T' = 0$. Then $\Lambda^{p,q} = 0$ as soon as $p > 0$, and

$$\Lambda^{0,q} = \Lambda^q\mathbb{C}T^*\mathcal{M},$$

$$d'^{0,q} = \text{exterior derivative acting on } q\text{-forms}.$$

All the differential complexes associated with T′ vanish if their degree is ≥ 1. The complex of degree zero is the *De Rham complex* over $\mathcal{M}$. ■

EXAMPLE I.6.2. Take now $T' = \mathbb{C}T^*\mathcal{M}$. Then $T'^{p,q} = \Lambda^{p+q}\mathbb{C}T^*\mathcal{M}$ and therefore $\Lambda^{p,q} = 0$ as soon as $q \geq 1$, while

$$\Lambda^{p,0} = \Lambda^p\mathbb{C}T^*\mathcal{M},\ p = 0,1,\ldots$$

All the maps $d'^{p,q}$ are zero. In this case $H'^{p,q}(\Omega) = 0$ if $q > 0$, and

$$H'^{p,0}(\Omega) = \mathscr{C}^\infty(\Omega;\Lambda^p). \blacksquare$$

EXAMPLE I.6.3. Let T′ define a complex structure on $\mathcal{M}$ (thus dim $\mathcal{M} = 2m$). If we use the direct sum decomposition $\mathbb{C}T^*\mathcal{M} = T'\otimes\overline{T}'$, we see that there is a natural vector bundle isomorphism

$$\mathbb{C}T^*\mathcal{M}/T' \cong \overline{T}'. \tag{I.6.24}$$

By virtue of (I.6.10) we derive

$$\Lambda^{p,q} \cong \Lambda^p T'\otimes\Lambda^q\overline{T}'. \tag{I.6.25}$$

In other words, we may identify any local section of $\Lambda^{p,q}$ to a linear combination of exterior products

$$\varphi_1\wedge\cdots\wedge\varphi_p\wedge\overline{\psi}_1\wedge\cdots\wedge\overline{\psi}_q$$

where the φ_j and the ψ_k are sections of T′. With this identification the differential operator d′ gets identified to the *Cauchy-Riemann operator* $\bar{\partial}$. The associated complex (I.6.16) will be called the *Cauchy-Riemann complex* (of degree p). It is sometimes called the *Dolbeault complex*.

Note the direct sum decomposition

$$\Lambda^r\mathbb{C}T^*\mathcal{M} = \bigoplus_{p+q=r} \Lambda^{p,q}. \blacksquare \tag{I.6.26}$$

The Vertical Complex

The differential complex (I.6.16) leads to the construction of a second differential complex associated with the formally integrable structure of $\mathcal{M}$. The construction is a standard one in the theory of spectral sequences (see Godement [1], chap. 1, 4.7). Here we shall describe it directly, without invoking general homological algebra.

Let $f \in \mathscr{C}^\infty(\Omega;T'^{p,q})$ be a representative of $\dot{f} \in \mathscr{C}^\infty(\Omega;\Lambda^{p,q})$. (Such a representative always exists if Ω is sufficiently small. In a large open set Ω the local representatives can be patched up together by means of $\mathscr{C}^\infty$ partitions of unity.) To say that $d'\dot{f} \equiv 0$ is equivalent to saying that $df \in \mathscr{C}^\infty(\Omega;T'^{p+1,q})$. In turn df is a representative of an element of $\mathscr{C}^\infty(\Omega;\Lambda^{p+1,q})$, which is obviously a cocycle. Thus the exterior derivative induces a linear map

$$Ker\, d'^{p,q} \to Ker\, d'^{p+1,q}. \tag{I.6.27}$$

Assume now $q \geq 1$ and $\dot{f} = d'\dot{g}$ with $\dot{g} \in \mathscr{C}^\infty(\Omega;\Lambda^{p,q-1})$. If $g \in \mathscr{C}^\infty(\Omega; T'^{p,q-1})$ is a representative of $\dot{g}$, then dg is a representative of $\dot{f}$. Any other representative of $\dot{f}$ is of the kind $f = h + dg$ with $h \in \mathscr{C}^\infty(\Omega;T'^{p+1,q-1})$ so that $df = dh$. Thus

$$(df)^{\boldsymbol{\cdot}} \in Im\, d'^{p+1,q-1}.$$

The exterior derivative induces a linear map

$$Im\, \mathrm{d}'^{p,q-1} \rightarrow Im\, \mathrm{d}'^{p+1,q-1}. \tag{I.6.28}$$

Taking the quotient of Equation (I.6.27) by Equation (I6.28), and letting p vary while keeping q fixed, yields a sequence of linear maps

$$\mathrm{d}''^{p,q} : \mathrm{H}'^{p,q}(\Omega) \rightarrow \mathrm{H}'^{p+1,q}(\Omega),\, p = 0,1,\ldots \tag{I.6.29}$$

When there is no risk of confusion we write d″ rather than $\mathrm{d}''^{p,q}$; we have $\mathrm{d}''^2 = 0$.

DEFINITION I.6.2. *We shall refer to the sequence of differential operators* (I.6.29) *as the* vertical differential complex *of degree q over Ω, associated with the formally integrable structure of $\mathcal{M}$.*

The cohomology spaces of the vertical complex will be denoted by $\mathrm{H}''^{p,q}(\Omega)$. Sometimes, in order to avoid misunderstanding, we shall refer to (I.6.16) as the *horizontal differential complex* of degree q over Ω, associated with the involutive structure of $\mathcal{M}$.

EXAMPLE I.6.4. Take $\mathrm{T}' = 0$ as in Example I.6.1. Then $\mathrm{H}'^{p,q}(\Omega) = 0$ if $p \geq 1$, and $\mathrm{H}'^{0,q}(\Omega)$ is the q-th De Rham cohomology space of $\mathcal{M}$. We see that $\mathrm{d}'' = 0$ for all p, q. ■

EXAMPLE I.6.5. Take $\mathrm{T}' = \mathbb{C}\mathrm{T}^*\mathcal{M}$ as in Example I.6.2. Then, if $q = 0$, (I.6.29) is the De Rham complex and d″ is the exterior derivative.

The cases $\mathrm{T}' = 0$ and $\mathrm{T}' = \mathbb{C}\mathrm{T}^*\mathcal{M}$ are mirror images of each other with respect to the diagonal in $\mathbb{Z}_+ \times \mathbb{Z}_+$ (where (p,q) varies); the vertical complex in one is the horizontal complex in the other, and vice versa. ■

EXAMPLE I.6.6. Let T′ define a complex structure on $\mathcal{M}$, as in Example I.6.3. By availing ourselves of the direct sum decomposition we see that each element of $\mathrm{H}'^{p,0}(\Omega)$ can be identified to a differential form of bidegree $(p,0)$ whose coefficients are holomorphic. If $q \geq 1$ each element of $\mathrm{H}'^{p,q}(\Omega)$ can be identified to an equivalence class of $\bar{\partial}$-closed forms of bidegree (p,q), which differ by a $\bar{\partial}$-exact form. The differential operator d″ can be regarded as the extension of the action of the *anti–Cauchy-Riemann operator* ∂ on such equivalence classes. Indeed, it is checked at once that $\partial\bar{\partial} + \bar{\partial}\partial = 0$ and therefore ∂ transforms $\bar{\partial}$-closed (resp., $\bar{\partial}$-exact) forms into $\bar{\partial}$-closed (resp., $\bar{\partial}$-exact) ones. ■

I.7. Local Representations in Locally Integrable Structures

Throughout this section we assume that $\mathcal{M}$ is equipped with a *locally integrable* structure. We take up, under this new hypothesis, the considerations of section I.5. As before the structure bundles are denoted by T′ and $\mathcal{V}$.

Let $\varphi_1,\ldots,\varphi_\nu,\varpi_1,\ldots,\varpi_d$ be the same smooth basis of T′ over the open neighborhood U of 0, as in section I.5. By hypothesis there are m $\mathcal{C}^\infty$ functions $Z_1,\ldots,Z_m$ whose differentials also make up a basis of T′ over some open neighborhood of 0. (We shall often refer to such functions as "*first integrals*.") We may and shall assume that all the functions Z_i vanish at the origin. There is a nonsingular complex matrix $(\gamma_{jk})_{1\le j,k\le m}$ such that, *at the origin*,

$$\varphi_j = \sum_{i=1}^{m} \gamma_{ji}\, dZ_i,\ \varpi_k = \sum_{i=1}^{m} \gamma_{(\nu+k)i}\, dZ_i,$$
$$j = 1,\ldots,\nu,\ k = 1,\ldots,d.$$

Keeping in mind that the entries γ_{ji} are *constant*, we may substitute $\sum_{i=1}^{m} \gamma_{ji} Z_i$ for Z_j $(j = 1,\ldots,m)$. In other words, we may assume that

$$\varphi_j = dZ_j,\ \varpi_k = dZ_{\nu+k} \textit{ at the origin } (j = 1,\ldots,\nu,\ k = 1,\ldots,d). \tag{I.7.1}$$

This implies that

$$dZ_1\wedge\cdots\wedge dZ_\nu\wedge d\bar{Z}_1\wedge\cdots\wedge d\bar{Z}_\nu\wedge dZ_{\nu+1}\wedge\cdots\wedge dZ_m \neq 0 \tag{I.7.2}$$

in a neighborhood of 0, which we take to be U; and that

$$dZ_{\nu+1},\ldots,dZ_m \textit{ are real at } 0. \tag{I.7.3}$$

But then we may regard

$$x_j = \mathcal{R}e Z_j,\ y_j = \mathcal{I}m\, Z_j\ (j = 1,\ldots,\nu),$$
$$s_k = \mathcal{R}e Z_{\nu+k} (k = 1,\ldots,d),$$

as part of a system of local coordinates in U (possible after having contracted U about 0). We shall denote by $t_1,\ldots,t_{n-\nu}$ the remaining coordinates in that system. Henceforth we write $z_j = x_j + \imath y_j$ instead of Z_j for $1 \le j \le \nu$ and w_k instead of $Z_{\nu+k}$. Note that

$$w_k = s_k + \imath\varphi_k(x,y,s,t),\ k = 1,\ldots,d. \tag{I.7.4}$$

Recall that the functions w_k all vanish at 0. Taking into account (I.7.3) shows that

$$\varphi_k|_0 = 0,\ d\varphi_k|_0 = 0,\ k = 1,\ldots,d. \tag{I.7.5}$$

Of the locally integrable structure of $\mathcal{M}$ restricted to the open set U we may now say the following: it is *real* if and only if $\nu = 0$ and all the functions φ_k ($k = 1,\dots,d$) vanish identically. It is a *complex structure* if and only if $\nu = n$, $d = 0$, i.e., there are no coordinates s_k and t_ℓ, and no functions w_k. It is *elliptic* if and only if $d = 0$, i.e., $\nu = m \leq n$ and there are no w_k; the coordinate system consists only of the functions x_j, y_j and t_ℓ. It is a *CR structure* if and only if $\nu = n \leq m$, i.e., there are no coordinates t_ℓ.

REMARK I.7.1. It ought to be noted that, by virtue of (I.7.5), the differentials $\mathrm{d}w_k$ are real at the origin.

However, even when the characteristic set T° is a vector bundle, and the one-forms $\varpi_1,\dots,\varpi_d$ are real (and span T° over U), it is not true, in general, that one can find real functions whose differentials span T° over a neighborhood of a given point. This is seen on the following:

EXAMPLE I.7.1. Call x, y, s the coordinates in $\mathbb{R}^3$. Equip $\mathbb{R}^3$ with the *Lewy structure*. This is the CR structure defined by the functions $z = x + \imath y$, $w = s + \imath|z|^2$, i.e., the cotangent structure bundle T' is spanned by $\mathrm{d}z$ and $\mathrm{d}w$. The tangent structure bundle is spanned by the *Lewy vector field*:

$$\mathrm{L} = \partial/\partial\bar{z} - \imath z\partial/\partial s. \tag{I.7.6}$$

Set $\overline{\mathrm{L}} = \partial/\partial z + \imath\bar{z}\partial/\partial s$ and notice that we have

$$[\mathrm{L},\overline{\mathrm{L}}] = 2\imath\partial/\partial s. \tag{I.7.7}$$

The characteristic set, which here is a line bundle, is spanned at every point by the single real form

$$\varpi = \mathrm{d}s + \imath(z\mathrm{d}\bar{z} - \bar{z}\mathrm{d}z). \tag{I.7.8}$$

Suppose there were a smooth real-valued function f in some open subset $\Omega \subset \mathbb{R}^3$ such that $\mathrm{d}f$ were also to span T° there. We would necessarily have $\mathrm{L}f \equiv 0$ in Ω, and since f is real, also $\overline{\mathrm{L}}f \equiv 0$. By letting $\overline{\mathrm{L}}$ act on $\mathrm{L}f$ and L on $\overline{\mathrm{L}}f$, and by subtracting, we obtain $[\mathrm{L},\overline{\mathrm{L}}]f \equiv 0$. According to (I.7.6) this would mean that $\partial f/\partial s \equiv 0$ in Ω. But then, again by the equation $\mathrm{L}f \equiv 0$ we conclude that $\partial f/\partial\bar{z} = \partial f/\partial z$ vanishes identically, which means $\mathrm{d}f \equiv 0$, a contradiction. ■

We return to the general case and we define the vector fields L_j by the following relations (cf. (I.5.4) to (I.5.6)):

$$\mathrm{L}_j z_i = \mathrm{L}_j w_k = 0,\quad i = 1,\dots,\nu,\ j = 1,\dots,n,\ k = 1,\dots,d; \tag{I.7.9}$$

$$\mathrm{L}_j \bar{z}_i = \delta_{ij},\ \mathrm{L}_j t_\ell = 0,\quad i,j = 1,\dots,\nu,\ \ell = 1,\dots,n-\nu; \tag{I.7.10}$$

$$L_j \bar{z}_i = 0, \; L_j t_\ell = \delta_{(j-\nu)\ell}, \quad i = 1,\ldots,\nu, \; j = \nu+1,\ldots,n, \; \ell = 1,\ldots,n-\nu. \tag{I.7.11}$$

As for the vector fields denoted by M_h in section 5, we define them now by the following relations (cf. (I.5.7) to (I.5.9)):

$$M_h \bar{z}_i = M_h t_\ell = 0, \quad h = 1,\ldots,m, \; i = 1,\ldots,\nu, \; \ell = 1,\ldots,n-\nu; \tag{I.7.12}$$

$$M_h z_i = \delta_{hi}, \; M_h w_k = 0, \quad h, i = 1,\ldots,\nu, \; k = 1,\ldots,d; \tag{I.7.13}$$

$$M_h z_i = 0, \; M_h w_k = \delta_{(h-\nu)k}, \quad h = \nu+1,\ldots,m, \; i = 1,\ldots,\nu, \; k = 1,\ldots,d. \tag{I.7.14}$$

From there it is easy to derive expressions of these vector fields:

$$L_j = \partial/\partial\bar{z}_j - \imath \sum_{k=1}^{d} (\partial\varphi_k/\partial\bar{z}_j) N_k, \; j = 1,\ldots,\nu; \tag{I.7.15}$$

$$L_{\nu+\ell} = \partial/\partial t_\ell - \imath \sum_{k=1}^{d} (\partial\varphi_k/\partial t_\ell) N_k, \; \ell = 1,\ldots,n-\nu; \tag{I.7.16}$$

$$N_k = \sum_{k'=1}^{d} \mu_{kk'} \partial/\partial s_{k'}, \; k = 1,\ldots,d, \tag{I.7.17}$$

where the $d \times d$ matrix $(\mu_{kk'})_{1 \le k,k' \le d}$ is the inverse of the "Jacobian" matrix $\partial w/\partial s = (\partial w_k/\partial s_{k'})_{1 \le k,k' \le d}$. (Thanks to (I.7.5) the matrix $\partial w/\partial s$ is nonsingular in a suitably small neighborhood of 0—which we take to contain U.) Notice that (I.7.12), (1.7.13) and (I.7.14) imply at once

$$M_h = \partial/\partial z_h - \imath \sum_{k=1}^{d} (\partial\varphi_k/\partial z_h) N_k, \; h = 1,\ldots,\nu; \tag{I.7.18}$$

and $M_{\nu+k} = N_k$ for all $k = 1,\ldots,d$.

PROPOSITION I.7.1. *The vector fields* L_j, M_i, N_k $(j = 1,\ldots,n, i = 1,\ldots,\nu, k = 1,\ldots,d)$ *commute pairwise in* U.

PROOF. By virtue of (I.7.9) to (I.7.14) each commutator of any two of those vector fields annihilates all the functions $z_i, \bar{z}_j, w_k, t_\ell$ $(i,j = 1,\ldots,\nu, k = 1,\ldots,d, \ell = 1,\ldots,n-\nu)$, whose differentials span $\mathbb{C}T^*\mathcal{M}$ at every point of U. ∎

The system of functions $z_j = x_j + iy_j$ $(j = 1,\ldots,\nu)$, $w_k = s_k + i\varphi_k(x,y,s,t)$ $(k = 1,\ldots,d)$ reflects some of the local features of the locally integrable structure on $\mathcal{M}$. The submanifolds $t = const.$ inherit a CR structure from the ambient structure (cf. Proposition I.5.3).

The preimages of points under the map

$$\mathrm{U} \ni (x,y,s,t) \rightarrow (z,w) \in \mathbb{C}^{\nu+d} \tag{I.7.19}$$

are the sets defined by

$$(x,y,s) = const., \ \varphi(x,y,s,t) = const., \tag{I.7.20}$$

where $\varphi = (\varphi_1,\ldots,\varphi_d)$. We shall refer to those preimages as the *fibres* of the map (z,w) in U.

EXAMPLES I.7.2. Recalling that, when the structure of $\mathcal{M}$ is a *CR structure*, we have $\nu = n$ and there are no t variables, we may state:

PROPOSITION I.7.2. *If the structure of $\mathcal{M}$ is a CR structure the fibres of the map (z,w) in* U *consist of single points.* ■

EXAMPLE I.7.3. When the structure of $\mathcal{M}$ is *elliptic* we have d $= 0$; there are no functions w_k and no variables s_k. Thus:

PROPOSITION I.7.3. *If the structure of $\mathcal{M}$ is elliptic the fibres of the map z in* U *consist of the intersections of* U *with the affine subspaces parallel to the t-coordinate subspace.*

Note also that the vector fields L_j and M_h have especially simple expressions:

$$\mathrm{L}_j = \partial/\partial\bar{z}_j, \ \mathrm{L}_{\nu+\ell} = \partial/\partial t_\ell, \ j = 1,\ldots,m, \ \ell = 1,\ldots,n-m; \tag{I.7.21}$$

$$\mathrm{M}_i = \partial/\partial z_i, \ i = 1,\ldots,m. \tag{I.7.22}$$

(Recall that, in an elliptic structure, $\nu = m \leq n$). ■

EXAMPLE I.7.4. When the structure of $\mathcal{M}$ is a *complex structure*, there are no coordinates s_k and t_ℓ, no functions w_k. The structure is defined by the *complex coordinates* z_j ($j = 1,\ldots,m$); the vector fields are given by $\mathrm{L}_j = \partial/\partial\bar{z}_j$, $\mathrm{M}_j = \partial/\partial z_j$. ■

EXAMPLE I.7.5. When the structure of $\mathcal{M}$ is *real* there are no coordinates x_j and y_j; and $\varphi_k \equiv 0$ for all $k = 1,\ldots,d$. The local coordinates are $s_1,\ldots,s_d$, $t_1,\ldots,t_n$ (since $\nu = 0$), and the vector fields are given by $\mathrm{L}_j = \partial/\partial t_j$, $\mathrm{M}_i = \partial/\partial s_i$. The fibres in U are the same as in the elliptic case (when $d = 0$, i.e., when there are no s variables, the structure is both real and elliptic). ■

Often one does not need all the detailed information provided by the functions z_j, w_k. It suffices to deal with a coarser local representation, one in which the first integrals $Z_1,\ldots,Z_m$, whose differentials make up a smooth basis of T′ over U, have the following simple form:

$$Z_j = x_j + \imath\Phi_j(x,t), j = 1,\ldots,m. \tag{I.7.23}$$

The local coordinates in U are denoted, here, by $x_1,\ldots,x_m,t_1,\ldots,t_n$. We may further assume

$$\Phi_j|_0 = 0,\ d_x\Phi_j|_0 = 0, j = 1,\ldots,m. \tag{I.7.24}$$

In (I.7.24) d_x stands for the differential with respect to x only. The functions z_j, w_k above are of the type Z_i in Eqq. (I.7.23) and (I.7.24), as one checks by setting $s_k = x_{\nu+k}$ and $y_j = t_{n-\nu+j}$.

Note that the pullbacks to any submanifold $t = const.$ of the differentials of the functions (I.7.23) span the full cotangent bundle of that submanifold: in the local representation (I.7.23), (I.7.24) the submanifolds $t = const.$ are maximally real.

When we make use of the functions Z_j we define the basic vector fields M_j and L_k by the following relations:

$$L_kZ_j = 0,\ L_kt_\ell = \delta_{k,\ell},\ M_iZ_j = \delta_{ij},\ M_it_\ell = 0,$$
$$i,j = 1,\ldots,m,\ k,\ell = 1,\ldots,n. \tag{I.7.25}$$

We derive at once, for $i = 1,\ldots,m$,

$$M_i = \sum_{j=1}^{m} \mu_{ij}\, \partial/\partial x_j, \tag{I.7.26}$$

where $(\mu_{ij})_{1\le i,j\le m}$ is the inverse of the Jacobian matrix $\partial Z/\partial x = (\partial Z_i/\partial x_j)_{1\le i,j\le m}$. Then, for $k = 1,\ldots,n$,

$$L_k = \partial/\partial t_k - \imath \sum_{j=1}^{m} (\partial\Phi_j/\partial t_k)M_j. \tag{I.7.27}$$

Here also the vector fields $L_1,\ldots,L_n,M_1,\ldots,M_m$ commute pairwise. Note that, because of (I.7.24), we have $M_i|_0 = \partial/\partial x_i$.

One can derive "concrete" local representations of the associated differential complex (I.6.16) from the choice of first integrals described above.

There are two types of local representations, in the open neighborhood U of 0, of the complex (I.I.16). One follows from the choice of the first integrals Z_j defined by (I.7.23). The other one, which is more refined, derives from the choice of the first integrals $z_j = x_j + iy_j$ $(j = 1,\ldots,\nu)$ and w_k $(k = 1,\ldots,d)$ defined by (I.7.4).

We begin by describing the representation gotten by making use of the functions (I.7.23). The differentials dZ_j ($j = 1,\dots,m$) span T′ over U and if we adjoin to them the differentials dt_k ($k = 1,\dots,n$), we get a smooth basis of $\mathbb{C}T^*\mathcal{M}$ over U. As a matter of fact, if F is any $\mathscr{C}^\infty$ function in U we have, according to (I.7.25),

$$dF = \sum_{j=1}^{m} M_j f dZ_j + \sum_{k=1}^{n} L_k F\, dt_k. \tag{I.7.28}$$

The differentials dt_k span a subbundle of $\mathbb{C}T^*\mathcal{M}$ over U that is supplementary to $T'|_U$; we may therefore identify $(\mathbb{C}T^*\mathcal{M}/T')|_U$ to the span of $dt_1,\dots,dt_n$. If we go back to the natural isomorphism (I.6.10), we conclude that any section $\dot{f}$ of $\Lambda^{p,q}$ over U has a unique representative of the kind

$$f = \sum_{|J|=p} \sum_{|K|=q} f_{J,K}(x,t)\, dZ_J \wedge dt_K, \tag{I.7.29}$$

with $dZ_J = dZ_{j_1}\wedge\cdots\wedge dZ_{j_p}$ assuming that $J = \{j_1,\dots,j_p\}$ (where $1 \le j_1 < \cdots < j_p \le \nu$) and with a similar meaning for dt_K. We shall refer to f as the *standard representative of* $\dot{f}$ in the *chart* $(U,Z_1,\dots,Z_m,t_1,\dots,t_n)$. To say that $\dot{f} \in \mathscr{C}^\infty(U;\Lambda^{p,q})$ is equivalent to saying that the coefficients $f_{J,K}$ belong to $\mathscr{C}^\infty(U)$; and $\dot{f} \in \mathscr{D}'(U;\Lambda^{p,q})$ means that those coefficients are distributions in U. In the absence of stricter specifications we shall always suppose the latter to be true.

The standard representative of $d'\dot{f}$ is

$$Lf = \sum_{|J|=p} \sum_{|K|=q} \sum_{\ell=1}^{n} L_\ell f_{J,K}(x,t)\, dt_\ell \wedge dZ_J \wedge dt_K. \tag{I.7.30}$$

Note that L is a true differential operator. We have, by virtue of (I.7.27),

$$Lf = d_t f - \iota \sum_{j=1}^{m} d_t\Phi_j \wedge M_j f, \tag{I.7.31}$$

with the M_j acting coefficientwise on f:

$$M_j f = \sum_{|J|=p} \sum_{|K|=q} M_j f_{J,K}(x,t) dZ_J \wedge dt_K. \tag{I.7.32}$$

We have $L^2 = 0$.

We go now to the local chart $(U,z_1,\dots,z_\nu,w_1,\dots,w_d,t_1,\dots,t_{n-\nu})$. Here if F is any $\mathscr{C}^\infty$ function in U, we have, according to the orthonormality relations (I.7.9) to (I.7.14), and to (I.7.17),

$$dF = \sum_{i=1}^{\nu} M_i F dz_i + \sum_{k=1}^{d} N_k F dw_k + \sum_{j=1}^{\nu} L_j F d\bar{z}_j + \sum_{\ell=1}^{n-\nu} L_{\nu+\ell} F dt_\ell. \tag{I.7.33}$$

The standard representative of a section $\dot{f}$ of $\Lambda^{p,q}$ over U has an expression of the kind

$$f = \sum_{|J|+|J'|=p} \sum_{|K|+|K'|=q} f_{J,J',K,K'}(x,y,s,t)\mathrm{d}z_J \wedge \mathrm{d}w_{J'} \wedge \mathrm{d}\bar{z}_K \wedge \mathrm{d}t_{K'} \qquad (\text{I.7.34})$$

and the standard representative of $\mathrm{d}'^{p,q}\dot{f}$ is

$$\mathrm{L}f = \sum_{|J|+|J'|=p} \sum_{|K|+|K'|=q} \left\{ \sum_{j=1}^{\nu} \mathrm{L}_j f_{J,J',K,K'}(x,y,s,t)\mathrm{d}\bar{z}_j \wedge \mathrm{d}z_J \wedge \mathrm{d}w_{J'} \wedge \mathrm{d}\bar{z}_{K'} \wedge \mathrm{d}t_{K'} \right.$$
$$\left. + \sum_{\ell=1}^{n-\nu} \mathrm{L}_{\nu+\ell} f_{J,J',K,K'}(x,y,s,t)\mathrm{d}t_\ell \wedge \mathrm{d}z_J \wedge \mathrm{d}w_{J'} \wedge \mathrm{d}\bar{z}_K \wedge \mathrm{d}t_{K'} \right\}. \qquad (\text{I.7.35})$$

By availing ourselves of the expressions (I.7.15) and (I.7.16) we see that

$$\mathrm{L}f = (\bar{\partial}_z + \mathrm{d}_t)f - \imath \sum_{k=1}^{d} [(\bar{\partial}_z + \mathrm{d}_t)\varphi_k] \wedge \mathrm{N}_k f, \qquad (\text{I.7.36})$$

with the vector fields N_k acting coefficientwise.

We can also obtain concrete representations of the vertical complex (see the end of section 6). The first representation is based on the functions (I.7.23). Let f be given by (I.7.29). Introduce the differential operator

$$\mathrm{M}f = \sum_{|J|=p} \sum_{|K|=q} \sum_{i=1}^{m} \mathrm{M}_i f_{J,K}(x,t)\ \mathrm{d}Z_i \wedge \mathrm{d}Z_J \wedge \mathrm{d}t_K. \qquad (\text{I.7.37})$$

If we take into account the fact that the vector fields L_j and M_i all commute pairwise, we see that

$$\mathrm{LM} + \mathrm{ML} = 0. \qquad (\text{I.7.38})$$

Thus, if $\mathrm{L}f = 0$, we also have $\mathrm{M}(\mathrm{L}f) = 0$ and if $f = \mathrm{L}u$ we have $\mathrm{M}f = -\mathrm{LM}u$, which shows that M can be made to act on a cohomology class $[F] \in \mathrm{H}'^{p,q}(\Omega)$: select a cocycle $\dot{f} \in \mathscr{C}^\infty(\Omega;\Lambda^{p,q})$ which belongs to the class and then let M act on the standard representative f (see (I.7.29)) of the cocycle $\dot{f}$; $\mathrm{M}f$ is the standard representative of a cocycle $(\mathrm{M}f)^{\bullet} \in \mathscr{C}^\infty(\Omega;\Lambda^{p+1,q})$ whose cohomology class is $\mathrm{d}''[F]$.

If we make use of the functions z_j $(j = 1,\ldots,\nu)$, w_k $(k = 1,\ldots,d)$, the differential operator M must be defined as follows. Suppose that f is given by (I.7.34). Then

$$\mathrm{M}f = \sum_{|J|+|J'|=p} \sum_{|K|+|K'|=q} \left\{ \sum_{i=1}^{\nu} \mathrm{M}_i f_{J,J',K,K'} \mathrm{d}z_i \wedge \mathrm{d}z_J \wedge \mathrm{d}w_{J'} \wedge \mathrm{d}\bar{z}_K \wedge \mathrm{d}t_{K'} \right.$$
$$\left. + \sum_{k=1}^{d} \mathrm{N}_k f_{J,J',K,K'} \mathrm{d}w_k \wedge \mathrm{d}z_J \wedge \mathrm{d}w_{J'} \wedge \mathrm{d}\bar{z}_K \wedge \mathrm{d}t_{K'} \right\}. \qquad (\text{I.7.39})$$

In other words, if we define the operator N in the obvious way, we have now

$$\mathrm{M}f = (\partial_z + \mathrm{N})f - \imath \sum_{k=1}^{d} \partial_z\varphi_k \wedge \mathrm{N}_k f, \tag{I.7.40}$$

where, in the sum with respect to k, N_k acts on f *coefficientwise*.

I.8. The Levi Form in a Formally Integrable Structure

We return to the study of a *formally* integrable structure on the manifold $\mathcal{M}$. Let x_0 be an arbitrary point of $\mathcal{M}$, L_1 and L_2 two smooth sections of $\mathcal{V}$ in some open neighborhood U of x_0, ω_0 a nonzero element of the characteristic set at x_0, $\mathrm{T}^0_{x_0}$.

LEMMA I.8.1. *The value of the complex number*

$$\langle \omega_0, (2\imath)^{-1}[\mathrm{L}_1,\bar{\mathrm{L}}_2]|_{x_0}\rangle \tag{I.8.1}$$

only depends on the values $\mathrm{L}_1|_{x_0}$, $\mathrm{L}_2|_{x_0}$ *of the sections* L_1, L_2 *at the point* x_0.

PROOF. Because ω_0 is a real covector, taking the complex conjugate of (I.8.1) amounts to exchanging L_1 and L_2. For this reason it suffices to show that (I.8.1) remains unchanged if we replace L_1 by another smooth section of $\mathcal{V}$ in U whose value at x_0 is equal to that of L_1. By subtraction this is the same as proving

$$\mathrm{L}_1|_{x_0} = 0 \Rightarrow \langle \omega_0, (2\imath)^{-1}[\mathrm{L}_1,\bar{\mathrm{L}}_2]|_{x_0}\rangle = 0. \tag{I.8.2}$$

Possibly after contracting U about x_0 we may assume that there is a smooth basis $\{\tilde{\mathrm{L}}_1,\dots,\tilde{\mathrm{L}}_n\}$ of $\mathcal{V}$ over the neighborhood U and we may write $\mathrm{L}_1 = \sum_{j=1}^{n} c_j(x)\tilde{\mathrm{L}}_j$ with $c_j \in \mathcal{C}^\infty(\mathrm{U})$. If $\mathrm{L}_1|_{x_0} = 0$, we must have $c_j(x_0) = 0$ for all $j = 1,\dots,n$; then, perforce,

$$[\mathrm{L}_1,\bar{\mathrm{L}}_2]|_{x_0} = -\sum_{j=1}^{n} (\bar{\mathrm{L}}_2 c_j(x_0))\tilde{\mathrm{L}}_j$$

is orthogonal to T'_{x_0}, which proves (I.8.2). ■

Thus we may define the *hermitian sesquilinear form* on $\mathcal{V}_{x_0} \times \mathcal{V}_{x_0}$,

$$\mathcal{B}_{(x_0,\omega_0)}(\mathbf{v}_1,\mathbf{v}_2), \tag{I.8.3}$$

as the form whose value at $(\mathbf{v}_1,\mathbf{v}_2)$ is equal to the number (I.8.1) with sections

L_1 and L_2 that take, at the point x_0, the values $\mathbf{v}_1$ and $\mathbf{v}_2$ respectively. We denote by $\mathcal{Q}_{(x_0,\omega_0)}(\mathbf{v})$ the associated *quadratic form on* $\mathcal{V}_{x_0}$:

$$\mathcal{Q}_{(x_0,\omega_0)}(\mathbf{v}) = \mathcal{B}_{(x_0,\omega_0)}(\mathbf{v},\mathbf{v}). \tag{I.8.4}$$

DEFINITION I.8.1. *The quadratic form* (I.8.4) (and sometimes also the hermitian form (I.8.3)) *will be called the* Levi form *at the characteristic point* (x_0,ω_0) *of the formally integrable structure of* $\mathcal{M}$.

We are going to express the Levi form in special bases of T′ and of $\mathbb{C}T^*\mathcal{M}$ over the neighborhood U of x_0. We select a smooth basis $\varphi_1,\dots,\varphi_\nu$, $\varpi_1,\dots,\varpi_d$ of T′ over U and local coordinates in U, x_j, y_j $(1 \le j \le \nu)$, s_k $(1 \le k \le d)$, t_ℓ $(1 \le \ell \le n-\nu)$, all vanishing at x_0, such that (I.5.3) holds. We shall assume that the one-forms φ_j and ϖ_k have the expressions (I.5.21), (I.5.22) (with (I.5.23) holding).

Let, as before, L_1, L_2 be two smooth sections of $\mathcal{V}$ over U. We have, according to (I.5.22) where we write $B_{kj} = b_{(\nu+k)j}$, $C_{k\ell} = c_{(\nu+k)\ell}$,

$$d\varpi_k = \sum_{j=1}^{\nu} dB_{kj}\wedge d\bar{z}_j + \sum_{\ell=1}^{n-\nu} dC_{k\ell}\wedge dt_\ell, \tag{I.8.5}$$

whence

$$\langle d\varpi_k, L_1\wedge\bar{L}_2\rangle = \sum_{j=1}^{\nu} (L_1B_{kj})\bar{L}_2\bar{z}_j - (\bar{L}_2B_{kj})L_1\bar{z}_j + \sum_{\ell=1}^{n-\nu} (L_1C_{k\ell})\bar{L}_2t_\ell - (\bar{L}_2C_{k\ell})L_1t_\ell. \tag{I.8.6}$$

On the other hand,

$$L_1\langle\varpi_k,\bar{L}_2\rangle = L_1\bar{L}_2s_k + \sum_{j=1}^{\nu} B_{kj}L_1\bar{L}_2\bar{z}_j + \sum_{j=1}^{\nu} (L_1B_{kj})\bar{L}_2\bar{z}_j + \sum_{\ell=1}^{n-\nu} [C_{k\ell}L_1\bar{L}_2t_\ell + (L_1C_{k\ell})\,\bar{L}_2t_\ell],$$

and a similar expression for $\bar{L}_2\langle\varpi_k,L_1\rangle$ (we disregard momentarily the fact that $\langle\varpi_k,L_1\rangle \equiv 0$). By subtraction we obtain

$$L_1\langle\varpi_k,\bar{L}_2\rangle - \bar{L}_2\langle\varpi_k,L_1\rangle = [L_1,\bar{L}_2]s_k + \sum_{j=1}^{\nu} B_{kj}[L_1,\bar{L}_2]\bar{z}_j + \sum_{\ell=1}^{n-\nu} C_{k\ell}[L_1,\bar{L}_2]t_\ell + \sum_{j=1}^{\nu} [(L_1B_{kj})\bar{L}_2\bar{z}_j - (\bar{L}_2B_{kj})L_1\bar{z}_j] + \sum_{\ell=1}^{n-\nu} [(L_1C_{k\ell})\bar{L}_2t_\ell - (\bar{L}_2C_{k\ell})L_1t_\ell]. \tag{I.8.7}$$

We apply (I.1.13) with $\varphi = \varpi_k$, $\mathbf{v}_1 = L_1$, $\mathbf{v}_2 = \overline{L}_2$. By taking into account (I.8.5), (I.8.6), (I.8.7) we get

$$\langle \varpi_k, [L_1, \overline{L}_2] \rangle = [L_1, \overline{L}_2] s_k + \sum_{j=1}^{\nu} B_{kj} [L_1, \overline{L}_2] \bar{z}_j + \sum_{\ell=1}^{n-\nu} C_{k\ell} [L_1, \overline{L}_2] t_\ell. \tag{I.8.8}$$

If then we also take into account (I.5.23), we see that $\langle \varpi_k, [L_1, \overline{L}_2] \rangle = [L_1, \overline{L}_2] s_k$ at x_0. Suppose then that

$$\varpi_0 = \sum_{k=1}^{d} \sigma_k (\varpi_k|_{x_0}). \tag{I.8.9}$$

We reach the conclusion that

$$\mathscr{B}_{(x_0, \varpi_0)}(\mathbf{v}_1, \mathbf{v}_2) = (2\imath)^{-1} \sum_{k=1}^{d} \sigma_k ([L_1, \overline{L}_2] s_k)|_{x_0}, \tag{I.8.10}$$

where now $\mathbf{v}_1$ and $\mathbf{v}_2$ are the values of L_1 and L_2, respectively, at x_0.

CR Structures

When the involutive structure of $\mathcal{M}$ is a CR structure, another expression of the Levi form can be devised. In this case we use a smooth basis $\varphi_1, \ldots, \varphi_n, \varpi_1, \ldots, \varpi_d$ of T′ over U such that

$$\varphi_1, \ldots, \varphi_n, \overline{\varphi}_1, \ldots, \overline{\varphi}_n, \varpi_1, \ldots, \varpi_d \textit{ form a basis of } \mathbb{C}T^*\mathcal{M} \textit{ over } U; \tag{I.8.11}$$

$$\varpi_1, \ldots, \varpi_d \textit{ are real} \tag{I.8.12}$$

(cf. (I.5.28)).

Since $dT' \subset T^{1,1}$ we see that, for each $k = 1, \ldots, d$,

$$d\varpi_k \equiv 2\imath \sum_{i,j=1}^{n} g_{ij;k} \varphi_i \wedge \overline{\varphi}_j + \sum_{\ell=1}^{d} \gamma_{k\ell} \wedge \varpi_\ell \bmod T^{2,0}, \tag{I.8.13}$$

where $g_{ij;k} \in \mathscr{C}^\infty(U)$, $g_{ij;k} = \bar{g}_{ji;k}$, and the $\gamma_{k\ell}$ are smooth one-forms. We apply (I.1.13). Since ϖ_k is real, we have $\langle \varpi_k, L \rangle = \langle \varpi_k, \overline{L} \rangle = 0$ whatever the smooth section L of $\mathcal{V}$ over U. We derive

$$\langle \varpi_k, [L_1, \overline{L}_2] \rangle = -\langle d\varpi_k, L_1 \wedge \overline{L}_2 \rangle. \tag{I.8.14}$$

If we then take (I.8.13) into account, we get

$$\langle \varpi_k, [L_1, \overline{L}_2] \rangle = 2\imath \sum_{i,j=1}^{n} g_{i,j;k} \langle \varphi_i, \overline{L}_1 \rangle \langle \overline{\varphi}_j, L_2 \rangle. \tag{I.8.15}$$

This has the following meaning. Assume that the covector ϖ_0 is given by (I.8.9). Let us use the coordinates ζ_j ($j = 1, \ldots, n$) in $\mathcal{V}$ with respect to the basis $\{L_j\}_{1 \le j \le n}$ of $\mathcal{V}$ defined by

$$\langle \overline{\varphi}_i, \mathrm{L}_j \rangle = \delta_{ij},\ i, j = 1,\ldots,n. \tag{I.8.16}$$

Then the quadratic form (I.8.4) can be written as

$$\sum_{i,j=1}^{n} \sum_{k=1}^{d} (g_{i,j;k}\sigma_k)\zeta_i\overline{\zeta}_j. \tag{I.8.17}$$

I.9. The Levi Form in a Locally Integrable Structure

When the manifold $\mathcal{M}$ is equipped with a *locally integrable* structure, the Levi form at a characteristic point can be given a more concrete expression. We make use of the first integrals $z_j = x_j + \imath y_j$ $(1 \le j \le \nu)$, $w_k = s_k + \imath\varphi_k(x,y,s,t)$ $(1 \le k \le d)$ and assume that (I.7.5) holds (the point 0 is the origin of the coordinates). We are going to apply (I.8.10). By making use of the basis (I.7.15), (I.7.16) of $\mathcal{V}$ over U we see that we may write

$$\mathrm{L}_1 = \delta_1 - \imath \sum_{k=1}^{d} (\delta_1\varphi_k)\mathrm{N}_k,$$

where δ_1 is a linear combination, with $\mathcal{C}^\infty$ coefficients, of partial differentiations $\partial/\partial\overline{z}_j$, $\partial/\partial t_\ell$; L_2 has the analogous expression. According to (I.7.5) we have

$$\delta_1\varphi_k|_0 = \overline{\delta}_2\varphi_k|_0 = 0,\ k = 1,\ldots,d. \tag{I.9.1}$$

Thus $\delta_1 = \mathrm{L}_1|_0$, $\delta_2 = \mathrm{L}_2|_0$ and

$$\mathrm{L}_1 s_{k'} = -\imath \sum_{k=1}^{d} \mu_{kk'}\delta_1\varphi_k,\quad \overline{\mathrm{L}}_2 s_{k'} = \imath \sum_{k=1}^{d} \overline{\mu}_{kk'}\overline{\delta}_2\varphi_k.$$

Thus, at the point 0,

$$\mathrm{L}_1\overline{\mathrm{L}}_2 s_{k'} = \imath \sum_{k=1}^{d} \overline{\mu}_{kk'}\delta_1\overline{\delta}_2\varphi_k,\quad \overline{\mathrm{L}}_2\mathrm{L}_1 s_{k'} = -\imath \sum_{k=1}^{d} \mu_{kk'}\overline{\delta}_2\delta_1\varphi_k.$$

But, still because of (I.7.5), we have $\mu_{kk'}|_0 = \delta_{kk'}$ (Kronecker's index) and

$$\delta_1\overline{\delta}_2\varphi_k|_0 = [(\delta_1|_0)(\overline{\delta}_2|_0)\varphi_k]|_0 = \overline{\delta}_2\delta_1\varphi_k|_0.$$

We reach the following conclusion:

$$\mathcal{B}_{(0,\omega_0)}(\mathbf{v}_1,\mathbf{v}_2) = \sum_{k=1}^{d} \sigma_k(\mathbf{v}_1\overline{\mathbf{v}}_2\varphi_k)|_0. \tag{I.9.2}$$

REMARK I.9.1. The reader should keep in mind that, in general, expressions such as (I.9.2) or (I.8.10) are only valid at the point 0. The bases of T′ that have been used are all "centered" at 0. This limited validity can easily be illustrated by examples, such as the following one.

EXAMPLE I.9.1. Consider a locally integrable structure on $\mathbb{R}^2$ (where the coordinates are denoted by s and t) defined by the single function $w = s + i\varphi(s,t)$ with φ a real-valued $\mathscr{C}^\infty$ function that vanishes, and whose first partial derivatives vanish, at the origin. The cotangent structure bundle is spanned by

$$\varpi = \mathrm{d}s + \imath[1+\imath\varphi_s(s,t)]^{-1}\varphi_t(s,t)\mathrm{d}t$$

and the tangent structure bundle is spanned by the single vector field $\mathrm{L} = \partial/\partial t - \imath(1+\imath\varphi_s)^{-1}\varphi_t\partial/\partial_s$. The characteristic set T° is equal to 0 at any point where $\varphi_t \neq 0$. At those points, among them the origin, where $\varphi_t = 0$, T° is the line spanned by $\mathrm{d}s$. On the other hand, we have

$$(2\imath)^{-1}\langle \mathrm{d}s,[\mathrm{L},\overline{\mathrm{L}}]\rangle = (1+\varphi_s^2)^{-2}(\varphi_{tt} + \varphi_{tt}\varphi_s^2 - 2\varphi_{st}\varphi_s\varphi_t + \varphi_{ss}\varphi_t^2). \quad \text{(I.9.3)}$$

It is clear that (I.9.3) is identical with (I.9.2) only at those points where both φ_s and φ_t vanish. ■

We continue to deal with the locally integrable structure on $\mathcal{M}$ defined near the point 0 by the functions z_j, w_k. Let us make use of the basis $\{\mathrm{L}_1,\ldots,\mathrm{L}_n\}$ of $\mathcal{V}$ over the neighborhood U given by (I.7.15) and (I.7.16). The latter expressions tell us that

$$\begin{aligned} \mathrm{L}_j|_0 &= \partial/\partial\bar{z}_j,\ 1 \leq j \leq \nu; \\ \mathrm{L}_{\nu+\ell}|_0 &= \partial/\partial t_\ell,\ 1 \leq \ell \leq n-\nu. \end{aligned} \quad \text{(I.9.4)}$$

For brevity let us write $\Phi = \sigma_1\varphi_1 + \cdots + \sigma_d\varphi_d$. According to (I.9.2) the Levi form at 0 can be identified to the following quadratic form on $\mathbb{C}^\nu \times \mathbb{C}^{n-\nu}$ (where the variable is denoted by (ζ,τ)),

$$\mathcal{Q}(\zeta,\tau) = \sum_{i,j=1}^{\nu} (\partial^2\Phi/\partial\bar{z}_i\partial z_j)(0)\zeta_i\bar{\zeta}_j +$$

$$2\,\mathcal{R}e\left[\sum_{i=1}^{\nu}\sum_{k=1}^{n-\nu} (\partial^2\Phi/\partial\bar{z}_i\partial t_k)(0)\zeta_i\bar{\tau}_k\right] + \sum_{k,\ell=1}^{n-\nu} (\partial^2\Phi/\partial t_k\partial t_\ell)(0)\tau_k\bar{\tau}_\ell.$$

Let us shorten the notation and rewrite this formula as

$$\mathcal{Q}(\zeta,\tau) = \Phi_{zz}(0)\zeta\cdot\bar{\zeta} + 2\,\mathcal{R}e[\Phi_{zt}(0)\tau\cdot\bar{\zeta}] + \Phi_{tt}(0)\tau\cdot\bar{\tau}. \quad \text{(I.9.5)}$$

We can represent $\mathbb{C}^{n-\nu}$ as the Hilbert sum (for the standard hermitian product $\tau\cdot\bar{\tau}'$) $\mathcal{N}^\perp \oplus \mathcal{N}$, with $\mathcal{N}$ the null-space of $\Phi_{tt}(0)$; we denote by α (resp., β) the component of τ in $\mathcal{N}^\perp$ (resp., $\mathcal{N}$). With this decomposition the right-hand side in (I.9.5) reads

$$\Phi_{zz}(0)\zeta\cdot\bar{\zeta} + 2\,\mathcal{R}e[\alpha\cdot\Phi_{zt}(0)\bar{\zeta}] + 2\,\mathcal{R}e[\beta\cdot\Phi_{zt}(0)\bar{\zeta}] + \Phi_{tt}(0)\alpha\cdot\bar{\alpha}. \quad \text{(I.9.6)}$$

Let us now take full advantage of the linear changes of variables that are allowed. They are of the following kind:

$$z' = \mathrm{A}z,\ t' = \mathrm{B}t + \mathrm{E}x + \mathrm{F}y, \quad \text{(I.9.7)}$$

with $A : \mathbb{C}^\nu \to \mathbb{C}^\nu$, $B = \mathbb{R}^{n-\nu} \to \mathbb{R}^{n-\nu}$ invertible linear maps and E, F linear maps $\mathbb{R}^\nu \to \mathbb{R}^{n-\nu}$. This yields the following substitutions in the complex tangent space

$$\partial/\partial z = {}^t A\partial/\partial z' + {}^t G\partial/\partial t', \; \partial/\partial t = {}^t B\partial/\partial t', \tag{I.9.8}$$

where $G = \frac{1}{2}(E - \imath F)$. Let ζ_j, τ_k denote the complex coordinates in $\mathcal{V}_0$ with respect to the basis $\{\partial/\partial\bar{z}_1,\ldots,\partial/\partial\bar{z}_\nu,\ \partial/\partial t_1,\ldots,\partial/\partial t_{n-\nu}\}$ and ζ_j', τ_k' the analogues in the basis $\{\partial/\partial\bar{z}_1',\ldots,\partial/\partial\bar{z}_\nu',\ \partial/\partial t_1',\ldots,\partial/\partial t_{n-\nu}'\}$. We have the right to make the substitutions

$$\zeta' = \overline{A}\zeta, \; \tau' = B\tau + \overline{G}\zeta. \tag{I.9.9}$$

We wish to select A, B, and G to achieve a simpler expression of the quadratic form (I.9.6) in the new coordinates. We shall decompose the transformation into several steps. In the first step we take $G = 0$, $A = \mathrm{Id}$; we select B to be the identity on $\mathcal{N}$ and to be such, on $\mathcal{N}^\perp$, that

$${}^t B^{-1}\Phi_{tt}(0)B^{-1} = \Pi_+ - \Pi_-,$$

where Π_+ (resp., Π_-) is the orthogonal projector of $\mathbb{C}^{n-\nu}$ onto the orthogonal sum of the eigenspaces of $\Phi_{tt}(0)$ on which $\Phi_{tt}(0)$ is definite-positive (resp., definite-negative). Setting $\alpha' = B\alpha$, we get

$$\begin{gathered}\Phi_{tt}(0)\alpha\cdot\alpha + 2\,\mathcal{R}e[\alpha\cdot\Phi_{zt}(0)\bar{\zeta}] = \\ |\Pi_+\alpha'|^2 - |\Pi_-\alpha'|^2 + 2\,\mathcal{R}e[\alpha'\cdot{}^t B^{-1}\Phi_{zt}(0)\bar{\zeta}] = \\ |\Pi_+[\alpha' + {}^t B^{-1}\Phi_{\bar{z}t}(0)\zeta]|^2 - \Pi_-[\alpha' - {}^t B^{-1}\Phi_{\bar{z}t}(0)\zeta]|^2 - \\ (\Pi_+ - \Pi_-){}^t B^{-1}\Phi_{\bar{z}t}(0)\zeta\cdot{}^t B^{-1}\Phi_{zt}(0)\bar{\zeta}.\end{gathered}$$

After deleting the primes the expression of (I.9.6) becomes

$$\begin{gathered}\Phi_{z\bar{z}}(0)\zeta\cdot\bar{\zeta} + 2\,\mathcal{R}e[\beta\cdot\Phi_{zt}(0)\bar{\zeta}] + |\Pi_+[\alpha + {}^t B^{-1}\Phi_{\bar{z}t}(0)\zeta]|^2 - \\ |\Pi_-[\alpha - {}^t B^{-1}\Phi_{\bar{z}t}(0)\zeta]|^2 - (\Pi_+ - \Pi_-){}^t B^{-1}\Phi_{\bar{z}t}(0)\zeta\cdot{}^t B^{-1}\Phi_{zt}(0)\bar{\zeta}.\end{gathered}$$

In the second step we let A and B be the identity and take

$$G = (\Pi_+ - \Pi_-){}^t B^{-1}\Phi_{zt}(0).$$

This means that we are reduced to dealing with a quadratic form

$$\Gamma\zeta\cdot\bar{\zeta} + 2\,\mathcal{R}e[B\cdot\Phi_{\bar{z}t}(0)\zeta] + |\Pi_+\alpha|^2 - |\Pi_-\alpha|^2, \tag{I.9.10}$$

where $\Gamma: \mathbb{C}^\nu \to \mathbb{C}^\nu$ is self-adjoint. We denote by P_+ (resp., P_-) the orthogonal projector of $\mathbb{C}^\nu$ onto the orthogonal sum of the eigenspaces of Γ in which Γ is definite-positive (resp., definite-negative). The third step does for Γ what the first step did for $\Phi_{tt}(0)$. We end up with a quadratic form

$$|P_+\zeta|^2 - |P_-\zeta|^2 + 2\,\mathcal{R}e(\mathcal{T}\beta\cdot\bar{\zeta}) + |\Pi_+\alpha|^2 - |\Pi_-\alpha|^2, \tag{I.9.11}$$

with $\mathcal{T}$ a linear map $\mathcal{N} \to \mathbb{C}^\nu$.

From this formula we derive a few consequences:

Proposition I.9.1. *Suppose that at least one of the following conditions is satisfied: (i)* $\nu = n$; *(ii)* $\nu = 0$; *(iii) the restriction of the Levi form at* $(0,\varpi_0)$ *to* τ*—space is nondegenerate. Then there is a change of variables* (I.9.7) *that transforms the Levi form at* $(0,\varpi_0)$ *into a sum of positive and negative squares.*

Proof. Indeed, under hypotheses (*i*) or (*iii*) there are no nonzero vectors β in (I.9.11). Under hypothesis (*ii*) there are no vectors ζ. ■

Remark I.9.2. Notice that the subspace $\zeta = 0$ of $\mathcal{V}_0$ is invariant; it is the intersection $\mathcal{V}_0 \cap \overline{\mathcal{V}}_0$. The restriction of the Levi form at $(0,\varpi_0)$ to $\mathcal{V}_0 \cap \overline{\mathcal{V}}_0$ is invariant under the admissible transformations (I.9.7). ■

Proposition I.9.2. *In order for the Levi form at* $(0,\varpi_0)$ *not to be semidefinite-positive, it is necessary and sufficient that at least one of the three operators,* P_-, Π_-, *or* $\mathcal{T}$ *be different from zero.*

Proof. It is clear that, if the Levi form at $(0,\varpi_0)$ is to take at least one value < 0, P_-, Π_- and $\mathcal{T}$ cannot all vanish. When $\Pi_- \neq 0$, in order to check that the Levi form at $(0,\varpi_0)$ takes some value < 0, it suffices to evaluate (I.9.11) at a point (ζ,α,β) where $\zeta = 0$, $\Pi_+\alpha = 0$, $\Pi_-\alpha \neq 0$. If $\mathrm{P}_- \neq 0$, one takes $\mathrm{P}_+\zeta = 0$, $\mathrm{P}_-\zeta \neq 0$, $\alpha = 0$, $\beta = 0$. Last, if $\mathcal{T} \neq 0$, one can take $\alpha = 0$, $\mathcal{T}\beta \neq 0$, and $\zeta = -\varepsilon\mathcal{T}\beta$. We see that (I.9.11) is then equal to $-2\varepsilon|\mathcal{T}\beta|^2 + O(\varepsilon^2)$ and it suffices to take $\varepsilon > 0$ sufficiently small. ■

Next we relate the expression (I.9.11) of the Levi form at $(0,\varpi_0)$ to the Taylor expansion of Φ about 0. After redefining the z_j and the t_k we may assume that

$$\partial^2\Phi/\partial z_i\partial \bar{z}_j|_0 = 0 \textit{ if } i, j = 1,\dots,\nu,\ i \neq j; \tag{I.9.12}$$

$$\partial^2\Phi/\partial z_j\partial \bar{z}_j|_0 = \begin{cases} 1 \textit{ if } 1 \leq j \leq p, \\ -1 \textit{ if } p < j \leq p+q, \\ 0 \textit{ if } p+q < j \leq n; \end{cases} \tag{I.9.13}$$

$$\partial^2\Phi/\partial t_k\partial t_\ell|_0 = 0 \textit{ if } k, \ell = 1,\dots,n-\nu,\ k \neq \ell; \tag{I.9.14}$$

$$\partial^2\Phi/\partial t_\ell^2|_0 = \begin{cases} 1 \textit{ if } 1 \leq \ell \leq p', \\ -1 \textit{ if } p' < \ell \leq p'+q', \\ 0 \textit{ if } p'+q' < \ell \leq n-\nu. \end{cases} \tag{I.9.15}$$

By (I.9.11) we shall have

$$\Phi|_{s=0} = \sum_{j=1}^{p}|z_j|^2 - \sum_{j=p+1}^{p+q}|z_j|^2 + \tfrac{1}{2}\sum_{\ell=1}^{p'} t_\ell^2 - \tfrac{1}{2}\sum_{\ell=p'+1}^{p'+q'} t_\ell^2 + \mathcal{R}(z,t) +$$
$$2\,\mathcal{R}e\left[\sum_{i,j=1}^{\nu}(\partial^2\Phi/\partial z_i\partial z_j)(0)z_iz_j\right] + O(|z|^3+|t|^3), \tag{I.9.16}$$

where we have used the notation

$$\mathcal{R}(z,t) = 2\,\mathcal{R}e\left[\sum_{j=1}^{\nu}\sum_{\ell=p'+q'+1}^{n-\nu} a_{j\ell}z_j t_\ell\right].$$

But we may replace the function w_k by the functions

$$\tilde{w}_k = w_k - 2\imath \sum_{i,j=1}^{\nu} (\partial^2\varphi_k/\partial z_i \partial z_j)(0) z_i z_j, \tag{I.9.17}$$

for each $k = 1,\ldots,d$; setting $\tilde{s}_k = \mathcal{R}e\tilde{w}_k$, we see that

$$\tilde{s}_k = s_k + O(|z|^2). \tag{I.9.18}$$

If we define $\tilde{\Phi} = \Sigma_{k=1}^{d}\,\sigma_k(\mathrm{Im}\,\tilde{w}_k)$ and recall that $d\varphi_k|_0 = 0$, we may assert, thanks to (I.9.16) and (I.9.18) and after deleting the tildes:

$$\Phi|_{s=0} = \sum_{j=1}^{p}|z_j|^2 - \sum_{j=p+1}^{p+q}|z_j|^2 + \frac{1}{2}\left\{\sum_{\ell=1}^{p'} t_\ell^2 - \sum_{\ell=p'+1}^{p'+q'} t_\ell^2\right\} +$$

$$\mathcal{R}(z,t) + O(|z|^3 + |t|^3). \tag{I.9.19}$$

In the above the cases $p = 0$, $q = 0$, $p' = 0$ or $q' = 0$ are not precluded; if for instance $p = 0$, the first sum at the right in (I.9.19) will not be present. We also remind the reader that

$$\Phi = \Phi|_{s=0} + O(|s|(|z| + |t| + |s|)). \tag{I.9.20}$$

Formula (I.9.19) simplifies when the restriction of the Levi form at $(0,\varpi_0)$ to $\mathcal{V}_0\cap\overline{\mathcal{V}}_0$ is nondegenerate. In that case $p' + q' = n - \nu$ and $\mathcal{R} \equiv 0$. It likewise simplifies in the following important particular cases.

EXAMPLE I.9.2. *Case* $\nu = 0$. There are no variables x_j, y_j. Then

$$\Phi = \frac{1}{2}\left\{\sum_{\ell=1}^{p'} t_\ell^2 - \sum_{\ell=p'+1}^{p'+q'} t_\ell^2\right\} + O(|s|^2 + |s||t| + |t|^3). \tag{I.9.21}$$

EXAMPLE I.9.3. *CR structures.* In this case there are no variables t_k and (I.9.19)–(I.9.20) read

$$\Phi(x,y,s) = \sum_{j=1}^{p}|z_j|^2 - \sum_{j=p+1}^{p+q}|z_j|^2 + O(|s||z| + |s|^2 + |z|^3). \tag{I.9.22}$$

Generic Submanifolds

The preceding computations are based on the data of the first integrals z_j and w_k. But when dealing with *embedded* CR manifolds many authors prefer to make use of the equations defining (locally or globally) the submanifold. Thus let us look at a generic submanifold $\mathcal{M}$ of $\mathbb{C}^{n+d}$ (see Proposition I.3.7 and subsequent remark). We assume that $d = \text{codim}\,\mathcal{M}$, and that $\mathcal{M}$ is defined by equations

$$\rho_k(z) = 0,\ k = 1,\ldots,d, \tag{I.9.23}$$

where the functions $\rho_k \in \mathscr{C}^\infty(\mathbb{C}^{n+d})$ are real-valued and

$$\partial\rho_1\wedge\cdots\wedge\partial\rho_d \neq 0 \tag{I.9.24}$$

in a full neighborhood of $\mathcal{M}$.

The complex conormal bundle of $\mathcal{M}$ in $\mathbb{C}^{n+d}$, $\mathbb{C}N^*\mathcal{M}$, is spanned by the (real) differentials $d\rho_k$. Since $\imath\partial\rho_k = -\imath\bar{\partial}\rho_k + \imath d\rho_k$ we see that the pullback to $\mathcal{M}$ of $\imath\partial\rho_k$ is a real one-form. It is obviously orthogonal to the tangential Cauchy-Riemann vector fields, i.e., to $\mathcal{V} = \mathbb{C}T\mathcal{M} \cap T^{0,1}$. It is therefore a section of the characteristic set T° of $\mathcal{M}$. By virtue of (I.9.24) we reach the following conclusion:

PROPOSITION I.9.3. *The one-forms on* $\mathcal{M}$, $\imath\partial\rho_k$ $(k = 1,\ldots,d)$, *make up a smooth basis of the characteristic set* T° *over the whole of* $\mathcal{M}$.

Assume that the coordinates in $\mathbb{C}^{n+d}$ are z_j, w_k $(1 \leq j \leq n,\ 1 \leq k \leq d)$ and that $0 \in \mathcal{M}$. Suppose furthermore that, near zero, we have the following defining equations for $\mathcal{M}$:

$$\mathscr{I}m\,w_k - \varphi_k(z,\mathscr{R}e\,w_k) = 0,\ k = 1,\ldots,d. \tag{I.9.25}$$

Call ρ_k the left-hand sides in (I.9.25). Then

$$2\imath\partial\rho_k = dw_k - 2\imath\partial\varphi_k. \tag{I.9.26}$$

Set $s = \mathscr{R}e\,w$ $(\in \mathbb{R}^d)$. The pullback of $2\imath\partial\rho_k$ to $\mathcal{M}$ is equal to

$$\sum_{\ell=1}^{d} (\delta_{k\ell} + \imath\partial\varphi_k/\partial s_\ell)ds_\ell + \imath\sum_{j=1}^{n} [(\partial\varphi_k/\partial z_j)dz_j + (\partial\varphi_k/\partial\bar{z}_j)d\bar{z}_j] -$$

$$2\imath\left\{\sum_{j=1}^{n} (\partial\varphi_k/\partial z_j)dz_j + \frac{1}{2}\sum_{\ell=1}^{d} (\partial\varphi_k/\partial s_\ell)dw_\ell\right\} =$$

$$\sum_{\ell=1}^{d} (\delta_{k\ell} + \imath\partial\varphi_k/\partial s_\ell)ds_\ell - \imath\sum_{j=1}^{n} [(\partial\varphi_k/\partial z_j)dz_j - (\partial\varphi_k/\partial\bar{z}_j)d\bar{z}_j] -$$

$$\imath\sum_{\ell,\ell'=1}^{d} (\partial\varphi_k/\partial s_\ell)(\delta_{\ell\ell'} + \imath\partial\varphi_\ell/\partial s_{\ell'})ds_{\ell'}.$$

We obtain that *the pullback of* $\imath\partial\rho_k$ *to* $\mathcal{M}$ *is equal to*

$$ds_k + \sum_{\ell,\ell'=1}^{d} (\partial\varphi_k/\partial s_\ell)(\partial\varphi_\ell/\partial s_{\ell'})ds_{\ell'} -$$

$$\imath \sum_{j=1}^{n} [(\partial\varphi_k/\partial z_j)dz_j - (\partial\varphi_k/\partial\bar{z}_j)d\bar{z}_j]. \qquad (I.9.27)$$

The forms (I.9.27) (for $k = 1,\dots,d$) form a basis of T° in the neighborhood of 0 in $\mathcal{M}$.

Also, according to (I.9.26) we have $\partial\bar{\partial}\rho_k = -\partial\bar{\partial}\varphi_k$. The pullback to $\mathcal{M}$ of $\bar{\partial}\partial\rho_k$ is given by

$$\frac{1}{4}\sum_{\ell,\ell'=1}^{d} (\partial^2\varphi_k/\partial s_\ell\partial s_{\ell'})dw_\ell \wedge d\bar{w}_{\ell'} +$$

$$\mathcal{I}m \sum_{\ell=1}^{d}\sum_{j=1}^{n}(\partial^2\varphi_k/\partial z_j\partial s_\ell)dz_j\wedge d\bar{w}_\ell + \sum_{j,j'=1}^{n} (\partial^2\varphi_k/\partial z_j\partial\bar{z}_{j'})dz_j\wedge d\bar{z}_{j'}. \qquad (I.9.28)$$

The pullback to $\mathcal{M}$ of $\bar{\partial}\partial\rho_k$ defines a hermitian quadratic form $\tilde{\mathfrak{Q}}_k$ on the fibre of $\mathbb{C}T\mathcal{M}$ at 0. Suppose that (I.7.5) holds. Then the value at 0 of the pullbacks to $\mathcal{M}$ of dw_ℓ and $d\bar{w}_\ell$ are both equal to ds_ℓ and are thus orthogonal to $\mathcal{V}_0$. It follows that, if **v** is the value at 0 of a smooth section L of $\mathcal{V}$, then

$$\tilde{\mathfrak{Q}}_k(\mathbf{v}) = \sum_{j,j'=1}^{n} (\partial^2\varphi_k/\partial z_j\partial\bar{z}_{j'})(\mathbf{v}\bar{z}_j)(\bar{\mathbf{v}}z_{j'}).$$

Let us make use of a smooth basis $\{L_1,\dots,L_n\}$ of $\mathcal{V}$ in an open neighborhood of 0, such that $L_j|_0 = \partial/\partial\bar{z}_j$, $j = 1,\dots,n$, and write

$$\mathbf{v} = \sum_{j=1}^{n} \zeta_j(L_j|_0).$$

We find

$$\tilde{\mathfrak{Q}}_k(\mathbf{v}) = \sum_{j,j'=1}^{n} (\partial^2\varphi_k/\partial\bar{z}_j\partial z_{j'})(0)\zeta_j\bar{\zeta}_{j'}.$$

Set $\sigma\cdot\rho = \sigma_1\rho_1 + \cdots + \sigma_d\rho_d$; then the quadratic form on $\mathcal{V}_0$ defined by $\partial\bar{\partial}(\sigma\cdot\rho)$ is equal to

$$\sigma\cdot\tilde{\mathfrak{Q}}(\mathbf{v}) = \sigma_1\tilde{\mathfrak{Q}}_1(\mathbf{v}) + \cdots + \sigma_d\tilde{\mathfrak{Q}}_d(\mathbf{v}) =$$

$$\sum_{j,j'=1}^{n} (\partial^2\Phi/\partial\bar{z}_j\partial z_{j'})(0)\zeta_j\bar{\zeta}_{j'}, \qquad (I.9.29)$$

in the notation of (I.9.5). If we compare with the latter formula (where there are no variables t_ℓ and τ_ℓ) we conclude that the quadratic form (I.9.29), defined by $\partial\bar{\partial}(\sigma\cdot\rho)$ on $\mathcal{V}_0$, is equal to the Levi form (I.9.5) at the origin.

Suppose now that we change defining equations for $\mathcal{M}$ near the origin. This means that we substitute for the functions ρ_k a set of functions

$$\rho_k^{\#} = \sum_{\ell=1}^{d} \gamma_{k\ell}\rho_\ell \ (1 \le k \le d),$$

where the $d \times d$ matrix $\gamma = (\gamma_{k\ell})_{1 \le k, \ell \le n}$ is real, nonsingular, and $\mathscr{C}^\infty$. The pullback to $\mathcal{M}$ of the two-form $\partial\bar{\partial}\rho_k^\#$ is congruent, modulo linear combinations of $\partial\rho_\ell$ and $\bar{\partial}\rho_{\ell'}$, to that of

$$\sum_{\ell=1}^{d} \gamma_{k\ell}\partial\bar{\partial}\rho_\ell. \tag{I.9.30}$$

But we know that the pullbacks of $\partial\rho_\ell$ and $\bar{\partial}\rho_\ell$ ($\approx - \partial\rho_\ell$) are orthogonal to $\mathcal{V}$ and therefore the quadratic form on $\mathcal{V}_0$ defined by $\partial\bar{\partial}\rho_k^\#$ is equal to that defined by (I.9.30). We reach the following conclusion:

> *Replacing the multiplet* $\rho = (\rho_1,\ldots,\rho_d)$ *by the multiplet* $\gamma\rho$ *leads to replacing the quadratic form* $\sigma \cdot \tilde{\mathfrak{Q}}$ *by the quadratic form* $({}^t\gamma\sigma)\cdot\tilde{\mathfrak{Q}}$.

The subsitution $\rho \to \gamma\rho$ does not change the *conjugacy class* of the quadratic map

$$\tilde{\mathfrak{Q}} = (\tilde{\mathfrak{Q}}_1,\ldots,\tilde{\mathfrak{Q}}_d) : \mathcal{V} \to \mathbb{R}^d.$$

This is why it is convenient to speak of the pullback to $\mathcal{M}$ of $\partial\bar{\partial}\rho$ as "the Levi form" on $\mathcal{M}$. Such a definition of the Levi form (as a conjugacy class of vector-valued quadratic forms on $\mathcal{V}$) is global, since it does not matter what set of equations we use to define $\mathcal{M}$ near any one of its points. When $\mathcal{M}$ is a hypersurface, i.e., when $d = 1$, the two-form $\partial\bar{\partial}\rho$ actually defines a scalar quadratic form on $\mathcal{V}$.

Inspection of the preceding remarks shows that we could have assumed $\mathcal{M}$ to be a generic submanifold of any complex manifold (of complex dimension $n+d$), not necessarily one of complex space $\mathbb{C}^{n+d}$.

I.10. Characteristics in Real and in Analytic Structures

The Real Frobenius Theorem

The following statement is one of the standard versions of the *Frobenius theorem*:

THEOREM I.10.1. *Any real involutive structure on* $\mathcal{M}$ *is locally integrable.*

PROOF. Thus let $\mathcal{M}$ be equipped with a real involutive structure. As usual, let n denote the rank of the tangent structure bundle $\mathcal{V}$. We shall reason by ascending induction on n. The assertion is trivially true when $n = 0$.

Let p be an arbitrary point of $\mathcal{M}$. We can find an open neighborhood U of p in which there is a smooth basis of $\mathcal{V}$ consisting of pairwise commuting vector fields $L_1,\ldots,L_n$ (Proposition I.5.1). Inspection of the argument that led to Proposition I.5.1 shows that, when the structure is real, the vector fields L_k

can be taken to be real. Let then $\mathcal{V}_1$ denote the involutive structure on U defined by the vector fields $L_2,\ldots,L_n$. By the induction hypothesis $\mathcal{V}_1$ is locally integrable. After contracting U about p we may assume that there are $m + 1$ smooth real-valued functions $x_0,\ldots,x_m$ in U, which are equal to zero at p and satisfy $L_k x_j = 0$ for all $k = 2,\ldots,n$, and whose differentials are linearly independent. To the x_j's we adjoin $n - 1$ functions $t_2,\ldots,t_n$, also $\mathcal{C}^\infty$ and real and vanishing at p, to obtain a complete system of local coordinates in U (possibly contracted about p). The vector fields $\partial/\partial t_k$ $(2 \le k \le n)$ span $\mathcal{V}_1$ over U. The coefficient of some partial differentiation $\partial/\partial x_j$ $(0 \le j \le m)$ in the expression of L_1 must be different from zero at the origin, and therefore in U if the latter is small enough. After relabeling the x_j's if need be, we may assume that it is the coefficient of $\partial/\partial x_0$. In the sequel we write t_1 instead of x_0. After dividing L_1 by the coefficient we may assume that

$$L_1 = \partial/\partial t_1 + \sum_{j=1}^{m} a_j(x,t)\partial/\partial x_j + \sum_{k=2}^{n} b_k(x,t)\partial/\partial t_k.$$

We conclude that

$$L = \partial/\partial t_1 + \sum_{j=1}^{m} a_j(x,t)\partial/\partial x_j$$

is a smooth section of $\mathcal{V}$ over U, and therefore L, $\partial/\partial t_2,\ldots,\partial/\partial t_n$ form a smooth (real) basis of $\mathcal{V}$ over U. Since $\mathcal{V}$ is involutive, the commutation brackets $[\partial/\partial t_k, L]$ are sections of $\mathcal{V}$, which is only possible if they vanish identically. We reach the conclusion that the coefficients $a_j(x,t)$ are independent of $t_2,\ldots,t_n$. Let then $F(y,t_1)$ denote the unique solution of the initial value problem

$$\dot{F} = \mathbf{a}(F,t_1), \quad F|_{t_1=0} = y. \tag{I.10.1}$$

Here $\dot{F}$ stands for the derivative of the function F (valued in $\mathbb{R}^m$) with respect to t_1; and $\mathbf{a} = (a_1,\ldots,a_m)$. The Jacobian determinant of F with respect to y is equal to 1 at $t = 0$ and therefore is $\neq 0$ in a neighborhood of p. We thus get a new system of local coordinates $y_1,\ldots,y_m,t_1,\ldots,t_n$, in which the expression of the vector field L is none other than $\partial/\partial t_1$; and thus $\mathcal{V}$ itself is spanned over a neighborhood of p by $\partial/\partial t_1,\ldots,\partial/\partial t_n$. The cotangent structure bundle T′ is spanned by $dy_1,\ldots,dy_m$. ■

We may rephrase Theorem I.10.1 in more geometrical terms. We see that $\mathcal{M}$ can be covered by coordinates patches $(U,x_1,\ldots,x_m,t_1,\ldots,t_n)$ with the following property: the complex tangent space to any submanifold $x = x_0$ of U, at any one of its points, is equal to the fibre of $\mathcal{V}$ at that point. We obtain a partition of the manifold $\mathcal{M}$ into connected subsets Σ that have the following property: an arbitrary point $p \in \Sigma$ can be viewed as the center of a local chart $(U,x_1,\ldots,x_m,t_1,\ldots,t_n)$ of the kind just considered, and $\Sigma \cap U$ is the union of a (countable) family of submanifolds $x = x_0$ of U. We may equip any such subset Σ with the manifold structure defined by the atlas made up of the local charts $(\Lambda_p, t_1, \dots, t_n)$ extracted from the charts $(U, x_1, \dots, x_m, t_1, \dots, t_n)$ considered above. One can take Λ_p to be the connected component of p in $\Sigma \cap U$. It is obvious that, when Σ is equipped with this manifold structure, the natural injection of Σ into $\mathcal{M}$ is an immersion. Its differential at any point $p \in \Sigma$ is a linear bijection of the tangent space $\mathbb{C}T_p\Sigma$ onto $\mathcal{V}|_\Sigma$. Because of this, Σ is called an *integral manifold* of $\mathcal{V}$. It is an immersed manifold without self-intersections. In general it will *not* be an embedded submanifold of $\mathcal{M}$.

It is convenient to take Σ to be *maximal*: any other integral manifold of $\mathcal{V}$ that intersects Σ must be an open subset of Σ. Any such maximal integral manifold of $\mathcal{V}$ will be called a *leaf* of $\mathcal{V}$, or of the real structure defined by $\mathcal{V}$, in $\mathcal{M}$. The family of all the leaves of $\mathcal{V}$ will be called the *foliation* defined by $\mathcal{V}$ on $\mathcal{M}$. The leaves of $\mathcal{V}$ are also called *characteristics of* $\mathcal{V}$.

If Ω is an open subset of $\mathcal{M}$, by a *leaf of* $\mathcal{V}$ *in* Ω we mean any connected component of the intersection with Ω of a leaf of $\mathcal{V}$ in $\mathcal{M}$.

THEOREM I.10.2. *Let* $\mathcal{V}$ *define a real involutive structure on* $\mathcal{M}$ *and let* Ω *be an open subset of* $\mathcal{M}$. *Any continuous solution in* Ω *is constant on any leaf of* $\mathcal{V}$ *in* Ω.

PROOF. Let Σ be a leaf of $\mathcal{V}$ in $\mathcal{M}$; denote by Σ_0 a connected component of $\Sigma \cap \Omega$. An arbitrary point $p \in \Sigma_0$ has an open neighborhood U in Ω such that, if $\mathbf{v}$ is a smooth vector field on the manifold Σ_0, the restriction of $\mathbf{v}$ to the connected component of p in $U \cap \Sigma_0$ extends as a $\mathcal{C}^\infty$ section L of $\mathcal{V}$ in U. If now $h \in \mathcal{C}^0(\Omega)$ is a solution, we have $Lh = 0$ in U, hence $\mathbf{v}(h|_{U \cap \Sigma_0}) \equiv 0$, which proves that $h = const.$ in $U \cap \Sigma_0$. Thus h is locally constant in Σ_0; but Σ_0 is connected. ■

The Holomorphic Frobenius Theorem

Let $\mathcal{M}$ now be a *complex manifold*. We denote by $T^{1,0}$ the holomorphic tangent bundle of $\mathcal{M}$. If $(U, z_1, \dots, z_N)$ is a complex coordinate patch in $\mathcal{M}$ ($N = \dim_{\mathbb{C}} \mathcal{M}$) a section of $T^{1,0}$ in U is a linear combination of the partial differentiations $\partial/\partial z_1, \dots, \partial/\partial z_N$. We denote by $T'^{1,0}$ the holomorphic cotangent bundle of $\mathcal{M}$; any section of $T'^{1,0}$ over the domain U is a linear combination of $dz_1, \dots, dz_N$.

By a *holomorphic involutive structure* on $\mathcal{M}$ we shall mean the datum of a *holomorphic* vector subbundle $\mathcal{V}$ of the holomorphic tangent bundle $T^{1,0}$ of $\mathcal{M}$ such that $[\mathcal{V}, \mathcal{V}] \subset \mathcal{V}$. That $\mathcal{V}$ is a holomorphic vector subbundle of $T^{1,0}$ means that $\mathcal{V}$ is a complex vector subbundle of $T^{1,0}$ and also a holomorphic submanifold of $T^{1,0}$ when the latter is equipped with its natural complex structure (defined by the local charts $(z_1,\ldots,z_N,\partial/\partial z_1,\ldots,\partial/\partial z_N)$). We may also define the holomorphic involutive structure of $\mathcal{M}$ by means of the vector subbundle T′ of $T'^{1,0}$ which is the orthogonal of $\mathcal{V}$ for the natural duality between $T^{1,0}$ and $T'^{1,0}$. As before we shall call n and m the fibre dimensions (over $\mathbb{C}$) of $\mathcal{V}$ and T′ respectively. Here also we have $n + m = N$.

The proof of Theorem I.10.1 can be directly translated into the holomorphic category (by replacing everywhere "real" by "holomorphic"):

THEOREM I.10.3. *Any holomorphic involutive structure on the complex manifold $\mathcal{M}$ is locally integrable, in the sense that the cotangent structure bundle* T′ *is locally spanned by the differentials of holomorphic functions.*

Likewise we may define the *holomorphic foliation* of $\mathcal{M}$ defined by $\mathcal{V}$, or by the holomorphic involutive structure of $\mathcal{M}$, exactly as in the real case. We shall also refer to the leaves in this foliation as the *complex characteristics* of $\mathcal{V}$. A leaf of $\mathcal{V}$ in an open subset Ω of $\mathcal{M}$ is any connected component of the intersection of Ω with a leaf of $\mathcal{V}$ in $\mathcal{M}$.

A *distribution solution* in Ω is a distribution u in Ω such that $\mathrm{L}u \equiv 0$ whatever the $\mathcal{C}^\infty$ section L of $\mathcal{V}$ over Ω. The proof of Theorem I.10.2 can be translated in holomorphic terms to yield:

THEOREM I.10.4. *Let the subbundle $\mathcal{V}$ of* $T^{1,0}$ *define a holomorphic involutive structure on the complex manifold $\mathcal{M}$, and let Ω be an open subset of $\mathcal{M}$. Any holomorphic solution in Ω is constant on any leaf of $\mathcal{V}$ in Ω.*

Complex Characteristics in Real-Analytic Structures

We go back to the case where $\mathcal{M}$ is a real manifold but now we assume that $\mathcal{M}$ is of class $\mathcal{C}^\omega$. By using the fact that $\mathcal{M}$ is countable at infinity one can construct a complex manifold $\hat{\mathcal{M}}$ and an embedding $\iota : \mathcal{M} \to \hat{\mathcal{M}}$, of class $\mathcal{C}^\omega$, such that $\iota(\mathcal{M})$ is a maximally real submanifold of $\hat{\mathcal{M}}$. The pair $(\hat{\mathcal{M}},\iota)$ is called a *complexification of $\mathcal{M}$*.

Assume that $\mathcal{M}$ is equipped with an involutive structure of class $\mathcal{C}^\omega$, and as usual let $\mathcal{V}$ and T′ denote the structure bundles. Let $(\hat{\mathcal{M}},\iota)$ be a complexification of $\mathcal{M}$; for the sake of simplicity we shall identify $\mathcal{M}$ to $\iota(\mathcal{M})$ and thus assume $\mathcal{M} \subset \hat{\mathcal{M}}$. It is clear that the real-analytic structure bundles $\mathcal{V}$ and T′ extend holomorphically to some open neighborhood of $\mathcal{M}$ in $\hat{\mathcal{M}}$. Let $(\hat{\mathrm{U}},z_1,\ldots,z_N)$ be a complex coordinate patch in $\hat{\mathcal{M}}$ such that $\mathrm{U} = \hat{\mathrm{U}}\cap\mathcal{M} \neq \emptyset$. Denote by x_j the restriction of z_j to U; $(\mathrm{U},x_1,\ldots,x_N)$ is an analytic coordinate patch in $\mathcal{M}$. Let

$$\mathrm{L} = \sum_{j=1}^{N} c_j(x)\partial/\partial x_j$$

be an analytic section of $\mathcal{V}$ over U. Then the coefficients c_j extend as holomorphic functions in some open neighborhood $\mathcal{O}$ of U in $\hat{\mathrm{U}}$ and we may consider the holomorphic vector field in $\mathcal{O}$,

$$\hat{\mathrm{L}} = \sum_{j=1}^{N} c_j(z)\partial/\partial z_j.$$

The union of all such open neighborhoods $\mathcal{O}$ is an open neighborhood of $\mathcal{M}$ in $\hat{\mathcal{M}}$; by contracting $\hat{\mathcal{M}}$ about $\mathcal{M}$, we may assume that $\hat{\mathcal{M}}$ is equal to the neighborhood in question. By freezing at arbitrary points of their domains of definition all the holomorphic vector fields $\hat{\mathrm{L}}$, we obtain the holomorphic vector subbundle $\hat{\mathcal{V}}$ of $\mathrm{T}^{1,0}$ ($\mathrm{T}^{1,0}$ is the holomorphic tangent bundle of $\hat{\mathcal{M}}$). This defines a holomorphic involutive structure on $\hat{\mathcal{M}}$, with a cotangent structure bundle $\hat{\mathrm{T}}' \subset \mathrm{T}'^{1,0}$.

We derive easily from Theorem I.10.3:

THEOREM I.10.5. *Any analytic involutive structure on the real-analytic manifold $\mathcal{M}$ is locally integrable.*

PROOF. Each point $p \in \mathcal{M}$ has an open neighborhood $\hat{\mathrm{U}}$ in $\hat{\mathcal{M}}$ in which there are m holomorphic functions $h_1,\ldots,h_m$ whose differentials ∂h_j form a basis of $\hat{\mathrm{T}}'$ in $\hat{\mathrm{U}}$. The restrictions of the h_j to $\mathrm{U} = \hat{\mathrm{U}} \cap \mathcal{M}$ are real-analytic functions whose differentials dh_j span T′ over U. ∎

To those leaves of the holomorphic structure bundle $\hat{\mathcal{V}}$ that intersect $\mathcal{M}$ we shall refer as the *complex characteristics* of the real-analytic structure bundle $\mathcal{V}$ (or of the analytic involutive structure on $\mathcal{M}$) in the complexification $\hat{\mathcal{M}}$. Note that any complex characteristic of $\mathcal{V}$ is a holomorphic submanifold of $\hat{\mathcal{M}}$, of complex dimension equal to n, whose holomorphic tangent bundle is equal to the pullback of $\hat{\mathcal{V}}$.

Let U be an open set in which there are m real-analytic (but, in general, complex-valued) functions $Z_1,\ldots,Z_m$ whose differentials span T′ over U. We may apply to these functions the reasoning of section I.7. For instance, after composing the map $Z\colon \mathrm{U} \to \mathbb{C}^m$ with an automorphism of $\mathbb{C}^m$, we may assume that the differentials of the functions $\mathcal{R}e Z_j$ are linearly independent and take those functions as part of a local coordinate system in U, $\{x_1,\ldots,x_m,t_1,\ldots,t_n\}$ (centered at some point $0 \in \mathrm{U}$). We may in fact assume that (I.7.23) and (I.7.24) hold, here with the added property that the functions $\Phi_j(x,t)$ are real-analytic (and real-valued). This, of course, allows us to solve with respect to x, in a neighborhood $\hat{\mathrm{U}}$ of 0 in $\hat{\mathcal{M}}$, the holomorphic equations

$$Z(x,t) = z. \qquad \text{(I.10.2)}$$

Here, x, z, and t are regarded as complex variables in some neighborhood of

0 in $\mathbb{C}^m$, $\mathbb{C}^m$, and $\mathbb{C}^n$ respectively. Let $x = H(z,t)$ be the unique solution of (I.10.2) that vanishes when $z = 0$, $t = 0$. Note then that, for $t' \in \mathbb{R}^n$ sufficiently near the origin, the functions of (x,t),

$$H_j(Z(x,t),t') \ (j = 1,\ldots,m),$$

are analytic solutions in a suitable neighborhood of the origin in $\mathcal{M}$. We have

$$H(Z(x,t),t) \equiv x. \tag{I.10.3}$$

In particular, (I.10.3) shows that if we substitute the solutions $H_i(Z,0)$ for the solutions Z_i in an open neighborhood U of 0 in $\mathcal{M}$, we may assume that the latter satisfy

$$Z_i(x,0) = x_i, \ i = 1,\ldots,m. \tag{I.10.4}$$

This observation allows us to prove the version of the Cauchy-Kovalevska theorem relevant to the present setup. Take $\mathrm{U} = \mathrm{V} \times \mathrm{W}$ with V an open neighborhood of 0 in x-space $\mathbb{R}^m$ and W one in t-space $\mathbb{R}^n$.

THEOREM I.10.6. *Suppose* (I.10.4) *holds. Let h_0 denote an analytic function of x in the open neighborhood* V. *There is a unique analytic solution h in an open neighborhood of the origin in $\mathcal{M}$,* $\mathrm{U}' \subset \mathrm{U}$, *such that $h(x,0) = h_0(x)$ if* $(x,0) \in \mathrm{U}'$.

PROOF. The existence follows by taking $h(x,t) = h_0(Z(x,t))$. To prove the uniqueness we must show that if an analytic solution h vanishes at $t = 0$, it vanishes identically (in a neighborhood of the origin). It suffices to use vector fields L_j as in (I.7.27) and expand their coefficients as series in powers of t with coefficients that are analytic functions of x. By equating to zero the coefficients of each power of t in the equations $\mathrm{L}_j h = 0$ $(j = 1,\ldots,n)$, one easily concludes that $h \equiv 0$. ∎

The Jacobian determinant of $Z(x,t)$ with respect to x is equal to 1 at $t = 0$, which shows that, indeed, we may use the complex coordinate system $\{z_1,\ldots,z_m,t_1,\ldots,t_n\}$ in a neighborhood of 0 in $\hat{\mathcal{M}}$ that we take to be $\hat{\mathrm{U}}$. In these coordinates, the vector fields $\partial/\partial t_1,\ldots,\partial/\partial t_n$ form a basis of $\hat{\mathcal{V}}$ over $\hat{\mathrm{U}}$; an integral manifold of $\hat{\mathcal{V}}$ in $\hat{\mathrm{U}}$ is defined by an equation $z = z_0$. In the coordinates x_j, t_k this submanifold is defined by the equation $x = H(z_0,t)$. Let then (x_0,t_0) be a point in $\mathrm{U} \subset \mathcal{M}$ suitably close to 0. The image of the map

$$t \to (H(Z(x_0,t_0),t),t)$$

is a piece of the complex characteristic of $\mathcal{V}$ that passes through $(x_0,t_0) \in \mathcal{M}$.

Let us call *fibres* of the map Z in a subset S of U, the preimages of points under the restriction $\mathrm{Z}|_{\mathrm{S}} : \mathrm{S} \to \mathbb{C}^m$ (cf. (I.7.20)). The fibres of Z, say in U, need not be connected (cf. Example I.10.1 below).

PROPOSITION I.10.1. *There is a subneighborhood* $U' \subset U$ *of* 0 *such that every fibre of the map* Z *in* U′ *is contained in one complex characteristic of* $\mathcal{V}$.

PROOF. Let z_j and t_k denote the complex coordinates in the open neighborhood $\hat{U}$ of 0 in $\hat{\mathcal{M}}$ introduced above; the integral manifolds of $\hat{\mathcal{V}}$ in $\hat{U}$ are defined by the equations $z = z_0$. Take $\hat{U} = \hat{V} \times \hat{W}$ with $\hat{V}$ (resp., $\hat{W}$) a polydisk in z-space $\mathbb{C}^m$ (resp., in t-space $\mathbb{C}^n$) centered at the origin. With such a choice those integral manifolds are connected and, thus, each is a piece of a single characteristic of $\hat{\mathcal{V}}$ in $\hat{\mathcal{M}}$. Since a fibre of the map Z is precisely the set of points (x,t) such that $Z(x,t) = z_0$ for some $z_0 \in \mathbb{C}^m$, we see that the choice $U' = \hat{U} \cap \mathcal{M}$ satisfies the requirements in Proposition I.10.1. ■

PROPOSITION I.10.2. *Let h denote an analytic solution in an open set* $\Omega \subset \mathcal{M}$ *containing* 0. *There is an open neighborhood* $\mathcal{N}_h \subset U$ *of* 0 *such that h is constant on the fibres of* Z *in* $\mathcal{N}_h$.

PROOF. Let $\hat{U} = \hat{V} \times \hat{W}$ be as in the proof of Proposition I.10.1. We may as well take $\mathcal{N}_h \subset U' = \hat{U} \cap \mathcal{M}$. Since h is analytic it extends as a holomorphic function $\hat{h}$ in an open neighborhood $\hat{U}_1 \subset \hat{U}$ of 0; take $\hat{U}_1 = \hat{V}_1 \times \hat{W}_1$ with $\hat{V}_1$ and $\hat{W}_1$ open polydisks centered at 0 in $\mathbb{C}^m$ and $\mathbb{C}^n$ respectively. By Theorem I.10.4 we know that $\hat{h} = const.$ on each submanifold $z = z_0 \in \hat{V}_1$. It suffices therefore to take $\mathcal{N}_h = \hat{U}_1 \cap \mathcal{M}$. ■

It follows from Proposition I.10.2 that the germs of sets at the point 0, defined by the fibres of the map Z, are invariants attached to the locally integrable structure of $\mathcal{M}$. Indeed, if $(U^\#, Z^\#)$ is another "chart" analogous to (U,Z) (i.e., each function $Z_j^\#$ is analytic and $dZ_1^\#, \dots, dZ_m^\#$ span T′ over $U^\#$) then there is an open neighborhood of 0, $\mathcal{N} \subset U \cap U^\#$, such that each function $Z_j^\#$, and therefore also the map $Z^\#$, is constant on the fibres of Z in $\mathcal{N}$, while Z is constant on the fibres of $Z^\#$ in $\mathcal{N}$. In other words, the fibres of the maps Z and $Z^\#$ in $\mathcal{N}$ are the same.

EXAMPLE I.10.1. Consider the *Mizohata structure* on $\mathbb{R}^2$ (where the coordinates are denoted by x and t) defined by the vector field $L = \partial/\partial t - \imath t \partial/\partial x$ (Example I.2.7).

We may extend L as a holomorphic vector field in $\mathbb{C}^2$, simply by regarding x and t as the complex coordinates. This vector field annihilates the holomorphic function $Z(x,t) = x + \imath t^2/2$. Here the solution of (I.10.2) is $H(z,t) = z - \imath t^2/2$. The complex characteristic in $\mathbb{C}^2$ of the Mizohata structure that passes through (x_0,t_0) is the complex curve

$$\mathbb{C} \ni t \to (x_0 + \imath(t_0^2 - t^2)/2, t) \in \mathbb{C}^2. \tag{I.10.5}$$

Notice that the curve (I.10.5) intersects the real space $\mathbb{R}^2$ at the points $(x_0,$

$\pm t_0$). The fibres of the map Z in $\mathbb{R}^2$ consist precisely of such pairs of points. Notice that, in every neighborhood of any point $(x,0)$, there are fibres of the map Z that consist of two distinct points. ∎

The next example shows that in general a result such as Proposition I.10.2 cannot be "globalized."

EXAMPLE I.10.2. Take $\mathcal{M} = \Delta \times \mathbb{R}$ where Δ is the open unit disk in $\mathbb{C}$; let $z = x + \imath y$ denote the complex coordinate in Δ and t the real one in $\mathbb{R}$. Then take

$$Z_1 = x + \imath(\cos t - 1),\ Z_2 = y + \imath \sin t.$$

The differentials dZ_1, dZ_2 span the cotangent structure bundle in an analytic involutive structure on $\mathcal{M}$. Consider the function

$$h(x,y,t) = e^{\imath t/2}(1 - ze^{-\imath t})^{\frac{1}{2}},$$

where $(\ \ )^{\frac{1}{2}}$ stands for the main branch of the square root. The function h is everywhere defined and analytic in $\mathcal{M}$. Locally it is equal to some determination of the square root of $1 - \imath(Z_1 + \imath Z_2)$ and therefore it is a solution. The fibre of $Z = (Z_1,Z_2)$ at the origin is the set of points $z = 0$, $t = 2k\pi$, $k \in \mathbb{Z}$. At such points $h = (-1)^k$.

We view $\mathbb{C}\times\mathbb{C}\times\mathbb{C}^1$ (with complex coordinates z, ζ, t) as a complexification of $\Delta\times\mathbb{R}^1$ (by identifying Δ to the open unit disk in the antidiagonal $\zeta = \bar{z}$ in $\mathbb{C}\times\mathbb{C}$). The complex characteristics of the structure are the holomorphic curves defined by the parametric equations

$$z = z_0 - \imath(e^{\imath t} - 1),\ \zeta = \zeta_0 - \imath(e^{-\imath t} - 1). \tag{I.10.6}$$

The curve (I.10.6) intersects the real space $\Delta \times \mathbb{R}^1$ at the points $(z_0, \bar{z}_0, 2k\pi)$. ∎

Whenever the (analytic) locally integrable structure of $\mathcal{M}$ is a CR structure, we can find an analytic "chart" (U,Z) such that the map $Z : U \to \mathbb{C}^m$ is injective (cf. Proposition I.7.2).

I.11. Orbits and Leaves. Involutive Structures of Finite Type

Let Ω be an open subset of the $\mathcal{C}^\infty$ manifold $\mathcal{M}$. If $\mathbf{v}$ is a $\mathcal{C}^\infty$ real vector field in Ω denote by $\Phi_{\mathbf{v}}(t)$ its *flow*: given any point $\mathfrak{p}_0 \in \Omega$, $t \to \mathfrak{p}(t) = \Phi_{\mathbf{v}}(t)\mathfrak{p}_0$ is a $\mathcal{C}^\infty$ map from an open interval $\mathcal{J}(\mathbf{v},\mathfrak{p}_0) \subset \mathbb{R}^1$, containing zero, into Ω; it is the unique solution of the initial value problem

$$\dot{p}(t) = \mathbf{v}(p(t)),\ p(0) = p_0. \tag{I.11.1}$$

The image of $\mathcal{J}(\mathbf{v},p_0)$ under the map $t \to \Phi_{\mathbf{v}}(t)p_0$ is called an *integral curve* of $\mathbf{v}$ through p_0. Note that such an integral curve is connected. If $\mathbf{v}|_{p_0} = 0$ then $p(t) \equiv p_0$ for all $t \in \mathbb{R}^1$ and any integral curve of $\mathbf{v}$ through p_0 reduces to the single point $\{p_0\}$.

It is convenient to reserve the notation $\mathcal{J}(\mathbf{v},p_0)$ for the largest open interval in $\mathbb{R}^1$ containing zero in which the solution $p(t)$ of (I.11.1) is defined and valued in Ω. The image of the map $\mathcal{J}(\mathbf{v},p_0) \ni t \to p(t)$ is then called *the* integral curve of $\mathbf{v}$ in Ω through p_0.

To any compact subset K of Ω there is an open interval $\mathcal{J}(\mathbf{v},\mathrm{K}) \subset \mathbb{R}^1$, containing zero, such that, for each $t \in \mathcal{J}(\mathbf{v},\mathrm{K})$, $p \to \Phi_{\mathbf{v}}(t)p$ is a diffeomorphism of some open neighborhood Ω' of K in Ω onto another open subset of Ω. One expresses this property by saying that $\Phi_{\mathbf{v}}(t)$ is a *local diffeomorphism* on Ω. Note that $(t,p) \to \Phi_{\mathbf{v}}(t)p$ is a $\mathscr{C}^\infty$ map of $\mathcal{J}(\mathbf{v},\mathrm{K})\times\Omega'$ into Ω. Of course $\Phi_{\mathbf{v}}(0) = \mathrm{Id}$.

One can compose and invert local diffeomorphisms provided one keeps track of the intervals of definition. Let $\mathbf{v}_1$, $\mathbf{v}_2$ be two $\mathscr{C}^\infty$ real vector fields in Ω and p_0 a point of Ω. If $t_1 \in \mathcal{J}(\mathbf{v}_1,p_0)$ and $t_2 \in \mathcal{J}(\mathbf{v}_2,\Phi_{\mathbf{v}_1}(t_1)p_0)$, then we may form $\Phi_{\mathbf{v}_2}(t_2)\Phi_{\mathbf{v}_1}(t_1)p_0$. If $\mathbf{v} = \mathbf{v}_1 = -\mathbf{v}_2$, and if we have $t_2 \in \mathcal{J}(-\mathbf{v},\Phi_{\mathbf{v}}(t_1)p_0)$, then

$$\Phi_{-\mathbf{v}}(t_2)\Phi_{\mathbf{v}}(t_1)p_0 = \Phi_{\mathbf{v}}(t_1 - t_2)p_0. \tag{I.11.2}$$

Let now $\mathcal{M}$ be equipped with a formally integrable structure, and let $\mathcal{V}$ stand for the tangent structure bundle. We shall denote by $\mathfrak{g}(\mathcal{V})$ the *Lie algebra*, for the commutation bracket, generated by the vector fields $\mathcal{R}e\mathrm{L}$, where L is a $\mathscr{C}^\infty$ section of $\mathcal{V}$ over $\mathcal{M}$: $\mathfrak{g}(\mathcal{V})$ is the smallest linear space (over $\mathbb{R}$) of $\mathscr{C}^\infty$ vector fields in $\mathcal{M}$ having the following two properties: (*i*) the vector fields $\mathcal{R}e\mathrm{L}$, $\mathrm{L} \in \mathscr{C}^\infty(\mathcal{M};\mathcal{V})$, belong to it; (*ii*) if two vector fields $\mathbf{v}_1$, $\mathbf{v}_2$ belong to it their bracket $[\mathbf{v}_1,\mathbf{v}_2] = \mathbf{v}_1\mathbf{v}_2 - \mathbf{v}_2\mathbf{v}_1$ also does. Note that $\mathfrak{g}(\mathcal{V})$ can be regarded as a (left-) module over the ring of real-valued $\mathscr{C}^\infty$ functions in $\mathcal{M}$.

If p is any point of $\mathcal{M}$ we shall denote by $\mathfrak{g}(\mathcal{V})|_p$ the set of tangent vectors to $\mathcal{M}$ at p obtained by freezing at p all the vector fields that belong to $\mathfrak{g}(\mathcal{V})$; $\mathfrak{g}(\mathcal{V})|_p$ is a linear subspace of $\mathrm{T}_p\mathcal{M}$; in general its dimension varies with p.

EXAMPLE I.11.1. In the plane $\mathbb{R}^2$ (with coordinates t and x) let $\mathcal{V}$ be spanned by the vector field $\mathrm{L} = \partial/\partial t - \imath x\partial/\partial x$ (*cf.* Example I.4.2). Then $\mathfrak{g}(\mathcal{V})|_p$ is one-dimensional if the point p lies on the t-axis and is two-dimensional otherwise. ■

We define an equivalence relation between pairs of points p, q of Ω by means of the Lie algebra $\mathfrak{g}(\mathcal{V})$: Let us write $p \underset{\Omega}{\approx} q$ if there is a finite set of points p_j ($j = 0,\ldots,r$) in Ω such that $p_0 = p$, $p_r = q$ and such that the following is true:

for each $j = 1,\ldots,r$, *the points* $\mathfrak{p}_{j-1}$ *and* $\mathfrak{p}_j$ *lie on an integral curve, entirely contained in* Ω, *of a vector field belonging to* $\mathfrak{g}(\mathcal{V})$. (I.11.3)

DEFINITION I.11.1. *Any equivalence class for the relation* $\mathfrak{p} \underset{\Omega}{\approx} \mathfrak{q}$ *will be called an* orbit *of* $\mathcal{V}$ (or of the formally integrable structure of $\mathcal{M}$) *in* Ω.

EXAMPLE I.11.2. Let the vector bundle $\mathcal{V} \subset \mathbb{C}T\mathbb{R}^2$ be defined as in Example I.11.1. There are three orbits in $\mathbb{R}^2$: the t-axis and each one of the open half-planes $x > 0$ and $x < 0$. ■

Let $\Omega_1 \subset \Omega_2$ be two open subsets of $\mathcal{M}$. Two points $\mathfrak{p}$, $\mathfrak{q}$ may belong to the same orbit in Ω_2 and yet belong to distinct orbits in Ω_1:

EXAMPLE I.11.3. Take $\mathcal{M} = \mathbb{R}^2$. Assume that $\mathcal{V}$ is spanned by the vector field

$$\mathrm{L} = \partial/\partial t + \imath b(x,t)\partial/\partial x,$$

where b is a real-valued $\mathscr{C}^\infty$ function in $\mathbb{R}^2$. Suppose $b \equiv 0$ in the open unit disk $\Omega = \{(x,t) \in \mathbb{R}^2;\, x^2+t^2 < 1\}$ and $b > 0$ if $x^2+t^2 > 1$. The orbits in Ω are the vertical segments $t^2 < 1 - x_0^2$, $x = x_0$ $(x_0^2 < 1)$, whereas in $\mathbb{R}^2$ there is only one orbit, the whole plane. ■

Let $\mathfrak{p}$ be any point in Ω. Select arbitrarily a finite number of vector fields $\mathbf{v}_j \in \mathfrak{g}(\mathcal{V})$ $(0 \leq j \leq r)$. Select any number $t_0 \in \mathcal{J}(\mathbf{v}_0,\mathfrak{p})$; after this use induction on $j = 1,\ldots,r$, to select any number $t_j \in \mathcal{J}(\mathbf{v}_j,[\Phi_{\mathbf{v}_{j-1}}(t_{j-1})\circ \cdots \circ\Phi_{\mathbf{v}_0}(t_0)]\mathfrak{p})$. The compose $\Phi_{\mathbf{v}_r}(t_r)\circ \cdots \circ\Phi_{\mathbf{v}_0}(t_0)$ is a diffeomorphism of an open neighborhood of $\mathfrak{p}$ in Ω onto one of $[\Phi_{\mathbf{v}_r}(t_r)\circ \cdots \circ\Phi_{\mathbf{v}_0}(t_0)]\mathfrak{p}$; we may let its differential act on the tangent space to $\mathcal{M}$ at $\mathfrak{p}$. This describes the action of the lie algebra $\mathfrak{g}(\mathcal{V})$ on $\mathrm{T}\mathcal{M}|_\Omega$, and allows us to talk of sets in $\mathrm{T}\mathcal{M}|_\Omega$ that are $\mathfrak{g}(\mathcal{V})$-*invariant*. In particular we may introduce the smallest $\mathfrak{g}(\mathcal{V})$-invariant subset of $\mathrm{T}\mathcal{M}|_\Omega$ which, given any point $\mathfrak{p}$ of Ω, contains $\mathfrak{g}(\mathcal{V})|_\mathfrak{p}$. We shall denote it by $\hat{\mathscr{G}}(\mathcal{V},\Omega)$. We call $\hat{\mathscr{G}}(\mathcal{V},\Omega)|_\mathfrak{p}$ the fibre of $\hat{\mathscr{G}}(\mathcal{V},\Omega)$ at $\mathfrak{p}$.

EXAMPLE I.11.4. Let $\mathcal{V}$ be the vector subbundle of $\mathbb{C}T\mathbb{R}^2$ defined in Example I.11.3. We see that $\hat{\mathscr{G}}(\mathcal{V},\mathbb{R}^2)|_\mathfrak{p} = \mathrm{T}_\mathfrak{p}\mathbb{R}^2$ for each $\mathfrak{p}$ and thus $\mathfrak{g}(\mathcal{V})|_\mathfrak{p} \neq \hat{\mathscr{G}}(\mathcal{V},\mathbb{R}^2)|_\mathfrak{p}$ when $|\mathfrak{p}| < 1$. On the other hand, if Ω is the open unit disk and if $\mathfrak{p} \in \Omega$, $\hat{\mathscr{G}}(\mathcal{V},\Omega)|_\mathfrak{p} = \mathfrak{g}(\mathcal{V})|_\mathfrak{p}$ is the line spanned by $\partial/\partial t$. ■

The following statement is a special case of Theorem 4.1 in Sussman [1].

THEOREM I.11.1. *Each orbit* Σ *of* $\mathcal{V}$ *in the open set* $\Omega \subset \mathcal{M}$ *is an immersed submanifold of* Ω *without self-intersections. The tangent space to* Σ *at any one of its points,* $\mathfrak{p}$, *is equal to* $\hat{\mathscr{G}}(\mathcal{V},\Omega)|_\mathfrak{p}$.

There are important particular situations where the preceding result can be improved, as shown in the following result of Hermann [1], also proved in Sussmann [1]:

THEOREM I.11.2. *Suppose the Lie algebra $\mathfrak{g}(\mathcal{V})$ has the following property:*

To each point $\mathfrak{p} \in \mathcal{M}$ there are vector fields $\mathbf{v}_1,\dots,\mathbf{v}_r$ in $\mathfrak{g}(\mathcal{V})$ (with $r \geq 0$ depending on the point $\mathfrak{p}$) *such that, for all points $\mathfrak{q}$ in some neighborhood of $\mathfrak{p}$,* (I.11.4)

$$\mathbf{v}_1|_{\mathfrak{q}},\dots,\mathbf{v}_r|_{\mathfrak{q}} \text{ span } \mathfrak{g}(\mathcal{V})|_{\mathfrak{q}}; \tag{I.11.5}$$

to every $\mathbf{v} \in \mathfrak{g}(\mathcal{V})$ there exist an open neighborhood U of *$\mathfrak{p}$ and $\mathcal{C}^\infty$ functions c_{ij} in* U *($1 \leq i, j \leq r$) such that, in* U, (I.11.6)

$$[\mathbf{v},\mathbf{v}_i] = \sum_{j=1}^{r} c_{ij}\mathbf{v}_j.$$

Then, given any open subset Ω of $\mathcal{M}$ and any point $\mathfrak{p} \in \Omega$, the tangent space at $\mathfrak{p}$ to the orbit of $\mathcal{V}$ in Ω (passing through $\mathfrak{p}$) *is equal to $\mathfrak{g}(\mathcal{V})|_{\mathfrak{p}}$.*

The conclusion in Theorem I.11.2 is that, whatever $\mathfrak{p} \in \mathcal{M}$, $\mathfrak{g}(\mathcal{V})|_{\mathfrak{p}} = \hat{\mathcal{G}}(\mathcal{V},\Omega)|_{\mathfrak{p}} = \hat{\mathcal{G}}(\mathcal{V},\mathcal{M})|_{\mathfrak{p}}$. An immersed submanifold $\Sigma \subset \mathcal{M}$ (with no self-intersections), such that $T_{\mathfrak{p}}\Sigma = \mathfrak{g}(\mathcal{V})|_{\mathfrak{p}}$ for all $\mathfrak{p} \in \Sigma$, is called an *integral manifold of* $\mathfrak{g}(\mathcal{V})$. By its very definition, an orbit Σ which is an integral manifold of $\mathfrak{g}(\mathcal{V})$ is a *maximal* integral manifold of $\mathfrak{g}(\mathcal{V})$: any other *connected* integral submanifold of $\mathfrak{g}(\mathcal{V})$ intersecting Σ must be an open subset of Σ.

Often, in the sequel, when each orbit of $\mathcal{V}$ is an integral manifold of $\mathfrak{g}(\mathcal{V})$, we shall refer to it as a *leaf* of $\mathcal{V}$ (or of the involutive structure of $\mathcal{M}$). The set of all leaves of $\mathcal{V}$ will be called the *foliation* on $\mathcal{M}$ defined by $\mathcal{V}$ (or by the involutive structure of $\mathcal{M}$).

The reader will notice that Condition (I.11.5) does not require that the tangent vectors $\mathbf{v}_j|_{\mathfrak{p}}$ ($j = 1,\dots,r$) be linearly independent. Note also that Condition (I.11.4) is satisfied when, in the neighborhood of each point, $\mathfrak{g}(\mathcal{V})$ is finitely generated as a module over the ring of real-valued $\mathcal{C}^\infty$ functions. There are two important cases in which this happens.

Real structures. An arbitrary point $\mathfrak{p}$ of $\mathcal{M}$ has an open neighborhood U in which there is a smooth basis of $\mathcal{V}$ consisting of real vector fields $L_1,\dots,L_n$. In U any element of $\mathfrak{g}(\mathcal{V})$ is a linear combination of the L_j with $\mathcal{C}^\infty$ coefficients. Thus $\mathfrak{g}(\mathcal{V})|_{\mathfrak{p}} = T_{\mathfrak{p}}\mathcal{M} \cap \mathcal{V}_{\mathfrak{p}}$ is n-dimensional whatever $\mathfrak{p}$ and the leaves are immersed n-dimensional submanifolds of $\mathcal{M}$. This is one of the standard versions of the Frobenius theorem (see section 10).

Analytic structures. Here we assume that $\mathcal{M}$ is a real-analytic manifold and that $\mathcal{V}$ is a real-analytic vector subbundle of $\mathbb{C}T\mathcal{M}$. But we continue to denote by $\mathfrak{g}(\mathcal{V})$ the Lie algebra generated by the vector fields. $\mathcal{R}e\mathrm{L}$ where L is any $\mathcal{C}^\infty$ section of $\mathcal{V}$ over $\mathcal{M}$.

PROPOSITION I.11.1. *Let $\mathcal{M}$ be a real-analytic manifold and $\mathcal{V}$ a real-analytic vector subbundle of $\mathbb{C}T\mathcal{M}$. Then Property* (I.11.4) *holds.*

PROOF. If Ω is any open subset of $\mathcal{M}$ we denote by $\mathfrak{g}^\omega(\mathcal{V},\Omega)$ the Lie subalgebra of $\mathfrak{g}(\mathcal{V})$ consisting of those vector fields that are analytic in Ω. Let p be an arbitrary point of $\mathcal{M}$. It will suffice to show that there is an open neighborhood U of p and vector fields $\mathbf{v}_1,\ldots,\mathbf{v}_r$ in $\mathfrak{g}^\omega(\mathcal{V},\mathrm{U})$ such that the following is true: if $\mathbf{v} \in \mathfrak{g}^\omega(\mathcal{V},\Omega)$ for some open set $\Omega \ni p$, then $\mathbf{v}$ is equal, in some open neighborhood of p in $\mathrm{U}\cap\Omega$, to a linear combination, with real analytic (and real-valued) coefficients, of $\mathbf{v}_1,\ldots,\mathbf{v}_r$.

For then, possibly after contracting U about p, we may select a basis $\mathrm{L}_1,\ldots,\mathrm{L}_n$ of $\mathcal{V}$ in U made up of $\mathcal{C}^\omega$ vector fields in U. After some further contraction of U and multiplication of each vector field L_j by a cutoff function we may assume that the real and imaginary parts of the vector fields L_j are elements of $\mathfrak{g}^\omega(\mathcal{V},\mathrm{U})$. Let now $\mathbf{v}$ be an arbitrary element of $\mathfrak{g}(\mathcal{V})$, not necessarily analytic in U. Then $\mathbf{v}$ is a finite sum of multibrackets $[\theta_1,[\theta_2,\ldots,[\theta_{q-1},\theta_q]\cdots]$ in which the vector fields θ_j are of the form $\mathcal{R}e\mathrm{L}$, $\mathrm{L} \in \mathcal{C}^\infty(\mathcal{M};\mathcal{V})$. But in U L is a linear combination with $\mathcal{C}^\infty$ coefficients of the analytic vector fields L_j. We conclude that, in U, $\mathbf{v}$ is a linear combination with $\mathcal{C}^\infty$ coefficients of elements of $\mathfrak{g}^\omega(\mathcal{V},\mathrm{U})$. By way of consequence, in some subneighborhood of p, it is a linear combination with $\mathcal{C}^\infty$ coefficients of $\mathbf{v}_1,\ldots,\mathbf{v}_r$.

Now, let $\mathcal{A}_p$ denote the ring of germs of real-analytic functions at a point p of $\mathcal{M}$. By means of an analytic basis of $T\mathcal{M}$ over an open neighborhood of p the space of germs of analytic vector fields at p can be identified to $\mathcal{A}_p^N$ ($N = \dim \mathcal{M}$). Since the ring $\mathcal{A}_p$ is noetherian any submodule of $\mathcal{A}_p^N$, such as the set of germs at p of elements of $\mathfrak{g}^\omega(\mathcal{V},\mathrm{U})$, is finitely generated over $\mathcal{A}_p$. This implies at once our claim. ∎

By combining Theorem I.11.2 and Proposition I.11.1 we derive the following special case of the classical result of Nagano [1]:

THEOREM I.11.3. *Suppose that $\mathcal{M}$ is real-analytic and that $\mathcal{V}$ is a real-analytic vector subbundle of $\mathbb{C}T\mathcal{M}$. The orbits of $\mathcal{V}$ are analytic immersed submanifolds of $\mathcal{M}$ without self-intersections; they are integral manifolds of* $\mathfrak{g}(\mathcal{V})$.

We return to the general case of a $\mathcal{C}^\infty$ manifold $\mathcal{M}$ equipped with an involutive structure.

DEFINITION I.11.2. *The involutive structure of $\mathcal{M}$ is said to be* of finite type *at a point $p \in \mathcal{M}$ if $\mathfrak{g}(\mathcal{V})|_p = T\mathcal{M}|_p$.*

The involutive structure of $\mathcal{M}$ is said to be of finite type if it is of finite type at every point of $\mathcal{M}$.

If the involutive structure of $\mathcal{M}$ is of finite type at p, the orbit of $\mathcal{V}$ through p must be an open subset of $\mathcal{M}$. If the structure of $\mathcal{M}$ is of finite type everywhere, the orbits of $\mathcal{V}$ are the connected components of $\mathcal{M}$.

There exist involutive structures that are not of finite type and in which the whole manifold $\mathcal{M}$ is the only orbit of $\mathcal{V}$. This is what happens in Example I.11.3.

EXAMPLE I.11.5. In order that a real structure be of finite type it is necessary and sufficient that $\mathcal{V} = \mathbb{C}T\mathcal{M}$. ■

EXAMPLE I.11.6. Every elliptic structure, in particular every complex structure, is of finite type. ■

EXAMPLE I.11.7. There are CR structures that are not of finite type. The simplest example is that of $\mathbb{C}\times\mathbb{R}$ (with coordinates z, s) equipped with the structure defined by the vector field $\partial/\partial\bar{z}$. The orbits, or rather, the leaves, in this structure, are the complex curves $s = const.$ ■

EXAMPLE I.11.8. The CR structure defined on $\mathbb{R}^3$ (with coordinates x, y, s) by the Lewy vector field $\mathrm{L} = \partial/\partial\bar{z} - \imath z\partial/\partial s$ is of finite type. In this case the Lie algebra $\mathfrak{g}(\mathcal{V})$ contains the vector fields $\partial/\partial x + 2y\partial/\partial s$, $\partial/\partial y - 2x\partial/\partial s$ and $(2\imath)^{-1}[\mathrm{L},\bar{\mathrm{L}}] = \partial/\partial s$. ■

I.12. A Model Case: Tube Structures

Throughout the present section the variable in $\mathbb{R}^m$ will be denoted by $x = (x_1,\dots,x_m)$, the one in $\mathbb{R}^n$ by $t = (t_1,\dots,t_n)$; Ω will be a nonempty open subset of $\mathbb{R}^n$.

DEFINITION I.12.1. *By a* tube structure *on $\mathbb{R}^m\times\Omega$ we shall mean the datum of m functions*

$$Z_j = x_j + \imath\Phi_j(t),\ j = 1,\dots,m, \tag{I.12.1}$$

where the Φ_j are real-valued $\mathcal{C}^\infty$ functions in Ω.

We may also consider tube structures of class $\mathcal{C}^r$ for $r < +\infty$ or $r = \omega$. It simply means that the map

$$\Phi = (\Phi_1,\dots,\Phi_m) : \Omega \to \mathbb{R}^m$$

is of class $\mathcal{C}^r$. When $r = \omega$ we say that the tube structure is analytic.

The tube structure on $\mathbb{R}^m \times \Omega$ defined by the functions (I.12.1) may, and shall, be identified to the locally integrable structure on $\mathbb{R}^m \times \Omega$ in which the cotangent structure bundle T′ is spanned by the differentials $dZ_j = dx_j + \imath d\Phi_j$. The tangent structure bundle $\mathcal{V}$ is spanned by the vector fields

$$L_k = \partial/\partial t_k - \imath \sum_{j=1}^{m} (\partial\Phi_j/\partial t_k)\partial/\partial x_j,\ k = 1,\dots,n, \qquad \text{(I.12.2)}$$

which obviously commute.

The reason for the name "tube structure" comes from the following:

EXAMPLE I.12.1. Take $m = n$. The complex structure on $\mathbb{R}^n \times \Omega$ defined by the functions $Z_j = x_j + \imath t_j$, $j = 1,\dots,n$, is a tube structure. It is the complex structure inherited from $\mathbb{C}^n$ (where the complex variable is $z = x + \imath t$); and $\mathbb{R}^n \times \Omega$ is the *tube with base* Ω; it is often denoted by $\mathbb{R}^n + \imath\Omega$. ∎

PROPOSITION I.12.1. *In order that the tube structure on* $\mathbb{R}^m \times \Omega$ *defined by the functions* (I.12.1) *be a real structure it is necessary and sufficient that* $\Phi \equiv 0$.

Evident.

PROPOSITION I.12.2. *In order that the tube structure on* $\mathbb{R}^m \times \Omega$ *defined by the functions* (I.12.1) *be a* CR *structure it is necessary and sufficient that the rank of* Φ *be equal to n at every point of* Ω.

PROOF. To say that the rank of Φ is equal to n is to say that the differentials $d\Phi_1,\dots,d\Phi_m$ have the same span as $dt_1,\dots,dt_n$; it is equivalent to saying that $dZ_1,\dots,dZ_m,d\bar{Z}_1,\dots,d\bar{Z}_m$ span the whole cotangent space to $\mathbb{R}^m \times \Omega$. ∎

EXAMPLE I.12.2. The tube structure on $\mathbb{R}^3$ defined by the functions $Z_1 = x_1 + \imath t$, $Z_2 = x_2 + \imath t^2$ is a CR structure. ∎

Here is an example of a tube structure that is not CR:

EXAMPLE I.12.3. The Mizohata structure on $\mathbb{R}^2$ (see Example I.2.7) is a tube structure, defined by the function $Z = x + \imath t^2/2$. ∎

Let us denote by ξ_j and τ_k the coordinates in the fibres of the cotangent bundle $T^*(\mathbb{R}^m \times \Omega)$ that are associated with the coordinates x_j and t_k in the base. The symbols of the vector fields (I.12.2) are given by

$$\sigma(L_k) = \tau_k - \imath \sum_{j=1}^{m} (\partial\Phi_j/\partial t_k)\xi_j. \qquad \text{(I.12.3)}$$

It follows at once that the characteristic set T° (Definition I.2.2) in the tube structure defined by the functions (I.12.1) is the subset of $T^*(\mathbb{R}^m \times \Omega)$ defined by the equations

$$d_t[\Phi(t)\cdot\xi] = 0, \tau = 0. \tag{I.12.4}$$

We see that T° is invariant under translations in x-space, as should be expected. If then we disregard the variables x and τ, we see that we may identify T° to the subset of (t,ξ)-space $\Omega \times \mathbb{R}^m$ consisting of the *critical points* of the function of t, $\Phi(t)\cdot\xi$.

EXAMPLE I.12.4. Suppose $m = 1$ (and $n \geq 1$ arbitrary). The tube structure on $\mathbb{R}^1 \times \Omega$ is defined by a single function, $Z = x + \imath\Phi(t)$. The characteristic set T° consists of all the points (x,t,ξ,τ) such that

$$d\Phi(t) = \tau = 0. \blacksquare \tag{I.12.5}$$

PROPOSITION I.12.3. *In order that the tube structure on $\mathbb{R}^m \times \Omega$ defined by the functions* (I.12.1) *be elliptic it is necessary and sufficient that the rank of Φ be equal to m at every point of Ω.*

PROOF. In order that the structure be elliptic it is necessary and sufficient that, whatever $t \in \Omega$, $d\Phi(t)\cdot\xi = 0 \Rightarrow \xi = 0$. This means exactly that the rank of the $n \times m$ real matrix $(\partial\Phi_j/\partial t_k)_{1\leq j\leq m, 1\leq k\leq n}$ is equal to m. ■

Let $t_0 \in R^n$ and $\xi \in \mathbb{R}^m\backslash\{0\}$ be such that $d\Phi(t_0)\cdot\xi = 0$. The differential form $\xi\cdot dx = \sum_{j=1}^m \xi_j dx_j$ is then an element of the charactertistic set at any point (x,t_0), $x \in \mathbb{R}^m$. Note that

$$(2\imath)^{-1}\langle\xi\cdot dx,[L_k,\overline{L}_\ell]\rangle = \sum_{j=1}^m \xi_j\partial^2\Phi_j/\partial t_k\partial t_\ell. \tag{I.12.6}$$

We see that the Levi form at the characteristic point $((x,t_0),\xi\cdot dx)$ (Definition I.8.1) can be identified to the Hessian at the critical point $t_0 \in \Omega$ of the function $\Phi\cdot\xi$.

In particular we see that if the Levi form at the characteristic point under consideration is nondegenerate, then t_0 is a nondegenerate (and therefore isolated) critical point of $\Phi\cdot\xi$. By virtue of the Morse lemma, near such a point there is a smooth change of variables t (possibly depending on ξ) which transforms $\Phi\cdot\xi$ into $c_0 + \mathfrak{Q}(t-t_0)$ with $\mathfrak{Q}$ a quadratic form in $\mathbb{R}^n$. This is what lends interest to the next example:

EXAMPLE I.12.5. Assume $m = 1$, $n \geq 1$, $\Omega = \mathbb{R}^n$ and let the tube structure on $\mathbb{R}^{n+1}$ be defined by the function $Z = x + \imath\mathfrak{Q}(t)/2$ with $\mathfrak{Q}(t)$ a quadratic form on $\mathbb{R}^n$. In order that the Levi form be nondegenerate at the points $((x,0),(\xi,0))$, it is necessary and sufficient that $\mathcal{Q}$ be a nondegenerate quadratic form. In this case the characteristic set T° is equal to zero in the region $t \neq 0$; on the line $t = 0$ it is spanned by dx. The Levi form at the characteristic points may be identified to $\mathcal{Q}(t)$.

We can always perform a linear change of variables in t-space so that $\mathcal{Q}(t)$ becomes a sum of positive or negative squares. When $\mathcal{Q}$ is nondegenerate we may assume that

$$\mathcal{Q}(t) = -t_1^2 - \cdots - t_\nu^2 + t_{\nu+1}^2 + \cdots + t_n^2.$$

The integer ν is the *index* of the quadratic form $\mathcal{Q}$; it is equal to the number of negative eigenvalues of the Levi form at points $((x,0),(\xi,0))$ with $\xi > 0$. ■

In a tube structure the vector field M_i (see (I.7.25), (I.7.26)) is simply $\partial/\partial x_i$. The associated differential complex admits a simple description. The standard representative of a section $\dot{f}$ of $\Lambda^{p,q}$ is a form f of the kind (I.7.29). That of $\mathrm{d}'\dot{f}$ is then

$$\mathrm{L}f = \mathrm{d}_t f - \imath \sum_{j=1}^{m} \mathrm{d}_t\Phi_j \wedge (\partial f/\partial x_j) \tag{I.12.7}$$

with

$$\partial f/\partial x_j = \sum_{|J|=p} \sum_{|K|=q} (\partial f_{J,K}/\partial x_j)\, \mathrm{d}Z_J \wedge \mathrm{d}t_K. \tag{I.12.8}$$

We close this section by pointing out that a tube structure on $\mathbb{R}^m \times \Omega$ is analytic with respect to the variable x. It is therefore natural to complexify it with respect to x alone. We embed $\mathbb{R}^m \times \Omega$ into $\mathbb{C}^m \times \Omega$ and replace x_k by the complex coordinate z_k. The "complex" characteristics of the structure (*cf.* section I.10) are then the *real* submanifolds of $\mathbb{C}^m \times \Omega$ consisting of the points $(z_0 - \imath\Phi(t), t)$ ($z_0 = x_0 + \imath y_0$ is fixed in $\mathbb{C}^m$). Such a submanifold intersects real space $\mathbb{R}^m \times \Omega$ at the points (x_0, t) with $t \in \Omega$ such that $\Phi(t) = y_0$, which is to say, at the points in the fibre $\overset{-1}{Z}(z_0)$.

Notes

Except for changes in terminology, the presentation in sections I.1 to I.7 follows closely that in Treves [7], chap. 1. Involutive structures, as well as locally integrable structures, in the sense given to these terms in the present book, seem to have been considered for the first time in Andreotti and Hill [1]. The concept of a general CR structure appears first in Greenfield [1]. The Levi form in several complex variable theory was introduced in Levi [1]. The expressions of the Levi matrix in sections I.8 and I.9 are taken form Baouendi,

Chang, and Treves [1]. According to Lawson [1], the Frobenius theorem is due to A. Clebsch and F. Deahna. The local integrability of the involutive structures of class $\mathcal{C}^\omega$ was first established, in full generality, in Andreotti and Hill [1]. It is derived, just as we have done here, from the holomorphic version of the Frobenius theorem (ibid., 314). The result had been preceded (Libermann [1], Eckmann and Frölicher [1]) by the proof of the local integrability of complex structures (which were, and often still are, called *almost-complex*) of class $\mathcal{C}^\omega$. The theory of characteristics, at least over the real field, has a long history, going back to Hamilton and Jacobi. The role of orbits of a locally integrable structure was first pointed out in Treves [1, chap. 2]; their existence follows from a general result of Sussmann [1]. The notion of finite type, in PDE theory, appears in Hörmander [2]; and in the context of several complex variables in Kohn [1]. The name of tube structures originates with *tubes domains* in complex space (see Bochner and Martin [1]); involutive tube structures were considered in Treves [6] and Baouendi and Treves [3].

II

Local Approximation and Representation in Locally Integrable Structures

Definite advantages accrue from local integrability: this is the message of chapter II. They flow from two sources: the Approximation Formula (section II.2) and its generalization to differential forms (and currents), the Approximate Poincaré Lemma (section II.6); the local structure of distribution solutions, and its generalization to currents that are closed in the differential complex of the structure (section II.5). Locally any distribution is the finite sum of derivatives of continuous functions. The statement remains valid if we insert the adjective "solution," meaning that both the distribution and the continuous functions are annihilated by all smooth sections L of the tangent structure bundle $\mathcal{V}$ (over their domain of definition).

Thus the starting point is the existence of m ($= \dim \mathcal{M} - \operatorname{rank} \mathcal{V}$) first integrals Z_i in the open neighborhood Ω of a central point $0 \in \mathcal{M}$. Let then $\mathcal{X}$ be any maximally real submanifold (Definition I.3.3) passing through 0. After an affine substitution one may assume that the functions Z_i are nearly real-valued; more precisely, that they have the expressions (I.7.23) as well as property (I.7.24); and that the submanifold $\mathcal{X}$ is defined in Ω by the equations $t_1 = \cdots = t_n = 0$. In such a setup the tangent structure bundle $\mathcal{V}$ is spanned, over a neighborhood of 0, by the vector fields $L_1,\ldots,L_n$ given in (I.7.27). A radial integration with respect to t (starting at 0) combined with a kind of convolution with a Gaussian in the "variables" Z, leads to a generalization of the Weierstrass Approximation Theorem (Theorems II.2.1, II.3.1).

The approximation, in a neighborhood of 0, by polynomials $P(Z_1,\ldots,Z_m)$, of any solution of the equations $L_j h = 0$ ($j = 1,\ldots,n$), is not the only consequence of the Approximation Formula that is worth noting. The integrand in the formula is the trace of the solution h on the "initial" submanifold $\mathcal{X}$ multiplied by the Gaussian alluded to above (the "volume form" is the pullback to $\mathscr{X}$ of $dZ_1\wedge\cdots\wedge dZ_m$). If the trace of h vanishes identically on $\mathscr{X}$, the same is true of h itself in a full neighborhood of $\mathscr{X}$. In other words we get, at no great expense, a result of uniqueness in the Cauchy problem when the Cauchy data are carried by a maximally real submanifold (Corollaries II.3.6, II.3.7). In turn this uniqueness entails that the support of any distribution solution is a union of orbits of the structure (Theorem II.3.3).

The local approximation of any solution by polynomials $P(Z_1,\ldots,Z_m)$ shows that the germs of sets, at the point 0, on which the (germ of) map $Z = (Z_1,\ldots,Z_m)$: $(\mathcal{M},0)\to(\mathbb{C}^m,0)$ is constant, are also those on which all the (germs of) continuous solutions are constant: these germs of sets are invariants of the structure. From there on we refer to them as the *fibres* of the structure. The reader ought to keep in mind that they can be highly singular sets; after all, the map Z is merely assumed to be smooth. Even when the fibres are smooth they need not be connected, and thus, although it is quite evident that the restriction to a smooth fibre of any continuous solution is locally constant, this by itself would not ensure that the solution is constant on the whole fibre.

After this we turn to an arbitrary L-closed current f; L stands for the "concrete" differential operator, built from the vector fields L_j, which locally represents the differential complex of the structure under study (see (I.7.30)). A consequence of the Approximate Poincaré Lemma is that the pullback of f to a smooth fibre is not just closed, it is exact (for the exterior derivative). This is an important piece of information if we are to solve (locally) the equation $Lu = f$. Another application of the Approximate Poincaré Lemma to local solvability is by way of the Mittag-Leffler process: if one now assumes that f is smooth and if one can show that, to each integer $k = 1,2,\ldots$, there is a form of class $\mathscr{C}^k$ in Ω, u_k, such that $Lu_k = f$, then one is permitted to conclude that the equation $Lu = f$ has a $\mathscr{C}^\infty$ solution in some neighborhood of the central point 0. Indeed, the Approximate Poincaré Lemma enables one to select the corrective terms Lh_k in the series representation

$$u = u_1 + \sum_{k=2}^{+\infty}(u_k - u_{k-1} - Lh_k)$$

that make it converge in the $\mathscr{C}^\infty$ sense. In passing we note that the existence of the solutions u_k will follow from the existence, for any smooth f, of a current v such that $Lv = f$ and whose coefficients lie in a Sobolev space $H^\sigma(\Omega)$ with $\sigma \in \mathbb{Z}$ independent of f. For then we can commute the differential operator L with any power (≥ 0 or < 0) of the natural "Laplacian" $\Delta_{L,\kappa M}$ (see end of section II.4).

The coordinates x_i, t_j, together with the first integrals $Z_i = x_i + \sqrt{-1}\Phi_i(x,t)$ ($1 \leq i \leq m$, $1 \leq j \leq n$), provide us with a local embedding of the base manifold $\mathcal{M}$ into $\mathbb{C}^m \times \mathbb{R}^n$, namely the mapping $(x,t) \to (Z(x,t),t)$. The space $\mathbb{C}^m \times \mathbb{R}^n$ carries a natural *elliptic* structure, in which the basic vector fields are

$$\partial/\partial\bar{z}_1,\ldots,\partial/\partial\bar{z}_m,\ \partial/\partial t_1,\ldots,\partial/\partial t_n, \tag{1}$$

and the first integrals are the functions $z_1,\ldots,z_m$. The map $(x,t) \to (Z,t)$ is a diffeomorphism of Ω onto a $\mathscr{C}^\infty$ submanifold of $\mathbb{C}^m \times \mathbb{R}^n$, Σ. The linear combinations of the vector fields (1) that are tangent to Σ define on Σ the locally integrable structure inherited from the elliptic structure of $\mathbb{C}^m \times \mathbb{R}^n$; one could call it the *tangential elliptic structure* on Σ. The first integrals are simply the restrictions to Σ of the complex coordinates z_i. The tangential elliptic structure on Σ is the same thing as the transfer of the locally integrable structure of $\mathcal{M}$ from Ω via the map $(x,t) \to (Z,t)$.

But one can also try to fit the proof of the Approximation Formula to a finer local embedding, mapping diffeomorphically Ω onto a submanifold Σ of $\mathbb{C}^\nu \times \mathbb{C}^d \times \mathbb{R}^{n'}$ (section II.7). We are talking here of the map $(x,y,s,t) \to (z,w,t)$ where $z_i = x_i + \sqrt{-1}y_i$ $(1 \le i \le \nu)$, $w_k = s_k + \sqrt{-1}\varphi_k(x,y,s,t)$ $(1 \le k \le d)$ are the first integrals introduced at the beginning of section I.7, and $t = (t_1,\ldots,t_{n'})$ (here $m = \nu + d$ is the rank of the cotangent structure bundle T', $n = \nu + n'$ is that of the tangent structure bundle $\mathcal{V}$). In the special case when there are no variables t this is the local realization of a CR structure as the tangential Cauchy-Riemann structure of a generic submanifold of $\mathbb{C}^m$ (of real codimension d). Indeed, the Approximation Formula and the local structure of solutions based on the fine local embedding are particularly suited to the study of locally integrable CR structures. The role played by the variables t_j $(1 \le j \le n)$ in the coarse embedding is played, in the fine one, by the variables z_i and t_ℓ $(1 \le i \le \nu,\ 1 \le \ell \le n')$. However, one cannot expect a solution to vanish identically in a full neighborhood of 0 when its trace on the submanifold γ, defined in Ω by the equations $z = 0$, $t = 0$, vanishes. But by analogy with what happens to holomorphic functions in the complex plane one might ask whether the solution h does vanish identically in a full neighborhood of 0 if it vanishes to *infinite order* on γ. This is shown (in section II.8) to be true under the *compact cycle* hypothesis (Definition II.8.1), a hypothesis that is always satisfied when the structure is of the *hypersurface type*, meaning that $d = 1$, or when the submanifold γ is real-analytic.

The dual track approach, based sometimes on the coarse local embedding $(x,t) \to (Z,t)$ and at other times on the fine local embedding $(x,y,s,t) \to (z,w,t)$, will also be followed in the future, for instance in our use of the Fourier (or of the FBI) transform in chapter IX—leading to the maxi- or to the minitransform. Continuity between these and the Approximation Formula should not come as a surprise, if one recalls the significance of convolution with the Gaussian $(\nu/\pi)^{\frac{1}{2}m}\exp(-\nu|x|^2)$ $(x \in \mathbb{R}^m,\ \nu \sim +\infty)$ in the theory of the Fourier transform.

II.1. The Coarse Local Embedding

Throughout the present chapter the manifold $\mathcal{M}$ will be equipped with a locally integrable structure; the tangent and cotangent structure bundles will be denoted by $\mathcal{V}$ and T' respectively. As usual the fibre dimension of T' is equal to m, that of $\mathcal{V}$ is equal to n, and thus $\dim \mathcal{M} = m + n$. The analysis will be carried out in the neighborhood of a point in $\mathcal{M}$ to which we refer as the *origin* and which we denote by 0.

Let $\mathcal{X}$ be a maximally real submanifold of $\mathcal{M}$ (Definition I.3.3) passing through 0. We introduce a coordinate chart $(U,x_1,\dots,x_m,t_1,\dots,t_n)$ *centered at* 0: $0 \in U$ and all the coordinates x_j and t_k vanish at 0. We shall further require that $\mathcal{X}\cap U$ be defined by $t_1 = \cdots = t_n = 0$; $dt_1,\dots,dt_n$ span the conormal bundle of $\mathcal{X}\cap U$.

We may also assume that there are m $\mathcal{C}^\infty$ functions in U, $Z_1,\dots,Z_m$, whose differentials span the structure bundle T' at each point of U. As usual we write $Z = (Z_1,\dots,Z_m)$: $U \to \mathbb{C}^m$. Since $T'\cap \mathbb{C}N^*\mathcal{X} = 0$ we must have $dZ_1\wedge\cdots\wedge dZ_m\wedge dt_1\wedge\cdots\wedge dt_n \neq 0$, which in turn demands that the Jacobian matrix $Z_x = (\partial Z_i/\partial x_j)_{1\le i,j\le m}$ be nonsingular at points of $\mathcal{X}\cap U$. After replacing Z by $Z_x^{-1}(0)Z$ and taking the function $\mathcal{R}e Z_j$ as coordinate x_j for $j = 1,\dots,m$, we may assume that the Z_j have the expressions (I.7.23) and that (I.7.24) holds.

We shall take the neighborhood U to be a product set

$$U = V\times W \tag{II.1.1}$$

with V (resp., W) an open neighborhood of the origin in x-space $\mathbb{R}^m$ (resp., in t-space $\mathbb{R}^n$). We shall take both V and W to be *convex*.

We recall that the subsets $V\times\{t\}$ $(t \in W)$ are *maximally real* submanifolds of U (Definition I.3.3); also, $\mathcal{X}\cap U = V\times\{0\}$.

We now have $Z_x|_0 = I$, the $m\times m$ identity matrix. As a consequence, given any $\varepsilon > 0$ there is $\delta > 0$ such that, if $\operatorname{diam} U < \delta$, then

$$\sup_U \|Z_x - I\| < \varepsilon. \tag{II.1.2}$$

Needless to say we shall always take ε small in comparison to 1.

Since V is convex we may write, for fixed $t \in W$,

$$Z(x,t) - Z(y,t) = \int_0^1 Z_x(\lambda x+(1-\lambda)y,t)(x-y)\, d\lambda. \tag{II.1.3}$$

From (II.1.2) and (II.1.3) we derive

$$|Z(x,t) - Z(y,t) - (x - y)| \le \varepsilon|x - y|. \tag{II.1.4}$$

We reach the conclusion that the map

$$U \ni (x,t) \to (Z(x,t),t) \in \mathbb{C}^m\times\mathbb{R}^n \tag{II.1.5}$$

is a diffeomorphism of U onto a $\mathscr{C}^\infty$ submanifold Σ of $\mathbb{C}^m \times \mathbb{R}^n$. Thus $\dim_{\mathbb{R}} \Sigma = m + n$. For each $t \in$ W the map V $\ni x \to Z(x,t)$ is a diffeomorphism of V onto a $\mathscr{C}^\infty$ submanifold Σ_t of $\mathbb{C}^m$ which is maximally real, i.e., it is totally real of (real) dimension m.

We shall make use of the vector fields L_j $(1 \le j \le n)$ and M_i $(1 \le i \le m)$ in U defined by the orthonormality relations (I.7.25) and whose expressions are found in (I.7.26) and (I.7.27).

We may view the map (II.1.5) as an embedding of U into the *model* manifold $\mathbb{C}^m \times \mathbb{R}^n$. The latter is equipped with the involutive structure defined by the vector fields

$$\partial/\partial\bar{z}_j \ (1 \le j \le m), \ \partial/\partial t_k \ (1 \le k \le n). \tag{II.1.6}$$

This is an elliptic structure (Definition I.2.3); it is locally integrable: the cotangent structure bundle is spanned by the differentials of the functions $z_1,\dots,z_m$ (these are the complex coordinates in $\mathbb{C}^m$ regarded as functions in $\mathbb{C}^m \times \mathbb{R}^n$). We shall refer to it as the *standard elliptic structure* on $\mathbb{C}^m \times \mathbb{R}^n$. The solutions, in this structure, are the functions that are holomorphic with respect to z and independent of t.

If we pullback, by means of the map (II.1.5), the restriction to Σ of the function z_j, we get the function $Z_j(x,t)$. This implies that the map (II.1.5) is an isomorphism of U, equipped with the locally integrable structure inherited from $\mathcal{M}$, onto Σ equipped with the structure inherited from the standard structure on $\mathbb{C}^m \times \mathbb{R}^n$.

By pushing forward, via (II.1.5), the vector fields M_i and L_j we get vector fields $\tilde{M}_i$ and $\tilde{L}_j$ on Σ, which can be described as follows. For each $i = 1,\dots,m$ (resp., $j = 1,\dots,n$) and each $k = 1,\dots,m$, select at random a $\mathscr{C}^\infty$ function α_{ik} (resp., β_{jk}) in an open neighborhood of Σ which extends the pushforward, under the map (II.1.5), of $M_i\bar{Z}_k$ (resp., of $L_j\bar{Z}_k$). Let then $\tilde{f}(z,\bar{z},t)$ be any $\mathscr{C}^\infty$ function in some open neighborhood $\mathcal{O}$ of Σ in $\mathbb{C}^m \times \mathbb{R}^n$. (In the present context it is convenient to represent the variable point in $\mathbb{R}^{2m} \cong \mathbb{C}^m$ by $(z,\bar{z})$ rather than by (x,y).) Then

$$\tilde{M}_i(\tilde{f}|_\Sigma) \text{ is equal to the restriction to } \Sigma \text{ of} \tag{II.1.7}$$

$$\partial\tilde{f}/\partial z_i + \sum_{k=1}^{m} \alpha_{ik}\partial\tilde{f}/\partial\bar{z}_k;$$

$$\tilde{L}_j(\tilde{f}|_\Sigma) \text{ is equal to the restriction to } \Sigma \text{ of} \tag{II.1.8}$$

$$\partial\tilde{f}/\partial t_j + \sum_{k=1}^{m} \beta_{jk}\partial\tilde{f}/\partial\bar{z}_k.$$

Note that when $\tilde{f}$ is *holomorphic with respect to* z we have

$$\tilde{\mathrm{M}}_i(\tilde{f}|_\Sigma) = \partial\tilde{f}/\partial z_i|_\Sigma,\ \tilde{\mathrm{L}}_j(\tilde{f}|_\Sigma) = \partial\tilde{f}/\partial t_j|_\Sigma. \tag{II.1.9}$$

Property (II.1.8) shows that the pushforward to Σ, under the map (II.1.5), of the tangent structure bundle over U, $\mathcal{V}|_{\mathrm{U}}$, is the vector subbundle of $\mathbb{C}T\Sigma$ whose sections are the linear combinations of the vector fields (II.1.6) that are tangent to Σ.

Let us now look at the differential complexes associated with the locally integrable structures under consideration (cf. section I.7). The differential complex associated with the standard locally integrable structure on $\mathbb{C}^m \times \mathbb{R}^n$ can simply be denoted by $\bar{\partial}_z + \mathrm{d}_t$: it can be viewed as the "sum" of the Cauchy-Riemann complex with respect to z and of the De Rham complex with respect to t. Let $\mathcal{O}$ be an open subset of $\mathbb{C}^m \times \mathbb{R}^n$. The standard representative of a class $\dot{\tilde{f}} \in \mathcal{C}^\infty(\mathcal{O};\Lambda^{p,q})$ (see (I.7.34)) will have an expression

$$\tilde{f} = \sum_{|I|=p} \sum_{|J|+|K|=q} \tilde{f}_{I,J,K}(z,\bar{z},t)\mathrm{d}z_I\wedge \mathrm{d}\bar{z}_J\wedge \mathrm{d}t_K. \tag{II.1.10}$$

Let us suppose now that $\Sigma \subset \mathcal{O}$ and compute the pullback of the differential form $\tilde{f}$ to Σ. Apply (I.7.28) to $\bar{Z}_k$:

$$\mathrm{d}\bar{Z}_k = \sum_{i=1}^{m}\mathrm{M}_i\bar{Z}_k\,\mathrm{d}Z_i + \sum_{j=1}^{n}\mathrm{L}_j\bar{Z}_k\,\mathrm{d}t_j. \tag{II.1.11}$$

If thus we put $z = Z(x,t)$ in (II.1.10) and if we reason modulo $\Lambda^{p+1,q-1}$, i.e., if we neglect any linear combination of exterior products $\mathrm{d}Z_J\wedge\mathrm{d}t_K$ with $|J| > p$ (and $|K| = p + q - |J|$), we obtain the standard representative of the pullback of $\tilde{f}$ to Σ. The pullback of the latter to U, under the map (II.1.5), is given by

$$f = \sum_{|I|=p} \sum_{|J|+|K|=q} \sum_{|J'|=|J|} \tilde{f}_{I,J,K}(Z(x,t),\bar{Z}(x,t),t)\cdot$$
$$\det[(\mathrm{L}_{j'}\bar{Z}_j)_{j\in J,j'\in J'}]\ \mathrm{d}Z_I\wedge\mathrm{d}t_{J'}\wedge\mathrm{d}t_K. \tag{II.1.12}$$

This is the standard representative of the pullback to U of a section $\dot{f} \in \mathcal{C}^\infty(\mathrm{U};\Lambda^{p,q})$ in the locally integrable structure on $\mathcal{M}$; one may say that $\dot{f}$ is the pullback to U of the section $\dot{\tilde{f}}$.

The reader will notice that the form (II.1.12) is also the pullback to U under the map (II.1.5) of a form in $\mathcal{O}$ of the kind (II.1.10) but which does not involve any differential $\mathrm{d}\bar{z}_j$. To see this it suffices to note that the pullback of $\tilde{f}$, given by (II.1.10), to Σ is equal to that of a form similar to $\tilde{f}$ but which does not involve any $\mathrm{d}\bar{z}_j$. Or else one can pushforward to Σ the coefficients of f,

$$\tilde{f}_{I,J',K}(Z(x,t),\bar{Z}(x,t),t)\det[(\mathrm{L}_j\bar{Z}_{j'})_{j\in J,j'\in J'}],$$

and extend them as $\mathcal{C}^\infty$ functions in $\mathcal{O}$, then replace $\mathrm{d}Z_I$ by $\mathrm{d}z_I$ in (II.1.12).

Thus, in the present setup, we may restrict our attention to forms (II.1.10) in which $J = 0$.

Actually, in the present chapter we shall somewhat change the meaning of the notation $\mathcal{C}^\infty(U;\Lambda^{p,q})$. Generally, given any integer $r \geq 0$ or $r = +\infty$, by $\mathcal{C}^r(U;\Lambda^{p,q})$ we shall mean the space of standard differential forms in U of bidegree (p,q) (rather than the equivalence classes they represent). Likewise, by $\mathcal{D}'(U;\Lambda^{p,q})$ we shall mean the space of standard currents of bidegree (p,q), in U.

It is convenient to express the duality between test-functions and distributions in U by means of the "duality bracket"

$$(u,v) \to \int u\, v\, dZ \wedge dt \tag{II.1.13}$$

where $dZ = dZ_1 \wedge \cdots \wedge dZ_m$, $dt = dt_1 \wedge \cdots \wedge dt_n$. The integral in (II.1.13) makes sense when both u and v are $\mathcal{C}^\infty$ functions in U and (supp u)$\cap$(supp v) is a compact subset of U. We extend the bilinear functional (II.1.13) by continuity to pairs $(u,v) \in \mathcal{D}'(U) \times \mathcal{C}^\infty(U)$ again with the proviso that (supp u)$\cap$(supp v) $\subset\subset$ U. When dealing with functions and distributions in V (resp., in W) we rely on the duality bracket

$$\int u\, v\, dZ \quad (\text{resp., } \int u\, v\, dt).$$

One of the advantages in making use of the duality bracket (II.1.13) is that the transpose of the vector field L_j (resp., M_i) is equal to $-L_j$ (resp., $-M_i$). It suffices to prove this fact when u and v are smooth:

Lemma II.1.1. *If $u, v \in \mathcal{C}^\infty(U)$ are such that* (supp u)$\cap$(supp v) *is a compact subset of* U, *then, for all $i = 1,\ldots,m$ and all $j = 1,\ldots,n$, we have:*

$$\int u\, L_j v\, dZ \wedge dt = -\int v\, L_j u\, dZ \wedge dt, \tag{II.1.14}$$

$$\int u\, M_i v\, dZ \wedge dt = -\int v\, M_i u\, dZ \wedge dt. \tag{II.1.15}$$

Proof. We derive from Formula (I.7.28):

$$d(uv\, dZ \wedge dt_1 \wedge \cdots \wedge \widehat{dt_j} \wedge \cdots \wedge dt_n) = (-1)^{m+j-1}(uL_j v + vL_j u)\, dZ \wedge dt; \tag{II.1.16}$$

$$d(uv\, dZ_1 \wedge \cdots \wedge \widehat{dZ_i} \wedge \cdots \wedge dZ_n \wedge dt) = (-1)^{i-1}(uM_i v + vM_i u)\, dZ \wedge dt. \tag{II.1.17}$$

(In the exterior products, the hatted factors must be omitted.) Integrate both sides in (II.1.16) and (II.1.17) over U. Stokes' theorem shows that the left-hand side integrals vanish, whence the assertion. ■

REMARK II.1.1. Lemma II.1.1 may be restated as follows:

The divergence (in the coordinates x_i and t_j) *of each vector field* (II.1.18)

$$(\det Z_x)M_i,\ (\det Z_x)L_k\ (1 \le i \le m,\ 1 \le k \le n),$$

is identically equal to zero in U.

Indeed, observe that

$$\int u\, L_j v\, dZ \wedge dt = \int u\, [(\det Z_x) L_j v]\, dx \wedge dt =$$
$$- \int v\, (\det Z_x) L_j u\, dx \wedge dt - \int u\, v\, \mathrm{div}[(\det Z_x) L_j]\, dx \wedge dt,$$

and likewise for M_i. ■

We shall make use of special spaces of distributions in U, V, W and, first of all, of the space of compactly supported distributions in V, $\mathcal{E}'(V)$. It is the dual of $\mathcal{C}^\infty(V)$, and it carries the natural dual topology (see Schwartz [1], chap. 3, sec. 7; also Treves [2], sec. 24). In general the weak dual topology will suffice for our needs. We also make use of the space $\mathcal{C}^\infty(W;\mathcal{D}'(V))$ (resp., $\mathcal{C}^\infty(W;\mathcal{E}'(V))$) of $\mathcal{C}^\infty$ functions of $t \in W$ valued in the space $\mathcal{D}'(V)$ (resp., $\mathcal{E}'(V)$). That a distribution $u \in \mathcal{D}'(U)$ belongs to $\mathcal{C}^\infty(W;\mathcal{D}'(V))$ simply means that, whatever the test-function $f \in \mathcal{C}_c^\infty(V)$, the duality bracket

$$\int u(x,t)\, f(x)\, [\det Z_x(x,t)]\, dx$$

is a $\mathcal{C}^\infty$ function of t in W. If $u \in \mathcal{C}^\infty(W;\mathcal{E}'(V))$ we may let f vary in $\mathcal{C}^\infty(V)$.

In the present setup the *structure theorem* for distributions (Schwartz [1], chap. 3, sec. 6, 7; Treves [2], sec. 24) can be stated as follows: if $u \in \mathcal{D}'(U)$ then, given any open set $\Omega \subset\subset U$, there is a finite family of continuous functions $f_{\alpha,\beta}$ ($\alpha \in \mathbb{Z}_+^n$, $\beta \in \mathbb{Z}_+^m$, $|\alpha| + |\beta| \le r$) in U such that

$$u = \sum_{|\alpha|+|\beta| \le r} L^\alpha M^\beta f_{\alpha,\beta} \tag{II.1.19}$$

in Ω. We have used the notation

$$L^\alpha = L_1^{\alpha_1} \cdots L_n^{\alpha_n},\quad M^\beta = M_1^{\beta_1} \cdots M_m^{\beta_m}. \tag{II.1.20}$$

Recall that the vector fields L_j and M_i commute. One may take all the functions $f_{\alpha,\beta}$ with their support contained in a compact subset of U.

If we know that u belongs to $\mathcal{C}^\infty(W;\mathcal{D}'(V))$ then we may assert the following: to every open subset $\Omega \subset\subset \mathrm{U}$ and every integer $r \geq 0$ there is a finite family of functions $f_\alpha \in \mathscr{C}^r(\mathrm{U})$ ($\alpha \in \mathbb{Z}_+^m$, $|\alpha| \leq r'$) such that, in Ω,

$$u = \sum_{|\alpha| \leq r'} \mathrm{M}^\alpha f_\alpha. \tag{II.1.21}$$

II.2. The Approximation Formula

We reason in the local chart (U, Z) of section II.1. We shall assume that all the conditions in section II.1, among them (I.7.23), (I.7.24), and (II.1.2), are satisfied. For technical reasons it is convenient to make the hypothesis that

$$Z_j \in \mathscr{C}^\infty(\mathscr{C}\ell\mathrm{U}),\ j = 1, \ldots, m. \tag{II.2.1}$$

Let us use the notation, for any $z \in \mathbb{C}^m$,

$$\langle z \rangle^2 = z \cdot z = z_1^2 + \cdots + z_m^2. \tag{II.2.2}$$

We define the following function in $\mathbb{C}^m \times (\mathscr{C}\ell\mathrm{U})$:

$$E_\tau(z; x, t) = (\tau/\pi)^{\frac{1}{2}m}\, \mathrm{e}^{-\tau\langle z - Z(x,t)\rangle^2}. \tag{II.2.3}$$

Then, given any $f \in \mathscr{C}^0(\mathscr{C}\ell\mathrm{U})$, we define

$$\tilde{\mathscr{E}}_\tau f(z,t) = \int_{\mathrm{V}} E_\tau(z; y, t)\, f(y,t)\, [\det Z_y(y,t)]\, \mathrm{d}y. \tag{II.2.4}$$

One can interpret the integral at the right in (II.2.4) as an integral on the slice Σ_t of the submanifold Σ of $\mathbb{C}^m \times \mathbb{R}^n$ (see section II.1): let us use the diffeomorphism (II.1.5) to transfer the function f to Σ; let $\tilde{f}(z,t)$ denote its push-forward; $\tilde{f} \in \mathscr{C}^0(\mathscr{C}\ell\Sigma)$. Then

$$\tilde{\mathscr{E}}_\tau f(z,t) = (\tau/\pi)^{m/2} \int_{\Sigma_t} \mathrm{e}^{-\tau\langle z - \tilde{z}\rangle^2}\, \tilde{f}(\tilde{z}, t)\, \mathrm{d}\tilde{z}, \tag{II.2.5}$$

where we view $\mathrm{d}\tilde{z} = \mathrm{d}\tilde{z}_1 \wedge \cdots \wedge \mathrm{d}\tilde{z}_m$ as a smooth m-form pulled back to Σ.

It is clear, on inspection of (II.2.4), that $\tilde{\mathscr{E}}_\tau f$ is a continuous function of (z,t) in $\mathbb{C}^m \times (\mathscr{C}\ell\mathrm{W})$, entire holomorphic with respect to z. We shall also make use of the pullback to U of $\tilde{\mathscr{E}}_\tau f$:

$$\mathscr{E}_\tau f(x,t) = \tilde{\mathscr{E}}_\tau f(Z(x,t), t). \tag{II.2.6}$$

LEMMA II.2.1. *Suppose that the support of $f \in \mathscr{C}^0(\mathscr{C}\ell\mathrm{U})$ is contained in a product set* $\mathrm{K} \times \mathrm{W}$ *for some compact subset* K *of* V. *Then, as $\tau \to +\infty$, $\mathscr{E}_\tau f$ converges uniformly to f in* U.

PROOF. We apply (II.1.4) and derive

$$\mathcal{R}e\langle Z(x,t)-Z(y,t)\rangle^2 = |x-y|^2 - |\Phi(x,t)-\Phi(y,t)|^2 \geq (1-\varepsilon)|x-y|^2. \tag{II.2.7}$$

In the integral at the right in (II.2.4) make the change of variables $y = x + y'/\sqrt{\tau}$. Thanks to the fact that $\operatorname{supp} f \subset K \times W$ we see that, as $\tau \to +\infty$, $\mathscr{E}_\tau f(x,t)$ converges uniformly to the product of $f(x,t)$ with the integral

$$\pi^{-\frac{1}{2}m}\int e^{-\langle Z_x(x,t)y'\rangle^2}\,[\det Z_x(x,t)]\,dy'.$$

Thanks to (II.1.2) we see that this integral is equal to 1. ∎

LEMMA II.2.2. *Assume the support of* $f \in \mathscr{C}^1(U)$ *is contained in a product set* $K \times W$ *for some compact subset* K *of* V. *Then, for every* $\tau > 0$, $\tilde{\mathscr{E}}_\tau f$ *belongs to* $\mathscr{C}^1(\mathbb{C}^m \times W)$ *and*

$$(\partial/\partial z_i)\tilde{\mathscr{E}}_\tau f = \tilde{\mathscr{E}}_\tau M_i f, \tag{II.2.8}$$

$$(\partial/\partial t_j)\tilde{\mathscr{E}}_\tau f = \tilde{\mathscr{E}}_\tau L_j f, \tag{II.2.9}$$

whatever $i = 1,\ldots,m$, $j = 1,\ldots,n$.

PROOF. In order to prove (II.2.8) it suffices to show that

$$\int \chi(t)\,(\partial/\partial z_i)\tilde{\mathscr{E}}_\tau f(z,t)\,dt = \int \chi(t)\tilde{\mathscr{E}}_\tau M_i f(z,t)\,dt \tag{II.2.10}$$

is valid whatever the function $\chi \in \mathscr{C}_c^\infty(W)$. But

$$(\partial/\partial z_i)E_\tau(z;x,t) = -M_i E_\tau(z;x,t)$$

with M_i acting in the variables x, t. Thus (II.2.10) reads

$$\int \chi(t)f(x,t)\,M_i E_\tau(z;x,t)\,dZ\wedge dt = -\int E_\tau(z;x,t)\,M_i[\chi(t)f(x,t)]\,dZ\wedge dt,$$

which is an immediate consequence of (II.1.15).

Next we prove (II.2.9). Let us write

$$L_j = \partial/\partial t_j + \sum_{i=1}^{m}\lambda_{ij}(x,t)\partial/\partial x_i,\ j = 1,\ldots,n \tag{II.2.11}$$

(cf. (I.7.26), (I.7.27)). Recalling the hypothesis about the support of f, we observe that

$$\int (\partial/\partial t_j)(E_\tau f \det Z_x)\, dx = \int L_j(E_\tau f)\, (\det Z_x)\, dx -$$

$$\int (\det Z_x) \sum_{i=1}^{m} \lambda_{ij} (\partial/\partial x_i)(E_\tau f)\, dx + \int E_\tau f\, (\partial/\partial t_j)(\det Z_x)\, dx.$$

If we integrate by parts in the second integral at the right and if we avail ourselves of (II.1.18), we obtain

$$\int (\partial/\partial t_j)(E_\tau f \det Z_x)\, dx = \int L_j(E_\tau f)\, (\det Z_x)\, dx.$$

Since $L_j E_\tau \equiv 0$ this equality is the same as (II.2.9). ■

COROLLARY II.2.1. *Let the support of $f \in \mathscr{C}^1(U)$ be contained in a product set $K \times W$ for some compact subset K of V. Then, for every $\tau > 0$, $\mathscr{E}_\tau f$ belongs to $\mathscr{C}^1(U)$ and*

$$M_i \mathscr{E}_\tau f = \mathscr{E}_\tau M_i f, \tag{II.2.12}$$

$$L_j \mathscr{E}_\tau f = \mathscr{E}_\tau L_j f, \tag{II.2.13}$$

whatever $i = 1,\ldots,m$, $j = 1,\ldots,n$.

PROOF. Take the restrictions to Σ of both sides in (II.2.8) and (II.2.9). Since $\tilde{\mathscr{E}}_\tau f$ is entire holomorphic with respect to z, according to (II.1.9) we may substitute $\tilde{M}_i(\tilde{\mathscr{E}}_\tau f|_\Sigma)$ for $(\partial/\partial z_i)(\tilde{\mathscr{E}}_\tau f)|_\Sigma$ and $\tilde{L}_j(\tilde{\mathscr{E}}_\tau f|_\Sigma)$ for $(\partial/\partial t_j)(\tilde{\mathscr{E}}_\tau f)|_\Sigma$. Pulling back to U the resulting identities yields (II.2.12) and (II.2.13). ■

COROLLARY II.2.2. *Let the support of $f \in \mathscr{C}^r(U)$ ($r \in \mathbb{Z}_+$) be contained in a product set $K \times W$ with K a compact subset of V. Then, for every $\tau > 0$, $\mathscr{E}_\tau f$ belongs to $\mathscr{C}^r(U)$ and we have, whatever $a \in \mathbb{Z}_+^m$, $\beta \in \mathbb{Z}_+^n$, $|\alpha| + |\beta| \leq r$,*

$$M^\alpha L^\beta \mathscr{E}_\tau f = \mathscr{E}_\tau M^\alpha L^\beta f. \tag{II.2.14}$$

We have used the notation (II.1.20).

LEMMA II.2.3. *Let the support of $f \in \mathscr{C}^r(U)$ ($r \in \mathbb{Z}_+$ or $r = +\infty$) be contained in a product set $K \times W$ for some compact subset K of V. Then, as $\tau \to +\infty$, $\mathscr{E}_\tau f$ converges to f in $\mathscr{C}^r(U)$.*

PROOF. We know that, for every pair of multiplets $\alpha \in \mathbb{Z}_+^m$, $\beta \in \mathbb{Z}_+^n$ such that $|\alpha| + |\beta| \leq r$, $M^\alpha L^\beta f$ is a $\mathscr{C}^0$ function in U. By Lemma II.2.1 and Corollary II.2.2, it is the uniform limit in U of $M^\alpha L^\beta \mathscr{E}_\tau f$. To reach the desired conclusion it suffices to recall that the vector fields $L_1,\ldots,L_n$, $M_1,\ldots,M_m$ span $\mathbb{C}T\mathcal{M}$ over U. ■

Henceforth the number ε in (II.1.2), (II.1.4) will be chosen $\leq 1/10$.

LEMMA II.2.4. *To each neighborhood of the origin* $V' \subset V$ *there are open neighborhoods of the origin* $\mathcal{O} \subset\subset \mathbb{C}^m$ *and* $W' \subset\subset W$, *and a constant* $\kappa > 0$ *such that, whatever* $r \in \mathbb{Z}_+$ *and whatever the function* $f \in \mathscr{C}^r(\mathscr{C}\ell U)$ *which vanishes identically in* $V' \times W$, *the norm in* $\mathscr{C}^r(\mathscr{C}\ell(\mathcal{O} \times W'))$ *of* $\tilde{\mathscr{E}}_\tau f$ *does not exceed* const. $e^{-\kappa\tau}$.

PROOF. We have

$$\mathscr{R}e\langle z - Z(x,t)\rangle^2 = |\mathscr{R}e z - x|^2 - |\mathscr{I}m z - \Phi(x,t)|^2 \geq$$
$$|\mathscr{R}e z - x|^2 - 2|\mathscr{I}m z - \Phi(0,t)|^2 - 2|\Phi(x,t) - \Phi(0,t)|^2.$$

We know that $|x| \geq \delta > 0$ on $V \backslash V'$. We can therefore choose $\mathcal{O}$ and W' small enough that

$$|\mathscr{R}e z - x|^2 - 2|\mathscr{I}m z - \Phi(0,t)|^2 \geq (1-\varepsilon)|x|^2$$

for all $x \in V\backslash V'$, $t \in W'$, and $z \in \mathcal{O}$. On the other hand we derive from (II.1.4)

$$|\Phi(x,t) - \Phi(0,t)|^2 \leq \varepsilon^2|x|^2.$$

We see thus that

$$\mathscr{R}e\langle z - Z(x,t)\rangle^2 \geq (1-\varepsilon-2\varepsilon^2)|x|^2,$$

and therefore, for some constant $\kappa > 0$ and for all $z \in \mathcal{O}$, $x \in V\backslash V'$, $t \in W_0$,

$$\mathscr{R}e\langle z - Z(x,t)\rangle^2 \geq 2\kappa, \tag{II.2.15}$$

$$|E_\tau(z;x,t)| \leq (\tau/\pi)^{\frac{1}{2}m} e^{-2\kappa\tau}, \tag{II.2.16}$$

whence the desired estimate for the norm in $\mathscr{C}^0(\mathscr{C}\ell(\mathcal{O} \times W'))$ of $\tilde{\mathscr{E}}_\tau f$. If $|\alpha| + |\beta| \leq r$, we may let $(\partial/\partial z)^\alpha(\partial/\partial t)^\beta$ act on $\tilde{\mathscr{E}}_\tau f$ (see (II.2.4)) under the integral sign. We get an integral of the same type as (II.2.4) but with f replaced by a different function $f_{\alpha,\beta}$, which depends polynomially on τ and on z. We have $\operatorname{supp} f_{\alpha,\beta} \subset \operatorname{supp} f$. The same reasoning as above applies to this integral with $f_{\alpha,\beta}$. Because of (II.2.16) its norm in $\mathscr{C}^0(\mathscr{C}\ell(\mathcal{O} \times W'))$ is $\leq$ *const.* $e^{-\kappa\tau}$. ∎

In the next statement we deal with *solutions* (Definition I.1.3) that are of class $\mathscr{C}^r$ $(0 \leq r \leq +\infty)$ in the closure of U.

THEOREM II.2.1. *If the open neighborhood* $U_0 \subset U$ *of the origin is sufficiently small, then, for all* $r \in \mathbb{Z}_+$ *or* $r = +\infty$, *and whatever the solution* $h \in \mathscr{C}^r(\mathscr{C}\ell U)$, *we have, in* $\mathscr{C}^r(U_0)$,

$$h(x,t) = \lim_{\tau \to +\infty} \int_V E_\tau(Z(x,t);y,0)\, h(y,0)\, [\det Z_x(y,0)]\, dy. \tag{II.2.17}$$

PROOF. It suffices to consider the case $r < +\infty$. Let $\chi \in \mathscr{C}_c^\infty(\mathrm{V})$ be equal to 1 in some open neighborhood V′ of the origin. Consider

$$\tilde{\mathscr{E}}_\tau(\chi h)(z,t) = \int E_\tau(z;y,t)\chi(y)h(y,t) \det Z_y(y,t)\, dy.$$

The fact that h is a solution implies

$$(\partial/\partial t_j)\tilde{\mathscr{E}}_\tau(\chi h) = \tilde{\mathscr{E}}_\tau[(\mathrm{L}_j\chi)h]. \qquad \text{(II.2.18)}$$

When $h \in \mathscr{C}^1(\mathscr{U}\mathrm{U})$ (II.2.18) is true thanks to (II.2.9). When $r = 0$ we derive from the equations

$$\partial h/\partial t_j = -\sum_{i=1}^{m} \lambda_{ij}\partial h/\partial x_i,\ j = 1,\ldots,n \text{ (cf. (II.2.11)},$$

that h is a $\mathscr{C}^1$ function of $t \in \mathrm{W}$ valued in the space of distributions of $x \in \mathrm{V}$ (cf. Proposition I.4.3). Clearly (II.2.9) and (II.2.18) remain valid if we interpret $\tilde{\mathscr{E}}_\tau(\chi h)$ as a duality bracket in x-space, with χ in the role of the test-function, depending in $\mathscr{C}^1$ fashion on the parameter t.

Whatever $r \geq 0$, Corollary II.2.1 entails that the right-hand side in (II.2.18) is a function of class $\mathscr{C}^r$ in $\mathbb{C}^m \times \mathrm{W}$, entire holomorphic with respect to z. We apply Lemma II.2.4, in which we take the neighborhood W′ to be *convex*:

$$\|d_t\tilde{\mathscr{E}}_\tau(\chi h)\|_{\mathscr{C}^r(\mathscr{U}(\mathcal{O}\times \mathrm{W}'))} \leq const.\ e^{-\kappa\tau}. \qquad \text{(II.2.19)}$$

As a consequence we can integrate with respect to t and find a function $v_\tau(z,t) \in \mathscr{C}^r(\mathbb{C}^m \times \mathrm{W})$, entire holomorphic with respect to z, such that

$$d_t v_\tau = d_t\tilde{\mathscr{E}}_\tau(\chi h), \qquad \text{(II.2.20)}$$

$$\|v_\tau\|_{\mathscr{C}^r(\mathscr{U}(\mathcal{O}\times \mathrm{W}'))} \leq const.\ e^{-\kappa\tau}. \qquad \text{(II.2.21)}$$

We derive from (II.2.20):

$$\tilde{\mathscr{E}}_\tau(\chi h)(z,t) - \tilde{\mathscr{E}}_\tau(\chi h)(z,0) = v_\tau(z,t) - v_\tau(z,0). \qquad \text{(II.2.22)}$$

It follows then from (II.2.21) that the limit, as $\tau \to +\infty$, in $\mathscr{C}^r(\mathscr{U}(\mathcal{O}\times \mathrm{W}_0))$ of the left-hand side in (II.2.22) is equal to zero.

Define U_0 then as the product of open neighborhoods of 0 in V and W respectively,

$$\mathrm{U}_0 = \mathrm{V}_0 \times \mathrm{W}_0. \qquad \text{(II.2.23)}$$

We select $\mathrm{V}_0 \subset \mathrm{V}'$ and $\mathrm{W}_0 \subset \mathrm{W}'$ small enough that $Z(\mathrm{U}_0) \subset \mathcal{O}$. By Lemmas II.2.3 and II.2.4 we know then that $\mathscr{E}_\tau(\chi h) \to h$ and $\mathscr{E}_\tau[(1-\chi)h] \to 0$ in $\mathscr{C}^r(\mathrm{U}_0)$, whence the sought conclusion. ∎

II.3. Consequences and Generalizations

Solutions That Are Functions

The first consequence of Theorem II.3.1 is the approximation of solutions by polynomials in the Z_j:

THEOREM II.3.1. *Let* $r \in \mathbb{Z}_+$ *or* $r = +\infty$. *Any solution* $h \in \mathcal{C}^r(\mathcal{C}\ell \mathrm{U})$ *is the limit, in* $\mathcal{C}^r(\mathrm{U}_0)$, *of a sequence of polynomials with respect to* $Z_1(x,t),\ldots,Z_m(x,t)$.

PROOF. It suffices to consider the case $r < +\infty$. Whatever the integer $\nu \geq 1$, we can find another integer $\nu' \geq 0$ such that

$$\|E_\nu(x,t;y,0) - (\nu/\pi)^{\frac{1}{2}m}\sum_{k=0}^{\nu'}\frac{(-\nu)^k}{k!}\langle Z(x,t)-Z(y,0)\rangle^{2k}\|_{\mathcal{C}^r(\mathcal{C}\ell \mathrm{U})} \leq 1/\nu.$$

We reach the conclusion that h is the limit, in $\mathcal{C}^r(\mathrm{U}_0)$, of the sequence of polynomials $\mathcal{P}_{h;\nu}(Z(x,t))$ ($\nu = 1,2,\ldots$) defined as follows:

$$\mathcal{P}_{h;\nu}(z) = (\nu/\pi)^{\frac{1}{2}m}\sum_{k=0}^{\nu'}\frac{(-\nu)^k}{k!}\int_{\mathrm{V}}\langle z-Z(y,0)\rangle^{2k}\, h(y,0)\, \mathrm{d}Z(y,0),$$

with $\nu' \in \mathbb{Z}_+$ depending on ν in a suitable manner. ■

EXAMPLE II.3.1. Consider the case $n = 0$ (which is not precluded), i.e., $\dim \mathcal{M} = m$ and there are no vector fields L_j. Select arbitrarily m complex valued $\mathcal{C}^\infty$ functions Z_j such that $\mathrm{d}Z_1,\ldots,\mathrm{d}Z_m$ span $\mathbb{C}T^*\mathcal{M}$ in some open neighborhood Ω of 0. Theorem II.3.1 implies that every continuous function in Ω is the uniform limit, in a smaller neighbohood of 0, of a sequence of polynomials $P_\nu(Z_1,\ldots,Z_m)$. This can be regarded as a local version of the Weierstrass Approximation Theorem. ■

EXAMPLE II.3.2. Let $\mathcal{M}$ be a generic (i.e., strongly noncharacteristic) submanifold of $\mathbb{C}^m$ equipped with the CR structure inherited from the ambient space $\mathbb{C}^m$. The solutions in the structure are the CR distributions. Theorem II.3.1 implies that, locally, each continuous CR function in an open subset Ω of $\mathcal{M}$ is the uniform limit of (the restrictions to $\mathcal{M}$ of) a sequence of holomorphic polynomials $P_\nu(z)$. ■

Recall that the *fibres* of the map Z in the subset U_0 of U are the preimages, under the restriction of Z to U_0, of single points in $\mathbb{C}^m$. The next statement is an immediate consequence of Theorem II.3.1:

COROLLARY II.3.1. *Every solution $h \in \mathcal{C}^0(\mathcal{C}\ell\mathrm{U})$ is constant on each fibre of the map Z in* U_0.

Corollary II.3.1 shows that the *germs of sets* at 0 defined by the fibres of the map Z are invariants of the locally integrable structure of $\mathcal{M}$: replacing the functions Z_j by another system of m solutions $Z_j^\#$ such that $dZ_1^\# \wedge \cdots \wedge dZ_m^\# \neq 0$ does not change any of those germs of sets (although, of course, the values that the maps Z and $Z^\#$ take on any one of them might differ). This generalizes the property of analytic structures established earlier (see the remarks following Proposition I.10.2).

EXAMPLE II.3.3. Let Ω be an open subset of $\mathbb{R}^n$ (with coordinates $t_1,\dots,t_n$) and Φ a $\mathcal{C}^\infty$ map $\Omega \to \mathbb{R}^n$. Equip $\mathbb{R}^m \times \Omega$ with the tube structure defined by the functions $Z_j = x_j + \imath\Phi_j(t)$ (see section I.12). To any open neighborhood U of a point $(x_0,t_0) \in \mathbb{R}^m \times \Omega$ there is a subneighborhood $\mathrm{U}_0 = \mathrm{V}_0 \times \mathrm{W}_0$ such that, if $h \in \mathcal{C}^0(\mathrm{U})$ is a solution, then, for each $x \in \mathrm{V}_0$, the function $\mathrm{W}_0 \ni t \to h(x,t)$ is constant on each level set $\Phi(t) = const.$

For instance, in the Mizohata structure on $\mathbb{R}^2$ (defined by $Z = x + \imath t^2/2$, see Example I.2.7), all germs of continuous solutions at a point $(x_0,0)$ are even with respect to t. ■

It should also be recalled that, in general, Corollary II.3.1 cannot be globalized, as shown in Example I.10.2 (which looks at a special *tube* structure).

COROLLARY II.3.2. *Let* K *be any compact subset of* U_0, *h any continuous solution in* $\mathcal{C}\ell\mathrm{U}$. *There exists a continuous function $\tilde{h}$ in* Z(K) *such that* $h = \tilde{h}\circ Z$ *in* K.

PROOF. For any $z_0 \in Z(K)$ define $\tilde{h}(z_0)$ to be the value of h on the intersection of the fibre $\overset{-1}{Z}(z_0)$ with U_0. It is an elementary result of point-set topology that $\tilde{h}$ is continuous on the compact set Z(K). ■

REMARK II.3.1. It follows from Theorem II.3.1 that the function $\tilde{h}$ in Corollary II.3.2 must have the following property:

$\tilde{h}$ is the uniform limit in Z(K) *of a sequence of holomorphic polynomials $P_\nu(z)$.* (II.3.1)

Conversely, take $\mathrm{K} \subset\subset \mathrm{U}$ to be the closure of some open neighborhood of the origin U* and let $\tilde{h}$ be any continuous function in Z(K) that has property (II.3.1). Then $h = \tilde{h}\circ Z$ is a solution in U*, since it is the uniform limit of the smooth solutions $P_\nu \circ Z$.

Suppose that $\tilde{h}$ extends as a $\mathcal{C}^1$ function to some open subset of $\mathbb{C}^m$ containing Z(K). Property (II.3.1) demands that $\tilde{h}$ be annihilated by the tangent vectors of type (0,1), i.e., its differential must be of type (1,0). Indeed its pullback $\tilde{h}\circ Z$, if it is to be a solution, must be annihilated by the tangent vectors L_j; its differential must belong to the span of $dZ_1,\ldots,dZ_m$. Thus, given any point z_0 of Z(K) and any tangent vector **v** to $\mathbb{C}^m$ at z_0 that is a linear combination of $\partial/\partial\bar{z}_1,\ldots,\partial/\partial\bar{z}_m$, we must have $\mathbf{v}\tilde{h}(z_0) = 0$. ■

COROLLARY II.3.3. *The set of continuous solutions in an open subset Ω of $\mathcal{M}$ form a ring for ordinary addition and multiplication.*

PROOF. The continuous solutions form a linear subspace of $\mathscr{C}^0(\Omega)$. Each point $p \in \Omega$ has an open neighborhood U_p such that the following is true: there is a $\mathscr{C}^\infty$ map $Z\colon U_p \to \mathbb{C}^m$ whose components Z_j are solutions; any two solutions f, $g \in \mathscr{C}^0(\Omega)$ are uniform limits in U_p of sequences of polynomials $\mathscr{P}_\nu\circ Z$, $\mathscr{Q}_\nu\circ Z$ ($\nu = 1,2,\ldots$), respectively. But then the products $(\mathscr{P}_\nu\mathscr{Q}_\nu)\circ Z$ converge uniformly to fg in U_p, perforce a distribution solution. ■

Next we shift the focus to distributions, specifically to those that belong to the spaces introduced at the end of section II.1:

Distribution Solutions

It follows at once from (II.2.4) that $\tilde{\mathscr{E}}_\tau$ extends as a bounded linear operator $\mathscr{C}^\infty(W;\mathscr{E}'(V)) \to \mathscr{C}^\infty(W;Hol(\mathbb{C}^m)) \cong Hol(\mathbb{C}^m;\mathscr{C}^\infty(W))$, the subspace of $\mathscr{C}^\infty(\mathbb{C}^m \times W)$ made up of the functions that are holomorphic with respect to z. And also that $\tilde{\mathscr{E}}_\tau$ acts on the space of distributions in U whose support is contained in a product set $K\times W$ with $K \subset\subset V$; it maps this space into $Hol(\mathbb{C}^m;\mathscr{D}'(W)) \cong \mathscr{D}'(W;\mathrm{Hol}(\mathbb{C}^m))$. Likewise we see easily that the operator $\mathscr{E}_\tau$ defined in (II.2.6) extends as a continuous linear operator $\mathscr{C}^\infty(W;\mathscr{E}'(V)) \to \mathscr{C}^\infty(U)$ and maps the space of distributions u in U such that supp $u \subset K\times W$, $K \subset\subset V$, into $\mathscr{C}^\infty(V;\mathscr{D}'(W)) \cong \mathscr{D}'(W;\mathscr{C}^\infty(V))$.

One could also define the linear operator $\mathscr{E}_\tau$ by transposition:

$$\int \mathscr{E}_\tau u \; v \; dZ\wedge dt = \int u \; \mathscr{E}_\tau v \; dZ\wedge dt \qquad \text{(II.3.2)}$$

for u and v in the appropriate distribution spaces: (supp u)$\cup$(supp v) $\subset K\times W$ with $K \subset\subset V$; the projection into t-space W of at least one of the two sets, supp u or supp v, must be compact; and if either u or v is an arbitrary distribution with respect to t, the other one must be a $\mathscr{C}^\infty$ function of t in W.

As stated in Lemma II.1.1, the transpose of L_j (resp., M_i) for the duality bracket (II.1.13) is equal to $-L_j$ (resp., $-M_i$). As a consequence, the transposition formula (II.3.2) allows us to extend (II.2.14) to those distributions; in general we may state

$$[L^\alpha M^\beta, \mathscr{E}_\tau] = 0. \qquad \text{(II.3.3)}$$

There are two ways of proving the next statement: one can either apply Lemma II.2.3 (for $r = +\infty$) and the transposition formula (II.3.2); or else one can make use of the local structure formulas (II.1.19) and (II.1.21). We leave the details to the reader.

LEMMA II.3.1. *If the support of* $f \in \mathscr{D}'(U)$ *is contained in a product set* $K \times W$ *with* $K \subset\subset V$, *then, as* $\tau \to +\infty$, $\mathscr{E}_\tau f$ *converges to* f *in* $\mathscr{D}'(U)$. *If* $f \in \mathscr{C}^\infty(W;\mathscr{E}'(V))$ *then* $\mathscr{E}_\tau f \to f$ *in* $\mathscr{C}^\infty(W;\mathscr{D}'(V))$.

Lemma II.2.4 can also easily be generalized to the same distribution spaces:

LEMMA II.3.2. *Let* V', W', *and* $\mathbb{O}$ *be the neighborhoods and* κ *the positive constant in Lemma* II.2.4. *Let* $r \in \mathbb{Z}_+$ *be arbitrary and the distribution* $f \in \mathscr{C}^\infty(W;\mathscr{D}'(V))$ *have its support in a product set* $K \times W$ *with* $K \subset\subset V$ *and vanish identically in* $V' \times W$. *Then the norm in* $\mathscr{C}^r(\mathscr{Cl}(\mathbb{O} \times W'))$ *of* $\mathscr{E}_\tau f$ *does not exceed* const. $e^{-\kappa\tau}$.

To state the analogue of Theorem II.2.1 for distribution solutions we avail ourselves of the property that a submanifold of U that is defined by an equation $t = const.$ is maximally real and therefore noncharacteristic (Definition I.4.1). We apply Proposition I.4.3: *every distribution solution in* U *is a* $\mathscr{C}^\infty$ *function of* t *in* W *valued in* $\mathscr{D}'(V)$.

We introduce the same cutoff function $\chi \in \mathscr{C}_c^\infty(V)$ and the same open neighborhood $U_0 \subset U$ of 0 as specified in the proof of Theorem II.2.4; in particular we assume that (II.2.23) holds.

THEOREM II.3.2. *Given any distribution solution* h *in* U, *we have, in* $\mathscr{C}^\infty(W_0;\mathscr{D}'(V_0))$,

$$h(x,t) = \lim_{\tau \to +\infty} \int E_\tau(x,t;y,0) h(y,0) \chi(y,0) [\det Z_x(y,0)]\, dy. \qquad \text{(II.3.4)}$$

PROOF. It is almost exactly the same as that of Theorem II.2.1; the only difference is that we must use Lemma II.3.1 in the place of Lemma II.2.3 and Lemma II.3.2 in that of Lemma II.2.4. ■

COROLLARY II.3.4. *Every distribution solution* h *in* U *is the limit, in* $\mathscr{C}^\infty(W_0;\mathscr{D}'(V_0))$, *of a sequence of solutions* $h_\tau \in \mathscr{C}^\infty(U)$ $(\tau = 1,2,\ldots)$.

PROOF. Note indeed that for any $\tau > 0$ the function

$$\int E_\tau(x,t;y,0)h(y,0)\chi(y,0)[\det Z_x(y,0)]\, dy \qquad \text{(II.3.5)}$$

is a $\mathscr{C}^\infty$ solution in U. ■

REMARK II.3.2. The distribution solution h is also the limit, in $\mathscr{C}^\infty(W_0;\mathscr{D}'(V_0))$, of the sequence of polynomials $\mathscr{P}_{h,\nu}(Z(x,t))$ with $\mathscr{P}_{h,\nu}$ as in the proof of Theorem II.3.1—possibly with an increased bound ν' on the summation. ■

The next corollary states a property of the type "uniqueness in the Cauchy problem":

COROLLARY II.3.5. *Any solution $h \in \mathscr{D}'(U)$ such that $h|_{t=0} \equiv 0$ vanishes identically in* U_0.

The reader is now referred to the beginning of section II.1. We recall that the equation $t = 0$ defines $\mathscr{X}\cap U$ in U, and that $\mathscr{X}$ is an arbitrary maximally real submanifold of $\mathcal{M}$. Since $\mathscr{X}$ is noncharacteristic the trace on $\mathscr{X}$ of any distribution solution defined in some open neighborhood of $\mathscr{X}$ is well defined (Corollary I.4.1). In view of Corollary II.3.5 we may state:

COROLLARY II.3.6. *Let Ω be an open subset of $\mathcal{M}$ containing the maximally real submanifold $\mathscr{X}$. Any distribution solution in Ω whose trace on $\mathscr{X}$ vanishes must vanish identically in an open neighborhood of $\mathscr{X}$ in Ω.*

COROLLARY II.3.7. *The conclusion in Corollary* II.3.6 *remains valid if $\mathscr{X}$ is a strongly noncharacteristic submanifold or if $\mathscr{X}$ is a noncharacteristic hypersurface.*

PROOF OF COROLLARY II.3.7. It suffices to prove the part of the statement concerning strongly noncharacteristic submanifolds, since every noncharacteristic hypersurface is strongly noncharacteristic (Proposition I.4.2). Through every point of $\mathscr{X}$ there passes a submanifold $\mathscr{Y}$ of $\mathscr{X}$ which is maximally real for the pullback to $\mathscr{X}$ of the involutive structure of $\mathcal{M}$ (Definition I.3.2). But it follows from Definition I.3.3 that $\mathscr{Y}$ is also a maximally real submanifold of $\mathcal{M}$. By Corollary II.3.5 any distribution solution whose trace on $\mathscr{X}$ vanishes identically must vanish identically in some open neighborhood of $\mathscr{Y}$ in $\mathcal{M}$. ■

COROLLARY II.3.8. *Assume that $\mathcal{M}$ is a complex manifold of complex dimension n, and let $\mathscr{X}$ be a totally real submanifold of $\mathcal{M}$ of real dimension n. Let Ω be a connected open subset of $\mathcal{M}$ containing $\mathscr{X}$. Any holomorphic function in Ω whose trace on $\mathscr{X}$ vanishes must vanish identically in Ω.*

Indeed, the maximally real submanifolds of the complex manifold $\mathcal{M}$ are the totally real submanifolds of $\mathcal{M}$ of maximum dimension (cf. Corollary I.3.2 and following remarks).

THEOREM II.3.3. *Let h be a distribution solution in an open subset Ω of $\mathcal{M}$. Any orbit* (Definition I.11.1) *of $\mathcal{V}$ in Ω that intersects* supp h *is entirely contained in* supp h.

PROOF. It will suffice to show that if a $\mathcal{C}^\infty$ curve $\gamma\colon [0,1] \to \Omega$ is an integral curve of $\mathcal{R}e\mathrm{L}$, with L a smooth section of $\mathcal{V}$ in Ω, and if $h \equiv 0$ in an open neighborhood of $\gamma(0)$, then $h \equiv 0$ in an open neighborhood of $\gamma(1)$. Actually we may subdivide the segment $[0,1]$ into a finite number of segments as short as we wish and prove the analogous result for each one of those segments. In other words we may assume that the image of γ (which is compact) is contained in an open neighborhood $\mathcal{N}$ of $\gamma(0)$ as small as needed. The only proviso is that the size of $\mathcal{N}$ not depend on the choice of the solution h. Thus we may assume that there are $\mathcal{C}^\infty$ coordinates in $\mathcal{N}$, $s_1,\ldots,s_{m+n-1}$, t, vanishing at $\gamma(0)$, and such that $\mathcal{R}e\mathrm{L} = \partial/\partial t$ in $\mathcal{N}$. We write $s = (s_1,\ldots,s_{m+n-1})$ and take $\mathcal{N}$ to be the cylinder $|s| < \rho$, $|t| < \delta$ with ρ, $\delta > 0$ suitably small.

We start from the hypothesis that, for some number λ_0, $0 < \lambda_0 < 1$, $h \equiv 0$ in the subcylinder

$$\Gamma_0 = \{ (s,t) \in \mathcal{N};\ |s| < \rho,\ |t| < \lambda_0\delta \}.$$

For any integer $k \geq 1$ and any number λ, $0 < \lambda \leq 1$, define

$$\Gamma_\lambda^{(k)} = \{ (s,t) \in \mathcal{N};\ (|s|/\rho)^{2k} + (|t|/\delta)^2 < \lambda^2 \}.$$

We are going to make use of the following properties:

$$\textit{if } \lambda < \lambda_0 \textit{ the closure of } \Gamma_\lambda^{(k)} \textit{ in } \mathcal{N} \textit{ is contained in } \Gamma_0; \tag{II.3.6}$$

$$\Gamma_\lambda^{(k)} \textit{ is equal to the union of all sets } \Gamma_{\lambda'}^{(k')} \textit{ with } k' < k,\ \lambda' < \lambda; \tag{II.3.7}$$

$$\textit{the union of the open sets } \Gamma_1^{(k)} \textit{ is equal to } \mathcal{N}. \tag{II.3.8}$$

Last we observe that the vector field $\mathcal{R}e\mathrm{L}$ is tangent to the boundary $\partial\Gamma_\lambda^{(k)}$ of $\Gamma_\lambda^{(k)}$ only at the "equator" $t = 0$. This entails that the portion of $\partial\Gamma_\lambda^{(k)}$ defined by $t \neq 0$ is a noncharacteristic hypersurface in $\mathcal{M}$.

Fix k arbitrarily. We know that $h \equiv 0$ in Γ_0, i.e., in some open neighborhood of the equator $t = 0$ in $\partial\Gamma_\lambda^{(k)}$. Suppose $h \equiv 0$ in $\Gamma_\lambda^{(k)}$; since h is a $\mathcal{C}^\infty$ function of t with values in the space of distributions with respect to s (in the ball $|s| < \rho$), we see that $h \equiv 0$ on the portion of $\partial\Gamma_\lambda^{(k)}$ off the equator (this makes sense since, by Corollary I.4.1, h has a well-defined trace on that portion of $\partial\Gamma_\lambda^{(k)}$). By Corollary II.3.6 we conclude that $h \equiv 0$ in some open neighborhood of the closure of $\Gamma_\lambda^{(k)}$. It follows from this that the set of numbers, λ,

$0 < \lambda \leq 1$, such that $h \equiv 0$ in $\Gamma_\lambda^{(k)}$ is open. By (II.3.7) it is closed and therefore it must be equal to the whole interval $]0,1]$: $h \equiv 0$ in $\Gamma_1^{(k)}$. By letting k go to $+\infty$ and by taking advantage of (II.3.8) we conclude that $h \equiv 0$ in $\mathcal{N}$. ■

COROLLARY II.3.9. *Suppose that the locally integrable structure of $\mathcal{M}$ is of finite type* (Definition I.11.2). *Then the support of any distribution solution in a connected and open subset Ω of $\mathcal{M}$ that does not vanish identically is equal to the whole of Ω.*

PROOF. If the structure of $\mathcal{M}$ is of finite type and if Ω is connected the only orbit of $\mathcal{V}$ in Ω is the whole of Ω. ■

Consider an embedded submanifold $\mathcal{N}$ of $\mathcal{M}$ such that $\mathcal{V}|_{\mathcal{N}} \subset \mathbb{C}T\mathcal{N}$; we have $\ell = \operatorname{codim} \mathcal{N} \leq m$. Let 0 be an arbitrary point of $\mathcal{N}$ and U an open neighborhood of 0 in which there are coordinates $x_1,\ldots,x_m$, $t_1,\ldots,t_n$ and $\mathcal{C}^\infty$ solutions Z_j such that (I.7.23) and (I.7.24) hold. We take U in the product form (II.1.1) and small enough that the $(m+n)$-form $dZ \wedge dt$ ($dZ = dZ_1 \wedge \cdots \wedge dZ_m$, $dt = dt_1 \wedge \cdots \wedge dt_n$) is nowhere zero in U. We further select U in such a way that $\mathcal{N} \cap U$ is closed in U. Because $\mathcal{V}|_{\mathcal{N}} \subset \mathbb{C}T\mathcal{N}$ each orbit of $\mathcal{V}$ in U through any point of $\mathcal{N} \cap U$ is entirely contained in $\mathcal{N} \cap U$: $\mathcal{N} \cap U$ is a union of orbits of $\mathcal{V}$ in U. Below we denote by $\pi_{\mathcal{N}}^*$ the pullback to $\mathcal{N} \cap U$ of covectors and differential forms in U. The fact that $\mathcal{V}|_{\mathcal{N}} \subset \mathbb{C}T\mathcal{N}$ entails that there are exactly $m-\ell$ differentials $\pi_{\mathcal{N}}^* dZ_1,\ldots,\pi_{\mathcal{N}}^* dZ_m$ that span $T'_{\mathcal{N}}$ over U ($T'_{\mathcal{N}}$: pullback to $\mathcal{N}$ of $T'|_{\mathcal{N}}$). After relabeling we may as well assume that they are $\pi_{\mathcal{N}}^* dZ_{\ell+1},\ldots,$ $\pi_{\mathcal{N}}^* dZ_m$. Keep in mind that, whatever j, $1 \leq j \leq \ell$,

$$\pi_{\mathcal{N}}^*(dZ_j \wedge dZ_{\ell+1} \wedge \cdots \wedge dZ_m) \equiv 0. \tag{II.3.9}$$

The case $\ell = m$ is not precluded; it simply means that $T'_{\mathcal{N}} = 0$, i.e., that $\pi_{\mathcal{N}}^* dZ_j \equiv 0$ for all $j = 1,\ldots,m$.

Possibly after some further contracting of U about 0 we may assume that the $(m+n-\ell)$-form $\pi_{\mathcal{N}}^*(dZ_{\ell+1} \wedge \cdots \wedge dZ_m \wedge dt)$ does not vanish at any point of $\mathcal{N} \cap U$ (it is equal to $\pi_{\mathcal{N}}^*(dx_{\ell+1} \wedge \cdots \wedge dx_m \wedge dt)$ at 0; when $\ell = m$ it is the form $\pi_{\mathcal{N}}^* dt$). Consequently, given any open set $\mathcal{O} \subset U$ such that $\mathcal{N} \cap \mathcal{O} \neq \emptyset$, there is $\psi \in \mathcal{C}_c^\infty(\mathcal{O})$ such that

$$\int_{\mathcal{N}} \pi_{\mathcal{N}}^*(\psi\, dZ_{\ell+1} \wedge \cdots \wedge dZ_m \wedge dt) \neq 0. \tag{II.3.10}$$

This said we define a distribution (i.e., a zero-current) h in U by the formula

$$\mathcal{C}_c^\infty(U;\Lambda^{m,n}) \ni \psi\, dZ \wedge dt \rightarrow$$

$$\langle h, \psi\, dZ \wedge dt \rangle = \int_{\mathcal{N} \cap U} \pi_{\mathcal{N}}^*(\psi\, dZ_{\ell+1} \wedge \cdots \wedge dZ_m \wedge dt). \tag{II.3.11}$$

Obviously supp $h \subset \mathcal{N} \cap U$; and choosing ψ as in (II.3.10) shows that supp $h = U \cap \mathcal{N}$.

Now let L_k be the vector fields defined by (I.7.25) and suppose that $\psi = L_k u$ for some k, $1 \leq k \leq n$, and some $u \in \mathscr{C}_c^\infty(U)$. It follows at once from (II.3.9) that

$$\pi_{\mathcal{N}}^*(\psi\, dZ_{\ell+1} \wedge \cdots \wedge dZ_m \wedge dt) =$$
$$\pi_{\mathcal{N}}^*[d(u\, dZ_{\ell+1} \wedge \cdots \wedge dZ_m \wedge dt_1 \wedge \cdots \wedge \widehat{dt_k} \wedge \cdots \wedge dt_n)] =$$
$$d[\pi_{\mathcal{N}}^*(u\, dZ_{\ell+1} \wedge \cdots \wedge dZ_m \wedge dt_1 \wedge \cdots \wedge \widehat{dt_k} \wedge \cdots \wedge dt_n)],$$

hence (cf. Lemma II.1.1)

$$\langle L_k h, u\, dZ \wedge dt \rangle = -\langle h, L_k u\, dZ \wedge dt \rangle = 0.$$

We reach the conclusion that, in U, $L_k h \equiv 0$, $k = 1, \ldots, n$. We may state:

THEOREM II.3.4. *Let $\mathcal{N}$ be a submanifold of $\mathcal{M}$ such that $\mathcal{V}|_{\mathcal{N}} \subset \mathbb{C}T\mathcal{N}$. An arbitrary point* 0 *of $\mathcal{N}$ has an open neighborhood* U *in which there is a distribution solution whose support is exactly equal to $\mathcal{N} \cap U$.*

The $\mathscr{C}^1$ Version of the Approximation Formula

One might want to apply the approximation formula (II.2.17), and its consequences, in particular Theorem II.3.2, to a generic submanifold $\mathcal{M}$ of $\mathbb{C}^m$ (or of any complex manifold) which is of class $\mathscr{C}^1$ only. In our setup this means that we must weaken the hypothesis (II.2.1) and merely assume that

$$Z_j \in \mathscr{C}^1(\mathscr{C}\ell U),\ j = 1, \ldots, m. \qquad \text{(II.3.12)}$$

Then we have indeed:

THEOREM II.3.5. *If the open neighborhood* $U_0 \subset U$ *of the origin is sufficiently small, then, whatever the solution $h \in \mathscr{C}^1(\mathscr{C}\ell U)$, Formula* (II.2.17) *is valid in* $\mathscr{C}^1(U_0)$.

PROOF. It is the same as that of Theorem II.2.1 in the case $r = 1$. One must be careful to check that the preliminary results on which the latter is based are still valid under the weaker hypothesis (II.3.12). Actually, such a generalization of Lemmas II.1.1 and II.2.2, and of Corollary II.2.1 follows by approximating the functions Z_j by $\mathscr{C}^\infty$ functions with similar properties. The stronger hypothesis (II.2.1) is not needed in the proof of Lemma II.2.1. The proofs of Lemmas II.2.3 and II.2.4, where we assume that (II.3.12) holds and take $r = 1$, do not require any modification (except that Corollary II.2.2 is not needed). ■

II.4. Analytic Vectors

Let $\mathbf{v}_1,\ldots,\mathbf{v}_r$ be smooth vector fields in an open subset Ω of $\mathcal{M}$ that commute pairwise. We shall use the notation

$$\mathbf{v}^\alpha = \mathbf{v}_1^{\alpha_1}\cdots\mathbf{v}_r^{\alpha_r},\ \alpha = (\alpha_1,\ldots,\alpha_r) \in \mathbb{Z}_+^r,$$

and adopt the following terminology (Nelson [1]):

DEFINITION II.4.1. *A continuous function f in Ω will be called an* analytic vector *of the system of vector fields $\{\mathbf{v}_1,\ldots,\mathbf{v}_r\}$ if $\mathbf{v}^\alpha f \in \mathcal{C}^0(\Omega)$ for every $\alpha \in \mathbb{Z}_+^r$, and if to every compact subset* K *of Ω there is a constant $\rho > 0$ such that, in* K,

$$\sup_{\alpha\in\mathbb{Z}_+^r} (\rho^{|\alpha|}|\mathbf{v}^\alpha f|/\alpha!) < +\infty. \tag{II.4.1}$$

We refer the reader to section II.1 whose notation we adopt in toto.

PROPOSITION II.4.1. *In order that $f \in \mathcal{C}^0(\mathrm{U})$ be an analytic vector of the system of vector fields $\{\mathrm{M}_1,\ldots,\mathrm{M}_m\}$ in the neighborhood* U *of* 0 *it is necessary and sufficient that there be an open neighborhood $\mathcal{O}$ of Σ in $\mathbb{C}^m\times \mathrm{W}$* (see (II.1.1)) *and a continuous function $\tilde{f}(z,t)$ in $\mathcal{O}$, holomorphic with respect to z and such that $f(x,t) = \tilde{f}(\mathrm{Z}(x,t),t)$. If such a function $\tilde{f}$ exists its germ at Σ is unique.*

PROOF. Suppose that (II.4.1) holds for $\mathbf{v}_j = \mathrm{M}_j$ (and $r = m$, $\mathrm{K} \subset\subset \mathrm{U}$). Define, for any $(x,t) \in \mathrm{U}$,

$$F(z,x,t) = \sum_{\alpha\in\mathbb{Z}_+^m} [z-\mathrm{Z}(x,t)]^\alpha\, \mathrm{M}^\alpha f(x,t)/\alpha!.$$

According to (II.4.1), there is a continuous function $\rho > 0$ in U such that the series defining F converges uniformly and absolutely on each compact subset of the open set

$$\mathrm{A} = \{\, (z,x,t) \in \mathbb{C}^m\times \mathrm{V}\times \mathrm{W};\ z \in \Delta(x,t)\,\},$$

where we have used the notation

$$\Delta(x,t) = \{\, z \in \mathbb{C}^m;\ |z_j - \mathrm{Z}_j(x,t)| < \rho(x,t),\ j = 1,\ldots,m\,\}.$$

The set A is open and connected. The function $F(z,x,t)$ is continuous in A and for any fixed $(x,t) \in \mathrm{U}$, it is holomorphic with respect to z in $\Delta(x,t)$.

Recall that $\Sigma \subset \mathbb{C}^m\times\mathbb{R}^n$ is the image of U under the diffeomorphism $(x,t) \to (\mathrm{Z}(x,t),t)$. For each $t \in \mathrm{W}$, $x \to \mathrm{Z}(x,t)$ is a diffeomorphism of the convex and open subset V of $\mathbb{R}^m$ onto a maximally real submanifold of $\mathbb{C}^m$, Σ_t. After decreasing the function $\rho(x,t) > 0$ we may assume that the union

$$\Gamma_t = \bigcup_{x \in V} \Delta(x,t)$$

is *simply connected* (this means that its first homotopy group π_1 vanishes). And, of course, Γ_t is a connected open set.

Until specified otherwise $t \in W$ will remain fixed. Consider $x_0 \in V$ and $z \in \Delta(x_0,t)$. There is an open neighborhood $V_0 \subset V$ of x_0 such that $z \in \Delta(x,t)$ for all $x \in V_0$. We derive from (I.7.25) that $M_jF \equiv 0$ in V_0 $(j = 1,\ldots,m)$. (Notice that, by (I.7.26), M_j is a vector field in x-space V whose coefficients depend on t—which is fixed.) Since the M_j span the whole complex tangent bundle of the submanifold $V \times \{t\}$ we derive that $F(z,x,t)$ is independent of x in V_0.

We avail ourselves of the diffeomorphism $x \to Z(x,t)$ to transfer the function ρ from $V \times \{t\}$ to Σ_t. We shall denote by z' the variable point in Σ_t and write $\rho(z')$ for the value of the transfer of ρ. We define the polydisk

$$\Delta(z') = \{ z \in \mathbb{C}^m;\, |z_j - z_j'| < \rho(z'),\, j = 1,\ldots,m \}.$$

We transfer the function F to the set

$$\mathcal{A}_t = \{ (z,z') \in \mathbb{C}^m \times \mathbb{C}^m;\, z' \in \Sigma_t,\, z \in \Delta(z') \}.$$

We write $\mathcal{F}(z,z')$ for the transfer of F. We have at our disposal the following information:

$$\begin{aligned}&\mathcal{F}(z,z') \textit{ is a continuous function in the set } \mathcal{A}_t\textit{; for each } z' \in \Sigma_t \\ &\mathcal{F}(z,z') \textit{ is a holomorphic function of } z \textit{ in } \Delta(z');\end{aligned} \tag{II.4.2}$$

$$\begin{aligned}&\textit{if } (z_0,z_0') \in \mathcal{A}_t \textit{ there is a neighborhood } \mathcal{N}_0' \textit{ of } z_0' \textit{ in } \Sigma_t \textit{ such that} \\ &z_0 \in \Delta(z') \textit{ and } \mathcal{F}(z_0,z') = \mathcal{F}(z_0,z_0') \textit{ for all } z' \in \mathcal{N}_0'.\end{aligned} \tag{II.4.3}$$

By an obvious compactness argument we derive from (II.4.3) that there are neighborhoods $\mathcal{N}_0$ of z_0, $\mathcal{N}_0'$ of z_0' in $\mathbb{C}^m$ and in Σ_t respectively, such that

$$\forall\, z \in \mathcal{N}_0,\, z' \in \mathcal{N}_0',\, z \in \Delta(z') \textit{ and } \mathcal{F}(z,z') = \mathcal{F}(z,z_0'). \tag{II.4.4}$$

We say that the relation $\mathcal{R}$ holds between two points (z_1,z_1') and (z_2,z_2') of $\mathcal{A}_t$ if $z_1 = z_2$ and $\mathcal{F}(z_1,z_1') = \mathcal{F}(z_2,z_2')$. It is seen at once that $\mathcal{R}$ is an equivalence relation. The quotient map $\mathcal{A}_t \to \mathcal{A}_t/\mathcal{R}$ induces a bijection of the subset $\Delta(z') \times \{z'\}$ onto a subset $\dot{\Delta}(z')$ of $\mathcal{A}_t/\mathcal{R}$. When z' ranges over Σ_t the sets $\dot{\Delta}(z')$ make up a covering of $\mathcal{A}_t/\mathcal{R}$. We may transfer the holomorphic coordinates z_j from $\Delta(z')$ to $\dot{\Delta}(z')$; they agree on any overlap $\dot{\Delta}(z_1')\cap\dot{\Delta}(z_2')$. It follows that the multiplets $(\dot{\Delta}(z'),z_1,\ldots,z_m)$ can be regarded as holomorphic local charts in a complex structure on $\mathcal{A}_t/\mathcal{R}$.

If we follow up the map $\Sigma_t \ni z' \to (z',z') \in \mathcal{A}_t$ with the quotient map $\mathcal{A}_t \to \mathcal{A}_t/\mathcal{R}$ we obtain an embedding of Σ_t into $\mathcal{A}_t/\mathcal{R}$. Each point $z' \in \Sigma_t \subset \mathcal{A}_t/\mathcal{R}$ is the center of the domain $\dot{\Delta}(z') \cong \Delta(z')$; it follows that the complex manifold $\mathcal{A}_t/\mathcal{R}$ is connected.

The first-coordinate projection $(z,z') \to z$ from $\mathcal{A}_t$ to Γ_t is constant in each equivalence class mod $\mathcal{R}$ and therefore induces a map π: $\mathcal{A}_t/\mathcal{R} \to \Gamma_t$ that is trivially holomorphic. We claim that it is a local biholomorphism. Let $z_0 \in \Gamma_t$. Select at random a point z_0' of Σ_t such that $z_0 \in \Delta(z_0')$. Select open neighborhoods $\mathcal{N}_0$ of z_0, $\mathcal{N}_0'$ of z_0' as in (II.4.4). Define then a map of $\mathcal{N}_0$ into $\mathcal{A}_t/\mathcal{R}$ by assigning to each $z \in \mathcal{N}_0$ the coset mod $\mathcal{R}$ that contains the pairs (z,z'), $z' \in \mathcal{N}_0'$. By fixing $z' \in \mathcal{N}_0'$ and letting z vary we obtain a holomorphic map of $\mathcal{N}_0$ onto an open subset $\dot{\mathcal{N}}_0$ of $\dot{\Delta}(z')$ whose inverse is equal to the restriction of π to $\dot{\mathcal{N}}_0$, which is what we wanted to prove.

Since any holomorphic map of a connected complex manifold $\mathcal{M}$ onto a connected and simply connected open subset of $\mathbb{C}^m$ that is a local biholomorphism is perforce a (global) biholomorphism, we see that Γ_t and $\mathcal{A}_t/\mathcal{R}$ are isomorphic.

By construction of $\mathcal{A}_t/\mathcal{R}$ the pushforward of $\mathcal{F}(z,z')$ under the quotient map $\mathcal{A}_t \to \mathcal{A}_t/\mathcal{R}$ is a holomorphic function on $\mathcal{A}_t/\mathcal{R}$, which in turn can be transferred as a holomorphic function $\tilde{f}(z,t)$ on Γ_t. By observing that if $z = z' = Z(x,t)$ then $\mathcal{F}(z,z') = f(x,t)$, we derive easily that $\tilde{f}(Z(x,t),t) = f(x,t)$.

Finally, if $\tilde{f}_1$ is any other continuous function of (z,t) in some open neighborhood of Σ in $\mathbb{C}^m \times \mathbb{R}^n$ that is equal to $\tilde{f}$ on Σ, then necessarily $\tilde{f}_1 \equiv \tilde{f}$ in a neighborhood of Σ. Indeed, for each $t \in \mathrm{W}$, Σ_t is totally real in $\mathbb{C}^m$ and has real dimension m; it suffices to apply Corollary II.3.8 to $\tilde{f} - \tilde{f}_1$. ∎

Next we introduce the second-order differential operator in U,

$$\Delta_{\mathrm{M}} = \mathrm{M}_1^2 + \cdots + \mathrm{M}_m^2.$$

Notice that the symbol of Δ_{M} is $\langle {}^t\mathrm{Z}_x(x,t)\xi\rangle^2$; ${}^t\mathrm{Z}_x$ is the transpose of the Jacobian matrix Z_x. At the points (x,t,ξ) where the symbol of Δ_{M} vanishes we must have $|\xi| = |{}^t\Phi_x(x,t)\xi| \leq \varepsilon|\xi|$ by (II.1.2), whence $\xi = 0$. Thus, if we regard Δ_{M} as a differential operator on the maximally real submanifold $\mathrm{V} \times \{t\}$ of U, we see that Δ_{M} is *elliptic*.

A function $f \in \mathcal{C}^0(\mathrm{U})$ is called an *analytic vector of the operator* Δ_{M} if $\Delta_{\mathrm{M}}^k f \in \mathcal{C}^0(\mathrm{U})$ for every $k \in \mathbb{Z}_+$ and if to every compact subset K′ of Δ there are constants C_1', $C_2' > 0$ such that

$$\operatorname*{Max}_{\mathrm{K}'} |\Delta_{\mathrm{M}}^k f| \leq C_1' \, C_2'^k (2k)!, \; \forall \, k \in \mathbb{Z}_+. \tag{II.4.5}$$

Elliptic theory implies at once that f is a $\mathcal{C}^\infty$ function of x, i.e., $f \in \mathcal{C}^\infty(\mathrm{V};\mathcal{C}^0(\mathrm{W})) \cong \mathcal{C}^0(\mathrm{W};\mathcal{C}^\infty(\mathrm{V}))$.

It is readily seen that every analytic vector for the system $\{\mathrm{M}_1,\ldots,\mathrm{M}_m\}$ (Definition II.4.1) is an analytic vector for Δ_{M}. We are now going to show that the converse is true. We are going to need a slightly more precise result.

LEMMA II.4.1. *To each compact subset* K_1 *of* V *and each compact subset* K_2

of W *there is a compact subset* K_1' *of* V *such that whatever the function* $f \in \mathscr{C}^\infty(V;\mathscr{C}^0(W))$, *the following is true.*

If the estimates (II.4.5) *hold with* $K' = K_1' \times K_2$ *then there are constants* $C_1, C_2 > 0$ *such that*

$$\operatorname*{Max}_{K} |M^\alpha f| \leq C_1 C_2^{|\alpha|}\alpha!, \ \forall\, \alpha \in \mathbb{Z}_+^m, \tag{II.4.6}$$

where $K = K_1 \times K_2$.

Moreover, the constant C_2 *can be chosen independently of* C_1' (the constant in (II.4.5)).

PROOF. We shall reason with $t \in W$ fixed. It will be quite evident that the estimates carry over to the case where t varies, as long as t remains in a compact subset of W. Also, we shall make use of the L^2 norm with respect to the variables x, instead of the maximum norm, as it is better suited to handling operators such as Δ_M. The result for the maximum norm will then follow, say, by way of the Sobolev embedding theorem.

Let V′ be a relatively compact open neighborhood of K_1 in V and $\{\Omega_s\}_{0\leq s\leq 1}$ a "one-parameter" family of relatively compact open subsets of V such that, for some constant $c_0 > 0$ and any pair of numbers s, s', $0 \leq s \leq s' \leq 1$, the following is true:

$$K_1 \subset \Omega_0 \subset \Omega_s \subset \Omega_{s'} \subset V'; \tag{II.4.7}$$

$$\operatorname{dist}[\Omega_s, \partial\Omega_{s'}] \geq c_0(s' - s). \tag{II.4.8}$$

We shall take $K_1' = \mathscr{C}\ell V'$.

Let now $\chi \in \mathscr{C}_c^\infty(V')$. The ellipticity of Δ_M entails that there is a constant $C_0 > 0$ such that, for all $f \in \mathscr{C}^\infty(V)$ and for $\kappa = 1, 2$,

$$\|\chi f\|_{H^\kappa} \leq C_0\left\{ \|\Delta_M(\chi f)\|_{H^{\kappa-2}} + \|\chi f\|_{H^0} \right\}. \tag{II.4.9}$$

In (II.4.9) $\|\cdot\|_{H^\kappa}$ is the standard Sobolev norm in x-space $\mathbb{R}^m$. Let us first apply (II.4.9) with $\kappa = 1$:

$$\sum_{i=1}^{m} \|\chi M_i f\|_{H^0} \leq C_0' \left\{ \|\Delta_M f\|_{L^2(V')} + \|f\|_{L^2(V')} \right\}. \tag{II.4.10}$$

Substitute $\Delta_M^k f$ for f and apply (II.4.5) with $K' = K_1' \times K_2$. Here we select χ so that $\chi \equiv 1$ in some neighborhood of $\mathscr{C}\ell\Omega_1$. We get

$$\sum_{i=1}^{m} \|M_i \Delta_M^k f\|_{L^2(\Omega_1)} \leq BC_1' \, C_2'^k (2k)!. \tag{II.4.11}$$

In the remainder of the proof we use the notation $\|\cdot\|_{(s)}$ to mean the norm in

$L^2(\Omega_s)$. We are going to prove that there are positive constants B_0, B_1, B_2, with B_1 and B_2 independent of C_1', such that, for all $\alpha \in \mathbb{Z}_+^m$, $k \in \mathbb{Z}_+$, $s \in [0,1]$ and all $f \in \mathscr{C}^\infty(V;\mathscr{C}^0(W))$, we have

$$(1-s)^{|\alpha|} \, \|M^\alpha \Delta_M^k f\|_{(s)} \leq B_0 B_1{}^{|\alpha|} \, B_2^k \, (2k+|\alpha|)!. \tag{II.4.12}$$

And all the constants will remain bounded if instead of keeping t fixed we let t vary in K_2. By putting $k = s = 0$ in (II.4.12) we derive the sought L^2 estimates:

$$\|M^\alpha f\|_{L^2(\Omega_0)} \leq B_0' B_1{}^{|\alpha|} \alpha!. \tag{II.4.13}$$

As we have said, the Sobolev inequalities enable us to derive readily from (II.4.13) a similar estimate for the maximum norm on K_1.

Note that, when $\alpha = 0$, (II.4.12) is an immediate consequence of (II.4.5) and (II.4.8). When $|\alpha| = 1$ it is a consequence of (II.4.11) (we must take $B_0 \geq C_1'$ and $B_1 \geq B$). We shall reason by induction on $|\alpha| \geq 2$. Fix s, s', with $0 \leq s < s' \leq 1$ and choose the cutoff function $\chi \in \mathscr{C}_c^\infty(\Omega_{s'})$ to be equal to 1 in some neighborhood of $\mathscr{C}\ell\Omega_s$. Thanks to (II.4.8) it is possible to have

$$\operatorname{Max} |M^\alpha \chi| \leq \tilde{C}_a (s'-s)^{-|\alpha|} \tag{II.4.14}$$

with $\tilde{C}_a > 0$ independent of s and s'. Then apply (II.4.9) with $\kappa = 2$. We get, for a suitable constant $E_0 > 0$, independent of s, s', f:

$$\sum_{|\gamma|=2} \|M^\gamma f\|_{(s)} \leq$$

$$E_0 \left\{ \|\Delta_M f\|_{(s')} + (s'-s)^{-1} \sum_{i=1}^m \|M_i f\|_{(s')} + (s'-s)^{-2} \|f\|_{(s')} \right\}. \tag{II.4.15}$$

Select β and γ such that $\alpha = \beta + \gamma$; $|\beta| = |\alpha| - 2$. Substitute $M^\beta \Delta_M^k f$ for f and keep in mind that the vector fields M_i commute with each other and with Δ_M. By the induction hypothesis on $|\alpha|$ we get

$$(1-s')^{|\beta|} \, \|M^\alpha \Delta_M^k f\|_{(s)}/(2k+|\beta|)! \leq$$

$$E_0 B_1 B_2{}^{|\beta|} \, B_3 \left\{ B_3 + m B_2 (2k+|\beta|+1)/[(s'-s)(1-s')] + (s'-s)^{-2} \right\}. \tag{II.4.16}$$

Putting $s' = s + (1-s)/|\alpha|$ in (II.4.16) yields

$$(1-1/|\alpha|)^{|\beta|} (1-s)^{|\beta|+2} \|M^\alpha \Delta_M^k f\|_{(s)}/(2k+|\beta|)! \leq$$

$$E_0 B_1 B_2{}^{|\beta|} \, B_3^k \left\{ B_3 + m B_2 (2k+|\beta|+1)|\alpha| + |\alpha|^2 \right\}. \tag{II.4.17}$$

We note that $[|\alpha|/(|\alpha|-1)]^{|\alpha|-1} \leq e$ and thus $|\alpha|/(|\alpha|-1) \leq e(1-1/|\alpha|)^{|\beta|}$. We derive from (II.4.17):

$$(1-s)^{|\alpha|}\,\|M^{\alpha}\Delta_M^k f\|_{(s)}/(2k+|\alpha|)! \leq eE_0B_1B_2{}^{|\alpha|}\,B_3^k[(B_3+1)/B_2^2+m/B_2]. \tag{II.4.18}$$

It suffices then to take B_2 large enough that $eE_0[(B_3+1)/B_2^2+m/B_2] \leq 1$. ■

Suppose for a moment that we are dealing with the locally integrable structure in U defined by all the functions Z_i and t_j $(1 \leq i \leq m,\ 1 \leq j \leq n)$; the cotangent structure bundle is then $\mathbb{C}T^*\mathcal{M}|_U$. In such circumstances the system of all the vector fields M_i and L_j plays the role that is played in the standard situation (when only the differentials dZ_i span the structure bundle) by the M_i's alone. The estimate (II.4.6) must be now replaced by

$$\operatorname*{Max}_K |M^{\alpha}L^{\beta}f| \leq C_1C_2{}^{|\alpha|+|\beta|}\,\alpha!\beta!,\ \forall\ \alpha \in \mathbb{Z}_+^m,\ \beta \in \mathbb{Z}_+^n, \tag{II.4.19}$$

where K is an arbitrary compact subset of U. Since the M_i and the L_j span $\mathbb{C}T\mathcal{M}|_U$ we see that necessarily $f \in \mathscr{C}^\infty(U)$. We derive at once from Proposition II.4.1:

PROPOSITION II.4.2. *In order for $f \in \mathscr{C}^\infty(U)$ to be an analytic vector of the system $\{M_1,\dots,M_m,L_1,\dots,L_n\}$ it is necessary and sufficient that there be a holomorphic function $\tilde{f}(z,t)$ in an open neighborhood $\mathcal{O}$ of Σ in $\mathbb{C}^m \times \mathbb{C}^n$ such that $f(x,t) = \tilde{f}(Z(x,t),t)$ in* U.

Set, for $\kappa > 0$,

$$\Delta_{L,\kappa M} = L_1^2 + \cdots + L_n^2 + \kappa^2(M_1^2 + \cdots + M_m^2).$$

By making use of the expressions (I.7.27) and of the fact that Δ_M is elliptic as a differential operator in x-space V we see that $\Delta_{L,\kappa M}$ will be elliptic in (x,t)-space U as soon as κ is sufficiently large. We derive from Lemma II.4.1:

LEMMA II.4.2. *Let the number $\kappa > 0$ be large enough that $\Delta_{L,\kappa M}$ be elliptic in* U. *Then to each compact subset* K *of* U *and each pair of positive constants C_1', C_2' there is another compact subset* K′ *of* U *and a pair of positive constants C_1, C_2 such that, given any function $f \in \mathscr{C}^\infty(U)$, the inequalities*

$$\operatorname*{Max}_{K'} |\Delta_{L,\kappa M}^k f| \leq C_1'C_2'^k(2k)!,\ \forall\ k \in \mathbb{Z}_+, \tag{II.4.20}$$

imply the estimates (II.4.19).

II.5. Local Structure of Distribution Solutions and of L-closed Currents

We are going to use the second-order differential operators Δ_M and $\Delta_{L,\kappa M}$ introduced in section II.4; $\kappa > 0$ will be large enough that $\Delta_{L,\kappa M}$ will be elliptic in U, which will allow us to apply Lemma II.4.2.

THEOREM II.5.1. *There is an open neighborhood of the origin,* $U_0 \subset U$, *such that, given any distribution solution h in* U *and any integer* $\mu \geq 0$, *there is a* $\mathscr{C}^\mu$ *solution* h_1 *in* U_0, *a holomorphic function of z in some open neighborhood of* $Z(U_0)$ *in* $\mathbb{C}^m$, $\tilde{\Psi}$, *and an integer* $\nu \geq 0$ *such that, in* U_0,

$$h = \Delta_M^\nu h_1 + \tilde{\Psi}\circ Z. \tag{II.5.1}$$

PROOF. Select an open neighborhood $U' \subset\subset U$ of the origin so small that the transpose of $\Delta_{L,\kappa M}$ is injective when acting on $\mathscr{C}_c^\infty(U')$. Then (cf. Hörmander [4], Theorems 13.4.1, 13.5.2) whatever the real number s and the integer ν, $\Delta_{L,\kappa M}^\nu$ will map $H_{loc}^s(U')$ *onto* $H_{loc}^{s-2\nu}(U')$.

First we select s large enough that $H_{loc}^s(U') \subset \mathscr{C}^\mu(U')$ and then select ν large enough that $h|_{U'} \in H_{loc}^{s-2\nu}(U')$. We may as well choose $\nu \geq 1$. Then let $v \in \mathscr{C}^\mu(U')$ satisfy

$$\Delta_{L,\kappa M}^\nu v = h \tag{II.5.2}$$

in U'. Since every L_j commutes with $\Delta_{L,\kappa L}$ we derive from (II.5.2)

$$\Delta_{L,\kappa M}^\nu L_j v = 0,\ j = 1,\dots,n. \tag{II.5.3}$$

It is convenient to take $U' = V' \times W'$ with V' and W' open balls centered at 0 in V and W respectively. Call Σ' the image of U' under the map $(x,t) \to (Z(x,t),t)$. By (II.5.3) we know that each $L_j v$ is an analytic vector of $\Delta_{L,\kappa M}$. We apply Lemma II.4.2 and Proposition II.4.2 (in that order): there is a holomorphic function $\tilde{g}_j(z,t)$ in an open and connected neighborhood $\mathscr{O}'$ of Σ' in $\mathbb{C}^m \times \mathbb{C}^n$ such that, in U',

$$L_j v(x,t) = \tilde{g}_j(Z(x,t),t),\ j = 1,\dots,n. \tag{II.5.4}$$

Since $L_k L_j v = L_j L_k v$ for all j, k, we have

$$(\partial/\partial t_k)\tilde{g}_j = (\partial/\partial t_j)\tilde{g}_k,\ \forall\, j,k = 1,\dots,n, \tag{II.5.5}$$

on Σ'. Since Σ' is maximally real in $\mathbb{C}^m \times \mathbb{C}^n$, (II.5.5) must be true in $\mathscr{O}'$. Select a polydisk $\Delta \subset \mathbb{C}^m$ and an open ball $\mathscr{B} \subset \mathbb{C}^n$, both centered at the origin, and such that $\Delta \times \mathscr{B} \subset \mathscr{O}'$. Thanks to (II.5.5) we can find a holomorphic function $\tilde{w}$ in $\Delta \times \mathscr{B}$ such that

$$\partial\tilde{w}/\partial t_j = \tilde{g}_j,\ j = 1,\dots,n. \tag{II.5.6}$$

Select U_0 small enough that its image under the map $(x,t) \to (Z(x,t),t)$ is contained in $\Delta \times \mathscr{B}$ and define in U_0, $w(x,t) = \tilde{w}(Z(x,t),t)$. By virtue of (II.5.4) and (II.5.6) we see that, in U_0,

$$L_j(v-w) = 0, j = 1,\dots,n. \tag{II.5.7}$$

Of course $v-w \in \mathscr{C}^{\mu}(U_0)$. On the other hand, we derive from (II.5.3)

$$(\partial/\partial t_j)(\Delta_t + \kappa^2\Delta_z)^{\nu}\tilde{w} \equiv 0, j = 1,\dots,n, \tag{II.5.8}$$

in $\Delta \times \mathscr{B}$ $[\Delta_t = (\partial/\partial t_1)^2 + \cdots + (\partial/\partial t_n)^2$, $\Delta_z = (\partial/\partial z_1)^2 + \cdots + (\partial/\partial z_m)^2]$; (II.5.8) means that $(\Delta_t + \kappa^2\Delta_z)^{\nu}\tilde{w}$ is independent of t, i.e., it is a holomorphic function of z alone, in the polydisk Δ, $\tilde{\Psi}(z)$. By (II.5.2) we have

$$h = \tilde{\Psi}\circ Z + \Delta^{\nu}_{L,\kappa M}(v-w).$$

But because of (II.5.7) we have $\Delta_{L,\kappa M}(v-w) = \kappa^2\Delta_M(v-w)$ and therefore

$$h = \tilde{\Psi}\circ Z + \kappa^{2\nu}\Delta^{\nu}_{M}(v-w). \tag{II.5.9}$$

We may therefore take $h_1 = \kappa^{2\nu}(v-w)$. ∎

REMARK II.5.1. It is always possible to solve the equation

$$\Delta^{\nu}_{z}\tilde{\psi} = \tilde{\Psi} \tag{II.5.10}$$

in the space of holomorphic functions in the polydisk Δ (see Treves [1], Theorem 9.4). Formula (II.5.9) may therefore be rewritten as

$$h = \Delta^{\nu}_{M}h_2 \tag{II.5.11}$$

where $h_2 = \tilde{\psi}\circ Z + \kappa^{2\nu}(v-w)$. ∎

EXAMPLE II.5.1. Consider the structure on $\mathbb{R}^2$ (with coordinates x, t) defined by the vector field $L = \partial/\partial t - \imath x\partial/\partial x$ (cf. Examples I.4.2, I.11.1, I.11.2). The following distributions are solutions:

$$h_k(x,t) = e^{\imath kt}\mathscr{Y}(x)x^k/k! \tag{II.5.12}$$

($k = 0,1,\dots$) where $\mathscr{Y}$ is the Heaviside function: $\mathscr{Y}(x) = 1$ for $x > 0$, $\mathscr{Y}(x) = 0$ for $x < 0$. We get another sequence of distribution solutions

$$h_{-k}(x,t) = e^{-\imath kt}\delta^{(k-1)}(x) \tag{II.5.13}$$

($k = 1,2,\dots$) where $\delta^{(j)}(x)$ stands for the derivative of order j of the Dirac distribution $\delta(x)$.

The differential of the function $Z = xe^{\imath t}$ spans the cotangent structure bundle T' (although it does not satisfy (I.7.24) it can be put to the same use as a solution that does). The corresponding vector field M is $e^{-\imath t}\partial/\partial x$. We see that, for any pair of integers $j \geq 1$, $k \geq 0$, we have

$$h_{-j} = M^{j+k}h_k. \tag{II.5.14}$$

Formula (II.5.14) may be regarded as a particular case of (II.5.1) (or of (II.5.11)). ■

We look now at currents of the kind (I.7.29); they are the standard representatives of equivalence classes $f \in \mathcal{D}'(U;\Lambda^{p,q})$ and, below, we shall always refer to them as *standard currents* in U. We let the differential operator L act on them as described in (I.7.30) and (I.7.31). We shall also let the second-order (scalar) operators Δ_M and $\Delta_{L,\kappa M}$ act on them, coefficientwise.

THEOREM II.5.2. *There is an open neighborhood* $U_0 \subset U$ *of the origin such that, given any* L*-closed standard current* f *of bidegree* (p,q), *as in* (I.7.29), *and any integer* $\mu \geq 0$, *there are standard currents* f_1 *and* ψ *in* U_0, *of bidegrees* (p,q) *and* $(p,q-1)$ *respectively, both with* $\mathcal{C}^\mu$ *coefficients, and an integer* $\nu \geq 0$ *such that, in* U_0,

$$f = \Delta^\nu_{L,\kappa M} f_1 + L\psi. \tag{II.5.15}$$

Moreover, one can take f_1 *to be* L*-closed and the coefficients of* ψ *to be the pullback, under the map* $(x,t) \to (Z(x,t),t)$, *of holomorphic functions of* (z,t) *in some open neighborhood of the image of* U_0 *in* $\mathbb{C}^m \times \mathbb{C}^n$.

PROOF. We begin by looking at standard currents of bidegree $(0,q)$:

$$f = \sum_{|J|=q} f_J(x,t)dt_J \tag{II.5.16}$$

$(f_J \in \mathcal{D}'(U))$. We shall assume $1 \leq q \leq n$.

The argument follows closely the proof of Theorem II.5.1. We make use of the same neighborhoods of the origin, $U' = V' \times W'$, Δ and $\mathcal{B}$ as in that proof. Taking $\mu > 1$, if ν is large enough we can find a form

$$v = \sum_{|J|=q} v_J(x,t)dt_J \tag{II.5.17}$$

with coefficients $v_J \in \mathcal{C}^\mu(U')$, such that, in U',

$$\Delta^\nu_{L,\kappa M} v = f. \tag{II.5.18}$$

By using the commutation of the differential operator L with the action (coefficientwise) of $\Delta_{L,\kappa M}$ and the hypothesis that f is L-closed we derive from (II.5.18)

$$\Delta^\nu_{L,\kappa M} Lv = 0. \tag{II.5.19}$$

As a consequence of this we see that each coefficient of Lv is an analytic vector of $\Delta_{L,\kappa M}$. Applying Lemma II.4.2 and Proposition II.4.2 once again shows that, to each multi-index J, $|J| = q+1$, there is a holomorphic function $\tilde{g}_J$ in an open neighborhood $\mathcal{O}'$ of Σ' (the image of U' under the map (II.1.5)) such that, if we define the $(q+1)$-form in $\mathcal{O}'$,

$$\tilde{g} = \sum_{|J|=q+1} \tilde{g}_J(z,t)\mathrm{d}t_J \tag{II.5.20}$$

then

in U′, Lv *is equal to the pullback of* $\tilde{g}$ *under the map* (II.1.5) (II.5.21)
(regarded as a map into $\mathbb{C}^m \times \mathbb{C}^n$).

Applying L to the pullback of $\tilde{g}$ to U′ is the same as pulling back $\mathrm{d}_t\tilde{g}$ to U′. Since $\mathrm{L}(\mathrm{L}v) \equiv 0$ we derive that the pullback of $\mathrm{d}_t\tilde{g}$ to Σ' vanishes; since Σ' is maximally real in $\mathbb{C}^m \times \mathbb{C}^n$, we conclude that

$$\mathrm{d}_t\tilde{g} \equiv 0 \tag{II.5.22}$$

in $\Delta \times \mathcal{B}$. We derive at once that there exists a differential form

$$\tilde{w} = \sum_{|J|=q} \tilde{w}_J(z,t)\mathrm{d}t_J \tag{II.5.23}$$

with holomorphic coefficients in $\Delta \times \mathcal{B}$ such that, there,

$$\mathrm{d}_t\tilde{w} = \tilde{g}. \tag{II.5.24}$$

As in the proof of Theorem II.5.1 take $\mathrm{U}_0 \subset \mathrm{U}' = \mathrm{V}' \times \mathrm{W}'$ small enough that its image under the map (II.1.5) is contained in $\Delta \times \mathcal{B}$. Call w the pullback of $\tilde{w}$ to U_0; then the pullback of $\mathrm{d}_t\tilde{w}$ is equal to Lw. By virtue of (II.5.21) and (II.5.24) we see that, in U_0,

$$\mathrm{L}(v-w) = 0. \tag{II.5.25}$$

Set $\tilde{\varphi} = (\Delta_t + \kappa^2\Delta_z)^\nu \tilde{w}$. We have $\mathrm{d}_t\tilde{\varphi} = (\Delta_t + \kappa^2\Delta_z)^\nu \tilde{g}$ and the pullback to U′ of $\mathrm{d}_t\tilde{\varphi}$ vanishes identically, according to (II.5.19) and (II.5.21). Again by using the property that Σ' is maximally real we derive that, in $\Delta \times \mathcal{B}$,

$$\mathrm{d}_t\tilde{\varphi} \equiv 0, \tag{II.5.26}$$

and therefore that there exists a differential form

$$\tilde{\psi} = \sum_{|J|=q-1} \tilde{\psi}_J(z,t)\mathrm{d}t_J \tag{II.5.27}$$

with holomorphic coefficients in $\Delta \times \mathcal{B}$, such that

$$\mathrm{d}_t\tilde{\psi} = \tilde{\varphi} \tag{II.5.28}$$

in $\Delta \times \mathcal{B}$. We take ψ to be the pullback to U_0 of $\tilde{\psi}$ under the map (II.1.5) and $h_1 = v - w$. This completes the proof of Theorem II.5.2 for standard currents of bidegree $(0,q)$.

It is now easy to extend the result to standard currents of bidegree (p,q) (see (I.7.29)) with $p \geq 1$. Consider such a current, written as follows:

$$f = \sum_{|I|=p} \mathrm{d}Z_I \wedge f_I \tag{II.5.29}$$

with each f_I a current of bidegree $(0,q)$, i.e., of the kind (II.5.16). Note that

$$Lf = \sum_{|I|=p} dZ_I \wedge Lf_I \tag{II.5.30}$$

and that $Lf \equiv 0$ if and only if $Lf_I \equiv 0$ for every I, $|I| = p$. As a consequence, if we assume that f is L-closed in U, we may apply the result for standard currents of bidegree $(0,q)$ to each f_I and replace it, in (II.5.29), by a decomposition (II.5.15), to obtain a similar decomposition for f. ∎

II.6. The Approximate Poincaré Lemma

We shall make use of the following notation: let S be a subset of the interval $[1,\ldots,n] \subset \mathbb{Z}$, not necessarily ordered; $\epsilon(S)$ will denote the signum of the permutation of S that orders S: $\epsilon(S) = +1$ if the permutation is even, -1 if it is odd. If $S_1 = \{a_1,\ldots,a_{r_1}\}$ and $S_2 = \{b_1,\ldots,b_{r_2}\}$ are two subsets of $[1,\ldots,n]$ such that $S_1 \cap S_2 = \emptyset$ we denote by S_1S_2 the subset $\{a_1,\ldots,a_{r_1},b_1,\ldots,b_{r_2}\}$.

We shall make use of the linear operator $\tilde{\mathscr{E}}_\tau$ defined in (II.2.4); it acts on functions, that is to say, on zero-forms. We shall now extend its action to standard forms of bidegree (p,q) with $0 \leq p \leq m$, $0 \leq q \leq n$. We shall assume, throughout, that (II.2.1) holds. We consider a form

$$f = \sum_{|I|=p} \sum_{|J|=q} f_{I,J}(x,t)\, dZ_I \wedge dt_J, \tag{II.6.1}$$

with coefficients $f_{I,J} \in \mathscr{C}^0(\mathscr{C}\ell\, U)$. We define

$$\tilde{\mathscr{E}}_\tau f(z,t) = \sum_{|I|=p} \sum_{|J|=q} \tilde{\mathscr{E}}_\tau f_{I,J}(z,t)\, dz_I \wedge dt_J; \tag{II.6.2}$$

$\tilde{\mathscr{E}}_\tau f$ is a continuous differential form on $\mathbb{C}^m \times (\mathscr{C}\ell\, W)$ whose coefficients are holomorphic with respect to z. Furthermore it involves no differentials $d\bar{z}_j$. We shall say that $\tilde{\mathscr{E}}_\tau f$ has bidegree (p,q).

Now assume $q \geq 1$. For any multi-index J, $|J| = q$, consider the $(q-1)$-form in t-space,

$$\varpi_J = \sum_{j \in J} \epsilon(j(J \backslash j))\, t_j\, dt_{J \backslash j}. \tag{II.6.3}$$

Let

$$F = \sum_{|J|=q} F_J\, dt_J$$

be a q-form in t-space whose coefficients belong to $\mathscr{C}^0(\mathscr{C}\ell\, W)$. The convexity of the neighborhood W in t-space enables us to define (see E. Cartan [1], p. 41)

$$K^{(q)}F = \sum_{|J|=q} \left\{ \int_0^1 F_J(\lambda t)\lambda^{q-1}\, d\lambda \right\} \varpi_J; \tag{II.6.4}$$

$K^{(q)}F$ is a $(q-1)$-form. Direct computation shows easily that

$$F = d_t K^{(q)}F + K^{(q+1)}d_t F. \tag{II.6.5}$$

We return to the form f given by (II.6.1). We combine the operators $\tilde{\mathscr{E}}_\tau$ and $K^{(q)}$ to define

$$\tilde{\mathscr{K}}_\tau^{(p,q)}f = (-1)^p K^{(q)}(\tilde{\mathscr{E}}_\tau f). \tag{II.6.6}$$

More explicitly we may write

$$\tilde{\mathscr{K}}_\tau^{(p,q)}f = (-1)^p \sum_{|I|=p} \sum_{|J|=q} (\tilde{\mathscr{K}}_\tau^{(q)}f)_{I,J}(z,t)dz_I \wedge \varpi_J, \tag{II.6.7}$$

$$(\tilde{\mathscr{K}}_\tau^{(q)}f)_{I,J}(z,t) =$$

$$\int_V \int_0^1 E_\tau(z;y,\lambda t) f_{I,J}(y,\lambda t) \det[Z_y(y,\lambda t)] \lambda^{q-1} d\lambda dy. \tag{II.6.8}$$

Notice that the form $\tilde{\mathscr{K}}_\tau^{(p,q)}f$ in $\mathbb{C}^m \times (\mathscr{C}\ell W)$ has bidegree $(p,q-1)$. Notice also that if r is any integer ≥ 0 and if the coefficients of f belong to $\mathscr{C}^r(\mathscr{C}\ell U)$, those of $\tilde{\mathscr{K}}_\tau^{(p,q)}f$ belong to $\mathscr{C}^r(\mathbb{C}^m \times (\mathscr{C}\ell W))$; and, of course, the functions (II.6.8) are holomorphic with respect to z.

We may pull $\tilde{\mathscr{K}}_\tau^{(p,q)}f$ back to U, by means of the map (II.1.5). We shall denote by $\mathscr{K}_\tau^{(p,q)}f$ its pullback.

THEOREM II.6.1. *Assume $0 \leq p \leq m$, $1 \leq q \leq n$. There is an open neighborhood of the origin, $U_0 \subset U$, such that, if r is any integer ≥ 0 or if $r = +\infty$, then, given any L-closed standard form f of bidegree (p,q), with $\mathscr{C}^{r+1}$ coefficients, in $\mathscr{C}\ell U$, we have, in $\mathscr{C}^r(U_0;\Lambda^{p,q})$,*

$$f = \lim_{\tau \to +\infty} L(\mathscr{K}_\tau^{(p,q)}f). \tag{II.6.9}$$

PROOF. Let us introduce the same cutoff function χ as in the proof of Theorem II.2.1. Set $g = \chi f$ and apply (II.6.5):

$$\tilde{\mathscr{E}}_\tau g = (-1)^p d_t[K^{(q)}(\tilde{\mathscr{E}}_\tau g)] + (-1)^p K^{(q+1)}d_t(\tilde{\mathscr{E}}_\tau g). \tag{II.6.10}$$

Next we apply (II.2.9):

$$d_t \tilde{\mathscr{E}}_\tau g = \tilde{\mathscr{E}}_\tau Lg, \tag{II.6.11}$$

where L is defined as in (I.7.30). Since f is L-closed, we have $Lg = L\chi \wedge f$. The coefficients of Lg vanish identically in the neighborhood $V' \subset V$ of the origin. We may apply Lemma II.2.4. If we choose the neighborhood $U_0 \subset U$ exactly as in the proof of Theorem II.2.1, we conclude that the pullback under the map (II.1.5) of $\tilde{\mathscr{E}}_\tau Lg$ converges to zero in $\mathscr{C}^r(U_0;\Lambda^{p,q+1})$. We require that W_0 be convex; since the operator $K^{(q+1)}$ involves integration only with respect to t along the rays issued from 0, we derive at once that $K^{(q+1)}d_t(\tilde{\mathscr{E}}_\tau g)$ converges to zero in $\mathscr{C}^r(U_0;\Lambda^{p,q})$.

A similar reasoning applies to the pullback under the map (II.5.1) of the form

$$(-1)^p \mathrm{d}_t[\mathrm{K}^{(q)}\tilde{\mathscr{E}}_\tau((1-\chi)f)].$$

If we add it to the right-hand side of (II.6.10), we reach the conclusion that $\tilde{\mathscr{E}}_\tau(\chi f)$ and the pullback of $(-1)^p \mathrm{d}_t[\mathrm{K}^{(q)}(\tilde{\mathscr{E}}_\tau f)]$, which is equal to $\mathrm{L}\mathscr{K}_\tau^{(p,q)}f$, converge to the same limit in $\mathscr{C}^r(\mathrm{U}_0;\Lambda^{p,q})$. But it follows at once from Lemma II.2.3 that $\mathscr{E}_\tau(\chi f)$ converges to χf in $\mathscr{C}^r(\mathrm{U};\Lambda^{p,q})$; and $\chi f = f$ in U_0. ∎

REMARK II.6.1. It must be emphasized that, in general, the form $\mathscr{K}_\tau^{(p,q)}f$ in (II.6.9) does not converge as $\tau \to +\infty$, not even in the space of currents $\mathscr{D}'(\mathrm{U}_0;\Lambda^{p,q-1})$. For if it did, and if u were its limit, we would have

$$\mathrm{L}u = f \tag{II.6.12}$$

in U_0. As we shall eventually see (in chapter VI) such a "local solvability" property is not generally true.

The existence of a solution u of (II.6.12) for any form f defined in some open neighborhood of an arbitrary point of $\mathcal{M}$ is often referred to as the *Poincaré Lemma* for the differential complex (I.6.16) associated with the locally integrable structure of $\mathcal{M}$. It is for this reason that we shall refer to Theorem II.6.1 as the Approximate Poincaré Lemma. ∎

Combining Theorems II.5.2 and II.6.1 yields an approximate Poincaré Lemma for L-closed *currents*. Let $\mathrm{U}_0 \subset \mathrm{U}, f, f_1$ and ψ be as in Theorem II.5.2; apply Theorem II.6.1 with U_0 in the place of U. We see that there is an open neighborhood $\mathrm{U}_1 \subset \mathrm{U}_0$ of the origin such that we have, in $\mathscr{C}^{\mu-1}(\mathrm{U}_1;\Lambda^{p,q})$, $f_1 = \lim_{\tau\to+\infty} \mathrm{L}(\mathscr{K}_\tau^{(p,q)}f_1)$ (with $\mu \geq 1$). Since the differential operators L and $\Delta_{\mathrm{L},\kappa\mathrm{M}}$ commute, we derive from this and from (II.5.15):

$$f = \lim_{\tau\to+\infty} \mathrm{L}(\Delta^\nu_{\mathrm{L},\kappa\mathrm{M}}\mathscr{K}_\tau^{(p,q)}f_1 + \psi) \tag{II.6.13}$$

in $\mathscr{D}'(\mathrm{U}_1;\Lambda^{p,q})$.

From the Taylor expansion of the coefficients of $\tilde{\mathscr{K}}_\tau^{(p,q)}f$ with respect to z and their Weierstrass approximation as functions of t in W_0 we derive:

COROLLARY II.6.1. *Let* U_0 *be as in Theorem* II.6.1. *Given any* L*-closed standard form* $f \in \mathscr{C}^{r+1}(\mathrm{U};\Lambda^{p,q})$, *there is a sequence of* $(q-1)$*-forms*

$$\tilde{\mathscr{P}}_\nu = \sum_{|I|=p}\sum_{|J|=q} \tilde{\mathscr{P}}_{\nu,I,J}(z,t)\mathrm{d}z_I \wedge \mathrm{d}t_J \tag{II.6.14}$$

whose coefficients are polynomials with respect to (z,t), *such that* f *is the limit in* $\mathscr{C}^r(\mathrm{U}_0;\Lambda^{p,q})$, *as* $\nu \to +\infty$, *of the pullbacks to* U_0, *under the map* (II.1.5), *of the forms* $\mathrm{d}_t\tilde{\mathscr{P}}_\nu$.

Formula (II.6.13) enables us to extend the previous result to L-closed standard currents.

COROLLARY II.6.2. *Provided the neighborhood* U_0 *is sufficiently small, the statement in Corollary* II.6.1 *remains valid if we replace* $\mathscr{C}^r$ *by* $\mathscr{D}'$.

PROOF. Apply Corollary II.6.1 to the form f_1 in (II.6.13) and observe that letting $\Delta_{\mathrm{L},\kappa\mathrm{M}}$ act on the pullback to U of $\mathscr{P}_\nu$ is the same as pulling back $(\Delta_t + \kappa^2\Delta_z)\mathscr{P}_\nu$, which is a polynomial with respect to (z,t). ■

Corollary II.6.1 has implications for the integration of L-closed standard forms on certain cycles:

COROLLARY II.6.3. *Let* γ *be any smooth, compact cycle in* U_0, *of dimension* $p+q$, *whose image under the map* $(x,t) \to (\mathrm{Z}(x,t),t)$ *is a cycle*

$$\tilde{\gamma} = \tilde{\gamma}_1 \times \tilde{\gamma}_2 \tag{II.6.15}$$

with $\tilde{\gamma}_1$ *a p-cycle in* $\mathbb{C}^m$ *and* $\tilde{\gamma}_2$ *a q-cycle in* $\mathrm{W}_0 \subset \mathbb{R}^n$. *Then, whatever the* L-*closed standard form* $f \in \mathscr{C}^1(\mathrm{U};\Lambda^{p,q})$, *we have*

$$\int_\gamma f = 0. \tag{II.6.16}$$

PROOF. Suppose f has the expression (II.5.29). Each $f_I \in \mathscr{C}^1(\mathrm{U};\Lambda^{0,q})$ can be regarded as a differential form on W_0 depending on $x \in \mathrm{V}$. But by (I.7.23), fixing $\mathrm{Z}(x,t)$ entails fixing x. Therefore it makes sense to integrate over $\tilde{\gamma}_2$ the push-forward $\tilde{f}_I$ of f_I under the map (II.1.5). We have

$$\int_\gamma f = \sum_{|I|=p} \int_{\tilde{\gamma}_1} \left(\int_{\tilde{\gamma}_2} \tilde{f}_I \right) \mathrm{d}z_I \tag{II.6.17}$$

and it suffices to prove that, if f is L-closed, then

$$\int_{\tilde{\gamma}_2} \tilde{f}_I(z,t) = 0, \forall\, I, |I| = p, z \in \mathrm{Z}(\mathrm{U}_0).$$

Each f_I is L-closed and we know by Corollary II.6.1 that $\tilde{f}_I$ is the limit in $\mathscr{C}^0(\Sigma_0)$ (Σ_0 : image of U_0 under the map (II.1.5)) of a sequence of q-forms $\mathrm{d}_t\tilde{\mathscr{P}}_\nu$ in t-space whose coefficients are polynomials with respect to (z,t). The integral of each $\mathrm{d}_t\tilde{\mathscr{P}}_\nu$ over $\tilde{\gamma}_2$ is equal to zero and so is therefore the integral of $\tilde{f}_I$. ■

COROLLARY II.6.4. *Let* γ *be any smooth, compact q-cycle in* U_0 *entirely con-*

tained in a fibre of Z *in* U_0. *Then, whatever the* L*-closed standard form* $f \in \mathscr{C}^1(U;\Lambda^{0,q})$, *we have* $\int_\gamma f = 0$.

PROOF. Indeed, if $\gamma \subset U_0 \cap \bar{Z}^1(z_0)$, the image of γ under the map (II.1.5) is of the kind $\{z_0\} \times \tilde{\gamma}_2$ with $\tilde{\gamma}_2 \subset W_0$. ∎

COROLLARY II.6.5. *Let* Λ *be a submanifold of* U_0 *entirely contained in a fibre of the map* Z. *The pullback of an* L*-closed standard form* $f \in \mathscr{C}^1(U;\Lambda^{0,q})$ *to* Λ *is exact.*

PROOF. Indeed the integral of f on any smooth, compact q-cycle in Λ is equal to zero. It suffices therefore to apply Theorems 14 to 17 in De Rham [1]. ∎

EXAMPLE II.6.1. Let $\mathscr{I}$ be an open interval in $\mathbb{R}^1$, W be an open ball in $\mathbb{R}^n$, both centered at the origin. In $\mathscr{M} = \mathscr{I} \times W$ consider the function

$$Z = x + \imath\Phi(x,t) \tag{II.6.18}$$

with Φ a real-valued analytic function in $\mathscr{M}$. The fibres of Z are the analytic sets $x = x_0$, $\Phi(x_0,t) = y_0$. Because of the analyticity of Φ in any open set U $\subset\subset \mathscr{M}$, all the fibres of Z, except possibly a finite number of them, are *analytic submanifolds* (without singularities) of U. And any singular fibre is an analytic subvariety of U; it is the union of a finite number of pairwise disjoint analytic submanifolds.

The differential dZ spans the cotangent structure bundle in an analytic involutive structure on $\mathscr{M}$. The differential operator L in the associated differential complex is given by

$$Lf = d_t f - \imath(d_t\Phi)\wedge(\bar{Z}_x^1 \partial f/\partial x) \tag{II.6.19}$$

(cf. (I.7.31)). By Corollary II.6.5 we know that the pullback to the regular part Λ of a fibre of Z in $U_0 \subset U$ of any L-closed form $f \in \mathscr{C}^1(U;\Lambda^{0,q})$ (see (II.5.16)) is exact—i.e., there is a $(q-1)$-form u on Λ, of class $\mathscr{C}^1$, such that the pullback of f to Λ is equal to $d_t u$. ∎

II.7. Approximation and Local Structure of Solutions Based on the Fine Local Embedding

Rather than using, in U, the coordinates x_j and t_k and the functions Z_j ($1 \leq j \leq m$, $1 \leq k \leq n$) we may want to use the coordinates x_i, y_j, s_k, and t_ℓ, and the functions $z_j = x_j + \imath y_j$ and w_k given by (I.7.4) ($1 \leq i, j \leq \nu$, $1 \leq k \leq d$, $1 \leq \ell \leq n-\nu$). For the sake of brevity we shall denote the imaginary part of w_k by $\varphi_k(z,s,t)$ rather than $\varphi_k(x,y,s,t)$. We shall always assume that $\varphi_k \in$

$\mathscr{C}^\infty(\mathscr{C}\ell\mathrm{U})$ for all $k = 1,\ldots,d$. We shall make use of the basis of $\mathcal{V}$ over U (or over $\mathscr{C}\ell\mathrm{U}$) consisting of the vector fields L_j given by (I.7.15) and (I.7.16); also of the vector fields N_k given by (I.7.17) and M_i given by (I.7.18). We recall that the vector fields L_j $(1 \leq j \leq n)$, M_i $(1 \leq \mathrm{i} \leq \nu)$, N_k $(1 \leq k \leq d)$ commute pairwise and make up a basis of $\mathbb{C}\mathrm{T}\mathcal{M}$ over $\mathscr{C}\ell\mathrm{U}$.

In accordance with this we are led to make use of the map

$$\mathrm{U} \ni (x,y,s,t) \rightarrow (z,w,t) \in \mathbb{C}^\nu \times \mathbb{C}^d \times \mathbb{R}^{n-\nu}. \tag{II.7.1}$$

We shall always assume that (I.7.5) holds and that

$$\mathrm{U} = \Delta \times \mathrm{V} \times \mathrm{W} \tag{II.7.2}$$

where Δ is an open polydisk $\{ z \in \mathbb{C}^\nu;\ |z_j| < \rho, j = 1,\ldots,\nu \}$ and V (resp., W) is an open and convex neighborhood of 0 in s-space $\mathbb{R}^d$ (resp., t-space $\mathbb{R}^{n-\nu}$). Henceforth the variable point in Δ will be called z rather than (x,y); a notation such as $f(z)$ will stand for a function in Δ, even if it is not holomorphic.

In the sequel we are going to use w_s as a short notation for the Jacobian matrix $\partial w/\partial s = (\partial w_k/\partial s_\ell)_{1\leq k,\ell\leq d}$; by (I.7.5) we have $w_s|_0 = I$, the $d \times d$ identity matrix. We shall select U small enough that

$$\sup_{\mathrm{U}} \|w_s - I\| < \varepsilon \tag{II.7.3}$$

with $0 < \varepsilon < 1/10$ (cf. (II.1.2)). We may then write (cf. (II.1.4))

$$|\varphi(z,s,t) - \varphi(z,s',t)| \leq \varepsilon|s - s'|. \tag{II.7.4}$$

Thanks to this we see that the map (II.7.1) is injective. It is a diffeomorphism of U onto a smooth submanifold Σ of $\Delta \times \mathbb{C}^d \times \mathrm{W}$. For each $t \in \mathrm{W}$ let Σ_t denote the image of the "slice" $\Sigma \cap (\Delta \times \mathbb{C}^d \times \{t\})$ under the coordinate projection $(z,w,t) \rightarrow (z,w)$ into $\mathbb{C}^{\nu+d}$; Σ_t is a generic submanifold of $\mathbb{C}^{\nu+d}$.

We view (II.7.1) as an embedding of U into the model manifold $\mathbb{C}^\nu \times \mathbb{C}^d \times \mathbb{R}^{n-\nu}$ equipped with its standard elliptic structure, as defined in section II.1 (it suffices to put $m = \nu + d$ and to replace n by $n - \nu$). In the present notation, the standard elliptic structure is defined by the vector fields

$$\begin{gathered} \partial/\partial\bar{z}_j\ (1 \leq j \leq \nu),\ \partial/\partial\bar{w}_k\ (1 \leq k \leq d), \\ \partial/\partial t_\ell\ (1 \leq \ell \leq n-\nu). \end{gathered} \tag{II.7.5}$$

The cotangent structure bundle is spanned by the differentials of the functions $z_1,\ldots,z_\nu,w_1,\ldots,w_d$. If we pull back, by means of the map (II.7.1), the functions w_k we get the functions (I.7.4). In other words, the map (II.7.1) is an isomorphism of U, equipped with the locally integrable structure inherited from $\mathcal{M}$, onto Σ equipped with the one inherited from the standard elliptic structure on $\mathbb{C}^\nu \times \mathbb{C}^d \times \mathbb{R}^{n-\nu}$.

The construction in section II.1, based on the map (II.1.5), has an obvious parallel here, based on the map (II.7.1). By the same token all the results of

Sections II.2 to II.6 have their analogues here. The role of the coordinates x_j ($j = 1,\dots,m$) in those sections may now be played by the x_j ($j = 1,\dots,\nu$) and the s_k ($k = 1,\dots,d$); then the role of the t_ℓ ($\ell = 1,\dots,n$) will presently be played by the y_j ($j = 1,\dots,\nu$) and the t_ℓ ($\ell = 1,\dots,n-\nu$) (recall that $m = \nu+d$). Statements and proofs are routine variants of those of the corresponding results in sections II.1 to II.6 and we leave them to the reader.

Rather, what we are going to do is to sketch a slightly different approach whose effect is to cut down the number of integrations in the "Gaussian" integrals of the kind (II.2.4) and (II.2.5). In this approach the role of the variables x_j in the previous sections is played by the s_k alone; that of the t_ℓ is played by the z_i and the t_ℓ.

We introduce the function in $\mathbb{C}^d \times (\mathcal{C}\ell\mathrm{U})$,

$$F_\tau(w;z,s,t) = (\tau/\pi)^{\frac{1}{2}d}\, e^{-\tau\langle w-s-\iota\varphi(z,s,t)\rangle^2}, \tag{II.7.6}$$

and define, for any $f \in \mathscr{C}^0(\mathcal{C}\ell\mathrm{U})$,

$$\tilde{\mathscr{F}}_\tau f(z,w,t) = \int_{\mathrm{V}} F_\tau(w;z,s,t)\, f(z,s,t)\, [\det w_s(z,s,t)]\, ds, \tag{II.7.7}$$

$$\mathscr{F}_\tau f(z,s,t) = \tilde{\mathscr{F}}_\tau f(z,s+\iota\varphi(z,s,t),t). \tag{II.7.8}$$

Note that $\tilde{\mathscr{F}}_\tau f(z,w,t)$ is a continuous function in $\Delta \times \mathbb{C}^d \times \mathrm{W}$, entire holomorphic with respect to w. Outside special cases it is *not* holomorphic with respect to z.

We shall now state, without proofs, the analogues of some of the results in section II.2. Each follows directly from the corresponding result in section II.2 or from an obvious adaptation of the latter's proof.

LEMMA II.7.1. *Let the support of $f \in \mathscr{C}^1(\mathrm{U})$ be contained in a product set $\Delta \times \mathrm{K} \times \mathrm{W}$ for some compact subset* K *of* V. *Then, for every $\tau > 0$, $\tilde{\mathscr{F}}_\tau f$ belongs to $\mathscr{C}^1(\Delta \times \mathbb{C}^d \times \mathrm{W})$ and*

$$(\partial/\partial z_i)\tilde{\mathscr{F}}_\tau f = \tilde{\mathscr{F}}_\tau \mathrm{M}_i f,\ i = 1,\dots,\nu; \tag{II.7.9}$$

$$(\partial/\partial w_k)\tilde{\mathscr{F}}_\tau f = \tilde{\mathscr{F}}_\tau \mathrm{N}_k f,\ k = 1,\dots,d; \tag{II.7.10}$$

$$(\partial/\partial \bar{z}_j)\tilde{\mathscr{F}}_\tau f = \tilde{\mathscr{F}}_\tau \mathrm{L}_j f,\ j = 1,\dots,\nu; \tag{II.7.11}$$

$$(\partial/\partial t_\ell)\tilde{\mathscr{F}}_\tau f = \tilde{\mathscr{F}}_\tau \mathrm{L}_{\nu+\ell} f,\ \ell = 1,\dots,n-\nu. \tag{II.7.12}$$

LEMMA II.7.2. *Let $r \in \mathbb{Z}_+$. Let the support of $f \in \mathscr{C}^r(\mathrm{U})$ be contained in a product set $\Delta \times \mathrm{K} \times \mathrm{W}$ with* K *a compact subset of* V. *Then, for every $\tau > 0$, $\mathscr{F}_\tau f$ belongs to $\mathscr{C}^r(\mathrm{U})$ and we have, whatever $\alpha \in \mathbb{Z}_+^n$, $\beta \in \mathbb{Z}_+^\nu$, $\gamma \in \mathbb{Z}_+^d$, $|\alpha|+|\beta|+|\gamma| \le r$,*

$$L^{\alpha}M^{\beta}N^{\gamma}\mathcal{F}_{\tau}f = \mathcal{F}_{\tau}L^{\alpha}M^{\beta}N^{\gamma}f. \tag{II.7.13}$$

We have used the notation analogous to (II.1.20).

LEMMA II.7.3. *Let the support of $f \in \mathscr{C}^r(U)$ ($r \in \mathbb{Z}_+$ or $r = +\infty$) be contained in a product set $\Delta \times K \times W$ for some compact subset K of V. Then, as $\tau \to +\infty$, $\mathcal{F}_{\tau}f$ converges to f in $\mathscr{C}^r(U)$.*

LEMMA II.7.4. *To each neighborhood of the origin $V' \subset V$ there are open neighborhoods of the origin $\Delta' \subset\subset \Delta$, $\mathcal{O} \subset\subset \mathbb{C}^d$ and $W' \subset\subset W$, and a constant $\kappa > 0$ such that, whatever the integer $r \in \mathbb{Z}_+$ and the function $f \in \mathscr{C}^r(\mathscr{Cl}U)$ that vanishes identically in $\Delta \times V' \times W$, the norm in $\mathscr{C}^r(\mathscr{Cl}(\Delta' \times \mathcal{O} \times W'))$ of $\tilde{\mathcal{F}}_{\tau}f$ does not exceed* const. $e^{-\kappa\tau}$.

Below we take $\Delta' = \{ z \in \mathbb{C}^{\nu};\, |z_j| < \rho', j = 1,\dots,\nu \}$, and W′ to be convex.

Consider now a solution $h \in \mathscr{C}^r(\mathscr{Cl}U)$ (r an integer ≥ 0) and a cutoff function $\chi \in \mathscr{C}_c^{\infty}(V)$ equal to 1 in the open neighborhood $V' \subset V$ of 0 in $\mathbb{C}^d$. We apply (II.7.11) and (II.7.12) with $f(z,s,t) = \chi(s)h(z,s,t)$. We have $L_j h = (L_j\chi)h$, $j = 1,\dots,n$. Combining the inhomogeneous Cauchy formula with (II.7.11) gets us, in $\Delta' \times \mathcal{O} \times W'$,

$$\tilde{\mathcal{F}}_{\tau}(\chi h)(z,w,t) =$$

$$(2\iota\pi)^{-1} \oint_{|z_1'|=\rho'} \tilde{\mathcal{F}}_{\tau}(\chi h)(z_1',z_2,\dots,z_{\nu},w,t)\,(z_1'-z_1)^{-1}\,dz_1' +$$

$$(2\iota\pi)^{-1} \iint_{|z_1'|\leq\rho'} \tilde{\mathcal{F}}_{\tau}L_1(\chi h)(z_1',z_2,\dots,z_{\nu},w,t)\,(z_1'-z_1)^{-1}\,dz_1'\wedge d\bar{z}_1'.$$

Lemma II.7.4 implies that the norm in $\mathscr{C}^r(\mathscr{Cl}(\Delta' \times \mathcal{O} \times W'))$ of $\tilde{\mathcal{F}}_{\tau}L_1(\chi h)$ is $\leq$ *const.* $e^{-\kappa\tau}$. We derive from this (and from the classical estimates for the Cauchy integral) that, as $\tau \to +\infty$, the limit in $\mathscr{C}^r(\Delta' \times \mathcal{O} \times W')$ of $\tilde{\mathcal{F}}_{\tau}(\chi h)$ is equal to that of

$$(2\iota\pi)^{-1} \oint_{|z_1'|=\rho'} \tilde{\mathcal{F}}_{\tau}(\chi h)(z_1',z_2,\dots,z_{\nu},w,t)\,(z_1'-z_1)^{-1}\,dz_1'.$$

Repeat the same reasoning with $\tilde{\mathcal{F}}_{\tau}(\chi h)(z_1',z_2,\dots,z_{\nu},w,t)$ in the place of $\tilde{\mathcal{F}}_{\tau}(\chi h)(z,w,t)$. After $\nu-1$ such repetitions one reaches the conclusion that, when $\tau \to +\infty$, the limit of $\tilde{\mathcal{F}}_{\tau}(\chi h)$ in $\mathscr{C}^r(\Delta' \times \mathcal{O} \times W')$ is equal to that of

$$(2\iota\pi)^{-\nu} \int\cdots\int_{\dot{\Delta}'} \tilde{\mathcal{F}}_{\tau}(\chi h)(z',w,t)\prod_{j=1}^{\nu}(z_j'-z_j)^{-1}\,dz' \tag{II.7.14}$$

where $\dot{\Delta}' = \{ z \in \mathbb{C}^\nu; |z_j| = \rho'_j, j = 1,\dots,\nu \}$ and $dz' = dz'_1 \wedge \cdots \wedge dz'_\nu$.

Next we let $\partial/\partial t_\ell$ act on (II.7.14) and avail ourselves of (II.7.12). Here the argument runs parallel to that in the proof of Theorem II.2.1: we may find a function $v_\tau(z,w,t) \in \mathscr{C}^r(\Delta \times \mathbb{C}^d \times \mathrm{W})$, entire holomorphic with respect to w, such that

$$d_t v_\tau = d_t \tilde{\mathscr{F}}_\tau(\chi h), \tag{II.7.15}$$

$$\|v_\tau\|_{\mathscr{C}^r(\mathscr{O}\!\ell(\Delta' \times \mathbb{O} \times \mathrm{W}'))} \leq const.\ e^{-\kappa\tau}. \tag{II.7.16}$$

We derive from (II.7.15) that

$$\tilde{\mathscr{F}}_\tau(\chi h)(z,w,t) - \tilde{\mathscr{F}}_\tau(\chi h)(z,w,0) = v_\tau(z,w,t) - v_\tau(z,w,0) \tag{II.7.17}$$

and from (II.7.16) that the limit in $\mathscr{C}^r(\mathscr{O}\!\ell(\Delta' \times \mathbb{O} \times \mathrm{W}'))$ of the left-hand side in (II.7.17) is equal to zero.

By Lemma II.7.4 we also know that

$$\|\tilde{\mathscr{F}}_\tau[(1-\chi)h]\|_{\mathscr{C}^r(\mathscr{O}\!\ell(\Delta' \times \mathbb{O} \times \mathrm{W}'))} \leq const.\ e^{-\kappa\tau}. \tag{II.7.18}$$

As a consequence the limits in $\mathscr{C}^r(\mathscr{O}\!\ell(\Delta' \times \mathbb{O} \times \mathrm{W}'))$ of $\tilde{\mathscr{F}}_\tau(\chi h)(z,w,t)$ and of $\tilde{\mathscr{F}}_\tau h(z,w,0)$ are the same, and the limit of (II.7.14) in $\mathscr{C}^r(\Delta' \times \mathbb{O} \times \mathrm{W}')$ is equal to that of

$$(2\iota\pi)^{-\nu} \int_{\dot{\Delta}'} \cdots \int \tilde{\mathscr{F}}_\tau h(z',w,0) \prod_{j=1}^{\nu} (z'_j - z_j)^{-1}\, dz'. \tag{II.7.19}$$

We conclude that the limit of (II.7.19) in $\mathscr{C}^r(\Delta' \times \mathbb{O} \times \mathrm{W}')$ is equal to that of $\tilde{\mathscr{F}}_\tau(\chi h)$.

We select a polydisk $\Delta_0 \subset \Delta$ centered at 0, open neighborhoods of the origin $\mathrm{V}_0 \subset \mathrm{V}'$ and $\mathrm{W}_0 \subset \mathrm{W}$ in $\mathbb{R}^d$ and $\mathbb{R}^{n-\nu}$ respectively, such that the image of

$$\mathrm{U}_0 = \Delta_0 \times \mathrm{V}_0 \times \mathrm{W}_0 \tag{II.7.20}$$

under the map (II.7.1) is contained in $\Delta' \times \mathbb{O} \times \mathrm{W}'$. Then we pull back to U_0 both $\tilde{\mathscr{F}}_\tau(\chi h)$ and (II.7.19). We let τ go to $+\infty$. By Lemma II.7.3 we know that $\tilde{\mathscr{F}}_\tau(\chi h)$ converges to χh in $\mathscr{C}^r(\mathrm{U})$; but $\chi \equiv 1$ in U_0. Thus we may state:

THEOREM II.7.1. *Let r be a finite integer ≥ 0 or $r = +\infty$. If the open neighborhood $\mathrm{U}_0 \subset \mathrm{U}$ is sufficiently small, then whatever the solution $h \in \mathscr{C}^r(\mathscr{O}\!\ell\mathrm{U})$, we have, in $\mathscr{C}^r(\mathrm{U}_0)$,*

$$h(z,s,t) =$$

$$\lim_{\tau \to +\infty} (2\iota\pi)^{-\nu} (\tau/\pi)^{\frac{1}{2}d} \int_{z' \in \dot{\Delta}'} \cdots \int \int_{s' \in \mathrm{V}} e^{-\tau(s + \iota\varphi(z,s,t) - s' - \iota\varphi(z',s',0))^2} h(z',s',0) \cdot$$

$$[\det w_s(z',s',0)]\prod_{j=1}^{\nu}(z_j'-z_j)^{-1}dz'\wedge ds'. \qquad \text{(II.7.21)}$$

EXAMPLE II.7.1. Suppose that $\mathcal{M}$ is equipped with a complex structure. In this case $m = n = \nu$; there are no variables s or t and no functions w; the solutions are holomorphic functions. Formula (II.7.21) reduces to the Cauchy formula. ∎

EXAMPLE II.7.2. Suppose that $\mathcal{M}$ is a CR manifold. In this case $n = \nu \le m$; there are no variables t. Formula (II.7.21) reads

$$h(z,s) =$$

$$\lim_{\tau\to+\infty}(2\iota\pi)^{-n}(\tau/\pi)^{\frac{1}{2}d}\int\cdots\int_{z'\in\Delta'}\int_{s'\in V} e^{-\tau\langle s+\iota\varphi(z,s)-s'-\iota\varphi(z',s')\rangle^2}h(z',s')\cdot$$

$$[\det w_s(z',s')]\prod_{j=1}^{n}(z_j'-z_j)^{-1}dz'\wedge ds'. \ ∎ \qquad \text{(II.7.22)}$$

REMARK II.7.1. Formula (II.7.21) entails at once that h is the limit in $\mathscr{C}^r(U_0)$ of a sequence of polynomials $\mathscr{P}_\nu(z,s+\iota\varphi(z,s,t))$ with respect to the z_j and the w_k, which is the present version of Theorem II.3.1. It also entails that $h \equiv 0$ in U_0 if $h(z,s,0) \equiv 0$ in $\Delta\times V$, a property already known to us through Corollary II.3.6 since the submanifold $t = 0$ of U is strongly noncharacteristic. However, formula (II.7.21) does not imply directly that if the trace of f vanishes on a maximally real submanifold of U, such as that defined by $y = 0$, $t = 0$, then $h \equiv 0$ in U_0 (cf. Corollary II.3.4).

On the other hand, formula (II.7.21) raises the following interesting question: *Is it true that if h vanishes to infinite order on the submanifold* $\Gamma_0 = \{0\}\times V\times\{0\}$ *of* U *then perforce* $h \equiv 0$ *in some neighborhood of* Γ_0? In the next section we are going to see that in a number of important cases the answer to this question is yes. ∎

Formula (II.7.21) can be generalized in a natural manner to distribution solutions. For this it should be noticed that the submanifolds

$$\Gamma_{z_0,t_0} = \{(z,s,t) \in U;\ z = z_0,\ t = t_0\} \qquad \text{(II.7.23)}$$

are *noncharacteristic* (Definition I.4.1). We may apply Proposition I.4.3: *every distribution solution in* $U = \Delta\times V\times W$ *is a* $\mathscr{C}^\infty$ *function of* (z,t) in $\Delta\times W$ *valued in the space of distribution* $\mathscr{D}'(V)$ *with respect to* s (we shall denote by $\mathscr{C}^\infty(\Delta\times W;\mathscr{D}'(V))$ the space of such distributions in U).

We reintroduce the cutoff function $\chi(s)$ used above. The generalization of (II.7.21) reads:

$$h(z,s,t) =$$

$$\lim_{\tau\to+\infty} (2\iota\pi)^{-\nu}(\tau/\pi)^{\frac{1}{2}d}\int\cdots\int_{z'\in\dot{\Delta}'}\int_{s'\in V} e^{-\tau(s+\iota\varphi(z,s,t)-s'-\iota\varphi(z',s',0))^2}\chi(s')\cdot$$

$$h(z',s',0)\,[\det w_s(z',s',0)]\prod_{j=1}^{\nu}(z_j'-z_j)^{-1}dz'\wedge ds'. \qquad \text{(II.7.24)}$$

Then the right-hand side in (II.7.24) makes perfectly good sense when h is a distribution solution in U and if we interpret $h(z',s',0)$ in the integral as the trace of h on the submanifold $\dot{\Delta}'\times V\times\{0\}$. The integral with respect to s' must be viewed as a duality bracket between a distribution (i.e., a zero-current) and a $\mathscr{C}^\infty$ d-form. The proof is an adaptation of that of Theorem II.7.1, in the same way as the proof of Theorem II.3.2 is an adaptation of that of Theorem II.2.1.

Another way of proving Formula (II.7.24) is by exploiting a local structure theorem for distribution solutions, the analogue of Theorem II.5.1. Here we make use of the second-order differential operator

$$\Delta_N = N_1^2 + \cdots + N_d^2. \qquad \text{(II.7.25)}$$

THEOREM II.7.2. *There is an open neighborhood of the origin*, $U_0 \subset U$, *such that, given any distribution solution h in* U *and any integer* $\mu \geq 0$, *there is a* $\mathscr{C}^\mu$ *solution* h_1 *in* U_0, *a holomorphic function* $\tilde{\Psi}(z,w)$ *in some open neighborhood of* Σ_0 *in* $\mathbb{C}^m$ *and an integer* $p \geq 0$ *such that, in* U_0,

$$h = \Delta_N^p h_1 + \tilde{\Psi}(z,s+\iota\varphi(z,s,t)). \qquad \text{(II.7.26)}$$

PROOF. It is a slight (but not totally banal) modification of that of Theorem II.5.1. The fact that h is a $\mathscr{C}^\infty$ function of (z,t) valued in $\mathscr{D}'(V)$ enables us to find a neighborhood $U' = \Delta'\times V'\times W' \subset\subset U$ of 0 and an integer $p \geq 0$ such that there is $u \in \mathscr{C}^{\mu+1}(U')$ satisfying

$$\Delta_N^p u = h, \qquad \text{(II.7.27)}$$

whence

$$\Delta_N^p L_j u = 0,\; j = 1,\ldots,n. \qquad \text{(II.7.28)}$$

We conclude that each $L_j u$ is an analytic vector of Δ_N^p hence, by Lemmas II.4.2 and Proposition II.4.2, that there is a $\mathscr{C}^\mu$ function $\tilde{g}_j(z,w,t)$, holomorphic with respect to w, in $\Delta'\times\mathscr{O}'\times W'$, with $\mathscr{O}'$ an open and connected neighborhood of V' in $\mathbb{C}^d$, such that, in U',

$$\mathrm{L}_j u(z,s,t) = \tilde{g}_j(z,s+\iota\varphi(z,s,t),t),\ j = 1,\ldots,n. \tag{II.7.29}$$

We may as well assume $\mu \geq 1$. We derive at once from (II.7.29) (by taking advantage of Corollary II.3.7) that the differential form in $\Delta' \times \mathbb{O}' \times \mathrm{W}'$,

$$\tilde{g} = \sum_{j=1}^{\nu} \tilde{g}_j d\bar{z}_j + \sum_{\ell=1}^{n-\nu} \tilde{g}_{\nu+\ell} dt_\ell,$$

is $(\bar{\partial}_z + \mathrm{d}_t)$-closed (it depends holomorphically on w). We can therefore find a $\mathscr{C}^\mu$ function $\tilde{v}$ in $\Delta' \times \mathbb{O}' \times \mathrm{W}'$, holomorphic with respect to w, such that, in that set,

$$(\bar{\partial}_z + \mathrm{d}_t)\tilde{v} = \tilde{g}. \tag{II.7.30}$$

We select U_0 small enough that its image under the map (II.7.1) is contained in $\Delta' \times \mathbb{O}' \times \mathrm{W}'$ and call v the pullback of $\tilde{v}$ to U_0. It is a consequence of (II.7.29) and (II.7.30) that $\mathrm{L}_j(u - v) = 0$, $j = 1,\ldots,n$, in U_0. From this and (II.7.28) we derive

$$(\bar{\partial}_z + \mathrm{d}_t)\Delta_w^p \tilde{v} \equiv 0 \tag{II.7.31}$$

in $\Delta' \times \mathbb{O}' \times \mathrm{W}'$ (recall that this open set is connected); here $\Delta_w = (\partial/\partial w_1)^2 + \cdots + (\partial/\partial w_d)^2$. The meaning of (II.7.31) is that $\Delta_w^p v$ is holomorphic with respect to z and independent of t; we know it is holomorphic with respect to w; call it $\tilde{\Psi}(z,w)$. By (II.7.27) we have

$$h = \tilde{\Psi}(z,s+\iota\varphi(z,s,t)) + \Delta_{\mathrm{N}}^p(u-v)$$

and we may take $h_1 = u - v$. ∎

There is a version of the Approximate Poincaré Lemma that stands in relation to Theorem II.6.1 in the same manner as Theorem II.7.2 stands in relation to Theorem II.5.1. We leave its statement and proof to the reader.

II.8. Unique Continuation of Solutions

We continue to reason within the framework set down in section II.7. We denote by Σ_0 the image of the neighborhood V of 0 in $\mathbb{C}^d$ (see (II.7.2)) under the map $s \to s + \iota\varphi(0,s,0)$; Σ_0 is a maximally real submanifold of $\mathbb{C}^d$; it passes through the origin and, by virtue of (I.7.5), its tangent space at 0 is defined by $\mathscr{I}m\, w = 0$.

DEFINITION II.8.1. *Let Λ be a maximally real submanifold of $\mathbb{C}^d$, $\mathfrak{p}$ one of its points. We shall say that Λ has the* compact cycle property *at $\mathfrak{p}$ if there is a basis of open neighborhoods of $\mathfrak{p}$ in $\mathbb{C}^d$ such that, if $\mathcal{N}$ is any one of these neighborhoods, then there is a holomorphic function F in $\mathcal{N}$ such that $F(\mathfrak{p}) \neq$*

0 *and that the closure of the connected component of* $\mathfrak{p}$ *in the set* $\{ w \in \Lambda \cap \mathcal{N};$ $F(w) \neq 0 \}$ *is a compact subset of* $\Lambda \cap \mathcal{N}$.

The reason for introducing Definition II.8.1 lies in the following result:

THEOREM II.8.1. *Assume that* Σ_0 *has the compact cycle property at the origin. Any Lipschitz continuous solution h in* U *that vanishes to infinite order on the subset* $\{ (z,s,t) \in U;\ z = 0,\ t = 0 \}$ *must vanish identically in some open neighborhood* $U_0 \subset U$ *of the origin.*

That h vanishes to infinite order on the subset of U defined by $z = 0$, $t = 0$, means that to each integer $k \geq 0$ and to every compact subset K of U there is a constant $C > 0$ such that

$$|h(z,s,t)| \leq C(|z| + |t|)^k,\ \forall\ (z,s,t) \in K. \tag{II.8.1}$$

PROOF. It suffices to prove that the restriction of the solution h to the submanifold $\{ (z,s,t) \in U;\ t = 0 \}$ vanishes in some neighborhood of 0 in Λ. Indeed, that submanifold is strongly noncharacteristic and one may then apply Corollary II.3.6. In other words it suffices to prove Theorem II.8.1 when there are no variables t, i.e., when the structure of $\mathcal{M}$ is CR. We shall assume this to be so through the remainder of the proof. Call Σ the image of U under the map $(z,s) \to (z,w)$; Σ is a generic submanifold of $\mathbb{C}^n \times \mathbb{C}^d$ and the submanifold Σ_0 of $\mathbb{C}^d$ can be identified to the intersection of Σ with the subspace $z = 0$.

Our hypothesis is that Σ_0 has the compact cycle property at 0. Select a neighborhood $\mathcal{N}$ and a holomorphic function F as in Definition II.8.1 (in which we take $\Lambda = \Sigma_0$ and $\mathfrak{p} = 0$). We solve with respect to z the following equation

$$z - \zeta F(s + \imath\varphi(z,s)) = 0 \tag{II.8.2}$$

with $s + \imath\varphi(z,s) \in \mathcal{N}$. Let V_1 denote the image of $\Sigma_0 \cap \mathcal{N}$ under the map $w \to \mathcal{R}e\, w$. Notice that, when $\zeta = 0$, the Jacobian determinant of the left-hand side in (II.8.2) with respect to (x,y) is equal to 1 whatever $s \in V_1$. In the neighborhood, in $\mathbb{C}^n \times \mathbb{R}^d$, of each point $(0,s)$, $s \in V_1$, one can find a $\mathcal{C}^\infty$ solution $G(\zeta,s)$ of (II.8.2) such that

$$G|_{\zeta=0} = 0. \tag{II.8.3}$$

By the uniqueness in the implicit function theorem these solutions can be patched together to yield a unique solution G of (II.8.2), defined and $\mathcal{C}^\infty$ in some open neighborhood $\mathcal{U}$ of $\{0\} \times V_1$ in $\mathbb{C}^n \times \mathbb{R}^d$ (where the variable is (ζ,s)), and satisfying (II.8.3). We shall make use of the

LEMMA II.8.1. *If* $\mathcal{U}$ *is suitably contracted about* $(0) \times V_1$, *we have*

$$G(\zeta,s) = [F(s+\iota\varphi(0,s))(1+O(|\zeta|)) + \overline{F}(s-\iota\varphi(0,s))(1+O(|\zeta|))]\zeta. \quad \text{(II.8.4)}$$

PROOF OF LEMMA II.8.1. Write $F(s+\iota\varphi(z,s)) = F(s+\iota\varphi(0,s)) + A(z,s)\cdot z + B(z,s)\cdot\bar{z}$, with $A = (A_1,\ldots,A_n)$, $B = (B_1,\ldots,B_n) \in \mathscr{C}^\infty(\mathrm{U};\mathbb{C}^n)$. Call $\zeta\circ A$ (resp., $\bar{\zeta}\circ B$) the n-vector with components $\zeta_k A_k(G(\zeta,s),s)$ (resp., $\bar{\zeta}_k B_k(G(\zeta,s),s)$). By (II.8.2) G is collinear to ζ; set $\chi_i = G_i/\zeta_i$; (II.8.4) follows by solving in $\mathscr{U}$ (suitably contracted)

$$\chi_i - (\zeta\circ A)\cdot\chi - (\bar{\zeta}\circ B)\cdot\bar{\chi} = F(s+\iota\varphi(0,s)) \; (i = 1,\ldots,n)$$

viewed as a system of $2n$ linear equations in the unknowns χ_i and $\bar{\chi}_j$. ■

Denote by K the image under the map $w \to \mathscr{R}e\, w$ of the closure of the connected component of the origin in the set $\{ w \in \Sigma_0 \cap \mathscr{N};\ F(w) \neq 0 \}$; K is a compact subset of V_1 since this map is a diffeomorphism of $\Sigma_0 \cap \mathscr{N}$ onto V_1. Observe that K is equal to the closure of its interior, $\mathscr{I}nt\,\mathrm{K}$; this is due to the fact that, by virtue of Corollary II.3.7, F cannot vanish identically in any open subset of Σ_0.

We may find a polydisk $\Delta \subset \mathbb{C}^n$ centered at 0 and an open subset V′ of V such that $\mathrm{K} \subset \mathrm{V}'$ and $\mathrm{U}' = \Delta\times\mathrm{V}' \subset\subset \mathscr{U}$. Then $\mathrm{G}_j \in \mathscr{C}^\infty(\mathscr{C}\ell\,\mathrm{U}')$ and according to (II.8.4) we can, and shall, select Δ so small that

$$\forall\, (\zeta,s) \in \Delta\times\mathrm{K},\ G(\zeta,s) = 0 \Leftrightarrow \zeta = 0 \text{ or } s \in \partial\mathrm{K}. \quad \text{(II.8.5)}$$

The functions

$$\zeta_j = \xi_j + \iota\eta_j \; (1 \le j \le n), \qquad w_k^\# = s_k + \iota\varphi_k(G(\zeta,s),s) \; (1 \le k \le d), \quad \text{(II.8.6)}$$

define a CR structure on U′ which, in $\mathrm{U}'' = \Delta\times(\mathscr{I}nt\,\mathrm{K})$, coincides with the pullback to U″, under the map $(\zeta,s) \to (G(\zeta,s),s)$, of the original CR structure in U. Indeed, by (II.8.5), that map has an inverse, $(z,s) \to (z/F(s+\iota\varphi(z,s)),s)$; and $z/F(w)$ is holomorphic when $F(w) \neq 0$.

Define then the following function in U′:

$$H(\zeta,s) = \begin{cases} h(G(\zeta,s),s) \text{ if } (z,s) \in \Delta\times\mathrm{K}, \\ 0 \text{ if } (\zeta,s) \in \Delta\times(\mathrm{V}'\backslash\mathrm{K}). \end{cases}$$

We claim that *H is a Lipschitz continuous solution in the structure on* U′ *defined by* (II.8.6). Indeed, observe that this is so in U″ where $H(\zeta,s) = h(z,s)$. On the other hand, by the hypothesis on h and by (II.8.5) we see that $H(\zeta,s)$ vanishes to infinite order on $\Delta\times\partial\mathrm{K}$, easily implying the claim.

Next we use the linear operators $\mathscr{F}_\tau$ defined in (II.7.7) and (II.7.8). But we do so in (ζ,s)-space U′; since supp $\mathrm{H} \subset \Delta\times\mathrm{K}$ we may apply Lemmas II.7.1 and II.7.2 directly to $f = H$. By (II.7.11) we see that $\tilde{\mathscr{F}}_\tau H$ is a holomorphic function of (ζ,w) in $\Delta\times\mathbb{C}^d$. Obviously, by the definition (II.7.7) of $\tilde{\mathscr{F}}_\tau$, $\tilde{\mathscr{F}}_\tau H$ vanishes to infinite order when $z = 0$. By Lemma II.7.3 we conclude that H

$\equiv 0$ in U′, hence that $h(z,s) \equiv 0$ in the image of U″ via the map $(\zeta,s) \to (G(\zeta,s),s)$, which, as we have pointed out earlier, is a diffeomorphism of U″ onto some open neighborhood of 0 in U. ■

We now state and prove two sufficient conditions for the validity of the compact cycle property:

PROPOSITION II.8.1. *Any maximally real submanifold of* $\mathbb{C}^1$ *has the compact cycle property at any one of its points.*

PROOF. A maximally real submanifold of $\mathbb{C}^1$ is a smooth real curve γ in the plane, i.e., an embedded, connected, one-dimensional submanifold of $\mathbb{R}^2$. Let $\mathcal{N}$ be any open disk centered at $p \in \gamma$ such that $\gamma \cap \mathcal{N}$ is a connected arc of curve; select arbitrarily two points a, b on $\gamma \cap \mathcal{N}$, on opposite sides of p. Then $F(z) = (z-a)(z-b)$ fulfills the requirement in Definition II.8.1. ■

EXAMPLE II.8.1. Let $\mathcal{M}$ be a real hypersurface in $\mathbb{C}^{n+1}$ (with complex coordinates $z_1,\dots,z_n,w$) equipped with the CR structure inherited from the ambiant complex space. We may assume that in some open neighborhood of one of its points, which we may as well take to be the origin, $\mathcal{M}$ is defined by an equation $t = \varphi(z,s)$ with $\varphi|_0 = 0$, $d\varphi|_0 = 0$ $(w = s+it)$. Proposition II.8.1 applies: if a Lipschitz continuous CR function h in an open neighborhood U of 0 in $\mathcal{M}$ vanishes to infinite order on the intersection of $\mathcal{M}$ with the complex line $z = 0$ then h vanishes identically in some neighborhood of 0. ■

PROPOSITION II.8.2. *Any real-analytic maximally real submanifold* Λ *of* $\mathbb{C}^d$ (d ≥ 1) *has the compact cycle property at any one of its points.*

PROOF. Select the open neighborhood $\mathcal{N}$ of $p \in \Lambda$ in $\mathbb{C}^d$ to be such that there is a biholomorphic map χ of $\mathcal{N}$ onto the open unit ball $\mathcal{B}^{2d} \subset \mathbb{C}^d$ which maps $\mathcal{N} \cap \Lambda$ onto $\mathcal{B}^d \subset \mathbb{R}^d$. The pullback F under the map χ of the function $1 - 2(w_1^2 + \cdots + w_d^2)$ fulfills the requirements of Definition II.8.1. ■

COROLLARY II.8.1. *If the function* $\varphi(0,s)$ *is real-analytic in some open neighborhood of* 0 *then* Σ_0 *has the compact cycle property at* 0.

There are maximally real submanifolds of $\mathbb{C}^d$ $(d > 1)$ that have the compact cycle property at every point without being real-analytic:

EXAMPLE II.8.2. Let γ_k $(k = 1,\dots,d)$ be smooth real curves in $\mathbb{C}$. Then regard $\gamma = \gamma_1 \times \cdots \times \gamma_d$ as a maximally real submanifold of $\mathbb{C}^d$. If $p = (p_1,\dots,p_d) \in \gamma$ and if a_k, $b_k \in \gamma_k$ are points lying on opposite sides of p_k for each k, then polynomials such as

$$F(w) = \prod_{k=1}^{d}(w_k - a_k)(w_k - b_k)$$

will fulfill the requirements in Definition II.8.1 (cf. Proposition II.8.1). ■

There exist maximally real submanifolds of $\mathbb{C}^d$ that do not have the compact cycle property at some of their points (see Jacobowitz [1]). And there are situations in which the conclusion in Theorem II.8.1 is true, even though it is not known whether Σ_0 possesses the compact cycle property at 0 (see Example 5.2 in Baouendi and Treves [4]).

Theorem II.8.1 leads one to ask what are the submanifolds of $\mathcal{M}$ that can play the role of the submanifold $\{ (z,s,t) \in U; z = 0, t = 0 \}$. We shall take a closer look at this question in Chapter III, in the context of hypo-analytic structures.

Notes

The exposition in sections II.1 to II.5 follows closely that in Treves [7], chap. 2. The approximation formula in section II.2 was first proved in Baouendi and Treves [2], which also contained a proof of the representation formula for distribution solutions (see section II.6) under a special hypothesis on the locally integrable structure. The proof of the approximation formula is a modification of the original proof of the Weierstrass Approximation Theorem (first proved in 1867; see Weierstrass [1]). That the support of any distribution solution is a union of orbits was shown in Treves [7]. The result that there exist, locally, CR distributions whose support consists of a single orbit (cf. Theorem II.3.4) is due to Baouendi and Rothschild [5]. Whether such distributions exist in the large is unknown. The proof of the representation formula in the general case, as well as the Approximate Poincaré Lemma, can be found in Treves [7]. Use of the fine local embedding is closely related to the FBI minitransform (section IX.6). The unique continuation result in section II.8 was first proved, by the integral kernels method, for a hypersurface in $\mathbb{C}^n$ in Rosay [1]. The proof given here, and the extension to locally integrable structures under the compact cycle property (Definition II.8.1), appeared in Baouendi and Treves [4]. Whether unique continuation holds in the absence of the compact cycle property is an open problem.

III

Hypo-Analytic Structures. Hypocomplex Manifolds

Let $\mathcal{M}$ be a generic submanifold of a complex manifold $\hat{\mathcal{M}}$; "generic" means that the restrictions to $\mathcal{M}$ of any set of local coordinates $z_1,\ldots,z_m$ ($m = \dim_{\mathbb{C}} \hat{\mathcal{M}}$) have $\mathbb{C}$-linearly independent differentials (which requires $\dim_{\mathbb{R}} \mathcal{M} \geq m$). The restrictions of the functions z_i to $\mathcal{M}$ form a set of first integrals (in some open subset U of $\mathcal{M}$) in a locally integrable structure on $\mathcal{M}$, specifically, the CR structure inherited by $\mathcal{M}$ from the complex structure of $\hat{\mathcal{M}}$. Another set of first integrals in U (for the same CR structure), $w_1,\ldots,w_m$, could also be the restrictions of complex coordinates in $\hat{\mathcal{M}}$; but in general need not be.

Suppose now that $\mathcal{M}$ is generic and, moreover, that $\dim_{\mathbb{R}} \mathcal{M} = \dim_{\mathbb{C}} \hat{\mathcal{M}}$ (in other words, $\mathcal{M}$ is a maximally real submanifold of $\hat{\mathcal{M}}$); and also that each point of $\mathcal{M}$ has an open neighborhood $\hat{U}$ in $\hat{\mathcal{M}}$ in which there are complex coordinates $z_1,\ldots,z_m$ whose restrictions to $U = \hat{U}\cap\mathcal{M}$ are *real-valued*. This is only possible if $\mathcal{M}$ is a real-analytic submanifold of $\hat{\mathcal{M}}$; then $\mathcal{M}$ inherits an analytic structure from $\hat{\mathcal{M}}$. Conversely, any real-analytic manifold can be embedded in its complexification as a maximally real submanifold, of class $\mathscr{C}^\omega$.

Embedded CR manifolds and real-analytic manifolds are subclasses of the larger *hypo-analytic* class to which the present book is devoted. A smooth manifold $\mathcal{M}$, if it is equipped with a locally integrable structure, can be covered by open sets each of which is mapped into $\mathbb{C}^m$ by a $\mathscr{C}^\infty$ map $Z = (Z_1,\ldots,Z_m)$ such that $Z_1,\ldots,Z_m$ are solutions and $dZ_1\wedge\cdots\wedge dZ_m \neq 0$. Suppose such "local charts" (U,Z), $(U^\#,Z^\#)$,... can be selected with domains U, $U^\#$,..., that still make up a covering of $\mathcal{M}$ and, furthermore, in a manner that the pairs of maps Z and $Z^\#$ agree in the overlaps $U\cap U^\#$ up to a biholomorphism. The special "atlas" consisting of those charts defines, on $\mathcal{M}$, a hypo-analytic structure (section III.1). A hypo-analytic function in U is the pull-back, under the map Z, of a holomorphic function in an open subset of $\mathbb{C}^m$ containing Z(U). The properties of hypo-analytic functions in $\mathcal{M}$ closely match those of analytic functions in $\mathbb{R}^m$ (section III.2).

A hypo-analytic structure determines a unique locally integrable structure that underlies it: the cotangent structure bundle T′ is spanned by the differentials of the hypo-analytic functions. There could be different, inequivalent ways of effecting the selection of the defining atlas (U,Z), ($U^{\#}$,$Z^{\#}$),...: the same locally integrable structure might underlie different hypo-analytic structures. However, there exist involutive structures that underlie a unique hypo-analytic structure; they are those in which every solution in the neighborhood of a point can be represented as a holomorphic function of the first integrals. We call *hypocomplex* these involutive structures, for the obvious reason that their prototypes are the complex structures. The properties of hypo-analytic functions in an open subset of a hypocomplex manifold closely match those of holomorphic functions in a domain in $\mathbb{C}^m$ (section III.5). In two dimensions the hypocomplex structures can easily be characterized: either $T' = 0$ or else the tangent structure bundle $\mathcal{V}$ is locally generated by a vector field L which is *hypo-elliptic*—a property that can be read on the symbol of L (section III.6). In more than two dimensions there exist lots of interesting hypocomplex structures, for instance among CR structures. The foremost example of a hypocomplex CR manifold is that of a real hypersurface $\mathcal{M}$ in $\mathbb{C}^m$ whose Levi form at every point has at least one positive and one negative eigenvalue. For then every solution—i.e., every CR function—in an open subset Ω of $\mathcal{M}$ can be extended as a holomorphic function to a neighborhood of Ω in $\mathbb{C}^m$ (section III.7).

An interesting aspect of hypo-analyticity and hypocomplexity, the fact that both notions can be *microlocalized*, will not be discussed in the present volume.

III.1. Hypo-Analytic Structures

Let $\mathcal{M}$ be a $\mathcal{C}^\infty$ manifold, equipped with a locally integrable structure. There is a natural equivalence relation among the local charts described in section I.7: two charts on the same domain, say (U,Z_1) and (U,Z_2), will be equivalent if there is a biholomorphic map H of an open neighborhood of Z_1(U) onto an open neighborhood of Z_2(U) such that $Z_2 = H \circ Z_1$. This equivalence relation points to a new type of structure on $\mathcal{M}$, finer than the locally integrable ones. Indeed, the apportioning of local charts among different equivalence classes cannot be characterized by means of the structure bundles T′ or $\mathcal{V}$ (see Example III.1.3 below). This leads to the following:

DEFINITION III.1.1. *By a* hypo-analytic structure *on $\mathcal{M}$ we mean the data of an open covering $\{\Omega_\iota\}$ of $\mathcal{M}$ and, for each index ι, of a $\mathcal{C}^\infty$ map $Z_\iota = (Z_{\iota,1},\dots,Z_{\iota,m}) : \Omega_\iota \to \mathbb{C}^m$ with $m \geq 1$ independent of ι, such that the following is true:*

$$\mathrm{d}Z_{\iota,1},\dots,\mathrm{d}Z_{\iota,m} \text{ are } \mathbb{C}\text{-linearly independent at each point of } \Omega_\iota; \quad \text{(III.1.1)}$$

if $\iota \neq \kappa$, to each point $\mathfrak{p}$ of $\Omega_\iota \cap \Omega_\kappa$ there is a holomorphic map $F^\iota_{\kappa,\mathfrak{p}}$ of an open neighborhood of $Z_\iota(\mathfrak{p})$ in $\mathbb{C}^m$ into $\mathbb{C}^m$ such that $Z_\kappa = F^\iota_{\kappa,\mathfrak{p}} \circ Z_\iota$ in a neighborhood of $\mathfrak{p}$ in $\Omega_\iota \cap \Omega_\kappa$. (III.1.2)

It follows from (III.1.1) that the differential of the map $F^\iota_{\kappa,\mathfrak{p}}$ at the point $Z_\iota(\mathfrak{p})$ is injective. From the implicit function theorem we conclude that $F^\iota_{\kappa,\mathfrak{p}}$ is a biholomorphism of some open neighborhood of $Z_\iota(\mathfrak{p})$ in $\mathbb{C}^m$ onto one of $Z_\kappa(\mathfrak{p})$. Actually, if these neighborhoods are sufficiently small, the inverse of $F^\iota_{\kappa,\mathfrak{p}}$ must perforce be $F^\kappa_{\iota,\mathfrak{p}}$. Indeed, as we have seen in the preceding chapters (see, e.g., the remarks which follow (I.7.24)), there passes through $\mathfrak{p}$ an m-dimensional submanifold $\mathcal{X}$ of U_ι on which the pullbacks of the differentials $dZ_{\iota,j}$ ($1 \leq j \leq m$) are linearly independent. As a consequence, $Z_\iota(\mathcal{X})$ is an immersed maximally real submanifold of $\mathbb{C}^m$. We may apply Corollary II.3.7: since the identity map and the compose $F^\kappa_{\iota,\mathfrak{p}} \circ F^\iota_{\kappa,\mathfrak{p}}$ are equal on $Z_\iota(\mathcal{X})$, they must also be equal in a full neighborhood of this set.

When equipped with a hypo-analytic structure, $\mathcal{M}$ will be referred to as a *hypo-analytic manifold*.

Suppose that $\mathcal{M}$ is a real-analytic manifold and that it is equipped with a hypo-analytic structure defined by "charts" (U_ι, Z_ι) as in Definition III.1.1. Suppose, moreover, that for each index ι the map $Z_\iota : U_\iota \to \mathbb{C}^m$ is real-analytic. Then we say that the hypo-analytic structure of $\mathcal{M}$ is *real-analytic* (or of class $\mathcal{C}^\omega$), and that $\mathcal{M}$ is a *real-analytic hypo-analytic manifold*, or a hypo-analytic manifold of class $\mathcal{C}^\omega$.

Definition III.1.1 is motivated by the following:

Example III.1.1. A *real-analytic structure* on a manifold $\mathcal{M}$ is a hypo-analytic structure on $\mathcal{M}$ defined by "charts" (Z_ι, Ω_ι) as in Definition III.1.1, such that $Z_\iota(\Omega_\iota) \subset \mathbb{R}^m$ for every index ι, with $m = \dim \mathcal{M}$. ■

We also have the obvious

Example III.1.2. The holomorphic local charts in a complex manifold $\mathcal{M}$ define what we shall refer to as the *natural* hypo-analytic structure on $\mathcal{M}$. In particular, when referring to the hypo-analytic structure of $\mathbb{C}^m$, we shall always assume that it is the structure defined by the chart $(\mathbb{C}^m, z_1, \ldots, z_m)$ where the z_j are the standard complex coordinates. ■

The following can be regarded as a (local) generalization of the two preceding examples:

Example III.1.3. Let $\mathcal{M}$ be a *generic* submanifold of a complex manifold $\tilde{\mathcal{M}}$ (cf. Proposition I.3.7). Suppose $\dim_{\mathbb{C}} \tilde{\mathcal{M}} = m$; then $\dim_{\mathbb{R}} \mathcal{M} = m + n$ for some n, $0 \leq n \leq m$. Let $(\tilde{U}, z)$ be a holomorphic local chart in $\tilde{\mathcal{M}}$: $\tilde{U}$ is an open subset of $\tilde{\mathcal{M}}$ and z is a biholomorphism of $\tilde{\mathrm{U}}$ onto an open subset of $\mathbb{C}^m$. When $(\tilde{\mathrm{U}},z)$ ranges over the set of all holomorphic local charts in $\tilde{\mathcal{M}}$ such that $\mathrm{U} = \tilde{\mathrm{U}}\cap\mathcal{M} \neq \emptyset$, the charts $(\mathrm{U},z|_{\mathrm{U}})$ define a hypo-analytic structure on $\mathcal{M}$. ∎

In the sequel we assume that $\mathcal{M}$ is equipped with a hypo-analytic structure defined by charts $(\mathrm{U}_\iota,Z_\iota)$ as in Definition III.1.1.

DEFINITION III.1.2. *A complex-valued function f defined in some open neighborhood of a point $\mathfrak{p}$ of $\mathcal{M}$ is said to be* hypo-analytic at $\mathfrak{p}$ *if, for some (or, equivalently, for every) index ι such that $\mathfrak{p} \in \mathrm{U}_\iota$, there exists a holomorphic function $\tilde{f}_\iota$ in some open neighborhood of $Z_\iota(\mathfrak{p})$ in $\mathbb{C}^m$ such that $f = \tilde{f}_\iota\circ Z_\iota$ in a neighborhood of $\mathfrak{p}$.*

We say that a complex function f defined in a subset S *of $\mathcal{M}$ is* hypo-analytic in S *if f is hypo-analytic at every point of* S.

Clearly a hypo-analytic function in S extends as a hypo-analytic function in an open set containing S. Any hypo-analytic function in an open subset Ω of $\mathcal{M}$ is $\mathscr{C}^\infty$ in Ω; it follows at once from Proposition I.1.2 that it is a solution in Ω. It is of class $\mathscr{C}^\omega$ if $\mathcal{M}$ is a hypo-analytic manifold of class $\mathscr{C}^\omega$.

If $h_1,\ldots,h_r$ are hypo-analytic functions in Ω and if F is a holomorphic function in some neighborhood of $h(\Omega)$, where $h = (h_1,\ldots,h_r)$, then $F\circ h$ is hypo-analytic in Ω (cf. Proposition I.1.2). In particular the hypo-analytic functions in Ω form a ring. If a hypo-analytic function f in Ω does not vanish at any point of Ω, $1/f$ is hypo-analytic in Ω.

We say that a distribution u in an open subset Ω of $\mathcal{M}$ is hypo-analytic at a *point* $\mathfrak{p} \in \Omega$ if, in some open neighborhood of $\mathfrak{p}$ in Ω, u is equal to a function that is hypo-analytic at $\mathfrak{p}$.

DEFINITION III.1.3. *Let u be a distribution in an open set Ω of $\mathcal{M}$. The complement in Ω of the* (open) *set of points at which u is a hypo-analytic function will be called the* hypo-analytic singular support *of u* (in Ω) *and will be denoted by* $\operatorname{sing\,supp}_{ha} u$.

Let $\mathcal{M}'$ be another hypo-analytic manifold. A map $\chi : \mathcal{M} \to \mathcal{M}'$ will be called a *hypo-analytic map* (or a *morphism of hypo-analytic manifolds*), if, given any open subset Ω' of $\mathcal{M}'$ and any hypo-analytic function $f : \Omega' \to \mathbb{C}$, the pullback of $f\circ\chi$ is a hypo-analytic function in the preimage $\chi^{-1}(\Omega')$. We shall say that χ is a *hypo-analytic isomorphism* if there is a hypo-analytic map $\psi : \mathcal{M}' \to \mathcal{M}$ such that $\chi\circ\psi = \mathrm{Id}_{\mathcal{M}'}$ and $\psi\circ\chi = \mathrm{Id}_{\mathcal{M}}$.

By a *hypo-analytic chart* in $\mathcal{M}$ we shall mean a pair (U,Z) made up of an open subset U of $\mathcal{M}$ and of a hypo-analytic map $Z = (Z_1,\ldots,Z_m) : \mathrm{U} \to \mathbb{C}^m$ such that $dZ_1\wedge\cdots\wedge dZ_m \neq 0$ at every point of U. It should be underlined that the linear independence of $dZ_1,\ldots,dZ_m$ is far from implying that Z is injective.

In general this is not even true after contracting U, as shown in the following two examples. In both the hypo-analytic structures are of a type quite different from those in Examples III.1.2 and III.1.3. The first one is obtained by relaxing, in Example III.1.1, the requirement that $m = \dim \mathcal{M}$.

EXAMPLE III.4. Let $\mathcal{M}$ be equipped with a hypo-analytic structure in which every point belongs to the domain U of some *real* hypo-analytic chart, i.e., a hypo-analytic chart (U,Z) such that $Z(U) \subset \mathbb{R}^m$. Such a structure can be interpreted as a *partially analytic* structure. Indeed, possibly after contracting U, the functions Z_j ($j = 1,\dots,m$) can be regarded as real coordinates in U and we can adjoin to them n real coordinates $t_1,\dots,t_n$ to form a complete coordinate patch. Any other set of real coordinates in U, $\{Z_j^\#, t_k^\#\}_{1 \le j \le m, 1 \le k \le n}$, such that the $Z_j^\#$ are hypo-analytic functions in U, must satisfy the condition that the $Z_j^\#$ be analytic functions of Z and be independent of t. If $m < \dim \mathcal{M}$ the map $Z : U \to \mathbb{R}^m$ cannot be injective. ■

EXAMPLE III.1.5. The *Mizohata structure* on $\mathbb{R}^2$ (where the coordinates are x and t) is defined by the local chart $(\mathbb{R}^2, Z)$ with $Z = x + \imath t^2/2$. Any hypo-analytic function is a holomorphic function of Z and, as a consequence, it is an even function of t (cf. Example II.3.3). Such a function cannot be injective in any neighborhood of any point $(x,0)$. ■

The hypo-analytic structure of $\mathcal{M}$ can be defined by the *sheaf of germs of hypo-analytic functions* in $\mathcal{M}$. Below we denote by ${}_m\mathcal{O}$ the ring of germs at the origin of holomorphic functions in $\mathbb{C}^m$; ${}_m\mathcal{O}$ can also be viewed as the ring of convergent power series in m variables. Consider a sheaf $\mathcal{S}$ of rings of germs of $\mathcal{C}^\infty$ functions in $\mathcal{M}$: the stalk $\mathcal{S}_\mathfrak{p}$ of $\mathcal{S}$ at an arbitrary point $\mathfrak{p}$ of $\mathcal{M}$ is a subring of the ring of germs at $\mathfrak{p}$ of $\mathcal{C}^\infty$ functions in $\mathcal{M}$ (for ordinary multiplication). Let $J_\mathfrak{p}$ be a ring homomorphism of ${}_m\mathcal{O}$ into $\mathcal{S}_\mathfrak{p}$. It defines a $\mathbb{C}$-linear map $DJ_\mathfrak{p} : \mathbb{C}^m \to \mathbb{C}T^*_\mathfrak{p}\mathcal{M}$ in the following manner. Identify to $\mathbb{C}^m$ the cotangent space to $\mathbb{C}^m$ at 0; as a complex vector space the latter is spanned by $dz_1,\dots,dz_m$. Define $DJ_\mathfrak{p}$ by the equations

$$DJ_\mathfrak{p}(dz_j) = d(J_\mathfrak{p} z_j)|_\mathfrak{p},\ j = 1,\dots,m. \qquad \text{(III.1.3)}$$

(By definition, the differential at $\mathfrak{p}$ of the germ at $\mathfrak{p}$ of a function f, which is defined and smooth in some neighborhood of $\mathfrak{p}$, is equal to $df|_\mathfrak{p}$.)

PROPOSITION III.1.1. *In order that a sheaf $\mathcal{S}$ of rings of germs of $\mathcal{C}^\infty$ functions in $\mathcal{M}$ be the sheaf of germs of hypo-analytic functions in some hypo-analytic structure on $\mathcal{M}$, it is necessary and sufficient that, for each $\mathfrak{p} \in \mathcal{M}$, there be a ring isomorphism $J_\mathfrak{p}$ of ${}_m\mathcal{O}$ onto the stalk $\mathcal{S}_\mathfrak{p}$ that has the following two properties:*

$$J_\mathfrak{p} \textit{ preserves the germs of constant functions;} \qquad \text{(III.1.4)}$$

$$DJ_p : \mathbb{C}^m \to \mathbb{C}T_p^*\mathcal{M} \text{ is injective.} \tag{III.1.5}$$

Condition (III.1.4) can be restated as follows: the germs of the constant functions belong to $\mathcal{S}_p$, which is an algebra over $\mathbb{C}$ with a unit element, $\dot{1}$; J_p is an algebra isomorphism, in particular $J_p(\dot{1}) = \dot{1}$.

PROOF. Necessity of the condition. Let $\mathcal{M}$ be equipped with a hypo-analytic structure. Let p be an arbitrary point of $\mathcal{M}$ and let (U,Z) be a hypo-analytic chart centered at p (i.e., $p \in$ U and $Z(p) = 0$). The pullback map $\tilde{f} \to \tilde{f} \circ Z$ is an isomorphism J_p of ${}_m\mathcal{O}$ onto the ring of germs of hypo-analytic functions at p which preserves the germs of constant functions. The pullback of the coordinate function z_j is the j-th component Z_j of the map Z; since the differentials $dZ_1, \ldots, dZ_m$ are linearly independent, the map DJ_p is injective.

Let us show the sufficiency. For $p \in \mathcal{M}$ arbitrary let J_p be a ring isomorphism of ${}_m\mathcal{O}$ onto $\mathcal{S}_p$ that has properties (III.1.4) and (III.1.5). Clearly J_p must transform any germ of a holomorphic function that vanishes at 0 into the germ of a $\mathcal{C}^\infty$ function that vanishes at p.

There is an open neighborhood U_p of p in which the germ $J_p(z_j)$ has a representative $Z_{p,j} \in \mathcal{C}^\infty(U_p)$ for every $j = 1, \ldots, m$. From what was just said we know that $Z_p(p) = 0$ (as usual we write $Z_p = (Z_{p,1}, \ldots, Z_{p,m})$). From (III.1.5) we derive that $dZ_{p,1}, \ldots, dZ_{p,m}$ are linearly independent at p, hence in U_p if the latter is small enough.

Now, let f be a continuous section of $\mathcal{S}$ over some open neighborhood U′ $\subset U_p$ of p; we may regard f as a $\mathcal{C}^\infty$ function in U′. Let $\dot{\tilde{f}}$ be the germ of a holomorphic function $\tilde{f}$ in some open neighborhood of 0 in $\mathbb{C}^m$ such that $J_p\dot{\tilde{f}}$ is equal to the germ of f at p. The compose $\tilde{f} \circ Z_p$ is a $\mathcal{C}^\infty$ function in an open neighborhood of p whose germ at p is also equal to $J_p\dot{\tilde{f}}$. We conclude that $f = \tilde{f} \circ Z_p$ in a full neighborhood of p in U_p.

Let then $r \in U_p$, $r \neq p$; call Z_r the analogue for r of the map Z_p. From what we have just said it follows that there is a holomorphic map of an open neighborhood $\mathcal{U}_r$ of $Z_r(r)$ onto one of $Z_p(r)$ such that $Z_p = F_{p,r} \circ Z_r$ in some open neighborhood of r; and $F_{p,r}$ is a biholomorphism provided $\mathcal{U}_r$ is small enough (since its differential at the point r must be injective).

Now consider three points in $\mathcal{M}$, p, $q \neq p$, and $r \in U_p \cap U_q$. In some open neighborhood of $Z_q(r)$ we may write

$$Z_p = F_{p,r} \circ \overset{-1}{F}_{q,r} \circ Z_q.$$

This shows that the charts (U_p, Z_p) ($p \in \mathcal{M}$) define a hypo-analytic structure on $\mathcal{M}$ as prescribed in Definition III.1.1; the compose $F_{p,r} \circ \overset{-1}{F}_{q,r}$ plays the role of the "transition map" F_κ^ι. ■

It is convenient to complete the set of hypo-analytic structures on $\mathcal{M}$ by adjoining to those defined in Definition III.1.1 (which requires $m \geq 1$) the *zero structure* : this is the structure defined by the sheaf of germs of *constant* functions. In this case, of course, we have $m = 0$.

The differentials of the hypo-analytic functions at a point $\mathfrak{p} \in \mathcal{M}$ span an m-dimensional subspace $T'_{\mathfrak{p}}$ of $\mathbb{C}T^*_{\mathfrak{p}}\mathcal{M}$; $T'_{\mathfrak{p}}$ is the fibre at $\mathfrak{p}$ of a vector subbundle of $\mathbb{C}T^*\mathcal{M}$, T', which defines a locally integrable structure on $\mathcal{M}$. We refer to the latter as the *locally integrable* (or the *involutive*) *structure underlying the hypo-analytic structure of* $\mathcal{M}$.

By virtue of (III.1.3) and (III.1.5) *the differential map* $DJ_{\mathfrak{p}}$ *is a linear bijection of* $\mathbb{C}^m$ *onto the fibre at* $\mathfrak{p}$ *of the cotangent structure bundle*, $T'_{\mathfrak{p}}$.

One and the same locally integrable structure may underlie two distinct hypo-analytic structures:

EXAMPLE III.1.6. Whatever the $\mathscr{C}^\infty$ function Z on $\mathbb{R}^1$ such that $dZ \neq 0$ at every point, the vector bundle spanned by dZ is equal to $\mathbb{C}T^*\mathbb{R}^1$. Two such functions Z_1 and Z_2 will define the same hypo-analytic structure on $\mathbb{R}^1$ if and only if, locally, they are holomorphic functions of each other. It is clear that this is not the case when $Z_1 = x$ (the coordinate in $\mathbb{R}^1$) and Z_2 is not analytic. ∎

PROPOSITION III.1.2. *Let* $\mathscr{A}_1$ *and* $\mathscr{A}_2$ *be two hypo-analytic structures on the manifold* $\mathcal{M}$ *such that the following holds:*

any function defined in an arbitrary open subset Ω *of* $\mathcal{M}$ *that is hypo-analytic in* Ω *in the sense of* $\mathscr{A}_1$ *is also hypo-analytic in* Ω *in the sense of* $\mathscr{A}_2$. (III.1.6)

Then the following two properties are true:

every solution in an open subset Ω *of* $\mathcal{M}$, *for the locally integrable structure underlying* $\mathscr{A}_1$, *is a solution for the structure underlying* $\mathscr{A}_2$; (III.1.7)

if a solution in Ω, *for the structure underlying* $\mathscr{A}_1$, *is hypo-analytic for* $\mathscr{A}_2$, *then it is hypo-analytic for* $\mathscr{A}_1$. (III.1.8)

PROOF. Call T'_1 (resp., T'_2) the cotangent structure bundle in $\mathscr{A}_1$ (resp., $\mathscr{A}_2$); below we call m_1 (resp., m_2) the rank of T'_1 (resp., T'_2). Property (III.1.6) entails $T'_1 \subset T'_2$ whence (III.1.7). Let $(U, Z_1, \ldots, Z_{m_1})$ be a hypo-analytic chart in Ω, in the sense of $\mathscr{A}_1$. By virtue of (III.1.6) and possibly after contracting U about one of its points, $\mathfrak{p}$, we may find $m_2 - m_1$ functions $Z_{m_1+1}, \ldots, Z_{m_2}$ such that (U, Z), with $Z = (Z_1, \ldots, Z_{m_2})$, is a hypo-analytic chart for $\mathscr{A}_2$. Let h be a function in U which is hypo-analytic at $\mathfrak{p}$ for $\mathscr{A}_2$. There is a holomorphic function $\tilde{h}$ in some open neighborhood of $Z(\mathfrak{p})$ in $\mathbb{C}^{m_2}$ such that $h = \tilde{h} \circ Z$ in a neighborhood of $\mathfrak{p}$ in U. Suppose that, in U, h is a solution in the sense of $\mathscr{A}_1$, i.e., dh belongs to the span of $dZ_1, \ldots, dZ_{m_1}$. It implies that the partial derivatives of $\tilde{h}$ with respect to the variables z_j ($m_1 < j \leq m_2$) vanish identically on Z(U). Corollary II.3.7 entails that, in a neighborhood of $Z(\mathfrak{p})$, $\tilde{h}$ is independent of z_j whatever j, $m_1 < j \leq m_2$, whence (III.1.8). ■

Corollary II.3.7 allows us also to prove the following:

THEOREM III.1.1. *Suppose $\mathcal{M}$ is a real-analytic manifold equipped with an involutive structure of class $\mathscr{C}^\omega$. There exists a unique hypo-analytic structure on $\mathcal{M}$, of class $\mathscr{C}^\omega$, whose underlying involutive structure is the same as that of $\mathcal{M}$.*

The sheaf of germs of hypo-analytic functions in this structure is equal to the sheaf of germs of analytic solutions.

PROOF. The proof of Theorem I.10.5 shows that there is a covering of $\mathcal{M}$ by open sets U in each of which there is a map $Z = (Z_1,\ldots,Z_m)$: $U \to \mathbb{C}^m$ whose components Z_j are $\mathscr{C}^\omega$ *solutions* and satisfy $dZ_1 \wedge \cdots \wedge dZ_m \neq 0$. Possibly after contracting U about one of its points we can find a real-analytic map $W : U \to \mathbb{C}^n$ such that (Z,W) is an analytic diffeomorphism of U onto a totally real submanifold $\mathscr{X}$ of $\mathbb{C}^{m+n}$. Of course $\mathscr{X}$ is a real-analytic submanifold of $\mathbb{C}^{m+n}$ and $\dim_{\mathbb{R}} \mathscr{X} = m + n$.

Now let $\mathscr{A}$ be a hypo-analytic structure on $\mathcal{M}$ of class $\mathscr{C}^\omega$ whose underlying involutive structure is equal to the one given on $\mathcal{M}$. Let f be a hypo-analytic function in U for $\mathscr{A}$. Since f is real-analytic, its push-forward to $\mathscr{X}$ under the map (Z,W) extends as a holomorphic function $\tilde{f}$ in an open neighborhood of $\mathscr{X}$ in $\mathbb{C}^{m+n}$. By hypothesis df belongs to the span of $dZ_1,\ldots,dZ_m$ over U; this means that the restrictions to $\mathscr{X}$ of the partial derivatives of $\tilde{f}$ with respect to $W_1,\ldots,W_n$ (regarded as coordinates in $\mathbb{C}^{m+n}$) vanish identically. By virtue of Corollary II.3.7 we conclude that $\tilde{f}$ is independent of $W_1,\ldots,W_n$ in a neighborhood of $\mathscr{X}$, which proves that $f = \tilde{f} \circ Z$.

The property just established implies at once that, if there is a hypo-analytic structure on $\mathcal{M}$ of class $\mathscr{C}^\omega$ with the same underlying involutive structure as $\mathcal{M}$, then it is unique; and that the sheaf of germs of hypo-analytic functions is equal to the sheaf of germs of analytic solutions, as asserted.

Actually, the same property also implies the existence of a $\mathscr{C}^\omega$ hypo-analytic structure on $\mathcal{M}$ whose underlying involutive structure is that of $\mathcal{M}$. Indeed, take as (Ω_ι, Z_ι) in Definition III.1.1 those charts in which the components of the map Z_ι are analytic solutions in Ω_ι and satisfy (III.1.1). The map Z_ι defines a hypo-analytic structure on Ω_ι for each ι; its underlying involutive structure is the same as that induced on Ω_ι by $\mathcal{M}$. According to what was said above, if $\Omega_\iota \cap \Omega_\kappa \neq \emptyset$ each component of the map Z_κ is hypo-analytic in $\Omega_\iota \cap \Omega_\kappa$ for the structure defined by Z_ι, which is a way of rephrasing (III.1.2). ■

We close the section with the following definition, valid in the general $\mathscr{C}^\infty$ case:

DEFINITION III.1.4. *A hypo-analytic structure on the manifold $\mathcal{M}$ will be called* maximal *if the underlying cotangent structure bundle is equal to* $\mathbb{C}T^*\mathcal{M}$.

It follows from Theorem III.1.1 that on a real-analytic manifold $\mathcal{M}$ there is only one maximal hypo-analytic structure of class $\mathscr{C}^\omega$, namely the real-analytic structure of $\mathcal{M}$.

III.2. Properties of Hypo-Analytic Functions

Let $\mathcal{M}$ be a manifold equipped with a hypo-analytic structure, of class $\mathscr{C}^\infty$, and let (U,Z) be a hypo-analytic chart in $\mathcal{M}$ centered at a point 0 (referred to as *the origin*). We may assume the situation to be that described at the start of section II.1: U is the domain of local coordinates x_j, t_k ($1 \leq j \leq m$, $1 \leq k \leq n$); the components Z_j have the expressions (I.7.23); (I.7.24) holds. Indeed, the substitutions of the Z_j and the changes of variables that are needed to bring us into such a situation do not violate the hypo-analyticity of the Z_j: the substitutions of the Z_j are $\mathbb{C}$-linear (and nonsingular). We may even (as we have done in section II.1) start from an arbitrary maximally real submanifold $\mathscr{X}$ of $\mathcal{M}$ passing through 0 and select the coordinates t_k in such a way that $\mathscr{X} \cap U$ is defined in U by the equations $t = 0$. After contracting U we may, and shall, assume that the map (II.1.5), $(x,t) \to (Z(x,t),t)$, is a diffeomorphism of U onto a $\mathscr{C}^\infty$ submanifold Σ of $\mathbb{C}^m \times \mathbb{R}^n$. We shall make use of the vector fields L_j and M_i defined in (I.7.25) (see (I.7.26) and (I.7.27)).

PROPOSITION III.2.1. *Let h be a continuous function in* U. *The following properties are equivalent:*

h is a hypo-analytic function in U; (III.2.1)

h is a $\mathscr{C}^\infty$ solution in U *and an analytic vector for the system of vector fields* $M_1, \ldots, M_m$; (III.2.2)

there is a $\mathscr{C}^\infty$ function $\tilde{h}(z,t)$ in an open neighborhood of Σ in $\mathbb{C}^m \times \mathbb{R}^n$, which is holomorphic with respect to z and locally constant with respect to t, and is such that $h(x,t) = \tilde{h}(Z(x,t),t)$ in U. (III.2.3)

PROOF. (III.2.1) $\Rightarrow$ (III.2.2). To each point p of U there is an open polydisk $\Delta_p \subset \mathbb{C}^m$ centered at $Z(p)$ and a holomorphic function $\tilde{h}_p$ in Δ_p such that $h = \tilde{h}_p \circ Z$ in $U_p = \overset{-1}{Z}(\Delta_p)$. This implies at once that h is a solution in U_p. According to (II.1.9), we have, in U_p:

$$M^\alpha h = \tilde{h}^{(\alpha)} \circ Z, \ \forall\, \alpha \in \mathbb{Z}_+^m, \tag{III.2.4}$$

where $\tilde{h}^{(\alpha)} = (\partial/\partial z)^\alpha \tilde{h}$. After contracting U_p about p we derive from the Cauchy inequalities for $\tilde{h}$:

$$\sup_{U_p} |\tilde{h}^{(\alpha)} \circ Z\,| \leq C_p^{|\alpha|+1}\, \alpha!. \tag{III.2.5}$$

By using a locally finite refinement of the covering $\{U_p\}$ of U we get at once the inequalities (II.4.6) for h (and arbitrary $K \subset\subset U$).

(III.2.2) $\Rightarrow$ (III.2.3). We apply Proposition II.4.1: there is a continuous function $\tilde{h}(z,t)$ holomorphic with respect to z, in an open neighborhood Ω of Σ in $\mathbb{C}^m \times \mathbb{R}^n$, such that $h(x,t) = \tilde{h}(Z(x,t),t)$ for all $(x,t) \in U$. We avail ourselves once again of (II.1.9):

$$(\partial\tilde{h}/\partial t_j)(Z(x,t),t) = L_j h(x,t) = 0.$$

This means that, for each $t \in W$, $d_t \tilde{h}|_{\Sigma_t} = 0$; since Σ_t is maximally real in $\mathbb{C}^m$ we have $d_t\tilde{h} \equiv 0$ in Ω_t and therefore everywhere in Ω.

(III.2.3) $\Rightarrow$ (III.2.1). To any point $(x_0,t_0) \in U$ there is an open neighborhood of $(Z(x_0,t_0),t_0)$, $\mathcal{O}_0 = \Delta_0 \times \mathcal{B}_0$ with Δ_0 an open polydisk in $\mathbb{C}^m$ centered at $Z(x_0,t_0)$ and $\mathcal{B}_0$ an open ball in $\mathbb{R}^n$ centered at t_0, in which the function $\tilde{h}(z,t)$ is defined, holomorphic with respect to z and independent of t. Consequently it defines a holomorphic function $\tilde{h}$ in Δ_0 such that $h = \tilde{h} \circ Z$ in $\overset{-1}{Z}(\Delta_0)$. ∎

The next statement is self-evident:

PROPOSITION III.2.2. *In order that a function h in* U *be hypo-analytic at* 0 *it is necessary and sufficient that, in some open neighborhood* $U_0 \subset U$ *of* 0, *h be equal to the sum of a convergent power series* $\sum_{\alpha \in \mathbb{Z}_+^m} c_\alpha Z^\alpha$ $(c_\alpha \in \mathbb{C})$.

Now let the function h be hypo-analytic in U. According to Proposition III.2.2 each point (x_0,t_0) of U has an open neighborhood U in which

$$h(x,t) = \sum_{\alpha \in \mathbb{Z}_+^m} c_\alpha(x_0,t_0)[Z(x,t) - Z(x_0,t_0)]^\alpha. \tag{III.2.6}$$

The convergence of the series at the right is uniformly absolute and, in fact, valid in $\mathscr{C}^\infty$. If then $\tilde{h}$ is a holomorphic function in some open neighborhood of $Z(x_0,t_0)$ such that $h = \tilde{h} \circ Z$ in a neighborhood of (x_0,t_0), we must necessarily have, in some polydisk centered at $Z(x_0,t_0)$,

$$\tilde{h}(z) = \sum_{\alpha \in \mathbb{Z}_+^m} c_\alpha(x_0,t_0)[z - Z(x_0,t_0)]^\alpha, \tag{III.2.7}$$

$$c_\alpha(x_0,t_0) = \tilde{h}^{(\alpha)}(Z(x_0,t_0))/\alpha!. \tag{III.2.8}$$

Indeed, the two sides in (III.2.7) are well defined and holomorphic in some open neighborhood of $Z(x_0,t_0)$. By (II.2.6) they are equal when $z = Z(x,t_0)$ for all x sufficiently close to x_0; these points z make up a maximally real submanifold of $\mathbb{C}^m$. Once again apply Corollary II.3.7. By (III.2.4) we have

$$c_\alpha(x_0,t_0) = \mathrm{M}^\alpha h(x_0,t_0)/\alpha!. \tag{III.2.9}$$

PROPOSITION III.2.3. *If a hypo-analytic function h in a connected open subset Ω of $\mathcal{M}$ vanishes to infinite order at a point of Ω it vanishes identically in Ω.*

PROOF. Suppose that h vanishes to infinite order at $p \in \Omega$ and let (U,Z) be a hypo-analytic chart in Ω centered at p. From (III.2.6) and (III.2.9) it follows at once that $h \equiv 0$ in some open neighborhood of p in U. We conclude that the subset of Ω consisting of the points at which h vanishes to infinite order is open; since it is obviously closed it must be identical to Ω. ∎

In certain hypo-analytic structures there exist solutions in an open connected subset Ω that vanish to infinite order at a point of Ω without vanishing identically in Ω.

EXAMPLE III.2.1. Let $\mathbb{R}^2$ be equipped with the hypo-analytic structure defined by the single hypo-analytic chart (U,Z) with $Z = x + \imath t^2/2$. The underlying locally integrable structure is the Mizohata structure. Using the main square-root branch we define

$$h(x,t) = \exp[-1/(t^2 - 2\imath x)^{1/2}]; \tag{III.2.10}$$

h is a $\mathcal{C}^\infty$ solution in $\mathbb{R}^2$ that vanishes to infinite order at (0,0). ∎

We may rephrase as follows the last part in the statement of Theorem III.1.1:

PROPOSITION III.2.4. *Let $\mathcal{M}$ be an analytic manifold and assume that the hypo-analytic structure of $\mathcal{M}$ is analytic. In order that a function h in some open subset Ω of $\mathcal{M}$ be hypo-analytic it is necessary and sufficient that h be an analytic solution in Ω.*

III.3. Submanifolds Compatible with the Hypo-Analytic Structure

Let $\mathcal{M}$ be a hypo-analytic manifold, $\mathcal{X}$ an embedded submanifold of $\mathcal{M}$. Denote by $\mathcal{S}$ the sheaf of germs of hypo-analytic functions in $\mathcal{M}$; its stalk at $p \in \mathcal{M}$ is the ring $\mathcal{S}_p$ of germs at p of the hypo-analytic functions in $\mathcal{M}$. If $p \in \mathcal{X}$ we denote by $\mathcal{I}_{\mathcal{X},p}$ the *ideal* of $\mathcal{S}_p$ consisting of those germs that vanish on (the germ at p of) $\mathcal{X}$. As p ranges over $\mathcal{X}$ the quotient rings

$$\mathcal{O}_{\mathcal{X},p} = \mathcal{O}_p / \mathcal{I}_{\mathcal{X},p}$$

make up a sheaf $\mathcal{O}_{\mathcal{X}}$ of germs of $\mathcal{C}^\infty$ functions in $\mathcal{X}$.

DEFINITION III.3.1. *We say that the submanifold $\mathcal{X}$ is* compatible *with the hypo-analytic structure of $\mathcal{M}$ if the sheaf $\mathcal{O}_{\mathcal{X}}$ defines a hypo-analytic structure on $\mathcal{X}$, to which we then refer as the hypo-analytic structure* induced by, *or* inherited from, *that of $\mathcal{M}$.*

If $\mathcal{X}$ is compatible with the hypo-analytic structure of $\mathcal{M}$ it is also compatible with the underlying locally integrable structure of $\mathcal{M}$ (Definition I.3.1). The converse is not true as shown by the following:

EXAMPLE III.3.1. Let $\mathbb{R}^2$ be equipped with its standard analytic structure (defined by the coordinate functions x, y); we have $T' = \mathbb{C}T^*\mathbb{R}^2$. Let $\mathcal{X}$ be the curve defined by the equation $y = e^{-1/x^2}$; the pullback of T' to $\mathcal{X}$ is equal to $\mathbb{C}T^*\mathcal{X}$. When p is the origin the ideal $\mathcal{I}_{\mathcal{X},p}$ is equal to 0, and thus $\mathcal{O}_{\mathcal{X},0}$ is equal to the ring of germs at 0 of the analytic functions in $\mathbb{R}^2$. The pullback of dx is $\neq 0$ at every point of $\mathcal{X}$. If $\mathcal{O}_{\mathcal{X}}$ were to define a hypo-analytic structure on $\mathcal{X}$ the restriction to $\mathcal{X}$ of any germ of analytic function at the origin should be equal to the restriction of the germ of some analytic function of x alone. This is certainly not the case for y (cf. Proposition III.2.3). ■

PROPOSITION III.3.1. *Let $\mathcal{X} \subset \mathcal{Y}$ be two embedded submanifolds of $\mathcal{M}$. If $\mathcal{Y}$ is compatible with the hypo-analytic structure of $\mathcal{M}$ and $\mathcal{X}$ is compatible with the hypo-analytic structure of $\mathcal{Y}$ inherited from that of $\mathcal{M}$ then $\mathcal{X}$ is compatible with the hypo-analytic structure of $\mathcal{M}$.*

PROOF. Let p be an arbitrary point of $\mathcal{X}$; denote by $\mathcal{I}_{\mathcal{X},\mathcal{Y},p}$ the ideal of germs at p of hypo-analytic functions in $\mathcal{Y}$ that vanish on $\mathcal{X}$ and by $\mathcal{O}_{\mathcal{X},\mathcal{Y},p}$ the quotient ring $\mathcal{O}_{\mathcal{Y},p}/\mathcal{I}_{\mathcal{X},\mathcal{Y},p}$. We have the commutative diagram

$$\begin{array}{ccccccccc}
 & & 0 & & 0 & & & & \\
 & & \downarrow & & \downarrow & & & & \\
0 & \to & \mathcal{I}_{\mathcal{Y},p} & \to & \mathcal{I}_{\mathcal{Y},p} & \to & 0 & & \\
 & & \downarrow & & \downarrow & & \downarrow & & \\
0 & \to & \mathcal{I}_{\mathcal{X},p} & \to & \mathcal{O}_p & \to & \mathcal{O}_{\mathcal{X},p} & \to & 0 \\
 & & \downarrow & & \downarrow & & \downarrow & & \\
0 & \to & \mathcal{I}_{\mathcal{X},\mathcal{Y},p} & \to & \mathcal{O}_{\mathcal{Y},p} & \to & \mathcal{O}_{\mathcal{X},\mathcal{Y},p} & \to & 0 \\
 & & \downarrow & & \downarrow & & \downarrow & & \\
 & & 0 & & 0 & & 0 & &
\end{array}$$

which shows that $\mathcal{O}_{\mathcal{X},p} \cong \mathcal{O}_{\mathcal{X},\mathcal{Y},p}$. ■

We reintroduce the ring homomorphism J_p of Proposition III.1.1. The preimage $\tilde{\mathcal{J}}_{\mathcal{X},p}$ of the ideal $\mathcal{J}_{\mathcal{X},p}$ under the ring homomorphism J_p is a proper ideal in ${}_m\mathcal{O}$. We have the commutative diagram

$$\begin{array}{ccccccccc} 0 & \to & \tilde{\mathcal{J}}_{\mathcal{X},p} & \to & {}_m\mathcal{O} & \to & {}_m\mathcal{O}/\tilde{\mathcal{J}}_{\mathcal{X},p} & \to & 0 \\ & & \downarrow J_p & & \downarrow J_p & & \downarrow & & \\ 0 & \to & \mathcal{J}_{\mathcal{X},p} & \to & \mathcal{S}_p & \to & \mathcal{S}_{\mathcal{X},p} & \to & 0 \end{array} \tag{III.3.1}$$

(of course, the two horizontal sequences are exact). By commutativity the last vertical arrow is an isomorphism. We know, by Proposition III.1.1, that if $\mathcal{S}_{\mathcal{X}}$ defines a hypo-analytic structure on $\mathcal{X}$ then, for some integer μ, $0 \leq \mu \leq \dim \mathcal{X}$, there is an isomorphism $\mathcal{S}_{\mathcal{X},p} \cong {}_\mu\mathcal{O}$. According to the diagram (III.3.1) this produces an isomorphism (for the structures of commutative ring with a unit element)

$$ {}_m\mathcal{O}/\tilde{\mathcal{J}}_{\mathcal{X},p} \cong {}_\mu\mathcal{O}, \tag{III.3.2}$$

with interesting consequences for the ideal $\tilde{\mathcal{J}}_{\mathcal{X},p}$:

LEMMA III.3.1. *The following properties of an ideal $\mathcal{J}$ of ${}_m\mathcal{O}$ are equivalent:*

$\mathcal{J}$ is the kernel of a surjective homomorphism $\tau : {}_m\mathcal{O} \to {}_\mu\mathcal{O}$; (III.3.3)

after a holomorphic change of variables in $\mathbb{C}^m$ the germs at the origin of the last $m-\mu$ coordinate functions, $z_{\mu+1},\ldots,z_m$, span the ${}_m\mathcal{O}$—module $\mathcal{J}$; (III.3.4)

$\mathcal{J}$ is the ideal of germs at 0 *of the holomorphic functions that vanish on the germ* (at 0) *of a μ-dimensional holomorphic submanifold of $\mathbb{C}^m$.* (III.3.5)

PROOF OF LEMMA III.3.1. For simplicity we shall not distinguish between functions and sets and their germs at 0. Suppose (III.3.3) holds, and let $u_j \in {}_m\mathcal{O}$ be such that $\tau(u_j) = z_j$, $j = 1,\ldots,\mu$. We claim that the differentials at 0 of the holomorphic functions (in m variables) u_j are linearly independent. Indeed, for each $i = 1,\ldots,m$, there is a holomorphic function at the origin in $\mathbb{C}^\mu$, q_i, such that $\tau(z_i) = q_i(\tau(u))$, with $q_i(0) = 0$; this means that the holomorphic function in $\mathbb{C}^m$, $h_i(z) = z_i - q_i(u)$, belongs to the ideal $\mathcal{J}$. If $(\mathrm{d}u_1|_0)\wedge\cdots\wedge(\mathrm{d}u_\mu|_0)$ were equal to zero the dimension (over $\mathbb{C}$) of the span of $(\mathrm{d}h_1|_0),\ldots,(\mathrm{d}h_m|_0)$ would be $> m - \mu$ and therefore we could effect a change of variables near 0 in $\mathbb{C}^m$ in such a way that at least $m - \mu + 1$ coordinate functions z_i be elements of $\mathcal{J}$. But then there would be an injective homomorphism ${}_\mu\mathcal{O} \cong {}_m\mathcal{O}/\mathcal{J} \to {}_{\mu-1}\mathcal{O}$, which is absurd.

Once we know that the differentials $\mathrm{d}u_j|_0$ are linearly independent we can take $u_1,\ldots,u_\mu$ as part of a complex coordinate system in $\mathbb{C}^m$. Call $u_{\mu+1},\ldots,u_m$ the remaining coordinates; $\mathcal{J}$ contain $m - \mu$ functions $u_k - p_k(u_1,\ldots,u_\mu)$, with

$p_k|_0 = 0$ $(k = \mu+1,\ldots,m)$. We now take as complex coordinates in $\mathbb{C}^m$, $z_j = u_j$ for $1 \le j \le \mu$ and $z_k = u_k - p_k(u_1,\ldots,u_\mu)$ for $\mu < k \le m$. Thus $z_{\mu+1},\ldots,z_m$ belong to $\mathcal{I}$; suppose they did not span the ${}_m\mathcal{O}$-module $\mathcal{I}$. There would be an element of $\mathcal{I}$ independent of the last $m-\mu$ coordinates, say $\psi(z_1,\ldots,z_\mu)$. The homomorphism τ would map ${}_m\mathcal{O}$ into the quotient of ${}_\mu\mathcal{O}$ modulo the principal ideal generated by ψ, and would definitely not be a surjection onto ${}_\mu\mathcal{O}$.

If (III.3.4) holds, the ideal $\mathcal{I}$ is exactly equal to the set of holomorphic functions that vanish when $z_{\mu+1} = \cdots = z_m = 0$. But any μ-dimensional submanifold of $\mathbb{C}^m$ passing through 0 can be transformed by a holomorphic change of coordinates, in some neighborhood of the origin, into that linear subspace. Any holomorphic function that vanishes on it is a linear combination of $z_{\mu+1},\ldots,z_m$ with coefficients in ${}_m\mathcal{O}$. We have thus shown that (III.3.5) $\Leftrightarrow$ (III.3.4).

Suppose (III.3.4) holds. The map

$$h(z_1,\ldots,z_m) \to h(z_1,\ldots,z_\mu,0,\ldots,0) \tag{III.3.6}$$

is a homomorphism ${}_m\mathcal{O} \to {}_\mu\mathcal{O}$ whose kernel is $\mathcal{I}$. ∎

PROPOSITION III.3.2. *In order that $\mathcal{X}$ be compatible with the hypo-analytic structure of $\mathcal{M}$ it is necessary and sufficient that there be an integer μ, $0 \le \mu \le \mathrm{Min}(m, \dim \mathcal{X})$, such that the following conditions are both satisfied:*

$$\mathrm{T}'|_{\mathcal{X}} \cap \mathbb{C}\mathrm{N}^*\mathcal{X} \text{ is a vector bundle over } \mathcal{X} \text{ of rank } m-\mu; \tag{III.3.7}$$

for all $\mathfrak{p} \in \mathcal{X}$ the ideal $\tilde{\mathcal{I}}_{\mathcal{X},\mathfrak{p}}$ is the kernel of a surjective homomorphism $\tau_{\mathcal{X},\mathfrak{p}} : {}_m\mathcal{O} \to {}_\mu\mathcal{O}$. (III.3.8)

The meaning of (III.3.7) is that the natural pullback map

$$\mathrm{T}'|_{\mathcal{X}} \to \mathbb{C}\mathrm{T}^*\mathcal{X} \tag{III.3.9}$$

maps $\mathrm{T}'|_{\mathcal{X}}$ onto a vector subbundle of $\mathbb{C}\mathrm{T}^*\mathcal{X}$ whose fibre dimension is equal to μ. It implies that $\mathcal{X}$ is compatible with the involutive structure of $\mathcal{M}$ (Definition I.3.1).

PROOF. We have already seen that the conditions are necessary; we show their sufficiency. We continue to identify the functions and their germs at the central point. Lemma III.3.1 allows us to choose the complex coordinates in $\mathbb{C}^m$ in such a way that the ${}_m\mathcal{O}$-module $\tilde{\mathcal{I}}_{\mathcal{X},\mathfrak{p}}$ is spanned by the functions $z_{\mu+1},\ldots,z_m$. We now have the diagram

$$\begin{array}{ccccccccc} 0 & \to & \tilde{\mathcal{I}}_{\mathcal{X},\mathfrak{p}} & \to & {}_m\mathcal{O} & \to & {}_u\mathcal{O} & \to & 0 \\ & & \downarrow & & J_\mathfrak{p} \downarrow & & J_\mathfrak{p} \downarrow & & \\ 0 & \to & \mathcal{I}_{\mathcal{X},\mathfrak{p}} & \to & \mathcal{S}_\mathfrak{p} & \to & \mathcal{S}_{\mathcal{X},\mathfrak{p}} & \to & 0 \end{array} \tag{III.3.10}$$

where, once again, the third vertical arrow stands for an isomorphism. The horizontal arrow ${}_m\mathcal{O} \to {}_\mu\mathcal{O}$ is now simply the map (III.3.6).

Set $Z_j = J_p z_j$ ($j = 1,\dots,m$). We know that $\mathcal{I}_{\mathcal{X},p}$ is spanned, over $\mathcal{S}_p$, by $Z_{\mu+1},\dots,Z_m$; it follows that the differentials at 0, $dZ_k|_0$ ($\mu < k \leq m$), belong to $\mathbb{C}N_p^*\mathcal{X}$ and therefore span $T_p' \cap \mathbb{C}N_p^*\mathcal{X}$. As for the elements of $\mathcal{S}_{\mathcal{X},p}$, they are the holomorphic functions of $Z_1,\dots,Z_\mu$. No linear combination of the differentials $dZ_j|_p$ ($1 \leq j \leq \mu$) can belong to $\mathbb{C}N_p^*\mathcal{X}$, otherwise (III.3.7) would be contradicted (keep in mind that $dZ_1|_p,\dots,dZ_m|_p$ are linearly independent). Therefore the differential at 0 of the third vertical arrow in Diagram (III.3.10) is injective. To reach the desired conclusion it suffices to apply Proposition III.1.1 to $\mathcal{X}$. ■

We get an important family of submanifolds of $\mathcal{M}$ that are compatible with the hypo-analytic structure of $\mathcal{M}$ by taking $\mu = m$: in this case the validity of (III.3.7) means that the submanifold $\mathcal{X}$ is *strongly noncharacteristic* (Proposition I.3.4); $\mathcal{I}_{\mathcal{X},p} = 0$ for all $p \in \mathcal{M}$ and Condition (III.3.8) is trivially satisfied ($\tau_{\mathcal{X},p} =$ Identity).

PROPOSITION III.3.3. *Let $\mathcal{X}$ be a strongly noncharacteristic submanifold of $\mathcal{M}$. Then $\mathcal{X}$ is compatible with the hypo-analytic structure of $\mathcal{M}$ and the restriction mapping to $\mathcal{X}$ is a sheaf isomorphism of $\mathcal{S}|_{\mathcal{X}}$ onto $\mathcal{S}_{\mathcal{X}}$.*

The next definition introduces a different class of submanifolds that are compatible with the hypo-analytic structure of $\mathcal{M}$.

DEFINITION III.3.2. *We say that an embedded submanifold $\mathcal{X}$ of $\mathcal{M}$ is a* hypo-analytic submanifold *of $\mathcal{X}$ if there is an integer μ, $0 \leq \mu \leq m$, such that* (III.3.7) *holds and every point p of $\mathcal{X}$ has an open neighborhood* U *in $\mathcal{M}$ in which there are $m-\mu$ hypo-analytic functions $h_1,\dots,h_{m-\mu}$ endowed with the following properties:*

$$dh_1 \wedge \cdots \wedge dh_{m-\mu} \neq 0 \text{ at every point of } U; \tag{III.3.11}$$

$$\mathcal{X} \cap U = \{q \in U;\ h_j(q) = 0,\ j = 1,\dots,m-\mu\}. \tag{III.3.12}$$

When $\mu = m$, a hypo-analytic submanifold is simply an open subset of $\mathcal{M}$. The only submanifolds of $\mathcal{M}$ that are both hypo-analytic and strongly noncharacteristic are the open subsets of $\mathcal{M}$. Indeed, the pullback to a strongly noncharacteristic submanifold $\mathcal{X}$ of $\mathcal{M}$ of a hypo-analytic function h in an open subset U of $\mathcal{M}$ that intersects $\mathcal{X}$, such that $dh \neq 0$ at every point of U, cannot vanish identically in any open subset of $U \cap \mathcal{X}$. Therefore, if $\mathcal{X}$ is to be hypo-analytic, the number μ in Definition III.3.2 is necessarily equal to m.

Keep in mind that, in general, the functions h_j in Definition III.3.2 are complex-valued. When (III.3.11) and (III.3.12) hold, the conormal bundle of $\mathcal{X}$ over $\mathcal{X} \cap U$ is spanned by the differentials of the functions $\mathcal{R}e\, h_j$ and $\mathcal{I}m\, h_j$; these differentials need not be linearly independent. Possibly after contracting U about p we can find μ hypo-analytic functions $Z_1,\dots,Z_\mu$ in U such that

$$dZ_1 \wedge \cdots \wedge dZ_\mu \wedge dh_1 \wedge \cdots \wedge dh_{m-\mu} \neq 0 \qquad \text{(III.3.13)}$$

at every point of U. Condition (III.3.7) implies that the pullback to $\mathcal{X}$ of any nontrivial linear combination of $dZ_1|_p, \ldots, dZ_\mu|_p$ is $\neq 0$, which in turn demands

$$\mu \leq \dim \mathcal{X}. \qquad \text{(III.3.14)}$$

The ideal $\mathcal{I}_{\mathcal{X},p}$ is generated by $h_1, \ldots, h_{m-\mu}$, the module $\mathcal{S}_{\mathcal{X},p}$ is generated by $Z_1, \ldots, Z_\mu$, and thus Condition (III.3.8) is satisfied. We can state:

PROPOSITION III.3.4. *Any hypo-analytic submanifold of $\mathcal{M}$ is compatible with the hypo-analytic structure of $\mathcal{M}$.*

Condition (III.3.7) in Definition III.3.2 is not redundant, as shown in the following:

EXAMPLE III.3.2. In $\mathbb{R}^3$, where the coordinates are x,y,t, consider the hypo-analytic structure defined by the functions $z = x + \imath y$, $w = x - \imath y + t^2$. The t-axis is the zero-set of the hypo-analytic function z, but the pullback to $z = 0$ of the differential dw vanishes at the point $t = 0$ (and only at that point). ■

It may happen that, through a given point, there does not pass any hypo-analytic submanifold $\mathcal{X}$ with $0 < \dim \mathcal{X} < \dim \mathcal{M}$:

EXAMPLE III.3.3. In the Mizohata structure on $\mathbb{R}^2$, defined by the function $Z = x + \frac{1}{2}\imath t^2$, the only hypo-analytic submanifolds that contain the origin are the plane $\mathbb{R}^2$ and the set $\{0\}$. ■

It might also happen that a subset $\mathcal{X}$ of $\mathcal{M}$ is not a submanifold of $\mathcal{M}$ and yet is equal to the zero set of $m - \mu$ hypo-analytic functions in $\mathcal{M}$ whose differentials are linearly independent at every point of $\mathcal{X}$:

EXAMPLE III.3.4. Consider the hypo-analytic structure on $\mathbb{R}^{n+1}$ (coordinates: $x, t_1, \ldots, t_n$) defined by the single function $Z = x + \imath\Phi(t)$. Unless the zero set of Φ in $\mathbb{R}^n$ is a submanifold, the zero set in $\mathbb{R}^{n+1}$ of the function Z will not be a submanifold of $\mathbb{R}^{n+1}$. ■

By combining Propositions III.3.1, III.3.3, and III.3.4 we get

PROPOSITION III.3.5. *Let $\mathcal{X} \subset \mathcal{Y}$ be two embedded submanifolds of $\mathcal{M}$. Suppose that $\mathcal{Y}$ is compatible with the hypo-analytic structure of $\mathcal{M}$ and that $\mathcal{X}$ is either a strongly noncharacteristic or else a hypo-analytic submanifold of $\mathcal{Y}$ for the structure inherited from that of $\mathcal{M}$. Then $\mathcal{X}$ is compatible with the hypo-analytic structure of $\mathcal{M}$.*

There are hypo-analytic manifolds $\mathcal{M}$ in which the nesting of submanifolds considered in Proposition III.3.5 produces all the submanifolds that are compatible with the hypo-analytic structure of $\mathcal{M}$:

EXAMPLE III.3.5. Let $\mathcal{M}$ be a generic submanifold of a complex manifold $\tilde{\mathcal{M}}$ (see end of section I.3). Set $\dim_{\mathbb{C}} \tilde{\mathcal{M}} = m$, $\operatorname{codim}_{\mathbb{R}} \mathcal{M} = d$; we have $0 \leq d \leq m$. In what follows the extreme cases are not precluded: when $d = m$, $\mathcal{M}$ is an open subset of $\tilde{\mathcal{M}}$; when $d = 0$, $\mathcal{M}$ is totally real of dimension m. We equip $\mathcal{M}$ with the CR structure inherited from $\tilde{\mathcal{M}}$; of course, this is a hypo-analytic structure on $\mathcal{M}$ (Example III.1.3).

By Proposition III.3.3 we know that the strongly noncharacteristic submanifolds of $\mathcal{M}$ are compatible with its hypo-analytic structure. By virtue of Proposition III.3.4 the same is true of any hypo-analytic submanifold of $\mathcal{M}$ (Definition III.3.2). Locally a hypo-analytic submanifold of $\mathcal{M}$ is equal to the intersection of $\mathcal{M}$ with a holomorphic submanifold of $\tilde{\mathcal{M}}$ (see Proposition III.3.7). More generally we may state:

PROPOSITION III.3.6. *In order that an embedded submanifold $\mathcal{X}$ of the CR manifold $\mathcal{M}$ be compatible with the hypo-analytic structure of $\mathcal{M}$ it is necessary and sufficient that through each point $\mathfrak{p} \in \mathcal{X}$ there pass a holomorphic submanifold $\mathcal{H}_\mathfrak{p}$ of $\tilde{\mathcal{M}}$ such that $\mathcal{X} \cap U_\mathfrak{p}$ is a generic submanifold of $\mathcal{H}_\mathfrak{p}$ for some open neighborhood $U_\mathfrak{p}$ of $\mathfrak{p}$ in $\mathcal{M}$.*

PROOF. It suffices to consider the case $\tilde{\mathcal{M}} = \mathbb{C}^m$, and to take $\mathfrak{p} \in \mathcal{M}$ to be the origin. The space $\mathcal{S}_0$ of germs at 0 of hypo-analytic functions in $\mathcal{M}$ is identical to ${}_m\mathcal{O}$. As always we denote by T$'$ the cotangent structure bundle of $\mathcal{M}$; it is spanned by the pullbacks to $\mathcal{M}$ of $dz_1,\dots,dz_m$.

Assume that $\mathcal{X}$ is compatible with the CR structure of $\mathcal{M}$. By Proposition III.3.1 (cf. also Lemma III.3.1) the coordinates z_i in (an open neighborhood of 0 in) $\mathbb{C}^m$ may be chosen in such a way that the ideal $\mathcal{I}_{\mathcal{X},0}$, which consists of those germs that vanish on $\mathcal{X}$, is spanned, as an ${}_m\mathcal{O}$-module, by $z_{\mu+1},\dots,z_m$: in a neighborhood of 0, $\mathcal{X}$ is contained in the subspace $\mathcal{H} = \{z \in \mathbb{C}^m;\ z_{\mu+1} = \cdots = z_m = 0\}$; the dz_j are sections of the conormal bundle of $\mathcal{X}$ in $\mathcal{M}$, $\mathbb{C}N^*\mathcal{X}$, over $\mathcal{X} \cap \mathcal{H}$. Since $\dim_{\mathbb{C}} T'_0 \cap \mathbb{C}N^*_0\mathcal{X} = m - \mu$ the pullbacks to $\mathcal{X} \cap \mathcal{H}$ of $dz_1,\dots,dz_\mu$ must be linearly independent in a neighborhood of 0. These differentials span the complex structure bundle of $\mathcal{H}$, whence the necessity of the condition.

To prove its sufficiency observe that if $\mathcal{X} \subset \mathcal{H}$, with $\mathcal{H}$ as above, then the functions z_j ($\mu < j \leq m$) vanish identically on $\mathcal{X}$ and $dz_{\mu+1},\dots,dz_m$ are linearly independent sections of $\mathbb{C}N^*\mathcal{X}$. Let ${}_\mu T'$ denote the span of the differentials dz_j ($1 \leq j \leq \mu$); the pullback map $\mathbb{C}T^*\mathcal{M}|_{\mathcal{H}} \to \mathbb{C}T^*\mathcal{H}$ induces an automorphism of ${}_\mu T'|_{\mathcal{H}}$. If we assume that $\mathcal{X}$ is a generic submanifold of $\mathcal{H}$, the pullback map ${}_\mu T'|_{\mathcal{H}} \to {}_\mu T'|_{\mathcal{X}}$ is bijective and perforce $({}_\mu T'|_{\mathcal{X}}) \cap \mathbb{C}N^*\mathcal{X} = 0$. As a consequence, $dz_{\mu+1},\ldots,dz_m$ make up a basis of $T'\cap\mathbb{C}N^*\mathcal{X}$ over such a neighborhood. This ensures the validity of (III.3.7). On the other hand, again due to the fact that $\mathcal{X}$ is generic in $\mathcal{H}$, no holomorphic function that is independent of $z_{\mu+1},\ldots,z_m$ can vanish identically in an open neighborhood of 0 in $\mathcal{X}$ without vanishing identically in an open neighborhood of 0 in $\mathcal{M}$; this means that the ideal $\mathcal{I}_{\mathcal{X},0}$ is generated by $z_{\mu+1},\ldots,z_m$. It suffices to apply Proposition III.3.2 to conclude that, near the origin, $\mathcal{X}$ is compatible with the hypo-analytic structure of $\mathcal{M}$. ■

PROPOSITION III.3.7. *In order that an embedded submanifold $\mathcal{X}$ of the CR manifold $\mathcal{M}$ be a hypo-analytic submanifold of $\mathcal{M}$ it is necessary and sufficient that $\mathcal{X}$ be compatible with the hypo-analytic structure of $\mathcal{M}$ and that, through each point $p \in \mathcal{X}$, there pass a holomorphic submanifold $\mathcal{H}_p$ of $\tilde{\mathcal{M}}$ such that $\mathcal{X}\cap U_p = \mathcal{H}_p\cap U_p$ for some open neighborhood U_p of p in $\mathcal{M}$.*

PROOF. The necessity of the conditions follows at one from Definition III.3.2 and Proposition III.3.6. Conversely, since $\mathcal{X}$ is compatible with the hypo-analytic structure of $\mathcal{M}$, condition (III.3.7) is satisfied. Thanks to the fact that $\mathcal{X}\cap U_p = \mathcal{H}_p\cap U_p$ we know that there are functions $h_1,\ldots,h_{m-\mu}$ in an open neighborhood of p, $U \subset U_p$, satisfying (III.3.11) and (III.3.12). ■

When $\mathcal{X}$ is compatible with the hypo-analytic structure of $\mathcal{M}$, the structure it inherits from $\mathcal{M}$ is CR. ■

EXAMPLE III.3.6. In $\tilde{\mathcal{M}} = \mathbb{C}^3$ (with complex coordinates z_1, z_2, z_3) consider the real hypersurface $\mathcal{M}$ defined by the equation $\mathcal{I}m z_3 = (x_2 - |z_1|^2)^2 + y_2^2$. Let $\mathcal{X}$ be the real surface in (z_1,z_2) – space $\mathbb{C}^2$ defined by the equations $x_2 = |z_1|^2$, $y_2 = 0$; $\mathcal{X}$ can be regarded as the (nontransversal) intersection of $\mathcal{M}$ with the complex hyperplane $z_3 = 0$. Since the pullbacks to $\mathcal{X}$ of dz_1 and dz_2 are linearly dependent at the origin but linearly independent at every other point of $\mathcal{X}$, Condition (III.3.7) is not satisfied: the submanifold $\mathcal{X}$ is not compatible with the hypo-analytic structure of $\mathcal{M}$. ■

III.4. Unique Continuation of Solutions in a Hypo-Analytic Manifold

Let $\mathcal{H}$ be a hypo-analytic submanifold (Definition III.3.2) of the hypo-analytic manifold $\mathcal{M}$ passing through a point $0 \in \mathcal{M}$ (called the origin). We can select hypo-analytic functions Z_j $(j = 1,\ldots,m)$ in an open neighborhood U of 0 such that, if $Z = (Z_1,\ldots,Z_m)$: $U \to \mathbb{C}^m$, then (U,Z) is a hypo-analytic chart in $\mathcal{M}$ and

$$\mathcal{H}\cap U = \{p \in U;\ Z_j(p) = 0,\ j = 1,\ldots,\nu\} \qquad \text{(III.4.1)}$$

($0 \le \nu \le m$; if $\nu = 0$, $\mathcal{H} \cap \mathrm{U} = \mathrm{U}$). Moreover (cf. (III.3.7)):

$$\mathrm{dZ}_1,\dots,\mathrm{dZ}_\nu \textit{ form a basis of } \mathrm{T}' \cap \mathbb{C}\mathrm{N}^*\mathcal{H} \textit{ over } \mathcal{H} \cap \mathrm{U}. \qquad \text{(III.4.2)}$$

We shall require that $\mathcal{H}$ be *noncharacteristic* (Definition I.4.1). This means that $\mathrm{T}' \cap \mathrm{N}^*\mathcal{H} = 0$, i.e.,

no $\mathbb{C}$*-linear combination of* $\mathrm{dZ}_1,\dots,\mathrm{dZ}_\nu$ *is real at any point of* $\mathcal{H} \cap \mathrm{U}$ *unless it is identically equal to zero.* (III.4.3)

We shall also take $\mathcal{H}$ to be *minimal* for all these properties, in the following sense:

Any noncharacteristic hypo-analytic submanifold of $\mathcal{M}$ *that is contained in* $\mathcal{H}$ *and passes through* 0 *must be equal to an open subset of* $\mathcal{H}$. (III.4.4)

Property (III.4.4) has the following consequence:

if ϖ *belongs to the* $\mathbb{C}$*-linear span of* $\mathrm{dZ}_{\nu+1}|_0,\dots,\mathrm{dZ}_m|_0$ *the pullback of* $\varpi \wedge \bar{\varpi}$ *to* $\mathcal{H}$ *vanishes.* (III.4.5)

Suppose the pullback to $\mathcal{H}$ of $\varpi \wedge \bar{\varpi}$ were $\neq 0$ for some $\varpi \in$ *Span*$(\mathrm{dZ}_{\nu+1},\dots,\mathrm{dZ}_m)$. We could write $\varpi = \mathrm{d}h|_0$ with $h = \tilde{h}(\mathrm{Z}_{\nu+1},\dots,\mathrm{Z}_m)$ and $\tilde{h} \in {}_m\mathcal{O}$. Then the equation $\tilde{h} = 0$ would define a $\mathcal{C}^\infty$ submanifold of $\mathcal{M}$ whose intersection with $\mathcal{H}$ would be a submanifold $\mathcal{H}_1$ of $\mathcal{M}$ such that $\dim \mathcal{H}_1 < \dim \mathcal{H}$. Properties (III.4.2) and (III.4.3) would remain true after substitution of $\mathcal{H}_1$ for $\mathcal{H}$ and $\nu+1$ for ν. This would contradict the minimality hypothesis (III.4.4).

By (III.4.3), and possibly after contracting U about 0, we can take the functions $\mathcal{R}e Z_j$, $\mathcal{I}m Z_j$ as coordinates in U; we call them x_j and y_j respectively ($1 \le j \le \nu$). After a $\mathbb{C}$-linear substitution of $Z_{\nu+1},\dots,Z_m$ we may add to those the coordinates $s_{k-\nu} = \mathcal{R}e Z_k$ ($\nu < k \le m$) and, lastly, complete this set of coordinates by selecting the residual number of coordinates, $t_1,\dots,t_{n-\nu}$. We shall then write z_j instead of Z_j if $j \le \nu$, $w_{k-\nu}$ instead of Z_k if $k > \nu$ (we set $d = m-\nu$). We now have a local chart of the kind considered in the first part of section I.7; we may assume that the w_k have the expressions (I.7.4). We notice, however, that (I.7.5) does not necessarily hold. The manifold $\mathcal{H}$ is defined by the equations $z_j = 0, j = 1,\dots,\nu$; and $\dim \mathcal{H} = n-\nu+d$. Condition (III.4.4) demands

$$\varphi_k = \sum_{j=1}^{\nu} (A_{jk}x_j + B_{jk}y_j) + \sum_{\ell=1}^{d} C_{k\ell}s_\ell + O(|z|^2+|s|^2+|t|^2)$$

with $A_{jk}, B_{jk}, C_{k\ell} \in \mathbb{R}$. Then replace w_k by

$$\tilde{w}_k = w_k - \iota \sum_{j=1}^{\nu} \zeta_{jk} z_j - \iota \sum_{\ell=1}^{d} C_{k\ell} w_\ell$$

for a suitable choice of the complex numbers ζ_{jk}; and take $\mathcal{R}e\tilde{w}_k$ to be the new coordinate s_k (for each $k = 1,\dots,d$), to find ourselves in the precise situation in which (I.7.4) and (I.7.5) are valid.

We are now in a position to apply the results of section II.8.

Let $\mathcal{X}$ be a maximally real submanifold of $\mathcal{H}$ for the involutive structure induced by $\mathcal{M}$; assume that $0 \in \mathcal{X}$. In view of (I.7.4) and (I.7.5) the pullbacks to $\mathcal{X}$ of $ds_1,\dots,ds_d$ must be a basis of $T^*\mathcal{X}$ in some neighborhood of 0 in $\mathcal{X}$. This means that there exists a $\mathcal{C}^\infty$ equation of $\mathcal{X}$, in some open neighborhood $\mathcal{N}$ of 0 in $\mathcal{H}$, of the kind

$$t = t(s). \tag{III.4.6}$$

Then, if $\mathcal{N}$ is sufficiently small, $\mathcal{X}\cap\mathcal{N}$ can be coordinatized by $s_1,\dots,s_d$ and the map $s \to s + \iota\varphi(0,s,t(s))$ induces a diffeomorphism of $\mathcal{X}\cap\mathcal{N}$ onto a maximally real submanifold $\tilde{\mathcal{X}}$ of $\mathbb{C}^d$. We shall say that $\mathcal{X}$ has the *compact cycle property at* 0 if this is true of the submanifold $\tilde{\mathcal{X}}$ of $\mathbb{C}^d$ (Definition II.8.1). It is quite clear that this property does not depend on the choice of the hypo-analytic functions w_k.

THEOREM III.4.1. *Let $\mathcal{H}$ be a hypo-analytic submanifold of $\mathcal{M}$ and* 0 *a point of $\mathcal{H}$ at which $\mathcal{H}$ is minimal noncharacteristic. Let $\mathcal{X}$ be a submanifold of $\mathcal{H}$ passing through* 0, *maximally real for the involutive structure on $\mathcal{H}$ inherited from $\mathcal{M}$. Assume that $\mathcal{X}$ has the compact cycle property at* 0.

If a Lipschitz-continuous solution in an open subset Ω of $\mathcal{M}$ that contains $\mathcal{X}$ vanishes to infinite order on $\mathcal{X}$ it must vanish identically in some neighborhood of 0 *in* Ω.

PROOF. We avail ourselves of the choice of the hypo-analytic functions z_j and w_k, and of the coordinates x_i, y_j, s_k, t_ℓ in the local description that precedes the statement of Theorem III.4.1. In particular we may assume that $\mathcal{X}$ is defined, in the neighborhood $\mathcal{N}$ of 0 in $\mathcal{H}$, by Equation (III.4.6). Actually we may even take $t_\ell - t_\ell(s)$ as new coordinate t_ℓ for each $\ell = 1,\dots,n-\nu$. We find ourselves exactly in the conditions of application of Theorem II.8.1; it implies that $h \equiv 0$ in some neighborhood of 0 in $\mathcal{M}$. ■

Proposition II.8.2 implies:

COROLLARY III.4.1. *Suppose that the manifold $\mathcal{M}$ and its hypo-analytic structure are of class $\mathcal{C}^\omega$. Let $\mathcal{H}$ be a hypo-analytic submanifold of $\mathcal{M}$, minimal noncharacteristic at* 0. *Let $\mathcal{X}$ be a real-analytic maximally real submanifold of $\mathcal{H}$ for the involutive structure on $\mathcal{H}$ inherited from $\mathcal{M}$. Then the conclusion in Theorem* III.4.1 *is valid.*

We have seen above that through each point of $\mathcal{M}$ there passes at least one hypo-analytic submanifold that is minimal noncharacteristic at that point. It might happen that the dimension of such a submanifold is equal to dim $\mathcal{M}$.

EXAMPLE III.4.1. Take the hypo-analytic structure in $\mathbb{R}^2$ (coordinates s, t) defined by the single function $Z = se^{\imath t}$ (cf. Example I.4.2). The only germs of hypo-analytic submanifolds that pass through the origin are the germ of $\mathbb{R}^2$ itself, or else the germ of the t-axis (defined by $Z = 0$). The latter is characteristic at 0. ■

Unique Continuation in Hypo-Analytic CR Manifolds

We shall assume that the involutive structure of the hypo-analytic manifold $\mathcal{M}$ is CR (Definition I.2.3). As usual we write $\dim_{\mathbb{R}} \mathcal{M} = m+n$ and $d = m-n$ $(0 \leq d \leq m)$. The extreme cases $n = 0$ and $n = m$ are not precluded. When $n = 0$ the hypo-analytic structure of $\mathcal{M}$ is maximal (Definition III.1.4); when $n = m$ it is complex (Definition I.2.3). The characteristic set T° of $\mathcal{M}$ is a vector bundle with fibre dimension d (cf. Proposition I.2.1).

It follows at once from (III.4.3) and (III.4.5) that *any hypo-analytic submanifold of $\mathcal{M}$ that is minimal noncharacteristic at one of its points has real dimension equal to d.*

An important feature of a hypo-analytic CR manifold is that through each of its points there passes a hypo-analytic submanifold that is minimal noncharacteristic at every one of its points. This prompts us to introduce the following:

DEFINITION III.4.1. *We shall say that a submanifold $\mathcal{X}$ of the CR manifold $\mathcal{M}$ is* minimal noncharacteristic *if it is noncharacteristic at every one of its points and if its dimension is equal to the fibre dimension of the characteristic set of $\mathcal{M}$.*

The local charts of the kind (U,z,w) with $z = (z_1,\ldots,z_n)$, $w = (w_1,\ldots,w_d)$, $z_j = x_j + \imath y_j$, $w_k = s_k + \imath\varphi_k(x,y,s)$, show that every point of the CR manifold $\mathcal{M}$ has an open neighborhood entirely fibred by hypo-analytic submanifolds that are minimal noncharacteristic. In the above local chart we take these to be the submanifolds $z = const.$ That they are noncharacteristic at every one of their points is due to the fact that, in the present situation, there are no variables t_k. And as their dimension is equal to d, they are perforce minimal noncharacteristic.

Furthermore, as a particular case of what has been shown at the beginning of the present section, we see that if $\mathcal{X}$ is any minimal noncharacteristic hypo-analytic submanifold of $\mathcal{M}$ passing through 0 the chart (U,z,w) can be so devised that $\mathcal{X} \cap \mathrm{U} = \{(x,y,s) \in \mathrm{U};\, z = 0\}$.

Note that the hypo-analytic structure induced by $\mathcal{M}$ on a minimal noncharacteristic submanifold $\mathcal{X}$ is maximal (Definition III.1.4). It follows that the only maximally real submanifolds of $\mathcal{X}$ are open subsets of $\mathcal{X}$. As a consequence, in the present situation, the statement of Theorem III.4.1 can be simplified.

THEOREM III.4.2. *Let $\mathcal{M}$ be a hypo-analytic CR manifold and $\mathcal{X}$ a minimal noncharacteristic hypo-analytic submanifold of $\mathcal{M}$. Assume that $\mathcal{X}$ has the compact cycle property at every one of its points.*

If a Lipschitz-continuous solution in an open subset Ω of $\mathcal{M}$ that contains $\mathcal{X}$ vanishes to infinite order on $\mathcal{X}$, it must vanish identically in some neighborhood of $\mathcal{X}$ in Ω.

In the present setup we can take advantage of Proposition II.8.1.

COROLLARY III.4.2. *The conclusion in* Theorem III.4.2 *is valid if $d = 1$.*

When $d = 1$ $\mathcal{M}$ is often said to be *of hypersurface type*. The reason is that, when $\mathcal{M}$ is a generic submanifold of a complex manifold $\tilde{\mathcal{M}}$, d is equal to the real codimension of $\mathcal{M}$ in $\tilde{\mathcal{M}}$.

We can also take advantage of Proposition II.8.2 and add precision to Corollary III.4.1. For this we observe that, in a hypo-analytic CR manifold, the notion of an *analytic* submanifold makes good sense.

DEFINITION III.4.2. *A submanifold $\mathcal{X}$ of the hypo-analytic CR manifold $\mathcal{M}$ will be called* analytic *if every point $\mathfrak{p}$ of $\mathcal{X}$ is the center of a hypo-analytic chart* (U,Z) *in $\mathcal{M}$ such that the restriction of the map Z to $\mathcal{X} \cap U$ is a diffeomorphism onto a real-analytic submanifold of $\mathbb{C}^m$.*

If $\mathcal{X} \subset \mathcal{M}$ is analytic, then whatever the hypo-analytic chart (U,Z) such that the restriction of Z to $\mathcal{X} \cap U$ is a diffeomorphism onto a submanifold $\hat{\mathcal{X}}$ of $\mathbb{C}^m$, the latter will perforce be a real-analytic submanifold of $\mathbb{C}^m$.

When $\mathcal{M}$ is a real-analytic CR manifold, Definition III.4.2 agrees with the standard definition; this is due to the fact that the hypo-analytic charts (U,Z) can be chosen in such a way that $Z\colon U \to \mathbb{C}^m$ is an analytic diffeomorphism of U onto an open subset of $\mathbb{C}^m$.

Return to the general case of a $\mathcal{C}^\infty$ hypo-analytic CR manifold $\mathcal{M}$. To say that the minimal noncharacteristic submanifold $\mathcal{X}$ is analytic is the same as saying that each point $\mathfrak{p}$ of $\mathcal{X}$ is the center of a hypo-analytic chart (U,z,w) of the type considered at the beginning such that $\mathcal{X} \cap U$ is defined in U by $z = 0$, and such that $s \to s + \imath\varphi(0,s)$ is a diffeomorphism of $\mathcal{X} \cap U$ onto an analytic submanifold $\hat{\mathcal{X}}$ of $\mathbb{C}^d$. This is of course equivalent to saying that the function $s \to \varphi(0,s)$ is real-analytic. By Proposition II.8.2 $\hat{\mathcal{X}}$ has the compact cycle property at every one of its points, and so therefore does $\mathcal{X}$. We may state:

COROLLARY III.4.3. *The conclusion in Theorem* III.4.2 *is valid if* $\mathcal{X}$ *is an analytic (minimal noncharacteristic hypo-analytic) submanifold of* $\mathcal{M}$.

We may interpret the hypothesis, in Theorem III.4.2, that $\mathcal{X}$ is minimal noncharacteristic in terms of the local embeddings provided by the hypo-analytic charts (U,z,w). To do this we may as well assume that $\mathcal{M}$ is a generic submanifold of $\mathbb{C}^m$ defined, near the origin, by the equations

$$t = \varphi(x,y,s) \tag{III.4.7}$$

(we have written $s = \mathcal{R}e\, w$, $t = \mathcal{I}m\, w$). We may also assume that (I.7.5) holds. Near 0 the submanifold $\mathcal{X}$ of $\mathcal{M}$ is equal to the intersection of $\mathcal{M}$ with the subspace $z = 0$. This observation will now enable us to describe the minimal noncharacteristic hypo-analytic submanifolds of any *embedded* CR manifold.

Let $\tilde{\mathcal{M}}$ be a complex manifold and J_z the complex structure map of $\tilde{\mathcal{M}}$ (see (I.3.15)); J_z is an automorphism of $T_z\tilde{\mathcal{M}}$ such that $J_z^2 = -\,\mathrm{Id}$. We define the *holomorphic tangent space* to a submanifold Λ of $\tilde{\mathcal{M}}$ at a point $z \in \Lambda$, as the intersection

$$\mathcal{T}_z\Lambda = T_z\Lambda \cap J_z T_z\Lambda; \tag{III.4.8}$$

$\mathcal{T}_z\Lambda$ is the largest complex subspace of $T_z\tilde{\mathcal{M}}$ (equipped with the complex structure defined by J_z) that is tangent to Λ at z. The map $\mathcal{R}e$: $\mathbb{C}T_z\tilde{\mathcal{M}} \to T_z\tilde{\mathcal{M}}$ induces a bijection of $T_z^{1,0} \cap \mathbb{C}T_z\Lambda$ onto $\mathcal{T}_z\Lambda$ (Proposition I.3.6; $T^{1,0}$: tangent structure bundle of the complex manifold $\tilde{\mathcal{M}}$).

If Λ is a holomorphic submanifold of $\tilde{\mathcal{M}}$, then $\mathcal{T}_z\Lambda = T_z\Lambda$; if Λ is totally real, $\mathcal{T}_z\Lambda = 0$ (Corollary I.3.2). On the other hand, let $\mathcal{M}$ be a *generic* submanifold (see end of Section I.3) of $\tilde{\mathcal{M}}$; $\dim_{\mathbb{R}} \mathcal{M} = m + n$ ($m = \dim_{\mathbb{C}} \tilde{\mathcal{M}}$, $m - n = \mathrm{codim}_{\mathbb{R}} \mathcal{M}$). If $z \in \mathcal{M}$ we have

$$\dim_{\mathbb{C}} \mathcal{T}_z\mathcal{M} = n. \tag{III.4.9}$$

Indeed, $T^{1,0} \cap \mathbb{C}T\mathcal{M} = \mathcal{V}$, the tangent structure bundle of $\mathcal{M}$.

We may also introduce the holomorphic conormal space of Λ at the point $z \in \Lambda$, $\mathcal{N}_z^*\Lambda$: it is the projection of $\mathbb{C}N_z^*\Lambda$ into $T_z'^{1,0}$, in the direct sum decomposition $\mathbb{C}T_z^*\tilde{\mathcal{M}} = T_z'^{1,0} \oplus T_z'^{0,1}$ ($T_z'^{1,0}$: cotangent structure bundle of the complex manifold $\tilde{\mathcal{M}}$; $T_z'^{0,1} = \overline{T}_z'^{1,0}$). Suppose that, in a neighborhood of the point z, Λ is defined as the set of common zeros of real-valued $\mathcal{C}^\infty$ functions f_j, $j = 1,\ldots,r$, with $\mathrm{d}f_1 \wedge \cdots \wedge \mathrm{d}f_r \neq 0$. Then $\mathcal{N}_z^*\Lambda$ is spanned by $\partial f_1,\ldots,\partial f_r$; keep in mind, however, that the one-forms ∂f_j may not be linearly independent.

If we regard $T_z^{1,0}$ and $T_z'^{1,0}$ as (complex) duals of each other, then $\mathcal{N}_z^*\Lambda$ may be viewed as the orthogonal of $T_z^{1,0} \cap \mathbb{C}T_z\Lambda$. This remark, combined with Proposition I.3.6, yields:

PROPOSITION III.4.1. *Let* Λ_1 *and* Λ_2 *be two submanifolds of* $\tilde{\mathcal{M}}$ *and* z *a point in* $\Lambda_1 \cap \Lambda_2$. *The following two properties are equivalent:*

$$\mathrm{T}_z\tilde{\mathcal{M}} = \mathcal{T}_z\Lambda_1 + \mathcal{T}_z\Lambda_2; \tag{III.4.10}$$

$$\mathcal{N}_z^*\Lambda_1 \cap \mathcal{N}_z^*\Lambda_2 = 0. \tag{III.4.11}$$

DEFINITION III.4.2. *We say that the submanifolds Λ_1 and Λ_2 of $\tilde{\mathcal{M}}$ have a* holomorphic transverse intersection, *or that they are* holomorphic transverse, *at the point $z \in \Lambda_1 \cap \Lambda_2$, if the equivalent properties* (III.4.10), (III.4.11) *hold.*

The implication (III.4.10) $\Rightarrow \mathrm{T}_z\tilde{\mathcal{M}} = \mathrm{T}_z\Lambda_1 + \mathrm{T}_z\Lambda_2$ is trivial, whence:

PROPOSITION III.4.2. *If Λ_1 and Λ_2 are holomorphic transverse at z, they are transverse at that point.*

In general the converse is not true.

EXAMPLE III.4.2. Let the coordinates in $\mathbb{C}^4$ be z_j, w_k ($j, k = 1,2$) and let $\mathcal{M} \subset \mathbb{C}^4$ be defined by the equations $\mathcal{I}m\, w_k = 0$, $k = 1,2$. Let $\mathcal{H}$ be defined by the equations $w_1 + z_1 = 0$, $w_2 + \imath z_1 = 0$. The intersection $\mathcal{M} \cap \mathcal{H}$ is equal to the z_2—plane $z_1 = w_1 = w_2 = 0$. It is transverse but not holomorphic transverse. ■

PROPOSITION III.4.3. *Let $\mathcal{M}$ be a real hypersurface and $\mathcal{H}$ a complex curve* (i.e., a holomorphic submanifold of complex dimension one) *in $\tilde{\mathcal{M}}$. If $\mathcal{M}$ and $\mathcal{H}$ intersect transversally at a point z, they are holomorphic transverse at z.*

PROOF. Suppose that, near z, $\rho = 0$ is a real equation defining $\mathcal{M}$ and $h_1 = \cdots = h_{m-1} = 0$ are holomorphic equations defining $\mathcal{H}$. That $\mathcal{M}$ and $\mathcal{H}$ intersect transversally at z means that, at the point z,

$$dh_1 \wedge \cdots \wedge dh_{m-1} \wedge d\bar{h}_1 \wedge \cdots \wedge d\bar{h}_{m-1} \wedge d\rho \neq 0, \tag{III.4.12}$$

which demands

$$\partial h_1 \wedge \cdots \wedge \partial h_{m-1} \wedge \overline{\partial h}_1 \wedge \cdots \wedge \overline{\partial h}_{m-1} \wedge \partial\rho \neq 0, \tag{III.4.13}$$

since the left-hand side in (III.4.12) is equal to the sum of the left-hand side in (III.4.13) and of the complex conjugate of the latter multiplied by $(-1)^{m-1}$. But (III.4.13) implies trivially

$$\partial h_1 \wedge \cdots \wedge \partial h_{m-1} \wedge \partial\rho \neq 0, \tag{III.4.14}$$

which (at the point z) is a restatement of $\mathcal{N}_z^*\mathcal{H} \cap \mathcal{N}_z^*\mathcal{M} = 0$. ■

PROPOSITION III.4.4. *Let $\mathcal{M}$ be a generic submanifold and $\mathcal{H}$ a holomorphic submanifold of $\tilde{\mathcal{M}}$. Assume that* $\operatorname{codim}_{\mathbb{R}} \mathcal{M} = \dim_{\mathbb{C}} \mathcal{H} = d$ *and let* 0 *be a point in $\mathcal{M} \cap \mathcal{H}$. The following properties are equivalent:*

$\mathcal{M}$ and $\mathcal{H}$ are holomorphic transverse at the point 0; (III.4.15)

$\mathcal{M}$ and $\mathcal{H}$ intersect transversally at 0 *and, in a neighborhood of* 0, *$\mathcal{M}\cap\mathcal{H}$ is a noncharacteristic submanifold of $\mathcal{M}$*; (III.4.16)

there are complex coordinates in some open neighborhood U *of* 0 *in $\tilde{\mathcal{M}}$, $z_1,\ldots,z_n,w_1,\ldots,w_d$, vanishing at* 0 *and such that, in* U, *$\mathcal{H}\cap$U is defined by the equation* $z = 0$ *and $\mathcal{M}\cap$U by the equation* (III.4.7). (III.4.17)

The integer n in Property (III.4.17) is necessarily equal to $m-d$.

PROOF. We may in any case assume that local coordinates z_j and w_k ($1 \le j \le n$, $1 \le k \le d$) can be chosen, in some open neighborhood $\hat{\mathrm{U}}$ of 0 in $\tilde{\mathcal{M}}$, in such a way that $\mathcal{H}\cap\hat{\mathrm{U}}$ is defined, in $\hat{\mathrm{U}}$, by $z = 0$, while $\mathcal{M}$ is defined by real equations $\rho_1 = \cdots = \rho_d = 0$, with $\rho_k \in \mathcal{C}^\infty(\hat{\mathrm{U}};\mathbb{R})$ and $d\rho_1\wedge\cdots\wedge d\rho_d \neq 0$ in $\hat{\mathrm{U}}$. The property that $\mathcal{M}$ and $\mathcal{H}$ are transverse at 0 ensures that, in a neighborhood of that point, $\mathcal{X} = \mathcal{H}\cap\mathcal{M}$ is an embedded submanifold of $\mathcal{M}$; the conormal bundle of $\mathcal{X}$ with respect to $\mathcal{M}$, $\mathrm{N}^*\mathcal{X}$, is spanned, near 0, by the pullbacks to $\mathcal{M}$ of $dx_1,\ldots,dx_n,dy_1,\ldots,dy_n$. To say that $\mathcal{X}$ is characteristic at 0 in $\mathcal{M}$ is to say that there is a linear combination

$$\sum_{j=1}^{n} \gamma_j dz_j + \bar{\gamma}_j d\bar{z}_j \quad (\gamma_j \in \mathbb{C})$$

whose pullback to $\mathcal{M}$ belongs to T_0°, i.e., there are complex numbers A_j, B_j and real numbers C_k such that

$$\sum_{j=1}^{n} (\gamma_j dz_j + \bar{\gamma}_j d\bar{z}_j) - \sum_{k=1}^{d} C_k d\rho_k =$$

$$\sum_{j=1}^{n} A_j dz_j + \sum_{k=1}^{d} B_k dw_k \tag{III.4.18}$$

at the point 0 (this equality is valid in $\tilde{\mathcal{M}}$). Since in $\tilde{\mathcal{M}}$ the dz_i, $d\bar{z}_j$, dw_k, $d\bar{w}_\ell$ are $\mathbb{C}$-linearly independent and since the left-hand side in (III.4.18) is real, this equality cannot hold unless $A_j = B_k = 0$, in which case it reads

$$\sum_{k=1}^{d} C_k \partial\rho_k = \sum_{j=1}^{n} \gamma_j dz_j. \tag{III.4.19}$$

This shows the equivalence of the two properties: $\mathcal{X}$ is noncharacteristic in $\mathcal{M}$ at 0; $\mathcal{M}$ and $\mathcal{H}$ are holomorphic transverse at 0; and therefore it also shows the equivalence of (III.4.15) and (III.4.16).

When (III.4.17) holds we have $\rho_k = \mathcal{I}m\, w_k - \varphi_k(z,\mathcal{R}e\, w)$, $k = 1,\ldots,d$, and it is seen at once that (III.4.17) $\Rightarrow$ (III.4.15). The converse follows from the construction at the beginning of the present section. ■

III.5. Hypocomplex Manifolds. Basic Properties

Certain hypo-analytic manifolds have the remarkable property that each distribution solution is a hypo-analytic function in its domain of definition. We call them *hypocomplex manifolds*. The prime examples of hypocomplex manifolds are the complex manifolds. In dealing with a hypocomplex manifold there is no need to talk of its hypo-analytic structure. The latter is unambiguously determined: hypocomplexity is a feature of the underlying locally integrable structure.

Let the manifold $\mathcal{M}$ be equipped with a locally integrable structure; as usual the structure bundles on $\mathcal{M}$ will be denoted by T' and $\mathcal{V}$. The ranks of T' and $\mathcal{V}$ are called m and n respectively; $\dim \mathcal{M} = m + n$.

DEFINITION III.5.1. *We say that the locally integrable structure of* $\mathcal{M}$ *is* hypocomplex at a point $\mathfrak{p}$, *or that* $\mathcal{M}$ *is hypocomplex at* $\mathfrak{p}$, *if there is an open neighborhood* U *of* $\mathfrak{p}$ *and a* $\mathcal{C}^\infty$ *map* $Z = (Z_1,\dots,Z_m)$: $\mathrm{U} \to \mathbb{C}^m$ *whose components are solutions and such that the following is true:*

$$\begin{array}{l}\textit{to any distribution solution } h \textit{ in some open neighborhood}\\ \textit{of } \mathfrak{p} \textit{ there is a holomorphic function } \tilde{h} \textit{ in some open}\\ \textit{neighborhood of } Z(\mathfrak{p}) \textit{ in } \mathbb{C}^m \textit{ such that } h = \tilde{h}\circ Z \textit{ in some}\\ \textit{open neighborhood of } \mathfrak{p} \textit{ contained in } \mathrm{U}.\end{array} \qquad \text{(III.5.1)}$$

We say that the locally integrable structure of $\mathcal{M}$ *is hypocomplex if it is hypocomplex at every point; we then say that* $\mathcal{M}$ *is a* hypocomplex manifold.

In Definition III.5.1 the differentials $dZ_1,\dots,dZ_m$ must be linearly independent at $\mathfrak{p}$, otherwise there would not be m $\mathcal{C}^\infty$ solutions whose differentials span $\mathrm{T}'_{\mathfrak{p}}$. It is also clear that in (III.5.1) the chart (U,Z) could be replaced by any other chart $(\mathrm{U}^\#,Z^\#)$ consisting of an open neighborhood $\mathrm{U}^\#$ of $\mathfrak{p}$ and of a $\mathcal{C}^\infty$ map $Z^\#$: $\mathrm{U}^\# \to \mathbb{C}^m$ such that $dZ_1^\#,\dots,dZ_m^\#$ span T' over U.

There is a unique hypo-analytic structure on a hypocomplex manifold whose underlying locally integrable structure is the given one on the manifold. In it the sheaf of germs of hypo-analytic functions is equal to the sheaf of germs of solutions (cf. Proposition III.1.1).

Proposition III.5.1. *Every elliptic structure is hypocomplex.*

PROOF. The simplest way to prove the statement is to use hypo-analytic charts of the kind described in section I.7. If the structure is elliptic, we can find coordinates $x_1,\ldots,x_m,y_1,\ldots,y_m,t_1,\ldots,t_n$ in an open neighborhood U of an arbitrary point $\mathfrak{p}$ of $\mathcal{M}$ such that T′ is spanned over U by $dz_1,\ldots,dz_m$ ($z_j = x_j + \imath y_j$; see the remarks that follow (I.7.5)). Then the solutions in U are the functions that are holomorphic with respect to z and independent of t. ■

Later on we will encounter hypocomplex structures that are not elliptic.

PROPOSITION III.5.2. *Assume the manifold $\mathcal{M}$ and its locally integrable structure to be analytic. In order for this structure to be hypocomplex at a point $\mathfrak{p}$ it is necessary and sufficient that every germ of solution at $\mathfrak{p}$ be the germ of an analytic functions at $\mathfrak{p}$.*

Proposition III.5.2 follows at once from Theorem III.1.1.

DEFINITION III.5.2. *Assume that the manifold $\mathcal{M}$ is of class $\mathcal{C}^\infty$ (resp., $\mathcal{C}^\omega$). Let $\mathbb{D}$ be a family of differential operators with $\mathcal{C}^\infty$ (resp., $\mathcal{C}^\omega$) coefficients in an open subset Ω of $\mathcal{M}$. We say that $\mathbb{D}$ is* hypo-elliptic (*resp.*, analytic hypo-elliptic) *at a point $\mathfrak{p}$ of Ω if, given any distribution u in some open neighborhood* U $\subset \Omega$ *of $\mathfrak{p}$, the fact that, for every $\mathcal{L} \in \mathbb{D}$, $\mathcal{L}u$ is a $\mathcal{C}^\infty$(resp., $\mathcal{C}^\omega$) function in some open neighborhood of $\mathfrak{p}$ contained in* U, *implies that u itself is a $\mathcal{C}^\infty$ (resp., $\mathcal{C}^\omega$) function in such a neighborhood.*

It is established PDE terminology to say that the family $\mathbb{D}$ is hypo-elliptic (resp., analytic hypo-elliptic) in Ω if it is so at every point of Ω.

PROPOSITION III.5.3. *Let $\mathcal{M}$ and its locally integrable structure be of class $\mathcal{C}^\omega$. In order for $\mathcal{M}$ to be hypocomplex at a point $\mathfrak{p}$ it is necessary and sufficient that any basis of $\mathcal{V}$ over some open neighborhood of $\mathfrak{p}$, that consists of analytic vector fields* $L_1,\ldots,L_n$, *be analytic hypo-elliptic at $\mathfrak{p}$.*

PROOF. The sufficiency is evident since, if the condition holds, any solution (i.e., any solution of the equations $L_j u = 0, j = 1,\ldots,n$) in an open neighborhood of $\mathfrak{p}$ will automatically be analytic in a suitably small open subneighborhood.

To prove the necessity consider a distribution solution in an open neighborhood U′ ⊂ U of $\mathfrak{p}$ such that $L_j u = f_j$ is analytic in U′ for every $j = 1,\ldots,n$. We can find an analytic solution v of the system of equations $L_j v = f_j$ ($1 \le j \le n$) in some open neighborhood U″ ⊂ U, for instance by applying the Cauchy-Kovalevska theorem or, more simply, by complexifying the base space, extending holomorphically the vector fields L_j and the right-hand sides f_j, then choosing the complex coordinates in such a way that the holomorphic vector fields L_j become the partial derivatives $\partial/\partial z_j$, $j = 1,\ldots,n$. At any rate $u - v$ is a solution in U″ and therefore, if $\mathcal{M}$ is hypocomplex at $\mathfrak{p}$, $u-v$ must be analytic in some neighborhood of $\mathfrak{p}$. The same must be true of u. ■

We shall eventually encounter examples of analytic locally integrable structures that are hypocomplex at one and only one point.

We now return to the general case of a hypo-analytic manifold $\mathcal{M}$ of class $\mathcal{C}^\infty$.

LEMMA III.5.1. *Suppose that $\mathcal{M}$ is hypocomplex at the point $\mathfrak{p}$ and let* (U,Z) *be any hypo-analytic chart such that* $\mathfrak{p} \in$ U. *Then to each open neighborhood $\mathcal{N} \subset$ U of $\mathfrak{p}$ there is an open neighborhood $\mathbb{O}$ of Z($\mathfrak{p}$) in $\mathbb{C}^m$ with the following property:*

Every function $\tilde{h}$ in Z($\mathcal{N}$) such that $\tilde{h}\circ$Z is a $\mathcal{C}^1$ solution in $\mathcal{N}$ can be extended, in a unique manner, as a holomorphic function in $\mathbb{O}$. (III.5.2)

PROOF. Let U′ be an open neighborhood of $\mathfrak{p}$ whose closure is compact and contained in $\mathcal{N}$. We denote by E(U′) the complex vector space consisting of those functions $\tilde{h}$ on Z(U′) such that $\tilde{h}\circ$Z is a $\mathcal{C}^1$ solution in U′ and such, moreover, that $\tilde{h}\circ$Z and its first partial derivatives extend as continuous functions in $\mathcal{C}\ell$ U′. We define the norm of an arbitrary element $\tilde{h}$ in E(U′) as the norm of $\tilde{h}\circ$Z in $\mathcal{C}^1(\mathcal{C}\ell\,$U′). Let $\{\tilde{h}_\nu\}$ (ν = 1,2,...) be a Cauchy sequence for that norm: on the one hand, the $\tilde{h}_\nu$ converge uniformly on Z(U′) to a function $\tilde{h}$; on the other hand, the functions $\tilde{h}_\nu\circ$Z converge in $\mathcal{C}^1(\mathcal{C}\ell\,$U′) to a limit necessarily equal, in U′, to $\tilde{h}\circ$Z. This proves that E(U′) is a Banach space.

Notice that restriction to Z(U′) transforms any function $\tilde{h}$ on Z($\mathcal{N}$) such that $\tilde{h}\circ$Z is a $\mathcal{C}^1$ solution in $\mathcal{N}$, into an element of E(U′). We are going to show that the hypothesis that $\mathcal{M}$ is hypocomplex at the point $\mathfrak{p}$ entails that

there is an open ball $\mathcal{B}$ centered at Z($\mathfrak{p}$) *in $\mathbb{C}^m$ such that, to every $\tilde{h} \in$* E(U′), *there is a unique holomorphic function in $\mathcal{B}$ equal to $\tilde{h}$ on $\mathcal{B}\cap$Z(U′).* (III.5.3)

For r = 1,2,... call Γ_r the subset of E(U′) made up of those functions $\tilde{h}$ that have a holomorphic extension to the open ball $\mathcal{B}_r = \{z \in \mathbb{C}^m;\ |z-Z(\mathfrak{p})| < 1/r\}$ with absolute value $\leq r$. The hypothesis implies that E(U′) is equal to the union of the sets Γ_r as $r \to +\infty$. Let $\{\tilde{h}_\nu\}_{\nu=1,2,\dots}$ be a sequence in Γ_r converging in E(U′) to some function $\tilde{h}$. The Montel theorem allows one to assume, after replacing $\{\tilde{h}_\nu\}_{\nu=1,2,\dots}$ by a subsequence, that the $\tilde{h}_\nu$ have holomorphic extensions to $\mathcal{B}_r$ that converge, uniformly on compact subsets of $\mathcal{B}_r$, to some holomorphic function $\tilde{h}_0$. Clearly $|\tilde{h}_0| \leq r$ in $\mathcal{B}_r$ and $\tilde{h}_0 = \tilde{h}$ on $\mathcal{B}_r\cap$Z(U′). This proves that $\tilde{h} \in \Gamma_r$, and thus that Γ_r is closed. The Baire category theorem demands that the interior of Γ_r be nonempty for some $r < +\infty$. Since Γ_r is clearly convex and stable under the symmetry $\tilde{h} \to -\tilde{h}$, its interior must contain the zero function. This means that every element of E(U′) is of the kind $\rho\tilde{h}$ with $\rho \geq 0$ and $\tilde{h} \in \Gamma_r$.

In order to prove (III.5.3) it remains to show that the extension of $\tilde{h} \in$ E(U′) to $\mathcal{B}_r$ is unique. This follows from the oft-used fact that the image, under the map Z, of any maximally real submanifold of U′ passing through $\mathfrak{p}$, is an immersed maximally real submanifold of $\mathbb{C}^m$ passing through $Z(\mathfrak{p})$. Any two holomorphic functions that agree on it must be equal in a full neighborhood of $Z(\mathfrak{p})$ (Corollary II.3.7) and therefore also in the whole of $\mathcal{B}_r$. ∎

Let us denote by $\mathcal{S}ol_{\mathfrak{p}}(\mathcal{M})$ the space of germs at $\mathfrak{p}$ of distribution solutions in $\mathcal{M}$. As we have done earlier, we call ${}_m\mathcal{O}$ the ring of germs at 0 of holomorphic functions in $\mathbb{C}^m$.

Corollary III.5.1. *Let $\mathcal{M}$ be hypocomplex at $\mathfrak{p}$, and let* (U,Z) *be a hypo-analytic chart in $\mathcal{M}$ centered at $\mathfrak{p}$* (i.e., $\mathfrak{p} \in$ U and Z($\mathfrak{p}$) = 0). *Then the pullback map $\tilde{h} \to \tilde{h}\circ Z$ is a linear bijection of ${}_m\mathcal{O}$ onto $\mathcal{S}ol_{\mathfrak{p}}(\mathcal{M})$.*

Of course, if $\mathcal{M}$ is hypocomplex at $\mathfrak{p}$, $\mathcal{S}ol_{\mathfrak{p}}(\mathcal{M})$ is equal to the ring of germs at $\mathfrak{p}$ of hypo-analytic functions in $\mathcal{M}$, and the pullback map $\tilde{h} \to \tilde{h}\circ Z$ is a ring isomorphism.

Theorem III.5.1. *Let* (Ω,z) *be a hypo-analytic chart in $\mathcal{M}$, centered at a point $\mathfrak{p}$. In order that $\mathcal{M}$ be hypocomplex at $\mathfrak{p}$ it is necessary and sufficient that, to every open neighborhood $\mathcal{N} \subset \Omega$ of $\mathfrak{p}$, there be an open neighborhood $\mathcal{O}$ of $Z(\mathfrak{p})$ in $\mathbb{C}^m$ such that, given any polynomial $P \in \mathbb{C}[z_1,\ldots,z_m]$, any one of the following equivalent properties holds true:*

$$P(\mathcal{O}) \subset P(Z(\mathcal{N})); \tag{III.5.4}$$

if the zero set of P intersects $\mathcal{O}$ it also intersects $Z(\mathcal{N})$; (III.5.5)

$$\sup_{\mathcal{O}}|P| \leq \sup_{Z(\mathcal{N})}|P|. \tag{III.5.6}$$

Proof. The equivalence (III.5.4) ⇔ (III.5.5) is evident: to say that $P(\mathcal{O}) \subset P(Z(\mathcal{N}))$ is the same as saying that to any $z_0 \in \mathcal{O}$ there is $\mathfrak{q} \in \mathcal{N}$ such that $P(z) - P(z_0) = 0$ when $z = Z(\mathfrak{q})$. The entailment (III.5.4) ⇒ (III.5.6) is trivial.

Let $\mathcal{M}$ be hypocomplex at $\mathfrak{p}$. Suppose there were a sequence of points $z_\nu \to Z(\mathfrak{p})$ ($\nu = 1,2,\ldots$) and a sequence of polynomials P_ν such that $P_\nu(z_\nu) \notin P_\nu(Z(\mathcal{N}))$ for every ν. The function $[P_\nu(z) - P_\nu(z_\nu)]^{-1}$ would be holomorphic in some open neighborhood of $Z(\mathcal{N})$. Lemma III.5.1 demands that there be an open neighborhood of $Z(\mathfrak{p})$ to which all those functions extend holomorphically, which is absurd.

Finally, we show that (III.5.6) implies that $\mathcal{M}$ is hypocomplex at $\mathfrak{p}$. We apply Theorem II.3.1 to $\mathfrak{p}$ as the origin and to an open neighborhood $U \subset\subset \Omega$ of $\mathfrak{p}$; we take $\mathcal{N} \subset U_0$. Any $\mathscr{C}^1$ solution h in Ω is the uniform limit of a sequence of solutions $P_\nu \circ Z$ in $\mathcal{N}$. It follows from (III.5.6) that the polynomials P_ν converge uniformly in $\mathcal{O}$, to a holomorphic function $\tilde{h}$; clearly $h = \tilde{h} \circ Z$ in $\overset{-1}{Z}(\mathcal{O}) \cap \mathcal{N}$. To see that every distribution solution has the same property it suffices to apply Theorem II.5.1. ■

In the above statements "polynomial" may be everywhere replaced by "entire function."

Corollary III.5.2. *If* $\mathcal{M}$ *is hypocomplex at* $\mathfrak{p}$, *then whatever the open neighborhood* U *of* $\mathfrak{p}$ *and the nowhere constant solution* $h \in \mathscr{C}^0(U)$, *the set* $h(U)$ *is a neighborhood of* $h(\mathfrak{p})$ *in* $\mathbb{C}$.

Proof. Select a hypo-analytic chart (Ω, Z) centered at $\mathfrak{p}$, with $\Omega \subset U$, and an open neighborhood $\mathcal{N} \subset \Omega$ such that h is the uniform limit, in $\mathcal{N}$, of a sequence of polynomials $P_\nu \circ Z$. Afterward select the open neighborhood $\mathcal{O}$ of $Z(\mathfrak{p})$ as prescribed by Theorem III.5.1; if $\tilde{h}$ is the uniform limit, in $\mathcal{O}$, of the polynomials P_ν, we have $\tilde{h}(\mathcal{O}) \subset h(\mathcal{N})$ by (III.5.4). But $\tilde{h}(\mathcal{O})$ is a neighborhood of $h(\mathfrak{p})$. ■

Corollary III.5.3. *Let* $\mathcal{M}$, $\mathfrak{p}$, (Ω, Z) *be as in Theorem* III.5.1; *but now suppose also that* $m = 1$. *In order for* $\mathcal{M}$ *to be hypocomplex at* $\mathfrak{p}$ *it is necessary and sufficient that the map* Z *be open at* $\mathfrak{p}$.

Proof. Indeed, when $m = 1$, every point is the zero set of some polynomial and thus, in this case, (III.5.5) reads $\mathcal{O} \subset Z(\mathcal{N})$. ■

In the remainder of the section we reason under the hypothesis that

$$\mathcal{M} \text{ is a hypocomplex manifold.} \tag{III.5.7}$$

Let Ω be an open subset of $\mathcal{M}$. We shall denote by $\mathcal{S}ol(\Omega)$ the space of solutions in Ω. By (III.5.7) we have $\mathcal{S}ol(\Omega) \subset \mathscr{C}^\infty(\Omega)$. We equip $\mathcal{S}ol(\Omega)$ with the topology of uniform convergence on the compact subsets of Ω; it is then a Fréchet space, and by the closed graph theorem the natural injection $\mathcal{S}ol(\Omega) \to \mathscr{C}^\infty(\Omega)$ is continuous. It can also be shown that $\mathcal{S}ol(\Omega)$ is a *Montel space* (see, e.g., Treves [2], p. 356). It is an algebra for ordinary multiplication; multiplication is obviously a continuous bilinear map $\mathcal{S}ol(\Omega) \times \mathcal{S}ol(\Omega) \to \mathcal{S}ol(\Omega)$. (Below we refer to this kind of algebra as a Fréchet-Montel algebra.)

Let (U, Z) be a hypo-analytic chart in $\mathcal{M}$. For any point $\mathfrak{p} \in U$ and any $\rho > 0$ we shall use the following notation:

$$\mathcal{B}_\rho(\mathfrak{p}) = \{z \in \mathbb{C}^m;\ |z - Z(\mathfrak{p})| < \rho\}.$$

LEMMA III.5.2. *Let $\mathcal{M}$ be a hypocomplex manifold and let* (U,Z) *be a hypo-analytic chart in $\mathcal{M}$. To each point $\mathfrak{p} \in$ U there is a number $\rho_{\mathfrak{p}} > 0$ such that the following is true:*

$$\text{to each } h \in \mathcal{S}ol(\mathrm{U}) \text{ there is a unique holomorphic function}$$
$$\tilde{h}_{\mathfrak{p}} \text{ in } \mathcal{B}_{\rho_{\mathfrak{p}}}(\mathfrak{p}) \text{ such that } h = \tilde{h}_{\mathfrak{p}} \circ Z \text{ in } \overset{-1}{Z}(\mathcal{B}_{\rho_{\mathfrak{p}}}(\mathfrak{p})). \tag{III.5.8}$$

Follows at once from Lemma III.5.1.

Now define the following open subset of $\mathbb{C}^m$:

$$\hat{\mathrm{U}} = \bigcup_{\mathfrak{p} \in \mathrm{U}} \mathcal{B}_{\rho_{\mathfrak{p}}/4}(\mathfrak{p}). \tag{III.5.9}$$

Of course $Z(\mathrm{U}) \subset \hat{\mathrm{U}}$. We denote by $\mathcal{O}(\hat{\mathrm{U}})$ the Fréchet-Montel algebra of holomorphic functions in $\hat{\mathrm{U}}$ (equipped with the topology of uniform convergence on the compact subsets of $\hat{\mathrm{U}}$).

PROPOSITION III.5.4. *Let $\mathcal{M}$ and* (U,Z) *be as in Lemma* III.5.2. *The pullback map $\tilde{h} \to \tilde{h} \circ Z$ is an isomorphism, for the structures of Fréchet-Montel algebra, of $\mathcal{O}(\hat{\mathrm{U}})$ onto $\mathcal{S}ol(\mathrm{U})$.*

PROOF. Suppose $\mathcal{B}_{\rho_{\mathfrak{p}}/4}(\mathfrak{p}) \cap \mathcal{B}_{\rho_{\mathfrak{p}'}/4}(\mathfrak{p}') \neq \emptyset$; this is only possible if

$$|Z(\mathfrak{p}) - Z(\mathfrak{p}')| < (\rho_{\mathfrak{p}} + \rho_{\mathfrak{p}'})/4.$$

Suppose $\rho_{\mathfrak{p}} \geq \rho_{\mathfrak{p}'}$. Then $\mathcal{B}_{\rho_{\mathfrak{p}}}(\mathfrak{p}) \cap \mathcal{B}_{\rho_{\mathfrak{p}'}}(\mathfrak{p}')$ contains $\mathcal{B}_{\rho_{\mathfrak{p}'}/2}(\mathfrak{p}')$ and thus also $Z(\mathcal{X})$, with $\mathcal{X} \subset$ U a maximally real submanifold of $\mathcal{M}$ such that $\mathfrak{p}' \in \mathcal{X} \subset \overset{-1}{Z}(\mathcal{B}_{\rho_{\mathfrak{p}'}/2}(\mathfrak{p}'))$. Let $\tilde{h}_{\mathfrak{p}} \in \mathcal{O}(\mathcal{B}_{\rho_{\mathfrak{p}}}(\mathfrak{p}))$ be related to $h \in \mathcal{S}ol(\mathrm{U})$ as in Lemma III.5.2; we must have $\tilde{h}_{\mathfrak{p}} = \tilde{h}_{\mathfrak{p}'}$ on $Z(\mathcal{X})$ and therefore also in the whole of $\mathcal{B}_{\rho_{\mathfrak{p}}}(\mathfrak{p}) \cap \mathcal{B}_{\rho_{\mathfrak{p}'}}(\mathfrak{p}')$. This implies that there is a holomorphic function in the union $\mathcal{B}_{\rho_{\mathfrak{p}}}(\mathfrak{p}) \cup \mathcal{B}_{\rho_{\mathfrak{p}'}}(\mathfrak{p}')$, which is equal to $\tilde{h}_{\mathfrak{p}}$ (resp., $\tilde{h}_{\mathfrak{p}'}$) in $\mathcal{B}_{\rho_{\mathfrak{p}}}(\mathfrak{p})$ (resp., $\mathcal{B}_{\rho_{\mathfrak{p}'}}(\mathfrak{p}')$), and therefore that there is $\tilde{h} \in \mathcal{O}(\hat{\mathrm{U}})$, whose restriction to $\mathcal{B}_{\rho_{\mathfrak{p}}/4}(\mathfrak{p})$ is equal to $\tilde{h}_{\mathfrak{p}}$ for each $\mathfrak{p} \in$ U. We have $h = \tilde{h} \circ Z$ in U; and $\tilde{h}$ is unique, by the uniqueness part in Lemma III.5.2. In other words, the pullback map $\tilde{h} \to \tilde{h} \circ Z$ from $\mathcal{O}(\hat{\mathrm{U}})$ to $\mathcal{S}ol(\mathrm{U})$ is a bijection. It is clearly continuous and an algebra homomorphism. It follows from the closed graph theorem that it is a homeomorphism. ■

COROLLARY III.5.4. *Let $\mathcal{M}$ and* (U,Z) *be as in Proposition* III.5.4. *To each compact subset $\tilde{\mathrm{K}}$ of $\hat{\mathrm{U}}$ there is a compact subset* K *of* U *such that*

$$\operatorname*{Max}_{\tilde{\mathrm{K}}} |\tilde{h}| \leq \operatorname*{Max}_{Z(\mathrm{K})} |\tilde{h}|, \ \forall\, \tilde{h} \in \mathcal{O}(\hat{\mathrm{U}}). \tag{III.5.10}$$

PROOF. Since the pullback map $\tilde{h} \to \tilde{h} \circ Z$ is an isomorphism, to each compact subset $\tilde{K}$ of $\hat{U}$ there is a compact subset K of U and a constant $C > 0$ such that, for all $\tilde{h} \in \mathcal{O}(\hat{U})$,

$$\operatorname*{Max}_{\tilde{K}} |\tilde{h}| \leq C \operatorname*{Max}_{Z(K)} |\tilde{h}|.$$

Apply this to $\tilde{h}^k$ ($k = 1,2\ldots$) in the place of $\tilde{h}$ and extract the k-th root of both sides; (III.5.10) follows by letting k go to $+\infty$. ■

One might rephrase the conclusion in Corollary III.5.4 by saying that *each compact subset of* $\hat{U}$ *is contained in the holomorphic convex hull, in* $\hat{U}$, *of the image via Z of some compact subset of* U.

Suppose now that the components Z_j of the map Z are given by (I.7.23) and let M_i be the vector field (I.7.26) ($1 \leq i, j \leq m$).

COROLLARY III.5.5. *Same hypotheses as in Corollary* III.5.4. *To each compact subset* K *of* U *there is another compact subset* K′ *of* U *and a constant* $C > 0$ *such that, whatever the solution h in* U *and* $\alpha \in \mathbb{Z}_+^m$,

$$\operatorname*{Max}_{K} |M^\alpha h| \leq C^{|\alpha|}\, \alpha!\, \operatorname*{Max}_{K'} |h|. \qquad \text{(III.5.11)}$$

PROOF. The image Z(K) is covered by a finite number of open balls $\mathcal{B}_{\rho_p/8}(p)$, $p \in U$. Call $\tilde{K}$ the union of the closures of those balls; $\tilde{K}$ is a compact subset of $\hat{U}$. Corollary III.5.4 allows us to select a compact subset K′ of U such that the holomorphic convex hull of Z(K′) in $\hat{U}$ contains $\tilde{K}$. It suffices then to apply (III.5.10) in conjunction with the Cauchy inequalities. ■

THEOREM III.5.2. *Let* $\mathcal{M}$ *be a hypocomplex manifold and let* Ω *be an open subset of* $\mathcal{M}$. *Given any nowhere constant solution h in* Ω *and any point* $p \in \Omega$, *we have*

$$|h(p)| < \sup_{\Omega} |h|. \qquad \text{(III.5.12)}$$

PROOF. It suffices to prove the assertion when Ω is replaced by some open neighborhood of p. We take U to be the domain of a hypo-analytic chart (U,Z) as before, and apply Proposition III.5.4. Let $\tilde{h} \in \mathcal{O}(\hat{U})$ be such that $h = \tilde{h} \circ Z$ in U. Since $\tilde{h}$ is nonconstant there is a compact neighborhood $\tilde{K}$ of $Z(p)$ in $\hat{U}$ such that

$$|\tilde{h}(Z(p))| < \operatorname*{Max}_{\tilde{K}} |\tilde{h}|.$$

It suffices then to apply Corollary III.5.4. ■

COROLLARY III.5.6. *If $\mathcal{M}$ is a compact and connected hypocomplex manifold every solution in $\mathcal{M}$ is a constant function.*

Theorem III.5.2 is the *maximum principle* for solutions in the hypocomplex manifold $\mathcal{M}$.

In general, without the hypothesis that $\mathcal{M}$ is hypocomplex, the maximum principle is not valid, as shown in the following:

EXAMPLE III.5.1. Consider the *Lewy structure* on $\mathbb{R}^3$ (coordinates: x, y, s). It is the hypo-analytic structure defined by the single chart $(\mathbb{R}^3, z, w)$ where $z = x + \imath y$ and

$$w = s + \imath|z|^2. \tag{III.5.13}$$

Set $g = w + \imath w^2 = s(1+2|z|^2) + \imath(|z|^2 + s^2 - |z|^4)$. Then $e^{\imath g}$ is a hypo-analytic function in $\mathbb{R}^3$ and has a strict local maximum at the origin. ∎

Actually, when $\mathcal{M}$ is hypocomplex, the maximum principle also applies to the *real parts* of solutions.

THEOREM III.5.3. *Let $\mathcal{M}$ be a hypocomplex manifold and let Ω be an open subset of $\mathcal{M}$. Given any nowhere constant solution h in Ω, we have*

$$\inf_{\Omega} \mathcal{R}e\, h < \mathcal{R}e\, h(\mathfrak{p}) < \sup_{\Omega} \mathcal{R}e\, h, \ \forall\ \mathfrak{p} \in \Omega. \tag{III.5.14}$$

PROOF. If (III.5.14) were not true, $h(\Omega)$ would not be an open subset of $\mathbb{C}$, contradicting Corollary III.5.2. Another proof derives (III.5.14) directly from (III.5.12) by substituting $e^{\pm h}$ for h in the latter inequality. ∎

III.6. Two-Dimensional Hypocomplex Manifolds

Throughout this section we shall reason under the hypothesis that

$$\dim \mathcal{M} = 2. \tag{III.6.1}$$

If $\mathcal{M}$ is a hypocomplex manifold (Definition III.5.1), the rank of the cotangent structure bundle T', m, cannot be equal to *two*. More generally, no maximal hypo-analytic structure (Definition III.1.4) can be hypocomplex (regardless of the dimension of the base manifold—provided the dimension is ≥ 1). For in such a structure any distribution in an open subset Ω of $\mathcal{M}$ is a solution in Ω.

When $T' = 0$ one could say that $\mathcal{M}$ is hypocomplex, but this is obviously a case devoid of interest. We may therefore limit ourselves to reasoning under the following hypothesis:

$$\text{\textit{The tangent and cotangent structure bundles of } } \mathcal{M} \text{ \textit{are complex line bundles.}} \tag{III.6.2}$$

THEOREM III.6.1. *Let $\mathcal{M}$ be a two-dimensional hypo-analytic manifold such that* (III.6.2) *holds. In order for $\mathcal{M}$ to be hypocomplex it is necessary and sufficient that each point $\mathfrak{p} \in \mathcal{M}$ have an open neighborhood mapped homeomorphically onto an open subset of $\mathbb{C}$ by a hypo-analytic function.*

PROOF. I. *Sufficiency of the condition.* Let h be a hypo-analytic function in some open neighborhood U of $\mathfrak{p}$ such that $h : \mathrm{U} \to h(\mathrm{U}) \subset \mathbb{C}$ is a homeomorphism. After contracting U about $\mathfrak{p}$ we may assume that there is a hypo-analytic function Z in U such that $\mathrm{dZ} \neq 0$ at every point of U. After further contracting of U we may assume that there is a holomorphic function $\tilde{h}$ in some open neighborhood of Z(U) in $\mathbb{C}$ such that $h = \tilde{h} \circ \mathrm{Z}$. But, then, if h is a homeomorphism so is Z. By Corollary III.5.3 we conclude that $\mathcal{M}$ is hypocomplex at the point $\mathfrak{p}$.

II. *Necessity of the condition.* Again by Corollary III.5.3 we know that if (U,Z) is a hypo-analytic chart in the hypocomplex manifold $\mathcal{M}$, the map $\mathrm{Z} : \mathrm{U} \to \mathbb{C}^1$ must be open. We claim that Z will be injective, after we contract U if need be. Indeed, take U to be the domain of local coordinates x and y such that $\mathrm{Z} = x + \imath\Phi(x,y)$; assume U is a square $\mathcal{I} \times \mathcal{I}$ in (x,y)-space. If there were points $x_0 \in \mathcal{I}$, $y_0, y_0' \in \mathcal{I}$, $y_0 < y_0'$ such that $\Phi(x_0,y_0) = \Phi(x_0,y_0')$ then Φ would have an extremum (not necessarily strict) at a point $y_1 \in \,]y_0,y_0'[$. There would be a neighborhood $\mathrm{U}_1 \subset \mathrm{U}$ of (x_0,y_1) such that $\mathrm{Z}(\mathrm{U}_1)$ would not intersect at least one of the two vertical segments originating at $\mathrm{Z}(x_0,y_1) = x_0 + \imath\Phi(x_0,y_1)$, contrary to the fact that Z is open. ∎

Thus, if (III.6.1) and (III.6.2) hold and if the manifold $\mathcal{M}$ is hypocomplex, there are hypo-analytic charts (U,Z) in $\mathcal{M}$ such that $\mathrm{Z} : \mathrm{U} \to \mathrm{Z}(\mathrm{U})$ is a homeomorphism and whose domains U cover $\mathcal{M}$. They define on $\mathcal{M}$ the structure of a *Riemann surface* (i.e., a one-dimensional complex manifold). In particular $\mathcal{M}$ must be *orientable*, and even more: there is a preferred orientation on $\mathcal{M}$, namely the orientation defined in any hypo-analytic chart (U,Z) by the two-form $\mathrm{d}\overline{\mathrm{Z}} \wedge \mathrm{dZ}/2\imath$. Although this two-form might vanish in U it has a well-defined sign there (cf. proof of Theorem III.6.2).

Two hypocomplex manifolds can be distinct while being isomorphic as Riemann surfaces.

EXAMPLE III.6.1. Let $\mathcal{M} = \mathbb{R}^2$ be equipped with the hypo-analytic structure defined by the function $Z = x + \imath y^3$; Z is a homeomorphism of the plane onto itself. According to Theorem III.6.1, $\mathcal{M}$ is a hypocomplex manifold; as a Riemann surface it is isomorphic to the complex plane. ■

Let $\mathcal{M}$ be a two-dimensional manifold equipped with an involutive structure such that (III.6.2) holds. We do not assume that $\mathcal{M}$ is hypocomplex, but we do assume that $\mathcal{M}$ is *orientable*, i.e., there exists a $\mathcal{C}^\infty$ *real* two-form Θ in $\mathcal{M}$ that does not vanish at any point. Then let ϖ be a $\mathcal{C}^\infty$ section of T' over some open set $\Omega \subset \mathcal{M}$, nowhere zero in Ω. There is a smooth, real-valued function g in Ω such that

$$\varpi \wedge \bar{\varpi}/2\imath = g\Theta \tag{III.6.3}$$

in Ω. The sign of g does not change if we multiply ϖ by a complex-valued $\mathcal{C}^\infty$ function in Ω. If $g \geq 0$ in Ω, we say that $\mathrm{T}'\wedge\overline{\mathrm{T}}'/2\imath$ is *nonnegative relative to* Θ. The property that g does not change sign in Ω remains valid if we replace Θ by any other nowhere zero, smooth real two-form: it is a property of the involutive structure of T'. We express it by saying that $\mathrm{T}'\wedge\overline{\mathrm{T}}'/2\imath$ *does not change sign in* Ω.

DEFINITION III.6.1. *Suppose* (III.6.2) *holds and that* $\mathcal{M}$ *is orientable. We shall say that the involutive structure of* $\mathcal{M}$ does not change sign in $\mathcal{M}$ *if there is a nowhere zero, real-valued* $\mathcal{C}^\infty$ *two-form* Θ *in* $\mathcal{M}$ *such that every point of* $\mathcal{M}$ *has an open neighborhood in which* $\mathrm{T}'\wedge\overline{\mathrm{T}}'/2\imath$ *is nonnegative relative to* Θ.

If the involutive structure of an orientable hypo-analytic manifold $\mathcal{M}$ (satisfying (III.6.2)) is *real* we shall have $\varpi\wedge\bar{\varpi} \equiv 0$ whatever the section ϖ of T' and thus $\mathrm{T}'\wedge\overline{\mathrm{T}}'/2\imath$ will not change sign in $\mathcal{M}$.

THEOREM III.6.2. *Let* $\mathcal{M}$ *be a two-dimensional manifold. In order that an involutive structure on* $\mathcal{M}$*, in which the ranks of the structure bundles* T' *and* $\mathcal{V}$ *are equal to one, underlie a hypocomplex structure on* $\mathcal{M}$ *it is necessary and sufficient that the following two conditions be satisfied:*

the pullback of T' *to any smooth curve* (i.e., any embedded one-dimensional submanifold) *cannot vanish identically;* (III.6.4)

$\mathcal{M}$ *is orientable and* $\mathrm{T}'\wedge\overline{\mathrm{T}}'/2\imath$ *does not change sign in* $\mathcal{M}$. (III.6.5)

PROOF. I. *Necessity of the conditions.* Suppose $\mathcal{M}$ is hypocomplex. Let (U,Z) be any hypo-analytic chart. If the pullback of dZ to a curve $\gamma \subset$ U were to vanish identically $Z(\gamma)$ would be a single point, contradicting Theorem III.6.1. This proves (III.6.4).

Now let (U,Z) be the chart in Part II of the proof of Theorem III.6.1. Then we may put $\varpi = \mathrm{d}Z = \mathrm{d}x + \imath\mathrm{d}\Phi$ in (III.6.3); we get

$$\mathrm{d}Z\wedge\mathrm{d}\bar{Z}/2\imath = \Phi_y\mathrm{d}x\wedge\mathrm{d}y. \tag{III.6.6}$$

As we have seen in the proof of Theorem III.6.1 the function $y \to \Phi(x,y)$ in $\mathcal{J}$ must be either strictly increasing or strictly decreasing whatever $x \in \mathcal{J}$. It follows at once that Φ_y cannot vanish on any open interval contained in $\{x_0\}\times\mathcal{J}$ and must keep the same sign throughout the square $\mathcal{J}\times\mathcal{J}$ (= U).

Among other things this shows that $\mathrm{d}Z\wedge\mathrm{d}\bar{Z}/2\imath$ does not vanish in an open and dense subset of U. On the other hand, let ($\mathrm{U}^\#$,$Z^\#$) be another hypo-analytic chart like (U,Z), such that $\mathrm{U}\cap\mathrm{U}^\# \neq \emptyset$; then in $\mathrm{U}\cap\mathrm{U}^\#$,

$$\mathrm{d}Z^\#\wedge\mathrm{d}\bar{Z}^\# = f\,\mathrm{d}Z\wedge\mathrm{d}\bar{Z},$$

with $f > 0$ at every point of $\mathrm{U}\cap\mathrm{U}^\#$. It follows from this that we may use the real two-forms $\mathrm{d}Z\wedge\mathrm{d}\bar{Z}/2\imath$ to define an orientation on $\mathcal{M}$, and that $\mathrm{T}'\wedge\bar{\mathrm{T}}'/2\imath$ does not change sign in $\mathcal{M}$.

II. *Sufficiency of the conditions*. We shall begin by reasoning under the hypothesis that the involutive structure of $\mathcal{M}$ is locally integrable, in the following sense:

To every integer $r \geq 1$ and every point $\mathfrak{p}$ of $\mathcal{M}$ there is an open neighborhood U_r of $\mathfrak{p}$ in which is defined a $\mathcal{C}^r$ function Z_r such that $\mathrm{d}Z_r$ spans T' over U_r. (III.6.7)

Let U_r and Z_r be as in (III.6.7). After contracting U_r about $\mathfrak{p}$ and multiplying Z_r by a complex number, we may select $\mathcal{C}^r$ coordinates x and y in U_r such that $Z_r = x + \imath\Phi(x,y)$. It is convenient to take U_r to be a rectangle $\mathcal{J}\times\mathcal{J}$. Since $\mathrm{d}Z_r$ spans T' and $\mathrm{T}'\wedge\bar{\mathrm{T}}'$ is orientable over U, we see that Φ_y keeps the same sign in U; since the pullback of $\mathrm{d}Z_r$ to any segment $x = x_0 \in \mathcal{J}$, $y \in [y_1,y_2] \subset \mathcal{J}$ (with $y_1 < y_2$) cannot vanish identically, we see that the function $y \to \Phi(x_0,y)$ is either strictly increasing for all $x_0 \in \mathcal{J}$, or else it is strictly decreasing for all $x_0 \in \mathcal{J}$. We get at once that the map Z_r is a homeomorphism of U_r onto an open subset $\hat{\mathrm{U}}_r$ of $\mathbb{C}$. Now consider any $\mathcal{C}^1$ solution h in U_r, and let $\tilde{h}$ denote its pushforward to $\hat{\mathrm{U}}_r$ under the map Z_r. We apply the $\mathcal{C}^1$ version of the approximation formula, Theorem II.3.4, and conclude that, locally in U_r, h is the uniform limit of a sequence of polynomials $\mathcal{P}_\nu(Z_r)$. It follows that, locally in $\hat{\mathrm{U}}_r$, $\tilde{h}$ is the uniform limit of the sequence of polynomials $\mathcal{P}_\nu(z)$, whence it follows that $\tilde{h}$ is holomorphic in $\hat{\mathrm{U}}_r$. Since $h = \tilde{h}\circ Z_r$ we see that h is a $\mathcal{C}^r$ solution. Since the integer $r \geq 1$ is arbitrary, we conclude that any $\mathcal{C}^1$ solution in an open subset Ω of $\mathcal{M}$ is in fact a $\mathcal{C}^\infty$ function in Ω. This is true, in particular, for Z_r. By once again using the local structure of distribution solutions (Theorem II.5.1) we also conclude that any distribution solution in U_r is a holomorphic function of Z_r.

Thus Theorem III.6.2 will be proved if we prove the following "weak local integrability" result:

LEMMA III.6.1. *Suppose dim* $\mathcal{M} = 2$ *and the involutive structure of* $\mathcal{M}$, *in which the structure bundles* T′ *and* $\mathcal{V}$ *have rank one, satisfies* (III.6.5). *Then* (III.6.7) *is true.*

This result can be deduced from results in section VIII.8, but since its proof is fairly simple we give it here.

PROOF. Let (U,x,y) be a coordinate patch in $\mathcal{M}$ and $\varpi = Adx + Bdy$ a $\mathcal{C}^\infty$ one-form in U that spans T′ over U. For simplicity assume that x and y vanish at a point $0 \in U$ (which we call the origin) and $A \neq 0$ at 0. After contracting U if needed we may even assume that $A \neq 0$ at every point of U. There is a real-valued $\mathcal{C}^\infty$ function μ, nowhere zero in a neighborhood of 0, which we take once again to be U, such that $\mu[dx + \mathcal{Re}(B/A)dy]$ is exact, i.e., is equal to dx' with x' also vanishing at (0,0). In other words, a change of coordinates brings us into the situation where T′ is spanned over U by the form $\varpi = dx + \imath b(x,y)dy$ with b real-valued. Then (III.6.5) implies

$$b \text{ does not change sign in } U. \tag{III.6.8}$$

Over U the tangent structure bundle $\mathcal{V}$ is generated by

$$L = \partial/\partial y - \imath b(x,y)\partial/\partial x. \tag{III.6.9}$$

It is convenient to take U to be a square: $U = \{(x,y) \in \mathbb{R}^2;\ |x| < \rho,\ |y| < \rho\}$ $(\rho > 0)$. Lemma III.6.1 will be a consequence of

LEMMA III.6.2. *If* (III.6.8) *and* (III.6.9) *hold, then to each integer* $r \geq 0$ *there is an open square* $U_r = \{(x,y) \in \mathbb{R}^2;\ |x| < \rho_r,\ |y| < \rho_r\} \subset U$ $(0 < \rho_r < \rho)$ *such that*

$$\forall f \in \mathcal{C}^\infty(U),\ \exists\, u \in \mathcal{C}^r(U_r) \text{ such that } Lu = f \text{ in } U_r. \tag{III.6.10}$$

PROOF THAT LEMMA III.6.2 ENTAILS LEMMA III.6.1. Lemma III.6.2 implies that there is a solution $u \in \mathcal{C}^r(U_r)$ of the equation in U_r,

$$Lu = \imath b_x \tag{III.6.11}$$

(subscripts stand for partial derivatives); (III.6.11) may be rewritten as

$$(e^u)_y - \imath(be^u)_x = 0. \tag{III.6.12}$$

If we then define

$$Z_r(x,y) = \int_0^x e^{u(s,y)}ds + \imath \int_0^y b(0,t)e^{u(0,t)}dt,$$

it is an immediate consequence of (III.6.12) that $LZ_r = 0$ in U_r. On the other hand, $(\partial/\partial x)Z_r = e^u \neq 0$ at every point of U_r. ■

PROOF OF LEMMA III.6.2. It is based on the results in the Appendix to the present section and on the following:

LEMMA III.6.3. *Same hypotheses as in Lemma* III.6.2. *There is $C > 0$ such that, for every suitably small $\varepsilon > 0$, if $\mathcal{R}_\varepsilon = \{ (x,y) \in \mathbb{R}^2; |x| < \rho, |y| < \varepsilon \} \subset$ U, then*

$$\|u\|_0 \leq C\varepsilon\|{}^t\mathrm{L}u\|_0,\ \forall\ u \in \mathcal{C}_c^\infty(\mathcal{R}_\varepsilon). \tag{III.6.13}$$

Here $\|\cdot\|_0$ is the norm in $L^2(\mathbb{R}^2)$ and ^{t}L is the transpose of L. The inner product in $L^2(\mathbb{R}^2)$ will be denoted by (,).

We combine Lemma III.6.3 with Lemmas III.6.A2 and III.6.A3 (see Appendix to present section) to reach the sought conclusion.

PROOF OF LEMMA III. 6.3. We shall assume $b \geq 0$ everywhere in U. Introduce the following pseudodifferential operators of order zero on the real line:

$$\mathbf{P}^+v(x) = \int_0^{+\infty} e^{\imath x\xi}\, \hat{v}(\xi)d\xi/2\pi,\ \mathbf{P}^-v(x) = \int_{-\infty}^{0} e^{\imath x\xi}\, \hat{v}(\xi)d\xi/2\pi,$$

where $v \in \mathcal{C}_c^\infty(\mathbb{R})$ and $\hat{v}$ is the Fourier transform of v; $\mathbf{P}^+$ and $\mathbf{P}^-$ are orthogonal projections in $\mathbf{L}^2(\mathbb{R})$, such that $\mathbf{P}^+ + \mathbf{P}^- = \mathrm{Id}$. We define

$$\mathbf{P}_\lambda = e^{-\lambda y}\mathbf{P}^+ - e^{\lambda y}\mathbf{P}^- \ (\lambda > 0).$$

Given any $u \in \mathcal{C}_c^\infty(\mathrm{U})$ consider

$$({}^t\mathrm{L}u,\mathbf{P}_\lambda u) = -(u_y,\mathbf{P}_\lambda u) + \imath((bu)_x,\mathbf{P}_\lambda u).$$

By exploiting the fact that $\mathbf{P}^\pm = (\mathbf{P}^\pm)^2$ commutes with $\partial/\partial y$ we get

$$(u_y,\mathbf{P}_\lambda u) = ((\mathbf{P}^+u)_y,e^{-\lambda y}\mathbf{P}^+u) - ((\mathbf{P}^-u)_y,e^{\lambda y}\mathbf{P}^-u).$$

Integration by parts yields, for any $v \in \mathcal{C}_c^\infty(\mathbb{R})$ and any $\alpha \in \mathbb{R}$,

$$2\mathcal{R}e\int v'(y)\bar{v}(y)e^{\alpha y}dy = -\alpha\int |v(y)|^2e^{\alpha y}dy,$$

whence

$$-2\mathcal{R}e({}^t\mathrm{L}u,\mathbf{P}_\lambda u) = \lambda\left\{\iint e^{-\lambda y}|\mathbf{P}^+u|^2dxdy + \iint e^{\lambda y}|\mathbf{P}^-u|^2dxdy\right\} + 2\mathcal{R}e(bu,\mathrm{D}_x\mathbf{P}_\lambda u), \tag{III.6.14}$$

where $\mathbf{D}_x = -\imath\partial/\partial x$. Set $\mathbf{D}^+ = \mathbf{D}_x\mathbf{P}^+$, $\mathbf{D}^- = -\mathbf{D}_x\mathbf{P}^-$; $\mathbf{D}^+$ and $\mathbf{D}^-$ are classical pseudodifferential operators of order *one* on the real line, both positive convolution operators on $L^2(\mathbb{R})$. Their square roots are readily defined:

$$(\mathbf{D}^+)^{1/2}v(x) = \int_0^{+\infty} e^{\imath x\xi}|\xi|^{1/2}\hat{v}(\xi)d\xi/2\pi,$$

and likewise for $(\mathbf{D}^-)^{1/2}$. We may write

$$(bu,\mathbf{D}_x\mathbf{P}_\lambda u) = (e^{-\lambda y}(\mathbf{D}^+)^{1/2}(bu),(\mathbf{D}^+)^{1/2}u) + (e^{\lambda y}(\mathbf{D}^-)^{1/2}(bu),(\mathbf{D}^-)^{1/2}u).$$

But $(\mathbf{D}^\pm)^{1/2}[(\mathbf{D}^\pm)^{1/2},b]$ is a pseudodifferential operator of order *zero* and, as a consequence, there is $C > 0$ such that

$$2\,\mathcal{R}e(bu,\mathbf{D}_x\mathbf{P}_\lambda u) \geq 2\left\{(e^{-\lambda y}b(\mathbf{D}^+)^{1/2}u,(\mathbf{D}^+)^{1/2}u) + (e^{\lambda y}b(\mathbf{D}^-)^{1/2}u,(\mathbf{D}^-)^{1/2}u)\right\} - C\iint cosh(\lambda y)|u|^2dxdy$$

(for a more precise result see Lemma VIII.8.1). Since $b \geq 0$ we get

$$2\mathcal{R}e(bu,\mathbf{D}_x\mathbf{P}_\lambda u) \geq -C\iint cosh(\lambda y)|u|^2dxdy. \qquad \text{(III.6.15)}$$

Require $|y| < 1/\lambda$ in $\mathcal{R}_\varepsilon$. Combining (III.6.14) and (III.6.15) yields

$$-2\mathcal{R}e({}^t\mathrm{L}u,\mathbf{P}_\lambda u) \geq (\lambda/e - Ce)\iint |u|^2dxdy, \qquad \text{(III.6.16)}$$

$$\iint |\mathbf{P}_\lambda u|^2dxdy \leq 2e^2\iint |u|^2dxdy. \qquad \text{(III.6.17)}$$

By the Cauchy-Schwarz inequality, requiring $2Ce^2 \leq \lambda \leq 1/\varepsilon$ yields (III.6.13). ■

The lemmas used to prove Theorem III.6.2 lead to another interesting characterization of the two-dimensional hypocomplex manifolds.

THEOREM III.6.3. *Let $\mathcal{M}$ be a two-dimensional manifold. In order for a line bundle $\mathcal{V} \subset \mathbb{C}T\mathcal{M}$ to be the tangent structure bundle in a hypocomplex structure on $\mathcal{M}$ it is necessary and sufficient that each point of $\mathcal{M}$ have an open neighborhood in which $\mathcal{V}$ is spanned by a $\mathcal{C}^\infty$ vector field that is, and whose transpose is, hypo-elliptic* (Definition III.5.2).

PROOF. Let Ω be an open subset of $\mathcal{M}$ in which there is a nowhere vanishing $\mathcal{C}^\infty$ section L of $\mathcal{V}$. We reason in the neighborhood of an arbitrary point $\mathfrak{p}$ of Ω.

Suppose the structure of $\mathcal{M}$ is hypocomplex. We may assume that the situation is the same as in the proof of Theorem III.6.2: p is the center of a local coordinate patch (U,x,y), with $U \subset \Omega$; in U L is given by (III.6.9) and (III.6.8) holds. If $u_0 \in \mathcal{D}'(\Omega)$ and $Lu_0 \in \mathcal{C}^\infty(\Omega)$, then, by Lemma III.6.2, given any integer $r \geq 0$, in some open neighborhood $U_r \subset U$ of 0, there is a $\mathcal{C}^r$ function u such that $Lu = Lu_0$. The solution $u - u_0$ belongs to $\mathcal{C}^\infty(U_r)$ by hypocomplexity, and therefore $u_0 \in \mathcal{C}^r(U_r)$. Repeating the same reasoning at every point of Ω shows that $u_0 \in \mathcal{C}^r(\Omega)$ and, since r is as large as we wish, $u_0 \in \mathcal{C}^\infty(\Omega)$. This means that L is hypo-elliptic.

Possibly after contracting U we may assume that there is $Z \in \mathcal{C}^\infty(U)$ such that $LZ = 0$, $dZ \neq 0$. In view of (III.6.9) this demands $Z_x \neq 0$ and $b = -\imath Z_y/Z_x$; therefore $Z_x L = Z_x\partial/\partial y - Z_y\partial/\partial x$. The transpose of $Z_x L$ is equal to $-Z_x L$ and we know that the latter is hypo-elliptic.

Suppose now ${}^t L$ is hypo-elliptic. We apply Lemma III.6.A4 and then Lemma III.6.A3. We conclude that each point p of Ω has an open neighborhood $U \subset \Omega$ such that to any $f \in \mathcal{C}^\infty(\Omega)$ there is a distribution u in U solution of $Lu = f$ in U. If then we suppose that L is hypo-elliptic, we must have $u \in \mathcal{C}^\infty(U)$. We may repeat the argument in the proof that Lemma III.6.2 $\Rightarrow$ Lemma III.6.1. We conclude that there is a function $Z \in \mathcal{C}^\infty(U)$ such that $dZ \neq 0$ and $LZ = 0$ in U. We may select local coordinates x, y in U such that $Z = x + \imath\Phi(x,y)$, with $x = y = Z = 0$ at p. The image $Z(\mathcal{Q})$ of $\mathcal{Q} = \{ (x,y) \in \mathbb{R}^2; |x| < \delta, |y| < \delta \} \subset U$ is a union of "vertical segments" $\mathcal{R}e z = const.$, $\rho_1 < \mathcal{I}m z < \rho_2$, and thus, if $Z(\mathcal{Q})$ contains some segment $\mathcal{R}e z = 0$, $-\rho < \mathcal{I}m\, z < \rho$ $(\rho > 0)$, then perforce $Z(\mathcal{Q})$ is a neighborhood of 0. If Z were not open at p there would be an open neighborhood $U' \subset \mathcal{Q}$ of p such that $Z(U')$ would not intersect at least one of the two vertical half-lines $\mathcal{R}e z = 0$, $\mathcal{I}m z > 0$ or $\mathcal{I}m z < 0$. But then there would be a branch of the square-root function defined on $Z(U')$; its pullback to U' under the map Z would be a nonsmooth solution, contradicting the hypo-ellipticity of L. ■

REMARK III.6.1. It can be proved that if L is a $\mathcal{C}^\infty$ vector field in an open subset $\Omega \subset \mathbb{R}^2$, nowhere zero in Ω, it is equivalent to say that L is hypo-elliptic or that ${}^t L$ is hypo-elliptic in Ω (see Treves [4]). It follows from this and from Theorem III.6.3 that the line bundle $\mathcal{V} \subset \mathbb{C}T\mathcal{M}$ is the tangent structure bundle in a hypocomplex structure on $\mathcal{M}$ if and only if the family of all $\mathcal{C}^\infty$ sections of $\mathcal{V}$ is hypo-elliptic. ■

Appendix to Section III.6: Some Lemmas about First-Order Differential Operators

We shall deal with a first-order differential operator in an open neighborhood Ω of the origin in $\mathbb{R}^{m+1}$ (coordinates: $x_1,\ldots,x_m,t$), of the following kind:

$$\mathcal{L} = \partial/\partial t + \sum_{j=1}^{m} c_j(x,t)\partial/\partial x_j + c_0(x,t).$$

We assume that all the coefficients c_0, c_j $(1 \leq j \leq m)$ are smooth in the open set Ω.

LEMMA III.6.A1. *Any distribution h in Ω that is a solution of the equation $\mathcal{L}h = 0$ and whose support is contained in the submanifold $S_0 = \{(x,t) \in \Omega; t = 0\}$ vanishes identically.*

PROOF. If supp $h \subset S_0$ then h is a linear combination of derivatives of the Dirac distribution $\delta(t)$ with coefficients that are distributions with respect to x:

$$h = \sum_{\nu=0}^{N} h_\nu(x) \otimes \delta^{(\nu)}(t),$$

whence

$$\mathcal{L}h = \sum_{\nu=0}^{N+1} g_\nu(x) \otimes \delta^{(\nu)}(t).$$

We have $\mathcal{L}h = 0$ if and only if each coefficient g_ν vanishes identically. But $g_{N+1} = h_N$; descending induction on N shows that $h \equiv 0$. ■

We denote by Δ the Laplace operator in (x,t)—space, and by $\| \ \|_s$ the standard Sobolev norm in the Sobolev space $H^s = H^s(\mathbb{R}^{m+1})$ $(s \in \mathbb{R})$:

$$\|u\|_s = \|(1 - \Delta)^{s/2}u\|_{L^2(\mathbb{R}^{m+1})}.$$

LEMMA III.6.A2. *Suppose that to each $\varepsilon > 0$ there is an open ball $\mathcal{B}_\varepsilon \subset \Omega$ centered at the origin, such that*

$$\|u\|_0 \leq \varepsilon \|\mathcal{L}u\|_0, \ \forall \ u \in \mathcal{C}_c^\infty(\mathcal{B}_\varepsilon). \tag{III.6.A1}$$

Then, to each real number s there is an open neighborhood of the origin, $U_s \subset \Omega$, and a constant $C_s > 0$ such that

$$\|u\|_s \leq C_s \|\mathcal{L}u\|_s, \ \forall \ u \in C_c^\infty(U_s). \tag{III.6.A2}$$

PROOF. We shall extend the coefficients of $\mathcal{L}$ to the whole space $\mathbb{R}^{m+1}$ as compactly supported $\mathcal{C}^\infty$ functions; this can always be achieved without modifying them in a neighborhood of the origin. We keep ε fixed in (III.6.A1) (with $0 < \varepsilon \leq 1$) and take the radius of the ball $\mathcal{B}_\varepsilon$ to be a monotone increasing function of ε, equal to zero when $\varepsilon = 0$. Select $g \in \mathcal{C}_c^\infty(\mathcal{B}_\varepsilon)$ equal to 1 in some open neighborhood Ω_ε of 0 and $g_1 \in \mathcal{C}_c^\infty(\Omega_\varepsilon)$ equal to 1 in an open neighborhood of 0, $U_s' \subset\subset \Omega_\varepsilon$. Put, in (III.6.A1), $u = g(1 - \Delta)^{s/2}v$, $v \in \mathcal{C}_c^\infty(U_s')$. Since $gv \equiv g_1v \equiv v$, we have

$$u = (1 - \Delta)^{s/2}v - \mathcal{R}_\varepsilon v,$$

where $\mathcal{R}_\varepsilon v = (1 - \Delta)^{s/2}(gg_1 v) - g(1 - \Delta)^{s/2}(g_1 v)$; $\mathcal{R}_\varepsilon$ is a pseudodifferential operator of order $-\infty$ (its total symbol is clearly identically equal to zero since $g_1 dg \equiv 0$); in other words, $\mathcal{R}_\varepsilon$ is smoothing. We have

$$\mathcal{L}u = (1 - \Delta)^{s/2}\mathcal{L}v - [(1 - \Delta)^{s/2},\mathcal{L}]v - \mathcal{L}\mathcal{R}_\varepsilon v.$$

Since $[(1 - \Delta)^{s/2},\mathcal{L}]$ is a pseudodifferential operator of order $\leq s$ (independent of ε), we derive from (III.6.A1):

$$\|v\|_s \leq \varepsilon\|\mathcal{L}v\|_s + C_0\varepsilon\|v\|_s + C(\varepsilon)\|v\|_{s-1}$$

with $C_0 > 0$ independent of ε of v, and $C(\varepsilon) > 0$ independent of v. If we take $\varepsilon < 1/2C_0$ we get

$$\|v\|_s \leq 2\varepsilon\|\mathcal{L}v\|_s + 2C(\varepsilon)\|v\|_{s-1}, \ \forall\, v \in C_c^\infty(U'_s). \qquad \text{(III.6.A3)}$$

Suppose the conclusion in Lemma III.6.A2 were not true for a given s. There would be a sequence of functions v_ν such that $|x|^2 + t^2 < 1/\nu$ on supp v_ν ($\nu = 1,2,\ldots$) and such that

$$\|v_\nu\|_s = 1 \text{ for all } \nu,\ \|\mathcal{L}v_\nu\|_s \to 0 \text{ as } \nu \to +\infty. \qquad \text{(III.6.A4)}$$

For any compact $K \subset \mathbb{R}^{m+1}$ call $H^s(K)$ the (closed) subspace of H^s consisting of those distributions whose support is contained in K. The natural embedding $H^s(K) \to H^{s-1}(K)$ is compact (see, e.g., Treves [5], Proposition 25.5). Therefore, possibly after replacing the sequence $\{v_\nu\}$ by a subsequence, we may assume that v_ν converges in H^{s-1} to an element w. We must have $\mathcal{L}w = 0$ and supp $w \subset \{0\}$; by Lemma III.6.A1 this entails $w \equiv 0$ and means that $\|v_\nu\|_{s-1} \to 0$. From (III.6.A3) it would follow that $\|v_\nu\|_s \to 0$ contradicting (III.6.A4). ■

Call ${}^t\mathcal{L}$ the transpose of $\mathcal{L}$:

$${}^t\mathcal{L} = -\mathcal{L} + 2c_0 - \sum_{j=1}^{m} \partial c_j/\partial x_j.$$

Lemma III.6.A3. *Let s, s' be two real numbers. Suppose there is an open neighborhood* $U \subset \Omega$ *of* 0 *and a constant* $C > 0$ *such that*

$$\|u\|_s \leq C\|{}^t\mathcal{L}u\|_{s'}, \ \forall\, u \in \mathscr{C}_c^\infty(U). \qquad \text{(III.6.A5)}$$

Then there is a continuous linear operator $\mathbf{G} : H^{-s} \to H^{-s'}$ *such that, in the open set* U,

$$\mathcal{L}\mathbf{G}f = f,\ \forall\, f \in H^{-s}. \qquad \text{(III.6.A6)}$$

Proof. Call $\mathcal{H}$ the linear subspace of $H^{s'}$ consisting of the distributions that are equal to ${}^t\mathcal{L}u$ for some $u \in \mathcal{C}_c^\infty(\mathrm{U})$. Then set $u = {}^t\mathbf{G}{}^t\mathcal{L}u$; this is a bona fide definition of a linear map ${}^t\mathbf{G} : \mathcal{H} \to \mathrm{H}^s$ since ${}^t\mathcal{L}u \equiv 0$ implies $u \equiv 0$ by (III.6.A5), which also shows that ${}^t\mathbf{G}$ is bounded for the $\mathrm{H}^{s'}$ norm on $\mathcal{H}$. As a consequence ${}^t\mathbf{G}$ extends by continuity to the closure of $\mathcal{H}$ in $\mathrm{H}^{s'}$, and then to the whole of $\mathrm{H}^{s'}$ if we set ${}^t\mathbf{G} \equiv 0$ on the orthogonal of $\mathcal{H}$. Define $\mathbf{G} : \mathrm{H}^{-s} \to \mathrm{H}^{-s'}$ as the transpose of ${}^t\mathbf{G}$ for the natural duality between H^s and H^{-s}. If $f \in \mathrm{H}^{-s}$, $u \in \mathcal{C}_c^\infty(\mathrm{U})$, we have $\langle f,u\rangle = \langle f,{}^t\mathbf{G}{}^t\mathcal{L}u\rangle = \langle \mathcal{L}\mathbf{G}f,u\rangle$. ∎

Lemma III.6.A4. *If ${}^t\mathcal{L}$ is hypo-elliptic in* Ω (Definition III.5.2), *then to every $s \in \mathbb{R}$ and $p \in \Omega$ there is $s' \in \mathbb{R}$ and an open neighborhood* $\mathrm{U} \subset \Omega$ *of p in which* (III.6.A5) *holds true.*

Proof. We may assume $s \geq 0$. The natural topology of $\mathcal{C}^\infty(\Omega)$ is equal to the intersection of all the topologies induced on $\mathcal{C}^\infty(\Omega)$ by the local Sobolev spaces $\mathrm{H}^\sigma_{loc}(\Omega)$, $\sigma \in \mathbb{R}$ (see e.g., Treves [5], p. 228). We may also equip $\mathcal{C}^\infty(\Omega)$ with the coarsest locally convex topology $\mathcal{T}$ for which the natural injection $\mathcal{C}^\infty(\Omega) \to \mathrm{H}^s_{loc}(\Omega)$ as well as all the mappings $v \to {}^t\mathcal{L}v$ from $\mathcal{C}^\infty(\Omega)$ into $\mathrm{H}^\sigma_{loc}(\Omega)$ ($\sigma \in \mathbb{R}$) are continuous. Clearly the topology $\mathcal{T}$ can be defined by a sequence of seminorms and is weaker than the natural $\mathcal{C}^\infty$ topology. A Cauchy sequence for $\mathcal{T}$, $\{u_\nu\}_{\nu=1,2,\ldots}$, must converge in $\mathrm{H}^s_{loc}(\Omega)$ to some distribution u; and ${}^t\mathcal{L}u_\nu$ must converge in $\mathcal{C}^\infty(\Omega)$, perforce to ${}^t\mathcal{L}u$. Since ${}^t\mathcal{L}$ is hypo-elliptic we must have $u \in \mathcal{C}^\infty(\Omega)$ and then u_ν converges to u in the sense of the topology $\mathcal{T}$. This proves that the locally convex space $\mathcal{C}^\infty(\Omega)$ equipped with $\mathcal{T}$ is a Fréchet space. The open mapping theorem shows that $\mathcal{T}$ must be identical to the standard $\mathcal{C}^\infty$ topology. If therefore Ω' is any open subset of Ω, with $\mathcal{Cl}\,\Omega' \subset\subset \Omega$, there exist a real number s' and a constant $C > 0$ such that

$$\|u\|_{s+1} \leq C(\|u\|_s + \|{}^t\mathcal{L}u\|_{s'}),\ \forall\, u \in \mathcal{C}_c^\infty(\Omega'). \tag{III.6.A7}$$

Take $p \in \Omega'$. To any $\varepsilon > 0$ there is an open neighborhood $\mathrm{U} \subset \Omega'$ of p such that $\|u\|_s \leq \varepsilon\|u\|_{s+1}$, $\forall\, u \in \mathcal{C}_c^\infty(\mathrm{U})$ (see e.g. Treves [5], Proposition 25.6). Combining this with (III.6.A7) and taking $\varepsilon < 1/2C$ yields what we wanted. ∎

III.7. A Class of Hypocomplex CR Manifolds

In the present section $\mathcal{M}$ will be a hypo-analytic CR manifold of the hypersurface type. If, as usual, n denotes the rank of the tangent structure bundle $\mathcal{V}$, then

$$\dim \mathcal{M} = 2n + 1, \tag{III.7.1}$$

and $\mathcal{V} \cap \overline{\mathcal{V}} = 0$; the rank of the cotangent structure bundle T′ will be equal to $n+1$. We shall reason within a hypo-analytic chart (U,z,w) centered at a point $0 \in \mathcal{M}$ (as before, called the *origin*). Here $z = (z_1,\ldots,z_n)$, $z_j = x_j + \imath y_j$ ($1 \leq j \leq n$), and

$$w = s + \imath\varphi(z,s), \tag{III.7.2}$$

with φ a real-valued $\mathscr{C}^\infty$ function in U, satisfying

$$\varphi|_0 = d\varphi|_0 = 0 \tag{III.7.3}$$

(cf. section I.7). It is convenient to take

$$U = \Delta \times \mathscr{I} \tag{III.7.4}$$

with Δ a polydisk $\{ z \in \mathbb{C}^n; |z_j| < \rho, j = 1,\ldots,n \}$ $(\rho > 0)$ and $\mathscr{I}$ an open interval centered at zero in the real line. We may view the map

$$U \ni (z,s) \to (z,w) \in \mathbb{C}^{n+1} \tag{III.7.5}$$

as a diffeomorphism of U onto a real hypersurface Σ of $\mathbb{C}^{n+1}$, specifically the hypersurface defined by the equation

$$\mathscr{I}m\, w = \varphi(z, \mathscr{R}e\, w) \tag{III.7.6}$$

where $z \in \Delta$, $\mathscr{R}e\, w \in \mathscr{I}$.

The characteristic set T° is a line bundle over $\mathcal{M}$. Over U it is spanned by the real one-form

$$\varpi = ds + \imath \sum_{j=1}^{n} (b_j d\bar{z}_j - \bar{b}_j dz_j), \tag{III.7.7}$$

with $b_j = (1 + \imath\varphi_s)^{-1}\partial\varphi/\partial\bar{z}_j$. We have $\varpi|_0 = ds|_0$.

The conditions, in the forthcoming statements, bear on the Levi form of $\mathcal{M}$ at the characteristic point $(0, \varpi|_0)$ (see section I.9). For the sake of brevity we shall allow ourselves a slight abuse of language and refer to it as the "Levi form at 0." We shall assume that

$$\varphi(z,s) = \sum_{j=1}^{p} |z_j|^2 - \sum_{j=p+1}^{p+q} |z_j|^2 + O(|s||z| + |s|^2 + |z|^3) \tag{III.7.8}$$

(cf. (I.9.22)). We shall then say that the Levi form at 0 has p eigenvalues > 0 and q eigenvalues < 0. The cases where either p or q, or both, are equal to zero are not a priori precluded.

THEOREM III.7.1. *Suppose the Levi form at* 0 *is definite* (i.e., $p = n$ or $q = n$). *Then there is an open neighborhood* $U' \subset U$ *of* 0 *such that the following properties hold:*

To each integer $k \geq 0$ there is a CR function of class $\mathscr{C}^k$ in U' but not of class $\mathscr{C}^{k+1}$ in any subneighborhood of 0. (III.7.9)

There is a CR function in U', *of class* $\mathscr{C}^\infty$, *nowhere zero in* $U' \setminus \{0\}$ *and vanishing to infinite order at* 0. (III.7.10)

PROOF. Let us suppose $p = n$. We derive from (III.7.8):

$$\varphi(z,s) \geq \frac{1}{2}|z|^2$$

in a sufficiently small open neighborhood of 0, $U' \subset U$. The functions $f_k = (\iota w)^{k+1/2}$ (with the main branch of the square root) satisfy the requirement (III.7.9). The function

$$g = e^{-1/h},$$

with $h = (\iota w)^{1/3}$ and the main branch of the cubic root, satisfies (III.7.10). ∎

COROLLARY III.7.1. *If the Levi form of $\mathcal{M}$ at 0 is definite, the CR structure of $\mathcal{M}$ is not hypocomplex at 0.*

REMARK III.7.1. Let Σ' denote the image of U' under the map (III.7.5) and let $\tilde{f}$ (resp., $\tilde{g}$) denote the pushforward to Σ', under the same map, of the function f (resp., g) in the proof of Theorem III.7.1. Both $\tilde{f}$ and $\tilde{g}$ extend as holomorphic functions of (z,w) in a neighborhood of $\Sigma' \backslash \{0\}$ in $\mathbb{C}^{n+1}$. ∎

Let Σ be the hypersurface defined by (III.7.6). The following classical extension result of Lewy [1] yields a sufficient condition of hypocomplexity:

THEOREM III.7.2. *Suppose the Levi form of $\mathcal{M}$ at 0 has at least one eigenvalue > 0 (i.e., $p \geq 1$). Then there is a compact neighborhood $\mathcal{K}$ of the origin in $\mathbb{C}^{n+1}$ such that an arbitrary continuous CR function h in Σ is equal, in $\Sigma \cap \mathcal{K}$, to the restriction of a continuous function $\tilde{h}$ in the set*

$$\mathcal{K}^+ = \{ (z,w) \in \mathcal{K};\ \mathcal{I}m\, w \geq \varphi(z, \mathcal{R}e\, w) \},$$

holomorphic in the interior of $\mathcal{K}^+$.

PROOF. We select an open neighborhood $U_0 \subset U$ of 0 in which the arbitrary CR function $h \in \mathcal{C}^0(U)$ is the uniform limit of a sequence of polynomials $P_\nu(z,w)$ (Theorem II.3.1). Call Σ_0 the image of U_0 under the map (III.7.5).

By virtue of (III.7.8) we may write

$$\varphi(z,s) = |z_1|^2 + O(|z'|^2 + |s|^2 + |z_1|^3), \tag{III.7.11}$$

where $z' = (z_2,\ldots,z_n)$; (III.7.11) enables us to select first δ and then ε such that $0 < \delta, \varepsilon < \rho$, and that

$$\varphi(z,\mathcal{R}e\, w) > \mathcal{I}m\, w \tag{III.7.12}$$

if $|z_1| = \delta$ and (z',w) remains in the closed polydisk

$$|z_2| \leq \varepsilon, \ldots, |z_n| \leq \varepsilon,\ |w| \leq \varepsilon. \tag{III.7.13}$$

Actually we take δ and ε small enough that the intersection of the hypersurface Σ with the set

$$\mathcal{K} = \{ (z,w) \in \mathbb{C}^{n+1};\ |z_1| \leq \delta,\ |z_2| \leq \varepsilon ,\dots,\ |z_n| \leq \varepsilon,\ |w| \leq \varepsilon \}$$

will be a compact subset of Σ_0.

Fix (z',w) in the polydisk (III.7.13). From what was said above the inequalities

$$|z_1| \leq \delta,\ \varphi(z,\mathcal{R}e\, w) \leq \mathcal{I}m\, w$$

define a *compact* subset $\mathcal{K}(z',w)$ of $\mathcal{K}$ that lies on a complex line $z' = const.$, $w = const.$, and whose boundary with respect to this line lies entirely on Σ_0. As (z',w) ranges over the whole set (III.7.13) the subsets $\mathcal{K}(z',w)$ cover the totality of $\mathcal{K}^+$ (as defined in the statement of Theorem III.7.2). Now, by the maximum principle, since the polynomials $P_\nu(z,w)$ converge uniformly in Σ_0, they must also converge in $\mathcal{K}^+$, to $\tilde{h}$. ∎

COROLLARY III.7.2. *If the Levi form of $\mathcal{M}$ at 0 has at least one eigenvalue > 0 and at least one eigenvalue < 0, then $\mathcal{M}$ is hypocomplex at 0.*

PROOF. By Theorem III.7.2, if the Levi form at 0 has at least one eigenvalue > 0, the continuous CR functions in Σ have continuous extensions to $\mathcal{K}^+$ that are holomorphic in $\mathcal{I}nt\,\mathcal{K}^+$. Likewise, by symmetry $w \to -w$, we see that if the Levi form has at least one eigenvalue < 0, the CR functions in Σ have continuous extension to $\mathcal{K}^- = \{ (z,w) \in \mathcal{K};\ \mathcal{I}m\, w \leq \varphi(z, \mathcal{R}e\, w) \}$ that are holomorphic in $\mathcal{I}nt\,\mathcal{K}^-$. A continuous function $\tilde{h}$ in $\mathcal{K} = \mathcal{K}^+ \cup \mathcal{K}^-$ that is holomorphic in $(\mathcal{I}nt\,\mathcal{K}^+) \cup (\mathcal{I}nt\,\mathcal{K}^-)$ is holomorphic in $\mathcal{I}nt\,\mathcal{K}$, since $\bar{\partial}\tilde{h}$, in the distribution sense, vanishes identically "across" $\mathcal{K} \cap \Sigma$. Thus all continuous CR functions are hypo-analytic at 0. It follows from the local structure Theorem II.5.1 that this is also true of all distribution solutions. ∎

COROLLARY III.7.3. *Suppose that the Levi form of $\mathcal{M}$ at 0 is nondegenerate. Then in order for $\mathcal{M}$ to be hypocomplex at 0, it is necessary and sufficient that the Levi form of $\mathcal{M}$ at 0 not be definite.*

Combine Corollaries III.7.1 and III.7.2.

REMARK III.7.2. The preceding results, at least insofar as they concern continuous CR functions, remain valid if we only assume that the function φ or, which amount to the same, the hypersurface Σ, is of class $\mathcal{C}^r$ with $r \geq 2$. Note that $\mathcal{C}^2$ is needed if we are to make use of the Taylor expansion of order two of φ about 0, and of the Levi form at that point. In (III.7.9) the smoothness degree k cannot, then, exceed $r-1$; and in (III.7.10) the function that vanishes to infinite order at 0 can only be of class $\mathcal{C}^r$ (it still vanishes to infinite order; cf. (II.8.1)). The proof of Theorem III.7.2 makes use of the approximation formulas (II.2.17) or (II.7.21). It is quite clear that such formulas remain valid under the hypothesis $\varphi \in \mathcal{C}^r$ $(r \geq 2)$ provided the solution and the convergence are taken in the $\mathcal{C}^{r'}$ sense, with $r' \leq r$. ∎

Notes

Hypo-analytic structures were defined in Baouendi, Chang, and Treves [1]. An error in Example 1.4 of that article raises the question of the conditions under which a submanifold inherits a hypo-analytic structure from the ambient manifold; the question is addressed in section III.3. Following Baouendi and Treves [4], section III.4 interprets from the viewpoint of hypo-analytic manifolds the unique continuation result in section II.8. In this connection we mention an open problem: Is it true that, on a connected hypersurface $\Sigma \subset \mathbb{C}^n$, CR functions that vanish to infinite order on a curve γ transverse at every point to the holomorphic directions in Σ vanish identically? Such a curve γ is minimal noncharacteristic; but it is of course not, in general, equal to the intersection of Σ with a holomorphic curve in $\mathbb{C}^n$, a property that is required if we are to apply Corollary III.4.2.

The concept of a hypocomplex manifold was introduced and studied in Treves [7]; sections III.5, III.6, and III.7 are adaptations of chapter II, section 6 of that monograph. The example in section III.7 is a direct consequence of the classical extension result in Lewy [1].

IV

Integrable Formal Structures. Normal Forms

The quest for local invariants of an involutive structure goes back to the early days of the theory of several complex variables, when H. Poincaré and E. Cartan first looked at the boundary of a strongly pseudoconvex domain in $\mathbb{C}^2$. The standpoint, in the present book, is less focused. In this fourth chapter we consider a general involutive structure and extract new invariants from the Taylor expansion (at a given point 0) of the coefficients of the vector fields L_j ($j = 1,\ldots,n$) that span the tangent structure bundle $\mathcal{V}$ over a neighborhood of 0. In such a study one may as well assume that the coefficients themselves are power series that need not converge. In other words, one deals with *formal* power series, say in N indeterminates $X_1,\ldots,X_N$, and with formal vector fields, i.e., linear combinations of $\partial/\partial X_1,\ldots,\partial/\partial X_N$ whose coefficients are formal power series.

The hypotheses on the system of (now formal) vector fields $\mathrm{L}_1,\ldots,\mathrm{L}_n$ must reflect what we always require from the tangent structure bundle: that it is indeed a vector bundle, and that it satisfies the Frobenius condition. The constant terms of the L_j must be linearly independent and their commutation brackets $[\mathrm{L}_j,\mathrm{L}_k]$ must belong to the span of $\mathrm{L}_1,\ldots,\mathrm{L}_n$ over the ring $\mathbb{C}[[X_1,\ldots,X_N]]$. This defines an involutive formal structure. But the first observation is that such a structure is always *integrable*: there always exist formal power series $Z_1,\ldots,Z_m \in \mathbb{C}[[X_1,\ldots,X_N]]$ such that $\mathrm{L}_jZ_i = 0$ for all i and j, and such that the constant terms of the formal differentials $\mathrm{d}Z_1,\ldots,\mathrm{d}Z_m$ are $\mathbb{C}$-linearly independent. In view of this we talk of *integrable formal structures* (section IV.1). As a matter of fact, the formal Cauchy-Kovalevska theorem allows one to select the indeterminates, then called x_i and t_j, and the first integrals Z_i so that $Z_i|_{t=0} = x_i$ ($1 \leq i \leq m$, $1 \leq j \leq n$). But there is more: every *solution* $h \in \mathbb{C}[[X_1,\ldots,X_N]]$ (of the equations $\mathrm{L}_jh = 0$, $j = 1,\ldots,n$) is a formal power series in $Z_1,\ldots,Z_m$ (and, of course, vice versa: every power series in $Z_1,\ldots,Z_m$ is a solution). One could say that every integrable formal structure is *formally hypocomplex* (cf. Definition III.5.1).

The approach followed here is the standard one, via the Lie algebra $\mathfrak{g}(\mathcal{V})$ generated by the real parts of the formal vector fields $\mathrm{L} \in \mathcal{V}$. The analogous Lie algebra in a locally integrable structure led us to the notion of orbit (Definition I.11.1); here it leads us to *normal forms*. One uses the natural filtration of $\mathfrak{g}(\mathcal{V})$, $\mathfrak{g}_1 \subset \cdots \subset \mathfrak{g}_i \subset \mathfrak{g}_{i+1} \subset \cdots$: $\mathfrak{g}_1$ consists of the formal vector fields $\mathcal{R}_e\mathrm{L}$, $\mathrm{L} \in \mathcal{V}$; if $i > 1$, $\mathfrak{g}_i$ is spanned by $\mathfrak{g}_{i-1}$ and by the brackets $[\mathbf{v},\mathbf{w}]$ with $\mathbf{v} \in \mathfrak{g}_1$, $\mathbf{w} \in \mathfrak{g}_{i-1}$. The dimension of the "freezing at 0" of $\mathfrak{g}_i$ might remain constant over certain intervals of the range of i, as i increases, and then jump up at certain values of i, called the *Hörmander numbers* (section IV.2). The structure has *finite type* if that dimension reaches the value N at a finite value of i. Our aim is to select the indeterminates (called x_i, y_j, s_k, t_ℓ) and the first integrals $z_j = x_j + \imath y_j$, $w_k = s_k + \imath\varphi_k(x,y,s,t)$ ($1 \leq j \leq \nu$, $1 \leq k \leq d$; cf. (I.7.4)), in such a way that, if we assign a *weight* to each indeterminate, specifically the weight 1 to x_i, y_j, t_ℓ and ϖ_k to s_k, and define the weights of monomials and of power series accordingly (see section IV.2), then the weight of φ_k is also equal to ϖ_k. (Each weight ϖ_k is equal to some Hörmander number; the number of indices k equal to one and the same Hörmander number m_ι is called the *multiplicity* of m_i.) The weights of the variables determine the weight of a formal vector field; the selection of the first integrals as indicated entails that $\mathcal{V}$ is spanned by n formal vector fields $\mathrm{L}_1,\ldots,\mathrm{L}_n$ (with expressions (I.7.15), (I.7.16)) whose weight is exactly equal to -1 (section IV.4). Such a "presentation" of $\mathcal{V}$ is what is called a normal form. The existence of normal forms is established in section IV.5.

The normal form can be further refined: the pluriharmonic parts of the series φ_k can be eliminated (section IV.5). Once this is done we call P_k the homogeneous part of lowest degree in the series φ_k; the system of "leading parts" $(P_1,\ldots,P_d)$ is essentially unique: it is unique up to the obvious substitutions of the indeterminates and of the first integrals that exchange those systems. The Hörmander numbers, their multiplicities, and the (equivalence classes of) systems of "leading parts" are invariants of the integrable formal structure.

The scope of chapter IV is limited to formal structures. But it must be pointed out that all the definitions and all the results extend routinely to *analytic structures*: it suffices to replace formal power series by convergent power series.

IV.1. Integrable Formal Structures

We begin by setting down some of the notation used throughout the present chapter.

We denote by $\mathbb{C}[[X]]$ the ring of formal power series in N indeterminates X_i ($i = 1,\ldots,N$). Later on the indeterminates will be chosen in a manner especially suited to the structures we shall be studying, and will be denoted by x_i,

y_j, s_k, t_ℓ, etc. The partial derivatives $\partial/\partial X_j$ ($j = 1,\ldots,N$) define *derivations* of the ring $\mathbb{C}[[X]]$, i.e., they are linear operators and the partial derivative of a product is computed in accordance with Leibniz's rule.

By a *formal vector field* we shall mean a linear combination of the partial differentiations $\partial/\partial X_j$ with coefficients in $\mathbb{C}[[X]]$. Formal vector fields make up a Lie algebra $\mathbb{D}[[X]]$ with scalar ring $\mathbb{C}[[X]]$; the bracket is the standard commutation bracket $[\mathrm{L}_1,\mathrm{L}_2]$. If $\mathrm{L} \in \mathbb{D}[[X]]$ we denote by $\mathrm{L}|_0$ the formal vector field (with complex coefficients) obtained by replacing each coefficient of L by its zero-order (i.e., constant) term. Obviously $\mathrm{L}|_0$ can be identified to a vector field with constant coefficients in Euclidean space $\mathbb{R}^N$. We shall often refer to $\mathrm{L}|_0$ as the formal vector field L "frozen at 0," or as the "freezing at 0" of L. Any formal vector field L defines a derivation of the ring $\mathbb{C}[[X]]$.

We shall also deal with *formal differential forms*. The formal differential forms of degree one, or one-forms, are the linear combinations of $\mathrm{d}X_1,\ldots,\mathrm{d}X_N$ with coefficients in $\mathbb{C}[[X]]$. They make up a module over $\mathbb{C}[[X]]$, which we denote by $\Lambda^1\mathbb{C}[[X]]$. We shall also introduce the p-th exterior powers $\Lambda^p\mathbb{C}[[X]]$ ($p = 2,3,\ldots$), and the whole exterior algebra $\Lambda\mathbb{C}[[X]]$. Of course, $\Lambda^0\mathbb{C}[[X]] = \mathbb{C}[[X]]$ and $\Lambda^p\mathbb{C}[[X]] = 0$ if $p > N$. Elements of $\Lambda^p\mathbb{C}[[X]]$ are *formal p-forms*. The exterior derivative,

$$\mathrm{d} : \Lambda^p\mathbb{C}[[X]] \to \Lambda^{p+1}\mathbb{C}[[X]],$$

acts in the customary fashion, and $\mathrm{d}^2 = 0$. If $\omega \in \Lambda^p\mathbb{C}[[X]]$ we shall denote by $\omega|_0$ the form obtained by replacing each coefficient in ω by its constant term; $\omega|_0$ can be regarded as a differential form of degree p, with constant coefficients, in $\mathbb{R}^N$.

There is a natural pairing between $\mathbb{D}[[X]]$ & $\Lambda^1\mathbb{C}[[X]]$, i.e., a $\mathbb{C}[[X]]$-bilinear map

$$\mathbb{D}[[X]] \times \Lambda^1\mathbb{C}[[X]] \ni (\mathrm{L},\omega) \to \langle \mathrm{L},\omega\rangle \in \mathbb{C}[[X]].$$

It is defined as follows: if

$$\mathrm{L} = \sum_{i=1}^{N} a_i(X)\partial/\partial X_i, \quad \omega = \sum_{i=1}^{N} b_i(X)\mathrm{d}X_i,$$

then

$$\langle \mathrm{L},\omega\rangle = \sum_{i=1}^{N} a_i(X)b_i(X).$$

If $\omega = \mathrm{d}f$ with $f \in \mathbb{C}[[X]]$, then obviously $\langle \mathrm{L},\omega\rangle = \mathrm{L}f$.

Taylor expansion at the origin in $\mathbb{R}^N$ defines a homomorphism $\mathcal{T}$ of the ring of complex $\mathscr{C}^\infty$ functions in $\mathbb{R}^N$ into the ring $\mathbb{C}[[X]]$. This homomorphism is clearly not injective. But $\mathcal{T}$ is surjective, according to a classical theorem of

E. Borel. Applying $\mathcal{T}$ to the coefficients, either of vector fields or of differential forms, yields surjective homomorphisms (which we also denote by $\mathcal{T}$) of the Lie algebra of $\mathcal{C}^\infty$ complex vector fields in $\mathbb{R}^N$, onto $\mathbb{D}[[X]]$, on the one hand, and on the other hand, of the exterior algebra of $\mathcal{C}^\infty$ complex differential forms in $\mathbb{R}^N$, onto $\Lambda\mathbb{C}[[X]]$.

DEFINITION IV.1.1. *By an* integrable formal structure *in N indeterminates, we shall mean a Lie subalgebra* $\mathcal{V}$ *of* $\mathbb{D}[[X]]$ *that has the following property*:

As a module over the ring $\mathbb{C}[[X]]$, $\mathcal{V}$ *is spanned by formal vector fields* $L_1,\dots,L_n$ *such that* $L_1|_0,\dots,L_n|_0$ *are* $\mathbb{C}$*-linearly independent.* (IV.1.1)

We shall refer to the integer n as the *rank* of $\mathcal{V}$; of course, $n \le N$. The fact that $\mathcal{V}$ is a Lie subalgebra of $\mathbb{D}[[X]]$ means that

for each pair $j,\ k = 1,\dots,n$, *the commutation bracket* $[L_j,L_k]$ *is a linear combination of* $L_1,\dots,L_n$ *with coefficients in* $\mathbb{C}[[X]]$. (IV.1.2)

We call T′ the $\mathbb{C}[[X]]$-submodule of $\Lambda^1\mathbb{C}[[X]]$ consisting of those formal one-forms ω such that $\langle L,\omega\rangle = 0$ for all $L \in \mathcal{V}$. By a *formal solution* we shall mean any formal power series h such that $dh \in T'$, i.e., such that $Lh = 0$ for all $L \in \mathcal{V}$. The formal solutions make up a *subring* of $\mathbb{C}[[X]]$.

Now select a basis $\{L_1,\dots,L_n\}$ of $\mathcal{V}$ as in (IV.1.1). Also select $m = N-n$ vector fields M_i $(i = 1,\dots,m)$ belonging to $\mathbb{D}[[X]]$ but with constant coefficients (i.e., $M_i = M_i|_0$), such that

$$L_1|_0,\dots,L_n|_0,\ M_1,\dots,M_m,$$

are linearly independent (and therefore make up a basis of the space of all vector fields). We may then write, for $j = 1,\dots,n$,

$$L_j = \sum_{k=1}^{n} \gamma_{jk} L_k|_0 + \sum_{\ell=1}^{m} \lambda_{j\ell} M_\ell,$$

where the coefficients are formal power series. The matrix $\{\gamma_{jk}\}_{1\le j,k\le n}$ is invertible in the ring of $n\times n$ matrices with entries in $\mathbb{C}[[X]]$, since its constant term is the identity. Thus, after a $\mathbb{C}[[X]]$-linear substitution of the L_j, we may assume that

$$L_j = L_j|_0 + \sum_{\ell=1}^{m} \lambda_{j\ell} M_\ell,\ j = 1,\dots,n. \qquad \text{(IV.1.3)}$$

Note that the constant term of each formal power series λ_{jk} vanishes, and that the brackets $[L_j, L_k]$ belong to the span of the M_i. By (IV.1.2) the latter demands

$$[L_j, L_k] = 0,\ \forall\, j,\ k = 1,\dots,n. \qquad \text{(IV.1.4)}$$

The preceding considerations enable us to prove the following:

LEMMA IV.1.1. *Let $\mathcal{V} \subset \mathbb{D}[[X]]$ be an integrable formal structure. If $f_0(X)$ is a linear form in $X_1,\ldots,X_N$ satisfying*

$$(\mathrm{L}|_0)f_0 = 0, \forall \mathrm{L} \in \mathcal{V}, \tag{IV.1.5}$$

then there is a formal solution f such that all the terms of degree ≤ 1 in $f - f_0$ are equal to zero.

PROOF. By induction on the integer $\nu = 0,1,\ldots$, we define a polynomial of degree $\nu+1, f_\nu(X)$, satisfying

$$f_\nu - f_{\nu-1} = \mathrm{O}(|X|^{\nu+1}), \tag{IV.1.6}$$

$$\mathrm{L}_j f_\nu = \mathrm{O}(|X|^{\nu+1}), \tag{IV.1.7}$$

where $\mathrm{O}(|X|^{\nu+1})$ stand for some formal power series in which all terms of order $\leq \nu$ are equal to zero. We do this as follows: when $\nu = 0, f_0$ satisfies (IV.1.5). When $\nu \geq 1$ call $g_{\nu,j}$ the homogeneous part of degree ν (i.e., of lowest degree) of $\mathrm{L}_j f_{\nu-1}$. In view of (IV.1.4) and of the homogeneity of the $g_{\nu,j}$, we have

$$(\mathrm{L}_j|_0)g_{\nu,k} = (\mathrm{L}_k|_0)g_{\nu,j}, \quad j, k = 1,\ldots,n,$$

and consequently there is a homogeneous polynomial of degree $\nu+1$, h_ν, satisfying

$$(\mathrm{L}_j|_0)h_\nu = -g_{\nu,j}, \quad j = 1,\ldots,n. \tag{IV.1.8}$$

Then set $f_\nu = f_{\nu-1} + h_\nu$; (IV.1.6) is trivially satisfied, and

$$\mathrm{L}_j f_\nu = g_{\nu,j} + \mathrm{O}(|X|^{\nu+1}) + (\mathrm{L}_j|_0)h_\nu + [\mathrm{L}_j - (\mathrm{L}_j|_0)]h_\nu = \mathrm{O}(|X|^{\nu+1}),$$

by (IV.1.8).

The unique formal power series f such that $f - f_\nu = \mathrm{O}(|X|^{\nu+2})$ for each $\nu \in \mathbb{Z}_+$ satisfies the requirements in Lemma IV.1.1. ■

The following direct consequence of Lemma IV.1.1 embodies the "formal Frobenius theorem" (cf. Theorems I.10.1, I.10.3):

THEOREM IV.1.1. *Let $\mathcal{V} \subset \mathbb{D}[[X]]$ be an integrable formal structure of rank n. There exist $m = N-n$ formal solutions Z_i ($i = 1,\ldots,m$) whose zero-order terms are equal to zero and which are such that*

$$\mathrm{d}Z_1|_0,\ldots,\mathrm{d}Z_m|_0 \text{ are } \mathbb{C}\text{-linearly independent.} \tag{IV.1.9}$$

PROOF. We select m linear forms $\zeta_1(X),\ldots,\zeta_m(X)$ such that

$$(\mathrm{L}_j|_0)\zeta_i = 0, \mathrm{M}_h\zeta_i = \delta_{hi} \; (h, i = 1,\ldots,m, j = 1,\ldots,n). \tag{IV.1.10}$$

We apply Lemma IV.1.1 with $f_0 = \zeta_j$ and call Z_j the formal solution f such that f–f_0 vanishes to second-order. That (IV.1.9) holds follows from (IV.1.10). ■

We need to distinguish between real and complex indeterminates. If u is a formal power series with respect to $X_1,\ldots,X_N$, its complex conjugate, $\bar{u}$, is obtained by substituting each coefficient in u by its complex conjugate. Thus the X_i are viewed as *real* indeterminates. We set $\mathcal{R}e\, u = (u + \bar{u})/2$, $\mathcal{I}m\, u = (u - \bar{u})/2\imath$; we say that the formal power series u is *real* if $u = \bar{u}$. The same terminology and notation will be used in dealing with formal vector fields and formal differential forms.

Let then $Z_1,\ldots,Z_m$ be formal solutions like those in Theorem IV.1.1. After a $\mathbb{C}$-linear substitution of the Z_i and one of the indeterminates X_j we may assume that

$$\mathbb{C}[[X]] = \mathbb{C}[[\mathcal{R}e Z_1,\ldots,\mathcal{R}e Z_m, X_{m+1},\ldots,X_{m+n}]].$$

To stress the analogy with the coarse local representation of $\mathscr{C}^\infty$ locally integrable structures (see section I.7) we shall write $x_i = \mathcal{R}e Z_i$, $t_j = X_{m+j}$ ($1 \le i \le m$, $1 \le j \le n$), and then assume that the Z_i have the expressions (I.7.23) with the proviso that, here, Φ_i stands for a formal power series. Further $\mathbb{C}$-linear substitutions ensure that (I.7.24) holds. Actually, here (as in the analytic case) an additional substitution enables us to go further and achieve that $\Phi(x,0) \equiv 0$, i.e., (I.10.4) holds. Indeed, if (I.7.23) and (I.7.24) hold, we can apply the implicit function theorem in the ring of formal power series and solve the equation $Z = H + \imath\Phi(H,0)$ to get a unique power series $H(Z)$ with $H(0) = 0$. This means, of course, that $H(x+\imath\Phi(x,0)) = x$. If we then set $Z^\# = Z - \imath\Phi(H(Z),0) = x + \imath[\Phi(x,t) - \Phi(H(x+\imath\Phi(x,t)),0)]$, we see that putting $t = 0$ yields $Z^\# = x$. Henceforth, we shall hypothesize that (I.10.4) holds.

The preceding choice of "first integrals" Z_i allows us to complement Lemma IV.1.1 with the version of the formal Cauchy-Kovalevska theorem for integrable formal structures (cf. Theorem I.10.6):

THEOREM IV.1.2. *Let $\mathcal{V} \subset \mathbb{C}[[x,t]]$ be a formal integrable structure and let $Z_1,\ldots,Z_m$ be formal solutions such that* (I.10.4) *holds. Then the substitution $h_0(x) \to h_0(Z)$ is a ring isomorphism of $\mathbb{C}[[x]]$ onto the subring of $\mathbb{C}[[x,t]]$ consisting of the formal solutions. Its inverse is the initial value map $h(x,t) \to h(x,0)$.*

We are using the notation $\mathbb{C}[[x,t]] = \mathbb{C}[[x_1,\ldots,x_m,t_1,\ldots,t_n]]$.

PROOF. That the map $h_0(x) \to h_0(Z)$ is a homomorphism into the ring of formal solutions is evident. If $h \in \mathbb{C}[[x,t]]$, by (I.10.4) $h(x,t) - h(Z(x,t),0)$ is also a formal solution, which vanishes identically when $t = 0$. The formal analogue of the argument used to prove uniqueness in Theorem I.10.6 shows that $h(x,t) \equiv h(Z(x,t),0)$. ■

COROLLARY IV.1.1. *The formal solutions make up the subring* $\mathbb{C}[[Z]]$ *of* $\mathbb{C}[[x,t]]$.

We have used the notation $\mathbb{C}[[Z]] = \mathbb{C}[[Z_1,\dots,Z_m]]$. Loosely speaking, Theorem IV.1.1 states that every "involutive" formal structure is integrable; Corollary IV.1.1, that every integrable formal structure is "hypocomplex" (cf. Definition III.5.1).

Another immediate consequence of Theorem IV.1.1 is that any integrable formal structure can be "lifted" to a locally integrable structure:

THEOREM IV.1.3. Let $\mathcal{V} \subset \mathbb{C}[[X]]$ *be an integrable formal structure. There is a locally integrable structure* $\hat{\mathcal{V}}$ *defined on open neighborhood* Ω *of* 0 *in* $\mathbb{R}^{m+n}$ *whose Taylor expansion at the origin is equal to* $\mathcal{V}$.

PROOF. Let $Z_1,\dots,Z_m$ be "first integrals" of $\mathcal{V}$ as in Theorem IV.1.1. We take advantage of the Borel theorem to select m $\mathcal{C}^\infty$ functions $\hat{Z}_1,\dots,\hat{Z}_m$ in $\mathbb{R}^{m+n}$ whose Taylor expansions at the origin are given by $Z_1,\dots,Z_m$ respectively. Thanks to (IV.1.9) we see that the differentials $d\hat{Z}_i$ are linearly independent in some open neighborhood Ω of 0 and therefore span a vector subbundle $\hat{T}'$ of $\mathbb{C}T^*\Omega$. Let $\hat{\mathcal{V}} \subset \mathbb{C}T\Omega$ denote its orthogonal. It is checked at once that the Taylor expansion map $\mathcal{T}$ maps $\hat{\mathcal{V}}$ onto $\mathcal{V}$. ■

We extend to integrable formal structures much of the terminology used in the study, near a point, of locally integrable structures. Thus we shall think of the $\mathbb{R}$-linear span of the differentials dX_i as the cotangent space to $\mathbb{R}^N$ at the origin. It will be denoted by T_0^*. The $\mathbb{R}$-linear span of the partial differentiations $\partial/\partial X_i$ will be denoted by T_0 and will be thought of as the tangent space to $\mathbb{R}^N$ at the origin. We shall also make use of the complexifications of those spaces, $\mathbb{C}T_0^*$ and $\mathbb{C}T_0$.

We denote by T_0' the vector space consisting of the differentials $dh|_0$ frozen at the origin, when h ranges over all formal solutions; T_0' is an m-dimensional vector space. Likewise, we call $\mathcal{V}_0$ the vector space consisting of the vector fields $L|_0$ with $L \in \mathcal{V}$ arbitrary; $\mathcal{V}_0$ is an n-dimensional complex vector space. For obvious reasons we refer to the intersection $T_0^\circ = T_0' \cap T_0^*$ as the *characteristic set at the origin*.

We shall say (cf. Definition I.2.3 and remark following (I.7.5)) that the integrable formal structure $\mathcal{V}$ is *real* if the following equivalent conditions are satisfied: (i) $\mathcal{V}$ is spanned over $\mathbb{C}[[X]]$ by real formal vector fields; (ii) T' is spanned over $\mathbb{C}[[X]]$ by real differentials; (iii) the formal power series $Z_1,\dots,Z_m$ in Theorem IV.1.1 can be chosen to have real coefficients.

We shall say that the structure is *complex* if $\mathbb{C}T_0 = \mathcal{V}_0 \oplus \overline{\mathcal{V}}_0$; *elliptic* if $\mathbb{C}T_0 = \mathcal{V}_0 + \overline{\mathcal{V}}_0$ or, equivalently, $T_0' \cap \overline{T}_0' = 0$; *CR* (*Cauchy-Riemann*) if $\mathcal{V}_0 \cap \overline{\mathcal{V}}_0 = 0$ or, equivalently, $\mathbb{C}T_0^* = T_0' + \overline{T}_0'$.

The aim of the present chapter is to isolate some significant properties of the integrable formal structure under study that are "invariant." By this we mean properties of $\mathcal{V}$ (or T') that do not depend on the choice of the formal vector fields L_j ($j = 1,\ldots,m$) that span $\mathcal{V}$ over $\mathbb{C}[[X]]$, or on that of the formal power series $Z_1,\ldots,Z_m$ whose differentials span T'. The first of these invariants are the integers m, n, and $d = \dim T_0^\circ$.

IV.2. Hörmander Numbers, Multiplicities, Weights. Normal Forms

Let $\mathcal{V} \subset \mathbb{C}[[X]]$ be an integrable formal structure of rank n. We denote by $\mathfrak{g}(\mathcal{V})$ the Lie algebra over $\mathbb{R}$ generated by the formal vector fields $\mathcal{R}e L$, $L \in \mathcal{V}$. We introduce *the natural filtration* of $\mathfrak{g}(\mathcal{V})$, $\{\mathfrak{g}_1,\mathfrak{g}_2,\ldots\}$: $\mathfrak{g}_1$ is the set of formal vector fields $\mathcal{R}e L$, $L \in \mathcal{V}$; if $i \geq 2$, $\mathfrak{g}_i$ is the smallest linear subspace of $\mathfrak{g}(\mathcal{V})$ that contains both $\mathfrak{g}_{i-1}$ and $[\mathfrak{g}_1,\mathfrak{g}_{i-1}]$ (the latter is the set of commutators of an element of $\mathfrak{g}_1$ with an element of $\mathfrak{g}_{i-1}$). We have $\mathfrak{g}_i \subset \mathfrak{g}_{i+1}$, $i = 1,2,\ldots$ Note that $\mathfrak{g}(\mathcal{V})$ is a module over the ring $\mathbb{R}[[X]]$ and that each $\mathfrak{g}_i$ is a submodule of $\mathfrak{g}(\mathcal{V})$. We denote by $\mathfrak{g}_i|_0 \subset T_0$ the image of $\mathfrak{g}_i$ under the "freezing at 0" map $\mathbf{v} \rightarrow \mathbf{v}|_0$.

Clearly $\mathfrak{g}_1|_0$ and $T_0^\circ \subset T_0^*$ are the orthogonal of each other; thus $\dim(\mathfrak{g}_1|_0) = m+n-d$ (unless specified otherwise all dimensions are computed over the real field). Suppose there is a smallest integer $i \geq 2$ such that $\mathfrak{g}_i|_0 \neq \mathfrak{g}_1|_0$; we denote it by m_1; and call $\kappa_1 = \dim(\mathfrak{g}_{m_1}|_0) - \dim(\mathfrak{g}_1|_0)$ the *multiplicity* of m_1. By induction on $\iota \geq 2$ we define m_ι as the smallest integer $i > m_{\iota-1}$ such that $\mathfrak{g}_i|_0 \neq \mathfrak{g}_{m_{\iota-1}}|_0$; the integer $\kappa_\iota = \dim(\mathfrak{g}_{m_\iota}|_0) - \dim(\mathfrak{g}_{m_{\iota-1}}|_0)$ is called the *multiplicity* of m_ι. There is an integer $r \geq 0$ such that $\mathfrak{g}_i|_0 = \mathfrak{g}_{m_r}|_0$ for all $i > m_r$ (with the understanding that $m_0 = 1$ and $\mathfrak{g}_{m_0} = \mathfrak{g}_1$). If $\mathfrak{g}_{m_r}|_0 = T_0$ the integrable formal structure $\mathcal{V}$ is said to be of *finite type* (cf. Definition I.11.2); we set $m_{r+1} = +\infty$, $\kappa_{r+1} = \dim T_0 - \dim(\mathfrak{g}_{m_r}|_0)$. In all cases we have $\kappa_1 + \cdots + \kappa_r + \kappa_{r+1} = d$. The integers m_ι ($\iota = 1,\ldots,r$) are called the *Hörmander numbers* of the structure $\mathcal{V}$.

There is a dual way of looking at Hörmander numbers and multiplicities. As already pointed out, $T_0^\circ = (\mathfrak{g}_1|_0)^\perp$. Set then, for any $i \geq 1$, $\mathcal{T}_i = (\mathfrak{g}_i|_0)^\perp$; we have $\mathcal{T}_1 = T_0^\circ$ and $\mathcal{T}_{i+1} \subset \mathcal{T}_i$ for all $i = 1,2,\ldots$ Set $S_i = \mathcal{T}_i \backslash \mathcal{T}_{i+1}$: $\sigma \in S_i$ means that there is $\vartheta \in \mathfrak{g}_{i+1}|_0$ such that $\langle\sigma,\vartheta\rangle \neq 0$ but no $\vartheta \in \mathfrak{g}_i|_0$ such that the same is true. We have $S_i = \emptyset$ unless i is one of the Hörmander numbers. We also define $\mathcal{T}_\infty \subset T_0^\circ$ as the orthogonal in T_0^* of the whole space $\mathfrak{g}(\mathcal{V})|_0$; to say that $\mathcal{V}$ is of finite order is the same as saying that $\mathcal{T}_\infty = 0$. We have

$$\mathrm{T}_0^\circ = S_{m_1} \cup \cdots \cup S_{m_r} \cup \mathcal{T}_\infty;\ S_{m_\iota} \cap S_{m_{\iota'}} = \emptyset \textit{ if } m_\iota \neq m_{\iota'};\ S_{m_\iota} \cap \mathcal{T}_\infty = \emptyset;$$
$$\kappa_1 = d - \dim \mathcal{T}_{m_1};\ \textit{if } \iota \geq 2,\ \kappa_\iota = \dim \mathcal{T}_{m_{\iota-1}} - \dim \mathcal{T}_{m_\iota}.$$

EXAMPLE IV.2.1. We take $n = 1$. Let $\mathcal{V} \subset \mathbb{D}[[X]]$ be spanned over $\mathbb{C}[[X]]$ by the vector field

$$\mathrm{L} = \partial/\partial X_{m+1} + \iota \sum_{j=1}^{J} X_{m+1}^{p_j} \partial/\partial X_j,$$

with $0 \leq p_1 < \cdots < p_J$ ($J \leq m$). If $p_1 = 0$, $\mathfrak{g}_1|_0$ is spanned by $\partial/\partial X_{m+1}$ and $\partial/\partial X_1$; if $p_1 \geq 1$, $\mathfrak{g}_1|_0$ is spanned by $\partial/\partial X_{m+1}$. In the former case $\mathrm{L}|_0$ and $\overline{\mathrm{L}}|_0$ are linearly independent; in the latter, they are equal. In all cases the structure is of finite type if $J = m$; it is not if $J < m$; all multiplicities are equal to 1 except, possibly, the last one when the structure is not of finite type, in which case $\kappa_{r+1} = m - J$.

When $p_1 = 0$ we change indeterminates: we write $x = \frac{1}{2}X_{m+1}$, $y = \frac{1}{2}X_1$, $s_j = -X_{j+1}$ ($j = 1,\ldots,m-1 = d$). Then

$$\mathrm{L} = \partial/\partial \bar{z} - \iota \sum_{j=1}^{J-1} (2x)^{p_{j+1}} \partial/\partial s_j; \tag{IV.2.1}$$

$\mathfrak{g}_1|_0$ is spanned by $\partial/\partial x$ and $\partial/\partial y$. The Hörmander number m_j is equal to $p_{j+1}+1$.

Now suppose $p_1 \geq 1$. In this case it is more suggestive to switch to the notation $x_i = -X_i$ ($i = 1,\ldots,m$), $t = X_{m+1}$. Then $\mathfrak{g}_1|_0$ is spanned by $\partial/\partial t$. We have

$$\mathrm{L} = \partial/\partial t - \iota \sum_{j=1}^{J} t^{p_j} \partial/\partial x_j; \tag{IV.2.2}$$

the Hörmander number m_j is equal to $p_j + 1$. ■

EXAMPLE IV.2.2. Call the real indeterminates $x_1,\ldots,x_m$, $t_1,\ldots,t_m$ ($m \geq 1$ arbitrary). Set $Z_j = x_j + \iota t_j^{p_j}$ ($p_j \in \mathbb{Z}_+$, $j = 1,\ldots,m$). Let then $\mathcal{V} \subset \mathbb{D}[[x,t]]$ be spanned over $\mathbb{C}[[x,t]]$ by the vector fields

$$\mathrm{L}_j = \partial/\partial t_j - \iota p_j t_j^{p_j - 1} \partial/\partial x_j,\ j = 1,\ldots,m. \tag{IV.2.3}$$

It is clear that the orthonormality relations (I.7.25) (with $\mathrm{M}_i = \partial/\partial x_i$) are satisfied. The structure is of finite type if and only if $p_j \geq 1$ for all j. The Hörmander numbers m_ι ($1 \leq \iota \leq r$) are equal to the values of the integers $p_i \geq 2$; if $\iota \leq r$, the multiplicity κ_ι is equal to the number of integers $p_j = m_\iota$; κ_{r+1} is equal to the number of integers $p_j = 0$. ■

We shall make use of a *fine representation* of the integrable formal structure

$\mathcal{V}$ (cf. sections I.7, II.7). The selection of indeterminates x_i and t_j (see Theorem IV.1.2) and of first integrals $Z_1,\ldots,Z_m$ given by (I.7.23) and of the corresponding vector fields (I.7.27) can be regarded as the *coarse representation* of $\mathcal{V}$ (cf. section II.1). After a $\mathbb{C}$-linear substitution we may assume that (I.7.2) holds at 0, as well as (I.7.3), for $\nu = m-d$. We shall always write z_i instead of Z_i if $i \leq \nu$, w_k instead of $Z_{\nu+k}$. We take then, as *real* indeterminates, $x_i = \mathcal{R}e z_i$, $y_j = \mathcal{I}m z_j$ $(1 \leq i, j \leq \nu)$, $s_k = \mathcal{R}e w_k$ $(k = 1,\ldots,d)$ and $n-\nu$ additional ones, t_ℓ $(\ell = 1,\ldots,n-\nu)$. After appropriate substitutions we may assume that the w_k have the expressions (I.7.4), where now φ_k is a real formal power series, and that (I.7.5) holds. The characteristic set at 0, T_0°, is spanned by $ds_1,\ldots,ds_d$. We are also going to make use of the formal vector fields M_i, L_j, N_k: they are the formal analogues of the $\mathcal{C}^\infty$ vector fields defined in section I.7 (see (I.7.9) to (I.7.18)).

Properties such as "the structure is real (resp., complex; resp., elliptic; resp., CR)" can be read in the fine representation exactly as in the $\mathcal{C}^\infty$ case (see the remark that follows (I.7.5)).

Our purpose is to select new real indeterminates x_i, y_j, s_k, t_ℓ and new first integrals $w_k = s_k + \imath\varphi_k(x,y,s,t)$ that somehow reflect the filtration $\{\mathfrak{g}_1,\mathfrak{g}_{m_1},\ldots,\mathfrak{g}_{m_r}\}$ of $\mathfrak{g}(\mathcal{V})$. We shall attach different *weights* to those indeterminates and first integrals, as follows: the weight of x_i, y_j (and therefore that of z_i and $\bar{z}_i$), t_ℓ will be equal to 1. We denote by ϖ_k the weight of s_k: we shall require that

$$\begin{aligned} \varpi_k &= m_\iota \ \textit{if}\ d_{\iota-1} < k \leq d_\iota \ (1 \leq \iota \leq r), \\ \varpi_k &= +\infty \ \textit{if}\ d_r < k \leq d, \end{aligned} \tag{IV.2.4}$$

where we have used the notation (also used in the sequel)

$$d_0 = 0,\ d_\iota = \sum_{p=1}^{\iota} \kappa_p \ \textit{if}\ 1 \leq \iota \leq r.$$

The weight of any monomial

$$cx^\alpha y^\beta t^\gamma s^\lambda \ (c \in \mathbb{C},\ \alpha, \beta \in \mathbb{Z}_+^\nu,\ \gamma \in \mathbb{Z}_+^{n-\nu},\ \lambda \in \mathbb{Z}_+^d)$$

will be equal to $|\alpha| + |\beta| + |\gamma| + \sum_{k=1}^{d} \varpi_k\lambda_k$ if $c \neq 0$ and to $+\infty$ if $c = 0$; the weight of a formal power series will be the minimum weight of the monomials in the series. We shall require that the weight of $\varphi_k(x,y,s,t)$ (and therefore that of w_k) be equal to ϖ_k. This leads to weighting the vector fields: the weights of $\partial/\partial x_i$, $\partial/\partial y_j$, $\partial/\partial t_\ell$ are all equal to -1, and that of $\partial/\partial s_k$ to $-\varpi_k$. If the weight of a formal power series f is equal to p and that of a formal vector field $\mathbf{v}$ is equal to q then the weight of the formal vector field $f\mathbf{v}$ will be equal to $p + q$ with the understanding that if $p = +\infty$ and $q = -\infty$, then the weight of $f\mathbf{v}$ is equal to $+\infty$.

DEFINITION IV.2.1. *The data of indeterminates* x_i, y_j, t_ℓ *with weight* 1, s_k *with weight* ϖ_k *given by* (IV.2.4) *and of first integrals* $z_j = x_j + \imath y_j$, $w_k = s_k + \imath\varphi_k(x,y,s,t)$ *such that the weight of the real formal power series* φ_k *is exactly equal to* ϖ_k $(1 \le i, j \le \nu,\ 1 \le k \le d,\ 1 \le \ell \le n-\nu)$ *is called* a normal form *for the integrable formal structure* $\mathcal{V}$.

In the forthcoming sections we shall establish the existence of normal forms. Before proceeding we consider a system of indeterminates and of first integrals for the integrable formal structure $\mathcal{V}$:

$$\{x_i, y_j, t_\ell, s_k, z_j = x_j + \imath y_j, w_k = s_k + \imath\varphi_k(x,y,s,t)\}_{1 \le i,j \le \nu, 1 \le k \le d, 1 \le \ell \le n-\nu}. \tag{IV.2.5}$$

The indeterminates shall be given weights as in Definition IV.2.1, but we shall not hypothesize that (IV.2.5) is a normal form for $\mathcal{V}$. It is convenient to partition the indeterminates s_k into subsets $s^{(\iota)} = \{s_{d_{\iota-1}+1},\dots,s_{d_\iota}\}$ for $1 \le \iota \le r+1$. We partition the first integrals w_k into analogous subsets $w^{(\iota)}$ $(1 \le \iota \le r+1)$.

PROPOSITION IV.2.1. *If* $m_1 < +\infty$ *and if* $1 \le k \le d_1$ *the sum* P_k *of monomials in the series* φ_k *with weight* $\le m_1$ *belongs to* $\mathbb{R}[x,y,t]$.

Assume $2 \le \iota \le r$, $m_\iota < +\infty$. *If* $d_{\iota-1} < k \le d_\iota$, *the sum* P_k *of monomials in the series* φ_k *with weight* $\le m_\iota$ *belongs to* $\mathbb{R}[x,y,s^{(1)},\dots,s^{(\iota-1)},t]$.

If weight $\varphi_k = +\infty$ $(d_r < k \le d)$ *then*

$$\varphi_k = \sum_{\ell=1}^{d_r} s_\ell \psi_{k\ell}, \tag{IV.2.6}$$

with power series $\psi_{k\ell} \in \mathbb{R}[[x,y,s,t]]$ *whose zero-order term vanishes.*

PROOF. The first two assertions in Proposition IV.2.1 follow from the fact that if P_k $(d_{\iota-1} < k \le d_\iota)$ were to depend on $s^{(\iota)}$ it would have to be linearly, which is excluded by (I.7.5). The last assertion is self-evident. ■

We observe that the notion of weight depends on the choice of indeterminates. Suppose we make use, as new indeterminates, of x_i, y_j, w_k (replacing s_k), t_ℓ and we assign to them the weights 1, 1, ϖ_k, 1. This defines new weights for formal power series. We contend that it is the same as the one defined by means of the s_k provided the (original) weight of $w_k - s_k$ is $\ge \varpi_k$ for each k. This is a consequence of:

LEMMA IV.2.1. *Let* x_i, y_j $(1 \le i, j \le \nu)$, t_ℓ $(1 \le \ell \le n-\nu)$ *be indeterminates with weight* 1, s_k *indeterminates with weight* $\varpi_k \ge 2$ $(1 \le k \le d$; none of these indeterminates is required to be real). *For each* $k = 1,\dots,d$ *let* $\psi_k \in \mathbb{C}[[x,y,s,t]]$ *be a power series whose terms of degree* ≤ 1 *all vanish and whose weight is*

$\geq \varpi_k$; *set* $w_k = s_k + \psi_k(x,y,s,t)$. *Then, for each* $k = 1,\ldots,d$, s_k *is a series in the powers of* x, y, w, t, *whose weight is equal to* ϖ_k *if we assign weight* 1 *to* x_i, y_j, t_ℓ *and weight* $\varpi_{k'}$ *to* $w_{k'}$ *for every* $k' = 1,\ldots,d$.

PROOF. Let $H_k^{(p)} \in \mathbb{C}[[x,y,w,t]]$ be power series defined inductively by the formulas

$$H_k^{(0)} = w_k, \; H_k^{(p+1)} = w_k - \psi_k(x,y,H^{(p)},t)$$

($p \in \mathbb{Z}_+$, $k = 1,\ldots,d$). We have used the notation $H^{(p)} = (H_1^{(p)},\ldots,H_d^{(p)})$. We shall prove, by induction, that weight of $H_k^{(p)} \geq \varpi_k$ for each k. The claim is obvious when $p = 0$; suppose it has been proved up to, and inclusively, p. We have

$$H_k^{(p+1)} = w_k - \sum_{\alpha\in\mathbb{Z}_+^d} (H^{(p)})^\alpha (\partial/\partial s)^\alpha \psi_k(x,y,0,t)/\alpha!.$$

The weight of $(\partial/\partial s)^\alpha \psi_k(x,y,0,t)$ is $\geq \varpi_k - \sum_{\ell=1}^{d} \alpha_\ell \varpi_\ell$. By the induction hypothesis that of $(H^{(p)})^\alpha$ is $\geq \sum_{\ell=1}^{d} \alpha_\ell \varpi_\ell$, which proves our claim for $p+1$. Going to the limit as $p \to +\infty$ yields that weight of $s_k = H_k(x,y,w,t) \geq \varpi_k$. But the weight of s_k cannot be $> \varpi_k$ for any k: otherwise, by reverting from the indeterminates w to the indeterminates s, we would conclude that weight of $w_k > \varpi_k$ for some k. ∎

PROPOSITION IV.2.2. *Let* (IV.2.5) *be a system of indeterminates and first integrals for the integrable formal structure* $\mathcal{V}$. *Let the weight of* x_i, y_j $(1 \leq i, j \leq \nu)$, t_ℓ $(1 \leq \ell \leq n-\nu)$ *be equal to* 1 *and that of* s_k *to* $\varpi_k \geq 2$ *for each* $k = 1,\ldots,d$. *Assume that, for each k, the weight of the series* φ_k *is* $\geq \varpi_k$. *Then, for each k, the formal vector field* N_k *defined in* (I.7.17) *has weight* $-\varpi_k$; *all the formal vector fields* L_j *defined in* (I.7.15) *and* (I.7.16), *as well as all the formal vector fields* M_h *defined in* (I.7.18), *have weight* -1.

It should be emphasized that we are not hypothesizing, in Proposition IV.2.2, that (IV.2.5) is a normal form for the formal structure $\mathcal{V}$.

PROOF. It suffices to prove the assertion about N_k, since the weights of $\partial\varphi_k/\partial z_i$ and $\partial\varphi_k/\partial t_\ell$ are $\geq \varpi_k - 1$. But if we switch to the indeterminates x_i, y_j, w_k, t_ℓ, we have $\mathrm{N}_k = \partial/\partial w_k$ whose weight is $-\varpi_k$. ∎

IV.3. Lemmas about Weights and Vector Fields

Until specified otherwise the scalar field will be $\mathbb{R}$. We deal with two sets of indeterminates: $X = (X_1,\dots,X_p)$, $S = (S_1,\dots,S_q)$ and one set of formal vector fields $A_1,\dots,A_p \in \mathbb{D}[[X,S]]$ having the following expressions:

$$A_i = \partial/\partial X_i + \sum_{k=1}^{q} a_{ik}\partial/\partial S_k,$$

where none of the formal power series $a_{ik} \in \mathbb{R}[[X,S]]$ is invertible.

To each indeterminate X_i we assign a weight $\varpi_i \in \mathbb{Z}_+$, $\varpi_i \geq 1$, and to all the indeterminates S_k we assign one and the same weight $\varpi > \operatorname{Max}_{1 \leq i \leq p} \varpi_i$; $\varpi \in \mathbb{Z}_+$ or $\varpi = +\infty$. The weights ϖ_i, ϖ give a weight to every formal vector field; recall that the weight of a vector field $S_k\partial/\partial S_\ell$ is equal to 0 if $\varpi < +\infty$, to $+\infty$ if $\varpi = +\infty$.

For each $i = 1,\dots,p$ we introduce the unique vector field

$$A_i^\circ = \partial/\partial X_i + \sum_{k=1}^{q} a_{ik}^\circ\partial/\partial S_k,$$

whose coefficients a_{ik}° are sums of monomials with weights $< \varpi - \varpi_i$ and which is such that the vector field

$$R_i = A_i - A_i^\circ = \sum_{k=1}^{q} R_{ik}\partial/\partial S_k$$

has weight $\geq -\varpi_i$. Thus the weight of A_i° is $< -\varpi_i$ unless all the series a_{ik}° vanish identically, in which case the weight of A_i° is exactly equal to $-\varpi_i$. As a consequence, for all i, k, $a_{ik}^\circ \in \mathbb{R}[[X]]$ and the weight of $R_{ik} \in \mathbb{R}[[X,S]]$ is $\geq \varpi - \varpi_i$.

Let $J = \{j_1,\dots,j_\lambda\}$ be a set of λ positive integers $j \leq p$ (the same integer j may appear repeatedly in J); the case $\lambda = 0$, i.e., $J = \emptyset$, is not precluded. We refer to λ as the *length* of J and to the integer $\varpi_J = \varpi_{j_1} + \cdots + \varpi_{j_\lambda}$ as its *weighted length*. We define C_J as the identity operator if $\lambda = 0$; $C_J = A_j$ if $\lambda = 1$ and $J = \{j\}$; and if $\lambda \geq 2$,

$$C_J = [A_{j_1},[\cdots[A_{j_{\lambda-1}},A_{j_\lambda}]\cdots].$$

We shall also use the analogous notation C_J° when A_j° is substituted for A_j.

LEMMA IV.3.1. *Suppose the following condition is satisfied*:

whatever the set J of integers $j \in [1,\dots,p]$ of length $\lambda \geq 2$, of weighted length $\varpi_J < \varpi$, we have $C_J|_0 = 0$. (IV.3.1)

Then there is a change of indeterminates

$$(X,S) \to (X,G(X,S)) \text{ with } G(0,0) = 0, \tag{IV.3.2}$$

such that, in the new indeterminates, $A_i^\circ = \partial/\partial X_i$ for each $i = 1,\dots,p$.

It is important to note that property (IV.3.1) is not invalidated by any change of indeterminates (IV.3.2) although, in general, the latter will modify the vector fields A_i°. It is also to be understood that through any sequence of changes (IV.3.2) the weights of X_i remain equal to ϖ_i and that of every S_k is set equal to ϖ.

Proof. First let $G_k(X,S) \in \mathbb{R}[[X,S]]$ be the unique solution of the Cauchy problem

$$A_p G_k = 0, \; G_k|_{X=0} = S_k. \tag{IV.3.3}$$

The corresponding change of indeterminates (IV.3.2) transforms the vector field A_p into $\partial/\partial X_p$.

From there on we reason by induction on $p \geq 1$, assuming that $A_p = \partial/\partial X_p$. When $p = 1$ the conclusion in Lemma IV.3.1 is valid.

Henceforth take $p > 1$. For $1 \leq j < p$ consider the vector field $\tilde{A}_j$ obtained by putting $X_p = 0$ in the coefficients of the vector field A_j; call $\tilde{C}_J$ the commutators analogous to C_J but with J consisting solely of positive integers $j < p$. We derive at once from (IV.3.1) that $\tilde{C}_J|_0 = 0$ if $\varpi_J < \varpi$. We may avail ourselves of the induction on p: we conclude that there is a change of indeterminates

$$(X_1,\ldots,X_{p-1},S) \to (X_1,\ldots,X_{p-1},G(X_1,\ldots,X_{p-1},S)) \tag{IV.3.4}$$

such that, in the new indeterminates, $\tilde{A}_i^\circ = \partial/\partial X_i$ for each $i = 1,\ldots,p-1$. Note that the change (IV.3.4) has no effect on A_p. In other words we have the right to assume that

$$a_{ik}^\circ(X) = X_p c_{ik}^\circ(X) \text{ if } 1 \leq i < p, \; 1 \leq k \leq q. \tag{IV.3.5}$$

Lemma IV.3.1 will then follow from:

Lemma IV.3.2. *Suppose* $A_p = \partial/\partial X_p$ *and that* (IV.3.5) *holds. Then* (IV.3.1) *entails*:

$$[A_p, A_i^\circ] = 0 \text{ if } 1 \leq i < p. \tag{IV.3.6}$$

Indeed (IV.3.6) entails $c_{ik}^\circ \equiv 0$ for all $k = 1,\ldots,q$, which is to say, $A_i^\circ = \partial/\partial x_i$ for all $i < p$, hence for all $i \leq p$.

Proof of Lemma IV.3.2. In what follows i will always stand for a positive integer $< p$. We have

$$[A_p, A_i^\circ] = \sum_{k=1}^{q} (\partial a_{ik}^\circ/\partial X_p)\partial/\partial S_k.$$

If (IV.3.6) were false we could find a set $H = \{h_1,\ldots,h_\lambda\}$ of positive integers $\leq p$ such that the freezing at 0 of

$$[A^\circ_{h_1},[\cdots[A^\circ_{h_\lambda},[A_p,A^\circ_i]]\cdots]] = \sum_{k=1}^{q}\Big[(\partial/\partial X_{h_1})\cdots(\partial/\partial X_{h_\lambda})(\partial a^\circ_{ik}/\partial X_p)\Big]\partial/\partial S_k$$

would not vanish. But since the weight of every monomial in the series a°_{ik} is $< \varpi - \varpi_i$ this is impossible unless the weighted length of $H\cup\{p\}$ is $< \varpi - \varpi_i$. We are going to show that

$$[A^\circ_{h_1},[\cdots[A^\circ_{h_\lambda},[A^\circ_i,A_p]]\cdots]]|_0 = 0 \tag{IV.3.7}$$

for all sets H of positive integers $\leq p$ with weighted length $\varpi_H < \varpi - \varpi_i - \varpi_p$.

Let us call $\mathcal{L}$ the subset of $\mathbb{D}[[X,S]]$ consisting of the vector fields

$$\sum_{k=1}^{q} c_k\partial/\partial S_k; \tag{IV.3.8}$$

$\mathcal{L}$ is a Lie algebra for the commutation bracket. For any $\rho \in \mathbb{Z}_+$ we denote by $\mathcal{L}^\rho$ the set of vector fields (IV.3.8) whose coefficients c_k have weight $\geq \rho$. Note that if $\varpi_J \leq \rho$, then

$$R \in \mathcal{L}^\rho \Rightarrow [\partial/\partial X_{j_1},[\cdots[\partial/\partial X_{j_\lambda},R]\cdots] \in \mathcal{L}^{\rho-\varpi_J}. \tag{IV.3.9}$$

On the other hand, if $\varpi \leq \rho + \rho'$ we have

$$[\mathcal{L}^\rho,\mathcal{L}^{\rho'}] \subset \mathcal{L}^{\rho+\rho'-\varpi}. \tag{IV.3.10}$$

Let μ, ν be integers ≥ 0. We call $\mathcal{I}_{\mu,\nu}$ the ideal in $\mathbb{R}[[X,S]]$ spanned by the coefficients $a^\circ_{\ell k}$ $(1 \leq \ell < p,\ 1 \leq k \leq q)$ and by their derivatives

$$(\partial/\partial X_1)^{\gamma_1}\cdots(\partial/\partial X_p)^{\gamma_p}a^\circ_{\ell k},\ \gamma_1+\cdots+\gamma_{p-1} \leq \mu,\ \gamma_p \leq \nu.$$

We denote by $\mathbb{D}_{\mu,\nu}$ the set of vector fields (IV.3.8) with coefficients $c_k \in \mathcal{I}_{\mu,\nu}$. Since the coefficients $a^\circ_{\ell k}$ are independent of S, $\mathbb{D}_{\mu,\nu}$ is an ideal in the Lie algebra $\mathcal{L}$.

Then let $H = \{h_1,\ldots,h_\lambda\}$ be an arbitrary set of λ positive integers $\leq p$, of weighted length $\varpi_H < \varpi - \varpi_i - \varpi_p$. For the sake of brevity we shall write $J = \{h_1,\ldots,h_\lambda,i,p\}$; $\varpi_J < \varpi$. We associate a new "index" to the set J: the number μ of integers h_ι $(1 \leq \iota \leq \lambda)$ which are $< p$. We claim that

$$C^\circ_J - C_J \in \mathcal{L}^{\varpi-\varpi_J} + \sum_{\kappa=0}^{\mu}\mathbb{D}_{\kappa,\lambda-\kappa}. \tag{IV.3.11}$$

We shall prove (IV.3.11) by ascending induction on λ. We have

$$[A_i,A_p] - [A^\circ_i,A^\circ_p] = [R_i,\partial/\partial X_p];$$

as $[\partial/\partial X_p, R_i] \in \mathcal{L}^{\varpi-\varpi_i-\varpi_p}$ we obtain (IV.3.11) when $\lambda = \mu = 0$.

Next suppose that property (IV.3.11) has been proved for any set $H = \{h_1,\ldots,h_\lambda\}$ with exactly μ elements $< p$. We assume there is a positive integer $\ell \le p$ such that $\varpi_\ell + \varpi_H < \varpi - \varpi_i - \varpi_p$. We continue to write $J = \{h_1,\ldots,h_\lambda,i,p\}$; we have

$$[A_\ell^\circ, C_J^\circ] - [A_\ell, C_J] = [A_\ell, C_J^\circ - C_J] - [R_\ell, C_J^\circ] =$$

$$[\partial/\partial X_\ell, C_J^\circ - C_J] + [A_\ell^\circ - \partial/\partial X_\ell, C_J^\circ - C_J] + [R_\ell, C_J^\circ - C_J] - [R_\ell, C_J^\circ].$$

We note that the vector field $[A_\ell^\circ - \partial/\partial X_\ell, C_J^\circ - C_J]$ belongs to $\mathbb{D}_{0,0}$ if $\ell < p$ and vanishes when $\ell = p$ (since $A_p^\circ = \partial/\partial X_p$). We have, by (IV.3.11):

$$[\partial/\partial X_\ell, C_J^\circ - C_J] \in [\partial/\partial X_\ell, \mathcal{L}^{\varpi-\varpi_J}] + [\partial/\partial X_\ell, \sum_{\kappa=0}^{\mu} \mathbb{D}_{\kappa,\lambda-\kappa}]$$

$$\subset \mathcal{L}^{\varpi-\varpi_J-\varpi_\ell} + \sum_{\kappa=0}^{\mu} \mathbb{D}_{\kappa+1,\lambda-\kappa} \text{ if } \ell < p;$$

$$\subset \mathcal{L}^{\varpi-\varpi_J-\varpi_\ell} + \sum_{\kappa=0}^{\mu} \mathbb{D}_{\kappa,\lambda-\kappa+1} \text{ if } \ell = p.$$

But note that

$$\sum_{\kappa=0}^{\mu} \mathbb{D}_{\kappa+1,\lambda-\kappa} = \sum_{\kappa=1}^{\mu+1} \mathbb{D}_{\kappa,\lambda+1-\kappa}.$$

Next we look at

$$[R_\ell, C_J^\circ - C_J] \in [R_\ell, \mathcal{L}^{\varpi-\varpi_J}] + [R_\ell, \sum_{\kappa=0}^{\mu} \mathbb{D}_{\kappa,\lambda-\kappa}]$$

$$\subset \mathcal{L}^{\varpi-\varpi_J-\varpi_\ell} + \sum_{\kappa=0}^{\mu} \mathbb{D}_{\kappa,\lambda-\kappa}$$

by (IV.3.10) and since $\mathbb{D}_{\kappa,\lambda-\kappa}$ is an ideal. Last we observe that the coefficients of $[R_\ell, C_J^\circ]$ belong to the span of those of C_J°, hence to $\mathcal{I}_{\mu,\lambda+1-\mu}$. This completes the proof of (IV.3.11) with $\lambda+1$ substituted for λ.

We freeze at 0 both sides in (IV.3.11). By virtue of (IV.3.1) and of the fact that $\varpi_J < \varpi$ we get

$$C_J^\circ|_0 \in \sum_{\kappa=0}^{\mu} \mathbb{D}_{\kappa,\lambda-\kappa}|_0. \tag{IV.3.12}$$

The left-hand side in (IV.3.12) is equal to that in (IV.3.7). Keeping λ and μ fixed we let the set $\{h_1,\ldots,h_\lambda,i\}$ vary arbitrarily (but always with $i < p$). At this juncture we observe that the coefficients of the vector fields $C_J^\circ|_0$ span $\mathcal{I}_{0,\lambda+1}|_0$ modulo $\mathcal{I}_{0,\lambda}|_0$ when $\mu = 0$ and that they span $\mathcal{I}_{\mu,\lambda+1-\mu}|_0$ mod $\{\mathcal{I}_{\mu,\lambda-\mu} +$

$\mathcal{I}_{\mu-1,\lambda+1-\mu}\}|_0$ when $\mu \geq 1$. When $\mu = 0$ we derive from (IV.3.12), by induction on λ:

$$\mathcal{I}_{0,\lambda+1}|_0 \subset \mathcal{I}_{0,\lambda}|_0 \subset \cdots \subset \mathcal{I}_{0,0}|_0 = 0.$$

When $\mu \geq 1$ we use induction on μ also and assume that $\mathcal{I}_{\mu-1,\lambda+1-\mu}|_0 = 0$. We derive then from (IV.3.12) and by induction on λ

$$\mathcal{I}_{\mu,\lambda+1-\mu}|_0 \subset \mathcal{I}_{\lambda,0}|_0 + \sum_{\kappa=0}^{\inf(\lambda-1,\mu)} \mathcal{I}_{\kappa,\lambda-1-\kappa}|_0 \subset$$

$$\mathcal{I}_{\lambda,0}|_0 + \sum_{\kappa=0}^{\inf(\lambda-2,\mu)} \mathcal{I}_{\kappa,\lambda-2-\kappa}|_0 = \cdots = \mathcal{I}_{\lambda,0}|_0.$$

It follows at once from (IV.3.5) that $\mathcal{I}_{\lambda,0}|_0 = 0$ whatever $\lambda \in \mathbb{Z}_+$, whence (IV.3.7). ■

We introduce additional vector fields

$$\mathrm{B}_j = \sum_{k=1}^{q} b_{jk}\partial/\partial S_k,\ j = 1,\ldots,p',$$

with coefficients $b_{jk} \in \mathbb{R}[[X,S]]$ none of which is invertible. We extend the definition of the commutators C_J to allow substitution of B_{j_α} for A_{j_α} for any number of indices j_α $(1 \leq \alpha \leq \lambda)$, with the agreement that, when this happens, ϖ_{j_α} *must be taken equal to one*. With this proviso we still refer to ϖ_J as the weighted length of C_J, and to (IV.3.1) thus modified, as *the extended hypothesis* (IV.3.1).

For each $j = 1,\ldots,p'$ we define the unique vector field

$$\mathrm{B}_j^\circ = \sum_{k=1}^{q} b_{jk}^\circ\partial/\partial S_k,$$

with each b_{jk}° a sum of monomials of weight $< \varpi - 1$ (thus the weight of B_j° is < -1 unless all the coefficients b_{jk}° vanish identically), such that the weight of the vector field $\mathrm{B}_j - \mathrm{B}_j^\circ$ is ≥ -1. These requirements have the consequence that, for all j, k, $b_{jk}^\circ \in \mathbb{R}[X]$.

LEMMA IV.3.3. *Suppose the extended hypothesis* (IV.3.1) *is satisfied. There is a change of indeterminates* (IV.3.2) *such that, in the new indeterminates, the weight of* A_i *is* $\geq -\varpi_i$ $(1 \leq i \leq p)$ *and that of* B_j *is* ≥ -1 $(1 \leq j \leq p')$.

The conclusion in Lemma IV.3.3 is equivalent to the fact that, after a change of indeterminates (IV.3.2),

$$\mathrm{A}_i^\circ = \partial/\partial X_i,\ \mathrm{B}_j^\circ \equiv 0,\ 1 \leq i \leq p,\ 1 \leq j \leq p'. \qquad \text{(IV.3.13)}$$

PROOF. Since the unmodified hypothesis (IV.3.1) holds, we may carry out the change of indeterminates in Lemma IV.3.1. In other words, we may assume $A_i^\circ = \partial/\partial X_i$ $(1 \le i \le p)$. If $H = \{h_1,\ldots,h_\lambda\}$ $(1 \le h_\alpha \le p)$ we set $C_{H,j} = [A_{h_1}, [\cdots[A_{h_\lambda},B_j]]\cdots]]$, $C_{H,j}^\circ = [A_{h_1}^\circ,[\cdots[A_{h_\lambda}^\circ,B_j^\circ]]\cdots]]$. It suffices to prove, for any j, $1 \le j \le p'$,

$$C_{H,j}^\circ|_0 = 0 \tag{IV.3.14}$$

for all sets H of λ positive integers $h_a \le p$. But since every monomial in every coefficient b_{jk}° has weight $< \varpi - 1$, it is enough to consider sets H with weighted length $\varpi_H < \varpi - 1$. In what follows j is kept fixed; we reason by induction on λ.

We continue to use the notation $\mathcal{L}^\rho$ as in the proof of Lemma IV.3.2. If j, ν are integers ≥ 0, $1 \le j \le p'$, we call $\mathcal{I}_{j,\nu}'$ the ideal in $\mathbb{R}[[X,S]]$ spanned by the coefficients b_{jk}° $(1 \le k \le q)$ and by their derivatives

$$(\partial/\partial X_1)^{\gamma_1}\cdots(\partial/\partial X_p)^{\gamma_p} b_{jk}^\circ,\ \gamma_1 + \cdots + \gamma_p \le \nu.$$

We denote by $\mathbb{D}_{j,\nu}'$ the set of vector fields (IV.3.8) with coefficients $c_k \in \mathcal{I}_{j,\nu}'$. Since the coefficients b_{jk}° are independent of S, $\mathbb{D}_{j,\nu}'$ is an ideal in the Lie algebra $\mathcal{L}$. Then setting $J = \{h_1,\ldots,h_\lambda,j\}$ and noting that $\varpi_J = \varpi_H + 1 < \varpi$, we claim that

$$C_{H,j}^\circ - C_{H,j} \in \mathcal{L}^{\varpi-\varpi J} + \mathbb{D}_{j,\lambda-1}'. \tag{IV.3.15}$$

We prove (IV.3.15) by induction on λ. By the definition of B_j° and with the agreement that $\mathbb{D}_{j,-1}' = 0$ the claim is valid when $\lambda = 0$. Suppose it has been proved up to λ. Then, if $1 \le \ell \le p$,

$$[A_\ell^\circ,C_{H,j}^\circ] - [A_\ell,C_{H,j}] = [A_\ell^\circ,C_{H,j}^\circ - C_{H,j}] + [R_\ell,C_{H,j}^\circ - C_{H,j}] - [R_\ell,C_{H,j}^\circ].$$

We have

$$[A_\ell^\circ,C_{H,j}^\circ - C_{H,j}] \in [\partial/\partial X_\ell,\mathcal{L}^{\varpi-\varpi_J}] + [\partial/\partial X_\ell,\mathbb{D}_{j,\lambda-1}'] \in \mathcal{L}^{\varpi-\varpi_J-\varpi_\ell} + \mathbb{D}_{j,\lambda}'$$

by (IV.3.9) and (IV.3.15). Next we observe that

$$[R_\ell,\mathcal{L}^{\varpi-\varpi_J}] + [R_\ell,\mathbb{D}_{j,\lambda-1}'] \in \mathcal{L}^{\varpi-\varpi_J-\varpi_\ell} +$$
$$[R_\ell,\mathbb{D}_{j,\lambda-1}'] \in \mathcal{L}^{\varpi-\varpi_J-\varpi_\ell} + \mathbb{D}_{j,\lambda-1}'.$$

Finally we note that the coefficients of $[R_\ell,C_{H,j}^\circ]$ belong to the span of those of $C_{H,j}^\circ$, hence to $\mathcal{I}_{j,\lambda}'$, which completes the proof of (IV.3.15) with $\lambda+1$ substituted for λ. We derive, by virtue of the extended hypothesis (IV.3.1) and of the fact that $\varpi_J < \varpi$,

$$C_{H,j}^\circ|_0 \in \mathbb{D}_{j,\lambda-1}'|_0.$$

But the coefficients of $C_{H,j}^\circ$ span $\mathcal{I}_{j,\lambda}'$ as H varies over the family of all sets of λ positive integers $\le p$. We derive

$$\mathcal{I}'_{j,\lambda}|_0 \subset \mathcal{I}'_{j,\lambda-1}|_0 \subset \cdots \subset \mathcal{I}'_{j,0}|_0 = 0$$

whereby the proof of Lemma IV.3.3 is complete. ∎

IV.4. Existence of Basic Vector Fields of Weight -1

Given indeterminates x_i, y_j $(1 \leq j \leq \nu)$, s_k $(1 \leq k \leq d)$ (cf. (IV.2.5)) we associate to them the formal vector fields L_j given by (I.7.15), (I.7.16) (with coefficients that are formal power series).

THEOREM IV.4.1. *The indeterminates x_i, y_j $(j = 1,\ldots,\nu)$, s_k $(k = 1,\ldots,d)$, t_ℓ $(\ell = 1,\ldots,n-\nu)$ can be chosen in such a manner that the weight of all the formal vector fields L_j $(j = 1,\ldots,n)$ be equal to -1.*

PROOF. We shall use the notation

$$N_0 = n+\nu\ (= \dim(\mathfrak{g}_1|_0)),\ N_\alpha = \dim(\mathfrak{g}_{m_\alpha}|_0)\ (1 \leq \alpha \leq r),\ N = m+n.$$

We start from a system of indeterminates and first integrals (IV.2.5) that do not constitute a normal form for $\mathcal{V}$. However, it is convenient to modify, for the duration of the proof of Theorem IV.4.1, the notation for the indeterminates. We shall write

$$X_i = 2x_i,\ X_{\nu+j} = 2y_j\ (i,j = 1,\ldots,\nu),$$
$$X_{2\nu+\ell} = t_\ell\ (\ell = 1,\ldots,n-\nu),\ \mathrm{X}_{n+\nu+k} = s_k\ (k = 1,\ldots,d). \tag{IV.4.1}$$

Consider the following (formal) real vector fields:

$$\vartheta_i = \mathcal{R}e\mathrm{L}_i,\ \vartheta_{\nu+j} = \mathcal{I}m\mathrm{L}_j\ (1 \leq i,j \leq \nu),$$

$$\vartheta_{2\nu+k} = \mathcal{R}e\mathrm{L}_{\nu+k},\ \vartheta'_\ell = \mathcal{I}m\mathrm{L}_{\nu+\ell}\ (1 \leq k,\ell \leq n-\nu).$$

The real "tangent vectors" $\vartheta_i|_0 = \partial/\partial X_i$ $(1 \leq i \leq N_0)$ form a basis of $\mathfrak{g}_1|_0$. By virtue of (I.7.15) and (I.7.16), if $1 \leq i \leq N_0$,

$$\vartheta_i = \partial/\partial X_i + \sum_{k>N_0}^{N} a_{ik}\partial/\partial X_k\ (a_{ik}|_0 = 0,\ N_0 < k \leq N). \tag{IV.4.2}$$

We can select real formal vector fields ϑ_k, $k = N_0+1,\ldots,N_r$, such that

$$\vartheta_k \in \mathfrak{g}_{m_1}\backslash\mathfrak{g}_1 \ \textit{if}\ N_0 < k \leq N_1,$$

$$\vartheta_k \in \mathfrak{g}_{m_\alpha}\backslash\mathfrak{g}_{m_{\alpha-1}} \ \textit{if}\ N_{\alpha-1} < k \leq N_\alpha\ (2 \leq \alpha \leq r;\ \text{cf. (IV.2.4)}); \tag{IV.4.3}$$

the vector fields $\vartheta_k|_0$ $(1 \leq k \leq N_r)$ are linearly independent. (IV.4.4)

According to (I.7.15), (I.7.16) every formal vector field belonging to $\mathfrak{g}(\mathcal{V})\backslash\mathfrak{g}_1$ only involves partial differentiations with respect to the indeterminates X_k

$(N_0 < k \le N)$. After a linear change of these indeterminates we may as well assume that (IV.4.2) is also valid for $N_0 < i \le N_r$.

Finally, if $N_r < N$ (i.e., if the integrable formal algebra $\mathcal{V}$ is not of finite type), we set

$$\vartheta_k = \partial/\partial X_k,\ k = N_r+1,\dots,N. \tag{IV.4.5}$$

The real vector fields ϑ_k $(k = 1,\dots N)$ span $\mathbb{D}[[X]]$ over $\mathbb{C}[[X]]$.

After multiplying the system of vector fields $(\vartheta_{N_0+1},\dots,\vartheta_{N_1})$ by the inverse of the matrix $(\delta_{ij} + a_{ij})_{N_0<i,j\le N_1}$ we may as well assume that, for $i = N_0+1,\dots,N_1$,

$$\vartheta_i = \partial/\partial X_i + \sum_{k>N_1}^{N} a_{ik}\partial/\partial X_k\ (a_{ik}|_0 = 0,\ N_1 < k \le N). \tag{IV.4.6}$$

For $i > N_1$,

$$\vartheta_i = \partial/\partial X_i + \sum_{k>N_1}^{N}\left[a_{ik} - \sum_{\ell>N_0}^{N_1} a_{i\ell}a_{\ell k}\right]\partial/\partial X_k$$

mod $(\vartheta_{N_0+1},\dots,\vartheta_{N_1})$. We may therefore select each ϑ_i to have the expression (IV.4.6), even for $i > N_1$. But then we may multiply the system of vector fields $(\vartheta_{N_1+1},\dots,\vartheta_{N_2})$ by the inverse of the matrix $(\delta_{ij} + a_{ij})_{N_1<i,j\le N_2}$. Repeating r times the same procedure we end up, if $N_{\alpha-1} < i \le N_\alpha$, $1 \le \alpha \le r$, with formal vector fields

$$\vartheta_i = \partial/\partial X_i + \sum_{k>N_\alpha}^{N} a_{ik}\partial/\partial X_k\ (a_{ik}|_0 = 0,\ N_\alpha < k \le N). \tag{IV.4.7}$$

Comparing with (IV.4.2) we see that (IV.4.7) is also valid for $\alpha = 0$, $1 \le i \le N_0$.

Note also that, for $j = 1,\dots,n-\nu$:

$$\vartheta_j' = \sum_{k>N_0}^{N} b_{jk}\partial/\partial X_k\ (b_{jk} \in \mathbb{R}[[X]]). \tag{IV.4.8}$$

Although we are going to carry out a sequence of changes of the indeterminates X_i it is convenient to spell out right away what the final weights will be: at the end of the construction, and for each i, $1 \le i \le N$, the weight of X_i will be the number ϖ_i defined according to the following rule:

$$\varpi_i = 1 \textit{ if } 1 \le i \le N_0;$$
$$\varpi_i = m_\alpha \textit{ if } N_{\alpha-1} < i \le N_\alpha\ (1 \le \alpha \le r); \tag{IV.4.9}$$
$$\varpi_i = +\infty \textit{ if } N_r < i \le N.$$

Comparing with (IV.2.4) the reader will notice the change in the index notation (but not in the values of the weight) if we accept the correspondence (IV.4.1).

If $N_0 = N$, we have $\vartheta_i = \partial/\partial X_i$ for every $i = 1,\dots,N$, $\vartheta'_{j'} = 0$ for all $j' = 1,\dots,n-\nu$. We do not modify these vector fields any further and assign the weight $+1$ to every indeterminate X_i.

Suppose $N_0 < N$. We apply Lemma IV.3.3 with the choices $\mathrm{A}_i = \vartheta_i$, $i = 1,\dots,p = N_0$; $\mathrm{B}_j = \vartheta'_j$, $j = 1,\dots,p' = n-\nu$; $S_k = X_{N_0+k}$, $k = 1,\dots,q = N-N_0$; $\varpi_i = 1$, $\varpi = m_1$. The extended hypothesis (IV.3.1) is satisfied by the very definition of the Hörmander number m_1. We conclude that there is a change of indeterminates

$$(X_1,\dots,X_{N_0},X_{N_0+1},\dots,X_N) \to (X_1,\dots,X_{N_0},G_{N_0+1}(X),\dots,G_N(X))$$

such that the weight, in the new indeterminates system, of the vector fields ϑ_i $(1 \le i \le N_0)$ and ϑ'_j $(1 \le j \le n-\nu)$ is ≥ -1. This means that the coefficients a_{ik} and b_{jk} will have weight $\ge m_1 - 1$ if $1 \le i \le N_0 < k \le N$, $1 \le j \le n-\nu$. (With respect to the choice (IV.4.9) the weight of the indeterminates X_k, $k > N_1$, has been decreased from ϖ_k to $m_1 < \varpi_k$.)

Suppose $N_1 = N$; this means that $\mathfrak{g}_{m_1}|_0 = \mathrm{T}_0$. The argument stops: we adopt the rule (IV.4.9) and all vectors ϑ_i $(1 \le i \le N_0)$ and ϑ'_j $(1 \le j \le n-\nu)$ have weight -1. If $N_0 < i \le N$, $\vartheta_i = \partial/\partial X_i$ has weight $-m_1$. In particular, when $m_1 = +\infty$ the coefficients of $\vartheta_i - \partial/\partial X_i$ and ϑ'_j belong to the ideal in $\mathbb{R}[[X]]$ generated by $X_{N_0+1},\dots,X_N$.

If $N_1 < N$ the argument is repeated, with appropriate modifications that we now proceed to describe. Actually we shall reason in general: we suppose that we have reached the following stage, for some $\alpha \ge 1$. The formal vector fields (IV.4.7) and (IV.4.8) satisfy the following condition:

Suppose weight of $X_i = \varpi_i$ *if* $1 \le i \le N_\alpha$, weight of $X_k = m_\alpha$ *if* $N_\alpha < k \le N$; *then* weight of $a_{ik} \ge \inf(\varpi_k, m_\alpha) - \varpi_i$ *if* $N_{\beta-1} < i \le N_\beta < k \le N$ $(0 \le \beta < \alpha)$, weight of $b_{jk} \ge \inf(\varpi_k, m_\alpha) - 1$ *if* $1 \le j \le n-\nu$, $N_0 < k \le N$. (IV.4.10)

Note that (IV.4.10) describes the situation at stage $\alpha = 1$.

If $N_\alpha = N$ the argument stops. Henceforth we suppose $N_\alpha < N$.

We let the inverse of the matrix

$$(\delta_{jk} + a_{jk})_{1 \le j,k \le N_\alpha} \tag{IV.4.11}$$

act on the system of vector fields $(\vartheta_1,\dots,\vartheta_{N_\alpha})$, getting new vector fields

$$\mathrm{A}_i = \partial/\partial X_i + \sum_{k>N_\alpha}^{N} \tilde{a}_{ik}\partial/\partial X_k,\ i = 1,\dots,p = N_\alpha.$$

The matrix (IV.4.11) is "block triangular" in the sense that $a_{jk} \equiv 0$ if $k > N_{\beta-1}$, $j, k \le N_\beta$ $(1 \le \beta \le \alpha)$. We also introduce the unique vector field

$$\mathrm{B}_j = \sum_{k>N_\alpha}^{N} \tilde{b}_{jk}\partial/\partial X_k$$

such that $\vartheta'_j - \mathrm{B}_j \in \mathrm{Span}(\mathrm{A}_1,\ldots,\mathrm{A}_p)$ $(j = 1,\ldots,p' = n-\nu)$. Let us write, for $N_{\beta-1} < i \leq N_\beta$ $(0 \leq \beta \leq \alpha)$, $1 \leq j \leq n-\nu$,

$$\mathrm{A}_i = \vartheta_i + \sum_{k>N_\beta}^{N_\alpha} a'_{ik}\vartheta_k,\ \mathrm{B}_j = \vartheta'_j + \sum_{k>N_0}^{N_\alpha} b'_{jk}\vartheta_k. \tag{IV.4.12}$$

LEMMA IV.4.1. *Suppose* (IV.4.10) *holds. Then*

$$\text{weight of } a'_{ik} \geq \varpi_k - \varpi_i;\ \text{weight of } b'_{jk} \geq \varpi_k - 1.$$

PROOF OF LEMMA IV.4.1. Consider the following linear relations between formal vector fields ξ_i and η_j $(1 \leq i, j \leq N_\alpha)$ with coefficients in $\mathbb{R}[[X]]$,

$$\eta_i = \xi_i + \sum_{k>N_\beta}^{N_\alpha} a_{ik}\xi_k\ (N_{\beta-1} < i \leq N_\beta,\ 0 \leq \beta \leq \alpha).$$

Suppose that, for each $i = 1,\ldots,N_\alpha$, the weight of ξ_i is equal to $-\varpi_i$ (as, for instance, when $\xi_i = \partial/\partial X_i$). By virtue of (IV.2.10) this is equivalent to assigning the weight $-\varpi_i$ to η_i. We introduce a new indeterminate T with weight $+1$ and multiply both sides of the preceding equation by T^{ϖ_i}. Replacing ξ_i by $T^{\varpi_i}\xi_i$, η_j by $T^{\varpi_j}\eta_j$ and a_{ik} by $T^{\varpi_i-\varpi_k}a_{ik}$ (which belongs to the field of quotients of $\mathbb{R}[[X,T]]$) leads to similar equations but in which the weights of the coefficients are ≥ 0 (by virtue of (IV.4.10)). Solving the latter equations with respect to the $\mathrm{T}^{\varpi_i}\xi_i$ leads to equations

$$\xi_i = \eta_i + \sum_{k>N_\beta}^{N_\alpha} a^{\#}_{ik}T^{\varpi_k-\varpi_i}\eta_k$$

where $a^{\#}_{ik} = a'_{ik}T^{\varpi_i-\varpi_k}$ have weight ≥ 0; thus weight of $a'_{ik} \geq \varpi_k - \varpi_i$. To see that weight of $b'_{jk} \geq \varpi_k - 1$ it suffices to substitute the expressions of the ξ_k in the linear forms $\sum_{k>N_0}^{N_\alpha} b'_{jk}\xi_k$ (see (IV.4.8)). ■

We resume the proof of Theorem IV.4.1. The next step is to prove that the extended hypothesis (IV.3.1) is satisfied with $\varpi = m_{\alpha+1}$. Suppose it were not: there would be a commutator

$$\mathrm{C} = [\mathbf{v}_1,[\cdots[\mathbf{v}_{\lambda-1},\mathbf{v}_\lambda]\cdots]],$$

where each formal vector field $\mathbf{v}_\iota$ is equal to a vector field A_i or B_j, of length $\lambda \geq 2$ and weighted length $< m_{\alpha+1}$, such that $\mathrm{C}|_0 \neq 0$ (the weighted length of

C is equal to $\kappa_1 + \cdots + \kappa_\lambda$ with $\kappa_\iota = m_i$ if $\mathbf{v}_\iota = A_i$, $\kappa_\iota = 1$ if $\mathbf{v}_\iota = B_j$). In view of (IV.4.12) we see that C is a sum of commutators

$$\tilde{C} = [g_1\tilde{\mathbf{v}}_1,[\cdots[g_{\lambda-1}\tilde{\mathbf{v}}_{\lambda-1},g_\lambda\tilde{\mathbf{v}}_\lambda]\cdots]]$$

where, for each $\iota = 1,\ldots,\lambda$, $\tilde{\mathbf{v}}_\iota$ is one of the vector fields ϑ_i $(1 \le i \le N_\alpha)$ or ϑ'_j $(1 \le j \le n-\nu)$; $g_\iota \in \mathbb{R}[[X,S]]$ and either $g_\iota \equiv 1$ or else $\tilde{\mathbf{v}}_\iota = \vartheta_{k_\iota}$ for some k_ι, $N_0 < k_\iota \le N_\alpha$, and then weight of $g_\iota \ge \omega_\iota = \varpi_{k_\iota} - \varpi_{\ell_\iota}$ for some $\ell_\iota < k_\iota$. With the understanding that $\omega_\iota = 0$ when $g_\iota \equiv 1$ the constraint on the weighted length of C demands

$$\varpi_{k_1} - \omega_1 + \cdots + \varpi_{k_\lambda} - \omega_\lambda \le m_{\alpha+1} - 1. \tag{IV.4.13}$$

Lemma IV.4.2. *Suppose* (IV.4.10) *holds. The commutator* $\tilde{C}$ *is equal to a linear combination, with coefficients in* $\mathbb{R}[[X,S]]$, *of commutators of vector fields* $\tilde{\mathbf{v}}_\iota$ *whose weighted length is* $< m_{\alpha+1}$ (if we assign weight $-\varpi_i$ to ϑ_i, $1 \le i \le N_\alpha$, and weight -1 to ϑ'_j, $1 \le j \le n-\nu$) *and of formal vector fields with coefficients whose zero-order terms vanish.*

Proof of Lemma IV.4.2. We observe that $\tilde{C}$ is a sum of terms

$$D_1g_1\cdots D_\lambda g_\lambda[\tilde{\mathbf{v}}_{p_1},[\cdots[\tilde{\mathbf{v}}_{p_{\mu-1}},\tilde{\mathbf{v}}_{p_\mu}]\cdots]] \tag{IV.4.14}$$

with $1 \le p_1 < \cdots < p_\mu \le \lambda$ and differential operators D_ι that are either the identity or else differential monomials $\tilde{\mathbf{v}}_{q_{\iota 1}}\cdots\tilde{\mathbf{v}}_{q_{\iota\mu_\iota}}$ with $1 \le q_{\iota 1} < \cdots < q_{\iota\mu_\iota} \le \lambda$. Actually, the multi-indices $\{p_1,\ldots,p_\mu\}$, $\{q_{\iota 1},\ldots,q_{\iota\mu_\iota}\}$ $(\iota = 1,\ldots,\lambda)$ must form a partition of the integral interval $\{1,\ldots,\lambda\}$. The case $\mu = 1$ is not precluded; in this case the multiple bracket in (IV.4.14) must be read to mean $\tilde{\mathbf{v}}_{p_1}$. Note that for (IV.4.14) not to vanish identically, we must have D_ι = Identity whenever $g_\iota \equiv 1$ $(1 \le \iota \le \lambda)$.

We assign the weight ϖ_i to X_i if $1 \le i \le N_\alpha$ and m_α to X_k if $N_\alpha < k \le N$. It is a restatement of (IV.4.10) that, with this choice, weight of $\vartheta_i = -\varpi_i$ if $1 \le i \le N_\alpha$ and weight of $\vartheta'_j = -1$ if $1 \le j \le n-\nu$. We call $-\tilde{\varpi}_\ell$ the weight of $\tilde{\mathbf{v}}_\ell$, computed in accordance with this choice. If the weight of the formal power series g is $\ge \omega$, then that of $\tilde{\mathbf{v}}_\ell g$ is $\ge \omega - \tilde{\varpi}_\ell$. By induction on the length μ_ι of D_ι we conclude that

$$\text{weight of } D_\iota g_\iota \ge \omega_\iota - \sum_{k=1}^{\mu_\iota} \tilde{\varpi}_{q_{\iota k}}.$$

As a consequence, if the zero-order term of $D_\iota g_\iota$ is not to vanish we must have

$$\omega_\iota \le \sum_{k=1}^{\mu_\iota} \tilde{\varpi}_{q_{\iota k}}. \tag{IV.4.15}$$

Combining this (for $\iota = 1,\ldots,\lambda$) with (IV.4.13) yields

$$\tilde{\varpi}_{p_1} + \cdots + \tilde{\varpi}_{p_\mu} = \tilde{\varpi}_1 + \cdots + \tilde{\varpi}_\lambda - \sum_{\iota=1}^{\lambda}\sum_{k=1}^{\mu_\iota} \tilde{\varpi}_{q_{\iota k}} \leq m_{\alpha+1} - 1,$$

which proves our claim. ∎

We derive from Lemma IV.4.2 that $\tilde{C}|_0 \in \mathfrak{g}_{m_{\alpha+1}-1}|_0$, which is not possible unless $\tilde{C}|_0 = 0$, as $\tilde{C}$ is a linear combination of $\partial/\partial S_1,\ldots,\partial/\partial S_q$ (recall that $\lambda \geq 2$). This proves that $C|_0 = 0$.

It allows us to apply Lemma IV.3.3 to the formal vector fields A_i and B_j defined above, taking $S_k = X_{N_\alpha+k}$, $k = 1,\ldots,N-N_\alpha$, and the weights to be

$$\varpi_i = 1 \textit{ if } 1 \leq i \leq N_0,\ \varpi_i = m_\beta \textit{ if } N_{\beta-1} < i \leq N_\beta\ (1 \leq \beta \leq \alpha),\ \varpi = m_{\alpha+1}. \tag{IV.4.16}$$

Lemma IV.3.3 implies that there is a change of indeterminates

$$(X_1,\ldots,X_{N_\alpha},X_{N_\alpha+1},\ldots,X_N) \to (X_1,\ldots,X_{N_\alpha},G_{N_\alpha+1}(X),\ldots,G_N(X))$$

such that the weight, in the new indeterminates system, of the vector fields A_i $(1 \leq i \leq N_0)$ and B_j $(1 \leq j \leq n-\nu)$ is ≥ -1 while that of the vector fields A_i $(N_0 \leq i \leq N_\alpha)$ is equal to $-\varpi_i$ (as given in (IV.2.4)). This is valid when the weight $m_{\alpha+1}$ $(= \varpi)$ is assigned to every indeterminate X_k, $N_\alpha < k \leq N$.

We revert to the vector fields ϑ_i and ϑ_j':

$$\vartheta_i = A_i + \sum_{k>N_\beta}^{N_\alpha} a_{ik}A_k,\quad \vartheta_j' = B_j - \sum_{k>N_0}^{N_\alpha} b_{jk}'\vartheta_k.$$

$(N_{\beta-1} < i \leq N_\beta,\ 0 \leq \beta \leq \alpha,\ 1 \leq j \leq n-\nu)$. If $N_{\beta-1} < i \leq N_\beta < k \leq N_\alpha$, weight of $a_{ik} \geq \varpi_k - \varpi_i$, hence weight of $\vartheta_i' = -\varpi_i$. And since weight of $\vartheta_k = -\varpi_k$ $(1 \leq k \leq N_\alpha)$ we get, by Lemma IV.4.1, weight of $\vartheta_j' - B_j \geq -1$, whence weight of $\vartheta_j' \geq -1$. We translate the preceding results in terms of the coefficients: if $0 \leq \beta \leq \alpha$ and $N_{\beta-1} < i \leq N_\beta < k \leq N_\alpha$, weight of $a_{ik} \geq \varpi_k - \varpi_i$; if $N_\alpha < k \leq N$, weight of $a_{ik} \geq m_{\alpha+1} - \varpi_i$. In this argument we assign the weight $m_{\alpha+1}$ to the indeterminates X_k $(N_\alpha < k \leq N)$. Thus (IV.4.10) is proved for $\alpha+1$ in the place of α.

By the induction on $\alpha \geq 1$ we eventually reach the stage $N_\alpha = N$. It should be underlined that, if $N_r < N$, the preceding construction yields vector fields ϑ_i and ϑ_j' in which the coefficients of $\partial/\partial X_{N_r+1},\ldots,\partial/\partial X_N$ belong to the ideal in $\mathbb{R}[[X]]$ generated by $X_{N_r+1},\ldots,X_N$. At the end of the process we revert to the indeterminates $x_i = \frac{1}{2}X_i$, $y_j = \frac{1}{2}X_{\nu+j}$, $s_k = X_{n+\nu+k}$, $t_\ell = X_{2\nu+\ell}$ (cf. (IV.4.1)) and to the formal vector fields $L_j = \vartheta_j + \imath\vartheta_{\nu+j}$, $L_{\nu+k} = \vartheta_{2\nu+k} + \imath\vartheta_k'$ $(j = 1,\ldots,\nu,\ k = 1,\ldots,n-\nu)$. Keep in mind that these vector fields have the expressions (I.7.15), (I.7.16). The weights are defined according to (IV.2.4); it is understood, always, that weight 1 is assigned to x_i, y_j, t_ℓ. This completes the proof of Theorem IV.4.1. ■

REMARK IV.4.1. The construction in the present section has yielded indeterminates x_i, y_j, s_k, t_ℓ in which the vector fields L_j have weight -1 and, at the same time, the formal vector fields

$$\vartheta_{n+\nu+k} = \partial/\partial s_k + \sum_{\ell=1}^{d} a_{(n+\nu+k)\ell}\partial/\partial s_\ell \ (k = 1,\dots,d) \qquad \text{(IV.4.17)}$$

have weight $-\varpi_k$ (defined in (IV.2.4)). Recall that properties (IV.4.3), (IV.4.4) hold. In particular, the vector fields L_i, $\overline{L}_j$, $\vartheta_{n+\nu+k}$ $(1 \leq i, j \leq n,\ 1 \leq k \leq d)$ span $\mathbb{D}[[x,y,s,t]]$ over $\mathbb{C}[[x,y,s,t]]$. ■

IV.5. Existence of Normal Forms. Pluriharmonic Free Normal Forms. Rigid Structures

We continue to deal with an integrable formal structure $\mathcal{V} \subset \mathbb{D}[[x,y,s,t]]$; a basis of $\mathcal{V}$ is given by the formal vector fields $L_1,\dots,L_n$ (see (I.7.15), (I.7.16)); $z_j = x_j + \iota y_j$ $(1 \leq j \leq \nu)$ are obvious first integrals. We show that the remaining first integrals w_k $(1 \leq k \leq d)$ can be selected to have the expressions and the weights required in a normal form (see Definition IV.2.1). We remind the reader that, in every definition of the weight, the indeterminates x_i, y_j, t_ℓ are assigned weight 1.

THEOREM IV.5.1. *The indeterminates x_i, y_j, s_k, t_ℓ and the first integrals $w_k = s_k + \iota\varphi_k(x,y,s,t)$ $(i, j = 1,\dots,\nu,\ k = 1,\dots,d,\ \ell = 1,\dots,n-\nu)$ can be chosen in such a manner that the weight* (computed according to the rule (IV.2.4)) *of the formal power series φ_k be equal to ϖ_k for each $k = 1,\dots,d$.*

PROOF. At first we let the indeterminates be those in Theorem IV.4.1. Let L_j denote the same formal vector fields as in Theorem IV.4.1. For the sake of simplicity let us write here $\delta_j = \partial/\partial\bar{z}_j$ if $1 \leq j \leq \nu$, $\delta_j = \partial/\partial t_{j-\nu}$ if $\nu < j \leq n$, so that

$$L_j = \delta_j - \iota\sum_{k=1}^{d}(\delta_j\varphi_k)N_k.$$

For each $k = 1,\dots,d$ we solve the Cauchy problem

$$L_j h_k = 0,\ j = 1,\dots,n, \qquad \text{(IV.5.1)}$$

with prescribed data $h_k = s_k$ when we put $x_i = y_j = t_\ell = 0$ in the power series h_k.

Let us denote by h_k° the series of monomials in the series h_k whose weight is $< \varpi_1$ ($= m_1$). Necessarily h_k° is independent of s. We have

$$\mathrm{L}_j h_k^\circ = \delta_j h_k^\circ = -\mathrm{L}_j(h_k - h_k^\circ) \ (1 \leq j \leq n). \tag{IV.5.2}$$

The extreme right-hand sides in (IV.5.2) have weight $\geq \varpi_1 - 1$; this is true of the left-hand sides only if they vanish. We conclude that $h_k^\circ \in \mathbb{C}[[z]]$. We set

$$w_k = h_k - h_k^\circ, \ k = 1,\ldots,d_1$$

(for the definition of d_ι see (IV.2.4); $d_1 = \kappa_1$ is the multiplicity of the first Hörmander number m_1). Notice that, according to the initial conditions attached to (IV.5.1), putting $x_i = y_j = t_\ell = 0$ in the series w_k yields $w_k = s_k$.

If $d_1 = d$ the argument stops. Suppose $d_1 < d$. Since the zero-order term of the Jacobian matrix of $w_1,\ldots,w_{d_1}$ with respect to $s_1,\ldots,s_{d_1}$ is the identity matrix, we can solve the equations

$$w_k = w_k(x,y,s,t), \tag{IV.5.3}$$

where $k = 1,\ldots,d_1$, with respect to $s^{(1)} = (s_1,\ldots,s_{d_1})$ (the zero-order term of the solutions vanish). Substituting in each h_k, $d_1 < k \leq d$, gets us a formal power series in x, y, $w^{(1)}$, s', t, where $w^{(1)} = (w_1,\ldots,w_{d_1})$, $s' = (s_{d_1+1},\ldots,s_d)$. We assign the weight ϖ_1 to the indeterminates w_k. This does not modify the definition of the weight in the ring of formal power series, by Lemma IV.2.1.

Let then h_ℓ^1 denote the series of monomials in $h_\ell(x,y,w^{(1)},s',t)$ whose weight is $< \varpi_2 = m_2$; $h_\ell^1 \in \mathbb{C}[[x,y,w^{(1)},t]]$ $(d_1 < \ell \leq d)$. Due to the fact that $\mathrm{L}_j w_k = 0$, $j = 1,\ldots,n$, $k = 1,\ldots,d_1$, we have, for all $\ell = d_1+1,\ldots,d$,

$$\mathrm{L}_j h_\ell^1 = \delta_j h_\ell^1 = -\mathrm{L}_j(h_\ell - h_\ell^1) \ (1 \leq j \leq n).$$

We use the fact that the weights of the extreme right-hand sides are $\geq \varpi_2 - 1$ and that those of the left-hand sides are $\leq \varpi_2 - 2$ unless they vanish identically. We conclude that $h_\ell^1 \in \mathbb{C}[[z,w^{(1)}]]$. We define

$$w_k = h_k - h_k^1, \ k = d_1+1,\ldots,d_2.$$

If $d_2 = d$ the argument stops. If $d_2 < d$ we repeat the preceding construction: we note that putting $x_i = y_j = t_\ell = 0$ gets us $w_k = s_k$ $(1 \leq k \leq d_2)$. We can solve Equations (IV.5.3), now for $k = d_1+1,\ldots,d_2$, with respect to $s_{d_1+1},\ldots,s_{d_2}$; substituting yields formal power series

$$h_k(x,y,w^{(1)},w^{(2)},s'',t)$$

where $w^{(2)} = (w_{d_1+1},\ldots,w_{d_2})$, $s'' = (s_{d_2+1},\ldots,s_d)$. We then assign the weight ϖ_2 to the indeterminates w_k $(d_1 < k \leq d_2)$ and we show that the series of monomials in $h_k(x,y,w^{(1)},w^{(2)},s'',t)$ whose weight is $< \varpi_2$ belongs to $\mathbb{C}[[z,w^{(1)},w^{(2)}]]$. For $d_2 < k \leq d_3$ we define w_k as the series of monomials in

$h_k(x,y,w^{(1)},w^{(2)},s'',t)$ whose weight is $\geq \varpi_2$. And so forth. Eventually we reach the stage where $d_\iota = d$.

At this stage we have d power series $w_j \in \mathbb{C}[[x,y,s,t]]$ $(1 \leq j \leq d_1)$, $w_k \in \mathbb{C}[[x,y,w^{(1)},s',t]]$ $(d_1 < k \leq d_2)$, $w_\ell \in \mathbb{C}[[x,y,w^{(1)},w^{(2)},s'',t]]$ $(d_2 < \ell \leq d_3)$, etc. We put $w^{(1)} = w^{(1)}(x,y,s,t)$ into w_k for all $k > d_1$; after having done this we put $w^{(2)} = w^{(2)}(x,y,s,t)$ into w_k for all $k > d_2$; and so forth. We end up with d solutions $w_k \in \mathbb{C}[[x,y,s,t]]$ of Equations (IV.5.1), such that $w_k = s_k$ if we substitute 0 for all x_i, y_j, t_ℓ. Finally we make the change of indeterminates $s_k \to \mathcal{R}_e w_k$ $(1 \leq k \leq d)$. Since the weight of $\mathcal{R}_e w_k$ is exactly equal to ϖ_k we may apply Lemma IV.2.1. The weight in $\mathbb{C}[[x,y,s,t]]$ has not been modified, and the first integrals w_k have the expected weights.

This tells us only that the indeterminates and the first integrals can be chosen in such a way that $\varpi_k^\# =$ weight of $\varphi_k \geq \varpi_k$ for each k. We change the definition of the weights and assign the weight $\varpi_k^\#$ to s_k for each k. Then the new weight of each φ_k is equal to or larger than its old weight, and thus weight of $\varphi_k \geq \varpi_k^\#$. Note, in passing, that this does *not* mean that weight of $\varphi_k = \varpi_k^\#$ for all k. In any event we now assign the weight $\varpi_k^\#$ to w_k for each k.

We are in a position to apply Proposition IV.2.2, and avail ourselves of the property that the (new) weights of the vector fields L_j are -1, and the weight of N_k is equal to $-\varpi_k^\#$. Consider then a commutator of length $\lambda \geq 2$,

$$C = [\mathbf{v}_1,\ldots,[\mathbf{v}_{\lambda-1},\mathbf{v}_\lambda]\ldots]],$$

where each $\mathbf{v}_\alpha$ is one of the vector fields L_j or $\overline{L}_{j'}$ $(1 \leq j, j' \leq n)$; we know (see (I.7.15), (I.7.16)) that

$$C = \sum_{k=1}^{d} a_k \partial/\partial s_k.$$

Here we claim that

$$\text{weight of } a_k \geq \varpi_k^\# - \lambda. \tag{IV.5.4}$$

PROOF OF (IV.5.4). We use induction on λ. Suppose first $\lambda = 2$, i.e.,

$$\pm\, C = [L_i,\overline{L}_j] = \iota \sum_{k=1}^{d}\Big[(L_i\overline{\delta}_j\varphi_k)\overline{N}_k + (\overline{L}_j\delta_i\varphi_k)N_k\Big] +$$

$$\sum_{k,\ell=1}^{d} (\delta_i\varphi_k)(\overline{\delta}_j\varphi_\ell)[N_k,\overline{N}_\ell].$$

If we go back to the expression (I.7.17) of N_k and make use of the fact that weight of N_k is $\geq -\varpi_k^\#$ we see that

$$\text{weight of } \mu_{kk'} \geq \varpi_{k'}^\# - \varpi_k^\#. \tag{IV.5.5}$$

We get

$$[\mathrm{L}_i,\overline{\mathrm{L}}_j] = \iota\sum_{k,k'=1}^{d}\Big[(\mathrm{L}_i\overline{\delta}_j\varphi_{k'})\overline{\mu}_{k'k} + (\overline{\mathrm{L}}_j\delta_i\varphi_{k'})\mu_{k'k}\Big]\partial/\partial s_k +$$

$$\sum_{k,k',\ell=1}^{d}(\delta_i\varphi_{k'})(\overline{\delta}_j\varphi_\ell)(\mathrm{N}_{k'}\overline{\mu}_{\ell k}-\overline{\mathrm{N}}_\ell\mu_{k'k})\partial/\partial s_k.$$

The weight of $(\mathrm{L}_i\overline{\delta}_j\varphi_{k'})\overline{\mu}_{k'k}$ and that of $(\overline{\mathrm{L}}_j\delta_i\varphi_{k'})\mu_{k'k}$ are $\geq \varpi_{k'}^{\#}-2+(\varpi_k^{\#}-\varpi_{k'}^{\#}) = \varpi_k^{\#}-2$. The same is true of the weight of $(\delta_i\varphi_{k'})(\overline{\delta}_j\varphi_\ell)\mathrm{N}_{k'}\overline{\mu}_{\ell k}$ and of that of $(\delta_i\varphi_{k'})(\overline{\delta}_j\varphi_\ell)\overline{\mathrm{N}}_\ell\mu_{k'k}$, whence the claim when $\lambda = 2$.

Suppose the claim proved up to, and inclusively, λ. We have

$$[\mathrm{L}_j,\mathrm{C}] = \sum_{k=1}^{d}(\mathrm{L}_j a_k)\partial/\partial s_k + \iota\sum_{k,k',\ell=1}^{d} a_\ell(\partial/\partial s_\ell)(\mu_{k'k}\delta_j\varphi_{k'})\partial/\partial s_k.$$

The weight of $\mathrm{L}_j a_k$ is $\geq \varpi_k^{\#}-\lambda-1$; that of $a_\ell(\partial/\partial s_\ell)(\mu_{k'k}\delta_j\varphi_{k'})$ is $\geq \varpi_\ell^{\#}-\lambda+\varpi_k^{\#}-\varpi_{k'}^{\#}+\varpi_{k'}^{\#}-1-\varpi_\ell^{\#}=\varpi_k^{\#}-\lambda-1$. ■

It follows from (IV.5.4) that $\mathrm{C}|_0 = 0$ whenever $\lambda < \varpi_k^{\#}$. If we had $\varpi_k^{\#} > m_\iota$ for some ι and k, $1 \leq \iota \leq r$, $d_{\iota-1} < k \leq d_\iota$, it would contradict the definition of the Hörmander numbers and of their multiplicity. We have thus proved that the weight of φ_k is exactly equal to ϖ_k for each $k = 1,\ldots,d$. The proof of Theorem IV.5.1 is complete. ■

It is possible to further streamline the expressions of the first integrals w_k in the normal form (IV.2.5). To do this it is convenient to switch to the complex indeterminates z_i, $\bar{z}_j$ $(1 \leq j \leq \nu)$. We use the notation $\varphi = (\varphi_1,\ldots,\varphi_d)$ and $\varphi(z,\bar{z},s,t) \in \mathbb{C}^d$ (rather than $\varphi(x,y,s,t)$).

THEOREM IV.5.2. *In the normal form* (IV.2.5) *the indeterminates s_k and the first integrals $w_k = s_k + \iota\varphi_k(z,\bar{z},s,t)$ $(1 \leq k \leq d)$ can be modified to ensure*

$$\varphi(z,0,s,0) \equiv 0, \tag{IV.5.6}$$

while preserving the property that weight of $\varphi_k = \varpi_k$ *for each $k = 1,\ldots,d$.*

Since φ is a *real* power series (IV.5.6) is equivalent to

$$\varphi(0,\bar{z},s,0) = 0 \tag{IV.5.7}$$

As usual, the indeterminates z_i and t_ℓ are given the weight 1.

PROOF. Let $G(z,w) \in \mathbb{C}[[z,w]]\otimes\mathbb{C}^d$ be the unique formal power series whose zero-order term is zero and which satisfies

$$G + \iota\varphi(z,0,G,0) = w. \tag{IV.5.8}$$

The constant term in the Jacobian matrix of G with respect to w is equal to the identity $d\times d$ matrix. We derive from (IV.5.8):

$$G(z,s+\iota\varphi(z,0,s,0)) = s. \qquad \text{(IV.5.9)}$$

We shall replace w by the following formal power series (with coefficients in $\mathbb{C}^d$):

$$F(z,w) = \overline{G}(0,G(z,w)-\iota\varphi(z,0,G(z,w),0)). \qquad \text{(IV.5.10)}$$

The constant term in the Jacobian matrix of F with respect to w is also equal to the identity $d\times d$ matrix. We have

$$F(z,w) - \overline{F}(0,s+\iota\varphi(z,0,s,0)) = \overline{G}(0,G(z,w)-\iota\varphi(z,0,G(z,w),0)) -$$
$$G(0,\overline{G}(0,s-\iota\varphi(z,0,s,0))+\iota\varphi(0,0,\overline{G}(0,s-\iota\varphi(z,0,s,0)),0)). \qquad \text{(IV.5.11)}$$

From (IV.5.9) we get directly

$$\overline{G}(0,G(z,s+\iota\varphi(z,0,s)) - \iota\varphi(z,0,G(z,s+\iota\varphi(z,0,s,0)),0)) =$$
$$\overline{G}(0,s-\iota\varphi(0,0,s,0)).$$

We may also apply (IV.5.9) with $z = 0$ and $\overline{G}(0,s-\iota\varphi(z,0,s,0))$ substituted for s:

$$G(0,\overline{G}(0,s-\iota\varphi(z,0,s,0))+\iota\varphi(0,0,\overline{G}(0,s-\iota\varphi(z,0,s,0)),0)) =$$
$$\overline{G}(0,s-\iota\varphi(z,0,s,0)).$$

We conclude that the left-hand side in (IV.5.11) vanishes identically when we put $w = s+\iota\varphi(z,0,s,0)$. If we write $F = (F_1,\ldots,F_d)$, this shows that, to satisfy the requirements in Lemma IV.5.1, we may indeed take $F_k(z,w)$ as new solution w_k, and the formal power series $\frac{1}{2}[F_k(z,s+\iota\varphi(z,\bar{z},s,t)) + \overline{F}_k(\bar{z},s-\iota\varphi(z,\bar{z},s,t))]$ as new indeterminate s_k $(1\le k\le d)$.

Concerning weights we apply Lemma IV.2.1: the original weights (as defined in (IV.2.4)) are unchanged if we replace the indeterminates s with the indeterminates w, these with the indeterminates G and the latter with the indeterminates F. In other words, the transition from the indeterminates s to the indeterminates F does not affect the weights. This means that, for all $k = 1,\ldots,d$, $F_k(z,s+\iota\varphi(z,\bar{z},s,t)) = s_k + \psi_k(z,\bar{z},s,t)$ with ψ_k fulfilling the requirements in Lemma IV.2.1. But then, again by Lemma IV.2.1, the switch from s to $\mathcal{R}e F(z,s+\iota\varphi(z,\bar{z},s,t))$ preserves the weights. The changes of first integrals in the proof of Theorem IV.5.1 do not invalidate Property (IV.5.6) and enable us to obtain that weight of $\varphi_k = \varpi_k$.■

Definition IV.5.1. *We say that the normal form* (IV.2.5) *is* pluriharmonic free *if property* (IV.5.6) *holds.*

Theorem IV.5.2 states that *the normal form* (IV.2.5) *can be chosen to be pluriharmonic free.*

EXAMPLE IV.5.1. Suppose there are only three indeterminates x, y, s and that these, together with the two first integrals $z = x + \imath y$, $w = s + \imath\varphi(z,\bar{z},s)$, constitute a normal form for the structure under consideration (thus $m = 2$, $n = \nu = d = 1$). In order that the structure not be of finite type it is necessary and sufficient that $\varphi(z,\bar{z},0) \equiv 0$. This statement is true because we are dealing with a normal form: indeed, if the structure is not of finite type the weight of s and w must be $+\infty$.

Suppose the structure is of finite type, and let $\varphi_0(z,\bar{z})$ be the homogeneous part of lowest degree in the series $\varphi(z,\bar{z},0)$. Since it is always assumed that the zero-order terms of φ and of $d\varphi$ vanish the degree of φ_0 must be ≥ 2. It is the first (and here the only) Hörmander number m_1.

Now suppose the structure is "*rigid*"—i.e., φ is independent of s (see Definition IV.5.2 below)—and write

$$\varphi(z,\bar{z}) = \sum_{p,q=2}^{+\infty} c_{pq} z^p \bar{z}^q,$$

with $c_{pq} = \bar{c}_{qp}$. We see that the normal form is pluriharmonic free if and only if $c_{pq} = 0$ whenever $q = 0$. In order to change the given normal form into one that is pluriharmonic free it suffices to effect the change of first integral $w \to \tilde{w} = w - \imath\varphi(z,0)$, and the concomitant change of indeterminate $s \to \tilde{s} = \mathcal{R}e\,\tilde{w}$. The new normal form is rigid and pluriharmonic free.

When φ depends on s, one must go through the construction described in the proofs of Theorems IV.5.1 and IV.5.2. ∎

EXAMPLE IV.5.2. Suppose now $m = 3$, $n = 1$, $d = 2$, and let the formal structure $\mathcal{V} \subset \mathbb{C}[[x,y,s_1,s_2]]$ be defined by the first integrals $z = x + \imath y$ and

$$w_1 = s_1 + \imath\mathcal{Q}(z,\bar{z}),\ w_2 = s_2 + \imath s_1\mathcal{Q}(z,\bar{z}), \tag{IV.5.12}$$

where $\mathcal{Q}$ is a quadratic form. A substitution $\tilde{w}_2 = w_2 - \frac{1}{2}w_1^2$, $\tilde{s}_2 = s_2 - \frac{1}{2}(s_1^2 - \mathcal{Q}(z,\bar{z})^2)$ yields the set of first integrals

$$w_1 = s_1 + \imath\mathcal{Q}(z,\bar{z}),\ \tilde{w}_2 = \tilde{s}_2, \tag{IV.5.13}$$

which shows that the structure $\mathcal{V}$ is not of finite type. There are two cases, depending on whether $\mathcal{Q}$ is harmonic or not. If $\mathcal{Q}$ is harmonic, i.e., $\mathcal{Q}(z,\bar{z}) = \mathcal{R}e(cz^2)$ for some $c \in \mathbb{C}$, the substitution $\tilde{w}_1 = w_1 - \imath cz^2$, $\tilde{s}_1 = s_1 + \mathcal{I}m(cz^2)$ transforms the set of first integrals (IV.5.13) into the set $\tilde{w}_1 = \tilde{s}_1$, $\tilde{w}_2 = \tilde{s}_2$; in which case, obviously, $m_1 = +\infty$. If $\mathcal{Q}$ is not harmonic we may assume that $\mathcal{Q}(z,\bar{z}) = a|z|^2$ for some real number $a \neq 0$. After the substitution $s_1 \to as_1$, $w_1 \to aw_1$ we may assume that

$$w_1 = s_1 + \imath|z|^2,\ w_2 = s_2, \tag{IV.5.14}$$

which is obviously a pluriharmonic free normal form for $\mathcal{V}$; and shows that $m_1 = 2$, $m_2 = +\infty$. ∎

DEFINITION IV.5.2. *We say that the integrable formal structure* $\mathcal{V} \subset \mathbb{D}[[x,y,s,t]]$ *is* rigid *if the indeterminates* x_i, y_j, s_k, t_ℓ *and the first integrals* $z_j = x_j + \imath y_j$, $w_k = s_k + \imath\varphi_k$ $(1 \leq i, j \leq \nu,\ 1 \leq k \leq d,\ 1 \leq \ell \leq n-\nu)$ *can be chosen in such a way that all the series* φ_k *belong to* $\mathbb{R}[[x,y,t]]$, *i.e., are independent of* s.

Example IV.5.2 shows that a rigid structure may admit a set of first integrals (IV.5.2) with imaginary parts φ_k that are not "rigid," i.e., that depend on s.

THEOREM IV.5.3. *If the integrable formal structure* $\mathcal{V}$ *is rigid, there is a pluriharmonic free normal form* (IV.2.5) *such that* $\varphi_k \in \mathbb{R}[[x,y,t]]$ *for all* $k = 1,\dots,d$.

PROOF. Let $w_k = s_k + \imath\varphi_k(z,\bar{z},t)$ $(1 \leq k \leq d)$ be first integrals such that the weight of $\varphi_k \leq \varpi_k$ (ϖ_k as in (IV.2.4)). Let us set $\tilde{w} = w - 2\imath\varphi(z,0,0)$. Then

$$\tilde{s} = \mathcal{R}e\,\tilde{w} = s + 2\mathcal{I}m\,\varphi(z,0,0),\ \tilde{\varphi}(z,\bar{z},t) = \varphi(z,\bar{z},t) - \varphi(z,0,0) - \varphi(0,\bar{z},0).$$

It is clear that the first integrals $\tilde{w}_k$ are pluriharmonic free. After substituting $\tilde{w}$ for w we may as well assume that

$$\varphi(z,0,0) \equiv 0. \tag{IV.5.15}$$

Then consider a basis of $\mathcal{V}$ consisting of vector fields of weight -1:

$$\mathrm{L}_j = \delta_j + \sum_{k=1}^{d}\lambda_{jk}(z,\bar{z},s,t)\partial/\partial s_k,\ j = 1,\dots,n$$

(δ_j has the same meaning as in the proof of Theorem IV.5.1). The equation $\mathrm{L}_j w_k \equiv 0$ shows that

$$\lambda_{jk} = -\delta_j\varphi_k, \tag{IV.5.16}$$

which shows that the coefficients λ_{jk} are independent of s. Recall that weight of $\lambda_{jk} \geq \varpi_k - 1$. Equations (IV.5.16) entail that, if φ_k° denotes the sum of all monomials in φ_k with weight $< \varpi_k$, then

$$\delta_j\varphi_k^\circ \equiv 0,\ j = 1,\dots,n. \tag{IV.5.17}$$

But (IV.5.17) means that φ_k° is independent of $\bar{z}$ and of t; according to (IV.5.15) $\varphi_k^\circ \equiv 0$, which proves that weight of $\varphi_k \geq \varpi_k$. When $\varpi_k < +\infty$ we cannot have weight of $\varphi_k > \varpi_k$ otherwise (IV.5.16) would imply weight of $\lambda_{jk} \geq \varpi_k$ for all j, contrary to the choice (IV.2.4). ■

EXAMPLE IV.5.3. Suppose $m = 3$, $n = 1$, $d = 2$, and let the formal structure $\mathcal{V}$ be defined by

$$w_1 = s_1 + \imath|z|^2,\ w_2 = s_2 + \imath(|z|^4 + s_1|z|^2). \tag{IV.5.18}$$

The substitution $\tilde{w}_2 = w_2 - \frac{1}{2}w_1^2$, $\tilde{s}_2 = s_2 - \frac{1}{2}(s_1^2 - |z|^4)$ transforms it into the system of first integrals

$$w_1 = s_1 + \imath|z|^2, \; \tilde{w}_2 = \tilde{s}_2 + \imath|z|^4, \tag{IV.5.19}$$

which shows that the structure is rigid and that $m_1 = 2$, $m_2 = 4$. But then it follows that (IV.5.18) constitutes a pluriharmonic free normal form, though not a "rigid" one. ■

IV.6. Leading Parts

Suppose the normal form (IV.2.5) is pluriharmonic free (Definition IV.5.1). For any k such that $1 \leq k \leq d$ and $\varpi_k < +\infty$, we shall denote by P_k the sum of monomials in the power series φ_k whose weight is exactly equal to ϖ_k. We have

$$\varphi_k = P_k + R_k, \tag{IV.6.1}$$

with $R_k \in \mathbb{R}[[x,y,s,t]]$, weight of $R_k > \varpi_k$. In the sequel we refer to P_k as the *leading part* of φ_k. If $1 \leq k \leq d_1$, P_k is a homogeneous polynomial of degree m_1 in the indeterminates x_i, y_j, t_ℓ. If $2 \leq \iota \leq r$, $d_{\iota-1} < k \leq d_\iota$, P_k is a polynomial in the indeterminates x_i, y_j, t_ℓ as well as $s_1,\ldots,s_{d_{\iota-1}}$; it is semihomogeneous (all its monomials have the same weight, m_ι). We shall refer to the ordered set of polynomials $\{P_1,\ldots,P_{d_r}\}$ as the *system of leading parts* of the pluriharmonic free normal form (IV.2.5).

We now give a rough description of how the system of leading parts transforms under changes of the indeterminates and of the first integrals. Invariance, here, can only be expected after we "quotient out" the obvious changes of indeterminates and of first integrals that modify the system of leading parts. Keep in mind that we hypothesize that the original, as well as the transformed normal form, is pluriharmonic free.

First consider substitutions of the indeterminates x, y, t that preserve the integrable formal structure, i.e., that transform the first integrals into another system of first integrals. They must be perforce of the kind

$$(z,t) \rightarrow (f(z,w), g(x,y,s,t)),$$

with g real, and with $A = \partial f/\partial z|_0$ and $B = \partial g/\partial t|_0$ nonsingular matrices. Let us also set $C = \partial g/\partial z|_0$. Note that

$$P_k(f,\bar{f},s,g) = P_k(Az,\overline{Az},s,Bt+Cz+\overline{Cz}) + \text{terms of weight} > \varpi_k.$$

(In the present context it is convenient to reason in $\mathbb{C}[[z,\bar{z},s,t]]$.) In other words, when dealing with the leading parts, we may as well limit our attention to the linear transformations

$$(z,t) \rightarrow (Az, Bt+Cz+\overline{Cz}). \tag{IV.6.2}$$

We follow up the transformation (IV.6.2) with a change of the first integrals

$$w_k \to G_k(z,w) \ (1 \le k \le d), \tag{IV.6.3}$$

and a concomitant change of the indeterminates $s_k \to \mathcal{R}e\,G_k(z,w)$. We require that the weight be the same in the new normal form as it was in the original. It means that the weight of G_k must be exactly equal to ϖ_k and that for each $\iota = 1,\dots,r+1$, the zero-order terms of both the Jacobian matrix of $G^{(\iota)} = (G_{d_{\iota-1}+1},\dots,G_{d_\iota})$ with respect to $w^{(\iota)}$, $\Gamma^{(\iota)}$, and the Jacobian matrix of $\mathcal{R}e\,G^{(\iota)}$ with respect to $s^{(\iota)}$, must be nonsingular. Suppose the Hörmander number m_ι is finite; then, if $d_{\iota-1} < k \le d_\iota$,

$$G^{(\iota)}(z,s+\imath\varphi) = H^{(\iota)}(z,w^{(1)},\dots,w^{(\iota-1)}) + \Gamma^{(\iota)}w^{(\iota)} + \text{terms of weight} > m_\iota.$$

Thus, if we limit our attention to terms of weight m_ι ($< +\infty$), for $\iota \le r$ it suffices to consider transformations

$$w^{(\iota)} \to H^{(\iota)}(z,w^{(1)},\dots,w^{(\iota-1)}) + \Gamma^{(\iota)}w^{(\iota)}. \tag{IV.6.4}$$

Moreover, if $d_r < d$, one may substitute $G^{(r+1)}(z,w)$ for $w^{(r+1)}$ with the understanding that the components of $G^{(r+1)}$ belong to the ideal in $\mathbb{C}[[z,w]]$ generated by $w_{d_r+1},\dots,w_d$. Then we shall use the notation

$$H(z,w) = (H^{(1)}(z),\dots,H^{(r)}(z,w^{(1)},\dots,w^{(r-1)}),0,\dots,0),$$

$$\Gamma w = (\Gamma^{(1)}w^{(1)},\dots,\Gamma^{(r+1)}w^{(r+1)}).$$

Actually the effect on the first integrals of the linear automorphism Γ is evident; to simplify matters let us take Γ = Identity.

First consider the case $\iota = 1$, $m_1 < +\infty$. The change (IV.6.4) reduces to $w^{(1)} \to w^{(1)} + H^{(1)}(z)$. We focus on $\psi^{(1)}(z,\bar{z},S,t) = \varphi^{(1)}(z,\bar{z},s,t) + \mathcal{I}m\,H^{(1)}(z)$ where $S = s + \mathcal{R}e\,H(z,w)$. The property that the normal forms are pluriharmonic free means that

$$\psi^{(1)}(z,0,S,0) = \varphi^{(1)}(z,0,s,0) \equiv 0, \text{ i.e., } H^{(1)}(z) \equiv 0.$$

Thus we may as well assume that, for $\iota = 1$, the change of first integrals (IV.6.4) is the identity.

From then on we reason by induction on $\iota \ge 2$, assuming that $m_\iota < +\infty$; still under the hypothesis that Γ is the identity, (IV.6.4) reads

$$w^{(\iota)} \to w^{(\iota)} + H^{(\iota)}(z,w^{(1)},\dots,w^{(\iota-1)}).$$

We focus on $\psi^{(\iota)}(z,\bar{z},S,t) = \varphi^{(\iota)}(z,\bar{z},s,t) + \mathcal{I}m\,H^{(\iota)}(z,w^{(1)},\dots,w^{(\iota-1)})$. If $\psi^{(\iota)}$ is to be pluriharmonic free putting $\bar{z} = 0$, $t = 0$ must yield

$$H^{(\iota)}(z,w^{(1)},\dots,w^{(\iota-1)}) \equiv \overline{H}^{(\iota)}(0,\overline{w}^{(1)},\dots,\overline{w}^{(\iota-1)}). \tag{IV.6.5}$$

By the induction hypothesis, when $\bar{z} = 0$, $t = 0$ we have $w^{(\kappa)} = s^{(\kappa)}$ if $1 \le \kappa < \iota$. Thus (IV.6.5) means

$$H^{(\iota)} \in \mathbb{R}[[w^{(1)},\ldots,w^{(\iota-1)}]]. \tag{IV.6.6}$$

Of course, in order that the term $H^{(\iota)}$ nontrivially contribute to the leading part of $\psi^{(\iota)}$ its semihomogeneous part of weight m_ι must not vanish identically. This demands that there be nonnegative integers $q_1,\ldots,q_{\iota-1}$ satisfying

$$m_\iota = q_1 m_1 + \cdots + q_{\iota-1} m_{\iota-1}. \tag{IV.6.7}$$

In particular we conclude that, in the absence of any relation (IV.6.7), and modulo the obvious linear substitutions and higher weight terms, the only transformation that preserves the pluriharmonic free nature of the normal forms is the identity transformation.

Notes

The Hörmander numbers made their appearance in the theory of subelliptic differential operators, first in Hörmander [2]. They reappear, in essentially the same context as in the present book, in Bloom and Graham [1]. The only difference is that here we deal with involutive structures, whereas Bloom and Graham dealt with the CR structure inherited from complex space by a generic submanifold. Thus the main theorem of chapter IV, Theorem IV.5.1, is a routine extension of the results in Bloom and Graham [1].

The whole subject matter of chapter IV is closely related, on the one hand, to subellipticity, and on the other, to the approximation of general vector fields by left-invariant vector fields on nilpotent Lie groups, a topic left untouched here. The approach to subelliptic differential operators from the standpoint of nilpotent groups, embryonic in Hörmander [2], was carried out in full generality in Rothschild and Stein [1]. The same approach can be used to obtain normal forms in CR structures as shown in Baouendi and Rothschild [2] (the proof is based on the main theorem in Helffer and Nourrigat [1]).

In the case of strongly pseudoconvex hypersurfaces in complex space, and more generally of hypersurfaces whose Levi form is nondegenerate, "normal form" has a much stronger meaning than the one adopted here in the general case. In the case of a nondegenerate hypersurface one can identify, by means of the Taylor expansion of the defining equations, a set of CR invariants that determine the (formal or analytic) CR structure up to isomorphisms. For an exposition of the theory in the nondegenerate case and for historical and bibliographical information, we refer the reader to Jacobowitz [2].

Recently, still in the hypersurface case but now allowing degeneracy of the Levi form, a different approach to normal forms has been elaborated in D'Angelo [1].

V

Involutive Structures with Boundary

In the present chapter the concept of an involutive structure on an open manifold is extended to a manifold $\overline{\mathcal{M}}$ with boundary. What the extended concept ought to be is evident: one deals with two open manifolds, $\mathcal{M}$, the interior of $\overline{\mathcal{M}}$, and $\partial\mathcal{M}$, its boundary; each carries an involutive structure; the two structures "fit" on $\partial\mathcal{M}$. In more precise language, we are given a vector subbundle $\mathcal{V}$ of the complex tangent bundle $\mathbb{C}T\overline{\mathcal{M}}$ over the whole manifold $\overline{\mathcal{M}}$ (boundary included), which satisfies the Frobenius formal integrability condition and whose pullback to the boundary is a vector bundle. Since the conormal bundle of the boundary is a line bundle over $\partial\mathcal{M}$, there are only two possibilities: either $\mathcal{V}$ is completely tangent to the boundary or else the rank of the tangent vector bundle $\mathcal{V}_{\partial\mathcal{M}}$ (in the involutive structure of $\partial\mathcal{M}$) is equal to rank $\mathcal{V} - 1$. In accordance with conventional PDE terminology, if $\mathcal{V}|_{\partial\mathcal{M}} \subset \mathbb{C}T\partial\mathcal{M}$ we say that the boundary $\partial\mathcal{M}$ is *totally characteristic*; otherwise, that it is *noncharacteristic* (section V.1).

The involutive structure on $\overline{\mathcal{M}}$ gives rise to two differential complexes: the extension to $\overline{\mathcal{M}}$ of the complex over the interior $\mathcal{M}$ associated with the involutive structure given on $\mathcal{M}$, and the *boundary complex* on $\partial\mathcal{M}$. Here we are thinking of the "generalized exterior derivatives" as acting on *smooth* sections. The boundary complex is isomorphic to the quotient of the differential complex over $\overline{\mathcal{M}}$ modulo the subcomplex whose pullback to $\partial\mathcal{M}$ vanishes identically. When the boundary is noncharacteristic one can set up, within the framework of the differential complex over $\overline{\mathcal{M}}$, a natural Cauchy problem with data at the boundary. Existence and uniqueness of solutions hold, provided one reasons modulo smooth sections that vanish to infinite order at the boundary (section V.2).

Starting in section V.3 we limit our attention to a *locally integrable* structure on $\overline{\mathcal{M}}$. This means that in the neighborhood of every point of $\overline{\mathcal{M}}$ there are smooth functions $Z_1,\ldots,Z_m$ whose differential span (over that neighborhood) the orthogonal of $\mathcal{V}$, i.e., the cotangent structure bundle T'. The existence of such "first integrals" allows us to extend the involutive structure across the boundary, at least locally (which suffices for our needs in the remainder of the chapter). Again in the case where the boundary is noncharacteristic, this leads to the Mayer-Vietoris sequence, which relates the cohomology of the differential complex over a small piece of the boundary to that of the differential complexes over its two sides (in the differential complex associated with the extended locally integrable structure; see section V.3). This generalizes the familiar property that, locally, any $\bar{\partial}_b$-closed differential form on a hypersurface Σ in complex space can be represented by the "jump" between two $\bar{\partial}$-closed forms on each side of Σ.

The dichotomy between the case of a noncharacteristic boundary and that of a totally characteristic one runs throughout the theory. It is true that, in both cases, every classical solution—i.e., every $\mathscr{C}^1$ function in $\overline{\mathcal{M}}$ annihilated by all smooth sections of $\mathcal{V}$—is locally the limit of a sequence of polynomials in the first integrals $Z_1,\ldots,Z_m$. But the proof of the approximation formula, based on the "pseudoconvolution" with a Gaussian function, which leads to that conclusion, must take into account the differences in the two situations (section V.4).

The contrast becomes glaring when we try to spell out what we mean by a *distribution solution*. The dilemma is most clearly exemplified by the two elementary structures on the upper half-plane $\mathbb{R}^2_+ = \{ (x,y) \in \mathbb{R}^2; y \geq 0 \}$: in one, defined by the vector field $\partial/\partial y$, the boundary is noncharacteristic; in the other, defined by the vector field $\partial/\partial x$, it is totally characteristic. The solutions in the latter structure are all distributions of y in the closed half-line $y \geq 0$; they have no regularity whatsoever with respect to y. But their coboundaries (in the sense of currents on a manifold with boundary) are distribution sections of T′. In contrast, the solutions in the structure defined by $\partial/\partial y$, that is, the distributions of x in the whole real line, are smooth with respect to y, the variable transversal to the boundary. But in general their coboundaries are not distribution sections of T′. For instance, the coboundary of the function $\mathbf{1}$ is the current $\mathbf{1}(x)\otimes\delta(y)dy$, which does not belong to the span of dx. However, their exterior derivative, extended from its natural domain, the set of $\mathscr{C}^1$ functions, to the natural space $\mathscr{A}'(\overline{\mathcal{M}})$ of the distributions in $\overline{\mathcal{M}}$ that admit a trace on the boundary, is a section of T′. All this remains valid in the general case. Once the correct definitions are adopted, the approximation formula and the local representation of solutions (the latter, an extension of (II.5.1)) and their customary consequences are valid in both cases (sections V.5 and V.6).

In the last section of this chapter, section V.7, it is shown how the general notions and results apply to smoothly bounded domains in complex space.

V.1. Involutive Structures with Boundary

Throughout this chapter $\overline{\mathcal{M}}$ will denote a $\mathscr{C}^\infty$ manifold with boundary $\partial\mathcal{M}$;

$\mathcal{M}$ will denote its interior. Thus $\overline{\mathcal{M}} = \mathcal{M} \cup \partial\mathcal{M}$. We shall always assume that $\overline{\mathcal{M}}$ is Hausdorff and countable at infinity. Provisionally we write $N = \dim \mathcal{M}$; then $\partial\mathcal{M}$ is a smooth manifold without boundary, countable at infinity; $\dim \partial\mathcal{M} = N - 1$.

We denote respectively by $\mathrm{T}\overline{\mathcal{M}}$ and $\mathrm{T}^*\overline{\mathcal{M}}$ the (real) tangent and cotangent bundles of $\overline{\mathcal{M}}$. They are vector bundles of rank N over the topological space $\overline{\mathcal{M}}$ (including over its boundary). They are smooth, i.e., they themselves are $\mathcal{C}^\infty$ manifolds with boundaries. By $\mathbb{C}\mathrm{T}\overline{\mathcal{M}}$ and $\mathbb{C}\mathrm{T}^*\overline{\mathcal{M}}$ we denote their respective complexifications. (In the absence of a bar on top, each one of the preceding bundles has the standard meaning, relative to the manifold without boundary, $\mathcal{M}$.)

We shall regard the tangent bundle of the boundary, $\mathrm{T}\partial\mathcal{M}$, as a hyperplane subbundle of $\mathrm{T}\overline{\mathcal{M}}$ over $\partial\mathcal{M}$. Its orthogonal in $\mathrm{T}^*\mathcal{M}$ (for the duality between tangent and cotangent vectors) is the conormal bundle of the boundary, $\mathrm{N}^*\partial\mathcal{M}$. The latter is a line bundle over $\partial\mathcal{M}$. Likewise with prefixes $\mathbb{C}$, when the scalar field is extended to the complex numbers.

DEFINITION V.1.1. *An* involutive (*or* formally integrable) *structure on the manifold with boundary* $\overline{\mathcal{M}}$ *is the datum of a* $\mathcal{C}^\infty$ *vector subbundle* $\mathcal{V}$ *of* $\mathbb{C}\mathrm{T}\overline{\mathcal{M}}$ *endowed with the following two properties:*

$\mathcal{V}$ *induces an involutive structure on the interior manifold* $\mathcal{M}$ (section I.1); (V.1.1)

$$\mathcal{V} \cap \mathbb{C}\mathrm{T}\partial\mathcal{M} \textit{ is a vector bundle over } \partial\mathcal{M}. \qquad \text{(V.1.2)}$$

Henceforth $\mathcal{V}$ shall be the *tangent structure bundle* of an involutive structure on the manifold with boundary $\overline{\mathcal{M}}$.

PROPOSITION V.1.1. *If* L_1 *and* L_2 *are two smooth sections of* $\mathcal{V}$ *over some open subset of* $\overline{\mathcal{M}}$ *the same is true of their commutation bracket* $[\mathrm{L}_1, \mathrm{L}_2]$.

PROOF. Indeed, the assertion is true over $\Omega \cap \mathcal{M}$, hence, by continuity, in the whole of Ω. ∎

Notice that (V.1.2) is equivalent to the following property:

$$(\mathcal{V}|_{\partial\mathcal{M}}) + \mathbb{C}\mathrm{T}\partial\mathcal{M} \textit{ is a vector bundle over } \partial\mathcal{M}. \qquad \text{(V.1.3)}$$

PROPOSITION V.1.2. *One of the following two mutually exclusive properties holds true:*

$$(\mathcal{V}|_{\partial\mathcal{M}}) + \mathbb{C}\mathrm{T}\partial\mathcal{M} = \mathbb{C}\mathrm{T}\overline{\mathcal{M}}|_{\partial\mathcal{M}}; \qquad \text{(V.1.4)}$$

$$\mathcal{V}|_{\partial\mathcal{M}} \subset \mathbb{C}\mathrm{T}\partial\mathcal{M}. \qquad \text{(V.1.5)}$$

PROOF. Indeed, $\mathbb{C}\mathrm{T}\partial\mathcal{M}$ is a hyperplane subbundle of $\mathbb{C}\mathrm{T}\overline{\mathcal{M}}$ over $\partial\mathcal{M}$. If

(V.1.5) does not hold, the rank of $(\mathcal{V}|_{\partial\mathcal{M}}) + \mathbb{C}T\partial\mathcal{M}$ must be equal to dim $\mathcal{M}$. ∎

The orthogonal T′ of $\mathcal{V}$ in $\mathbb{C}T^*\overline{\mathcal{M}}$ will be referred to as the *cotangent structure bundle* of $\overline{\mathcal{M}}$. The fibre dimension of T′ will be m, that of $\mathcal{V}$, n; we have dim $\mathcal{M} = m + n$. The subbundle $T' \subset \mathbb{C}T\overline{\mathcal{M}}$ is closed (Definition I.1.1). By duality (V.1.2) and (V.1.3) are respectively equivalent to

$$(T'|_{\partial\mathcal{M}}) + \mathbb{C}N^*\partial\mathcal{M} \textit{ is a vector bundle over } \partial\mathcal{M}; \tag{V.1.6}$$

$$T'\cap\mathbb{C}N^*\partial\mathcal{M} \textit{ is a vector bundle over } \partial\mathcal{M}. \tag{V.1.7}$$

On the other hand, (V.1.4) is equivalent to

$$T'\cap\mathbb{C}N^*\partial\mathcal{M} = 0, \tag{V.1.8}$$

whereas (V.1.5) is equivalent to

$$\mathbb{C}N^*\partial\mathcal{M} \subset T'. \tag{V.1.9}$$

That either (V.1.8) or (V.1.9) holds true is evident, if we note that the rank (over $\mathbb{C}$) of $T'\cap\mathbb{C}N^*\partial\mathcal{M}$ is ≤ 1.

As in the preceding chapters we denote by T° the intersection of T′ with the real cotangent bundle $T^*\overline{\mathcal{M}}$; T° is the *characteristic set* of the structure (Definition I.2.2). We see that (V.1.8) is equivalent to

$$T^\circ\cap N^*\partial\mathcal{M} = 0, \tag{V.1.10}$$

whereas (V.1.9) is equivalent to

$$N^*\partial\mathcal{M} \subset T^\circ. \tag{V.1.11}$$

In accordance with Definition I.4.1 we say that $\partial\mathcal{M}$ is *noncharacteristic* if (V.1.10) (or (V.1.8)) holds. In accordance with Definition I.4.2 we say that $\partial\mathcal{M}$ is *totally characteristic* if (V.1.11) (or (V.1.9)) holds.

Observe now that, in both cases (V.1.8) or (V.1.9), the natural map

$$\mathbb{C}T^*\overline{\mathcal{M}}|_{\partial\mathcal{M}} \to \mathbb{C}T^*\partial\mathcal{M} \tag{V.1.12}$$

(whose kernel is $\mathbb{C}N^*\partial\mathcal{M}$) maps $T'|_{\partial\mathcal{M}}$ onto a vector subbundle $T'_{\partial\mathcal{M}}$ of $\mathbb{C}T^*\partial\mathcal{M}$; $T'_{\partial\mathcal{M}}$ defines an *involutive structure* on $\partial\mathcal{M}$ to which we refer as the involutive structure *inherited from* or *induced by* that of $\overline{\mathcal{M}}$.

The boundary $\partial\mathcal{M}$ is noncharacteristic if and only if the natural map (V.1.12) induces an isomorphism $T'_{\partial\mathcal{M}} \cong T'|_{\partial\mathcal{M}}$. It is totally characteristic if and only if the map (V.1.12) induces an isomorphism $T'_{\partial\mathcal{M}} \cong (T'|_{\partial\mathcal{M}})/N^*\partial\mathcal{M}$. When $\partial\mathcal{M}$ is noncharacteristic the rank of $T'_{\partial\mathcal{M}}$ is equal to m, whereas it is equal to $m-1$ when $\partial\mathcal{M}$ is totally characteristic.

Let ρ be a $\mathcal{C}^\infty$ function in some open subset U of $\overline{\mathcal{M}}$ such that $\partial\mathcal{M} \cap U \neq \emptyset$ is defined in U by $\rho = 0$, and that $d\rho \neq 0$ at every point of U. Then $d\rho$ spans

$\mathbb{C}N^*\partial\mathcal{M}$ over $\partial\mathcal{M}\cap U$. To say that the boundary $\partial\mathcal{M}$ is noncharacteristic is equivalent to saying that $d\rho \notin T'$ at any point of $\partial\mathcal{M} \cap U$, whereas to say that $\partial\mathcal{M}$ is totally characteristic is equivalent to saying that $d\rho$ is a section of T' over $\partial\mathcal{M}\cap U$.

Denote by $\mathcal{V}_{\partial\mathcal{M}}$ the tangent structure bundle in the involutive structure of $\partial\mathcal{M}$ inherited from $\overline{\mathcal{M}}$; $\mathcal{V}_{\partial\mathcal{M}}$ is the orthogonal of $T'_{\partial\mathcal{M}}$ in $\mathbb{C}T\partial\mathcal{M}$. When $\partial\mathcal{M}$ is noncharacteristic, the rank of $\mathcal{V}_{\partial\mathcal{M}}$ is equal to $n - 1$ (n = rank of $\mathcal{V}$); when $\partial\mathcal{M}$ is totally characteristic, it is equal to n. Over $U\cap\partial\mathcal{M}$ (see above) the sections of $\mathcal{V}_{\partial\mathcal{M}}$ can be regarded as those vector fields that are simultaneously orthogonal to $T'|_{\partial\mathcal{M}}$ and to $d\rho$.

EXAMPLE. V.1.1. Take $\mathcal{V} = \mathbb{C}T\overline{\mathcal{M}}$, $T' = 0$. Then $\partial\mathcal{M}$ is noncharacteristic. ■

EXAMPLE V.1.2. Take $T' = \mathbb{C}T^*\overline{\mathcal{M}}$, $\mathcal{V} = 0$. Then the boundary $\partial\mathcal{M}$ is totally characteristic. ■

EXAMPLE V.1.3. Let $\mathcal{M}$ be an open subset of a complex manifold $\hat{\mathcal{M}}$ (with $\dim_{\mathbb{C}} \hat{\mathcal{M}} = m$). Suppose the boundary $\partial\mathcal{M}$ is smooth and that $\mathcal{M}$ lies on one side of it. Then the union $\mathcal{M}\cup\partial\mathcal{M}$ can be regarded as a manifold with boundary, $\overline{\mathcal{M}}$. Let T' denote the restriction to $\overline{\mathcal{M}}$ of the complex structure bundle of $\hat{\mathcal{M}}$ (usually denoted by $T'^{1,0}$): for any $p \in \mathcal{M}\cup\partial\mathcal{M}$, T'_p is spanned by the differentials of the holomorphic functions in some open neighborhood of p in $\hat{\mathcal{M}}$. Since $T'_p \cap T^*\hat{\mathcal{M}} = 0$ property (V.1.10) must perforce be valid: the boundary of $\overline{\mathcal{M}}$ is noncharacteristic. ■

EXAMPLE V.1.4. Let x, y denote the coordinates in the plane and let $\overline{\mathcal{M}}$ be the closed upper half-plane $y \geq 0$. Take T' to be the vector bundle spanned over $\overline{\mathcal{M}}$ spanned by the one-form $dy + \imath y dx$. The pullback of T' to $\partial\mathcal{M}$, i.e., to the x-axis, vanishes identically: $\partial\mathcal{M}$ is totally characteristic. ■

It is perhaps worthwhile to give an example akin to Example V.1.4, of "something" that is *not* an involutive structure:

EXAMPLE V.1.5. Let $\overline{\mathcal{M}}$ denote the upper half-plane, as in Example V.1.4, but let T' now be spanned by the differential $dy + \imath x dx$. The pullback of T' to $\partial\mathcal{M}$ vanishes at $x = 0$ but not when $x \neq 0$ and thus it is not a vector bundle over $\partial\mathcal{M}$. ■

We look now at the pullback to $\partial\mathcal{M}$ of the "canonical bundle" of the structure of $\overline{\mathcal{M}}$, $\Lambda^m T'$. The pullback is a complex line bundle when the boundary is noncharacteristic. By (V.1.9) it vanishes identically when the boundary is totally characteristic. When $m = 1$ this is equivalent to saying that $T'_{\partial\mathcal{M}} = 0$. Let us point out the following consequence of this fact:

PROPOSITION V.1.3. *Suppose that $m = 1$ and that the boundary is totally characteristic. Then any $\mathscr{C}^1$ solution in an open subset Ω of $\overline{\mathcal{M}}$ is locally constant on $\Omega \cap \partial\mathcal{M}$.*

We shall extend the terminology of Definition I.2.3 to manifolds with boundary. The next two statements are evident:

If the structure of $\overline{\mathcal{M}}$ is real, then the structure induced on $\partial\mathcal{M}$ is also real. (V.1.13)

If the structure of $\overline{\mathcal{M}}$ is elliptic, then the boundary is noncharacteristic. (V.1.14)

PROPOSITION V.1.4. *If the structure of $\overline{\mathcal{M}}$ is complex, it induces on the boundary $\partial\mathcal{M}$ a CR structure. The characteristic set of $\partial\mathcal{M}$ is a real line bundle.*

PROOF. Call n the complex dimension of $\mathcal{M}$. Let $T'_{\partial\mathcal{M}}$ denote the pullback to $\partial\mathcal{M}$ of $T'^{1,0}$, the cotangent structure bundle of $\overline{\mathcal{M}}$. Necessarily $\mathbb{C}T^*\partial\mathcal{M} = T'_{\partial\mathcal{M}} + \overline{T}'_{\partial\mathcal{M}}$, which proves that the structure induced on $\partial\mathcal{M}$ is CR. Moreover, the complex fibre dimension of $T'_{\partial\mathcal{M}}$, which is $\leq n$, is $\geq \frac{1}{2}\dim \partial\mathcal{M} = n - 1/2$; it must be equal to n, and the dimension of $T'_{\partial\mathcal{M}} \cap \overline{T}'_{\partial\mathcal{M}}$ must be equal to one.■

The concepts introduced above have their counterparts in the real-analytic category: $\overline{\mathcal{M}}$ is a $\mathscr{C}^\omega$ manifold with boundary, which subsumes that $\partial\mathcal{M}$ is a $\mathscr{C}^\omega$ hypersurface in $\overline{\mathcal{M}}$; $\mathcal{V}$ (resp., T') is a $\mathscr{C}^\omega$ vector subbundle of $\mathbb{C}T\overline{\mathcal{M}}$ (resp., $\mathbb{C}T^*\overline{\mathcal{M}}$), etc.

Let us return to the general case: the $\mathscr{C}^\infty$ manifold with boundary $\overline{\mathcal{M}}$ is equipped with an involutive structure, of class $\mathscr{C}^\infty$ (with structure bundles $\mathcal{V}$ and T').

DEFINITION V.1.2. *We say that the involutive structure of $\overline{\mathcal{M}}$ is* locally integrable *if every point of $\overline{\mathcal{M}}$ has an open neighborhood in which T' is spanned by exact differentials* (cf. Definition I.1.2).

PROPOSITION V.1.5. *If the involutive structure of $\overline{\mathcal{M}}$ is locally integrable, the same is true of the involutive structure it induces on $\partial\mathcal{M}$.*

THEOREM V.1.1. *If the involutive structure of $\overline{\mathcal{M}}$ is either real or analytic, it is locally integrable.*

PROOF. We know (Theorems I.10.1, I.10.5) that the involutive structure of the interior manifold $\mathcal{M}$ is locally integrable. Then let p be an arbitrary point of $\partial\mathcal{M}$ and $(U, x_1,\dots,x_N)$ a local chart centered at p, such that

$$U\cap\mathcal{M} = \{x \in U;\, x_N > 0\},\quad U\cap\partial\mathcal{M} = \{x \in U;\, x_N = 0\}.$$

We shall write $x' = (x_1,\ldots,x_{N-1})$. In the analytic case we take the coordinates x_i to be real-analytic.

We introduce a $\mathscr{C}^\infty$ (resp., $\mathscr{C}^\omega$) basis of $\mathscr{V}$ over U, $L_1,\ldots,L_n$, consisting of pairwise commuting vector fields (cf. section I.5). In the $\mathscr{C}^\omega$ case the coefficients of the vector fields $L_1,\ldots,L_n$ extend analytically across the piece of boundary $U\cap\partial\mathcal{M}$ to values $x_N < 0$ and their extensions commute. We may then apply Theorem I.10.5 to reach the desired conclusion.

Suppose the vector fields $L_1,\ldots,L_n$ are real. We may assume that $L_1,\ldots,L_\nu$ are tangent to $\partial\mathcal{M}$ (on $U\cap\partial\mathcal{M}$), with $\nu = n - 1$ when $\partial\mathcal{M}$ is noncharacteristic or with $\nu = n$ when $\partial\mathcal{M}$ is totally characteristic.

We look first at a noncharacteristic boundary. In this case we may assume that

$$L_j = \partial/\partial x_j + \sum_{k=n}^{N-1} a_{j,k}(x)\partial/\partial x_k \ (1 \le j < n),\ L_n = \partial/\partial x_N + \sum_{k=n}^{N-1} a_{n,k}(x)\partial/\partial x_k.$$

We apply the Frobenius theorem (Theorem I.10.1) in x'-space, or rather a version of Theorem I.10.1 with parameter, here x_N, to the vector fields $L_1,\ldots,L_{n-1}$: there exists a $\mathscr{C}^\infty$ change of variables $(x',x_N) \to (F(x),x_N)$ that transforms their linear span into that of $\partial/\partial x_1,\ldots,\partial/\partial x_{n-1}$. To obtain a basis of $\mathscr{V}$ over U (suitably contracted) we may adjoin to the latter a vector field L_n whose expression (in the new coordinates) is of the same kind as above. The involution condition requires that $\partial/\partial x_1,\ldots,\partial/\partial x_{n-1}$ and L_n commute, which simply means that the coefficients $a_{n,k}$ must be independent of $x_1,\ldots,x_{n-1}$. Extend those coefficients to the region $x_N \le 0$ as $\mathscr{C}^\infty$ functions independent of $x_1,\ldots,x_{n-1}$. The system, consisting of $\partial/\partial x_1,\ldots,\partial/\partial x_{n-1}$ and of L_n thus extended, commutes and therefore defines a real structure in a full neighborhood of the origin in x-space. It suffices to apply Theorem I.10.1 to this system, to reach the desired conclusion.

Finally we look at a totally characteristic boundary. In this case, after applying the standard Frobenius theorem to the vector fields $L_j|_{\partial\mathcal{M}}$ we may assume that

$$L_j = \partial/\partial x_j + x_N \sum_{k=n+1}^{N} a_{j,k}(x)\partial/\partial x_k \ (1 \le j \le n).$$

The coefficients $a_{j,k}$ are real-valued. The special expressions of these commuting vector fields makes it possible to solve the boundary value problems

$$L_j\Psi_h = 0 \ \textit{if}\ x_N > 0,\ j = 1,\ldots,n, \tag{V.1.15}$$

$$\Psi_h|_{x_N=0} = x_h \tag{V.1.16}$$

($h = n+1,\ldots,N$) in the ring of formal series in the powers of x_N with coefficients that are *real*-valued functions of x': just write $\Psi_h = x_h + \sum_{p=1}^{\infty} x_N^p B_{h,p}(x')$

and determine the coefficients $B_{h,p}$ successively, by exploiting the differential equations (V.1.15) (the commutation of the vector fields L_j yields the needed compatibility conditions). We can then select m real-valued $\mathscr{C}^\infty$ functions $\hat{Z}_h$ whose Taylor expansions with respect to x_N about zero are equal to Ψ_h for each $h = n+1,\ldots,m+n$ (cf. Theorem IV.1.3). This entails that $L_j\hat{Z}_h$ vanishes to infinite order and that $\hat{Z}_h - x_h$ vanishes as $x_N \to +0$. If we take $\hat{Z}_h$ as new coordinate x_h for each $h = n+1,\ldots,N$ (in a suitably contracted neighborhood U), then in the new coordinates system we shall have

$$L_j = \partial/\partial x_j + \sum_{k=n+1}^{N} b_{j,k}(x)\partial/\partial x_k \ (1 \leq j \leq n)$$

with coefficients $b_{j,k}$ that vanish to infinite order at $x_N = 0$ and can therefore be extended to a full neighborhood of the origin in $\mathbb{R}^N$ by $b_{j,k}(x) \equiv 0$ if $x_N < 0$. Applying the standard Frobenius theorem to the system of vector fields $L_1,\ldots,L_n$ thus extended yields the desired conclusion. ■

Let $\mathcal{M}$ denote a $\mathscr{C}^\infty$ manifold without boundary, equipped with an involutive structure. The definitions just introduced lead to a new notion of local integrability for the involutive structure of $\mathcal{M}$.

DEFINITION V.1.3. *We shall say that the involutive structure of $\mathcal{M}$ is* locally integrable on one side, *at a point $\mathfrak{p} \in \mathcal{M}$, if there exists an isomorphism, for the induced involutive structures, of an open neighborhood of $\mathfrak{p}$ onto a relatively open subset of the boundary $\partial\mathcal{M}_1$ of a manifold with boundary $\overline{\mathcal{M}}_1$ endowed with an involutive structure which is locally integrable in the interior $\mathcal{M}_1$.*

PROPOSITION V.1.6. *Any $\mathscr{C}^\infty$ manifold $\mathcal{M}$ equipped with an involutive structure can be regarded as the noncharacteristic (resp., totally characteristic) boundary of a manifold $\overline{\mathcal{M}}_1$ with boundary, equipped with an involutive structure. The involutive structure of $\mathcal{M}$ is the same as that inherited from $\overline{\mathcal{M}}_1$. For the involutive structure of $\mathcal{M}$ to be locally integrable at a point $\mathfrak{p} \in \mathcal{M}$ it is necessary and sufficient that the same be true of that of $\overline{\mathcal{M}}_1$.*

PROOF. The manifold $\mathcal{M}$ can be regarded as the boundary $\partial\mathcal{M}_1$ of the "half-cylinder" $\overline{\mathcal{M}}_1 = \mathcal{M}\times\overline{\mathbb{R}}_+ = \{(\mathfrak{p},\tau) \in \mathcal{M}\times\mathbb{R}^1; \tau \geq 0\}$. If T′ denotes the cotangent structure bundle of $\mathcal{M}$, its pullback π^*T' under the "coordinate projection" $\pi : \mathcal{M}\times\overline{\mathbb{R}}_+ \to \mathcal{M}$ defines an involutive structure on $\overline{\mathcal{M}}_1$. The pullback map to $\partial\mathcal{M}_1$, $\mathbb{C}T^*\overline{\mathcal{M}}_1|_{\partial\mathcal{M}_1} \to \mathbb{C}T^*\partial\mathcal{M}_1 \cong \mathbb{C}T^*\mathcal{M}$, induces an isomorphism of $\pi^*T'|_{\partial\mathcal{M}_1}$ onto T′. In this setup $\partial\mathcal{M}_1$ is noncharacteristic. Instead, if we define the cotangent structure bundle T'_1 on $\overline{\mathcal{M}}_1$ as the Whitney sum of π^*T' and of the line bundle spanned by the one-form $d\tau$, the boundary $\partial\mathcal{M}_1$ will be totally characteristic. (In the former case the tangent structure bundle of $\overline{\mathcal{M}}_1$ is the direct sum of the "horizontal lift" of that of $\mathcal{M}$, $\mathcal{V}$, and of the line bundle spanned by $\partial/\partial\tau$, whereas in the latter case it is exactly the lift of $\mathcal{V}$.) If U is an open subset of $\mathcal{M}$ in which there exist $\mathcal{C}^\infty$ functions $Z_1,\ldots,Z_m$ whose differentials dZ_j span T′ over U, the pullbacks $d(Z_j\circ\pi)$ span π^*T' over $U\times\overline{\mathbb{R}}_+$ whereas to span T'_1 we must adjoin $d\tau$ to those pullbacks. ∎

Proposition V.1.7. *Let $\mathcal{M}$ be a CR manifold whose characteristic set is a line bundle. If the CR structure of $\mathcal{M}$ is locally integrable, $\mathcal{M}$ is locally isomorphic to the boundary of a complex domain in $\mathbb{C}^{n+1}$ (and* $\dim \mathcal{M} = 2n + 1$).

Proof. Let T′ be the cotangent structure bundle of $\mathcal{M}$ and m denote its rank. Since, by hypothesis, $\mathbb{C}T^*\mathcal{M} = T' + \overline{T}'$ and the rank of $T'\cap\overline{T}'$ is equal to one, we conclude that the rank of the tangent structure bundle $\mathcal{V}$ of $\mathcal{M}$ is equal to $n = m - 1$ and $\dim \mathcal{M} = 2n + 1$. Since the CR structure of $\mathcal{M}$ is locally integrable, over some open neighborhood U of an arbitrary point p of $\mathcal{M}$ T′ is spanned by the differentials of smooth functions $z_j = x_j + \imath y_j$, $j = 1,\ldots,n$, and $w = s + \imath\varphi(z,s)$. We may assume that the functions x_i, y_j $(1 \le i, j \le n)$ and s form a coordinate system in U. This means that $(x,y,s) \to (z,w)$ is a $\mathcal{C}^\infty$ diffeomorphism of U onto a hypersurface Σ in $\mathbb{C}^{n+1}$. This is an isomorphism for the CR structure induced on U by $\mathcal{M}$ and on Σ by $\mathbb{C}^{n+1}$. Furthermore, if we hypothesize $\varphi|_0 = 0$ and $d\varphi|_0 = 0$, and possibly contract U about p, we see that Σ is the boundary of the domain

$$\Omega^+ = \{ (z,w) \in \mathbb{C}^{n+1};\ (z,\mathcal{R}e\,w) \in U,\ \mathcal{I}m\,w \ge \varphi(z,s) \}$$

regarded as a manifold with boundary. ∎

In section VII.7 the reader will find an example of a CR structure that is locally integrable on one side but not locally integrable.

V.2. The Associated Differential Complex. The Boundary Complex

The differential complexes associated with the involutive structure of the manifold with boundary $\overline{\mathcal{M}}$ are defined exactly as for a manifold without boundary (see section I.6). Here we shall pay particular attention to the $\mathcal{C}^\infty$ complexes (for $p = 0,1,\ldots,m$)

$$d'^{p,q} : \mathcal{C}^\infty(\overline{\mathcal{M}};\Lambda^{p,q}) \to \mathcal{C}^\infty(\overline{\mathcal{M}};\Lambda^{p,q+1}),\ q = 0,1,\ldots, \qquad \text{(V.2.1)}$$

and their sheaf-theoretical analogues.

We may also consider the differential complex associated with the involutive structure inherited by $\partial\mathcal{M}$ from that of $\overline{\mathcal{M}}$. It is the *boundary complex*,

$$d_b'^{p,q} : \mathscr{C}^\infty(\partial\mathcal{M};\Lambda_b^{p,q}) \to \mathscr{C}^\infty(\partial\mathcal{M};\Lambda_b^{p,q+1}),\ q = 0,1,\ldots \tag{V.2.2}$$

Here the vector bundles $\Lambda_b^{p,q}$ are the quotient bundles (I.6.6) relative to the pullback $T'_{\partial\mathcal{M}}$ to $\partial\mathcal{M}$ of the cotangent structure bundle T′ of $\overline{\mathcal{M}}$. The reader will easily check that the pullback map $T'|_{\partial\mathcal{M}} \to T'_{\partial\mathcal{M}}$ induces a surjection $\Lambda^{p,q}|_{\partial\mathcal{M}} \to \Lambda_b^{p,q}$ to which we refer as the *pullback map* to $\partial\mathcal{M}$.

EXAMPLE V.2.1. Denote by x_i ($i = 1,\ldots,n$) and t the coordinates in $\mathbb{R}^{n+1}$ and take $\overline{\mathcal{M}} = \{ (x,t) \in \mathbb{R}^{n+1}; t \geq 0 \}$. Let the involutive structure of $\overline{\mathcal{M}}$ be defined by $\partial/\partial t$; the cotangent structure bundle T′ is spanned by $dx_1,\ldots,dx_n$. The boundary $\partial\mathcal{M}$ is noncharacteristic and $T'_{\partial\mathcal{M}} = \mathbb{C}T^*\partial\mathcal{M}$. When $q = 0$, $\Lambda^{p,q}$ is spanned by the exterior products dx_I, $|I| = p$; when $q = 1$, it is spanned by the products $dx_I \wedge dt$. The pullback to $\partial\mathcal{M}$ maps $\Lambda^{p,0}|_{\partial\mathcal{M}}$ onto $\Lambda^p\mathbb{C}T^*\partial\mathcal{M}$ and $\Lambda^{p,1}|_{\partial\mathcal{M}}$ onto $\{0\}$. ■

EXAMPLE V.2.2. Take $\overline{\mathcal{M}}$ as in Example V.2.1 but let its involutive structure be defined by the vector field $\partial/\partial x_1$. Now the boundary $\partial\mathcal{M}$ is totally characteristic. When $q = 0$ the bundle $\Lambda^{p,q}$ is spanned by the exterior products dx_I, $|I| = p$ and $1 \notin I$, and $dx_J \wedge dt$, $|J| = p-1$ and $1 \notin J$; when $q = 1$ it is spanned by the exterior products dx_I with $|I| = p+1$ and $1 \in I$, and $dx_J \wedge dt$ with $|J| = p$ and $1 \in J$. The pullback map $\Lambda^{p,q}|_{\partial\mathcal{M}} \to \Lambda_b^{p,q}$ consists in equating to zero all products that contain a factor dt; $\Lambda_b^{p,0}$ is spanned by the dx_I, $|I| = p$ and $1 \notin I$; $\Lambda_b^{p,1}$ by the dx_J, $|J| = p+1$ and $1 \in J$. ■

Back to the general case we have the natural pullback map

$$i_b^{*p,q} : \mathscr{C}^\infty(\overline{\mathcal{M}};\Lambda^{p,q}) \to \mathscr{C}^\infty(\partial\mathcal{M};\Lambda_b^{p,q}). \tag{V.2.3}$$

Below we write often i_b^* rather than $i_b^{*p,q}$.

LEMMA V.2.1. *The map* (V.2.3) *is surjective.*

PROOF. We select a locally finite covering $\{U_\iota\}$ of $\partial\mathcal{M}$ consisting of open subsets U_ι of $\mathcal{M}$ in each one of which there is a smooth basis of $\mathbb{C}T^*\overline{\mathcal{M}}$, $\{\varphi_{\iota 1},\ldots,\varphi_{\iota N}\}$, such that m of the forms $\varphi_{\iota j}$ make up a smooth basis of T′ over U_ι. We may, and shall, choose each U_ι in such a way that the restriction map $\mathscr{C}^\infty(U_\iota) \to \mathscr{C}^\infty(U_\iota \cap \partial\mathcal{M})$ is surjective.

Let $\dot{f} \in \mathscr{C}^\infty(\partial\mathcal{M};\Lambda_b^{p,q})$ be arbitrary; there is a representative of $\dot{f}$ over $U_\iota \cap \partial\mathcal{M}$ of the kind

$$f_\iota = \sum_{|J|=p+q}{}' f_{\iota,J}\, i_b^*\varphi_{\iota,J},$$

where the apostrophe means that the sum is to be restricted to those multi-indices J such that exactly p forms $\varphi_{\iota,j}$ ($j \in J$) are sections of T′ over U_ι while q are not. Extend each coefficient $f_{\iota,J}$ as a $\mathscr{C}^\infty$ function in the whole of U_ι, and define the $(p+q)$-form in U_ι,

$$F_\iota = \sum_{|J|=p+q}{}' f_{\iota,J}\varphi_{\iota,J}.$$

For each ι we select a function $\chi_\iota \in \mathscr{C}^\infty(\overline{\mathcal{M}})$ such that supp $\chi_\iota \subset U_\iota$, and that $\sum_\iota \chi_\iota \equiv 1$ in an open neighborhood of $\partial\mathcal{M}$ in $\overline{\mathcal{M}}$. Then $F = \sum_\iota \chi_\iota F_\iota$ is a $\mathscr{C}^\infty$ form in the whole of $\overline{\mathcal{M}}$. Its pullback to $\partial\mathcal{M}$ is a representative of f. ■

EXAMPLE V.2.3. On $\overline{\mathcal{M}} = \{ (x,t) \in \mathbb{R}^2; t \geq 0 \}$ consider the structure defined by the vector field $\partial/\partial t$. The elements of Ker i_b^* can be identified to the following $(p+q)$-forms: when $q = 0$, the zero-forms μ if $p = 0$, the one-forms μdx if $p = 1$, with $\mu \in \mathscr{C}^\infty(\overline{\mathcal{M}})$, $\mu|_{t=0} \equiv 0$; when $q = 1$, the one-forms $\mu_1 dt$ if $p = 0$ and the two-forms $\mu_1 dx \wedge dt$ if $p = 1$, with $\mu_1 \in \mathscr{C}^\infty(\overline{\mathcal{M}})$ arbitrary. ■

EXAMPLE V.2.4. Take $\overline{\mathcal{M}}$ as in Example V.2.3 but now with the structure defined by $\partial/\partial x$. In this case the elements of Ker i_b^* can be identified to the following $(p+q)$-forms: when $p = 0$, the zero-forms μ if $q = 0$, the one-forms μdx if $q = 1$, with $\mu \in \mathscr{C}^\infty(\overline{\mathcal{M}})$, $\mu|_{t=0} \equiv 0$; when $p = 1$, the one-forms $\mu_1 dt$ if $q = 0$ and the two-forms $\mu_1 dx \wedge dt$ if $q = 1$, with $\mu_1 \in \mathscr{C}^\infty(\overline{\mathcal{M}})$ arbitrary. ■

Next we introduce the quotient spaces

$$\mathscr{C}_*^\infty(\overline{\mathcal{M}};\Lambda^{p,q}) = \mathscr{C}^\infty(\overline{\mathcal{M}};\Lambda^{p,q})/\text{Ker } i_b^{*p,q}. \tag{V.2.4}$$

Since $d_b'^{p,q} i_b^{*p,q} = i_b^{*p,q+1} d'^{p,q}$ we may also introduce the differential complex

$$d_*'^{p,q} : \mathscr{C}_*^\infty(\overline{\mathcal{M}};\Lambda^{p,q}) \to \mathscr{C}_*^\infty(\overline{\mathcal{M}};\Lambda^{p,q+1}),\ q = 0,1,\ldots, \tag{V.2.5}$$

for each $p = 0,\ldots,m$, and its cohomology spaces, $H_*'^{p,q}(\overline{\mathcal{M}})$.

A representative of a class $[f]_* \in H_*'^{p,q}(\overline{\mathcal{M}})$ is the equivalence class mod Ker i_b^* of a $\mathscr{C}^\infty$ section f of $\Lambda^{p,q}$ over $\overline{\mathcal{M}}$ such that $i_b^*(d'f) \equiv 0$ in $\partial\mathcal{M}$. If $f_1 \in \mathscr{C}^\infty(\overline{\mathcal{M}};\Lambda^{p,q})$ defines the same class $[f]_*$, then the following must be true: when $q = 0$ the pullback of $f - f_1$ to $\partial\mathcal{M}$ vanishes identically; when $q > 1$ there is $v \in \mathscr{C}^\infty(\overline{\mathcal{M}};\Lambda^{p,q-1})$ such that the pullback to $\partial\mathcal{M}$ of $f - f_1 - d'^{p,q}v$ vanishes identically. In the latter case note that necessarily $i_b^*(d'f_1) \equiv 0$ on $\partial\mathcal{M}$.

By Lemma V.2.1 the pullback map (V.2.3) induces a linear *bijection*

$$\mathscr{C}_*^\infty(\overline{\mathcal{M}};\Lambda^{p,q}) \to \mathscr{C}^\infty(\partial\mathcal{M};\Lambda_b^{p,q}). \tag{V.2.6}$$

If the horizontal arrows are interpreted as maps (V.2.6), the diagram

$$\begin{array}{ccc} \mathscr{C}_*^\infty(\overline{\mathcal{M}};\Lambda^{p,q}) & \to & \mathscr{C}^\infty(\partial\mathcal{M};\Lambda_b^{p,q}) \\ d_*'^{p,q} \downarrow & & \downarrow d_b'^{p,q} \\ \mathscr{C}_*^\infty(\overline{\mathcal{M}};\Lambda^{p,q+1}) & \to & \mathscr{C}^\infty(\partial\mathcal{M};\Lambda_b^{p,q+1}) \end{array} \tag{V.2.7}$$

is commutative. Thus we may state

PROPOSITION V.2.1. *The boundary complex* (V.2.2) *and the differential complex* (V.2.5) *are isomorphic.*

To get a more meaningful connection between the differential complex (V.2.1) and the boundary complex (V.2.2) we must restrict our attention to the case where

$$\textit{the boundary } \partial\mathcal{M} \textit{ is noncharacteristic.} \tag{V.2.8}$$

We introduce

$\mathcal{C}_0^\infty(\overline{\mathcal{M}};\Lambda^{p,q})$, *the subspace of* $\mathcal{C}^\infty(\overline{\mathcal{M}};\Lambda^{p,q})$ *consisting of the sections that vanish to infinite order at the boundary,*

and the subcomplex of (V.2.1),

$$\mathrm{d}' : \mathcal{C}_0^\infty(\overline{\mathcal{M}};\Lambda^{p,q}) \to \mathcal{C}_0^\infty(\overline{\mathcal{M}};\Lambda^{p,q}),\ q = 0,1,\ldots, \tag{V.2.9}$$

We denote by $\mathrm{H}_0'^{p,q}(\overline{\mathcal{M}})$ the cohomology spaces of (V.2.9).

The following is an existence and uniqueness statement for solutions of what could be called the *Cauchy problem* at the boundary, for sections of $\Lambda^{p,q}$ *modulo* $\mathcal{C}_0^\infty(\overline{\mathcal{M}};\Lambda^{p,q})$:

THEOREM V.2.1. *Suppose* $\partial\mathcal{M}$ *noncharacteristic and* $q > 0$. *Let* $f \in \mathcal{C}^\infty(\overline{\mathcal{M}};\Lambda^{p,q})$ *and* $g \in \mathcal{C}^\infty(\partial\mathcal{M};\Lambda^{p,q-1})$ *have the following properties:*

$$\mathrm{d}'f \in \mathcal{C}_0^\infty(\overline{\mathcal{M}};\Lambda^{p,q+1}); \tag{V.2.10}$$

$$i_b^* f = \mathrm{d}_b' g. \tag{V.2.11}$$

Then there exists $u \in \mathcal{C}^\infty(\overline{\mathcal{M}};\Lambda^{p,q-1})$ *satisfying*

$$\mathrm{d}'u - f \in \mathcal{C}_0^\infty(\overline{\mathcal{M}};\Lambda^{p,q}), \tag{V.2.12}$$

$$i_b^* u = g. \tag{V.2.13}$$

Moreover, if $q = 1$ *and if* $u_1 \in \mathcal{C}^\infty(\overline{\mathcal{M}};\Lambda^{p,0})$ *also satisfies* (V.2.12) *and* (V.2.13) *then*

$$u - u_1 \in \mathcal{C}_0^\infty(\overline{\mathcal{M}};\Lambda^{p,0}). \tag{V.2.14}$$

REMARK V.2.1. The existence of the section $u \in \mathcal{C}^\infty(\overline{\mathcal{M}};\Lambda^{p,q-1})$ satisfying (V.2.12) and (V.2.13) implies (V.2.10) and (V.2.11). The latter are the natural *compatibility conditions* for the solvability of equations (V.2.12) and (V.2.13). ■

REMARK V.2.2. When $q \geq 2$ the *existence* result in Theorem V.2.1 automatically entails a *uniqueness* result:

Let $u_1 \in \mathcal{C}^\infty(\overline{\mathcal{M}};\Lambda^{p,q-1})$ also satisfy (V.2.12) and (V.2.13). By subtraction we get

$$\mathrm{d}'(u-u_1) \in \mathscr{C}_0^\infty(\overline{\mathcal{M}};\Lambda^{p,q}), \tag{V.2.15}$$

$$i_b^*(u-u_1) \equiv 0. \tag{V.2.16}$$

This means that $u-u_1$ satisfies conditions (V.2.10) and (V.2.11) with $g \equiv 0$ (and $q-1$ substituted for q). But then, according to Theorem V.2.1, there exists $v \in \mathscr{C}^\infty(\overline{\mathcal{M}};\Lambda^{p,q-2})$, satisfying

$$u-u_1-\mathrm{d}'v \in \mathscr{C}_0^\infty(\overline{\mathcal{M}};\Lambda^{p,q-1}), \tag{V.2.17}$$

$$i_b^*v \equiv 0. \blacksquare \tag{V.2.18}$$

Proof of Theorem V.2.1. A preliminary simplification is in order. Lemma V.2.1 enables us to find a section $G \in \mathscr{C}^\infty(\overline{\mathcal{M}};\Lambda^{p,q})$ such that $i_b^*G = g$. Define

$$F = f - \mathrm{d}'G.$$

It is obvious that F also satisfies (V.2.10). Furthermore, $i_b^*F \equiv 0$ in $\mathrm{U}\cap\partial\mathcal{M}$. Suppose we find v such that $\mathrm{d}'v - \mathrm{F} \in \mathscr{C}_0^\infty(\overline{\mathcal{M}};\Lambda^{p,q})$ and $i_b^*v = 0$. Then $u = v + \mathrm{G}$ satisfies (V.2.10) and (V.2.11). In other words we may as well assume $g \equiv 0$.

We begin by proving a local version of Theorem V.2.1. Actually we avail ourselves of (I.6.12) and limit our attention to the case $p = 0$. We reason in an open neighborhood U in $\overline{\mathcal{M}}$ of an arbitrary point $\mathfrak{p}$ of the boundary. We take U to be the domain of local coordinates $x_1,\dots,x_m$, $t_1,\dots,t_n$, vanishing at $\mathfrak{p}$ and such that $\mathrm{d}t_1,\dots,\mathrm{d}t_n$ span, at each point of U, a supplementary of T′ in $\mathbb{C}\mathrm{T}^*\overline{\mathcal{M}}$. Thanks to our hypothesis (V.2.8) we may even choose those coordinates in such a manner that $t_n = 0$ is a defining equation for $\partial\mathcal{M}$ in U. We shall use the notation $t' = (t_1,\dots,t_{n-1})$. It is convenient to assume that the coordinates x_i, t_j define a diffeomorphism

$$\mathrm{U} \cong \{\, (x,t) \in \mathbb{R}^N;\ (x,t') \in \mathbb{O},\ 0 \le t_n < T \,\} \tag{V.2.19}$$

with $\mathbb{O}$ an open neighborhood of the origin in $\mathbb{R}^{N-1}$.

Furthermore, possibly after contracting U about $\mathfrak{p}$, one can select a basis $\{\mathrm{L}_1,\dots,\mathrm{L}_n\}$ of $\mathcal{V}$ over U submitted to the requirement that $\mathrm{L}_j t_k = \delta_{jk}$, the Kronecker index, for all j, $k = 1,\dots,n$. We may also assume that the vector fields L_j commute (see section I.5). Note, for use below, that

$$\mathrm{L}_n = \partial/\partial t_n + \sum_{i=1}^{m} \lambda_i(x,t)\partial/\partial x_i. \tag{V.2.20}$$

Each class $\dot{h} \in \mathscr{C}^\infty(\mathrm{U};\Lambda^{0,q})$ has a unique standard representative

$$h = \sum_{|J|=q} h_J(x,t)\mathrm{d}t_J, \tag{V.2.21}$$

with $h_J \in \mathscr{C}^\infty(\mathrm{U})$. For the sake of simplicity we shall not distinguish between $\dot{h}$ and h; we shall identify sections of $\Lambda^{0,q}$ to differential forms in t-space whose coefficients are smooth functions in U, and we shall omit the dots on top. In the same vein $d'h$ will be identified to

$$\mathrm{L}h = \sum_{|J|=q} \sum_{k=1}^{n} \mathrm{L}_k h_J(x,t)\mathrm{d}t_k \wedge \mathrm{d}t_J. \tag{V.2.22}$$

We shall also use the notation

$$\mathrm{L}'h = \sum_{|J|=q} \sum_{k=1}^{n-1} \mathrm{L}_k h_J(x,t)\mathrm{d}t_k \wedge \mathrm{d}t_J. \tag{V.2.23}$$

Note that $\mathrm{L}'^2 = 0$ and that, along $\mathrm{U}\cap\partial\mathcal{M}$, L' is "tangential" to the boundary. Also, below, we shall let L_n act coefficientwise on differential forms that only involve differentials $\mathrm{d}t_k$, $k < n$. This action of L_n commutes with the operator L'.

Consider the section f in the statement. We may write

$$f = f_0 + \mathrm{d}t_n \wedge f_1,$$

where f_0 and f_1 are standard forms of respective degrees q and $q-1$, which do not involve $\mathrm{d}t_n$. We have

$$\mathrm{L}f = \mathrm{L}'f_0 + \mathrm{d}t_n \wedge (\mathrm{L}_n f_0 - \mathrm{L}'f_1),$$

and thus (V.2.10) reads

$$\mathrm{L}'f_0 \in \mathscr{C}_0^\infty(\mathrm{U};\Lambda^{0,q+1}),\ \mathrm{L}_n f_0 - \mathrm{L}'f_1 \in \mathscr{C}_0^\infty(\mathrm{U};\Lambda^{0,q}), \tag{V.2.24}$$

while (V.2.11) reads

the coefficients of f_0 vanish identically when $t_n = 0$. (V.2.25)

We introduce a form $v \in \mathscr{C}^\infty(\mathrm{U};\Lambda^{0,q-1})$ which does not involve $\mathrm{d}t_n$ and satisfies

$\mathrm{L}_n v - f_1$ *vanishes to infinite order at* $t_n = 0$; (V.2.26)

every coefficient of v vanishes at $t_n = 0$. (V.2.27)

Finding such a form v is straightforward: It suffices to construct, for each multi-index J, $|J| = q-1$ ($n \notin J$), a function $v_J \in \mathscr{C}^\infty(\mathrm{U})$ vanishing at $t_n = 0$, and such that $\mathrm{L}_n v_J - f_{1J}$ (the J-th coefficient of f_1) vanishes to infinite order at $t_n = 0$. For this we expand in powers of t_n each coefficient of L_n, the right-hand side f_J and the solution v_J. We determine, by induction on $k = 1,2,\ldots,$and by taking advantage of (V.2.20), the coefficient $v_{J,k}(x,t')$ of t_n^k in the expansion of v_J. We define

$$v_J(x,t) = \sum_{k=1}^{+\infty} \chi(t_n/R_k) v_{J,k}(x,t') t_n^k,$$

where $\chi \in \mathcal{C}^\infty(\mathbb{R})$, $\chi(\tau) = 1$ for $\tau < 1$, $\chi(\tau) = 0$ for $\tau > 2$ and the positive numbers R_k tend to $+\infty$ fast enough to ensure that the above series converges in $\mathcal{C}^\infty(\mathrm{U})$. We then set

$$v = \sum_{|J|=q-1} v_J(x,t)\mathrm{d}t_J.$$

With this choice of v we have

$$f - \mathrm{L}v = f_0 - \mathrm{L}'v + \mathrm{d}t_n \wedge (f_1 - \mathrm{L}_n v),$$

whence

$$(f-\mathrm{L}v) - (f_0-\mathrm{L}'v) \in \mathcal{C}_0^\infty(\mathrm{U};\Lambda^{0,q}). \tag{V.2.28}$$

From (V.2.24) and (V.2.26) it follows that

$$\mathrm{L}_n(f_0-\mathrm{L}'v) \in \mathcal{C}_0^\infty(\mathrm{U};\Lambda^{0,q}), \tag{V.2.29}$$

and from (V.2.25) and (V.2.27), that

$$\textit{every coefficient of } f_0-\mathrm{L}'v \textit{ vanishes at } t_n = 0. \tag{V.2.30}$$

Then let c be an arbitrary coefficient of $f_0-\mathrm{L}'v$; $c|_{t_n=0} = 0$ and $\mathrm{L}_n c$ vanishes to infinite order at t_n. If once again we avail ourselves of (V.2.20) and we consider the Taylor expansion of c in powers of t_n, we conclude that c itself vanishes to infinite order at $t_n = 0$. Combining this with (V.2.28) yields

$$f - \mathrm{L}v \in \mathcal{C}_0^\infty(\mathrm{U};\Lambda^{0,q}), \tag{V.2.31}$$

which implies the local existence result we were seeking.

When $q = 1$, if $v_1 \in \mathcal{C}^\infty(\mathrm{U})$ also satisfies (V.2.27) and (V.2.31), then $v-v_1|_{t_n=0} = 0$ and $\mathrm{L}_n(v-v_1)$ vanishes to infinite order at $t_n = 0$. The above argument about the coefficient c also shows that $v-v_1$ vanishes to infinite order at $t_n = 0$.

This completes the proof of the local version of Theorem V.2.1.

We now show how Theorem V.2.1 follows from its local version. This is obvious for the uniqueness statement, when $q = 1$. Let $\{\mathrm{U}_\iota\}$ be a locally finite open covering of $\partial\mathcal{M}$ and $\{\chi_\iota\}$ a $\mathcal{C}^\infty$ partition of unity in a neighborhood of $\partial\mathcal{M}$, like those introduced in the proof of Lemma V.2.1. We suppose that the local existence result established in the first part of the proof does apply to $\mathrm{U} = \mathrm{U}_\iota$, whatever the index ι: there is $u_\iota \in \mathcal{C}^\infty(\mathrm{U}_\iota;\Lambda^{p,q-1})$ satisfying

$$\mathrm{d}'u_\iota - f|_{\mathrm{U}_\iota} \in \mathcal{C}_0^\infty(\mathrm{U}_\iota;\Lambda^{p,q}), \tag{V.2.32}$$

$$i_b^* u_\iota = 0 \textit{ in } \mathrm{U}_\iota \cap \partial\mathcal{M}. \tag{V.2.33}$$

Define

$$v = \sum_\iota \chi_\iota u_\iota.$$

It follows at once from (V.2.33) that $i_b^* v = 0$. On the other hand,

$$d'v = f - R + S, \tag{V.2.34}$$

with $R = \sum_\iota \chi_\iota(f - d'u_\iota)$, $S = \sum_\iota d'\chi_\iota \wedge u_\iota$. Clearly $R \in \mathscr{C}_0^\infty(\overline{\mathcal{M}};\Lambda^{p,q})$.

Let Ω be an open subset of $\overline{\mathcal{M}}$ containing $\partial\mathcal{M}$ in which $\sum_\iota \chi_\iota \equiv 1$. In Ω,

$$S = \sum_{\iota,\iota'} \chi_{\iota'} d'\chi_\iota \wedge (u_\iota - u_{\iota'}). \tag{V.2.35}$$

Suppose first $q = 1$. We apply the local uniqueness result in U_ι: the sections $u_\iota - u_{\iota'}$ vanish to infinite order on $U_\iota \cap U_{\iota'} \cap \partial\mathcal{M}$ and we conclude, thanks to (V.2.35), that $S \in \mathscr{C}_0^\infty(\overline{\mathcal{M}};\Lambda^{p,1})$. By the same token, and still in the case $q = 1$, we note that property (V.2.13) (with $g \equiv 0$) is trivially local in nature.

Next suppose $q \geq 2$. We use induction on q. By the argument used in Remark V.2.2, there is $v_{\iota\iota'} \in \mathscr{C}^\infty(U_\iota \cap U_{\iota'};\Lambda^{p,q-2})$ satisfying

$$d'v_{\iota\iota'} - (u_\iota - u_{\iota'}) \in \mathscr{C}_0^\infty(U_\iota \cap U_{\iota'};\Lambda^{p,q-1}), \tag{V.2.36}$$

$$i_b^* v_{\iota\iota'} \equiv 0 \text{ in } U_\iota \cap U_{\iota'} \cap \partial\mathcal{M}. \tag{V.2.37}$$

From (V.2.35) and (V.2.36) we get

$$S - d'\Big(\sum_{\iota,\iota'} \chi_{\iota'} d'\chi_\iota \wedge v_{\iota\iota'}\Big) - S_1 \in \mathscr{C}_0^\infty(\Omega;\Lambda^{p,q}), \tag{V.2.38}$$

with $S_1 = \sum_{\iota,\iota'} d'\chi_\iota \wedge d'\chi_{\iota'} \wedge v_{\iota\iota'}$. But, still in Ω,

$$S_1 = \sum_{\iota,\iota',\iota''} \chi_{\iota''} d'\chi_\iota \wedge d'\chi_{\iota'} \wedge v_{\iota\iota'},$$

$$\sum_{\iota,\iota',\iota''} \chi_{\iota''} d'\chi_\iota \wedge d'\chi_{\iota'} \wedge v_{\iota'\iota''} \equiv \sum_{\iota,\iota',\iota''} \chi_{\iota''} d'\chi_\iota \wedge d'\chi_{\iota'} \wedge v_{\iota''\iota} \equiv 0,$$

whence

$$S_1 = \sum_{\iota,\iota',\iota''} \chi_{\iota''} d'\chi_\iota \wedge d'\chi_{\iota'} \wedge v_{\iota\iota'\iota''},$$

where we have used the notation

$$v_{\iota\iota'\iota''} = v_{\iota\iota'} + v_{\iota'\iota''} + v_{\iota''\iota}.$$

We derive at once from (V.2.36) and (V.2.37)

$$d'v_{\iota\iota'\iota''} \in \mathscr{C}_0^\infty(U_\iota \cap U_{\iota'} \cap U_{\iota''};\Lambda^{p,q-1}),$$

$$i_b^* v_{\iota\iota'\iota''} \equiv 0 \text{ in } U_\iota \cap U_{\iota'} \cap U_{\iota''}.$$

If $q = 2$ once again the uniqueness for one-forms enables us to conclude that $v_{\iota\iota'\iota''} \in \mathscr{C}_0^\infty(U_\iota \cap U_{\iota'} \cap U_{\iota''};\Lambda^{p,q-2})$, and therefore that S_1 vanishes to infinite order on $\partial\mathcal{M}$. By returning to (V.2.34) and (V.2.38) we conclude that, in this case,

$$u_0 = v + \sum_{\iota,\iota'} \chi_{\iota'} d'\chi_\iota \wedge v_{\iota\iota'} \in \mathscr{C}^\infty(\Omega;\Lambda^{p,q-1})$$

satisfies $d'u_0 - f \in \mathscr{C}_0^\infty(\Omega;\Lambda^{p,q})$. We also have $i_b^* u_0 \equiv 0$ by virtue of (V.2.37), and of the fact that $i_b^* v_0 \equiv 0$.

If $q > 2$ we apply the local result with $q - 2$ in the place of q and $v_{\iota\iota'\iota''}$ in that of f. We obtain that there is $w_{\iota\iota'\iota''} \in \mathscr{C}^\infty(U_\iota \cap U_{\iota'} \cap U_{\iota''};\Lambda^{p,q-3})$ satisfying

$$d'w_{\iota\iota'\iota''} - v_{\iota\iota'\iota''} \in \mathscr{C}_0^\infty(U_\iota \cap U_{\iota'} \cap U_{\iota''};\Lambda^{p,q-2}),$$

$$i_b^* w_{\iota\iota'\iota''} = 0 \textit{ in } U_\iota \cap U_{\iota'} \cap U_{\iota''} \cap \partial\mathcal{M}.$$

We get, in Ω,

$$S_1 - d'\Big(\sum_{\iota,\iota',\iota''} \chi_{\iota''} d'\chi_\iota \wedge d'\chi_{\iota'} \wedge w_{\iota\iota'\iota''}\Big) - S_2 \in \mathscr{C}_0^\infty(\Omega;\Lambda^{p,q}),$$

with

$$S_2 = \sum_{\iota,\iota',\iota'',\kappa} \chi_\kappa d'\chi_\iota \wedge d'\chi_{\iota'} \wedge d'\chi_{\iota''} \wedge w_{\iota\iota'\iota''\kappa},$$

$$w_{\iota\iota'\iota''\kappa} = w_{\iota\iota'\iota''} + w_{\iota'\iota''\kappa} + w_{\iota''\kappa\iota} + w_{\kappa\iota\iota'}.$$

At this point we repeat the argument used for $v_{\iota\iota'\iota''}$, and so forth, until we reach the stage where the uniqueness result for one-forms applies.

At the end of this process we get a form $u_0 \in \mathscr{C}^\infty(\Omega;\Lambda^{p,q-1})$ such that $d'u_0 - f \in \mathscr{C}_0^\infty(\Omega;\Lambda^{p,q})$ and $i_b^* u_0 \equiv 0$. Then select any function $\chi \in \mathscr{C}^\infty(\overline{\mathcal{M}})$ with supp $\chi \subset \Omega$ and $\chi \equiv 1$ in some neighborhood of $\partial\mathcal{M}$; $u = \chi u_0$ satisfies (V.2.12) and $i_b^* u \equiv 0$. ■

We introduce the quotient spaces

$$\mathscr{C}_\bullet^\infty(\overline{\mathcal{M}};\Lambda^{p,q}) = \mathscr{C}^\infty(\overline{\mathcal{M}};\Lambda^{p,q})/\mathscr{C}_0^\infty(\overline{\mathcal{M}};\Lambda^{p,q}). \tag{V.2.39}$$

Since (V.2.9) is a subcomplex of (V.2.1), we may also introduce the quotient complexes,

$$d_\bullet'^{p,q} : \mathscr{C}_\bullet^\infty(\overline{\mathcal{M}};\Lambda^{p,q}) \to \mathscr{C}_\bullet^\infty(\overline{\mathcal{M}};\Lambda^{p,q+1}),\ q = 0,1,\ldots, \tag{V.2.40}$$

and their cohomology spaces $H_\bullet'^{p,q}(\overline{\mathcal{M}})$.

A representative of a cohomology class $[f]_\bullet \in H_\bullet'^{p,q}(\overline{\mathcal{M}})$ is the equivalence class mod $\mathscr{C}_0^\infty(\overline{\mathcal{M}};\Lambda^{p,q})$ of a $\mathscr{C}^\infty$ section f of $\Lambda^{p,q}$ over $\overline{\mathcal{M}}$ such that $d'f \in \mathscr{C}_0^\infty(\overline{\mathcal{M}};\Lambda^{p,q+1})$. If $f_1 \in \mathscr{C}^\infty(\overline{\mathcal{M}};\Lambda^{p,q})$ defines the same class $[f]_\bullet$, then the following must be true: when $q = 0, f - f_1 \in \mathscr{C}_0^\infty(\overline{\mathcal{M}};\Lambda^{p,q})$; when $q \geq 1$ there exists $v \in \mathscr{C}^\infty(\overline{\mathcal{M}};\Lambda^{p,q-1})$ such that $f - f_1 - d'v \in \mathscr{C}_0^\infty(\overline{\mathcal{M}};\Lambda^{p,q})$. In the latter case note that necessarily $d'f_1 \in \mathscr{C}_0^\infty(\overline{\mathcal{M}};\Lambda^{p,q+1})$.

Since clearly $\mathcal{C}_0^\infty(\mathcal{M};\Lambda^{p,q}) \subset \operatorname{Ker} i_b^{*p,q}$ (see (V.2.3)), we have a natural surjection

$$\tau^{p,q} : \mathcal{C}_\bullet^\infty(\overline{\mathcal{M}};\Lambda^{p,q}) \to \mathcal{C}_*^\infty(\overline{\mathcal{M}};\Lambda^{p,q}), \tag{V.2.41}$$

clearly such that $\mathrm{d}_*'^{p,q}\tau^{p,q} = \tau^{p,q+1}\mathrm{d}_\bullet'^{p,q}$. Below we often write τ rather than $\tau^{p,q}$.

PROPOSITION V.2.2. *Suppose the boundary* $\partial\mathcal{M}$ *is noncharacteristic. The map* (V.2.41) *defines a bijection*

$$\mathrm{H}_\bullet'^{p,q}(\overline{\mathcal{M}}) \cong \mathrm{H}_*'^{p,q}(\overline{\mathcal{M}}). \tag{V.2.42}$$

PROOF. I. *The map* (V.2.41) *defines a surjection* $[\tau^{p,q}] : \mathrm{H}_\bullet'^{p,q}(\overline{\mathcal{M}}) \to \mathrm{H}_*'^{p,q}(\overline{\mathcal{M}})$. Indeed, let $h \in \mathcal{C}^\infty(\overline{\mathcal{M}};\Lambda^{p,q})$ be such that $i_b^*(\mathrm{d}'\, h) \equiv 0$. We apply Theorem 2.1 with $f \equiv 0$, $g = i_b^* h$ (and $q+1$ substituted for q). We obtain that $i_b^*(h-u) \equiv 0$ for some $u \in \mathcal{C}^\infty(\overline{\mathcal{M}};\Lambda^{p,q})$ such that $\mathrm{d}'u \in \mathcal{C}_0^\infty(\overline{\mathcal{M}};\Lambda^{p,q+1})$. The image via τ of the equivalence class of u mod $\mathcal{C}_0^\infty(\overline{\mathcal{M}};\Lambda^{p,q})$ is equal to the class of h mod $\operatorname{Ker} i_b^*$. Furthermore $\mathrm{d}_\bullet' u = 0$ and we may define

$$[\tau^{p,q}][u]_\bullet = [h]_*.$$

II. *The map* $[\tau^{p,q}]$ *is injective*. Indeed, let $f \in \mathcal{C}^\infty(\overline{\mathcal{M}};\Lambda^{p,q})$ satisfy (V.2.10) and (V.2.11). Property (V.2.10) means that the equivalence class of f mod $\mathcal{C}_0^\infty(\mathcal{M};\Lambda^{p,q})$ is a cocycle in the complex (V.2.40). Property (V.2.11) entails that the class of f mod $\operatorname{Ker} i_b^*$ is a coboundary in the complex (V.2.5), and therefore $[\tau^{p,q}][f] = 0$. By Theorem V.2.1, (V.2.12) means that the equivalence class of f mod $\mathcal{C}_0^\infty(\overline{\mathcal{M}};\Lambda^{p,q})$ is a coboundary in the complex (V.2.40). ■

By combining Propositions V.2.1 and V.2.2 we get

PROPOSITION V.2.3. *If the boundary* $\partial\mathcal{M}$ *is noncharacteristic there is a natural isomorphism*

$$\mathrm{H}_b'^{p,q}(\partial\mathcal{M}) \cong \mathrm{H}_\bullet'^{p,q}(\overline{\mathcal{M}}). \tag{V.2.43}$$

It is easy to describe directly the isomorphism (V.2.43): let $g \in \mathcal{C}^\infty(\partial\mathcal{M};\Lambda_b^{p,q})$ be such that $\mathrm{d}_b' g = 0$. By Theorem V.2.1 there is $u \in \mathcal{C}^\infty(\overline{\mathcal{M}};\Lambda^{p,q})$ such that $\mathrm{d}'u \in \mathcal{C}_0^\infty(\overline{\mathcal{M}};\Lambda^{p,q+1})$ and $i_b^* u = g$. Let $u_1 \in \mathcal{C}^\infty(\overline{\mathcal{M}};\Lambda^{p,q})$ be also such that $\mathrm{d}'u_1 \in \mathcal{C}_0^\infty(\overline{\mathcal{M}};\Lambda^{p,q})$ and $i_b^* u_1 = g$. If $q = 1$ we have $u - u_1 \in \mathcal{C}_0^\infty(\overline{\mathcal{M}};\Lambda^{p,q})$ and if $q > 1$, by Remark V.2.2, we have (V.2.17) for some $v \in \operatorname{Ker} i_b^{*p,q-1}$. It follows that u and u_1 define the same class $[u] \in \mathrm{H}_\bullet'^{p,q}(\overline{\mathcal{M}})$, which corresponds to the class $[g] \in \mathrm{H}_b'^{p,q}(\partial\mathcal{M})$ in the isomorphism (V.2.43).

COROLLARY V.2.1. *If the boundary* $\partial\mathcal{M}$ *is noncharacteristic, there exists, for each* p $(0 \leq p \leq m)$, *a natural long exact sequence*

$$0 \to \mathrm{H}_0'^{p,0}(\overline{\mathcal{M}}) \to \cdots \to \mathrm{H}_0'^{p,q}(\overline{\mathcal{M}}) \to \mathrm{H}'^{p,q}(\overline{\mathcal{M}}) \to \mathrm{H}_b'^{p,q}(\partial\mathcal{M}) \to \mathrm{H}_0'^{p,q+1}(\overline{\mathcal{M}}) \to \cdots \qquad \text{(V.2.44)}$$

PROOF. From the obvious short exact sequence

$$0 \to \mathscr{C}_0^\infty(\overline{\mathcal{M}};\Lambda^{p,q}) \to \mathscr{C}^\infty(\overline{\mathcal{M}};\Lambda^{p,q}) \to \mathscr{C}_\bullet^\infty(\overline{\mathcal{M}};\Lambda^{p,q}) \to 0$$

we draw a long exact sequence

$$0 \to \mathrm{H}_0'^{p,0}(\overline{\mathcal{M}}) \to \cdots \to \mathrm{H}_0'^{p,q}(\overline{\mathcal{M}}) \to \mathrm{H}'^{p,q}(\overline{\mathcal{M}}) \to \mathrm{H}_\bullet'^{p,q}(\overline{\mathcal{M}}) \to \mathrm{H}_0'^{p,q+1}(\overline{\mathcal{M}}) \to \cdots,$$

and it suffices to take (V.2.43) into account. ∎

COROLLARY V.2.2. *Suppose the boundary $\partial\mathcal{M}$ is noncharacteristic, and let p be any integer, $0 \le p \le m$. Suppose that*

$$\mathrm{H}_0'^{p,q}(\overline{\mathcal{M}}) = 0 \ \forall\, q,\ 0 \le q \le \nu\ (\le n). \qquad \text{(V.2.45)}$$

Then the pullback map i_b^* defines an isomorphism

$$\mathrm{H}'^{p,q}(\overline{\mathcal{M}}) \cong \mathrm{H}_b'^{p,q}(\partial\mathcal{M}),\ \forall\, q,\ 0 \le q \le \nu - 1. \qquad \text{(V.2.46)}$$

Let us show, by an example, that Proposition V.2.3 has no validity when the boundary is totally characteristic.

EXAMPLE V.2.5. Take $\overline{\mathcal{M}} = \{ (x,t) \in \mathbb{R}^2; t \ge 0 \}$ with the structure defined by $\partial/\partial x$. We have $\mathrm{H}_b'^{0,0}(\partial\mathcal{M}) \cong \mathbb{C}$ since the boundary complex is the De Rham complex on the x-axis. On the other hand, let $f \in \mathscr{C}^\infty(\overline{\mathcal{M}})$ be such that f_x vanishes to infinite order at $t = 0$. By integration with respect to x we see that there is a $\mathscr{C}^\infty$ function g in $\overline{\mathcal{M}}$ that vanishes to infinite order at $t = 0$ and is such that $(f-g)_x \equiv 0$. This means that $f(x,t) - g(x,t) = h(t)$, with $h \in \mathscr{C}^\infty(\overline{\mathbb{R}}_+)$, and therefore, that $\mathrm{H}_\bullet'^{0,0}(\overline{\mathcal{M}})$ is isomorphic to the quotient of $\mathscr{C}^\infty(\overline{\mathbb{R}}_+)$ modulo the subspace $\mathscr{C}_0^\infty(\mathbb{R}_+)$ made up of the $\mathscr{C}^\infty$ functions that vanish to infinite order at the origin. The quotient $\mathscr{C}^\infty(\overline{\mathbb{R}}_+)/\mathscr{C}_0^\infty(\mathbb{R}_+)$ is naturally isomorphic to the Taylor expansions of $\mathscr{C}^\infty$ functions at the origin (in the real line), and thus, by the classical Borel theorem, $\mathscr{C}^\infty(\overline{\mathbb{R}}_+)/\mathscr{C}_0^\infty(\mathbb{R}_+) \cong \mathbb{C}^{\mathbb{Z}_+}$. ∎

V.3. Locally Integrable Structures with Boundary. The Mayer-Vietoris Sequence

In the present section the manifold with boundary $\overline{\mathcal{M}}$ will be endowed with a locally integrable structure (Definition V.1.2); as usual the structure bundles will be called T′ and $\mathcal{V}$ and have respective ranks m and n. We shall be interested in local representations in some open neighborhood U of an arbitrary point p of the boundary. For the analogue at interior points we refer the reader to section I.7.

We take U small enough that there is a *nonnegative* function $\rho \in \mathscr{C}^\infty(U)$ such that $d\rho \neq 0$ at every point of U and that $\rho = 0$ be a defining equation for $U \cap \partial\mathcal{M}$ in U. Along $U \cap \partial\mathcal{M}$ the differential $d\rho$ spans the conormal bundle to $\partial\mathcal{M}$, $\mathbb{C}N^*\mathcal{M}$. We shall always assume that the geometry of U is suitably simple; in particular we shall assume that $U \cap \partial\mathcal{M}$ *is connected*.

Possibly after further shrinking of U we may also assume that T′ is spanned over U by the differentials of m $\mathscr{C}^\infty$ functions, $Z_1, \ldots, Z_m$; we take these functions to vanish at p. Below we shall select n additional $\mathscr{C}^\infty$ functions t_j, real-valued and vanishing at p, whose differentials span, over U, a supplementary of T′ in $\mathbb{C}T^*\overline{\mathcal{M}}$. We shall introduce vector fields L_j $(1 \leq j \leq n)$ and M_i $(1 \leq i \leq m)$ defined by the "orthonormality relations" (I.7.25). To tailor the choice of the "first integrals" more closely to our goals we must distinguish between the two basic situations.

Noncharacteristic boundary. Here, by (V.1.8), $d\rho \wedge dZ_1 \wedge \cdots \wedge dZ_m \neq 0$ over $U \cap \partial\mathcal{M}$. After a $\mathbb{C}$-linear substitution and further contracting of U about p we may assume that

$$d\rho \wedge d(\mathscr{R}e Z_1) \wedge \cdots \wedge (\mathscr{R}e Z_m) \neq 0,$$

and the functions $\mathscr{R}e Z_i$ $(i = 1, \ldots, m)$ as well as ρ can be taken as members of a coordinate system. We write $x_i = \mathscr{R}e Z_i$ and $t_n = \rho$. We may also arrange things in such a way that the differentials with respect to x of all the functions $\mathscr{I}m Z_i$ vanish at p. As in the case when there is no boundary, the rest of the coordinates will be denoted by $t_1, \ldots, t_{n-1}$; they shall all vanish at p. The differentials $dt_1, \ldots, dt_n$ span a supplementary of T′ in $\mathbb{C}T^*\mathcal{M}|_U$. Here also the formulas (I.7.23) and (I.7.24) are valid. We have $t_n \geq 0$ in U and

$$U \cap \partial\mathcal{M} = \{ (x,t) \in U;\, t_n = 0 \}. \tag{V.3.1}$$

Since the vector fields L_j $(1 \leq j < n)$ and M_i $(1 \leq i \leq m)$ all annihilate t_n, they are all tangent to $U \cap \partial\mathcal{M}$; L_n is transversal to $\partial\mathcal{M}$. It has an expression of the kind (V.2.20). ■

Totally characteristic boundary. Here $d\rho$ is a section of T′ over $U \cap \partial\mathcal{M}$, i.e.,

$$d\rho = c_1 dZ_1 + \cdots + c_m dZ_m \text{ on } U \cap \partial\mathcal{M}, \tag{V.3.2}$$

which implies (cf. the remark preceding Proposition V.1.3)

PROPOSITION V.3.1. *When the boundary is totally characteristic the pullback to* $U \cap \partial\mathcal{M}$ *of* $dZ_1 \wedge \cdots \wedge dZ_m$ *vanishes identically.*

When $m = 1$ we write simply Z instead of Z_1.

COROLLARY V.3.1. *When the boundary is totally characteristic and* $m = 1$ *we have* $Z \equiv 0$ *in* $U \cap \partial \mathcal{M}$.

PROOF. We know (Proposition V.1.3) that Z is locally constant in $U \cap \partial \mathcal{M}$. But on the other hand, $Z|_p = 0$ and $U \cap \partial \mathcal{M}$ is connected. ■

In (V.3.2) we may assume $c_m|_p \neq 0$. After substituting $(c_1|_p)Z_1 + \cdots + (c_m|_p)Z_m$ for Z_m we may assume $c_j|_p = 0$ for $j < m$, $c_m|_p = 1$. This implies that $dZ_1 \wedge \cdots \wedge dZ_{m-1} \wedge d\rho \neq 0$ at the point p and therefore in some neighborhood of p. After a contraction of U and a $\mathbb{C}$-linear substitution of the Z_j $(1 \leq j \leq m-1)$, we may assume that

$$d\rho \wedge d(\mathcal{Re} Z_1) \wedge \cdots \wedge d(\mathcal{Re} Z_{m-1}) \neq 0,$$

and take $\mathcal{Re} Z_1, .., \mathcal{Re} Z_{m-1}$ and ρ as coordinates. We write $x_i = \mathcal{Re} Z_i$ $(1 \leq i \leq m-1)$ and $x_m = \rho$. To these we adjoin n additional coordinates $t_1, \ldots, t_n$, which vanish at p and span a supplementary of T′ in $\mathbb{C}T^*\overline{\mathcal{M}}$, over U. We have $x_m \geq 0$ in U and

$$U \cap \partial \mathcal{M} = \{ (x,t) \in U; \; x_m = 0 \}. \tag{V.3.3}$$

We have, here,

$$Z_j = x_j + \imath \Phi_j(x,t), \; j = 1, \ldots, m-1. \tag{V.3.4}$$

We also know that

$$\mathcal{Re} Z_m = x_m + F(x,t), \; F(x,t) = O(|x|^2 + |t|^2).$$

Actually, we can write $\mathcal{Re} F(x,t) = A(x',t) + x_m B(x,t)$, where $x' = (x_1, \ldots, x_{m-1})$. We may take $x_m[1 + B(x,t)]$ as new variable x_m. We reach the conclusion that

$$\begin{gathered} Z_m = x_m + A(x',t) + \imath \Phi_m(x,t), \\ |A(x',t)| \leq const. \; (|x'|^2 + |t|^2). \end{gathered} \tag{V.3.5}$$

By letting the matrix $(\partial Z/\partial x)^{-1}|_p$ act on the m-vector $Z = (Z_1, \ldots, Z_m)$ we may assume that (I.7.24) holds.

In the present case, we may write

$$L_j = \partial/\partial t_j + \sum_{k=1}^{m-1} \lambda_{jk}(x,t)\partial/\partial x_k + \gamma_j(x,t) x_m \partial/\partial x_m. \; ■ \tag{V.3.6}$$

In all cases the coordinates x_i and t_j enable us to identify U to a subset of $\mathbb{R}^{m+n}$. When the boundary is noncharacteristic we may identify U to a subset of the half-space $t_n \geq 0$; when the boundary is totally characteristic, to a subset of the half-space $x_m \geq 0$. Let V (resp., W) be an open ball in x-space $\mathbb{R}^m$ (resp., t-space $\mathbb{R}^n$) centered at 0 and take, when the boundary $\partial\mathcal{M}$ is noncharacteristic,

$$\mathrm{U} = \{ (x,t) \in \mathrm{V}\times\mathrm{W};\ t_n \geq 0 \}, \tag{V.3.7}$$

whereas, when $\partial\mathcal{M}$ is totally characteristic,

$$\mathrm{U} = \{ (x,t) \in \mathrm{V}\times\mathrm{W};\ x_m \geq 0 \}. \tag{V.3.8}$$

Let us then define, in the former case,

$$\overline{\mathrm{U}}_- = \{ (x,t) \in \mathrm{V}\times\mathrm{W};\ t_n \leq 0 \}, \tag{V.3.9}$$

and in the latter,

$$\overline{\mathrm{U}}_- = \{ (x,t) \in \mathrm{V}\times\mathrm{W};\ x_m \leq 0 \}. \tag{V.3.10}$$

In the immediate sequel, in order to stress the contrast with $\overline{\mathrm{U}}_-$, we write $\overline{\mathrm{U}}_+$ in the place of U. We set $\mathrm{U}_0 = \overline{\mathrm{U}}_+\cap\partial\mathcal{M} = \overline{\mathrm{U}}_+\cap\overline{\mathrm{U}}_-$, $\hat{\mathrm{U}} = \overline{\mathrm{U}}_+\cup\overline{\mathrm{U}}_- = \mathrm{V}\times\mathrm{W}$.

Possibly after decreasing the radii of V and W we extend each first integral $Z_i(x,t)$ as a $\mathscr{C}^\infty$ function in $\hat{\mathrm{U}}$, in such a way that the differentials $dZ_1,\ldots,dZ_m$ define a locally integrable structure in $\hat{\mathrm{U}}$. We continue to denote by T′ (resp., $\mathcal{V}$) the cotangent (resp., tangent) structure bundle in the extended structure. We regard the subset U_0 as the boundary of the manifolds with boundary $\overline{\mathrm{U}}_+$ and $\overline{\mathrm{U}}_-$, and we equip it with the induced locally integrable structure. The bundles $\Lambda^{p,q}$ below must be understood in the sense of these locally integrable structures. In the remainder of the present section we are going to reason under the hypothesis that

the boundary $\partial\mathcal{M}$ is noncharacteristic.

Thus $\mathrm{U} = \overline{\mathrm{U}}_+$ is given by (V.3.7), and the vector field L_n by (V.2.20). In what follows p, $0 \leq p \leq m$, will be fixed. We introduce the following pullback maps:

$$i_+ : \mathscr{C}^\infty(\hat{\mathrm{U}};\Lambda^{p,q}) \to \mathscr{C}^\infty(\overline{\mathrm{U}}_+;\Lambda^{p,q}),\ i_- : \mathscr{C}^\infty(\hat{\mathrm{U}};\Lambda^{p,q}) \to \mathscr{C}^\infty(\overline{\mathrm{U}}_-;\Lambda^{p,q}),$$

$$i_0 : \mathscr{C}^\infty(\Omega;\Lambda^{p,q}) \to \mathscr{C}^\infty(\mathrm{U}_0;\Lambda^{p,q}),\ \Omega = \hat{\mathrm{U}},\ \overline{\mathrm{U}}_+ \text{ or } \overline{\mathrm{U}}_-,$$

and the linear maps

$$\mathrm{H}'^{p,q}(\hat{\mathrm{U}}) \xrightarrow{\alpha^q} \mathrm{H}'^{p,q}(\overline{\mathrm{U}}_+)\oplus\mathrm{H}'^{p,q}(\overline{\mathrm{U}}_-) \xrightarrow{\beta^q} \mathrm{H}_b'^{p,q}(\mathrm{U}_0)$$

defined by

$$\alpha^q([f]) = [i_+f]\oplus[i_-f], \tag{V.3.11}$$

$$\beta^q([f_+]\oplus[f_-]) = [i_0f_+ - i_0f_-]. \tag{V.3.12}$$

Lemma V.3.1. *We have $\beta^q\circ\alpha^q = 0$.*

PROOF. If $f \in \mathscr{C}^\infty(\hat{U};\Lambda^{p,q})$ we have $i_0 i_+ f = i_0 i_- f$. ■

We introduce two more linear maps:

$$\gamma^q : H_b'^{p,q}(U_0) \to H'^{p,q+1}(\hat{U}), \tag{V.3.13}$$

$$\delta^q : H_\bullet'^{p,q}(\overline{U}_+) \oplus H_\bullet'^{p,q}(\overline{U}_-) \to H'^{p,q+1}(\hat{U}). \tag{V.3.14}$$

Let us first describe how (V.3.14) is defined.

Let $f_\pm, g_\pm \in \mathscr{C}^\infty(\overline{U}_\pm;\Lambda^{p,q})$ be such that $f_\pm - g_\pm \in \mathscr{C}_0^\infty(\overline{U}_\pm;\Lambda^{p,q})$, i.e., $f_\pm$ and $g_\pm$ define the same equivalence class in $\mathscr{C}_\bullet^\infty(\overline{U}_\pm;\Lambda^{p,q})$. Assume that this class is a cocycle for $d_\bullet'$ (see (V.2.40), (V.2.41)): $d' f_\pm$ (and therefore $d' g_\pm$) belongs to $\mathscr{C}_0^\infty(\overline{U}_\pm;\Lambda^{p,q+1})$, i.e., vanishes to infinite order on U_0. We can define an element $F \in \mathscr{C}^\infty(\hat{U};\Lambda^{p,1})$ by setting

$$F = d' f_+ \ in\ U_+,\ F = -d' f_- \ in\ U_-. \tag{V.3.15}$$

Of course $d'F = 0$. Denote by G the analogue of F with g substituted for f.

First, suppose $q = 0$. Then set

$$u = f_+ - g_+ \ in\ U_+,\ u = g_- - f_- \ in\ U_-.$$

It is clear that $u \in \mathscr{C}^\infty(\hat{U};\Lambda^{p,0})$. We have

$$d'u = F - G, \tag{V.3.16}$$

i.e., F and G belong to the same cohomology class in $H'^{p,1}(\hat{U})$. By definition this class is set to be $\delta^0([f_+]\oplus[f_-])$.

Next suppose $q \geq 1$. We want to say that the cohomology classes defined by $f_\pm$ and $g_\pm$ in $H_\bullet'^{p,q}(\overline{U}_\pm)$ are the same: there exists $v_\pm \in \mathscr{C}^\infty(\overline{U}_\pm;\Lambda^{p,q-1})$ such that

$$f_\pm - g_\pm - d' v_\pm \in \mathscr{C}_0^\infty(\overline{U}_\pm;\Lambda^{p,q}).$$

If here we set

$$u = f_+ - g_+ - d' v_+ \ in\ \overline{U}_+,\ u = g_- - f_- + d' v_- \ in\ \overline{U}_-,$$

(V.3.16) is valid. Thus F and G belong to the same cohomology class in $H'_{p,q+1}(\hat{U})$. By definition, this class is taken to be $\delta^q([f_+]\oplus[f_-])$.

Finally, we define the map (V.3.13). We apply Proposition V.2.3 with $\overline{U}_\pm$ in the place of $\overline{\mathcal{M}}$. We get two bijections

$$\lambda_\pm^q : H_b'^{p,q}(U_0) \to H_\bullet'^{p,q}(\overline{U}_\pm),$$

and we set

$$\gamma^q([f_0]) = \delta^q([\lambda_+^q f_0]\oplus[\lambda_-^q f_0]). \tag{V.3.17}$$

It is not difficult to describe directly the map γ^q: Let $f_0 \in \mathscr{C}^\infty(U_0;\Lambda^{p,q})$ be such that $d_b' f_0 = 0$. Apply Theorem V.2.1 with $q+1$ in the place of q and with

$f \equiv 0$, $g = f_0$. Then there is $u_\pm \in \mathscr{C}^\infty(\overline{\mathrm{U}}_\pm;\Lambda^{p,q})$ such that $\mathrm{d}'u_\pm \in \mathscr{C}_0^\infty(\overline{\mathrm{U}}_\pm;\Lambda^{p,q+1})$ and $i_0 u_\pm = f_0$. We can define an element $F \in \mathscr{C}^\infty(\hat{\mathrm{U}};\Lambda^{p,1})$ by setting

$$F = \mathrm{d}'u_+ \ \mathit{in}\ \overline{\mathrm{U}}_+,\ F = -\mathrm{d}'u_- \ \mathit{in}\ \overline{\mathrm{U}}_-. \tag{V.3.18}$$

Of course $\mathrm{d}'F = 0$. Then $\gamma^q([f_0]) = [F]$.

LEMMA V.3.2. *We have* $\alpha^{q+1}\circ\gamma^q = 0$.

PROOF. Let $u_\pm$ and F be as in (V.3.18). We have $i_+F = \mathrm{d}'u_+$ and therefore $[i_+F] = 0$; likewise, $[i_-F] = 0$, whence $\alpha^{q+1}[F] = 0$. ∎

LEMMA V.3.3. *We have* $\gamma^q\circ\beta^q = 0$.

PROOF. Take $f_\pm \in \mathscr{C}^\infty(\overline{\mathrm{U}}_\pm;\Lambda^{p,q})$ such that $\mathrm{d}'f_\pm \equiv 0$, $f_0 = i_0 f_+ - i_0 f_-$. Let $\tilde{f}_\pm \in \mathscr{C}^\infty(\hat{\mathrm{U}};\Lambda^{p,q})$ extend $f_\pm$. Set $u_+ = \tilde{f}_+ - \tilde{f}_-$ in $\overline{\mathrm{U}}_+$, $u_- = \tilde{f}_+ - \tilde{f}_-$ in $\overline{\mathrm{U}}_-$. We have $\mathrm{d}'u_\pm \in \mathscr{C}_0^\infty(\overline{\mathrm{U}}_\pm;\Lambda^{p,q+1})$ and $i_0 u_\pm = f_0$. Then define F as in (V.3.18). Since $\mathrm{d}'\tilde{f}_+ = 0$ in U_+ and $\mathrm{d}'\tilde{f}_- = 0$ in U_- we have, in $\hat{\mathrm{U}}$, $F = -\mathrm{d}'(\tilde{f}_+ + \tilde{f}_-)$. This proves that $[F] = 0$, i.e., $\gamma^q([f_0]) = \gamma^q(\beta^q([f_+]\oplus[f_-])) = 0$. ∎

The following sequence can be regarded as the *Mayer-Vietoris sequence* of the pair $(\overline{\mathrm{U}}_+, \overline{\mathrm{U}}_-)$:

$$0 \to \mathrm{H}'^{p,0}(\hat{\mathrm{U}}) \to \cdots \to \mathrm{H}'^{p,q}(\hat{\mathrm{U}}) \xrightarrow{\alpha^q} \mathrm{H}'^{p,q}(\overline{\mathrm{U}}_+)\oplus\mathrm{H}'^{p,q}(\overline{\mathrm{U}}_-) \xrightarrow{\beta^q}$$

$$\mathrm{H}_b'^{p,q}(\mathrm{U}_0) \xrightarrow{\gamma^q} \mathrm{H}'^{p,q+1}(\hat{\mathrm{U}}) \to \cdots \tag{V.3.19}$$

THEOREM V.3.1. *Suppose the boundary $\partial\mathcal{M}$ is noncharacteristic. Then the Mayer-Vietoris sequence* (V.3.19) *is exact.*

PROOF. It is clear that the map α^q is injective when $q = 0$. Thanks to Lemmas V.3.1, V.3.2, and V.3.3 it suffices to prove

$$\mathrm{Ker}\ \alpha^{q+1} \subset \mathrm{Im}\ \gamma^q, \tag{V.3.20}$$

$$\mathrm{Ker}\ \beta^q \subset \mathrm{Im}\ \alpha^q, \tag{V.3.21}$$

$$\mathrm{Ker}\ \gamma^q \subset \mathrm{Im}\ \beta^q. \tag{V.3.22}$$

Below we use the notation $\mathscr{C}_0^\infty(\hat{\mathrm{U}};\Lambda^{p,q})$ to mean the subspace of $\mathscr{C}^\infty(\hat{\mathrm{U}};\Lambda^{p,q})$ made up of the sections that vanish to infinite order on U^0, i.e., at $t_n = 0$.

PROOF OF (V.3.20). Let $f \in \mathscr{C}^\infty(\hat{\mathrm{U}};\Lambda^{p,q+1})$ satisfy $\mathrm{d}'f \equiv 0$ and $\alpha^{q+1}([f]) = 0$. This means that there are forms $g_\pm \in \mathscr{C}^\infty(\overline{\mathrm{U}}_\pm;\Lambda^{p,q})$ such that $i_\pm f = \mathrm{d}'g_\pm$. Let $\tilde{g}_\pm \in \mathscr{C}^\infty(\hat{\mathrm{U}};\Lambda^{p,q})$ be arbitrary extensions of $g_\pm$ and define

$$h_1 = \tfrac{1}{2}(\tilde{g}_+ - \tilde{g}_-),\ h_2 = \tfrac{1}{2}(\tilde{g}_+ + \tilde{g}_-).$$

Since $d'h_1 = \frac{1}{2}(f - d'\tilde{g}_-) - \frac{1}{2}(f - d'\tilde{g}_+)$ and since f, $d'\tilde{g}_+$ and $d'\tilde{g}_-$ agree to infinite order when $t_n = 0$, we conclude that $d'h_1 \in \mathscr{C}_0^\infty(\hat{U};\Lambda^{p,q+1})$. Moreover,

$$f - d'h_2 = d'i_+h_1 \textit{ in } \overline{U}_+, f - d'h_2 = -\ d'i_-h_1 \textit{ in } \overline{U}_-.$$

This means that, in (V.3.18), we may take $f_0 = i_0h_1$, $F = f - d'h_2$, and $u_\pm = i_\pm h_1$. It proves that $[f] = \gamma^q([f_0])$.

Proof of (V.3.21). Let $f_\pm \in \mathscr{C}^\infty(\overline{U}_\pm;\Lambda^{p,q})$ satisfy $d'f_\pm = 0$ and $[i_0f_+] = [i_0f_-]$. Suppose first $q = 0$ and let $f_{\pm J}$ denote the coefficient of dZ_J, $|J| = p$, in $f_\pm$. We have

$$L_nf_{\pm J} \equiv 0 \tag{V.3.23}$$

in $\overline{U}_\pm$, and the restriction to U_0 of f_{+J} and f_{-J} are equal. This implies that all the tangential derivatives, i.e., the derivatives with respect to x_i $(1 \leq i \leq m)$ and t_j $(1 \leq j \leq n-1)$, of f_{+J} and f_{-J} are equal on U_0. But from (V.3.23), which entails

$$(\partial/\partial x)^\alpha(\partial/\partial t)^\beta L_nf_{\pm J} \equiv 0,\ \forall\ \alpha \in \mathbb{Z}_+^m,\ \beta \in \mathbb{Z}_+^n,$$

and from the expression (V.2.20) of L_n, we derive that all derivatives of f_{+J} and f_{-J} are equal on U_0. It follows that there is a form $f \in \mathscr{C}^\infty(\hat{U};\Lambda^{p,q})$ such that $i_\pm f = f_\pm$ and thus $[f_+]\oplus[f_-] = \alpha^0([f])$.

Suppose now $q \geq 1$. Let $\tilde{f}_+ \in \mathscr{C}^\infty(\hat{U};\Lambda^{p,q})$ extend f_+; then $d'\tilde{f}_+ \equiv 0$ in $\overline{U}_+$, hence $d'i_-\tilde{f}_+ \in \mathscr{C}_0^\infty(\overline{U}_-;\Lambda^{p,q+1})$; also, $i_0\tilde{f}_+ - i_0f_- = d_b'v_0$ with $v_0 \in \mathscr{C}^\infty(U_0;\Lambda^{p,q-1})$. Select arbitrarily a section $v_- \in \mathscr{C}^\infty(\overline{U}_-;\ \Lambda^{p,q-1})$ whose pullback to U_0, i_0v_-, is equal to v_0. Let us define

$$g_- = i_-\tilde{f}_+ - f_- - d'v_- \in \mathscr{C}^\infty(\overline{U}_-;\Lambda^{p,q}).$$

We have $d'g_- \in \mathscr{C}_0^\infty(\overline{U}_-;\Lambda^{p,q+1})$ and $i_0g_- = 0$. Apply Theorem V.2.1 with $\mathcal{M} = \overline{U}_-$, g_- in the place of f and zero in that of g: there exists $u_- \in \mathscr{C}_0^\infty(\overline{U}_-;\Lambda^{p,q-1})$ such that

$$g_- - d'u_- \in \mathscr{C}_0^\infty(\overline{U}_-;\Lambda^{p,q}).$$

Then define

$$f = f_+ \textit{ in } U_+, f = f_- + d'(u_- + v_-) \textit{ in } U_-.$$

In U_-, $i_-\tilde{f}_+ - f \in \mathscr{C}_0^\infty(\overline{U}_-;\Lambda^{p,q})$ and therefore f belongs to $\mathscr{C}^\infty(\hat{U};\Lambda^{p,q})$. Moreover, $d'f \equiv 0$ and $\alpha^q([f]) = [f_+]\oplus[f_-]$.

PROOF OF (V.3.22). Let $f_0 \in \mathscr{C}^\infty(U_0;\Lambda^{p,q})$ satisfy $d_b' f_0 = 0$ and $\gamma^q([f_0]) = 0$. Let $u_\pm \in \mathscr{C}^\infty(\overline{U}_\pm;\Lambda^{p,q})$ satisfy $d'u_\pm \in \mathscr{C}_0^\infty(\overline{U}_\pm;\Lambda^{p,q+1})$ and $i_0 u_\pm = f_0$. Let F be defined by (V.3.18). Our hypothesis is that there exists $G \in \mathscr{C}^\infty(\hat{U};\Lambda^{p,q})$ such that $F = dG$. Define

$$f_+ = -\tfrac{1}{2}(G-u_+) \text{ in } U_+, f_- = -\tfrac{1}{2}(G+u_-) \text{ in } U_-.$$

Since $d'f_\pm = 0$ the class $[f_\pm]$ is well defined; and $i_0 f_+ - i_0 f_- = f_0$, which implies $[f_0] = \beta^q([f_+]\oplus[f_-])$.

The proof of Theorem V.3.1 is complete. ■

COROLLARY V.3.2. *Suppose that the boundary $\partial\mathcal{M}$ is noncharacteristic and the extension to $\hat{U}$ of the functions Z_i is such that $H'^{p,q}(\hat{U}) = 0$ for all $q \geq 1$. Then, whatever $q \geq 0$, every section $f_0 \in \mathscr{C}^\infty(U_0;\Lambda^{p,q})$ such that $d_b' f_0 = 0$ can be written as a difference*

$$f_0 = i_0 f_+ - i_0 f_- \tag{V.3.24}$$

with $f_\pm \in \mathscr{C}^\infty(\overline{U}_\pm;\Lambda^{p,q})$, $d'f_\pm = 0$.

PROOF. By virtue of Theorem V.3.1 and of the hypothesis that $H^{p,q+1}(\hat{U}) = 0$, the map β^q is surjective. If we go back to the definition (V.3.12) of β^q, it means that there exist sections $f_\pm \in \mathscr{C}^\infty(\overline{U}_\pm;\Lambda^{p,q})$ such that, when $q \geq 1$, $i_0 f_+ - i_0 f_- - f_0 = d_b' g_0$ for some $g_0 \in \mathscr{C}^\infty(U_0;\Lambda^{p,q-1})$, or simply such that (V.3.24) holds when $q = 0$. When $q \geq 1$, let $\tilde{g} \in \mathscr{C}^\infty(\hat{U};\Lambda^{p,q})$ be an arbitrary extension of g_0. Substituting $f_\pm \mp \frac{1}{2} d' i_\pm \tilde{g}$ for $f_\pm$ yields (V.3.24). ■

V.4. Approximation of Classical Solutions in Locally Integrable Structures with Boundary

Throughout the remainder of the present chapter, $\overline{\mathcal{M}}$ will stand for a $\mathscr{C}^\infty$ manifold with boundary, equipped with a *locally integrable* structure. As usual, the tangent and cotangent bundles will be denoted by $\mathcal{V}$ and T′, their fibre dimensions by m and n.

In the present section the setup will exactly be the one described in the first part of section V.3: the analysis will take place in an open neighborhood U of an arbitrary point 0 of the boundary, to which we refer as the origin. Some open neighborhood $\mathcal{O}$ of $\mathcal{C}\ell U$ will be the domain of local coordinates x_i, t_j $(1 \leq i \leq m,\ 1 \leq j \leq n)$; U itself will be given by (V.3.7) when the boundary $\partial\mathcal{M}$ is noncharacteristic and by (V.3.8) when $\partial\mathcal{M}$ is totally characteristic. We assume that there are smooth functions Z_j in $\mathcal{O}$ whose differentials span T′ and which satisfy (I.7.23) and (I.7.24) when $\partial\mathcal{M}$ is noncharacteristic, (V.3.4), (V.3.5), and (I.7.24) when $\partial\mathcal{M}$ is totally characteristic.

All this allows us to introduce the function $E_\tau(z;x,t)$ in (II.2.3). Below we write

$$\Sigma = \{ x \in \mathbb{R}^m ;\ (x,0) \in U \}.$$

When $\partial\mathcal{M}$ is noncharacteristic, Σ is the neighborhood V in (V.3.7); when $\partial\mathcal{M}$ is totally characteristic, it is the "positive half" of V: $\Sigma = \{ x \in V;\ x_m \geq 0 \}$. We can then extend Theorem II.2.1 as follows:

THEOREM V.4.1. *There is an open neighborhood of the origin in* $\overline{\mathcal{M}}$, $U' \subset U$, *such that, given any solution* $h \in \mathcal{C}^1(\mathcal{O}/U)$, *we have, uniformly in* U',

$$h(x,t) = \lim_{\tau \to +\infty} \int_\Sigma E_\tau(Z(x,t);y,0)\, h(y,0)\, [\det Z_x(y,0)] dy. \qquad \text{(V.4.1)}$$

PROOF. Write $dZ = dZ_1 \wedge \cdots \wedge dZ_m$. We shall call Γ the portion of the boundary of Σ that lies on the sphere ∂V. When the boundary is noncharacteristic, $\Gamma = \partial V = \partial\Sigma$, whereas when the boundary is totally characteristic, Γ is the hemisphere $\{ x \in \partial V;\ x_m \geq 0 \}$. Let t_0 be a point of W, with $t_{0n} \geq 0$ if the boundary is noncharacteristic. Let $\ell(t_0)$ be the straight-line segment joining 0 to t_0 in $\mathbb{R}^n$, oriented from 0 to t_0. Whatever the solution $h \in \mathcal{C}^1(\mathcal{O}/U)$ we have, with a suitable orientation of Γ,

$$\int_{\partial[\Sigma \times \ell(t_0)]} hdZ = \int_{\Sigma \times \{t_0\}} hdZ - \int_{\Sigma \times \{0\}} hdZ - \int_{\Gamma \times \ell(t_0)} hdZ. \qquad \text{(V.4.2)}$$

This is obvious when $\partial\mathcal{M}$ is noncharacteristic. When $\partial\mathcal{M}$ is totally characteristic the pullback of dZ to the part of $\partial\Sigma$ that lies in $\partial\mathcal{M}$ vanishes identically (Proposition V.3.1). We apply the Stokes theorem. Since $d(hdZ) \equiv 0$ we derive from (V.4.2):

$$\int_{\Sigma \times \{t_0\}} hdZ = \int_{\Sigma \times \{0\}} hdZ + \int_{\Gamma \times \ell(t_0)} hdZ. \qquad \text{(V.4.3)}$$

We replace $h(x,t)$ by $E_\tau(Z(x_0,t_0);x,t)h(x,t)$ ($x_0 \in \Sigma$). Then Lemma II.2.1 implies that, as $\tau \to +\infty$, the left-hand side converges to $h(x_0,t_0)$, uniformly with respect to (x_0,t_0) in U. The argument used in the proof of Lemma II.2.4 shows that the integral over $\Gamma \times \ell(t_0)$ converges to zero, provided (x_0,t_0) stays in a suitably chosen neighborhood of 0 contained in U. ■

As consequences of Theorem V.4.1 we can state, for classical solutions, the analogues of Theorem II.3.1, Corollaries II.3.2 and II.3.5:

COROLLARY V.4.1. *Every solution* $h \in \mathcal{C}^1(\mathcal{O}/U)$ *is the uniform limit in* U' *of a sequence of polynomials, with complex coefficients, with respect to* $Z_1,\ldots,Z_m$.

COROLLARY V.4.2. *There is a compact neighborhood* $K \subset U$ *of* 0 *in* $\overline{\mathcal{M}}$ *with the property that, to every solution* $h \in \mathcal{C}^1(U)$, *there is a continuous function* $\tilde{h}$ *in* Z(K) *such that* $h = \tilde{h} \circ Z$ *in* K.

(We have written $Z = (Z_1, \ldots, Z_m)$: $\mathcal{O} \to \mathbb{C}^m$.)

COROLLARY V.4.3. *If the restriction to* $\{ (x,t) \in U; t = 0 \}$ *of any solution* $h \in \mathcal{C}^1(\mathcal{O}\!\!/U)$ *vanishes identically, then* $h \equiv 0$ *in* U′.

Thus, also in a manifold with boundary equipped with a locally integrable structure, we have *local approximation by polynomials in the first integrals, local constancy on the fibres* and *uniqueness in the Cauchy problem* with data on maximally real submanifolds (possibly contained in the boundary).

V.5. Distribution Solutions in a Manifold with Totally Characteristic Boundary

It is necessary to clarify what is meant by a "distribution solution." There is no ambiguity about what it means in the interior: let Ω be an open subset of $\overline{\mathcal{M}}$; a distribution h in the interior $\Omega \cap \mathcal{M}$ is a solution if, given any $\mathcal{C}^\infty$ section L of $\mathcal{V}$ over $\Omega \cap \mathcal{M}$, we have $Lh = 0$ in $\Omega \cap \mathcal{M}$. The difficulties arise at the boundary $\Omega \cap \partial\mathcal{M}$.

The cases of a noncharacteristic boundary and of a totally characteristic one must be treated differently. The present section is solely devoted to the latter case.

We begin by recalling some definitions and terminology pertaining to currents on a manifold with boundary. We limit our attention to orientable manifolds (with boundary) to avoid bringing in twisted forms (and sections of the density bundle). It is not difficult also to handle the case of a nonorientable manifold; we leave this as an exercize to the reader. On all this we refer to Schwartz [1], chap. 9.

Let p be an integer, $0 \le p \le N = \dim \overline{\mathcal{M}}$. In accordance with the notation adopted throughout this book, $\mathcal{C}^\infty(\overline{\mathcal{M}};\Lambda^p)$ will denote the space of $\mathcal{C}^\infty$ p-forms in $\overline{\mathcal{M}}$. The topology of $\mathcal{C}^\infty(\overline{\mathcal{M}};\Lambda^p)$ is that of convergence of the sections of $\Lambda^p = \Lambda^p\mathbb{C}T^*\overline{\mathcal{M}}$, and of each one of their derivatives, on every compact subset K of $\overline{\mathcal{M}}$. We denote by $\mathcal{C}_c^\infty(K;\Lambda^p)$ the subspace of $\mathcal{C}^\infty(\overline{\mathcal{M}};\Lambda^p)$ consisting of the sections whose support is contained in K; $\mathcal{C}_c^\infty(K;\Lambda^p)$ is equipped with the topology inherited from $\mathcal{C}^\infty(\overline{\mathcal{M}};\Lambda^p)$. We denote by $\mathcal{C}_c^\infty(\overline{\mathcal{M}};\Lambda^p)$ the space of the $\mathcal{C}^\infty$ p-forms that have compact support, that is to say, the union of the spaces $\mathcal{C}_c^\infty(K;\Lambda^p)$ as K ranges over the family of all compact subsets of $\overline{\mathcal{M}}$. The elements of $\mathcal{C}_c^\infty(\overline{\mathcal{M}};\Lambda^p)$ are often referred to as *test-forms*. Keep in mind that their support might well intersect $\partial\mathcal{M}$.

By definition, a *current of degree p* (what we also call a p-current) in $\overline{\mathcal{M}}$ is a linear functional on $\mathscr{C}_c^\infty(\overline{\mathscr{M}};\Lambda^{N-p})$ whose restriction to each subspace $\mathscr{C}_c^\infty(K;\Lambda^p)$, whatever the compact subset K of $\overline{\mathscr{M}}$, is continuous. A *distribution* in $\overline{\mathscr{M}}$ is a current of degree zero in $\overline{\mathscr{M}}$. The space of p-currents in $\overline{\mathscr{M}}$ will be denoted by $\mathscr{D}'(\overline{\mathscr{M}};\Lambda^p)$, the space of distributions by $\mathscr{D}'(\overline{\mathscr{M}})$. The analogous notation with $\overline{\mathscr{M}}$ replaced by one of its open subsets, Ω, will also be used.

Many of the concepts and operations associated with distributions (in a manifold without boundary) extend routinely to currents on a manifold with boundary, for instance the notion of *support*. Note also that, in a local chart $(\mathscr{O}, s_1,\ldots,s_N)$ in $\overline{\mathscr{M}}$, an arbitrary p-current u can be written as a sum

$$u = \sum_{|J|=p} u_J \mathrm{d}s_J,$$

with coefficients u_J that are distributions in the open set $\mathscr{O}$. This shows, among other things, that our notation is coherent: if Ω is an open subset of $\overline{\mathscr{M}}$, $\mathscr{D}'(\Omega;\Lambda^p)$ is indeed the space of distribution sections of Λ^p over Ω.

The subspace of $\mathscr{D}'(\Omega;\Lambda^p)$ consisting of the p-currents whose support is compact will be denoted by $\mathscr{E}'(\overline{\mathscr{M}};\Lambda^p)$; that of the compactly supported distributions, by $\mathscr{E}'(\overline{\mathscr{M}})$. The space $\mathscr{E}'(\overline{\mathscr{M}};\Lambda^p)$ can be identified to the dual of $\mathscr{C}^\infty(\overline{\mathscr{M}};\Lambda^{N-p})$.

Convergence of p-currents in $\overline{\mathscr{M}}$ is defined in standard fashion: a sequence of such currents will converge if its evaluation at an arbitrary test-form converges. In $\mathscr{E}'(\overline{\mathscr{M}};\Lambda^p)$ there is an additional requirement: the supports of the currents in the convergent sequence must remain in a fixed compact set.

There is a natural injection $\mathscr{C}^\infty(\overline{\mathscr{M}};\Lambda^p) \to \mathscr{D}'(\overline{\mathscr{M}};\Lambda^p)$, which induces a natural injection $\mathscr{C}_c^\infty(\overline{\mathscr{M}};\Lambda^p) \to \mathscr{E}'(\overline{\mathscr{M}};\Lambda^p)$: a p-form $f \in \mathscr{C}^\infty(\overline{\mathscr{M}};\Lambda^p)$ defines a p-current T_f by the formula

$$\langle T_f,\varphi\rangle = \int_{\mathscr{M}} f \wedge \varphi, \quad \varphi \in \mathscr{C}_c^\infty(\overline{\mathscr{M}};\Lambda^{N-p}). \tag{V.5.1}$$

We recall the notion of *coboundary operator* acting on p-currents in the manifold with boundary $\overline{\mathscr{M}}$. The *exterior derivative* acting on smooth p-forms in $\overline{\mathscr{M}}$ is the linear operator

$$\mathrm{d} : \mathscr{C}^\infty(\overline{\mathscr{M}};\Lambda^p) \to \mathscr{C}^\infty(\overline{\mathscr{M}};\Lambda^{p+1}) \tag{V.5.2}$$

such that, given any $\mathscr{C}^\infty$ p-form φ in $\overline{\mathscr{M}}$, the restriction (i.e., the pullback) of $\mathrm{d}\varphi$ to the interior $\mathscr{M}$ is equal to the (standard) exterior derivative of the pullback of φ. The linear operator (V.5.2) is a differential operator and therefore induces a continuous linear map $\mathscr{C}_c^\infty(\overline{\mathscr{M}};\Lambda^p) \to \mathscr{C}_c^\infty(\overline{\mathscr{M}};\Lambda^{p+1})$.

Consider $f \in \mathscr{C}^\infty(\overline{\mathscr{M}};\Lambda^p)$, $\varphi \in \mathscr{C}_c^\infty(\overline{\mathscr{M}};\Lambda^{N-p-1})$. By the Stokes theorem we have

$$\int_{\overline{\mathscr{M}}} \mathrm{d}(f\wedge\varphi) = \int_{\partial\mathscr{M}} f\wedge\varphi = \int_{\overline{\mathscr{M}}} \mathrm{d}f\wedge\varphi + (-1)^p \int_{\overline{\mathscr{M}}} f\wedge\mathrm{d}\varphi,$$

which, in the notation (V.5.1), can be rewritten as follows:

$$\langle T_{\mathrm{d}f},\varphi\rangle = \langle T_f,(-1)^{p-1}\mathrm{d}\varphi\rangle + \int_{\partial\mathcal{M}} f \wedge \varphi. \tag{V.5.3}$$

Consider then the continuous linear map

$$(-1)^{p-1}\mathrm{d} : \mathscr{C}_c^\infty(\overline{\mathcal{M}};\Lambda^{N-p-1}) \to \mathscr{C}_c^\infty(\overline{\mathcal{M}};\Lambda^{N-p}). \tag{V.5.4}$$

By definition, the transpose of (V.5.4) is the coboundary operator,

$$\mathrm{b} : \mathscr{D}'(\overline{\mathcal{M}};\Lambda^p) \to \mathscr{D}'(\overline{\mathcal{M}};\Lambda^{p+1}). \tag{V.5.5}$$

The analogous notation will be used with Ω, an open subset of $\overline{\mathcal{M}}$, substituted for $\overline{\mathcal{M}}$. Formula (V.5.3) implies that, when Ω does not intersect $\partial\mathcal{M}$ and when f is a smooth p-form in Ω, then $\mathrm{b}T_f = T_{\mathrm{d}f}$. But (V.5.3) also shows that when Ω does intersect $\partial\mathcal{M}$, even if f is a smooth p-form in Ω, we shall in general have $\mathrm{b}T_f \neq T_{\mathrm{d}f}$. (In the sequel we shall write f instead of T_f.)

What precedes can be made more precise. Let us introduce the following one-current in $\overline{\mathcal{M}}$:

$$\langle \delta_{\partial\mathcal{M}},\varphi\rangle = -\int_{\partial\mathcal{M}} \varphi, \quad \varphi \in \mathscr{C}_c^\infty(\overline{\mathcal{M}};\Lambda^{N-1}). \tag{V.5.6}$$

Observe that $\delta_{\partial\mathcal{M}}$ is continuous on $\mathscr{C}_c^\infty(\overline{\mathcal{M}};\Lambda^{N-1})$ for the topology of uniform convergence on the compact subsets of $\overline{\mathcal{M}}$. One could say that $\delta_{\partial\mathcal{M}}$ is a *Radon measure section* of Λ^1. It is also clear that the support of $\delta_{\partial\mathcal{M}}$ is contained in $\partial\mathcal{M}$. If g is a continuous function in $\Omega\cap\partial\mathcal{M}$, it defines a one-current in Ω, $g\delta_{\partial\mathcal{M}}$, by the formula

$$\langle g\delta_{\partial\mathcal{M}},\varphi\rangle = -\int_{\Omega\cap\partial\mathcal{M}} g\varphi, \quad \varphi \in \mathscr{C}_c^\infty(\Omega;\Lambda^{N-1}). \tag{V.5.7}$$

Then let $f \in \mathscr{C}^\infty(\Omega)$. If we apply (V.5.3) with $p = 0$, Ω in the place of $\overline{\mathcal{M}}$ and take the definition of the operator (V.5.5) into account, we get

$$\mathrm{b}f = \mathrm{d}f + (f|_{\Omega\cap\partial\mathcal{M}})\delta_{\partial\mathcal{M}}. \tag{V.5.8}$$

Clearly, (V.5.8) extends to functions $f \in \mathscr{C}^1(\Omega)$.

In (V.5.8) put $f = \mathit{1}$, the function identically equal to one in $\overline{\mathcal{M}}$. We get

$$\mathrm{b}\mathit{1} = \delta_{\partial\mathcal{M}}. \tag{V.5.9}$$

The justification for the choice of the minus sign in front of the integrals, in (V.5.6) and (V.5.7), lies in the following:

EXAMPLE V.5.1. Take $\overline{\mathcal{M}} = \overline{\mathbb{R}}_+$, the half-line $t \geq 0$. The restriction map

$$\mathscr{C}_c^\infty(\mathbb{R};\Lambda^1) \to \mathscr{C}_c^\infty(\overline{\mathbb{R}}_+;\Lambda^1)$$

is a surjection. As a consequence, its transpose is an injection $\mathcal{D}'(\overline{\mathbb{R}}_+) \to \mathcal{D}'(\mathbb{R})$: it extends an arbitrary distribution $u \in \mathcal{D}'(\overline{\mathbb{R}}_+)$ by setting $u = 0$ for $t < 0$.

Now consider $f \in \mathcal{C}^1(\overline{\mathbb{R}}_+)$, $\varphi \in \mathcal{C}_c^\infty(\overline{\mathbb{R}}_+)$. By definition, we have

$$\int_0^{+\infty} \varphi(\mathrm{b}f) = -\int_0^{+\infty} f\mathrm{d}\varphi. \tag{V.5.10}$$

Integration by parts in the right-hand side yields

$$\mathrm{b}f = \mathrm{d}f + f(0)\delta, \tag{V.5.11}$$

where δ is the Dirac measure at $t = 0$. ■

DEFINITION V.5.1. *Suppose that the boundary $\partial\mathcal{M}$ is totally characteristic and let Ω be an open subset of $\overline{\mathcal{M}}$. A distribution h in Ω will be called a* distribution solution *if its coboundary,* bh, *is a (distribution) section of* T′ *over* Ω.

The following example can be thought of as the model case for a locally integrable structure with totally characteristic boundary:

EXAMPLE V.5.2. Take $\overline{\mathcal{M}} = \{(x,t) \in \mathbb{R}^2; x \geq 0\}$ with the structure defined by the vector field $\mathrm{L} = \partial/\partial t - ix\partial/\partial x$ (see Example II.5.1). The cotangent structure bundle T′ is spanned by dZ with $Z = xe^{it}$. We have the sequence of distribution solutions h_k ($k \in \mathbb{Z}$) given by (II.5.12) and (II.5.13).

PROPOSITION V.5.1. *We have:*

$$\mathrm{b}h_k = h_{k-1}\mathrm{d}Z, \ \forall\, k \in \mathbb{Z}. \tag{V.5.12}$$

PROOF. Let $\varphi = \alpha\mathrm{d}x + \beta\mathrm{d}t \in \mathcal{C}_c^\infty(\overline{\mathcal{M}};\Lambda^1)$ be arbitrary. If $k \geq 1$ we have

$$\iint_{x\geq 0} (\mathrm{b}h_k)\wedge\varphi = -\iint_{x\geq 0} h_k\mathrm{d}\varphi = \tfrac{1}{k!}\iint_{x\geq 0} (e^{it}x)^k(\alpha_t - \beta_x)\mathrm{d}x\mathrm{d}t =$$

$$-\iint_{t\geq 0} (e^{it}x)^{k-1}(ix\alpha - \beta)e^{it}\mathrm{d}x\wedge\mathrm{d}t/(k-1)! = \iint_{t\geq 0} h_{k-1}\mathrm{d}(xe^{it})\wedge(\alpha\mathrm{d}x + \beta\mathrm{d}t),$$

which is what we wanted in this case. Likewise,

$$\iint_{x\geq 0} (\mathrm{b}h_0)\wedge\varphi = -\iint_{x\geq 0} \mathrm{d}\varphi = \iint_{x\geq 0} (\alpha_t - \beta_x)\mathrm{d}x\mathrm{d}t = \int_{\mathbb{R}} \beta(0,t)\mathrm{d}t$$

$$= \langle\delta(x)\mathrm{d}x,\varphi\rangle = \langle e^{-it}\delta(x)\mathrm{d}Z,\varphi\rangle = \langle h_{-1}\mathrm{d}Z,\varphi\rangle.$$

Suppose now $k \geq 1$ and consider

$$\langle \mathrm{b}h_{-k},\varphi\rangle = -\langle h_{-k},\mathrm{d}\varphi\rangle = -\langle e^{-\imath kt}\delta^{(k-1)}(x),(\beta_x-\alpha_t)\mathrm{d}x\wedge\mathrm{d}t\rangle =$$
$$\langle e^{-\imath kt}\delta^{(k)}(x),\ \beta\mathrm{d}x\wedge\mathrm{d}t\rangle + \imath k\langle e^{-\imath kt}\delta^{(k-1)}(x),\ \alpha\mathrm{d}x\wedge\mathrm{d}t\rangle =$$
$$\langle e^{-\imath(k+1)t}\delta^{(k)}(x),e^{\imath t}(\beta-\imath x\alpha)\mathrm{d}x\wedge\mathrm{d}t\rangle = \langle h_{-k-1},\mathrm{d}Z\wedge\varphi\rangle,$$

whence (V.5.12) in all cases. ∎

We return to the general case (of a manifold with totally characteristic boundary).

PROPOSITION V.5.2. *When the boundary $\partial\mathcal{M}$ is totally characteristic the function **1** is a distribution solution in $\overline{\mathcal{M}}$.*

PROOF. It suffices to show that an arbitrary point $\mathfrak{p}$ of the boundary $\partial\mathcal{M}$ has an open neighborhood U in which ***1*** is a distribution solution. Assume U is given by (V.3.8) and that there are first integrals Z_j in U as given by (V.3.4) and (V.3.5). It is convenient, as in section V.3, to introduce the "biball" $\hat{\mathrm{U}} = \overline{\mathrm{U}}_+ \cup \overline{\mathrm{U}}_- = \mathrm{V}\times\mathrm{W}$ in (x,t)-space, and to extend to it, as $\mathscr{C}^\infty$ functions, the Z_j. We may write, for any $\varphi \in \mathscr{C}_c^\infty(\hat{\mathrm{U}};\Lambda^{N-1})$,

$$\langle \mathrm{b}\mathbf{1},i_+\varphi\rangle = -\int_{\hat{\mathrm{U}}}\mathscr{Y}(x_m)\mathrm{d}\varphi, \tag{V.5.13}$$

where i_+ is the pullback map from $\hat{\mathrm{U}}$ to $\overline{\mathrm{U}}_+$ and $\mathscr{Y}$ is the Heaviside function. We have

$$-\int_{\hat{\mathrm{U}}}\mathscr{Y}(x_m)\mathrm{d}\varphi = \int_{\mathbb{R}^N}\mathrm{d}[\mathscr{Y}(x_m)]\wedge\varphi.$$

Since the differentials $\mathrm{d}Z_i$ and $\mathrm{d}t_j$ span $\mathbb{C}T^*\overline{\mathcal{M}}$ over U we may write (cf. (I.7.28))

$$\mathrm{d}[\mathscr{Y}(x_m)] = \sum_{i=1}^{m}\mathrm{M}_i[\mathscr{Y}(x_m)]\mathrm{d}Z_i + \sum_{j=1}^{n}\mathrm{L}_j[\mathscr{Y}(x_m)]\mathrm{d}t_j,$$

where the vector fields M_i and L_j are defined by (I.7.25). But we know that L_j is given by (V.3.6) and therefore

$$\mathrm{L}_j[\mathscr{Y}(x_m)] = 0,\ \forall\, j = 1,\ldots,n,$$

and thus $\mathrm{d}[\mathscr{Y}(x_m)]$ belongs to the span of $\mathrm{d}Z_1,\ldots,\mathrm{d}Z_m$. ∎

REMARK V.5.1. An argument similar to that used in the proof of Proposition V.5.2 shows that Definition V.5.1 is not acceptable when the boundary $\partial\mathcal{M}$ is noncharacteristic: Indeed, let U be given by (V.3.7) and the first integrals Z_j by (I.7.23) and (I.7.24). Here, instead of (V.5.13), we have

$$\langle \mathrm{b}\mathbf{1}, i_+\varphi\rangle = -\int_{\hat{U}} \mathcal{Y}(t_n)\mathrm{d}\varphi. \tag{V.5.14}$$

But now $\mathrm{d}\mathcal{Y}(t_n) = \mathrm{L}_n\mathcal{Y}(t_n)\,\mathrm{d}t_n = \delta(t_n)\,\mathrm{d}t_n$ according to (V.2.20). Thus $\mathrm{b}\mathbf{1}$ is not a section of T′. ■

Proposition V.5.3. *Suppose that the boundary $\partial\mathcal{M}$ is totally characteristic and let Ω be an open subset of $\overline{\mathcal{M}}$. If a function $h \in \mathcal{C}^1(\Omega)$ is a solution in the interior $\Omega\cap\mathcal{M}$, then h is a distribution solution in Ω.*

Proof. If $h \in \mathcal{C}^1(\Omega)$ its differential $\mathrm{d}h$ is a continuous section of $\mathbb{C}\mathrm{T}^*\overline{\mathcal{M}}$ over Ω. If moreover h is a solution in $\Omega\cap\mathcal{M}$ then $\mathrm{d}h$ is a section of T′ over $\Omega\cap\mathcal{M}$. By continuity it must be a section of T′ over the whole of Ω. Since, by (V.5.9) and Proposition V.5.2, $\delta_{\partial\mathcal{M}}$ is a distribution section of T′, the same is true of $h\delta_{\partial\mathcal{M}}$, whence the result, by (V.5.8). ■

Local Representation and Approximation of Distribution Solutions When the Boundary Is Totally Characteristic

We continue to reason within the framework set up in section V.3. Thus the neighborhood $\mathrm{U} = \overline{\mathrm{U}}_+$ is given by (V.3.8). We assume that the functions Z_i have been smoothly extended to the biball $\hat{\mathrm{U}} = \mathrm{V}\times\mathrm{W}$ as indicated in section V.3; that the differentials $\mathrm{d}Z_i$ span a vector subbundle of $\mathbb{C}\mathrm{T}^*\mathbb{R}^N$ over $\hat{\mathrm{U}}$, which we continue to call T′. We also extend to $\hat{\mathrm{U}}$ the vector fields L_j and M_i defined by the orthonormality relations (I.7.25). The vector fields $\mathrm{L}_1,\dots,\mathrm{L}_n$ span the vector subbundle $\mathcal{V} = \mathrm{T}'^{\perp}$ over $\hat{\mathrm{U}}$.

The pullback map

$$i_+ : \mathcal{C}_c^\infty(\hat{\mathrm{U}};\Lambda^{N-p}) \to \mathcal{C}_c^\infty(\overline{\mathrm{U}}_+;\Lambda^{N-p}) \tag{V.5.15}$$

is surjective. Its transpose is the map $u \to \tilde{u}$, with

$$\tilde{u} = u \textit{ in } \overline{\mathrm{U}}_+,\ \tilde{u} = 0 \textit{ in } \hat{\mathrm{U}}\backslash\overline{\mathrm{U}}_+. \tag{V.5.16}$$

The map (V.5.15) enables us to identify $\mathcal{D}'(\mathrm{U}_+;\Lambda^p)$ to the subspace of $\mathcal{D}'(\hat{\mathrm{U}};\Lambda^p)$ consisting of the currents whose support is contained in $\overline{\mathrm{U}}_+$.

Proposition V.5.4. *For any $u \in \mathcal{D}'(\overline{\mathrm{U}}_+;\Lambda^p)$ we have*

$$(\mathrm{b}u)^{\sim} = \mathrm{d}\tilde{u}. \tag{V.5.17}$$

Proof. In $\hat{\mathrm{U}}$, which has no boundary, $\mathrm{b}\tilde{u} = \mathrm{d}\tilde{u}$. If $\varphi \in \mathcal{C}_c^\infty(\hat{\mathrm{U}};\Lambda^{N-p-1})$ we have, by Stokes' theorem,

$$\int d\tilde{u}\wedge\varphi = (-1)^{p-1}\int \tilde{u}\wedge d\varphi = (-1)^{p-1}\int_U u\wedge d(i_+\varphi) =$$

$$\langle bu, i_+\varphi\rangle = \int (bu)^\sim \wedge\varphi. \ \blacksquare$$

Proposition V.5.4 has the following immediate consequence:

PROPOSITION V.5.5. *Assume that the boundary $\partial\mathcal{M}$ is totally characteristic. In order that a distribution u in* U *be a solution, it is necessary and sufficient that its extension $\tilde{u}$* (see (V.5.16)) *be a solution in* $\hat{U}$ *for the extended locally integrable structure.*

PROOF. To say that $\tilde{u}$ is a solution in $\hat{U}$ is to say that $d\tilde{u} = b\tilde{u}$ is a distribution section of T′ over $\hat{U}$, which is the same as saying that bu is a distribution section of T′ over $\overline{U}_+$, whence the result, by Definition V.5.1. ■

Proposition V.5.5 leads to the straightforward generalization to the present case of a number of properties that are valid in manifolds without boundary:

PROPOSITION V.5.6. *Suppose that $\partial\mathcal{M}$ is totally characteristic. Every distribution solution in* U (given by (V.3.8)) *can be regarded as a $\mathcal{C}^\infty$ function of t in* W *valued in the space of distributions in* V *that vanish identically in the open set $x_m < 0$.*

Another immediate consequence of Proposition V.5.5 is the generalization to the present situation of the Approximation Formula (II.3.4). Indeed, (II.3.4) holds in the extended structure in $\hat{U}$, i.e., with $\tilde{h}$ substituted for h. But then the "integration" at the right can be limited to $\overline{V}_+ = \{y \in V; y_m \geq 0\}$, and the variation of x can be limited to $\overline{V}_{0_+} = \{x \in V_0; x_m \geq 0\}$, which leads to the following conclusion:

THEOREM V.5.1. *If the open balls, centered at the origin, $V_0 \subset V$ in x-space $\mathbb{R}^m$ and $W_0 \subset W$ in t-space $\mathbb{R}^n$, are suitably chosen, then, given any distribution solution h in* U, *we have, in $\mathcal{C}^\infty(W_0;\mathcal{D}'(\overline{V}_{0_+}))$,*

$$h(x,t) = \lim_{\nu\to+\infty}\int_{V_+} E_\tau(x,t;y,0)h(y,0)\chi(y,0)[\det Z_x(y,0)]\, dy. \qquad (V.5.18)$$

We recall that $\chi \in \mathcal{C}_c^\infty(V)$ is identically equal to one in a suitable neighborhood of the closure of V_0. We should emphasize that, in the right-hand side of (V.5.18), the integral stands for the duality bracket between currents and smooth forms *in the manifold with boundary* $\overline{V}_+$.

We may now generalize Corollaries V.4.1 and V.4.3 to distribution solutions.

COROLLARY V.5.1. *Every distribution solution in* U *is the distribution limit, in* U′*, of a sequence of complex polynomials in* $Z_1,\ldots,Z_m$.

COROLLARY V.5.2. *If the trace on the submanifold* $\{ (x,t) \in U; t = 0 \}$ *of a distribution solution h in* U *vanishes identically, then* $h \equiv 0$ *in* U′.

Finally another obvious consequence of Proposition V.5.4. is that the local representation formula (II.5.1) in the extended locally integrable structure on Û yields the local representation in U.

V.6. Distribution Solutions in a Manifold with Noncharacteristic Boundary

The concept of a distribution solution, as laid out in Definition V.5.1 when the boundary of $\overline{\mathcal{M}}$ is totally characteristic, is not suited to noncharacteristic boundaries. One surely wants the constant function ***1*** to be a distribution solution, notwithstanding that, when the boundary is noncharacteristic, its coboundary b***1*** is not a distribution section of the cotangent structure bundle T′ (Remark V.5.1). This compels one to restrict the pool of distributions on $\overline{\mathcal{M}}$ out of which the *solutions* are extracted.

Throughout the present section, and without recalling it any more, we reason under the hypothesis that *the boundary* $\partial\mathcal{M}$ *is noncharacteristic*. Here also it is convenient to assume that the manifold $\overline{\mathcal{M}}$ is orientable. Let p, q be a pair of integers such that $0 \leq p \leq m$, $0 \leq q \leq n$. Choosing an orientation in $\overline{\mathcal{M}}$ allows us to identify the vector bundle $\Lambda^{p,q}$ to the dual of $\Lambda^{m-p,n-q}$ (the proof of this fact is the same as in a manifold without boundary; see Proposition VIII.1.1). With this identification in mind we define the space of currents $\mathcal{D}'(\overline{\mathcal{M}};\Lambda^{p,q})$ as the (strong) dual of the space of test-forms $\mathcal{C}_c^\infty(\overline{\mathcal{M}};\Lambda^{m-p,n-q})$.

Below we make frequent use of an open subset U of $\overline{\mathcal{M}}$ such that $U\cap\partial\mathcal{M} \neq \emptyset$. We shall always assume that U is the domain of local coordinates x_i, t_j ($1 \leq i \leq m$, $1 \leq j \leq n$) which define a diffeomorphism as in (V.2.19); $U\cap\partial\mathcal{M}$ is defined by $t_n = 0$. We shall also assume that there are first integrals $Z_i \in \mathcal{C}^\infty(U)$ ($i = 1,\ldots,m$) with the expressions (I.7.23). We extend each function $\Phi_i(x,t)$ in a $\mathcal{C}^\infty$ manner to $\hat{U} = \{ (x,t) \in \mathbb{R}^N; (x,t') \in \mathbb{O}, |t_n| < T\}$; we assume that (x,t), $\to (Z(x,t),t)$ is a diffeomorphism of Û onto a $\mathcal{C}^\infty$ submanifold of $\mathbb{C}^m \times \mathbb{R}^n$. We may then define the vector fields $L_1,\ldots,L_n$ by the orthonormality relations (I.7.25). We get thus a locally integrable structure on Û; below the vector bundles $\Lambda^{p,q}$ over Û are those associated with this structure. The restriction map to U induces a surjection of $\mathcal{C}_c^\infty(\hat{U};\Lambda^{m-p,n-q})$ onto $\mathcal{C}_c^\infty(U;\Lambda^{m-p,n-q})$ whose transpose is the injection of $\mathcal{D}'(U;\Lambda^{p,q})$ into $\mathcal{D}'(\hat{U};\Lambda^{p,q})$: we can identify the elements of $\mathcal{D}'(U;\Lambda^{p,q})$ to the distribution sections of $\Lambda^{p,q}$ whose support is contained in the relatively closed subset U.

It is convenient to make use of the local Sobolev spaces $\dot{H}^s_{loc}(\overline{\mathcal{M}};\Lambda^{p,q})$ ($s \in \mathbb{R}$). These are not the standard Sobolev spaces $H^s_{loc}(\overline{\mathcal{M}};\Lambda^{p,q})$ (which, due to the presence of the boundary, are "standard" only when s is an integer). The spaces $\dot{H}^s_{loc}(\overline{\mathcal{M}};\Lambda^{p,q})$ take that presence into account. It suffices to define the spaces $\dot{H}^s_{loc}(U;\Lambda^{p,q})$ with U a suitably small open subset of $\overline{\mathcal{M}}$: by definition a distribution u shall belong to $\dot{H}^s_{loc}(\overline{\mathcal{M}};\Lambda^{p,q})$ if its restriction to any such open set U belongs to $\dot{H}^s_{loc}(U;\Lambda^{p,q})$. When U is entirely contained in the interior $\mathcal{M}$ we set $\dot{H}^s_{loc}(U;\Lambda^{p,q}) = H^s_{loc}(U;\Lambda^{p,q})$. When $U \cap \partial\mathcal{M} \neq \emptyset$ and U is given as in (V.2.19) $\dot{H}^s_{loc}(U;\Lambda^{p,q})$ can be identified (through his natural injection into $\mathcal{D}'(\hat{U};\Lambda^{p,q})$) to the (closed) linear subspace of $H^s_{loc}(\hat{U};\Lambda^{p,q})$ consisting of the distribution sections with support contained in U. Whether U intersects the boundary or not $\dot{H}^s_{loc}(U;\Lambda^{p,q})$ can be equipped with a natural Fréchet space structure. In turn this makes a Fréchet space out of $\dot{H}^s_{loc}(\overline{\mathcal{M}};\Lambda^{p,q})$: its topology is the coarsest locally convex topology that renders all the restriction mappings $f \to f|_U$ continuous. Observe that

$$\dot{H}^0_{loc}(\overline{\mathcal{M}};\Lambda^{p,q}) = L^2_{loc}(\overline{\mathcal{M}};\Lambda^{p,q}); \tag{V.6.1}$$

$$\dot{H}^s_{loc}(\overline{\mathcal{M}};\Lambda^{p,q}) \subset \dot{H}^{s'}_{loc}(\overline{\mathcal{M}};\Lambda^{p,q}) \text{ if } s \geq s'. \tag{V.6.2}$$

Next we introduce the *totally characteristic differential operators* in $\overline{\mathcal{M}}$. Those are the differential operators P with coefficients in $\mathscr{C}^\infty(\overline{\mathcal{M}})$ whose total symbol, except for the zero order terms, vanishes identically on the conormal bundle of the boundary $\partial\mathcal{M}$. Of course the latter requirement entails no restriction in the interior $\mathcal{M}$. Let us once again consider an open set U as in (V.2.19). In the local coordinates x_i, t_j a totally characteristic differential operator P of order k will have an expression

$$P = \sum_{|\alpha|+|\beta'|+\ell\leq k} c_{\alpha,\beta',\ell}(x,t)(\partial/\partial x)^\alpha(\partial/\partial t')^{\beta'}(t_n\partial/\partial t_n)^\ell \tag{V.6.3}$$

where $t' = (t_1,\ldots,t_{n-1})$, $\beta' = (\beta_1,\ldots,\beta_{n-1})$, $c_{\alpha,\beta'\ell} \in \mathscr{C}^\infty(U)$.

PROPOSITION V.6.1. *The totally characteristic differential operators in $\overline{\mathcal{M}}$ form an algebra for composition. The transpose of a totally characteristic differential operator is a totally characteristic differential operator.*

PROOF. It suffices to deal with differential operators (V.6.3) in the open set U such that $U \cap \partial\mathcal{M} \neq \emptyset$ introduced above. The last claim in the statement follows from the fact that the transpose of the operator $t_n\partial/\partial t_n$ is $-t_n\partial/\partial t_n - 1$. In order to prove the first part it suffices to show that the commutator of the operator $t_n\partial/\partial t_n$ with multiplication by an element $g \in \mathscr{C}^\infty(U)$ is equal to multiplication by another such element. We have

$$\partial g/\partial t_n = [\partial g/\partial t_n] + g(x,t',0)\otimes\delta(t_n), \tag{V.6.4}$$

where $[\partial g/\partial t_n]$ is the element of $\mathscr{C}^\infty(\mathrm{U})$ equal to $\partial g/\partial t_n$ in $\mathrm{U}\cap\mathcal{M}$. We conclude that $t_n\partial g/\partial t_n = t_n[\partial g/\partial t_n] \in \mathscr{C}^\infty(\mathrm{U})$. ∎

We denote by $\dot{\mathscr{A}}^{(s)}(\overline{\mathcal{M}};\Lambda^{p,q})$ the subspace of $\dot{\mathrm{H}}^s_{loc}(\overline{\mathcal{M}};\Lambda^{p,q})$ consisting of all the distribution sections u such that, given any totally characteristic differential operator P in $\overline{\mathcal{M}}$, we have $Pu \in \dot{\mathrm{H}}^s_{loc}(\overline{\mathcal{M}};\Lambda^{p,q})$ (P acts on the current u coefficientwise). We equip $\dot{\mathscr{A}}^{(s)}(\overline{\mathcal{M}};\Lambda^{p,q})$ with the coarsest locally convex topology that ensures the continuity of all the linear maps $\dot{\mathscr{A}}^{(s)}(\overline{\mathcal{M}};\Lambda^{p,q}) \ni u \to Pu \in \dot{\mathrm{H}}^s_{loc}(\overline{\mathcal{M}};\Lambda^{p,q})$; $\dot{\mathscr{A}}^{(s)}(\overline{\mathcal{M}};\Lambda^{p,q})$ is a Fréchet space. Given any compact subset K of $\overline{\mathcal{M}}$ we denote by $\dot{\mathscr{A}}^{(s)}(\mathrm{K};\Lambda^{p,q})$ the closed linear subspace of $\dot{\mathscr{A}}^{(s)}(\overline{\mathcal{M}};\Lambda^{p,q})$ consisting of the sections u with $\operatorname{supp} u \subset \mathrm{K}$; $\dot{\mathscr{A}}^{(s)}(\mathrm{K};\Lambda^{p,q})$ will be equipped with the topology inherited from $\dot{\mathscr{A}}^{(s)}(\overline{\mathcal{M}};\Lambda^{p,q})$. We define the $\mathscr{LF}$ space $\dot{\mathscr{A}}^{(s)}_c(\overline{\mathcal{M}};\Lambda^{p,q})$ as the locally convex inductive limit of the spaces $\dot{\mathscr{A}}^{(s)}(\mathrm{K};\Lambda^{p,q})$, $\mathrm{K} \subset\subset \overline{\mathcal{M}}$. According to (V.6.2) we have

$$\dot{\mathscr{A}}^{(s)}(\overline{\mathcal{M}};\Lambda^{p,q}) \subset \dot{\mathscr{A}}^{(s')}(\overline{\mathcal{M}};\Lambda^{p,q}) \text{ if } s \geq s'. \tag{V.6.5}$$

We write $\dot{\mathscr{A}}^{(s)}(\overline{\mathcal{M}})$ rather than $\dot{\mathscr{A}}^{(s)}(\overline{\mathcal{M}};\Lambda^{0,0})$. Observe that the restriction map from $\overline{\mathcal{M}}$ to the interior $\mathcal{M}$ induces a continuous linear map of $\dot{\mathscr{A}}^{(s)}(\overline{\mathcal{M}};\Lambda^{p,q})$ into $\mathscr{C}^\infty(\mathcal{M};\Lambda^{p,q})$.

PROPOSITION V.6.2. *If $s \leq 0$, $\mathscr{C}^\infty(\overline{\mathcal{M}};\Lambda^{p,q}) \subset \dot{\mathscr{A}}^{(s)}(\overline{\mathcal{M}};\Lambda^{p,q})$ and $\mathscr{C}^\infty_c(\mathcal{M};\Lambda^{p,q})$ is dense in $\dot{\mathscr{A}}^{(s)}_c(\overline{\mathcal{M}};\Lambda^{p,q})$ and in $\dot{\mathscr{A}}^{(s)}(\overline{\mathcal{M}};\Lambda^{p,q})$.*

PROOF. Suppose $s \leq 0$ and $g \in \mathscr{C}^\infty(\overline{\mathcal{M}})$. According to (V.6.4), if U is defined according to (V.2.19) $(t_n\partial/\partial t_n)g = t_n[\partial g/\partial t_n] \in \mathrm{L}^2(\mathrm{U})$, which entails $g \in \dot{\mathscr{A}}^{(s)}(\overline{\mathcal{M}})$. By using local trivializations of $\Lambda^{p,q}$ one derives at once that $\mathscr{C}^\infty(\overline{\mathcal{M}};\Lambda^{p,q}) \subset \dot{\mathscr{A}}^{(s)}(\overline{\mathcal{M}};\Lambda^{p,q})$.

Use of a partition of unity in $\mathscr{C}^\infty_c(\overline{\mathcal{M}})$ shows that the density of $\mathscr{C}^\infty_c(\mathcal{M};\Lambda^{p,q})$ in $\dot{\mathscr{A}}^{(s)}_c(\overline{\mathcal{M}};\Lambda^{p,q})$ or in $\dot{\mathscr{A}}^{(s)}(\overline{\mathcal{M}};\Lambda^{p,q})$ will follow from that of $\mathscr{C}^\infty_c(\mathrm{U}\cap\mathcal{M};\Lambda^{p,q})$ in $\dot{\mathscr{A}}^{(s)}_c(\mathrm{U};\Lambda^{p,q})$ with U as in (V.2.19). By using a trivialization of $\Lambda^{p,q}$ in U it suffices to establish the density of $\mathscr{C}^\infty_c(\mathrm{U}\cap\mathcal{M})$ in $\dot{\mathscr{A}}^{(s)}_c(\mathrm{U})$.

Let $\rho \in \mathscr{C}^\infty(\mathrm{U})$ have compact support contained in the interior of U, $\mathrm{U}\cap\mathcal{M}$, and be such that $\int \rho dx dt = +1$; set $\rho_\varepsilon(x,t) = \varepsilon^{-N}\rho(x/\varepsilon,t/\varepsilon)$ for any $\varepsilon > 0$. If $u \in \dot{\mathscr{A}}^{(s)}_c(\mathrm{U})$ let $\tilde{u}$ be the function in $\hat{\mathrm{U}}$ equal to u in U and to zero in $\hat{\mathrm{U}}\backslash\mathrm{U}$; then $\rho_\varepsilon\star\tilde{u} \in \mathscr{C}^\infty_c(\mathrm{U}\cap\mathcal{M})$ as soon as ε is small enough. We have

$$(t_n\partial/\partial t_n)(\rho_\varepsilon\star\tilde{u}) = [(\partial/\partial t_n)(t_n\rho_\varepsilon)]\star\tilde{u} + \rho_\varepsilon\star(t_n\partial\tilde{u}/\partial t_n).$$

We know that $\tilde{u}$ and $t_n\partial\tilde{u}/\partial t_n$ both belong to $\mathrm{H}^s(\hat{\mathrm{U}})$. The convolution operator $\rho_\varepsilon\star$ acting on $\mathrm{H}^s(\mathbb{R}^N)$ converges to the identity as $\varepsilon \to +0$; on the other hand, as $\varepsilon \to +0$, the convolution operators $[(\partial/\partial t_n)(t_n\rho_\varepsilon)]\star$ form a bounded set of linear operators on the Hilbert space $H^s(\mathbb{R}^N)$; for each element $v \in \mathscr{C}_c^\infty(\mathbb{R}^N)$ $[(\partial/\partial t_n)(t_n\rho_\varepsilon)]\star v \to 0$ in $\mathscr{C}_c^\infty(\mathbb{R}^N)$. We conclude that $[(\partial/\partial t_n)(t_n\rho_\varepsilon)]\star\tilde{u}$ converges to zero in $H^s(\mathbb{R}^N)$ and therefore that $(t_n\partial/\partial t_n)(\rho_\varepsilon\star\tilde{u})$ converges to $t_n\partial\tilde{u}/\partial t_n$. It follows easily that $P(\rho_\varepsilon\star u)$ converges to Pu in $\dot{H}^s_{loc}(U)$ whatever the totally characteristic differential operator P given by (V.6.3). ∎

For any $u \in \mathscr{D}'(\overline{\mathscr{M}};\Lambda^{p,q})$ and any $\varphi \in \mathscr{C}^\infty(\overline{\mathscr{M}};\Lambda^{m-p,n-q-1})$ we define

$$\langle \mathrm{b}'^{p,q}u, \varphi\rangle = (-1)^{p-q-1}\langle u, \mathrm{d}'^{m-p,n-q-1}\varphi\rangle \tag{V.6.6}$$

(cf. (V.2.1), (V.5.3)). This leads us to the differential complex

$$\mathrm{b}'^{p,q}: \mathscr{D}'(\overline{\mathscr{M}};\Lambda^{p,q}) \to \mathscr{D}'(\overline{\mathscr{M}};\Lambda^{p,q+1}),\ q = 0,1,\ldots \tag{V.6.7}$$

Most of the time we shall write b' rather than $\mathrm{b}'^{p,q}$.

PROPOSITION V.6.3. *The differential operator* b' *defines a continuous linear map* $\dot{\mathscr{A}}^{(s)}(\overline{\mathscr{M}};\Lambda^{p,q}) \to \dot{\mathscr{A}}^{(s-1)}(\overline{\mathscr{M}};\Lambda^{p,q+1})$.

PROOF. Once again it suffices to limit our attention to an open subset U given by (V.2.19). If $u \in \dot{\mathscr{A}}^{(s)}(U;\Lambda^{p,q})$ and $\varphi \in \mathscr{C}_c^\infty(U;\Lambda^{m-p,n-q-1})$ the duality bracket $\langle u, \mathrm{d}'\varphi\rangle$ is a linear combination of integrals

$$\mathscr{I} = \int_{U\cap\mathscr{M}} u_J\, \mathrm{L}_j\varphi_K\, \mathrm{d}Z\wedge\mathrm{d}t.$$

If $j < n$ we have $\mathscr{I} = -\int_{U\cap\mathscr{M}} \mathrm{L}_j u_J\, \varphi_K\, \mathrm{d}Z\wedge\mathrm{d}t$. Suppose now $j = n$. Let $\tilde{\varphi}_K \in \mathscr{C}_c^\infty(\hat{U})$ be such that $\tilde{\varphi}_K = \varphi_K$ in U. Let P be the differential operator (V.6.3), ${}^tP$ its transpose with respect to the measure $(\det Z_x)\,\mathrm{d}x\mathrm{d}t$; we have

$$\int_{t_n>0}\int_{\mathbb{R}^{N-1}} \tilde{u}_J\, \mathrm{L}_n{}^tP\tilde{\varphi}_K\, \mathrm{d}Z\wedge\mathrm{d}t =$$

$$\int_{t_n>0}\int_{\mathbb{R}^{N-1}} Q\tilde{u}_J\,(\partial\tilde{\varphi}_K/\partial t_n)\, \mathrm{d}Z\wedge\mathrm{d}t + \int_{t_n>0}\int_{\mathbb{R}^{N-1}} R\tilde{u}_J\, \tilde{\varphi}_K\, \mathrm{d}Z\wedge\mathrm{d}t,$$

with Q and R totally characteristic differential operators. The assignment

$$\tilde{\varphi}_K \to \int_{t_n>0}\int_{\mathbb{R}^{N-1}} Q\tilde{u}_J\,(\partial\tilde{\varphi}_K/\partial t_n)\, \mathrm{d}Z\wedge\mathrm{d}t$$

defines an element of $H^{s-1}_{loc}(\hat{U})$ whose support is contained in U. The assignment

$$\tilde{\varphi}_K \to \int_{t_n>0}\int_{\mathbb{R}^{N-1}} R\tilde{u}_J\, \tilde{\varphi}_K \mathrm{d}Z\wedge\mathrm{d}t$$

defines an element of $H^s_{loc}(\hat{U})$, also with support contained in U. This proves that $Pb'u \in \dot{H}^{s-1}_{loc}(U;\Lambda^{p,q+1})$. ■

Let i_b^* be the pullback to the boundary (V.2.3). We define a linear map

$$E : \mathcal{D}'(\partial\mathcal{M};\Lambda_b^{p,q}) \to \mathcal{D}'(\overline{\mathcal{M}};\Lambda^{p,q+1})$$

by the formula, valid for all $u \in \mathcal{D}'(\partial\mathcal{M};\Lambda_b^{p,q})$, $\psi \in \mathcal{C}_c^\infty(\overline{\mathcal{M}};\Lambda^{m-p,n-q-1})$,

$$\langle Eu,\psi\rangle = \int_{\partial\mathcal{M}} u\wedge i_b^*\psi. \tag{V.6.8}$$

If $\psi \equiv 0$ in an open neighborhood of supp u, the same is true of $i_b^*\psi$; thus

$$\text{supp } Eu \subset \text{supp } u. \tag{V.6.9}$$

LEMMA V.6.1 *The linear operator E maps $\mathcal{C}^\infty(\partial\mathcal{M};\Lambda_b^{p,q})$ into $\dot{\mathcal{A}}^{(-1)}(\overline{\mathcal{M}};\Lambda^{p,q+1})$.*

PROOF. Let $\varphi \in \mathcal{C}^\infty(\partial\mathcal{M};\Lambda_b^{p,q})$; if U is our standard open set in (V.2.19) we have, for any $\psi \in \mathcal{C}_c^\infty(U;\Lambda^{m-p,n-q-1})$,

$$\langle(\partial/\partial x)^\alpha(\partial/\partial t')^{\beta'}(1+t_n\partial/\partial t_n)^\ell E\varphi,\psi\rangle =$$

$$(-1)^{|\alpha|+|\beta'|+\ell}\int_{U\cap\partial\mathcal{M}} \varphi\wedge i_b^*[(\partial/\partial x)^\alpha(\partial/\partial t')^{\beta'}(t_n\partial/\partial t_n)^\ell\psi],$$

which vanishes unless $\ell = 0$. On the other hand, by (V.6.8),

$$|\langle(\partial/\partial x)^\alpha(\partial/\partial t')^{\beta'}E\varphi,\psi\rangle| \le C_1\|i_b^*\psi\|_0.$$

Let $E\varphi$ define a current in $\hat{U}$ by setting $\langle E\varphi,\tilde{\varphi}\rangle = \langle E\varphi,i_+\tilde{\varphi}\rangle$ for any $\tilde{\varphi} \in \mathcal{C}_c^\infty(\hat{U};\Lambda^{m-p,n-q-1})$ (i_+ is the pullback map to U). Clearly $E\varphi \equiv 0$ in $\hat{U}\backslash U$; on the other hand, $\|i_b^*\tilde{\varphi}\|_0 \le C_2\|\tilde{\varphi}\|_1$. Conclusion: $(\partial/\partial x)^\alpha(\partial/\partial t')^{\beta'}(t_n\partial/\partial t_n)^\ell E\varphi$ defines an element of $\dot{H}^{-1}_{loc}(U;\Lambda^{p,q})$. ■

DEFINITION V.6.1. *We shall denote by $\mathcal{A}'(\overline{\mathcal{M}};\Lambda^{p,q})$ the space of distribution sections $u \in \mathcal{D}'(\overline{\mathcal{M}};\Lambda^{p,q})$ such that, given any real number $s \le 0$, the linear functional $\varphi \to \langle u,\varphi\rangle$ extends continuously from $\mathcal{C}_c^\infty(\overline{\mathcal{M}};\Lambda^{m-p,n-q})$ to $\mathcal{A}_c^{(s)}(\overline{\mathcal{M}};\Lambda^{m-p,n-q})$.*

Once again let the open set U be as in (V.2.19). If u is an arbitrary element of $\mathcal{A}'(\overline{\mathcal{M}};\Lambda^{p,q})$, to every compact subset K of U and to every real number $s \le 0$ there are constants $C > 0$, $k \in \mathbb{Z}_+$, such that, for all $\varphi \in \mathcal{C}_c^\infty(K;\Lambda^{m-p,n-q})$,

$$|\langle u,\varphi\rangle| \le C\sum_{|\alpha|+|\beta'|+\ell\le k}\|(\partial/\partial x)^\alpha(\partial/\partial t')^{\beta'}(t_n\partial/\partial t_n)^\ell\varphi\|_s. \tag{V.6.10}$$

We list a number of basic properties of the spaces $\mathcal{A}'(\overline{\mathcal{M}};\Lambda^{p,q})$.

First of all, the natural injection of $\mathscr{C}^\infty(\overline{\mathcal{M}};\Lambda^{p,q})$ into $\mathscr{D}'(\overline{\mathcal{M}};\Lambda^{p,q})$, i.e., the map $f \to T_f$ given by (V.5.1), identifies $\mathscr{C}^\infty(\overline{\mathcal{M}};\Lambda^{p,q})$ to a linear subspace of $\mathscr{A}'(\overline{\mathcal{M}};\Lambda^{p,q})$.

Also, $\mathscr{E}'(\mathcal{M};\Lambda^{p,q}) \subset \mathscr{A}'(\overline{\mathcal{M}};\Lambda^{p,q})$. Indeed, as noted earlier, the restriction mapping to $\mathcal{M}$ maps $\dot{\mathscr{A}}_c^{(s)}(\overline{\mathcal{M}};\Lambda^{m-p,n-q})$ into $\mathscr{C}^\infty(\mathcal{M};\Lambda^{m-p,n-q})$; the image is dense since it contains $\mathscr{C}_c^\infty(\mathcal{M};\Lambda^{m-p,n-q})$ whence the claim, by transposition.

Another property used below is the following:

$$\forall\, u \in \mathscr{A}'(\mathcal{M};\Lambda^{p,q}),\ \operatorname{supp} u \subset \partial\mathcal{M} \Rightarrow u \equiv 0. \qquad \text{(V.6.11)}$$

Indeed, if $\operatorname{supp} u \subset \partial\mathcal{M}$ then $\langle u,\psi\rangle = 0$ for all $\psi \in \mathscr{C}_c^\infty(\mathcal{M};\Lambda^{m-p,n-q})$ whence $u \equiv 0$ by Proposition V.6.2.

For our purposes here it suffices to equip the space $\mathscr{A}'(\overline{\mathcal{M}};\Lambda^{p,q})$ with the *weak dual topology*: a net (or a sequence) of distributions $u_\iota \in \mathscr{A}'(\overline{\mathcal{M}};\Lambda^{p,q})$ converges to 0 if, for all $s \leq 0$ and $f \in \dot{\mathscr{A}}_c^{(s)}(\overline{\mathcal{M}};\Lambda^{m-p,n-q})$, the complex numbers $\langle u_\iota,f\rangle$ converge to zero. The union $\dot{\mathscr{A}}_c(\overline{\mathcal{M}};\Lambda^{m-p,n-q})$ of all the spaces $\dot{\mathscr{A}}_c^{(s)}(\overline{\mathcal{M}};\Lambda^{m-p,n-q})$ $(s \leq 0)$ can be identified to the topological dual of $\mathscr{A}'(\overline{\mathcal{M}};\Lambda^{p,q})$. We contend that

$$\mathscr{C}_c^\infty(\overline{\mathcal{M}};\Lambda^{p,q}) \textit{ is dense in } \mathscr{A}'(\overline{\mathcal{M}};\Lambda^{p,q}). \qquad \text{(V.6.12)}$$

PROOF. Thanks to the Hahn-Banach theorem it suffices to prove that if any element $v \in \dot{\mathscr{A}}_c(\overline{\mathcal{M}};\Lambda^{m-p,n-q})$ is such that $\langle v,\varphi\rangle = 0$ for all $\varphi \in \mathscr{C}_c^\infty(\overline{\mathcal{M}};\Lambda^{p,q})$ then $v = 0$. But this is true when v is an arbitrary distribution section of $\Lambda^{m-p,n-q}$ over $\overline{\mathcal{M}}$. ∎

One of the main motivations for Definition V.6.1 lies in the next statement:

PROPOSITION V.6.4. *The pullback map* (V.2.3) *extends as a* (weakly continuous) *linear map* $i_b^* : \mathscr{A}'(\overline{\mathcal{M}};\Lambda^{p,q}) \to \mathscr{D}'(\partial\mathcal{M};\Lambda_b^{p,q})$.

PROOF. Consider $u \in \mathscr{A}'(\overline{\mathcal{M}};\Lambda^{p,q})$. The "extended" pullback i_b^*u is defined by the formula

$$\langle i_b^*u,\psi\rangle = (-1)^{p+q}\langle u,E\psi\rangle,\ \psi \in \mathscr{C}_c^\infty(\partial\mathcal{M};\Lambda_b^{m-p,n-q-1}), \qquad \text{(V.6.13)}$$

where E is the linear operator defined in (V.6.8). This makes sense since $E\psi \in \dot{\mathscr{A}}_c^{(-1)}(\mathcal{M};\Lambda^{m-p,n-q})$. When $u = \varphi \in \mathscr{C}^\infty(\overline{\mathcal{M}};\Lambda^{p,q})$, the right-hand side in (V.6.13) is equal to

$$(-1)^{(p+q)(N-1)}\int_{\mathcal{M}} E\psi\wedge\varphi \ \text{(cf. (V.5.1))}.$$

But $\langle E\psi,\varphi\rangle = \int_{\partial\mathcal{M}} \psi\wedge i_b^*\varphi = (-1)^{(p+q)N}\langle i_b^*\varphi,\ \psi\rangle$, which, in view of (V.6.13), shows that $i_b^*u = i_b^*\varphi$. ∎

PROPOSITION V.6.5. *The differential operator* $\mathrm{d}'\colon \mathscr{C}^\infty(\overline{\mathcal{M}};\Lambda^{p,q}) \to \mathscr{C}^\infty(\overline{\mathcal{M}};\Lambda^{p,q+1})$ *extends as a linear operator* $\mathrm{d}'\colon \mathscr{A}'(\overline{\mathcal{M}};\Lambda^{p,q}) \to \mathscr{A}'(\overline{\mathcal{M}};\Lambda^{p,q+1})$.

PROOF. If $\psi \in \mathscr{C}_c^\infty(\overline{\mathcal{M}};\Lambda^{m-p,n-q-1})$ and $u \in \mathscr{A}'(\overline{\mathcal{M}};\Lambda^{p,q})$ we may define

$$\langle \mathrm{d}'u,\psi\rangle = (-1)^{p+q}\langle u,\mathrm{b}'\psi\rangle. \tag{V.6.14}$$

It follows from Proposition V.6.3 that $\mathrm{d}'u \in \mathscr{A}'(\overline{\mathcal{M}};\Lambda^{p,q+1})$. Suppose $u = \varphi \in \mathscr{C}^\infty(\overline{\mathcal{M}};\Lambda^{p,q})$; then, by (V.6.6),

$$\langle \varphi,\mathrm{b}'\psi\rangle = (-1)^{(p+q)(N-1)}\langle \mathrm{b}'\psi,\varphi\rangle =$$
$$(-1)^{N-p-q-1}(-1)^{(p+q)(N-1)}\langle \psi,\mathrm{d}'\varphi\rangle =$$
$$(-1)^{N-p-q-1}(-1)^{(p+q+1)(N-1)}(-1)^{(p+q)(N-1)}\langle \mathrm{d}'\varphi,\psi\rangle =$$
$$(-1)^{p+q}\langle \mathrm{d}'u,\psi\rangle. \ \blacksquare$$

Thus we may consider the differential complex

$$\mathrm{d}' : \mathscr{A}'(\overline{\mathcal{M}};\Lambda^{p,q}) \to \mathscr{A}'(\overline{\mathcal{M}};\Lambda^{p,q+1}),\ q = 0,1,\ldots \tag{V.6.15}$$

PROPOSITION V.6.6. *Let* $f \in \mathscr{A}'(\overline{\mathcal{M}};\Lambda^{0,1})$. *If* $u \in \mathscr{D}'(\overline{\mathcal{M}})$ *satisfies the equation* $\mathrm{d}'u = f$ *in the interior* $\mathcal{M}$ *there exists a unique element* v *of* $\mathscr{A}'(\overline{\mathcal{M}})$ *equal to* u *in* $\mathcal{M}$. *We have* $\mathrm{d}'v = f$ *in* $\overline{\mathcal{M}}$.

PROOF. Since $\mathscr{C}_c^\infty(\mathcal{M};\Lambda^{m,n})$ is dense in $\dot{\mathscr{A}}_c^{(s)}(\overline{\mathcal{M}};\Lambda^{m,n})$ the fact that $v = u$ in $\mathcal{M}$ determines v uniquely. It also entails that the support of $f - \mathrm{d}'v \in \mathscr{A}'(\overline{\mathcal{M}};\Lambda^{0,1})$ is contained in $\partial\mathcal{M}$, hence $\mathrm{d}'v$ must be equal to f by (V.6.11). The existence of v will follow if we prove that the linear functional $\psi \to \langle u,\psi\rangle$ is continuous for the topology induced by $\dot{\mathscr{A}}_c^{(s)}(\overline{\mathcal{M}};\Lambda^{m,n})$ $(s \leq 0)$.

Once again we reason in the open set U given by (V.2.19). Select a cutoff function $\chi \in \mathscr{C}^\infty(\mathbb{R}_+)$, $\chi(\tau) = 1$ if $\tau < \varepsilon/2$, $\chi(\tau) = 0$ if $\tau > \varepsilon$. Then $\mathrm{d}'[\chi(t_n)u] = \chi(t_n)f + \chi'(t_n)u\mathrm{d}t_n \in \mathscr{A}'(\mathrm{U};\Lambda^{0,1})$. Let K be a compact subset of U; we can select $\varepsilon > 0$ sufficiently small that

$$\mathrm{K}_\varepsilon = \{ (x,t) \in \mathrm{U};\ \exists\ (x,t',\tau) \in \mathrm{K} \textit{ such that } \tau \leq t_n \leq \varepsilon \} \subset\subset \mathrm{U}.$$

It will suffice to show that the linear functional $\mathscr{C}_c^\infty(\mathrm{K};\Lambda^{m,n}) \ni \psi \to \langle \chi(t_n)u,\psi\rangle$ is continuous for the topology induced by $\dot{\mathscr{A}}^{(s)}(\mathrm{K};\Lambda^{m,n})$ since the analogous claim about $[1-\chi(t_n)]u$ is obvious. In other words, we may as well assume that $u \equiv 0$ in the region $t_n > \varepsilon$. We may write $\psi = \Psi \mathrm{d}x\wedge \mathrm{d}t$ and

$$\langle u,\psi\rangle = \langle u,\Psi\rangle,$$

where the bracket at the right is that of the duality between scalar distributions and scalar test-functions. We claim that, whatever $s \in \mathbb{R}$, there are constants $C > 0$, $k \in \mathbb{Z}_+$, such that, for all $\Psi \in \mathscr{C}_c^\infty(\mathrm{K})$,

$$|\langle u,\Psi\rangle| \leq C \sum_{|\alpha|+|\beta'|+\ell\leq k} \|(\partial/\partial x)^\alpha(\partial/\partial t')^{\beta'}(t_n\partial/\partial t_n)^\ell\Psi\|_s. \qquad (V.6.16)$$

Thanks to the local structure of distributions the claim is true for some value s_0 of s (with $k = 0$). We shall reason by descending induction on s. We suppose the claim true for some $s \leq s_0$. Using the notation

$$\partial_n^{-1}\Psi(x,t) = \int_0^{t_n} \Psi(x,t',s)ds,$$

we may write

$$\langle u,\Psi\rangle = - \langle \partial u/\partial t_n, \chi_1\partial_n^{-1}\Psi\rangle,$$

with $\chi_1 \in \mathscr{C}_c^\infty(U)$ equal to 1 in some neighborhood of the compact set K_ε. Since $d'u = f$ we have

$$\partial u/\partial t_n = f_n - \sum_{i=1}^n \lambda_{ni}\partial u/\partial x_i. \qquad (V.6.17)$$

in $U\cap\mathcal{M} = \{(x,t) \in U;\, t_n > 0\}$. There are constants $C_0 > 0$, $k \in \mathbb{Z}_+$ such that

$$|\langle f_n,\chi_1\partial_n\Psi\rangle| \leq$$
$$C_0 \sum_{|\alpha|+|\beta'|+\ell\leq k} \|(\partial/\partial x)^\alpha(\partial/\partial t')^{\beta'}(t_n\partial/\partial t_n)^\ell\Psi\|_{s-1}. \qquad (V.6.18)$$

On the other hand, by (V.6.16),

$$\sum_{i=1}^n |\langle \lambda_{ni}\partial u/\partial x_i, \chi_1\partial_n^{-1}\Psi\rangle| \leq$$
$$C_1 \sum_{|\alpha|+|\beta'|+\ell\leq k+1} \|(\partial/\partial x)^\alpha(\partial/\partial t')^{\beta'}(t_n\partial/\partial t_n)^\ell(\chi_1\partial_n^{-1}\Psi)\|_s.$$

It suffices to observe that

$$\|\chi_1\partial_n^{-1}\Psi\|_s \leq C_2 \sum_{|\alpha|+|\beta'|\leq 1} \|(\partial/\partial x)^\alpha(\partial/\partial t')^{\beta'}\Psi\|_{s-1}$$

to obtain

$$\sum_{k=1}^n |\langle \lambda_{nk}\partial u/\partial x_k, \chi_1\partial_n^{-1}\Psi\rangle| \leq$$
$$C_3 \sum_{|\alpha|+|\beta'|+\ell\leq k+2} \|(\partial/\partial x)^\alpha(\partial/\partial t')^{\beta'}(t_n\partial/\partial t_n)^\ell\Psi\|_{s-1}$$

(C_1, C_2, C_3 are suitably large positive constants). If we combine this last inequality with (V.6.18) we get what we wanted. ■

Corollary V.6.1. *If* $u \in \mathscr{D}'(\overline{\mathcal{M}})$ *satisfies the homogeneous equation* $d'u = 0$

in $\mathcal{M}$, there is a unique element v of $\mathcal{A}'(\overline{\mathcal{M}})$ equal to u in $\mathcal{M}$. We have $\mathrm{d}'v = 0$ *in $\overline{\mathcal{M}}$.*

DEFINITION V.6.2. *We shall say that a distribution section $u \in \mathcal{D}'(\overline{\mathcal{M}};\Lambda^{p,q})$ is* smooth transversally to the boundary *if every point of $\partial\mathcal{M}$ has an open neighborhood* U *of the kind* (V.2.19) *in which the coefficients of u in the basis* $\mathrm{d}Z_I \wedge \mathrm{d}t_J$ $(|I| = p,\ |J| = q)$ *are $\mathcal{C}^\infty$ function of $t_n \geq 0$ valued in the space of distributions of $(x,t') \in \mathcal{O}$.*

The statement and proof of the following result are similar to those of Proposition I.4.3.

PROPOSITION V.6.7. *Assume that the boundary $\partial\mathcal{M}$ is noncharacteristic. Let Ω be an open subset of $\overline{\mathcal{M}}$ and $u \in \mathcal{A}'(\Omega)$ be such that $f = \mathrm{d}'u \in \mathcal{A}'(\Omega;\Lambda^{0,1})$ is smooth transversally to the boundary. Then u is smooth transversally to the boundary.*

PROOF. Let $\mathrm{U} \subset \Omega$ be an open set of the kind (V.2.19). We avail ourselves of (V.6.17). We have

$$u(x,t) = u_0(x,t') + \mathcal{Y}(t_n)\star(f_n - \sum_{i=1}^{n} \lambda_{ni}\partial u/\partial x_i). \qquad \text{(V.6.19)}$$

Here $\mathcal{Y}$ is the Heaviside function and the star $\star$ stands for convolution with respect to the coordinate t_n. We are taking advantage of Proposition V.6.4: u_0 is the pullback, i.e., the *trace*, of u on the boundary piece $\mathrm{U}\cap\partial\mathcal{M}$ $(\cong\mathcal{O})$.

We may view u as a distribution in $[0,T[$ (where t_n varies) valued in the space of distributions of (x,t') in $\mathcal{O}$. Given any number $T' \in [0,T[$ and any open set $\mathcal{O}' \subset\subset \mathcal{O}$ there are numbers $\mu, \kappa \in \mathbb{Z}$ such that the restriction of u to $\mathcal{O}' \times [0,T'[$ defines an element of $\mathrm{H}^\mu([0,T'[;\mathrm{H}^\kappa(\mathcal{O}'))$ and that of u_0 to $\mathcal{O}'$ belongs to $\mathrm{H}^\kappa(\mathcal{O}')$. The right-hand side in (V.6.19) belongs to $\mathrm{H}^{\mu+1}([0,T'[;\mathrm{H}^{\kappa-1}(\mathcal{O}'))$, whence the assertion by ascending induction on μ. ■

DEFINITION V.6.3. *Suppose the manifold $\overline{\mathcal{M}}$ is equipped with a locally integrable structure such that the boundary $\partial\mathcal{M}$ is noncharacteristic. By a* distribution solution *in an open subset Ω of $\overline{\mathcal{M}}$ we shall mean an element h of $\mathcal{A}'(\Omega)$ satisfying $d'h = 0$ in Ω.*

In Definition V.6.3 the open subset Ω is viewed as a $\mathcal{C}^\infty$ manifold with boundary; the boundary of Ω is the submanifold $\Omega\cap\partial\mathcal{M}$.

REMARK V.6.1. It is instructive to compare the situation in Definition V.6.3 with that in Example V.5.2. Let the manifold with boundary $\overline{\mathcal{M}} = \{ (x,t) \in \mathbb{R}^2;\ x \geq 0 \}$ be equipped with the involutive structure defined by the vector field $L = \partial/\partial t - \imath x\partial/\partial x$ (with first integral $Z = xe^{\imath t}$). One can define the space $\mathcal{A}'(\overline{\mathcal{M}})$ exactly as in the case of a noncharacteristic boundary (of course, here L is totally characteristic). Here, however, $u = e^{-\imath t}\otimes\delta(x)$ is a solution of the homogeneous equation $Lu = 0$. But supp $u \subset \partial\mathcal{M}$ and thus $u \notin \mathcal{A}'(\overline{\mathcal{M}})$. ∎

We derive from Proposition V.6.7:

Corollary V.6.2. *Every distribution solution in an open subset Ω of $\overline{\mathcal{M}}$ is smooth transversally to the boundary.*

Corollary V.6.3. *Suppose a distribution h in Ω is smooth transversally to the boundary and satisfies $d'h = 0$ in the interior $\Omega\cap\mathcal{M}$. Then $h \in \mathcal{A}'(\Omega)$ and $d'h = 0$ in Ω.*

Proof. By Corollary V.6.1 (with Ω in the place of $\overline{\mathcal{M}}$) there is $v \in \mathcal{A}'(\Omega)$ such that $u = v$ in $\Omega\cap\mathcal{M}$, and $d'v = 0$ in Ω. By Proposition V.6.7 v and therefore also $h - v$ are smooth transversally to the boundary. But supp$(h-v) \subset \Omega\cap\partial\mathcal{M}$ which implies $h - v \equiv 0$. ∎

Local Representation and Approximation of Distribution Solutions When the Boundary Is Noncharacteristic

We can generalize to locally integrable structures with noncharacteristic boundary the local representation formula (II.5.1). Assume that U as in (V.2.19) is a neighborhood of a point $0 \in \partial\mathcal{M}$ to which we refer below as the origin.

Theorem V.6.1. *Suppose that the boundary $\partial\mathcal{M}$ is noncharacteristic. There is an open neighborhood of the origin in $\overline{\mathcal{M}}$, $U_1 \subset U$, such that, given any distribution solution h in U and any integer $\mu \geq 1$, there is a solution $h_1 \in \mathcal{C}^{\mu}(U_1)$ and an integer $\nu \geq 0$ such that, in U_1,*

$$h = \Delta_M^{\nu} h_1. \tag{V.6.20}$$

Here, as in section II.5, $\Delta_M = M_1^2 + \cdots + M_m^2$.

Proof. By Definition V.6.3 and Corollary V.6.2 h is a $\mathcal{C}^\infty$ function of $t_n \geq 0$ valued in the space of distributions with respect to (x,t'). Consider an open neighborhood of the origin $U' \subset\subset U$ of the form $U' \cong \mathcal{O}' \times [0,\varepsilon[$ with $\mathcal{O}' \subset\subset \mathcal{O}$ an open ball in (x,t')-space $\mathbb{R}^{N-1}$ centered at the origin, and $0 < \varepsilon < T$ (cf. (V.2.19)). We observe that h is a solution, in U', of the equations $L_1 h = \cdots = L_{n-1}h = 0$. If $\mathcal{O}'' \subset\subset \mathcal{O}'$ is another open ball in (x,t')-space with suitably small radius, thanks to (II.5.11), we may write, in $U_2 = \mathcal{O}'' \times [0,\varepsilon[$,

$$h = \Delta_M^{\nu} h_2, \tag{V.6.21}$$

where $h_2 \in \mathscr{C}^{\mu+1}(U_2)$ and

$$L_j h_2 = 0 \textit{ in } U_2, j = 1,\ldots,n-1. \tag{V.6.22}$$

(Actually, the manner in which the function h_2 is constructed, in the proof of Theorem II.5.1, where now t_n is regarded as a parameter, shows that h_2 depends in $\mathscr{C}^\infty$ fashion on t_n, $0 \leq t_n < \varepsilon$.) We let L_n act on both sides of (V.6.21), in the interior set $\mathbb{O}'' \times]0,\varepsilon[$. We get

$$\Delta_M^\nu L_n h_2 = 0, \tag{V.6.23}$$

which shows that $L_n h_2$ is an analytic vector of the system $\{M_1,\ldots,M_m\}$ (see section II.4). We reason as in the proof of Theorem II.5.1:

$$L_n h_2 = g(Z(x,t),t), \tag{V.6.24}$$

where $g(z,t)$ is a $\mathscr{C}^\mu$ function in some neighborhood of the origin in $\mathbb{C}^m \times \mathbb{R}^{n-1} \times [0,\varepsilon[$, holomorphic with respect to z. We derive form (V.6.23):

$$\Delta_z^\nu g \equiv 0, \tag{V.6.25}$$

where $\Delta_z = (\partial/\partial z_1)^2 + \cdots + (\partial/\partial z_m)^2$. Furthermore, here, by (V.6.22) and (V.6.24),

$$L_j[g(Z(x,t),t)] = (\partial g/\partial t_j)(Z(x,t),t) = 0, j = 1,\ldots,n-1.$$

By using the fact that, for each t, the image of the map $x \to Z(x,t)$ is a maximally real submanifold of $\mathbb{C}^m$, we conclude that g does not depend on $t_1,\ldots,t_{n-1}$. We write $g(z,t_n)$ rather than $g(z,t)$ and define

$$v(z,t_n) = \int_0^{t_n} g(z,\tau)d\tau, \; h_1(x,t) = h_2(x,t) - v(Z(x,t),t_n).$$

Then, by (V.6.24), we get $L_n h_1 = 0$, and by (V.6.25), $\Delta_z^\nu v \equiv 0$. We also have, by (V.6.22), $L_j h_1 = 0$ for $j = 1,\ldots,n-1$, whence the result. ■

By Definition V.6.3 and Proposition V.6.4 the trace h_0 on the boundary of a distribution solution h is well defined; h_0 is a distribution solution in the locally integrable structure induced on $\partial\mathcal{M}$ by that of $\overline{\mathcal{M}}$. If follows that the trace of h on any maximally real submanifold of $\partial\mathcal{M}$ is well defined. We can extend Theorem V.4.2 to distribution solutions. In this connection it is convenient to take U in the form (V.3.7).

THEOREM V.6.2. *There is an open neighborhood of the origin in $\overline{\mathcal{M}}$, $U' \subset U$, and a function $\chi \in \mathscr{C}_c^\infty(V)$ equal to one in an open neighborhood $V' \subset V$ of the origin in x-space, such that, given any distribution solution h in U, we have, in the sense of distributions in U′,*

$$h(x,t) = \lim_{\tau \to +\infty} \int_V E_\tau(Z(x,t);y,0)\chi(y)h(y,0)[\det Z_x(y,0)]dy. \tag{V.6.26}$$

At the right in (V.6.26) the integral signifies a duality bracket.

PROOF. If supp χ is sufficiently small we may write, according to (V.6.20) and to the integration by parts formula (II.1.15),

$$\int_V E_\tau(Z(x,t);y,0)\chi(y)h(y,0)[\det Z_x(y,0)]dy =$$

$$\int_V E_\tau(Z(x,t);y,0)\chi(y)\Delta_M^\nu h_1(y,0)[\det Z_x(y,0)]dy =$$

$$\int_V \left\{(\Delta_M^\nu)_{y,s}[\chi(y)E_\tau(Z(x,t);y,s)]\Big|_{s=0}\right\}h_1(y,0)[\det Z_x(y,0)]dy =$$

$$(\Delta_M^\nu)_{x,t}\left\{\int_V E_\tau(Z(x,t);y,0)h_1(y,0)[\det Z_x(y,0)]dy\right\} +$$

$$\sum_{j=1}^r \int_V [P_j E_\tau(Z(x,t);y,0)]\chi_j(y)h_1(y,0)[\det Z_x(y,0)]dy,$$

where P_j ($j = 1,\ldots,r$) is a differential operator in y-space, with $\mathscr{C}^\infty$ coefficients, and $\chi_j \in \mathscr{C}^\infty(V)$, $\chi_j \equiv 0$ in V′. It is immediately seen (cf. Lemma II.2.4) that, as $\tau \to +\infty$, there is a neighborhood of the origin, contained in U, in which

$$\sum_{j=1}^r \int_V [P_j E_\tau(Z(x,t);y,0)]\chi_j(y)h_1(y,0)[\det Z_x(y,0)]dy$$

converges uniformly to zero. On the other hand we know, by Theorem V.4.1, that there is a neighborhood of the origin, also contained in U, in which

$$\int_V E_\tau(Z(x,t);y,0)h_1(y,0)[\det Z_x(y,0)]dy$$

converges uniformly to $h_1(x,t)$, whence the result, by (V.6.20). ∎

Here also, as in the case of a totally characteristic boundary, Corollaries V.4.1 and V.4.3 generalize to distribution solutions: Corollaries V.5.1 and V.5.2 are also valid when the boundary is noncharacteristic.

V.7. Example: Domains in Complex Space

Let $\mathcal{M}$ be an open and connected subset of $\mathbb{C}^n$ that lies on one side of its boundary, $\partial\mathcal{M}$, assumed to be of class $\mathscr{C}^\infty$. We regard $\overline{\mathcal{M}} = \mathcal{M}\cup\partial\mathcal{M}$ as a $\mathscr{C}^\infty$ manifold with boundary and endow it with the involutive structure inherited from $\mathbb{C}^n$: the cotangent structure bundle T′ is spanned by the differentials dz_i

of the complex coordinates z_i in $\mathbb{C}^n$, the tangent structure bundle $\mathcal{V}$ by the Cauchy-Riemann vector fields $\partial/\partial\bar{z}_i$. At interior points the solutions are the holomorphic functions.

We shall now reason in an open neighborhood U in $\overline{\mathcal{M}}$ of a point of the boundary, point which we take to be the origin in $\mathbb{C}^n$. We have $\mathrm{U} = \hat{\mathrm{U}} \cap \overline{\mathcal{M}}$ with $\hat{\mathrm{U}}$ an open neighborhood of 0 in $\mathbb{C}^n$. We may, and shall, assume that the tangent hyperplane to $\partial\mathcal{M}$ at 0 is defined by the equation $y_n = 0$. In $\hat{\mathrm{U}}$, U will be defined by an inequality

$$\rho(z) = y_n - \varphi(z',x_n) \geq 0. \tag{V.7.1}$$

As usual, $x_i = \mathcal{Re} z_i$, $y_i = \mathcal{Im} z_i$ $(i = 1,\ldots,n)$, $z' = (z_1,\ldots,z_{n-1})$; φ is a real-valued $\mathscr{C}^\infty$ function in an open neighborhood of the origin in $\mathbb{R}^{2n-1}$, and satisfies

$$\varphi|_0 = 0, \ \mathrm{d}\varphi|_0 = 0. \tag{V.7.2}$$

If we make use of the local coordinate ρ we get the following set of first-integrals in the neighborhood U:

$$\mathrm{Z}_j = x_j + \imath y_j \ (= z_j) \ \textit{if} \ 1 \leq j < n, \ \mathrm{Z}_n = x_n + \imath\varphi(z',x_n) + \imath\rho. \tag{V.7.3}$$

Over U the vector bundle $\mathcal{V}$ is spanned by the vector fields

$$\begin{aligned} \mathrm{L}_k &= \partial/\partial y_k - \imath\partial/\partial x_k + \lambda_k \partial/\partial x_n \ (k = 1,\ldots,n-1), \\ \mathrm{L}_n &= \partial/\partial\rho - \imath\mu\partial/\partial x_n, \end{aligned} \tag{V.7.4}$$

where

$$\mu = (1 + \imath\partial\varphi/\partial x_n)^{-1}, \ \lambda_k = -\imath\mu(\partial\varphi/\partial y_k - \imath\partial\varphi/\partial x_k). \tag{V.7.5}$$

The vector fields M_i (see (I.7.25)) have the following expressions:

$$\mathrm{M}_j = \partial/\partial x_j - \imath\mu(\partial\varphi/\partial x_j)\partial/\partial x_n \ (j = 1,\ldots,n-1), \ \mathrm{M}_n = \mu\partial/\partial x_n. \tag{V.7.6}$$

As expected, $\mathrm{M}_1,\ldots,\mathrm{M}_n$, $\mathrm{L}_1,\ldots,\mathrm{L}_{n-1}$ are tangential to the boundary; L_n is transversal to it.

Next we define an operator of integration $\mathcal{I}$ acting on an arbitrary holomorphic function $h(z)$ in the interior $\mathrm{U}\cap\mathcal{M}$. For this it is convenient to take the neighborhood $\hat{\mathrm{U}}$ in the form

$$\hat{\mathrm{U}} = \{ z \in \mathbb{C}^n; z' \in \mathcal{O}', |x_n| < \delta, |y_n| < r \},$$

where $\mathcal{O}'$ is an open neighborhood of 0 in $\mathbb{C}^{n-1}$, and δ and r are small numbers > 0. We require $\mathcal{O}'$ and δ to be small enough that

$$|\varphi(z',x_n)| < r/2, \ \forall \ z' \in \mathcal{O}', \ x_n \in \,]-\delta,\delta[. \tag{V.7.7}$$

Let a be a number such that $r/2 < a < r$. We define

$$\mathcal{I}h(z) = -\int_{z_n}^{\imath a} h(z',\zeta)\mathrm{d}\zeta, \tag{V.7.8}$$

where the integration is carried out over the path that starts at the purely imaginary point $\imath a$, follows the horizontal line to $x_n + \imath a$, and then moves along the vertical line to $x_n + \imath y_n$. In all this $z' \in \mathcal{O}'$, $|x_n| < \delta$ and $\varphi(z',x_n) < y_n < r$. Obviously $\mathcal{I}h$ is holomorphic in $\mathrm{U} \cap \mathcal{M}$ and we have the identities:

$$\mathrm{M}_n \mathcal{I} h = h, \tag{V.7.9}$$

$$\mathcal{I}\mathrm{M}_n h(z) = h(z) - h(z,\imath a). \tag{V.7.10}$$

Let ν be an integer ≥ 0 and $\varepsilon(\nu) = 1$ if $\nu = 0$, $\varepsilon(\nu) = 0$ if $\nu \geq 1$. We shall denote by $\mathcal{H}_\nu(\mathrm{U})$ the space of holomorphic functions in $\mathrm{U} \cap \mathcal{M}$ such that

$$\sup_{\mathrm{U} \cap \mathcal{M}} \{\rho^\nu |\log \rho|^{-\varepsilon(\nu)} |h|\} < +\infty. \qquad \text{(V.7.11)}_\nu$$

The functional at the left in $(\text{V.7.11})_\nu$ is the natural norm on $\mathcal{H}_\nu(\mathrm{U})$; thus normed the latter is a Banach space.

Lemma V.7.1. *Whatever* $\nu \geq 1$, $\mathcal{I}$ *is a bounded linear operator* $\mathcal{H}_\nu(\mathrm{U}) \to \mathcal{H}_{\nu-1}(\mathrm{U})$.

It also induces a bounded linear operator from $\mathcal{H}_0(\mathrm{U})$ *into the Banach space of bounded and continuous functions in* U *that are holomorphic in the interior* $\mathrm{U} \cap \mathcal{M}$.

Proof. By the definition of $\mathcal{I}$ we have, for any $h \in \mathcal{H}_\nu(\mathrm{U})$,

$$\mathcal{I}h(z) = \int_0^{x_n} h(z',\xi + \imath a)d\xi - \imath \int_{y_n}^{a} h(z',x_n + \imath\eta)d\eta. \tag{V.7.12}$$

Given any $z' \in \mathcal{O}'$ and any $x_n \in]-\delta,\delta[$ we cannot possibly have $(z',x_n+\imath a) \in \partial\mathcal{M}$, for it would mean that $a = \varphi(z',x_n)$, which is precluded by (V.7.7) and our choice of a. It follows that the first integral, in the right-hand side of (V.7.12), is bounded in $\mathcal{O}' \times]-\delta,\delta[$. On the other hand, set, for $\eta > 0$,

$$\mathcal{P}(\eta) = \eta^{-\nu}|\log \eta|^{\varepsilon(\nu)}.$$

By virtue of (V.7.1) and $(\text{V.7.11})_\nu$ we have, for some $C > 0$,

$$\left|\int_{y_n}^{a} h(z',x_n+\imath\eta)d\eta\right| \leq C\int_{y_n}^{a} \mathcal{P}(\eta - \varphi(z',x_n))d\eta = C\int_{\rho(z)}^{a-\varphi(z',x_n)} \mathcal{P}(\eta)d\eta.$$

If $\nu \geq 2$, the integral at the far right is equal to

$$(\nu-1)^{-1}\left\{\rho(z)^{-\nu-1} - [a - \varphi(z',x_n]^{-\nu-1}\right\} \leq (\nu-1)^{-1}\rho(z)^{-\nu-1}.$$

If $\nu = 1$, it is equal to

$$|\log \rho(z)| - |\log[a - \varphi(z',x_n]| \leq |\log \rho(z)|,$$

and if $\nu = 0$, it is bounded independently of z, and continuous with respect to z. ■

PROPOSITION V.7.1. *Suppose that* $h \in \mathcal{H}_\nu(\mathrm{U})$ $(\nu \geq 0)$. *There is a bounded and continuous function* h_1 *in* U, *holomorphic in the interior* $\mathrm{U}\cap\mathcal{M}$, *such that, in* $\mathrm{U}\cap\mathcal{M}$,

$$h = \mathrm{M}_n^{\nu+1} h_1. \tag{V.7.13}$$

PROOF. Take $h_1 = \mathcal{J}^{\nu+1} h$ and combine (V.7.9) with Lemma V.7.1. ■

COROLLARY V.7.1. *Let h be as in Proposition* V.7.1. *Then h defines a distribution solution in* U (regarded as a manifold with boundary $\mathrm{U}\cap\partial\mathcal{M}$).

PROOF. Formula (V.7.13) shows that h is a continuous function of ρ with values in the space of distributions with respect to (x,y') (in a neighborhood of the origin). But the same applies to $\partial^\ell h/\partial z_n^\ell$ whatever $\ell = 1,2,\ldots$,which shows that h is smooth transversally to the boundary. It suffices then to apply Corollary V.6.3. ■

DEFINITION V.7.1. *We shall say that a holomorphic function h in* $\mathcal{M}$ grows slowly at the boundary *if, given any compact subset* K *of* $\overline{\mathcal{M}}$, *there is an integer* $\nu \geq 0$ *and a constant* $C > 0$ *such that*

$$|h(z)| < C[\mathrm{dist}(z,\partial\mathcal{M})]^{-\nu}, \ \forall\, z \in \mathrm{K}\cap\mathcal{M}. \tag{V.7.14}$$

The space of holomorphic functions in $\mathcal{M}$ that grow slowly at the boundary will be denoted by $\mathcal{H}_{\mathrm{slow}}(\mathcal{M})$. Clearly, the property that a given holomorphic function h in $\mathcal{M}$ grows slowly at the boundary is local: it suffices to check it in arbitrarily small neighborhoods of points of the boundary.

THEOREM V.7.1. *The restriction mapping* $h \to h|_{\mathcal{M}}$ *is a linear bijection of the space of distribution solutions in* $\overline{\mathcal{M}}$ *onto the space* $\mathcal{H}_{\mathrm{slow}}(\mathcal{M})$ *of holomorphic functions in the interior* $\mathcal{M}$ *that grow slowly at the boundary.*

PROOF. Let h be a distribution solution in $\overline{\mathcal{M}}$. An arbitrary point p of $\partial\mathcal{M}$ has an open neighborhood U in $\overline{\mathcal{M}}$, of the kind considered above, in which h has a representation (V.6.20); $h_1 \in \mathcal{C}^\mu(\mathrm{U})$ is holomorphic in $\mathrm{U}\cap\mathcal{M}$. The first integrals (V.7.3) are nothing else but the restrictions to $\overline{\mathcal{M}}$ of the complex coordinates z_i; and here also the orthonormality relations (I.7.25) are valid, whence $\mathrm{M}_i h_1 = \partial h_1/\partial z_i$ and $h = \Delta_z^\nu h_1$ in $\mathrm{U}\cap\mathcal{M}$. Set $d(z) = \mathrm{dist}[z,\mathrm{U}\cap\partial\mathcal{M}]$; select a compact subset K of $\mathbb{O}$ such that

$$d(z) = \mathrm{dist}(z,\mathbb{C}^n\backslash\mathrm{U}), \ \forall\, z \in \mathrm{K}. \tag{V.7.15}$$

(In other words, the points in the boundary of U that are closest to any point of K lie in $U \cap \partial \mathcal{M}$.) The Cauchy inequalities entail, for every $z \in K \cap \mathcal{M}$,

$$|\Delta_z^\nu h_1(z)| \leq C d(z)^{-2\nu} \operatorname*{Max}_{|\zeta - z| \leq d(z)/2} |h_1(\zeta)|. \tag{V.7.16}$$

But the set $K' = \{ \zeta \in U; \exists\, z \in K, |\zeta - z| \leq d(z)/2 \}$ is a compact subset of U and thus (V.7.16) implies (V.7.14). Since every compact subset of $\overline{\mathcal{M}}$ is a finite union of compact sets that are either contained in the interior or of the kind of the compact set K just considered, this shows that $h \to h|_{\mathcal{M}}$ transforms distribution solutions into holomorphic functions in $\mathcal{M}$ that grow slowly at the boundary. The map is trivially injective, due to the fact that every distribution solution is smooth transversally to the boundary, and therefore could not have its support contained in $\partial\mathcal{M}$. The map is also surjective since, by Corollary V.7.1, every function $h \in \mathcal{H}_{\text{slow}}(\mathcal{M})$ defines a distribution solution in some open neighborhood of every point of $\partial\mathcal{M}$, hence in the whole of $\overline{\mathcal{M}}$. ■

COROLLARY V.7.2. *Every function $h \in \mathcal{H}_{\text{slow}}(\mathcal{M})$ has a well-defined trace on the boundary; the trace is a distribution solution for the induced CR structure on $\partial\mathcal{M}$.*

PROOF. Every function $h \in \mathcal{H}_{\text{slow}}(\mathcal{M})$ is smooth transversally to the boundary. ■

The bijection $h \to h|_{\mathcal{M}}$ becomes a homeomorphism when the space $\mathcal{H}_{\text{slow}}(\mathcal{M})$ is equipped with the natural locally convex topology defined by means of the inequalities (V.7.14), and the space of distribution solutions is equipped with its own natural locally convex topology. These topologies are fairly complicated; we limit ourselves to pointing out that, in the interior $\mathcal{M}$, they coincide with the standard topology of uniform convergence on compact subsets, and that, near the boundary, they reflect, on one hand, the conditions of slow growth, on the other, the smoothness transversally to the boundary.

It is in the sense of these topologies (with $\overline{\mathcal{M}}$ replaced by one of its relatively open subsets, U) that the word "limit" is used in the following statement:

THEOREM V.7.2. *Every point $z_0 \in \overline{\mathcal{M}}$ has an open neighborhood* U *such that every holomorphic function h in $\mathcal{M}$ that grows slowly at the boundary is the limit, in $\mathcal{H}_{\text{slow}}(U)$, of a sequence of holomorphic polynomials.*

Theorem V.7.2 follows from Theorem V.7.1 and a variant of Corollary V.6.1 in which the meaning of limit is strengthened: here the convergence must take into account the smoothness transversally to the boundary (cf. Theorem II.3.2).

THEOREM V.7.3. *Let $\mathcal{X} \subset \overline{\mathcal{M}}$ be a totally real submanifold of $\mathbb{C}^n$, of real* *dimension n. If the trace of a distribution solution h on $\mathcal{X}$ vanishes identically, then $h \equiv 0$ in $\overline{\mathcal{M}}$.*

PROOF. Since the domain $\mathcal{M}$ is connected, it suffices to show that $h \equiv 0$ in some open set $\Omega \subset \mathcal{M}$, $\Omega \neq \emptyset$. If $\mathcal{X} \cap \mathcal{M} \neq \emptyset$, the result follows at once from Corollary II.3.8. Suppose then that $\mathcal{X} \subset \partial\mathcal{M}$. We can find a coordinate chart $(U, x_1, \ldots, x_m, t_1, \ldots, t_n)$ in $\overline{\mathcal{M}}$ of the kind used in section V.6 and such that $U \cap \mathcal{X} \neq \emptyset$ (here $m = n$ and x_i is not necessarily the real part of z_i). But we may choose those coordinates in such a way that $U \cap \mathcal{X}$ is defined, in U, by the equation $t = 0$ (cf. beginning of section II.1). Corollary V.5.2 entails that $h \equiv 0$ in some nonempty open set $\Omega \subset \mathcal{M}$. ∎

Notes

Chapter V is an elaboration of Treves [9]. The notion of boundary complex, when the boundary is noncharacteristic, is the natural generalization of the induced differential complex on the boundary of a complex manifold (with boundary). The latter was first formalized and studied in Kohn and Rossi [1]. The Mayer-Vietoris sequence in locally integrable structures with noncharacteristic boundary (section V.3) generalizes the analogous sequence in Andreotti and Hill [2], Part I. The invariant presentation in section V.6, in particular the use of the spaces $\mathcal{A}'$ introduced in Melrose [1], follows a suggestion of P. Cordaro. The notion of one-sided local integrability (Definition V.1.3) was first introduced, in the context of complex structures, in Hill [1].

VI

Local Integrability and Local Solvability in Elliptic Structures

Chapters VI and VII complement each other. Both are devoted to the study of noteworthy classes of involutive structures from the viewpoint of local solvability and local integrability. Chapter VI discusses the basic classes of structures in which both properties hold; chapter VII presents a number of examples in which at least one of them does not.

Local solvability usually concerns equations $\mathrm{L}u = f$ in which the data, i.e., the right-hand sides f, are L-closed one-forms. The concept can be generalized as local exactness, at any level, in the differential complex associated with the involutive structure under consideration (section I.6). The prototype of a locally integrable structure in which local exactness occurs at *every* level is the structure defined on the space $\mathbb{C}^\nu \times \mathbb{R}^r$ by the vector fields $\partial/\partial\bar{z}_1,\ldots,\partial/\partial\bar{z}_\nu$, $\partial/\partial t_1,\ldots,\partial/\partial t_{n-\nu}$ (the real coordinates in $\mathbb{C}^\nu \times \mathbb{R}^r$ are $x_i = \mathcal{R}e z_i$, $y_j = \mathcal{I}m z_j$, t_k $(1 \leq i, j \leq \nu,\ 1 \leq k \leq r;\ n - \nu \leq r)$). A set of first integrals consists of the functions $z_1,\ldots,z_\nu$, $t_{n-\nu+1},\ldots,t_r$. This structure is the *local* model of all real structures when $\nu = 0$; of all elliptic structures when $n - \nu = r$; of all complex structures when $n = \nu$, $r = 0$ (locally, the real structures can be classified according to the values of n and r; the elliptic structures, according to the values of n and ν; the complex structures, according to the values of n). The structure is CR when $\nu = n$; it is then the local model of all CR structures that are Levi flat (classified according to the values of n and r). Not only does the local cohomology in the associated differential complex vanish in all dimensions but there are relatively simple homotopy formulas from which the vanishing ensues. It suffices to establish those formulas in the elliptic case $r = n - \nu$; the other cases are then taken care of, either by ignoring some of the first integrals or else by treating some of the extra t variables, $t_{n-\nu+1},\ldots,t_r$, as parameters. At first (in sections VI.1 and VI.2) we restrict our attention to complex structures (there are only z's, no t's). We have chosen to construct the homotopy operators following the Bochner-Martinelli-Koppelman-Leray recipe (in convex domains). Sections VI.3 and VI.4 are devoted to estimating the norms of these homotopy operators when they act on Hölder-continuous functions or forms. These estimates, combined with the Newton method for solving nonlinear equations, enable one to prove the all important Newlander-Nirenberg theorem, that every complex structure is locally integrable (sections VI.5 and VI.6). In section VI.7 we return to general elliptic structures and deduce from the Newlander-Nirenberg theorem that they also are locally integrable, which means that locally they can be modeled after $\mathbb{C}^\nu \times \mathbb{R}^{n-\nu}$ as already said. Once this is known it is possible to combine the homotopy formula in $\mathbb{C}^\nu$ with the familiar one in $\mathbb{R}^{n-\nu}$ (based on radial integration) to build up a homotopy formula in $\mathbb{C}^\nu \times \mathbb{R}^{n-\nu}$.

Local integrability can follow from partial ellipticity, provided some algebraic constraint is imposed on the "nonelliptic directions." In the last two sections of the chapter, VI.8 and VI.9, a "transverse group action" is allowed. Roughly speaking, when the involutive structure is locally invariant under a transverse group action, the base manifold $\mathcal{M}$ is locally modeled after the product of a Lie group with a manifold onto which the involutive structure of $\mathcal{M}$ projects as an elliptic structure. The two simplest examples of this situation are the tube structures and the rigid CR structures: in the former, the basic vector fields are of the kind

$$L_j = \partial/\partial t_j + \sum_k \lambda_{jk}(t)\partial/\partial s_k;$$

in the latter,

$$L_j = \partial/\partial \bar{z}_j + \sum_k \lambda_{jk}(x,y)\partial/\partial s_k.$$

In both cases the Lie group is simply the group of translations in s-space $\mathbb{R}^d$. In tube structures the base manifold is locally the product $\mathbb{R}^n \times \mathbb{R}^d$, in rigid structures it is $\mathbb{C}^n \times \mathbb{R}^d$; the structure of $\mathcal{M}$ projects onto the first factor as the De Rham structure in the first case and as the Dolbeault one in the second case.

VI.1. The Bochner-Martinelli Formulas

The complex coordinates in $\mathbb{C}^n$ are $z_j = x_j + \imath y_j$ ($j = 1,\dots,n$) and the variable point is $z = (z_1,\dots,z_n)$. The orientation of $\mathbb{C}^n$ will be the one determined by the $2n$-form

$$dx_1\wedge\cdots\wedge dx_n\wedge dy_1\wedge\cdots\wedge dy_n = (2\imath)^{-n} d\bar{z}_1\wedge\cdots\wedge d\bar{z}_n\wedge dz_1\wedge\cdots\wedge dz_n.$$

This volume form will often be denoted by $d\bar{z}\wedge dz/(2\imath)^n$.

Below Ω denotes a bounded domain in $\mathbb{C}^n$, with piecewise $\mathscr{C}^1$ boundary $\partial\Omega$. We shall deal with forms f of bidegree (p,q) and of class $\mathscr{C}^k$ in the closure $\mathscr{C}\ell\Omega$ of Ω; these are differential forms

$$f = \sum_{|I|=p} \sum_{|J|=q} f_{I,J}(z)\, dz_I \wedge d\bar{z}_J \tag{VI.1.1}$$

with coefficients $f_{I,J}$ that are $\mathscr{C}^k$ functions in Ω whose derivatives of order $< k+1$ extend as continuous functions to $\mathscr{C}\ell\Omega$ $(0 \leq k \leq +\infty)$. We denote by $\mathscr{C}^k(\mathscr{C}\ell\Omega;\Lambda^{p,q})$ the space of such forms; it will be equipped with the topology of uniform convergence, on $\mathscr{C}\ell\Omega$, of all the derivatives of order $< k + 1$ of every coefficient.

The following consequence of the piecewise $\mathscr{C}^1$ nature of $\partial\Omega$ is, for us, of the essence: there exists a sequence of open sets Ω_ν $(\nu = 1,2,\ldots)$ such that $\Omega_\nu \subset\subset \Omega_{\nu+1} \to \Omega$ (i.e., whose union is equal to Ω), each with a $\mathscr{C}^\infty$ boundary $\partial\Omega_\nu$ (it will always be assumed that Ω_ν lies entirely on one side of $\partial\Omega_\nu$) such that the following is true: let $f \in \mathscr{C}^0(\mathscr{C}\ell\Omega;\Lambda^{n,n})$, $g \in \mathscr{C}^0(\mathscr{C}\ell\Omega;\Lambda^{n-q,n+q-1})$ $(q = 0, 1)$. Then

$$\int_{\mathscr{C}\ell\Omega} f = \lim_{\nu\to+\infty} \int_{\mathscr{C}\ell\Omega_\nu} f; \tag{VI.1.2}$$

$$\int_{\partial\Omega} g = \lim_{\nu\to+\infty} \int_{\partial\Omega_\nu} g. \tag{VI.1.3}$$

Actually, we are going to deal, below, with continuous differential forms on $\Omega\times(\mathscr{C}\ell\Omega)$. The variable in the first factor, Ω, will be denoted by z, while the one in the second factor, $\mathscr{C}\ell\Omega$, will be denoted by z'. The notation $f(z,z')$ will stand for such a form (in general not a homogeneous one), which we integrate either over Ω or else over $\partial\Omega$, *with respect to* z'. This has the following meaning. Write

$$f(z,z') = \sum_{p,q=0}^{n} \sum_{|I|=p,|J|=q} g_{I,J}(z,z') \wedge dz_I \wedge d\bar{z}_J$$

where the $g_{I,J}(z,z')$ are differential forms in $\mathscr{C}\ell\Omega$ regarded as a subset of z'-space, whose coefficients are continuous functions of $(z,z') \in \Omega\times(\mathscr{C}\ell\Omega)$. Then, for any r-chain $\mathrm{E} \subset \mathscr{C}\ell\Omega$, we define

$$\int_{\mathrm{E}} f(z,z') = \sum_{p,q=0}^{n} \sum_{|I|=p,|J|=q} \left\{ \int_{\mathrm{E}} g_{I,J}^{(r)}(z,z') \right\} dz_I \wedge d\bar{z}_J \tag{VI.1.4}$$

where $g_{I,J}^{(r)}(z,z')$ is the homogeneous component of degree r of the differential form $g_{I,J}(z,z')$; (VI.1.4) is a continuous differential form in Ω regarded as a subset of z-space. Below E will always be equal either to Ω or to $\partial\Omega$, and the degree r above will be equal to $2n$ or to $2n-1$ respectively.

We shall look at the Cauchy-Riemann (also known as the Dolbeault) complex

$$\bar{\partial} : \mathscr{C}^\infty(\mathscr{C}\ell\Omega;\Lambda^{p,q}) \to \mathscr{C}^\infty(\mathscr{C}\ell\Omega;\Lambda^{p,q+1}) \tag{VI.1.5}$$

($q = 0,1,\dots,n$). All these complexes, for the various values of $p = 0,1,\dots,n$, are isomorphic. It is convenient, from our viewpoint, to focus on the case $p = n$ (cf. Remark I.6.2). A typical form of bidegree (n,q) has an expression

$$f = f_0 \wedge \mathrm{d}z, \tag{VI.1.6}$$

where $\mathrm{d}z = \mathrm{d}z_1 \wedge \cdots \wedge \mathrm{d}z_n$ and f_0 is a $(0,q)$-form

$$f_0 = \sum_{|J|=q} f_J(z)\, \mathrm{d}\bar{z}_J. \tag{VI.1.7}$$

The advantage of dealing with (n,q)-forms is that the action of the $\bar{\partial}$-operator on such forms is equal to that of the exterior derivative d. Thus we are looking at the differential complex

$$\mathrm{d} : \mathscr{C}^\infty(\mathscr{C}\ell\Omega;\Lambda^{n,q}) \to \mathscr{C}^\infty(\mathscr{C}\ell\Omega;\Lambda^{n,q+1}). \tag{VI.1.8}$$

Next we introduce the following distributions in $\mathbb{R}^{2n}$:

$$\mathscr{E}_j = (n-1)!\, \bar{z}_j/|z|^{2n} \ (1 \le j \le n). \tag{VI.1.9}$$

PROPOSITION VI.1.1. *We have, in* $\mathbb{R}^{2n}$,

$$\sum_{j=1}^{n} \bar{\partial}_j \mathscr{E}_j = \pi^n \delta. \tag{VI.1.10}$$

In (VI.1.10) δ is the Dirac distribution in $\mathbb{R}^{2n}$ and $\bar{\partial}_j = \partial/\partial\bar{z}_j$.

PROOF. When $n = 1$ the formula (VI.1.10) is a restatement of the fact that $1/\pi z$ is a fundamental solution of $\partial/\partial\bar{z}$. When $n > 1$ it is an immediate consequence of the fact that $-(n-2)!/[4\pi^n|z|^{2(n-1)}]$ is a fundamental solution of the Laplacian $\Delta = 4(\bar{\partial}_1\partial_1 + \cdots + \bar{\partial}_n\partial_n)$ (see, e.g., Treves [5], sec. 5, 9). ∎

Introduce then the following $(n,n-1)$-form in $\mathbb{C}^n$:

$$\mathscr{E} = (2\iota\pi)^{-n}\sum_{j=1}^{n}(-1)^{j-1}\mathscr{E}_j \mathrm{d}\bar{z}_1 \wedge \cdots \wedge \widehat{\mathrm{d}\bar{z}_j} \wedge \cdots \wedge \mathrm{d}\bar{z}_n \wedge \mathrm{d}z,$$

where the hatted factor, in the exterior product, is to be omitted. Notice that the coefficients of $\mathscr{E}$ belong to $L^1_{loc}(\mathbb{R}^{2n})$.

COROLLARY VI.1.1. *We have, in* $\mathbb{R}^{2n}$,

$$\mathrm{d}\mathscr{E} = \delta\, \mathrm{d}\bar{z} \wedge \mathrm{d}z/(2\iota)^n. \tag{VI.1.11}$$

We shall denote by $\mathscr{E}(z-z')$ the pullback of the form $\mathscr{E}$ to $\mathbb{C}^n \times \mathbb{C}^n$ under the map $(z,z') \to z-z'$. Below we denote by d the exterior derivative in z-space, by d$'$ the one in z'-space (there is no danger here of misinterpreting d′ as the operator in (I.6.19)). If ω is a differential form in $\mathbb{R}^{2n}$ we denote by $\omega(z-z')$ its pullback to $\mathbb{R}^{2n}\times\mathbb{R}^{2n}$. Then, the pullback of its exterior derivative, dω, is equal to

$$(\mathrm{d} + \mathrm{d}')\omega(z-z').$$

COROLLARY VI.1.2. *We have, in* $\mathbb{R}^{2n}\times\mathbb{R}^{2n}$,

$$(\mathrm{d} + \mathrm{d}')\mathscr{E}(z-z') = \delta(z-z')\, \mathrm{d}(\bar{z}-\bar{z}')\wedge \mathrm{d}(z-z')/(2\iota)^n. \qquad \text{(VI.1.12)}$$

Define

$$\mathcal{K}_\Omega f(z) = \int_\Omega f(z')\wedge\mathscr{E}(z-z'), \quad f \in \mathscr{C}^0(\mathscr{C}\ell\,\Omega;\Lambda^{n,q}), \qquad \text{(VI.1.13)}$$

$$\mathcal{K}_{\partial\Omega} f(z) = \int_{\partial\Omega} f(z')\wedge\mathscr{E}(z-z'), \quad f \in \mathscr{C}^0(\partial\Omega;\Lambda^{n,q}). \qquad \text{(VI.1.14)}$$

These integrals (with respect to z') define forms in Ω. They are akin to standard convolution and the kernel $\mathscr{E}(z-z')$ is a $\mathscr{C}^\infty$ form off the diagonal. As a matter of fact (VI.1.13) is equal to the convolution of the form $\mathscr{E} \in L^1_{loc}(\mathbb{R}^{2n};\Lambda^{n,n-1})$ with the form equal to f in Ω and to 0 in $\mathbb{C}^n\backslash\Omega$. It is a simple matter of using the definition (VI.1.4) and of counting degrees, to check that $\mathcal{K}_\Omega f$ is an $(n,q-1)$-form and $\mathcal{K}_{\partial\Omega} f$ is an (n,q)-form. It follows easily that

$$\begin{aligned} \mathcal{K}_\Omega &: \mathscr{C}^0(\mathscr{C}\ell\,\Omega;\Lambda^{n,q}) \to \mathscr{C}^0(\Omega;\Lambda^{n,q-1}), \\ \mathcal{K}_\Omega &: \mathscr{C}^\infty(\mathscr{C}\ell\,\Omega;\Lambda^{n,q}) \to \mathscr{C}^\infty(\Omega;\Lambda^{n,q-1}), \end{aligned} \qquad \text{(VI.1.15)}$$

$$\mathcal{K}_{\partial\Omega} : \mathscr{C}^0(\mathscr{C}\ell\,\Omega;\Lambda^{n,q}) \to \mathscr{C}^\infty(\Omega;\Lambda^{n,q}), \qquad \text{(VI.1.16)}$$

are bounded linear operators. When $q = 0$, $\mathcal{K}_\Omega f \equiv 0$ (simply because the integral over Ω of an $(n,n-1)$-form must vanish). We shall refer to (VI.1.15) and (VI.1.16) as the *Bochner-Martinelli operators*.

THEOREM VI.1.1. *Given any* $f \in \mathscr{C}^1(\mathscr{C}\ell\,\Omega;\Lambda^{n,q})$ *we have, in* Ω,

$$f = \mathrm{d}[(-1)^{n-q}\mathcal{K}_\Omega f] + (-1)^{n-q-1}\mathcal{K}_\Omega \mathrm{d}f + (-1)^{n-q}\mathcal{K}_{\partial\Omega} f. \qquad \text{(VI.1.17)}$$

PROOF. By differentiating under the integral sign we get

$$\mathrm{d}\mathcal{K}_\Omega f(z) = (-1)^{n+q}\int_\Omega f(z')\wedge(\mathrm{d}+\mathrm{d}')\mathscr{E}(z-z') -$$

$$(-1)^{n+q}\int_\Omega f(z')\wedge \mathrm{d}'[\mathscr{E}(z-z')],$$

that is,

$$(-1)^{n+q}\mathrm{d}\mathcal{K}_\Omega f(z) = \int_\Omega \delta(z-z')\, f(z')\wedge \mathrm{d}(\bar{z}-\bar{z}')\wedge \mathrm{d}z/(2\imath)^n -$$

$$\int_\Omega f(z')\wedge \mathrm{d}'[\mathscr{E}(z-z')]. \tag{VI.1.18}$$

We make use of the expression (VI.1.6)–(VI.1.7):

$$\int_\Omega \delta(z-z')\, f_0(z')\wedge \mathrm{d}z'\wedge \mathrm{d}(\bar{z}-\bar{z}') =$$

$$\sum_{|J|=q}\int_\Omega \delta(z-z')\, f_J(z')\mathrm{d}\bar{z}'_J\wedge \mathrm{d}z'\wedge \mathrm{d}(\bar{z}-\bar{z}') =$$

$$\sum_{|J|=q}\int_\Omega \delta(z-z')\, f_J(z')\mathrm{d}\bar{z}'\wedge \mathrm{d}z'\wedge \mathrm{d}\bar{z}_J = (2\imath)^n f_0(z).$$

Thus, for $z \in \Omega$,

$$\int_\Omega \delta(z-z')\, f(z')\wedge \mathrm{d}(\bar{z}-\bar{z}')\wedge \mathrm{d}z/(2\imath)^n = f(z). \tag{VI.1.19}$$

On the other hand, by Stokes' theorem,

$$\int_{\partial\Omega} f(z')\wedge\mathscr{E}(z-z') = \int_\Omega \mathrm{d}f(z')\wedge\mathscr{E}(z-z') + (-1)^{n+q}\int_\Omega f(z')\wedge \mathrm{d}'[\mathscr{E}(z-z')],$$

i.e.,

$$\int_\Omega f(z')\wedge \mathrm{d}'[\mathscr{E}(z-z')] = (-1)^{n+q}\mathcal{K}_{\partial\Omega} f + (-1)^{n+q-1}\mathcal{K}_\Omega \mathrm{d}f.$$

Putting this, together with (VI.1.19), into (VI.1.18) yields (VI.1.17). ∎

COROLLARY VI.1.4. *Suppose* $f \in \mathscr{C}^1(\mathscr{C}\ell\,\Omega;\Lambda^{n,n})$. *We have, in* Ω,

$$f = \mathrm{d}\mathcal{K}_\Omega f. \tag{VI.1.20}$$

PROOF. Indeed, $\mathrm{d}f \equiv 0$ and $f(z')\wedge\mathscr{E}(z-z')$ has degree $2n$ with respect to z', hence its integral over $\partial\Omega$ must vanish. ∎

COROLLARY VI.1.5. *Suppose* $f \in \mathscr{C}^1(\mathscr{C}\ell\,\Omega;\Lambda^{n,0})$. *Then we have, in* Ω,

$$(-1)^n f = \mathcal{K}_\Omega \mathrm{d}f + \mathcal{K}_{\partial\Omega} f. \tag{VI.1.21}$$

PROOF. Indeed, $\mathcal{K}_\Omega f \equiv 0$. ∎

COROLLARY VI.1.6. *Let* $f \in \mathscr{C}^1(\mathscr{C}\ell\,\Omega;\Lambda^{n,0})$ *be such that* $\bar{\partial} f = 0$ *in* Ω. *Then we have, in* Ω:

$$f = (-1)^n \mathcal{K}_{\partial\Omega} f. \tag{VI.1.22}$$

If $f \in \mathcal{C}^\infty(\mathcal{Cl}\,\Omega;\Lambda^{n,0})$ and $\bar{\partial} f \equiv 0$, it means that $f = h(z)\mathrm{d}z$ with $h \in \mathcal{C}^\infty(\mathcal{Cl}\,\Omega)$ and h holomorphic in Ω. When $n = 1$ (VI.1.22) is essentially the Cauchy formula; (VI.1.21) is the inhomogeneous Cauchy formula. In general, these formulas are called the Bochner-Martinelli formulas.

REMARK VI.1.1. Corollary VI.1.5 can be applied to prove that all the classical solutions of a system of inhomogeneous Cauchy-Riemann equations with $\mathcal{C}^\infty$ right-hand sides are $\mathcal{C}^\infty$ functions. Indeed, let f be a $\mathcal{C}^1$ function in an open subset $\mathcal{O}$ of $\mathbb{C}^n$ such that $\bar{\partial} f \in \mathcal{C}^\infty(\mathcal{O};\Lambda^{0,1})$. Apply (VI.1.21) to $f\,\mathrm{d}z$ and to any open set $\Omega \subset\subset \mathcal{O}$ whose boundary is smooth, and note that the right-hand side is a $\mathcal{C}^\infty$ $(n,0)$-form in Ω (cf. (VI.1.15), (VI.1.16)). ■

VI.2. Homotopy Formulas for $\bar{\partial}$ in Convex and Bounded Domains

Throughout this section we shall reason under the hypothesis that

Ω is an open, bounded, and convex neighborhood of the origin in $\mathbb{C}^n$. (VI.2.1)

Now we shall also assume that the boundary $\partial\Omega$ of Ω is of class $\mathcal{C}^2$. This means that there is a real-valued $\mathcal{C}^2$ function ρ in some open neighborhood $\tilde{\Omega}$ of $\mathcal{Cl}\,\Omega$ in $\mathbb{C}^n$ such that $\Omega = \{ z \in \tilde{\Omega};\ \rho(z) < 0 \}$ and $\partial\rho \neq 0$ at every point of $\partial\Omega$. Hypothesis (VI.2.1) entails, then,

$$\mathcal{Re}[\partial\rho(z')\cdot(z-z')] < 0,\ \forall\ z \in \Omega,\ z' \in \partial\Omega. \tag{VI.2.2}$$

In (VI.2.2), and in the sequel as well, we use the notation

$$\partial\rho = (\partial_1\rho,\ldots,\partial_n\rho),\ \partial_j\rho = \partial\rho/\partial z_j.$$

We introduce an additional variable, t, which varies in the closed interval $[0,1]$. For the sake of simplicity we denote provisionally by D the exterior derivative in (z,z',t)-space $\mathbb{C}^n \times \mathbb{C}^n \times [0,1]$. Thus $\mathrm{D} = \mathrm{d} + \mathrm{d}' + \mathrm{d}_t$ where d and d′ have the meaning given to them in section VI.1. However, we continue to write $\mathrm{d}z$, $\mathrm{d}z'$, $\mathrm{d}t$.

We shall make use of the following function of $z \in \Omega$, $z' \in \partial\Omega$, and $t \in [0,1]$, of class $\mathcal{C}^1$, and valued in $\mathbb{C}^n$,

$$\bar{\zeta} = t\partial\rho(z')/[\partial\rho(z')\cdot(z-z')] + (1-t)(\bar{z}-\bar{z}')/|z-z'|^2.$$

Below, and until otherwise specified, the whole analysis will take place in the set $\Omega \times \partial\Omega \times [0,1]$, and we shall not constantly recall this fact. Notice that we have, in that set,

$$(z-z')\cdot\bar{\zeta} \equiv 1. \tag{VI.2.3}$$

LEMMA VI.2.1. *We have, in* $\Omega \times \partial\Omega \times [0,1]$,

$$\sum_{j=1}^{n}(z_j-z_j')\mathrm{D}\bar{\zeta}_j \wedge \mathrm{d}(z-z') \equiv 0. \tag{VI.2.4}$$

PROOF. From (VI.2.3) we derive

$$\sum_{j=1}^{n}(z_j-z_j')\mathrm{D}\bar{\zeta}_j \ + \ \bar{\zeta}_j\mathrm{d}(z_j-z_j') \equiv 0.$$

To get (VI.2.4) carry out the exterior product of the left-hand side of this identity with $\mathrm{d}(z-z') = \mathrm{d}(z_1-z_1')\wedge\cdots\wedge\mathrm{d}(z_n-z_n')$. ■

COROLLARY VI.2.1. *We have, in* $\Omega \times \partial\Omega \times [0,1]$,

$$\mathrm{D}\bar{\zeta}\wedge\mathrm{d}(z-z') \equiv 0. \tag{VI.2.5}$$

We are using the notation $\mathrm{D}\bar{\zeta} = \mathrm{D}\bar{\zeta}_1\wedge\cdots\wedge\mathrm{D}\bar{\zeta}_n$.

We define the following differential form of degree $2n-1$:

$$\mathscr{F} = (2\iota\pi)^{-n}(n-1)!\sum_{j=1}^{n}(-1)^{j-1}\bar{\zeta}_j\mathrm{D}\bar{\zeta}_1\wedge\cdots\wedge\widehat{\mathrm{D}\bar{\zeta}_j}\wedge\cdots\wedge\mathrm{D}\bar{\zeta}_n\wedge\mathrm{d}(z-z'). \tag{VI.2.6}$$

We derive from (VI.2.5):

COROLLARY VI.2.2. *We have, in* $\Omega \times \partial\Omega \times [0,1]$,

$$\mathrm{D}\mathscr{F} \equiv 0. \tag{VI.2.7}$$

LEMMA VI.2.2. *Let* $\psi, u_1,\ldots,u_n$ *be* $\mathscr{C}^\infty$ *functions in some* $\mathscr{C}^\infty$ *manifold; we have*

$$\sum_{j=1}^{n}(-1)^{j-1}\psi u_j\mathrm{d}(\psi u_1)\wedge\cdots\wedge\widehat{\mathrm{d}(\psi u_j)}\wedge\cdots\wedge\mathrm{d}(\psi u_n) =$$

$$\psi^n\sum_{j=1}^{n}(-1)^{j-1}u_j\mathrm{d}u_1\wedge\cdots\wedge\widehat{\mathrm{d}u_j}\wedge\cdots\wedge\mathrm{d}u_n. \tag{VI.2.8}$$

PROOF. The difference between the left-hand and right-hand sides in (VI.2.8) is equal to

$$\psi^{n-1}\mathrm{d}\psi\wedge\left\{\sum_{j=1}^{n}\sum_{k<j}(-1)^{j+k}u_ju_k\mathrm{d}u_1\wedge\cdots\wedge\widehat{\mathrm{d}u_k}\wedge\cdots\wedge\widehat{\mathrm{d}u_j}\wedge\cdots\wedge\mathrm{d}u_n \ -\right.$$

$$\sum_{j=1}^{n}\sum_{k>j}(-1)^{j+k}u_j u_k \mathrm{d}u_1\wedge\cdots\wedge\widehat{\mathrm{d}u_j}\wedge\cdots\wedge\widehat{\mathrm{d}u_k}\wedge\cdots\wedge\mathrm{d}u_n\Bigg\},$$

which vanishes identically as one sees by exchanging j and k in the second double sum. ■

COROLLARY VI.2.3. *The pullback of the form $\mathcal{F}$ to the subspace $t = 0$ is equal to the form $\mathscr{E}$* (defined in section VI.1).

PROOF. Apply Lemma VI.2.2 to $\psi = |z-z'|^{-2}$, $u_j = \bar{z}_j-\bar{z}'_j$. ■

COROLLARY VI.2.4. *The pullback of the form $\mathcal{F}$ to the subspace $t = 1$ is equal to the differential form in $\Omega\times\partial\Omega$,*

$$\mathscr{E}^{\#} = (n-1)![\partial\rho(z')\cdot(z-z')]^{-n}\cdot$$

$$\sum_{j=1}^{n}(-1)^{j-1}\partial_j\rho(z')\, \mathrm{d}[\partial_1\rho(z')]\wedge\cdots\wedge\widehat{\mathrm{d}[\partial_j\rho(z')]}\wedge\cdots\wedge\mathrm{d}[\partial_n\rho(z')]\wedge\mathrm{d}(z-z').$$

PROOF. Apply Lemma VI.2.2 to $\psi = [\partial\rho(z')\cdot(z-z')]^{-1}$, $u_j = \partial_j\rho(z')$. ■

COROLLARY VI.2.5. *Let q be an integer, $1\le q\le n$. Then, whatever the form $f\in\mathscr{C}^0(\partial\Omega;\Lambda^{n,q})$, we have*

$$\int_{\partial\Omega\times\{1\}} f(z')\wedge\mathcal{F}(z,z',t) \equiv 0. \tag{VI.2.9}$$

In (VI.2.9) we are using partial integration with respect to (z',t).

PROOF. It suffices to observe that if $q\ge 1$ the degree of $f\wedge\mathscr{E}^{\#}$ with respect to (x',y') is $\ge 2n$ and the integral in (VI.2.9) is carried out over a compact $(2n-1)$-dimensional submanifold, $\partial\Omega$. ■

Given any differential form $f\in\mathscr{C}^0(\partial\Omega;\Lambda^{n,q})$, with $1\le q\le n$, we define the following differential form in Ω,

$$\mathscr{B}_{\partial\Omega}f(z) = \int_{\partial\Omega\times[0,1]} f(z')\wedge\mathcal{F}(z,z',t). \tag{VI.2.10}$$

It is readily checked that $\mathscr{B}_{\partial\Omega}f\in\mathscr{C}^0(\Omega;\Lambda^{n,q-1})$; $\mathscr{B}_{\partial\Omega}f\equiv 0$ if $q = n$.

THEOREM VI.2.1. *Let the open set $\Omega\subset\mathbb{C}^n$ be bounded, convex, and have a $\mathscr{C}^2$ boundary. Then, for all $f\in\mathscr{C}^1(\mathcal{Cl}\,\Omega;\Lambda^{n,q})$ $(1\le q\le n)$, we have in Ω,*

$$\mathscr{K}_{\partial\Omega}f = \mathrm{d}(\mathscr{B}_{\partial\Omega}f) - \mathscr{B}_{\partial\Omega}\mathrm{d}f. \tag{VI.2.11}$$

PROOF. When $q = n$ both sides in (VI.2.11) vanish identically. Assume $1 \leq q \leq n-1$. By differentiating under the integral sign and by Corollary VI.2.2 we get

$$(-1)^{n+q}\mathrm{d}\mathcal{B}_{\partial\Omega}f(z) = \int_{\partial\Omega\times[0,1]} f(z')\wedge \mathrm{D}\mathcal{F}(z,z',t) -$$

$$\int_{\partial\Omega\times[0,1]} f(z')\wedge(\mathrm{d}'+\mathrm{d}_t)\mathcal{F}(z,z',t) =$$

$$(-1)^{n+q-1}\int_{\partial\Omega\times[0,1]} (\mathrm{d}'+\mathrm{d}_t)[f(z')\wedge\mathcal{F}(z,z',t)] +$$

$$(-1)^{n+q}\int_{\partial\Omega\times[0,1]} \mathrm{d}f(z')\wedge\mathcal{F}(z,z',t),$$

whence, by Stokes' theorem:

$$\mathrm{d}\mathcal{B}_{\partial\Omega}f - \mathcal{B}_{\partial\Omega}\mathrm{d}f = \int_{\partial\Omega\times\{0\}} f(z')\wedge\mathcal{F}(z,z',t) -$$

$$\int_{\partial\Omega\times\{1\}} f(z')\wedge\mathcal{F}(z,z',t).$$

Combining this identity with Corollaries VI.2.3 and VI.2.5 yields (VI.2.11). ■

If we combine Theorems VI.1.1 and VI.2.1, we obtain:

THEOREM VI.2.2. *Assume that the open set $\Omega \subset \mathbb{C}^n$ is bounded, convex, and has a $\mathcal{C}^2$ boundary; assume also that $1 \leq q \leq n$. Given any differential form $f \in \mathcal{C}^1(\mathcal{C}\ell\Omega;\Lambda^{n,q})$, we have then*

$$f = \mathrm{d}[(-1)^{n-q}(\mathcal{K}_\Omega + \mathcal{B}_{\partial\Omega})f] + (-1)^{n-q-1}(\mathcal{K}_\Omega + \mathcal{B}_{\partial\Omega})\mathrm{d}f. \quad \text{(VI.2.12)}$$

Some authors refer to (VI.2.12) as the *Koppelman-Leray formula*.

COROLLARY VI.2.6. *Let Ω and q be as in Theorem* VI.2.2. *Given any form $f \in \mathcal{C}^1(\mathcal{C}\ell\Omega;\Lambda^{n,q})$ such that* $\mathrm{d}f = 0$ *in* Ω, *we have*

$$f = \mathrm{d}[(-1)^{n-q}(\mathcal{K}_\Omega + \mathcal{B}_{\partial\Omega})f]. \quad \text{(VI.2.13)}$$

As a preparation for the next section we shall give a slightly more explicit expression for $\mathcal{F}$ (see (VI.2.6)) or, more precisely, for the terms in $\mathcal{F}$ that contain one factor dt; these are the only terms that make a contribution to the integral at the right, in (VI.2.10). Thus we write

$$\mathcal{F} = (\mathcal{F}_0 + dt\wedge\mathcal{F}_1)\wedge d(z-z')$$

where $\mathcal{F}_0$ contains no factor dt.

Since $\mathcal{B}_{\partial\Omega}f \equiv 0$ when $n = 1$, we assume throughout $n \geq 2$.

PROPOSITION VI.2.1. *We have*

$$(2\imath\pi)^n\mathcal{F}_1/(n-1)! =$$

$$\sum_{j=1}^{n}(-1)^{j-1}(\bar{z}_j-\bar{z}'_j)\Bigg\{\sum_{k<j}(-1)^{k-1}\partial_k\rho(z')\omega_1\wedge\cdots\wedge\widehat{\omega_k}\wedge\cdots\wedge\widehat{\omega_j}\wedge\cdots\wedge\omega_n -$$

$$\sum_{k>j}(-1)^{k-1}\partial_k\rho(z')\omega_1\wedge\cdots\wedge\widehat{\omega_j}\wedge\cdots\wedge\widehat{\omega_k}\wedge\cdots\wedge\omega_n\Bigg\}/[|z-z'|^2\partial\rho(z')\cdot(z-z')], \tag{VI.2.14}$$

where

$$\omega_j = (1-t)d(\bar{z}_j-\bar{z}'_j)/|z-z'|^2 + td[\partial_j\rho(z')]/[\partial\rho(z')\cdot(z-z')].$$

PROOF. Let $\varpi_1,\ldots,\varpi_n$ be smooth one-forms and J a multi-index of lenght q: $J = \{j_1,\ldots,j_q\}$ with $1 \leq j_1 <\cdots< j_q \leq n$. We shall make use of the following notation:

$$\varpi_{(J)} = (-1)^{j_1+\cdots+j_q-q}\varpi_{J^*} \tag{VI.2.15}$$

where J^* is the ordered multi-index of integers j such that $1 \leq j \leq n$ and $j \notin J$. If J is a set of distinct integers j, $1 \leq j \leq n$, not necessarily ordered, we call $[J]$ the multi-index obtained by ordering J and denote by $\epsilon(J)$ the signum of the permutation that orders J, equal to $+1$ if that permutation is even and to -1 if it is odd. We extend the notation (VI.2.15) by setting

$$\varpi_{(J)} = \epsilon(J)\varpi_{([J])}. \tag{VI.2.16}$$

Finally we set $\varpi_{(J)} = 0$ if two elements of J are equal.

Exchange of the summation indices j and k shows that, for any set of functions $u_1,\ldots,u_r$ and one-forms $\varpi_1,\ldots,\varpi_r$, we have

$$\sum_{j,k=1}^{r} u_ju_k\varpi_{(jk)} \equiv 0. \tag{VI.2.17}$$

In the present proof we shall also use the notation $a = \bar{z}-\bar{z}'$, $b = \partial\rho$, $A = \bar{a}\cdot a$ and $B = \bar{a}\cdot b$. Thus

$$\bar{\zeta} = a/A - t(a/A-b/B),\ \omega_j = (1-t)da_j/A + tdb_j/B.$$

In this notation we may write:

$$-(2\imath\pi)^n \mathcal{F}_1/(n-1)! = \sum_{j,k=1}^{n} \bar{\zeta}_j(a_k/A - b_k/B)\mathrm{d}\bar{\zeta}_{(jk)}.$$

From the definition of $\bar{\zeta}_j$ and from (VI.2.17) we derive

$$\mathcal{F}_1 = (n-1)!(2\imath\pi)^{-n}(AB)^{-1}\sum_{j,k=1}^{n} a_j b_k \mathrm{d}\bar{\zeta}_{(jk)}. \qquad \text{(VI.2.18)}$$

Let us also define

$$\mu_j = (1-t)a_j\mathrm{d}(1/A) + tb_j\mathrm{d}(1/B).$$

Thus $\mathrm{d}\bar{\zeta}_j = \mu_j + \omega_j$ $(1 \leq j \leq n)$. We observe that

$$\mathrm{d}\bar{\zeta}_{(jk)} = \omega_{(jk)} + \sum_{\ell=1}^{n} \mu_\ell \wedge \omega_{(jk\ell)} + \sum_{\ell,\ell'=1}^{n} \mu_\ell \wedge \mu_{\ell'} \wedge \omega_{(jk\ell\ell')}. \qquad \text{(VI.2.19)}$$

It is understood that the last sum is not present when $n < 4$, nor is the one before the last present when $n = 2$. In this last case $\mathrm{d}\bar{\zeta}_{(jk)} = \omega_{(jk)}$ is equal to 1 if $j < k$ and to -1 if $j > k$. We have

$$\sum_{\ell=1}^{n} \mu_\ell \wedge \omega_{(jk\ell)} = (1-t)\mathrm{d}(1/A) \wedge \sum_{\ell=1}^{n} a_\ell \omega_{(jk\ell)} + t\mathrm{d}(1/B) \sum_{\ell=1}^{n} b_\ell \omega_{(jk\ell)}.$$

We derive from (VI.2.17):

$$\sum_{j,\ell=1}^{n} a_j a_\ell \omega_{(jk\ell)} \equiv 0 \textit{ for each } k;$$

$$\sum_{k,\ell=1}^{n} b_k b_\ell \omega_{(jk\ell)} \equiv 0 \textit{ for each } j.$$

From this it follows straightaway that

$$\sum_{j,k=1}^{n} a_j b_k \sum_{\ell=1}^{n} \mu_\ell \wedge \omega_{(jk\ell)} \equiv 0. \qquad \text{(VI.2.20)}$$

On the other hand,

$$\sum_{\ell,\ell'=1}^{n} \mu_\ell \wedge \mu_{\ell'} \wedge \omega_{(jk\ell\ell')} =$$

$$2t(1-t)\mathrm{d}(1/A) \wedge \mathrm{d}(1/B) \wedge \sum_{\ell,\ell'=1}^{n} a_\ell b_{\ell'} \omega_{(jk\ell\ell')}.$$

Now we avail ourselves of the following facts:

$$\sum_{j,\ell=1}^{n} a_j a_\ell \omega_{(jk\ell\ell')} \equiv 0 \textit{ for each pair } k,\ell';$$

$$\sum_{k,\ell'=1}^{n} b_k b_{\ell'} \omega_{(jk\ell\ell')} \equiv 0 \textit{ for each pair } j,\ell.$$

We conclude that

$$\sum_{j,k=1}^{n} a_j b_k \sum_{\ell,\ell'=1}^{n} a_\ell b_{\ell'} \omega_{(jk\ell\ell')} \equiv 0.$$

If we combine this with (VI.2.19) and (VI.2.20) we get

$$\sum_{j,k=1}^{n} a_j b_k \mathrm{d}\bar{\zeta}_{(jk)} = \sum_{j,k=1}^{n} a_j b_k \omega_{(jk)}$$

which is equivalent to (VI.2.14). ■

VI.3. Estimating the Sup Norms of the Homotopy Operators

Consider two elements of the fibre of $\Lambda^{p,q}$ at some point of $\mathbb{C}^n$,

$$\varphi = \sum_{|I|=p,|J|=q} \varphi_{I,J} \mathrm{d}z_I \wedge \mathrm{d}\bar{z}_J, \ \psi = \sum_{|I|=p,|J|=q} \psi_{I,J} \mathrm{d}z_I \wedge \mathrm{d}\bar{z}_J.$$

Their inner product is defined to be the complex number

$$(\varphi,\psi) = \sum_{|I|=p,|J|=q} \varphi_{I,J} \bar{\psi}_{I,J}$$

and the norm $|\varphi|$ of φ is equal to $(\varphi,\varphi)^{1/2}$. This allows us to speak of the norm $|f(z)|$ of a form $f \in \mathscr{C}^0(\mathscr{Cl}\Omega;\Lambda^{p,q})$ at a point $z \in \mathscr{Cl}\Omega$.

We shall begin by estimating the sup norm, in Ω, of $\mathscr{K}_\Omega f$ and $\mathscr{B}_{\partial\Omega} f$ in terms of that of $f \in \mathscr{C}^0(\mathscr{Cl}\Omega;\Lambda^{n,q})$. In what follows the integer q will always be ≥ 1. Notice that, in the definition (VI.1.13), the only property of Ω that is needed is that Ω be open and bounded.

Proposition VI.3.1. *Let Ω be a bounded open subset of $\mathbb{C}^n$. There is a constant $C_n > 0$ that depends solely on the dimension n and is such that, for all $f \in \mathscr{C}^0(\mathscr{Cl}\Omega;\Lambda^{n,q})$,*

$$\sup_{\Omega} |\mathscr{K}_\Omega f| \leq C_n(\text{diam } \Omega) \operatorname*{Max}_{\mathscr{Cl}\Omega} |f|. \tag{VI.3.1}$$

Proof. Let $\mathscr{E}_j$ be the function defined in (VI.1.9). By using polar coordinates in $\mathbb{R}^{2n}$ one immediately gets

$$\sup_{z \in \Omega} \int_\Omega |\mathscr{E}_j(z-z')| \, \mathrm{d}\bar{z}' \wedge \mathrm{d}z'/(2\imath)^n \leq \textit{const.}(\text{diam } \Omega), \tag{VI.3.2}$$

whence (VI.3.1). ■

In order to estimate the norm of $\mathcal{B}_{\partial\Omega}f$ we shall strengthen our hypotheses on the boundary of Ω. We continue to assume that Ω is a bounded domain in $\mathbb{C}^n$. But now we assume that $\partial\Omega$ is of class $\mathcal{C}^\infty$ and that it is *strictly convex*. More explicitly, we shall reason under the following hypothesis:

There exists a function $\rho \in \mathcal{C}^\infty(\tilde{\Omega})$ *such that* $\Omega = \{ z \in \tilde{\Omega};\ \rho(z) < 0 \}$ *with* $d\rho \neq 0$ *in* $\partial\Omega$, *and that, for some* $c_0 > 0$ *and all* $z \in \Omega$, $z' \in \partial\Omega$, (VI.3.3)

$$\rho(z) - 2\mathcal{R}_e[\partial\rho(z')\cdot(z-z')] \geq c_0|z-z'|^2. \qquad \text{(VI.3.4)}$$

One could easily weaken the smoothness requirement on $\partial\Omega$, but this would be unimportant for the applications in which we shall be interested.

PROPOSITION VI.3.2. *Under hypothesis* (VI.3.3) *we have*

$$\sup_{\Omega} |\mathcal{B}_{\partial\Omega}f| \leq C \operatorname*{Max}_{\mathcal{C}\ell\Omega} |f|,\ \forall f \in \mathcal{C}^0(\mathcal{C}\ell\Omega;\Lambda^{n,q}). \qquad \text{(VI.3.5)}$$

PROOF. We may assume $n \geq 2$, otherwise $\mathcal{B}_{\partial\Omega}f \equiv 0$. Let us write, provisionally,

$$f = dz\wedge \sum_{|J|=q} f_J d\bar{z}_J,$$

$$\mathcal{B}_{\partial\Omega}f(z) = \sum_{|I|=q-1}\left\{\sum_{|J|=q}\int_{\partial\Omega\times[0,1]} B_{I,J}(z,z',t)f_J(z')dS(z')dt\right\} dz\wedge d\bar{z}_I,$$

where dS is the area element on $\partial\Omega$ defined by the measure $d\bar{z}\wedge dz/(2\iota)^n$ on $\mathbb{R}^{2n}$. The inequality to be proved, (VI.3.5), is a direct consequence of the existence of a constant $C > 0$, independent of $z \in \Omega$, such that

$$\int_{\partial\Omega\times[0,1]} |B_{I,J}(z,z',t)|dS(z')dt \leq C,$$
$$\forall\, I, J,\ |I| = q-1,\ |J| = q. \qquad \text{(VI.3.6)}$$

In order to prove (VI.3.6) it suffices to prove a similar estimate when the integration with respect to z' is carried out over some neighborhood $\mathcal{N}$, in $\partial\Omega$, of an arbitrary point of $\partial\Omega$. It is convenient to take this point to be the origin; after a change of the complex coordinates (cf. Example I.9.3) we may assume that, in some open neighborhood $\mathcal{U}$ of 0 in $\mathbb{C}^n$,

$$\rho(z) = -y_n + |z|^2 + \mathcal{R}(z) \qquad \text{(VI.3.7)}$$

with $|\mathcal{R}(z)| \leq const.|z|^3$. Since Ω is defined by $\rho < 0$, we see that $y_n > |z|^2/2$ in $\Omega\cap\mathcal{U}$ (possibly after some contracting of $\mathcal{U}$ about 0).

We also derive, from (VI.3.7),

$$\partial\rho(z')\cdot(z-z') = \imath[(x_n-x_n')+\imath(y_n-y_n')]/2 + \bar{z}'\cdot(z-z') +$$

$$\partial\mathcal{R}(z')\cdot(z-z') = \tfrac{\imath}{2}[(x_n-x_n')+2\mathcal{I}m(\bar{z}'\cdot z)] - \tfrac{1}{2}[|\rho(z)|+$$

$$|z-z'|^2+\mathcal{R}(z)-\mathcal{R}(z')] + \partial\mathcal{R}(z')\cdot(z-z'),$$

whence

$$\mathcal{I}m\partial\rho(z')\cdot(z-z') = \tfrac{1}{2}(x_n-x_n') + \mathcal{I}m(\bar{z}'\cdot z) + \mathcal{I}m[\partial\mathcal{R}(z')\cdot(z-z')]. \tag{VI.3.8}$$

In the integral with respect to z' we may use the coordinates $\tau_j = x_j-x_j'$, $\tau_{n+j-1} = y_j-y_j'$ $(1 \leq j \leq n-1)$, $\sigma = \mathcal{I}m\partial\rho(z')\cdot(z-z')$ in the neighborhood $\mathcal{N}$. Clearly, if $\mathcal{N}$ and $c_1 > 0$ are sufficiently small,

$$c_1 \leq |z-z'|/(|\rho(z)|+|\sigma|+|\tau|) \leq c_1^{-1}. \tag{VI.3.9}$$

We derive from (VI.3.4) and (VI.3.8):

$$c_2 \leq |\partial\rho(z')\cdot(z-z')|/(|\rho(z)|+|\sigma|+|\tau|^2) \leq c_2^{-1} \tag{VI.3.10}$$

for some $c_2 > 0$ and all $z \in \mathcal{U}$, $z' \in \mathcal{N}$.

At this stage we return to the expression (VI.2.14). Combining it with (VI.3.9) and (VI.3.10) yields an estimate of the following kind:

$$\int_0^1 |B_{I,J}(z,z',t)|dt \leq$$

$$const. \sum_{p=0}^{n-2} (|\rho|+|\sigma|+|\tau|)^{-2p-1}(|\rho|+|\sigma|+|\tau|^2)^{p+1-n} \tag{VI.3.11}$$

($\rho = \rho(z)$; recall that $n \geq 2$). The claim follows then from the fact that, for each p, $0 \leq p \leq n-2$, the integral

$$\mathrm{I}_p = \int_0^1\int_0^1 (s+t)^{-2p-1}(s+t^2)^{p+1-n}t^{2n-3}dsdt \leq \int_0^1\int_0^{1/t^2} (1+u)^{p+1-n}dudt$$

is finite. Indeed, if $p \leq n-2$, $\mathrm{I}_p \leq \mathrm{I}_{n-2} \leq \int_0^1 log(1+1/t^2)dt < +\infty$. ■

REMARK VI.3.1. The inequalities (VI.3.11) can be used to show that the *one-half Hölder norms* of $\mathcal{K}_\Omega f$ and $\mathcal{B}_{\partial\Omega} f$ do not exceed *const.* $\operatorname{Max}_{\mathcal{C}\ell\Omega} |f|$. On all this see Henkin and Leiterer [1], sec. 2.2. ■

Next we strengthen the regularity requirements on the boundary $\partial\Omega$ and on the form f: we hypothesize that both are $\mathcal{C}^\infty$. We are going to estimate the sup norms of the derivatives of $\mathcal{K}_\Omega f$ and of $\mathcal{B}_{\partial\Omega} f$ in terms of those of f. Let $\alpha = (\alpha_1,\ldots,\alpha_n)$, $\beta = (\beta_1,\ldots,\beta_n)$ belong to $\mathbb{Z}_+^n$ and the differential operator

$$\partial^\alpha\bar{\partial}^\beta = (\partial/\partial z_1)^{\alpha_1}\cdots(\partial/\partial z_n)^{\alpha_n}(\partial/\partial\bar{z}_1)^{\beta_1}\cdots(\partial/\partial\bar{z}_n)^{\beta_n} \tag{VI.3.12}$$

act coefficientwise on the form $f \in \mathscr{C}^\infty(\mathscr{C}\ell\Omega;\Lambda^{n,q})$: if f is given by (VI.1.1),

$$\partial^\alpha\bar{\partial}^\beta f = \sum_{|I|=p,|J|=q} \partial^\alpha\bar{\partial}^\beta f_{I,J}dz_I\wedge d\bar{z}_J. \tag{VI.3.13}$$

To any subset Γ of $\mathscr{C}\ell\Omega$ and any integer $r \geq 0$ we associate the following seminorm:

$$\|f\|_{\Gamma,r} = \sup_\Gamma \sum_{|\alpha|+|\beta|\leq r} |\partial^\alpha\bar{\partial}^\beta f(z)|. \tag{VI.3.14}$$

PROPOSITION VI.3.3. *Let Ω be a bounded open subset of $\mathbb{C}^n$ whose boundary $\partial\Omega$ is smooth. To every integer $r \geq 0$ there is a constant $C_{\Omega,r} > 0$ such that*

$$\|\mathscr{K}_\Omega f\|_{\Omega,r} \leq C_{\Omega,r}\|f\|_{\Omega,r},\ \forall\, f \in \mathscr{C}^\infty(\mathscr{C}\ell\Omega;\Lambda^{n,q}). \tag{VI.3.15}$$

PROOF. According to (VI.1.13) the coefficients of $\mathscr{K}_\Omega f$ are linear combinations (with constant coefficients) of integrals of the kind

$$\mathscr{I}(z) = \int_\Omega \mathscr{E}_j(z-z')\, f_J(z')\, dV(z')$$

(where $dV(z') = d\bar{z}'\wedge dz'/(2\imath)^n$). Denote by $\delta_1,\dots,\delta_r$ any one of the partial derivatives $\partial/\partial z_k$ or $\partial/\partial\bar{z}_\ell$ ($1 \leq k, \ell \leq n$); and by $\delta_1',\dots,\delta_r'$ the same operators but acting in the variables z'. We have:

$$\delta_1\cdots\delta_r\mathscr{I}(z) = (-1)^r\int_\Omega [\delta_1'\cdots\delta_r'\mathscr{E}_j(z-z')]\, f_J(z')\, dV(z') =$$

$$(-1)^{r-1}\int_\Omega [\delta_1'\cdots\delta_{r-1}'\mathscr{E}_j(z-z')]\ \delta_r' f_J(z')\, dV(z') +$$

$$\int_{\partial\Omega} [\delta_1'\cdots\delta_{r-1}'\mathscr{E}_j(z-z')]\, f_J(z')\, d\mu_r(z'),$$

where $d\mu_r$ is the appropriate $(2n-1)$-form on $\partial\Omega$: up to sign $d\mu_r$ is equal to the pullback to $\partial\Omega$ of the contraction of the vector field δ_r with the $2n$-form $d\bar{z}\wedge dz/(2\imath)^n$. Further integrations by parts in Ω yield:

$$\delta_1\cdots\delta_r\mathscr{I}(z) = \int_\Omega \mathscr{E}_j(z-z')\ (\delta_1\cdots\delta_r f_J)(z')\, dV(z') +$$

$$\sum_{s=1}^r \int_{\partial\Omega} [\delta_0'\cdots\delta_{s-1}'\mathscr{E}_j(z-z')]\ (\delta_{s+1}\cdots\delta_{r+1} f_J)(z')\, d\mu_s(z'),$$

with the understanding that δ_s is the identity operator if $s = 0$ or if $s = r+1$. We also integrate by parts in $\partial\Omega$, noting that $\mathscr{E}_j = \partial_j E$ with $E = -(n-2)!/|z|^{2n-2}$ if $n \geq 2$ and $E = 2\log|z|$ if $n = 1$. We get

$$\delta_1 \cdots \delta_r \mathcal{I}(z) = \int_\Omega \mathscr{E}_j(z-z')\,(\delta_1 \cdots \delta_r f_J)(z') +$$

$$\int_{\partial\Omega} E(z-z')\, \mathrm{P}_\delta f_J(z')\, dS(z'), \qquad \text{(VI.3.16)}$$

where P_δ is a linear partial differential operator with smooth coefficients in $\mathbb{R}^{2n}$, of order r (dS is the volume form on $\partial\Omega$). We use once again (VI.3.2) and the obvious fact that

$$\int_{\partial\Omega} |E(z-z')| dS(z') \le C_\Omega,$$

to obtain (VI.3.15). ■

PROPOSITION VI.3.4. *Assume $n \ge 2$ and let Ω be as in Proposition* VI.3.2. *To each integer $r \ge 0$ there is a constant $C_{\Omega,r} > 0$ such that*

$$\|\mathscr{B}_{\partial\Omega} f\|_{\Omega,r} \le C_{\Omega,r}\, \|f\|_{\Omega,r}, \ \forall\, f \in \mathscr{C}^\infty(\mathscr{C}\ell\,\Omega;\Lambda^{n,q}). \qquad \text{(VI.3.17)}$$

PROOF. Once again we make use of the expression (VI.2.14) of $\mathscr{F}_1$. The coefficients of $\mathscr{B}_{\partial\Omega} f$ are linear combinations of integrals

$$\mathcal{J}(z) = \int_{\partial\Omega} \mathscr{F}_{p,j}(z', z-z')\, f_J(z')\, g(z')\, dS(z')$$

where $g \in \mathscr{C}^\infty(\partial\Omega)$,

$$\mathscr{F}_{p,j}(z',z) = [\partial\rho(z')\cdot z]^{p+1-n} |z|^{-2(p+1)} \bar{z}_j$$

and $0 \le p \le n-2$. In the same notation as in the proof of Proposition VI.3.3 we may write

$$\delta_1 \cdots \delta_r \mathcal{J}(z) =$$

$$(-1)^r \int_{\partial\Omega} [\delta_1' \cdots \delta_r' \mathscr{F}_{p,j}(w, z-z')]|_{w=z'}\, f_J(z')\, g(z')\, dS(z').$$

Integration by parts shows that $\delta_1 \cdots \delta_r \mathcal{J}(z)$ is the sum of integrals

$$\int_{\partial\Omega} \delta_1'' \cdots \delta_\nu'' \mathscr{F}_{p,j}(w, z-z')|_{w=z'}\, \mathrm{P}_{\delta,\nu} f_J(z')\, dS(z') \qquad \text{(VI.3.18)}$$

where $\delta_1'', \ldots, \delta_\nu''$ are partial derivatives of the kind $\partial/\partial w_k$ or $\partial/\partial \bar{w}_\ell$ ($1 \le k, \ell \le n$; $0 \le \nu \le r$) and $\mathrm{P}_{\delta,\nu}$ is a linear partial differential operator with $\mathscr{C}^\infty$ coefficients in $\partial\Omega$, of order $\le r - \nu$. Let δ'' be any one of the operators δ_i'' ($1 \le i \le \nu$). We have

$$\delta'' \mathscr{F}_{p,j}(w,z) = -(n-1-p)[\partial\rho(w)\cdot z]^{p-n}[\delta''\partial\rho(w)\cdot z]|z|^{-2(p+1)}\bar{z}_j.$$

But recalling that $\partial\rho$ does not vanish at any point of $\partial\Omega$, we may write

$$-(n-1-p)[\partial\rho(w)\cdot z]^{p-n} = |\partial\rho(w)|^{-2}\sum_{k=1}^{n}\bar{\partial}_k\rho(w)\partial/\partial z_k\{[\partial\rho(w)\cdot z]^{p+1-n}\}.$$

As a consequence, more integration by parts will show that (VI.3.18) is a sum of integrals

$$\int_{\partial\Omega}[\partial\rho(z')\cdot(z-z')]^{p+1-n}\, Q_{\delta,\nu'}\left\{H_{\nu'}(z-z')P_{\delta,\nu}f_J(z')]\right\} dS(z'),$$

where $Q_{\delta,\nu'}$ is a differential operator with $\mathscr{C}^\infty$ coefficients in $\partial\Omega$, of order $\nu' \leq \nu$ (acting in the variable z') and

$$H_{\nu'}(z) = z_{i_1}\cdots z_{i_{\nu'}}z_j|z|^{-2(p+1)}.$$

We conclude that

$$\left|\int_{\partial\Omega}[\partial\rho(z')\cdot(z-z')]^{p+1-n}\, Q_{\delta,\nu'}\left\{H_{\nu'}(z-z')P_{\delta,\nu}f_J(z')]\right\} dS(z')\right| \leq$$

$$const.\left\{\int_{\partial\Omega}|\partial\rho(z')\cdot(z-z')|^{p+1-n}|z-z'|^{-2p-1}\, dS(z')\right\}\|f\|_{\partial\Omega,r},$$

and from there the proof is completed like that of Proposition VI.3.2. ∎

The conjunction of Propositions VI.3.3 and VI.3.4 implies

THEOREM VI.3.1. *Let Ω be a bounded and open subset of $\mathbb{C}^n$ that satisfies Condition* (VI.3.3). *Assume that* $1 \leq q \leq n$. *Then $\mathscr{K}_\Omega + \mathscr{B}_{\partial\Omega}$ defines a bounded linear operator $\mathscr{C}^\infty(\mathscr{C}\ell\Omega;\Lambda^{n,q}) \to \mathscr{C}^\infty(\mathscr{C}\ell\Omega;\Lambda^{n,q-1})$.*

If we combine Corollary VI.2.6 with Theorem VI.3.1 we obtain:

COROLLARY VI.3.1. *Let Ω and q be as in Theorem* VI.3.1. *Given any differential form $f \in \mathscr{C}^\infty(\mathscr{C}\ell\Omega;\Lambda^{n,q})$ such that* $\mathrm{d}f = 0$, *there is $u \in \mathscr{C}^\infty(\mathscr{C}\ell\Omega;\Lambda^{n,q-1})$ satisfying* $\mathrm{d}u = f$ *in Ω.*

VI.4. Hölder Estimates for the Homotopy Operators in Concentric Balls

We are now going to extend the estimates of section 3 to Hölder norms in *open balls* (centered at one and the same point, which we take to be the origin). We shall estimate the Hölder norm of order $r+1+\kappa$ ($r \in \mathbb{Z}_+$, $0 < \kappa < 1$) of $\mathscr{K}_\Omega f$ and of $\mathscr{B}_{\partial\Omega} f$ in an open ball of radius $R' > 0$ in terms of the Hölder norm of order $r+\kappa$ of f in an open ball of radius $R > R'$. We are going to show that the constants in the estimates can be taken equal to

$C_{r,\kappa}R^2/(R-R')$. These estimates will be used in the proof of the Newlander-Nirenberg theorem (see section VI.5).

Throughout the present section Ω_R shall denote the open ball of radius $R > 0$ in $\mathbb{C}^n$, centered at the origin. In the formulas of section VI.2 we take $\Omega = \Omega_R$ and we put $\rho(z) = |z|^2 - R^2$; thus $\partial\rho = \bar{z}$ and

$$\bar{\zeta} = t\bar{z}'/\bar{z}'\cdot(z-z') + (1-t)(\bar{z}-\bar{z}')/|z-z'|^2.$$

Of course condition (VI.3.3) is satisfied: if $z \in \Omega$, $z' \in \partial\Omega$ (i.e., $|z'| = R$),

$$\rho(z) - 2\mathcal{Re}[\partial\rho(z')\cdot(z-z')] = |z-z'|^2 \text{ (cf. (VI.3.4))}.$$

Let R and κ be positive numbers, with $\kappa < 1$; we shall make use of the *Hölder quotient*

$$\mathcal{H}_R^{r+\kappa}(f) = \sup_{z,z'\in\Omega_R}\left\{\sum_{|\alpha+\beta|=r} |\partial^\alpha\bar{\partial}^\beta f(z) - \partial^\alpha\bar{\partial}^\beta f(z')|/|z-z'|^\kappa\right\}$$

and of the *Hölder norm*

$$\mathrm{N}_R^{r+\kappa}(f) = \sum_{k=0}^{r} R^k\|f\|_{\Omega_R,k} + R^{r+\kappa}\mathcal{H}_R^{r+\kappa}(f).$$

Here f is either a complex-valued function or a (p,q)-form of the kind (VI.1.1), in Ω_R. We have defined the norm of a (p,q)-form at a point z in the beginning of section VI.3; $\partial^\alpha\bar{\partial}^\beta f$ is defined in (VI.3.13), $\|f\|_{\Omega_R,k}$ in (VI.3.14).

A preliminary remark concerning the effect of a dilation will be useful. Consider the pullback under the map $z \to Rz$ from Ω_1 to Ω_R,

$$\pi_R^* f = R^{p+q} \sum_{|I|=p,|J|=q} f_{I,J}(Rz)dz_I\wedge d\bar{z}_J,$$

of a form

$$f = \sum_{|I|=p,|J|=q} f_{I,J}(z)dz_I\wedge d\bar{z}_J \in \mathcal{C}^\infty(\mathcal{Cl}\,\Omega_R;\Lambda^{p,q}).$$

For any $r \in \mathbb{Z}_+$,

$$\sup_{\Omega_1}\sum_{|\alpha+\beta|=r} |\partial^\alpha\bar{\partial}^\beta(\pi_R^* f)| = R^{p+q+r}\sup_{\Omega_R}\sum_{|\alpha+\beta|=r}|\partial^\alpha\bar{\partial}^\beta f|.$$

Likewise,

$$\mathcal{H}_1^{r+\kappa}(\pi_R^* f) = R^{p+q+r+\kappa}\mathcal{H}_R^{r+\kappa}(f).$$

We conclude from all this that

$$\mathrm{N}_1^{r+\kappa}(\pi_R^* f) = R^{p+q}\mathrm{N}_R^{r+\kappa}(f). \tag{VI.4.1}$$

It is also an easy exercise to check that the expressions of the kernels $\mathcal{E}$ and $\mathcal{F}$

(see (VI.1.9), (VI.2.6)) and the definitions of the linear operators $\mathcal{K}_\Omega$, $\mathcal{B}_{\partial\Omega}$ (see (VI.1.13), (VI.2.10)) entail, for all $f \in \mathcal{C}^\infty(\mathcal{Cl}\Omega;\Lambda^{n,q})$,

$$\pi_R^*(\mathcal{K}_{\Omega_R} f) = \mathcal{K}_{\Omega_1}\pi_R^* f,\ \pi_R^*(\mathcal{B}_{\partial\Omega_R} f) = \mathcal{B}_{\partial\Omega_1}\pi_R^* f. \tag{VI.4.2}$$

In what follows q will always be an integer, $0 \le q \le n$.

THEOREM VI.4.1. *Let q be an integer, $0 \le q \le n$. To any number κ, $0 < \kappa < 1$, and any integer $r \ge 0$ there is a constant $C > 0$ such that, given any two numbers R, R', $0 < R' < R$,*

$$\mathrm{N}_{R'}^{r+1+\kappa}(\mathcal{K}_{\Omega_R} f) + \mathrm{N}_{R'}^{r+1+\kappa}(\mathcal{B}_{\partial\Omega_R} f) \le CR^2(R-R')^{-1}\mathrm{N}_R^{r+\kappa}(f),$$
$$\forall f \in \mathcal{C}^\infty(\mathcal{Cl}\Omega_R;\Lambda^{n,q}). \tag{VI.4.3}$$

PROOF. It suffices to prove (VI.4.3) when $R = 1$. For then, given any λ, $0 < \lambda < 1$, and any $f \in \mathcal{C}^\infty(\mathcal{Cl}\Omega_R;\Lambda^{n,q})$, one gets

$$\mathrm{N}_\lambda^{r+1+\kappa}(\mathcal{K}_{\Omega_1}\pi_R^* f) + \mathrm{N}_\lambda^{r+1+\kappa}(\mathcal{B}_{\partial\Omega_1}\pi_R^* f) \le C(1-\lambda)^{-1}\mathrm{N}_1^{r+\kappa}(\pi_R^* f).$$

Taking (VI.4.1) and (VI.4.2) into account and putting $R' = \lambda R$ yields precisely (VI.4.3).

In the remainder of the proof we assume $R = 1$ and we write Ω rather than Ω_1. We derive at once from (VI.3.15) and (VI.3.17):

$$\sum_{k=0}^{r}\lambda^k[\|\mathcal{K}_\Omega f\|_{\Omega_\lambda,k} + \|\mathcal{B}_{\partial\Omega} f\|_{\Omega_\lambda,k}] \le C_r\sum_{k=0}^{r}\|f\|_{\Omega,k}. \tag{VI.4.4}$$

We are left with estimating $\|u\|_{\Omega_\lambda,r+1}$ and $\mathcal{H}_\lambda^{r+1+\kappa}(u)$ when $u = \mathcal{K}_\Omega f$ or $u = \mathcal{B}_{\partial\Omega} f$. We shall say that a quantity $\mathcal{Q}(f,z,w)$ is *well bounded* if there is a constant $C > 0$ that depends solely on the dimension n and on the smoothness degrees r, κ, such that

$$\mathcal{Q}(f,z,w) \le C(1-\lambda)^{-1}\mathrm{N}_1^{r+\kappa}(f). \tag{VI.4.5}$$

I. *Estimate of* $\|\mathcal{K}_\Omega f\|_{\Omega_\lambda,r+1}$

We go back to the proofs of Propositions VI.3.3 and VI.3.4, and first of all, to (VI.3.16), where we substitute $r+1$ for r. We shall systematically write $\mathrm{D}^r f_J = \delta_1\cdots\delta_r f_J$. Integration by parts yields

$$\delta_1\cdots\delta_{r+1}\mathcal{I}(z) = \int_\Omega \mathcal{E}_j(z-z')\delta'_{r+1}[\mathrm{D}^r f_J(z') - \mathrm{D}^r f_J(z)]\mathrm{d}V(z') +$$
$$\int_{\partial\Omega}\mathcal{E}_j(z-z')\mathfrak{Q} f_j(z')\mathrm{dS}(z'),$$

where δ'_{r+1} is the operator δ_{r+1} acting in the variables z' and $\mathfrak{Q}$ is a linear partial differential operator of order r, with $\mathscr{C}^\infty$ coefficients, on $\partial\Omega$. In the first integral at the right we integrate by parts once, thus getting

$$\delta_1\cdots\delta_{r+1}\mathscr{I}(z) = \int_\Omega \delta_{r+1}\mathscr{E}_j(z-z')[D^r f_J(z') - D^r f_J(z)]dV(z') +$$

$$\int_{\partial\Omega}\mathscr{E}_j(z-z')\left\{\mathscr{Q}f_J(z') + \varphi(z')[D^r f_J(z') - D^r f_J(z)]\right\}dS(z'), \quad \text{(VI.4.6)}$$

with $\varphi \in \mathscr{C}^\infty(\partial\Omega)$. We get at once

$$|\delta_1\cdots\delta_{r+1}\mathscr{I}(z)| \le \int_\Omega |z-z'|^{\kappa-2n}[|D^r f_J(z') - D^r f_J(z)|/|z-z'|^\kappa]dV(z') +$$

$$\int_{\partial\Omega}|z-z'|^{1-2n}\left|\mathscr{Q}f_J(z') + \varphi(z')[D^r f_J(z') - D^r f_J(z)]\right|dS(z').$$

The standard estimates for convolution yields

$$\operatorname*{Max}_{z\in\Omega}\int_\Omega |z-z'|^{\kappa-2n}[|D^r f_J(z') - D^r f_J(z)|/|z-z'|^\kappa]dV(z') \le C_\kappa \mathscr{H}_1^{r+\kappa}(f).$$

On the other hand, if we restrict the variation of z to Ω_λ we get

$$\int_{\partial\Omega}|z-z'|^{1-2n}\left|\mathscr{Q}f_J(z') + \varphi(z')[D^r f_J(z') - D^r f_J(z)]\right|dS(z') \le$$

$$(1-\lambda)^{-1}\int_{\partial\Omega}|z-z'|^{2-2n}\left|\mathscr{Q}f_J(z') + \varphi(z')[D^r f_J(z') - D^r f_J(z)]\right|dS(z') \le$$

$$C(1-\lambda)^{-1}\|f\|_{\Omega,r}.$$

We conclude that $\|\mathscr{K}_\Omega f\|_{\Omega_\lambda,r+1}$ is well bounded.

II. *Estimate of* $\|\mathscr{B}_{\partial\Omega}f\|_{\Omega_\lambda,r+1}$

We recall that, in the study of $\mathscr{B}_{\partial\Omega}$, we assume $n \ge 2$; and that $0 \le p \le n-2$. In the present situation, as already noted, $\partial\rho(z') = \bar{z}'$. Now and later the reasoning will be based on the following inequality, valid if $|z'| = 1$ and $|z| \le \lambda < 1$,

$$|\bar{z}'\cdot(z-z')| \ge \tfrac{1}{4}[|\mathscr{Im}(\bar{z}'\cdot z)| + |z'-z|^2 + 1 - \lambda], \quad \text{(VI.4.7)}$$

which is an immediate consequence of the identity

$$-2\bar{z}'\cdot(z_* - z') = |z'-z_*|^2 + (1-|z_*|^2) - 2\imath\mathscr{Im}(\bar{z}'\cdot z_*).$$

We go back to the proof of Proposition VI.3.4, specifically to the study of the integral

$$\int_{\partial\Omega} [\bar{z}'\cdot(z-z')]^{p+1-n}\, Q_{\delta,\nu'}[H_{\nu'}(z-z')P_{\delta,\nu}f_J(z')]dS(z')$$

where now, however, the order of the differential operator $P_{\delta,\nu}$ is $\leq r+1-\nu$. Further integration by parts shows that

$$\left|\int_{\partial\Omega} [\bar{z}'\cdot(z-z')]^{p+1-n}\, Q_{\delta,\nu'}[H_{\nu'}(z-z')P_{\delta,\nu}f_J(z')]dS(z')\right| \leq$$

$$C_0\left\{\int_{\partial\Omega} |\bar{z}'\cdot(z-z')|^{p+1-n}|z-z'|^{-2p-1}[(1-|z|)^{-1} + |z-z'|^{-1}]dS(z')\right\} \|f\|_{\partial\Omega,r}$$

$$\leq 2C_0(1-\lambda)^{-1}\left\{\int_{\partial\Omega} |\bar{z}'\cdot(z-z')|^{p+1-n}|z-z'|^{-2p-1}dS(z')\right\} \|f\|_{\partial\Omega,r}$$

whatever $z \in \Omega_\lambda$. At the end of the proof of Proposition VI.3.2 it is shown that

$$\sup_{z\in\Omega}\int_{\partial\Omega} |\bar{z}'\cdot(z-z')|^{p+1-n}|z-z'|^{-2p-1}dS(z') < +\infty,$$

which allows us to conclude that $\|\mathcal{B}_{\partial\Omega}f\|_{\Omega_\lambda,r+1} \leq C(1-\lambda)^{-1}\|f\|_{\Omega,r}$.

III. *Estimate of* $\mathcal{H}_\lambda^{r+1+\kappa}(\mathcal{K}_\Omega f)$

Throughout the argument z and w will remain in Ω_λ. Call $\mathcal{I}_0(z)$ the first integral at the right in (VI.4.6). Let χ be a $\mathcal{C}^\infty$ function in the real line, $0 \leq \chi \leq 1$, $\chi(\tau) = 0$ for $\tau < 4$, $\chi(\tau) = 1$ for $\tau > 9$. We decompose $\mathcal{I}_0(z) - \mathcal{I}_0(w)$ into a sum $\mathcal{I}_1(z,w) + \mathcal{I}_2(z,w)$ with

$$\mathcal{I}_1(z,w) = \int_\Omega \Big\{[\delta_{r+1}\mathcal{E}_j(z-z')][D^rf_J(z')-D^rf_J(z)] -$$

$$[\delta_{r+1}\mathcal{E}_j(w-z')][D^rf_J(z')-D^rf_J(w)]\Big\}[1-\chi(|z-z'|^2/|z-w|^2)]\, dV(z'),$$

$$\mathcal{I}_2(z,w) = \int_\Omega \Big\{[\delta_{r+1}\mathcal{E}_j(z-z')][D^rf_J(z')-D^rf_J(z)] -$$

$$[\delta_{r+1}\mathcal{E}_j(w-z')][D^rf_J(z')-D^rf_J(w)]\Big\}\chi(|z-z'|^2/|z-w|^2)\, dV(z'),$$

Set $\Delta_1 = \{ z' \in \Omega; |z-z'| \leq 3|z-w| \}$. We have

$$|\mathcal{I}_1(z,w)| \leq C_0\mathcal{H}_1^{r+\kappa}(f)\int_{\Delta_1}[|z-z'|^{\kappa-2n} + |w-z'|^{\kappa-2n}]\, dV(z')$$

$$\leq C|z-w|^{\kappa}\mathscr{H}_1^{r+\kappa}(f),$$

which shows that $|\mathscr{I}_1(z,w)|/|z-w|^{\kappa}$ is well bounded.

In order to estimate $\mathscr{I}_2$ we select a cutoff function $g \in \mathscr{C}_c^{\infty}(\Omega)$, $0 \leq g \leq 1$, such that

$$g(z) = 1 \text{ if } |z| < (1+\lambda)/2, \tag{VI.4.8}$$

$$|dg| \leq const.(1-\lambda)^{-1}. \tag{VI.4.9}$$

We decompose $\mathscr{I}_2(z,w)$ into a sum $\mathscr{A}(z,w) + \mathscr{B}(z,w) + \mathscr{C}(z,w)$, with

$$\mathscr{A}(z,w) =$$

$$[D^r f_J(w) - D^r f_J(z)]\int_{\Omega}\mathscr{E}_j(z-z')\delta'_{r+1}[g(z')\chi(|z-z'|^2/|z-w|^2)]dV(z'),$$

$$\mathscr{B}(z,w) =$$

$$[D^r f_J(w) - D^r f_J(z)]\int_{\Omega}[1-g(z')]\chi(|z-z'|^2/|z-w|^2)]\ \delta_{r+1}\mathscr{E}_j(z-z')\ dV(z'),$$

$$\mathscr{C}(z,w) = \int_{\Omega}[\delta_{r+1}\mathscr{E}_j(z-z') - \delta_{r+1}\mathscr{E}_j(w-z')][D^r f_J(z') -$$

$$D^r f_J(w)]\ \chi(|z-z'|^2/|z-w|^2)\ dV(z').$$

We note that

$$|\delta'_{r+1}[g(z')\chi(|z-z'|^2/|z-w|^2)]| \leq$$

$$C_0[(1-\lambda)^{-1} + |\chi'(|z-z'|^2/|z-w|^2)|\ |z-z'|/|z-w|^2],$$

and also that $\chi'(|z-z'|^2/|z-w|^2) \neq 0$ entails $2 < |z-z'|/|z-w| < 3$, whence

$$|z-w|^{\kappa}\left|\int_{\Omega}\mathscr{E}_j(z-z')\ \delta'_{r+1}[g(z')\chi(|z-z'|^2/|z-w|^2)]dV(z')\right| \leq$$

$$C_1\left\{(1-\lambda)^{-1}|z-w|^{\kappa}\int_{\Omega}|z-z'|^{1-2n}dV(z') + \int_{\Delta_1}|z-z'|^{\kappa-2n}dV(z')\right\} \leq$$

$$C_2(1-\lambda)^{-1}|z-w|^{\kappa},$$

which shows that $|\mathscr{A}(z,w)|/|z-w|^{\kappa}$ is well bounded.

Since

$$|z'-z| \geq \tfrac{1}{2}(1-\lambda) \text{ when } |z| < \lambda,\ z' \in \operatorname{supp}(1-g).$$

we get at once

$$|\mathscr{B}(z,w)| \leq C_0|D^r f_J(z) - D^r f_J(w)|\int_{\Omega}[1-g(z')]|z-z'|^{-2n}dV(z') \leq$$

$$C_1(1-\lambda)^{-1}|D^r f_J(z) - D^r f_J(w)|,$$

which shows that $|\mathcal{B}(z,w)|/|z-w|^\kappa$ is well bounded.

Finally we estimate $\mathcal{C}(z,w)$. We note that there is a point z_* on the segment joining z to w such that

$$|\delta_{r+1}\mathcal{E}_j(z-z') - \delta_{r+1}\mathcal{E}_j(w-z')| \leq const.|z-w|/|z'-z_*|^{-2n-1}.$$

But if $\chi(|z-z'|^2/|z-w|^2) \neq 0$ then $|z-z'| > 2|z-w|$ and

$$|z-w| \leq |z'-w| \leq \tfrac{3}{2}|z'-z_*|,$$

whence

$$\Bigg|\int_\Omega [\delta_{r+1}\mathcal{E}_j(z-z') - \delta_{r+1}\mathcal{E}_j(w-z')]\chi(z')[D^r f_J(z') -$$

$$D^r f_J(w)]\ \chi(|z-z'|^2/|z-w|^2)\ dV(z')\Bigg| \leq$$

$$C_1\mathcal{H}_1^{r+\kappa}(f)|z-w| \int_{|z-w|<|z'-w|} |z'-w|^{\kappa-2n-1}dV(z') \leq C_2\mathcal{H}_1^{r+\kappa}(f)|z-w|^\kappa.$$

We conclude that $|\mathcal{C}(z,w)|/|z-w|^\kappa$ is well bounded, and consequently that the same is true of $|\mathcal{I}_2(z,w)|/|z-w|^\kappa$ and of $|\mathcal{I}_0(z) - \mathcal{I}_0(w)|/|z-w|^\kappa$.

Let $\mathcal{J}_\#(z)$ denote the second integral in the right-hand side of (VI.4.6). For any pair of points z, w in Ω_λ we decompose $\mathcal{J}_\#(z) - \mathcal{J}_\#(w)$ into the sum

$$[D^r f_J(z) - D^r f_J(w)]\int_{\partial\Omega} \mathcal{E}_j(w-z')\varphi(z')dS(z') + \mathcal{R}(z,w),$$

where

$$\mathcal{R}(z,w) = \int_{\partial\Omega} [\mathcal{E}_j(z-z') - \mathcal{E}_j(w-z')]\Big\{\mathcal{Q}f_J(z') +$$

$$\varphi(z')[D^r f_J(z') - D^r f_J(z)]\Big\}dS(z').$$

We have

$$\left|\int_{\partial\Omega} \mathcal{E}_j(w-z')\varphi(z')dS(z')\right| \leq C_0(1-\lambda)^{-1}\int_{\partial\Omega}|w-z'|^{2-2n}dS(z') \leq C(1-\lambda)^{-1}.$$

In order to estimate $\mathcal{R}(z,w)$ we partition the unit sphere $\partial\Omega$ into three parts: Γ_w, in which $|z'-z| \geq 2|z'-w|$; Γ_z, where $|z'-w| \geq 2|z'-z|$; Γ_0, in which

$$1/2 \leq |z-z'|/|w-z'| \leq 2.$$

In Γ_w we simply write

$$|\mathscr{E}_j(z-z')-\mathscr{E}_j(w-z')| \leq 2|z'-w|^{1-2n} \leq 2(1-\lambda)^{-1}|z'-w|^{2-2n}.$$

We note that $|z-w| \geq |z-z'|-|z'-w| \geq |z'-w|$, whence

$$|\mathscr{E}_j(z-z')-\mathscr{E}_j(w-z')| \leq C(1-\lambda)^{-1}|z-w|^{\kappa}|z'-w|^{2-2n-\kappa},$$

for all $z' \in \Gamma_w$. Likewise, when $z' \in \Gamma_z$, we get

$$|\mathscr{E}_j(z-z')-\mathscr{E}_j(w-z')| \leq C(1-\lambda)^{-1}|z-w|^{\kappa}|z'-z|^{2-2n-\kappa}. \quad \text{(VI.4.10)}$$

Suppose now z' belongs to Γ_0. There is a point z_* in the segment joining z to w such that

$$|\mathscr{E}_j(z'-z)-\mathscr{E}_j(z'-w)| \leq const.|z-w|/|z'-z_*|^{2n}.$$

But $|z-z'| \leq 2|z_*-z'|$ and $|z-w| \leq |z-z'|+|z'-w| \leq 3|z-z'|$, which shows that (VI.4.10) is also valid when $z' \in \Gamma_0$.

Combining all these inequalities shows that $|\mathscr{R}(z,w)|/|z-w|^{\kappa}$ is well bounded, and thus that this is also true of $|\mathscr{J}_{\#}(z)-\mathscr{J}_{\#}(w)|/|z-w|^{\kappa}$, and as a consequence, of $\mathscr{H}_1^{r+1+\kappa}(\mathscr{K}_\Omega f)$.

IV. *Estimate of* $\mathscr{H}_1^{r+1+\kappa}(\mathscr{B}_{\partial\Omega} f)$

The points z and w continue to vary in Ω_λ. We must estimate the quantities (cf. proof of Proposition VI.3.4)

$$\int_{\partial\Omega} [\bar{z}'\cdot(z-z')]^{p+1-n} Q_{\delta,\nu'}\left\{H_{\nu'}(z-z')P_{\delta,\nu}f_J(z')]\right\} dS(z') -$$

$$\int_{\partial\Omega} [\bar{z}'\cdot(w-z')]^{p+1-n} Q_{\delta,\nu'}\left\{H_{\nu'}(w-z')P_{\delta,\nu}f_J(z')]\right\} dS(z'),$$

where $P_{\delta,\nu}$ and $Q_{\delta,\nu'}$ are differential operators of orders $\leq r+1-\nu$ and $\leq \nu'$ ($\nu' \leq \nu$) respectively ($Q_{\delta,\nu'}$ acts in the variables z'). Recall that

$$H_{\nu'}(z) = z_{i_1}\cdots z_{i_{\nu'}} z_j|z|^{-2(p+1)}$$

and also that $n \geq 2$ and $0 \leq p \leq n-2$. Carrying out the differentiations, we see that we must deal with a sum of integrals, some of which are of the kind

$$\int_{\partial\Omega}\Big\{[\bar{z}'\cdot(z-z')]^{p+1-n}\, D^{\gamma}H_{\nu'}(z-z') -$$

$$[\bar{z}'\cdot(w-z')]^{p+1-n}\, D^{\gamma}H_{\nu'}(w-z')\Big\}\, \psi(z')D^{s}f_J(z')\, dS(z'), \quad \text{(VI.4.11)}$$

while others are of the kind

$$\int_{\partial\Omega}\Big\{[\bar{z}'\cdot(z-z')]^{p+1-n}H_{\nu'}(z-z') -$$

$$[\bar{z}'\cdot(w-z')]^{p+1-n}H_{\nu'}(w-z')\Big\}\ \delta'[\varphi(z')D^r f_J(z')]\ dS(z') \qquad \text{(VI.4.12)}$$

($\delta' = \partial/\partial z_j'$ or $\partial/\partial\bar{z}_j'$ for some j). In these formulas $D^s f_J$ stands for some partial derivative of f_J of order $s \le r$, $D^\gamma H_{\nu'}$ stands for a partial derivative of $H_{\nu'}$ of order $\gamma \le \nu'$; φ and ψ are $\mathscr{C}^\infty$ functions in $\partial\Omega$.

The estimate of (VI.4.11) is straightforward. Call $\mathscr{D}_\gamma(z,z')$ the gradient with respect to z of $[\bar{z}'\cdot(z-z')]^{p+1-n}\, D^\gamma H_{\nu'}(z-z')$; we write

$$[\bar{z}'\cdot(z-z')]^{p+1-n}D^\gamma H_{\nu'}(z-z') - [\bar{z}'\cdot(w-z')]^{p+1-n}D^\gamma H_{\nu'}(w-z') =$$

$$(z-w)\cdot\int_0^1 \mathscr{D}_\gamma(z_*,z')dt \qquad \text{(VI.4.13)}$$

with $z_* = tz + (1-t)w$. We have, by virtue of (VI.4.7),

$$|\mathscr{D}_\gamma(z_*,z')| \le C_0(1-\lambda)^{-1}|\bar{z}'\cdot(z'-z_*)|^{p+1-n}|z'-z_*|^{-2p-1}. \qquad \text{(VI.4.14)}$$

We conclude that the absolute value of (VI.4.11) does not exceed

$$C_0(1-\lambda)^{-1}|z-w|\ \|f\|_{\Omega,r}\int_0^1\Big\{\int_{\partial\Omega}|\bar{z}'\cdot(z'-z_*)|^{p+1-n}|z'-z_*|^{-2p-1}dS(z')\Big\}dt$$

and the proof of Proposition VI.3.2 shows that the integral inside $\{\cdots\}$ is bounded independently of z_*.

We focus our attention on (VI.4.12). Let $\chi \in \mathscr{C}^\infty(\mathbb{R})$ be the function introduced above and define $\chi_0(\tau) = 1 - \chi(\tau) - \chi(1/\tau)$. Observe that $\chi_0(\tau) = 0$ if $\tau < 1/9$ or $\tau > 9$. Note that (VI.4.12) is equal to the sum of the following three integrals:

$$\mathscr{L}_0(z,w) = \int_{\partial\Omega}\Big\{[\bar{z}'\cdot(z-z')]^{p+1-n}H_{\nu'}(z-z') - [\bar{z}'\cdot(w-z')]^{p+1-n}H_{\nu'}(w-z')\Big\}\cdot$$

$$\delta'[\chi_0(|z-z'|^2/|w-z'|^2)\varphi(z')D^r f_J(z')]\ dS(z'),$$

$$\mathscr{L}_1(z,w) = \int_{\partial\Omega}\Big\{[\bar{z}'\cdot(z-z')]^{p+1-n}H_{\nu'}(z-z') - [\bar{z}'\cdot(w-z')]^{p+1-n}H_{\nu'}(w-z')\Big\}\cdot$$

$$\delta'[\chi(|z-z'|^2/|w-z'|^2)\varphi(z')D^r f_J(z')]\ dS(z'),$$

$$\mathscr{L}_2(z,w) = \int_{\partial\Omega}\Big\{[\bar{z}'\cdot(z-z')]^{p+1-n}H_{\nu'}(z-z') - [\bar{z}'\cdot(w-z')]^{p+1-n}H_{\nu'}(w-z')\Big\}\cdot$$

$$\delta'[\chi(|w-z'|^2/|z-z'|^2)\varphi(z')D^r f_J(z')]\ dS(z').$$

We begin by estimating $\mathscr{L}_1(z,w)$. We write

$$\mathscr{L}_1(z,w) = \mathscr{M}_1(z,w) + [D^r f_J(w)][\mathscr{N}_1(z,w) + \mathscr{N}'_1(z,w)],$$

$$\mathscr{M}_1(z,w) = \int_{\partial\Omega}\Big\{[\bar{z}'\cdot(z-z')]^{p+1-n}H_{\nu'}(z-z') - [\bar{z}'\cdot(w-z')]^{p+1-n}H_{\nu'}(w-z')\Big\}\cdot$$

$$\delta'\Big\{\chi(|z-z'|^2/|w-z'|^2)\varphi(z')[D^r f_J(z') - D^r f_J(w)]\Big\}\, dS(z'),$$

$$\mathscr{N}_1(z,w) = \int_{\partial\Omega}\Big\{[\bar{z}'\cdot(z-z')]^{p+1-n}H_{\nu'}(z-z') - [\bar{z}'\cdot(w-z')]^{p+1-n}H_{\nu'}(w-z')\Big\}\cdot$$

$$\chi(|z-z'|^2/|w-z'|^2)\,\delta'\varphi(z')\, dS(z'),$$

$$\mathscr{N}'_1(z,w) = \int_{\partial\Omega}\Big\{[\bar{z}'\cdot(z-z')]^{p+1-n}H_{\nu'}(z-z') - [\bar{z}'\cdot(w-z')]^{p+1-n}H_{\nu'}(w-z')\Big\}\cdot$$

$$\varphi(z')\delta'[\chi(|z-z'|^2/|w-z'|^2)]\, dS(z').$$

First we estimate $\mathscr{M}_1$. Let δ'' denote the transpose of δ' with respect to the measure $dS(z')$. It suffices to apply the rough inequality

$$\left|\delta''\Big\{[\bar{z}'\cdot(z-z')]^{p+1-n}H_{\nu'}(z-z') - [\bar{z}'\cdot(w-z')]^{p+1-n}H_{\nu'}(w-z')\Big\}\right| \le$$

$$C_0(1-\lambda)^{-1}\Big\{|\bar{z}'\cdot(z-z')|^{p+1-n}|z-z'|^{-2p-1} + |\bar{z}'\cdot(w-z')|^{p+1-n}|w-z'|^{-2p-1}\Big\}$$

to obtain

$$|\mathscr{M}_1(z,w)| \le C_0(1-\lambda)^{-1}\mathscr{H}_1^{r+\kappa}(f)\int_{\partial\Omega}|z'-w|^{\kappa}\Big\{|\bar{z}'\cdot(z-z')|^{p+1-n}|z-z'|^{-2p-1} +$$

$$|\bar{z}'\cdot(w-z')|^{p+1-n}|w-z'|^{-2p-1}\Big\}\,\chi(|z-z'|^2/|w-z'|^2)\, dS(z').$$

We make then use of the fact that

$$\begin{gathered} w \in \Omega_\lambda \textit{ and } \chi(|z-z'|^2/|w-z'|^2) \neq 0 \Rightarrow \\ 1-\lambda < |w-z'| \le |z-z'| - |w-z'| \le |w-z|. \end{gathered} \tag{VI.4.15}$$

This yields

$$|\mathscr{M}_1(z,w)|/|z-w|^{\kappa} \le$$

$$C_0(1-\lambda)^{-1}\mathscr{H}_1^{r+\kappa}(f)\sup_{\zeta\in\Omega}\int_{\partial\Omega}|\bar{z}'\cdot(\zeta-z')|^{p+1-n}|\zeta-z'|^{-2p-1}dS(z')$$

and the estimates at the end of the proof of Proposition VI.3.2 allow us to conclude that $|\mathscr{M}_1(z,w)|/|z-w|^{\kappa}$ is well bounded.

The argument at the end of the proof of Proposition VI.3.2 also shows right away that $|\mathscr{N}_1(z,w)|$ is well bounded. By using (VI.4.15) once again we see that

$$|\mathscr{N}_1(z,w)|/|z-w|^{\kappa} \leq C(1-\lambda)^{-1}. \qquad \text{(VI.4.16)}$$

Next we estimate $\mathscr{N}_1'$. Observe that $d\chi(|w-z'|^2/|z-z'|^2) \neq 0$ entails $2 \leq |w-z'|/|z-z'| \leq 3$, and therefore, if we continue to use the notation $z_* = tz + (1-t)w$,

$$|z_*-z'|/|z-w| \leq 2, \ \forall\, t \in [0,1]. \qquad \text{(VI.4.17)}$$

On the other hand, we have

$$|\delta'[\chi(|w-z'|^2/|z-z'|^2)]| \leq C_1|z-z'|^{-1}.$$

When $d\chi(|w-z'|^2/|z-z'|^2) \neq 0$ we also have

$$|z-w| \leq |z-z'| + |z'-w| \leq 4|z-z'|.$$

By applying (VI.4.14) with $\gamma = 0$ we see thus that

$$|(z-w)\cdot\mathscr{D}_0(z_*,z')\delta'[\chi(|w-z'|^2/|z-z'|^2)]| \leq$$
$$C_1(1-\lambda)^{-1}|z-w|^{\kappa}|\bar{z}'\cdot(z'-z_*)|^{p+1-n}|z'-z_*|^{-2p-1-\kappa}. \qquad \text{(VI.4.18)}$$

The inequality

$$|\mathscr{N}_1'(z,w)|/|z-w|^{\kappa} \leq C(1-\lambda)^{-1} \qquad \text{(VI.4.19)}$$

will follow if we show that the integral

$$\int_{\partial\Omega} |\bar{z}'\cdot(z'-z_*)|^{p+1-n}|z'-z_*|^{-2p-1-\kappa}dS(z')$$

is bounded independently of z_*. By using local charts on the sphere $\partial\Omega$ as in the proof of Proposition VI.3.2, and the inequality (VI.4.7), one sees that this is a consequence of the finiteness of the integrals

$$J_p = \int_0^1\int_0^1 (s+t)^{-2p-1-\kappa}(s+t^2)^{p+1-n}t^{2n-3}ds dt \leq$$
$$\int_0^1\int_0^{1/t^2} (1+u)^{p+1-n}t^{-\kappa}du dt.$$

And indeed, if $p \leq n-2$, $J_p \leq J_{n-2} \leq \int_0^1 t^{-\kappa}\log(1+1/t^2)dt < +\infty$.

Combining (VI.4.16), (VI.4.19), and the fact that $|\mathscr{M}_1(z,w)|/|z-w|^{\kappa}$ is well bounded shows that $|\mathcal{L}_1(z,w)|/|z-w|^\kappa$ is also well bounded. The same is true of $|\mathcal{L}_2(z,w)|/|z-w|^\kappa$ since $\mathcal{L}_2(z,w) = -\mathcal{L}_1(w,z)$.

Last we estimate

$$\mathcal{L}_0(z,w) = (z-w)\cdot\int_0^1 \mathcal{M}_0(z,w,t)dt, \tag{VI.4.20}$$

where

$$\mathcal{M}_0(z,w,t) = \int_{\partial\Omega} \mathfrak{D}_0(z_*,z')\delta'[\chi_0(|z-z'|^2/|w-z'|^2)\varphi(z')D^r f_J(z')]dS(z')$$

$$= \int_{\partial\Omega} [\delta''\mathfrak{D}_0(z_*,z')]\chi_0(|z-z'|^2/|w-z'|^2)\varphi(z')[D^r f_J(z') - D^r f_J(z_*)]dS(z') +$$

$$[D^r f_J(z_*)]\int_{\partial\Omega} \mathfrak{D}_0(z_*,z')\delta'[\chi_0(|z-z'|^2/|w-z'|^2)\varphi(z')]dS(z'),$$

where $z_* = tz + (1-t)w$ as before. By reasoning more or less as in the proof of (VI.4.19) (in particular, by making use of (VI.4.18)), we get

$$\left|(z-w)\cdot\int_{\partial\Omega} \mathfrak{D}_0(z_*,z')\delta'[\chi_0(|z-z'|^2/|w-z'|^2)\varphi(z')]dS(z')\right| \le$$
$$C(1-\lambda)^{-1}|z-w|.$$

We now make use of the estimate

$$|\delta''\mathfrak{D}_0(z_*,z')| \le C_0|\bar{z}'\cdot(z'-z_*)|^{p-n-1}|z'-z_*|^{-2p-1}.$$

In the set $\Sigma = \{ z' \in \partial\Omega;\ |z-z'| \ge 2\,|z-w| \}$ we have $|z-w| \le |z'-z_*|$ and therefore

$$\left|(z-w)\cdot\int_{\Sigma} [\delta''\mathfrak{D}_0(z_*,z')]\chi_0(|z-z'|^2/|w-z'|^2)\varphi(z')\cdot\right.$$
$$\left.[D^r f_J(z') - D^r f_J(z_*)]dS(z')\right| \le$$
$$C_1|z-w|^\kappa \mathcal{H}_1^{r+\kappa}(f)\int_{\partial\Omega} |\bar{z}'\cdot(z'-z_*)|^{p-n-1}|z'-z_*|^{-2p}dS(z').$$

It will suffice to prove that

$$(1-\lambda)\int_{\partial\Omega} |\bar{z}'\cdot(z'-z_*)|^{p-n-1}|z'-z_*|^{-2p}dS(z')$$

is bounded independently of λ and of $z_* \in \Omega_\lambda$. We use local charts in the unit sphere $\partial\Omega$ as in the proof of Proposition VI.3.2 and we avail ourselves of the inequality (VI.4.7). This leads to proving the finiteness of integrals

$$\mathrm{K}_p(\varepsilon) = \varepsilon\int_0^1\int_0^1(\varepsilon+s+t^2)^{p-n-1}(s+t)^{-2p}t^{2n-3}\mathrm{d}s\mathrm{d}t,$$

where $\varepsilon = 1 - \lambda$. We have

$$\mathrm{K}_p(\varepsilon) \leq \varepsilon\int_0^1\int_0^1(\varepsilon+s+t^2)^{p-n-1}t^{2(n-p)-3}\mathrm{d}s\mathrm{d}t \leq$$

$$\frac{1}{2}\int_0^{+\infty}\int_0^{+\infty}(1+\sigma)^{p-n-1}(1+\tau)^{p-n}\tau^{n-p-2}\mathrm{d}\sigma\mathrm{d}\tau < +\infty$$

since $0 \leq p \leq n-2$. When $z' \notin \Sigma$ then $|z_* - z'| \leq 3|z-w|$; we do not rely on (IV.4.20) but rather on the rough estimate of $\delta''\ \{[\bar{z}'\cdot(z-z')]^{p-n+1}H_{v'}(z-z')\}$ used to estimate $\mathcal{M}_1$. We conclude that $|\mathcal{L}_0(z,w)|/|z-w|^\kappa$, and thus also $\mathcal{H}_1^{r+1+\kappa}(\mathcal{B}_{\partial\Omega}f)$, is well bounded.

This completes the proof of Theorem VI.4.1.

VI.5. The Newlander-Nirenberg Theorem

The present section will be devoted to the proof of the following statement (Newlander and Nirenberg [1]):

THEOREM VI.5.1. *Every complex structure* (Definition I.2.3) *on a $\mathcal{C}^\infty$ manifold $\mathcal{M}$ is locally integrable.*

The (real) dimension of $\mathcal{M}$ is even and will be denoted by $2n$. The result is purely local and we may take $\mathcal{M}$ to be the Euclidean space $\mathbb{R}^{2n}$ (coordinates x_i, y_j, $1 \leq i, j \leq n$). As always we write $z_j = x_j + \imath y_j$, $z = (z_1,\dots,z_n)$, etc. In the course of the proof the choice of these coordinates will be modified. Meanwhile we write $\partial_j = \partial/\partial z_j$, $\bar{\partial}_j = \partial/\partial\bar{z}_j$, $\partial f = \sum_{j=1}^n \partial_j f\mathrm{d}z_j$, and so forth.

At each step in the proof of Theorem VI.5.1 we shall be changing the complex coordinates z_j in an open neighborhood $\mathcal{O}$ of $\mathbb{C}^n$. To avoid changing frames, each time, in the bundles $\mathrm{T}'^{0,q}$, it is convenient to identify $\mathrm{T}'^{0,1}$ to $\mathbb{C}^n$ via the isomorphism that transforms the basis $\{\mathrm{d}\bar{z}_1,\dots,\mathrm{d}\bar{z}_n\}$ into the natural basis of $\mathbb{C}^n$. This equates the vector bundle $\mathrm{T}'^{0,q}$ over $\mathcal{O}$, to $\mathcal{O}\times\Lambda^q\mathbb{C}^n$ ($\Lambda^q\mathbb{C}^n$ is the q-th exterior power of $\mathbb{C}^n$). A (0,1)-form will then be a vector-valued function $f = (f_1,\dots,f_n)$, a (0,2)-form will be a function with values that are skew-symmetric covariant tensors of degree two, $F = (F_{jk})_{1\leq j<k\leq n}$, etc.

Suppose that the components f_j are $\mathcal{C}^\infty$ functions in $\mathcal{O}$ and consider a system of $\mathcal{C}^\infty$ complex vector fields $\{D_1,\dots,D_n\}$ in $\mathcal{O}$. They shall act on f as follows, giving rise to the (0,2)-form

$$Df = (D_j f_k - D_k f_j)_{1 \le j < k \le n}.$$

When acting on the (0,2)-form F the vector fields D_j define the (0,3)-form

$$DF = (D_j F_{k\ell} - D_k F_{j\ell} + D_\ell F_{jk})_{1 \le j < k < \ell \le n};$$

and so forth.

It will not always be the case that the vector fields D_j commute. When they do, the differential operators

$$D : \mathscr{C}^\infty(\mathcal{O};\Lambda^q\mathbb{C}^n) \to \mathscr{C}^\infty(\mathcal{O};\Lambda^{q+1}\mathbb{C}^n) \; (q = 0,1,\ldots) \qquad \text{(VI.5.1)}$$

make up a differential complex, i.e., we have $D^2 = 0$. We use standard terminology: $f \in \mathscr{C}^\infty(\mathcal{O};\Lambda^q\mathbb{C}^n)$ is D-closed if $Df = 0$; f is D-exact if $f = Du$ for some $u \in \mathscr{C}^\infty(\mathcal{O};\Lambda^{q-1}\mathbb{C}^n)$.

We begin by choosing the coordinates in such a way that the tangent structure bundle $\mathcal{V}$ (in the given complex structure on $\mathbb{R}^{2n}$) is spanned over some open ball U centered at the origin by n $\mathscr{C}^\infty$ vector fields

$$\mathrm{L}_j = \partial/\partial \bar{z}_j - \sum_{j,k=1}^{n} \gamma_{jk}\partial/\partial z_k \qquad \text{(VI.5.2)}$$

with $\gamma_{jk}|_0 = 0$ ($j, k = 1,\ldots,n$; cf. (I.5.23), (I.5.24)). In the ball U $\mathrm{L}_1,\ldots,\mathrm{L}_n$, $\bar{\mathrm{L}}_1,\ldots,\bar{\mathrm{L}}_n$ are linearly independent, and therefore form a smooth basis of $\mathbb{C}T\mathcal{M}$ over U. Let γ denote the $n \times n$ matrix with entries γ_{jk}; we shall assume that the matrix $\mathrm{I}+\gamma$ is nonsingular at every point of U. According to the identification of one-forms with n-vectors, $\mathrm{L}f = (\mathrm{L}_1 f,\ldots,\mathrm{L}_n f)$.

Lemma VI.5.1. *Given $h \in \mathscr{C}^\infty(\mathrm{U})$, there is $H \in \mathscr{C}^\infty(\mathrm{U})$ such that $\mathrm{L}H$ vanishes to infinite order, and $H - h$ vanishes, at $y = 0$.*

Proof. (VI.5.2) allows us to write

$$-2\imath \mathrm{L}_j = \partial/\partial y_j + \sum_{k=1}^{n} \gamma_{jk}\partial/\partial y_k - \imath\left\{\partial/\partial x_j - \sum_{k=1}^{n} \gamma_{jk}\partial/\partial x_k\right\}.$$

Letting the matrix $-2\imath(\mathrm{I}+\gamma)^{-1}$ act on the system $\{\mathrm{L}_1,\ldots,\mathrm{L}_n\}$ gets us a new system of vector fields

$$\mathrm{L}_j^{\#} = \partial/\partial y_j - \sum_{k=1}^{n} c_{jk}\partial/\partial x_k, \; j = 1,\ldots,n.$$

The vector fields $\mathrm{L}_j^{\#}$ commute since the commutation brackets $[\mathrm{L}_j^{\#},\mathrm{L}_k^{\#}]$ belong to the span of $\mathrm{L}_1^{\#},\ldots,\mathrm{L}_n^{\#}$ (in U). We write as follows the Taylor expansions with respect to y of c_{jk} and H respectively:

$$\sum_{\nu=0}^{+\infty} c_{jk;\nu}(x,y), \; \sum_{\nu=0}^{+\infty} H_\nu(x,y),$$

with $c_{jk;\nu}(x,y)$ and $H_\nu(x,y)$ homogeneous polynomials of degree ν with respect to y whose coefficients are $\mathscr{C}^\infty$ functions of x in the ball $\mathrm{U}_0 = \{x \in \mathbb{R}^n;\ (x,0) \in \mathrm{U}\}$. We take $H_0(x,y) = h(x,0)$ and solve inductively, for $\nu = 0,1,\ldots,$ the equations

$$(\partial/\partial y_j)H_{\nu+1} = \sum_{\mu=0}^{\nu}\sum_{k=1}^{n} c_{jk;\mu}\partial H_{\nu-\mu}/\partial x_k \ (j = 1,\ldots,n).$$

That the right-hand sides satisfy the required compatibility conditions follows from induction on ν and from the commutation of the vector fields $\mathrm{L}_j^\#$.

Let $\chi \in \mathscr{C}_c^\infty(\mathbb{R})$, $\chi(\tau) = 1$ for $|\tau| \leq 1$. We can select inductively a sequence of numbers $\varepsilon_\nu > 0$, $\varepsilon_\nu \to +0$, such that the series

$$\sum_{\nu=0}^{+\infty} \chi(|y|/\varepsilon_\nu)H_\nu(x,y)$$

converges in $\mathscr{C}^\infty(\mathrm{U}_0 \times \mathbb{R}^n)$ to a function H whose restriction to U satisfies the requirements in Lemma VI.5.1. ■

We apply Lemma VI.5.1 to obtain, for each $j = 1,\ldots,n$, a function $\mathrm{Z}_{0j} \in \mathscr{C}^\infty(\mathrm{U})$ satisfying the conditions:

whatever k, $1 \leq k \leq n$, $\mathrm{L}_k\mathrm{Z}_{0j}$ *vanishes to infinite order at* $y = 0$; (VI.5.3)

$$\mathrm{Z}_{0j}|_{y=0} = x_j. \tag{VI.5.4}$$

There is an open neighborhood $\mathscr{U}$ of the origin in $\mathbb{R}^{2n}$ such that the map $\mathrm{Z}_0 = (\mathrm{Z}_{01},\ldots,\mathrm{Z}_{0n})\colon \mathscr{U} \to \mathbb{C}^n$ is a diffeomorphism onto an open subset of $\mathbb{C}^n$. This means that we may use the functions Z_{0j} as complex coordinates in $\mathscr{U}$.

By induction on the integer $\nu \geq 0$ we select a sequence of open subsets Ω_ν, Ω'_ν of $\mathbb{C}^n$ and of functions $\mathrm{Z}_{\nu+1,j}$ ($j = 1,\ldots,n$) in Ω_ν as follows. Let us write $\mathrm{Z}_\nu = (\mathrm{Z}_{\nu,1},\ldots,\mathrm{Z}_{\nu,n})$ and agree that, when $\nu = -1$, $\Omega_\nu = \mathscr{U}$ and $\mathrm{Z}_{\nu+1} = \mathrm{Z}_0$. For each $\nu \geq 0$ we choose numbers δ_ν, δ'_ν, with $0 < \delta_{\nu+1} < \delta'_\nu < \delta_\nu$, in such a way that

the functions $\mathrm{Z}_{\nu,j}$ ($j = 1,\ldots,n$) *form a system of complex coordinates in* $\Omega'_{\nu-1}$; (VI.5.5)$_\nu$

the $n \times n$ *matrix* $\mathscr{L}\overline{\mathrm{Z}}_\nu = (\mathrm{L}_j\overline{\mathrm{Z}}_{\nu,k})_{1\leq j,k\leq n}$ *is nonsingular at every point of* $\Omega'_{\nu-1}$; (VI.5.6)$_\nu$

(below we also use the notation $\mathscr{L}\mathrm{Z}_\nu$ for the matrix with entries $\mathrm{L}_j\mathrm{Z}_{\nu,k}$);

the ball Ω_ν *in* Z_ν*-space, centered at the origin and having radius* δ_ν*, has its closure contained in the ball* $\Omega'_{\nu-1}$ *in* $\mathrm{Z}_{\nu-1}$*-space, centered at the origin and having radius* $\delta'_{\nu-1}$. (VI.5.7)$_\nu$

Provisionally call $\mu_{\nu,jk}$ the entries of the inverse matrix $(\mathcal{L}\bar{Z}_\nu)^{-1}$ and set, for each $j = 1,\ldots,n$,

$$L_{\nu,j} = \sum_{k=1}^{n} \mu_{\nu,jk} L_k.$$

We have $L_{\nu,j}\bar{Z}_{\nu,k} = \delta_{jk}$ (Kronecker's index), and therefore in the coordinates $Z_{\nu,j}$ the vector field $L_{\nu,j}$ has an expression

$$L_{\nu,j} = \partial/\partial\bar{Z}_{\nu,j} + \sum_{k=1}^{n}\left(\sum_{\ell=1}^{n} \mu_{\nu,j\ell} L_\ell Z_{\nu,k}\right)\partial/\partial Z_{\nu,k}. \tag{VI.5.8}$$

The vector fields $L_{\nu,j}$ are defined and smooth in $\Omega_{\nu-1}$, by virtue of (VI.5.5) and (VI.5.6). They make up an involutive system, which demands that they commute. We call L_ν the differential operator (VI.5.1) when $D_j = L_{\nu,j}$; of course, $L_\nu^2 = 0$. The expressions (VI.5.8) can be rewritten as

$$L_\nu = (\mathcal{L}\bar{Z}_\nu)^{-1}L = \bar{\partial}_\nu + (\mathcal{L}\bar{Z}_\nu)^{-1}(\mathcal{L}Z_\nu)\partial_\nu. \tag{VI.5.9}$$

We explain the notation: $\bar{\partial}_\nu$ stands for the Cauchy-Riemann operator in Z_ν-space, acting on a 0-form f according to the rule

$$\bar{\partial}_\nu f = (\partial f/\partial\bar{Z}_{\nu,j})_{1\le j\le n}; \tag{VI.5.10}$$

whereas the operators ∂_ν acts according to the rule

$$\partial_\nu f = (\partial f/\partial Z_{\nu,j})_{1\le j\le n}. \tag{VI.5.11}$$

The actions of $\bar{\partial}_\nu$ and ∂_ν on $(0,q)$-forms ($q \ge 1$) follow from these formulas.

Keep in mind that since we have identified any one-form to an n-vector, any $n\times n$ matrix, such as $\mathcal{L}Z_\nu$ or $(\mathcal{L}\bar{Z}_\nu)^{-1}$, acts on it in natural fashion.

For each $q = 1,2,\ldots$, the (Bochner-Martinelli-Leray-Koppelman) operators $(-1)^{n-q}(\mathcal{K}_{\Omega_\nu} + \mathcal{B}_{\partial\Omega_\nu})$ of sections VI.1 and VI.2 relative to the open ball Ω_ν in Z_ν-space can be transferred as bounded linear operators

$$G_\nu : \mathcal{C}^\infty(\mathcal{C}\ell\Omega_\nu;\Lambda^q\mathbb{C}^n) \to \mathcal{C}^\infty(\mathcal{C}\ell\Omega_\nu;\Lambda^{q-1}\mathbb{C}^n).$$

In order to carry out this transfer we compose the canonical isomorphism $T'^{n,q} \cong T'^{0,q}$ with the isomorphism $T'^{0,q} \cong \Lambda^q\mathbb{C}^n$ over $\Omega_{\nu-1}$ derived by transforming the basis $\{dZ_{\nu,1},\ldots,dZ_{\nu,n}\}$ into the canonical basis of $\mathbb{C}^n$. If $f \in \mathcal{C}^\infty(\mathcal{C}\ell\Omega_\nu)$ we have, in $\mathcal{C}\ell\Omega_\nu$,

$$f = G_\nu\bar{\partial}_\nu f, \tag{VI.5.12}$$

and if $F \in \mathcal{C}^\infty(\mathcal{C}\ell\Omega_\nu;\Lambda^q\mathbb{C}^n)$ with $q \ge 1$,

$$F = G_\nu\bar{\partial}_\nu F + \bar{\partial}_\nu G_\nu F \tag{VI.5.13}$$

(see Theorem VI.2.2).

In the sequel we shall deal with vector-valued forms of the kind $L_\nu Z_\nu =$

$\{L_\nu Z_{\nu,1},\ldots,L_\nu Z_{\nu,n}\}$: each component $L_\nu Z_{\nu,j}$ of $L_\nu Z_\nu$ is a one-form in $\Omega_{\nu-1}$, a fact we shall express by writing $L_\nu Z_\nu \in \mathscr{C}^\infty(\Omega_{\nu-1};\Lambda^1\mathbb{C}^n)\otimes\mathbb{C}^n$. We may also view each component $L_\nu Z_{\nu,j}$ as a function in $\Omega_{\nu-1}$ valued in $\mathbb{C}^n$; then $L_\nu Z_\nu$ can be viewed as a function in $\Omega_{\nu-1}$ valued in the space of $n\times n$ complex matrices. If we replace L_ν by L the last remark allows us to identify LZ_ν to $\mathscr{L}Z_\nu$.

We define

$$\begin{aligned} W_\nu &= G_\nu L_\nu Z_\nu = \{G_\nu L_\nu Z_{\nu,1},\ldots,G_\nu L_\nu Z_{\nu,n}\}, \\ R_\nu &= G_\nu[\bar\partial_\nu(L_\nu Z_\nu)]. \end{aligned} \tag{VI.5.14}$$

Note that

$$W_\nu \in \mathscr{C}^\infty(\mathscr{Cl}\Omega_\nu)\otimes\mathbb{C}^n\,(\cong \mathscr{C}^\infty(\mathscr{Cl}\Omega_\nu;\mathbb{C}^n)),$$

$$R_\nu \in \mathscr{C}^\infty(\mathscr{Cl}\Omega_\nu;\Lambda^1\mathbb{C}^n)\otimes\mathbb{C}^n\ (\cong \mathscr{C}^\infty(\mathscr{Cl}\Omega_\nu;\mathbb{C}^{n^2})).$$

By virtue of (VI.5.13) we have, in Ω_ν,

$$L_\nu Z_\nu = \bar\partial_\nu W_\nu + R_\nu. \tag{VI.5.15}$$

We set then, in Ω_ν,

$$Z_{\nu+1} = Z_\nu - W_\nu. \tag{VI.5.16}$$

The next step is to seek estimates for the Hölder norms of the functions and differential forms that have just been introduced. Since Ω_ν is the ball in $\mathbb{C}^n$ (where the coordinates are $Z_{\nu,j}$) centered at 0 with radius δ_ν, we may apply Theorem VI.4.1 in Z_ν-space. The reader ought to keep in mind that the Hölder norms $N_{\delta_\nu}^{m+\kappa}$ and $N_{\delta'_\nu}^{m+\kappa}$ used below are defined by means of the coordinates $\mathscr{Re}Z_{\nu,j}$, $\mathscr{Im}Z_{\nu,k}$. Later we shall compare them to the analogous Hölder norms defined by means of the original coordinates x_j, y_k.

We shall eventually choose the sequence $\{\delta_\nu\}_{\nu=0,1,\ldots}$ in such a way that

$$0 < c_0 < \delta_\nu \le 1$$

with c_0 henceforth kept fixed. The constants below will depend on the choice of c_0 although the notation will not indicate this fact. We shall denote by C various positive constants that have in common the property of being independent of ν.

Thus let m be an integer ≥ 1, κ a number such that $0 < \kappa < 1$. We get:

$$N_{\delta'_\nu}^{m+1+\kappa}(W_\nu) \le C(\delta_\nu - \delta'_\nu)^{-1}N_{\delta_\nu}^{m+\kappa}(L_\nu Z_\nu). \tag{VI.5.17}$$

We get at once, from (VI.5.16) and (VI.5.17),

$$N_{\delta'_\nu}^{m+1+\kappa}(Z_{\nu+1} - Z_\nu) \le C(\delta_\nu - \delta'_\nu)^{-1}N_{\delta_\nu}^{m+\kappa}(L_\nu Z_\nu). \tag{VI.5.18}$$

On the other hand, by (VI.5.9) we get

$$L_\nu Z_{\nu+1} - L_\nu Z_\nu = -L_\nu W_\nu = -L_\nu Z_\nu - (\mathscr{L}\bar Z_\nu)^{-1}(\mathscr{L}Z_\nu)\partial_\nu W_\nu + R_\nu,$$

whence

$$\mathrm{L}_\nu Z_{\nu+1} = R_\nu - (\mathscr{L}\overline{Z}_\nu)^{-1}(\mathscr{L}Z_\nu)\partial_\nu W_\nu. \tag{VI.5.19}$$

By (VI.5.9) and thanks to the fact that $\mathrm{L}_\nu^2 = 0$, we get

$$R_\nu = \mathrm{G}_\nu[\bar\partial_\nu(\mathrm{L}_\nu Z_\nu)] = -\mathrm{G}_\nu[(\mathscr{L}\overline{Z}_\nu)^{-1}(\mathscr{L}Z_\nu)\partial_\nu(\mathrm{L}_\nu Z_\nu)].$$

Thus, thanks to the property that $\mathrm{N}_\delta^{m+\kappa}(fg) \le C\mathrm{N}_\delta^{m+\kappa}(f)\mathrm{N}_\delta^{m+\kappa}(g)$ for all functions f, $g \in \mathscr{C}^\infty(\Omega_\delta)$, we get

$$\mathrm{N}_{\delta_\nu'}^{m+\kappa}(R_\nu) \le C(\delta_\nu - \delta_\nu')^{-1}\, \mathrm{N}_{\delta_\nu}^{m-1+\kappa}((\mathscr{L}\overline{Z}_\nu)^{-1})\, \mathrm{N}_{\delta_\nu}^{m-1+\kappa}(\mathscr{L}Z_\nu)\, \mathrm{N}_{\delta_\nu}^{m+\kappa}(\mathrm{L}_\nu Z_\nu).$$

Combining this with (VI.5.17) and (VI.5.19) yields

$$\mathrm{N}_{\delta_\nu'}^{m+\kappa}(\mathrm{L}_\nu Z_{\nu+1}) \le$$

$$C(\delta_\nu - \delta_\nu')^{-1}\, \mathrm{N}_{\delta_\nu}^{m+\kappa}((\mathscr{L}\overline{Z}_\nu)^{-1})\, \mathrm{N}_{\delta_\nu}^{m+\kappa}(\mathscr{L}Z_\nu)\, \mathrm{N}_{\delta_\nu}^{m+\kappa}(\mathrm{L}_\nu Z_\nu). \tag{VI.5.20}$$

We also have, according to (VI.5.18),

$$\mathrm{N}_{\delta_\nu'}^{m+\kappa}(\mathscr{L}\overline{Z}_{\nu+1} - \mathscr{L}\overline{Z}_\nu) \le C(\delta_\nu - \delta_\nu')^{-1}\mathrm{N}_{\delta_\nu}^{m+\kappa}(\mathrm{L}_\nu Z_\nu). \tag{VI.5.21}$$

The definition of L_ν entails, for any $f \in \mathscr{C}^\infty(\mathscr{C}\ell\Omega_\nu;\Lambda^1\mathbb{C}^n)$ $(\cong \mathscr{C}^\infty(\mathscr{C}\ell\Omega_\nu;\mathbb{C}^n))$,

$$\begin{aligned} \mathrm{N}_{\delta_\nu}^{m+\kappa}(\mathrm{L}_\nu f) &\le C\mathrm{N}_{\delta_\nu}^{m+\kappa}((\mathscr{L}\overline{Z}_\nu)^{-1})\mathrm{N}_{\delta_\nu}^{m+\kappa}(\mathrm{L}f), \\ \mathrm{N}_{\delta_\nu}^{m+\kappa}(\mathrm{L}f) &\le C\mathrm{N}_{\delta_\nu}^{m+\kappa}(\mathscr{L}\overline{Z}_\nu)\mathrm{N}_{\delta_\nu}^{m+\kappa}(\mathrm{L}_\nu f). \end{aligned} \tag{VI.5.22}$$

First we derive from (VI.5.18):

$$\mathrm{N}_{\delta_\nu'}^{m+1+\kappa}(Z_{\nu+1} - Z_\nu) \le C(\delta_\nu - \delta_\nu')^{-1}\mathrm{N}_{\delta_\nu}^{m+\kappa}((\mathscr{L}\overline{Z}_\nu)^{-1})\mathrm{N}_{\delta_\nu}^{m+\kappa}(\mathscr{L}Z_\nu). \tag{VI.5.23}$$

Next we derive from (VI.5.20) and the first line in (VI.5.22):

$$\mathrm{N}_{\delta_\nu'}^{m+\kappa}(\mathrm{L}_\nu Z_{\nu+1}) \le$$

$$C(\delta_\nu - \delta_\nu')^{-1}\left\{\mathrm{N}_{\delta_\nu}^{m+\kappa}((\mathscr{L}\overline{Z}_\nu)^{-1})\right\}^2\mathrm{N}_{\delta_\nu}^{m+\kappa}(\mathscr{L}\overline{Z}_\nu)\left\{\mathrm{N}_{\delta_\nu}^{m+\kappa}(\mathscr{L}Z_\nu)\right\}^2.$$

If we take into account the second line in (VI.5.22), we get

$$\mathrm{N}_{\delta_\nu'}^{m+\kappa}(\mathscr{L}Z_{\nu+1}) \le C(\delta_\nu - \delta_\nu')^{-1}\left\{\mathrm{N}_{\delta_\nu}^{m+\kappa}((\mathscr{L}\overline{Z}_\nu)^{-1})\mathrm{N}_{\delta_\nu}^{m+\kappa}(\mathscr{L}\overline{Z}_\nu)\mathrm{N}_{\delta_\nu}^{m+\kappa}(\mathscr{L}Z_\nu)\right\}^2. \tag{VI.5.24}$$

Third, we derive from (VI.5.21):

$$\mathrm{N}_{\delta_\nu'}^{m+\kappa}(\mathscr{L}\overline{Z}_{\nu+1} - \mathscr{L}\overline{Z}_\nu) \le C(\delta_\nu - \delta_\nu')^{-1}\mathrm{N}_{\delta_\nu}^{m+\kappa}((\mathscr{L}\overline{Z}_\nu)^{-1})\, \mathrm{N}_{\delta_\nu}^{m+\kappa}(\mathscr{L}Z_\nu). \tag{VI.5.25}$$

The proof of Theorem VI.5.1 will be completed in the next section.

VI.6. End of the Proof of the Newlander-Nirenberg Theorem

It is convenient to rewrite the estimates (VI.5.23), (VI.5.24), and (VI.5.25) in terms of the Hölder norms in the original coordinates. The norms $N^{m+\kappa}_{\delta_\nu}$ and $N^{m+\kappa}_{\delta'_\nu}$ are defined in the coordinates $\mathscr{R}e Z_{\nu,j}$, $\mathscr{I}m Z_{\nu,k}$; we now relate them to their analogues in the original coordinates. Let Ω be an open subset of $\mathbb{C}^n$ whose closure is compact and contained in U (see beginning of section VI.5); let f be an arbitrary element of $\mathscr{C}^\infty(\mathscr{C}\ell\Omega)$. We shall use the following notation:

$$\mathscr{H}N^{m+\kappa}_{\Omega}(f) = \|f\|_{\Omega,m} + \mathscr{H}^{m+\kappa}_{\Omega}(f),$$

where $\|f\|_{\Omega,m}$ is defined in (VI.3.14) and where

$$\mathscr{H}^{m+\kappa}_{\Omega}(f) = \sup_{z,z'\in\Omega}\left\{\sum_{|\alpha+\beta|=m}|\partial^\alpha\bar\partial^\beta f(z)-\partial^\alpha\bar\partial^\beta f(z')|/|z-z'|^\kappa\right\}.$$

In these definitions we are using the original coordinates x_j, y_k even when the open set Ω is one of the Ω_ν in the sequence we are currently using.

Since the vector fields L_j and $\overline{L}_k$ form a basis of the tangent bundle over U, there is a constant $C' > 0$ that depends solely on the system $\{L_1,\ldots,L_n\}$ (and therefore not on ν) such that, at every point of $\Omega_{\nu-1}$,

$$\|\partial Z_\nu\| + \|\bar\partial Z_\nu\| \leq C'(\|\mathscr{L}Z_\nu\| + \|\overline{\mathscr{L}}Z_\nu\|) \leq C'^2(\|\partial Z_\nu\| + \|\bar\partial Z_\nu\|). \qquad \text{(VI.6.1)}$$

Here $\partial Z_\nu = (\partial Z_{\nu,j}/\partial z_k)_{1\leq j,k\leq n}$, $\bar\partial Z_\nu = (\partial Z_{\nu,j}/\partial \bar z_k)_{1\leq j,k\leq n}$. For any $\nu\in\mathbb{Z}_+$ we have

$$\|\mathscr{L}Z_\nu\|\,\|(\mathscr{L}\overline{Z}_\nu)^{-1}\| \leq 1/2, \qquad \text{(VI.6.2)}$$

which entails, for $C'' > 0$ independent of ν,

$$\|(DZ_\nu)^{-1}\| \leq C''\|(\mathscr{L}\overline{Z}_\nu)^{-1}\| \qquad \text{(VI.6.3)}$$

(DZ_ν: differential of the map Z_ν).

Thanks to (VI.6.3), and by applying the chain rule and the mean value theorem in Z_ν-space (valid because Ω_ν is a ball; cf. (VI.6.19) below and subsequent remarks), we reach the following conclusion:

$$\mathscr{H}N^{m+\kappa}_{\Omega_\nu}(f) \leq$$
$$C[\mathscr{H}N^{m-1+\kappa}_{\Omega_\nu}(\mathscr{L}Z_\nu)+\mathscr{H}N^{m-1+\kappa}_{\Omega_\nu}(\mathscr{L}\overline{Z}_\nu)]^{m+\kappa}N^{m+\kappa}_{\delta_\nu}(f), \qquad \text{(VI.6.4)}$$

$$N^{m+\kappa}_{\delta_\nu}(f) \leq C[\mathscr{H}N^{m-1+\kappa}_{\Omega_\nu}((\mathscr{L}\overline{Z}_\nu)^{-1})]^{m+\kappa}\mathscr{H}N^{m+\kappa}_{\Omega_\nu}(f). \qquad \text{(VI.6.5)}$$

The same estimate is valid after we replace Ω_ν by Ω'_ν and δ_ν by δ'_ν.

We take (VI.6.4) and (VI.6.5) into account in connection with the estimates established earlier. In order to shorten the notation let us set

$$A_\nu = \mathscr{H}N^{m+\kappa}_{\Omega_\nu}(\mathscr{L}Z_\nu),\ B_\nu = \mathscr{H}N^{m+\kappa}_{\Omega_\nu}(I-\mathscr{L}\overline{Z}_\nu),\ B'_\nu = \mathscr{H}N^{m+\kappa}_{\Omega_\nu}((\mathscr{L}\overline{Z}_\nu)^{-1})$$

(I: $n\times n$ identity matrix).

By (VI.5.3) we know that, to each integer $N \geq 0$, there is $C_N > 0$ such that, whatever the open neighborhood of 0, $\Omega \subset\subset U$,

$$\mathcal{H}N_{\Omega}^{m+\kappa}(\mathcal{L}Z_0) \leq C_N(\operatorname{diam} \Omega)^N. \tag{VI.6.6}$$

In particular we have

$$A_0 \leq C_N \delta_0^N. \tag{VI.6.7}$$

On the other hand, by virtue of (VI.5.3) and (VI.5.4) we know that to each $\varepsilon > 0$, there is $\delta > 0$ such that, if $\operatorname{diam} \Omega < \delta$, then

$$\mathcal{H}N_{\Omega}^{m+\kappa}(I - \mathcal{L}\overline{Z}_0) < \varepsilon. \tag{VI.6.8}$$

It follows that, whatever the number ε_0, $0 < \varepsilon_0 < 1$, we can choose $\delta_0 > 0$ so small that, if $\operatorname{diam} \Omega_0 \leq \delta_0$, then

$$B_0 \leq \varepsilon_0,\ B_0' \leq (1-\varepsilon_0)^{-1}. \tag{VI.6.9}$$

We set, for any integer $\nu \geq 1$,

$$\varepsilon_\nu = \varepsilon_0^{(3/2)^\nu} \tag{VI.6.10}$$

and take

$$\delta_\nu' = \delta_\nu - \delta_0 \varepsilon_{\nu+1}. \tag{VI.6.11}$$

We shall use induction on ν to prove the following properties:

$$A_\nu \leq (CM)^{-1} \varpi^{6(m+\kappa)+2} \delta_0 \varepsilon_\nu^3, \qquad \text{(VI.6.12)}_\nu$$

$$(1 + A_\nu + B_\nu)^{m+\kappa+2} \leq M, \qquad \text{(VI.6.13)}_\nu$$

$$B_\nu' \leq 1/\varpi_\nu, \qquad \text{(VI.6.14)}_\nu$$

where

$$\varpi_\nu = (1-\varepsilon_0)(1-\varepsilon_1)\cdots(1-\varepsilon_\nu) \geq \varpi = \lim_{\nu \to +\infty} \varpi_\nu.$$

We note that, thanks to (VI.6.7), (VI.6.12)_0 will hold provided

$$C_N \delta_0^{N-1} \leq \varpi^{6(m+\kappa)} \varepsilon_0^3 / CM \tag{VI.6.15}$$

and by (VI.6.7) and (VI.6.9) the inequality (VI.6.13)_0 will hold as soon as

$$(1 + C_N \delta_0^N + \varepsilon_0)^{m+\kappa+2} \leq M. \tag{VI.6.16}$$

Clearly (VI.6.14)_0 is identical to the second inequality (VI.6.9).

We derive from (VI.5.23), (VI.6.4) and (VI.6.5):

$$N_{\delta_\nu'}^{m+1+\kappa}(Z_{\nu+1} - Z_\nu) \leq C(\delta_\nu - \delta_\nu')^{-1} B_\nu'^{2(m+\kappa)+1} A_\nu,$$

$$\mathcal{H}N_{\Omega_\nu'}^{m+1+\kappa}(Z_{\nu+1} - Z_\nu) \leq C(\delta_\nu - \delta_\nu')^{-1}(1 + A_\nu + B_\nu)^{m+\kappa} B_\nu'^{2(m+\kappa)+1} A_\nu.$$

But (VI.6.12)_ν, (VI.6.13)_ν and (VI.6.14)_ν show that

$$C(\delta_\nu-\delta'_\nu)^{-1}(1+A_\nu+B_\nu)^{m+\kappa}B'^{2(m+\kappa)+1}_\nu A_\nu \leq \varepsilon^{-1}_{\nu+1}\varpi^{4(m+\kappa)+1}\varepsilon^3_\nu.$$

And since $\varepsilon^3_\nu = \varepsilon^2_{\nu+1}$ (cf. (VI.6.10)), we get

$$N^{m+1+\kappa}_{\delta'_\nu}(Z_{\nu+1}-Z_\nu) \leq \varepsilon_{\nu+1}, \tag{VI.6.17}$$

$$\mathcal{H}N^{m+1+\kappa}_{\Omega'_\nu}(Z_{\nu+1}-Z_\nu) \leq \varepsilon_{\nu+1}. \tag{VI.6.18}$$

Denote by $J_\nu Z_{\nu+1}$ the Jacobian matrix of the map $Z_{\nu+1}$: $\Omega'_\nu \to \mathbb{C}^n$ in the coordinates $\xi_j = \mathcal{R}e Z_{\nu,j}$, $\xi_{n+k} = \mathcal{I}m Z_{\nu,k}$ $(1 \leq j,\ k \leq n)$. We derive from (VI.6.17)

$$\sup_{\Omega'_\nu} \|I-J_\nu Z_{\nu+1}\| \leq \varepsilon_{\nu+1} \tag{VI.6.19}$$

(I: $n\times n$ identity matrix) which implies that $J_\nu Z_{\nu+1}$ is nonsingular at each point of Ω'_ν. Furthermore, we may write, in the ball Ω'_ν,

$$\xi - \xi' - [Z_{\nu+1}(\xi)-Z_{\nu+1}(\xi')] = \left\{\int_0^1 [I-J_\nu Z_{\nu+1}(\xi+t(\xi-\xi'))]\mathrm{d}t\right\}(\xi-\xi').$$

As a consequence of (VI.6.19) we get

$$|\xi - \xi' - [Z_{\nu+1}(\xi)-Z_{\nu+1}(\xi')]| \leq \varepsilon_{\nu+1}|\xi-\xi'|,$$

which shows that the map $Z_{\nu+1}$ is injective. This proves $(\text{VI.5.5})_{\nu+1}$.

The estimate (VI.6.17) also implies $|Z_\nu| \leq \delta_{\nu+1}+\varepsilon_{\nu+1}$ in $\Omega_{\nu+1}$. Property $(\text{VI.5.7})_{\nu+1}$ will be satisfied, i.e., $\Omega_{\nu+1} \subset\subset \Omega'_\nu$, if we require

$$\delta_{\nu+1}+\varepsilon_{\nu+1} < \delta'_\nu. \tag{VI.6.20}$$

In order that there exists a number δ'_ν that satisfies both (VI.6.11) and (VI.6.20) we must have

$$\delta_{\nu+1} < \delta_\nu - (1+\delta_0)\varepsilon_{\nu+1}. \tag{VI.6.21}$$

Property $(\text{VI.5.7})_{\nu+1}$ enables us to derive from (VI.5.24):

$$A_{\nu+1} \leq C(\delta_\nu-\delta'_\nu)^{-1}(1+A_\nu+B_\nu)^{m+\kappa+2}B'^{6(m+\kappa)+2}_\nu A^2_\nu \leq$$

$$CM\delta_0^{-1}\varepsilon^{-1}_{\nu+1}\varpi^{-6(m+\kappa)-2}A^2_\nu \leq (CM)^{-1}\delta_0\varpi^{6(m+\kappa)+2}\varepsilon^{-1}_{\nu+1}\varepsilon^6_\nu,$$

and since $\varepsilon^{-1}_{\nu+1}\varepsilon^6_\nu = \varepsilon^3_{\nu+1}$ this proves $(\text{VI.6.12})_{\nu+1}$.

On the other hand, we derive from (VI.5.25):

$$\mathcal{H}N^{m+\kappa}_{\Omega'_\nu}(\mathcal{L}\bar{Z}_{\nu+1}-\mathcal{L}\bar{Z}_\nu) \leq C(\delta_\nu-\delta'_\nu)^{-1}(1+A_\nu+B'_\nu)^{m+\kappa}B'^{2(m+\kappa)+1}_\nu A_\nu \leq$$

$$CM\delta_0^{-1}\varpi^{-2(m+\kappa)-1}\varepsilon^{-1}_{\nu+1}A_\nu \leq \varpi^{4(m+\kappa)+1}\varepsilon^{-1}_{\nu+1}\varepsilon^3_\nu \leq \varpi^{4(m+\kappa)+1}\varepsilon_{\nu+1}$$

and, as a consequence,

$$\mathcal{H}N^{m+\kappa}_{\Omega'_\nu}(\mathcal{L}\bar{Z}_{\nu+1}-\mathcal{L}\bar{Z}_\nu) \leq \varpi_\nu\varepsilon_{\nu+1}. \tag{VI.6.22}$$

First we derive from (VI.6.22)

$$B_{\nu+1} \leq \sum_{j=1}^{\nu+1} \varepsilon_j. \qquad \text{(VI.6.23)}_{\nu+1}$$

By combining $(\text{VI.6.12})_{\nu+1}$ with $(\text{VI.6.23})_{\nu+1}$ we obtain $(\text{V.6.13})_{\nu+1}$, provided M is large enough to ensure

$$\left\{(CM)^{-1}\varpi^{6(m+\kappa)+2}\delta_0\varepsilon_0^3 + \sum_{\nu=0}^{+\infty}\varepsilon_\nu\right\}^{m+\kappa+2} \leq M. \qquad \text{(VI.6.24)}$$

Next we derive from (VI.6.22):

$$\sup_{\Omega_\nu'} \|\overline{\mathcal{L}}Z_{\nu+1} - \overline{\mathcal{L}}Z_0\| \leq \sum_{j=1}^{\nu+1} \varepsilon_j.$$

If we now avail ourselves of the first inequality (VI.6.9) and recall the choice (VI.6.10) we conclude that Property $(\text{VI.5.6})_{\nu+1}$ holds, i.e., $\overline{\mathcal{L}}Z_{\nu+1}$ is nonsingular at every point of Ω_ν', and that its inverse, $(\overline{\mathcal{L}}Z_{\nu+1})^{-1}$, is bounded in Ω_ν'. By the formula for the derivatives of the inverse and for their difference-quotients it follows that $B_{\nu+1}'$ is finite. We observe then that

$$\mathcal{H}N_{\Omega_\nu'}^{m+\kappa}((\mathcal{L}\overline{Z}_{\nu+1})^{-1} - (\mathcal{L}\overline{Z}_\nu)^{-1}) \leq B_\nu' B_{\nu+1}' \mathcal{H}N_{\Omega_\nu'}^{m+\kappa}(\mathcal{L}\overline{Z}_{\nu+1} - \mathcal{L}\overline{Z}_\nu),$$

and therefore (VI.6.22) entails

$$\mathcal{H}N_{\Omega_\nu'}^{m+\kappa}((\mathcal{L}\overline{Z}_{\nu+1})^{-1} - (\mathcal{L}\overline{Z}_\nu)^{-1}) \leq B_\nu' B_{\nu+1}' \varpi_\nu \varepsilon_{\nu+1}.$$

Taking into account $(\text{VI.6.14})_\nu$ we derive from the preceding inequality:

$$\mathcal{H}N_{\Omega_\nu'}^{m+\kappa}((\mathcal{L}\overline{Z}_{\nu+1})^{-1} - (\mathcal{L}\overline{Z}_\nu)^{-1}) \leq \varepsilon_{\nu+1} B_{\nu+1}', \qquad \text{(VI.6.25)}$$

and as a consequence

$$B_{\nu+1}' - B_\nu' \leq \varepsilon_{\nu+1} B_{\nu+1}',$$

which implies at once $(\text{VI.6.14})_{\nu+1}$.

We return to (VI.6.21) to get

$$\delta = \inf_\nu \delta_\nu < \delta_0 - (1+\delta_0)\sum_{\nu=1}^{+\infty}\varepsilon_\nu. \qquad \text{(VI.6.26)}$$

We require $\delta > \delta_0/2$ and therefore

$$\sum_{\nu=1}^{+\infty}\varepsilon_\nu < \tfrac{1}{2}\delta_0/(1+\delta_0). \qquad \text{(VI.6.27)}$$

Let us point out that (VI.6.15), (VI.6.16), and (VI.6.27) can be simultaneously satisfied by taking δ_0 sufficiently small, $\varepsilon_0 = \delta_0$, for instance, and $N \geq 5$.

From (VI.6.18) we derive, for all $\nu = 1,2,\ldots$,

$$|Z_\nu| \leq |Z_0| + \sum_{j=1}^{\nu}\varepsilon_j < |Z_0| + \tfrac{1}{2}\delta_0/(1+\delta_0).$$

This implies that the closure of the open set Ω defined by $|Z_0| < \frac{1}{2}\delta_0^2(1+\delta_0)$ is contained in the open set defined by $|Z_\nu| < \delta < \delta_\nu$, a fortiori in Ω_ν. From (VI.6.18) we conclude that the maps Z_ν converge in $\mathscr{C}^{m+1}(\mathscr{Cl}\Omega;\mathbb{C}^n)$ to a limit Z such that $\mathscr{L}Z = 0$ and $\mathscr{L}\bar{Z}$ is nonsingular at every point of Ω (by going to the limit, as $\nu \to +\infty$, in (VI.6.12)$_\nu$ and (VI.6.14)$_\nu$).

In order to complete the proof of Theorem VI.5.1 one must check that Z is a $\mathscr{C}^\infty$ map. The *ellipticity* of the second-order differential operator $\Delta_L = \bar{L}_1L_1 + \cdots + \bar{L}_nL_n$ and the fact that each component Z_j satisfies $\Delta_L Z_j = 0$ ensure that every Z_j belongs to $\mathscr{C}^\infty(\Omega)$. Another way of establishing the smoothness of Z is by exploiting fully what has been proved above, namely that each point p of $\mathcal{M}$ has a basis of open neighborhoods $U_m(p)$ ($m = 1,2,\ldots$; $U_{m+1}(p) \subset U_m(p)$) in each one of which there exists a $\mathscr{C}^m$ map $Z_{p,m}$: $U_m(p) \to \mathbb{C}^n$ whose components $Z_{p,m,j}$ ($1 \le j \le n$) are solutions and can be used as complex coordinates. The tangent structure bundle $\mathcal{V}$ is then spanned over $U_m(p)$ by the Cauchy-Riemann vector fields with respect to these coordinates. This entails that all other solutions in $U_m(p)$, such as the functions $Z_{p,m',j}$ for $m' < m$, are holomorphic functions of $Z_{p,m}$, and therefore are also of class $\mathscr{C}^m$. Letting m go to $+\infty$ proves that they are $\mathscr{C}^\infty$ functions. ∎

VI.7. Local Integrability and Local Solvability of Elliptic Structures. Levi Flat Structures

The Newlander-Nirenberg theorem has the following consequence:

THEOREM VI.7.1. *Any elliptic involutive structure* (Definition I.2.3) *on a* $\mathscr{C}^\infty$ *manifold* $\mathcal{M}$ *is locally integrable.*

PROOF. As usual we denote by m (resp., n) the fibre dimension (over $\mathbb{C}$) of the cotangent (resp., tangent) structure bundle of $\mathcal{M}$. The ellipticity of the structure of $\mathcal{M}$ demands $\mathbb{C}T\mathcal{M} = \mathcal{V} + \bar{\mathcal{V}}$ (and thus $m \le n$). It follows that $\mathcal{V} \cap \bar{\mathcal{V}}$ is a vector bundle over $\mathcal{M}$ of fibre dimension $n' = n - m$; it is spanned, over $\mathbb{C}$, by its intersection $\mathcal{V}_0$ with the real tangent bundle $T\mathcal{M}$. Of course $[\mathcal{V}_0,\mathcal{V}_0] \subset \mathcal{V}_0$ since both $\mathcal{V}$ and $\bar{\mathcal{V}}$ are involutive. We apply the Frobenius theorem to the real structure defined by $\mathcal{V}_0$. It follows that in some open neighborhood U of a point $0 \in \mathcal{M}$ we may choose the coordinates ξ_i ($1 \le i \le 2m$), t_k ($1 \le k \le n'$) in such a way that $\mathcal{V}_0$ will be spanned over U by the vector fields $\partial/\partial t_1,\ldots,\partial/\partial t_{n'}$. To obtain a full basis of $\mathcal{V}$ over U (possibly contracted about 0) we adjoin to them m additional $\mathscr{C}^\infty$ vector fields

$$L_i = \sum_{j=1}^{2m} a_{ij}(\xi,t)\partial/\partial\xi_j, \quad i = 1,\ldots,m.$$

The $m \times (2m)$ matrix $(a_{ij})_{1 \le i < m, 1 \le j \le 2m}$ must have rank m and, therefore, after a linear substitution of the L_i and a relabeling of the coordinates ξ_j we may assume that

$$L_i = \partial/\partial\xi_{m+i} + \sum_{j=1}^{m} a_{ij}\partial/\partial\xi_j,\ i = 1,\ldots,m.$$

That $\mathcal{V}$ is involutive demands that the vector fields L_j and $\partial/\partial t_k$ commute, which in turn entails that the coefficients $a_{ij} \in \mathcal{C}^\infty(U)$ be independent of t. The system $\{L_1,\ldots,L_m\}$ defines a complex structure on an open neighborhood V of the origin in $\mathbb{R}^{2m}$. According to Theorem VI.5.1 there exist $\mathcal{C}^\infty$ functions $Z_1(\xi),\ldots,Z_m(\xi)$ in V (possibly contracted about 0) satisfying the following conditions in V:

$$dZ_1\wedge\cdots\wedge dZ_m \neq 0, \tag{VI.7.1}$$

$$L_jZ_i = 0,\ i = 1,\ldots,m, \tag{VI.7.2}$$

for all $j = 1,\ldots,m$. If we view Z_i as a function of (ξ,t) in $V\times\mathbb{R}^{n'}$ the equations (VI.7.2) are trivially satisfied when L_j is replaced by $\partial/\partial t_k$, $k = 1,\ldots,n'$. ∎

Until specified otherwise we assume that $\mathcal{M}$ is equipped with an elliptic structure. The analysis takes place in the open neighborhood U of the point $0 \in \mathcal{M}$. We go on using the notation in the proof of Theorem VI.7.1. We have the right to take $\mathcal{R}e Z_i$ (resp., $\mathcal{I}m Z_j$) as coordinate x_i (resp., y_j) in U; henceforth we write z_j instead of Z_j. Thus T′ is spanned over U by $dz_1,\ldots,dz_m$, and $\mathcal{V}$ by $\partial/\partial\bar{z}_j$, $\partial/\partial t_k$ $(1 \leq j \leq m,\ 1 \leq k \leq n')$. It is convenient to take U to contain the closure of a product set $\Omega\times\mathcal{O}$ with Ω (resp., $\mathcal{O}$) an open ball centered at the origin in z-space $\mathbb{C}^m$ (resp., t-space $\mathbb{R}^{n'}$).

According to (I.7.34), the standard representative of a section $\dot{f} \in \mathcal{C}^\infty(U;\Lambda^{p,q})$ will be a form of the kind

$$f = \sum_{|I|=p}\ \sum_{|J|+|J'|=q} f_{I,J,J'}(x,y,t)\ dz_I\wedge d\bar{z}_J\wedge dt_{J'} \tag{VI.7.3}$$

with $f_{I,J,J'} \in \mathcal{C}^\infty(U)$. The standard representative of $d'\dot{f}$ is then $(\bar{\partial}_z + d_t)f$. We are going to construct a bounded linear operator

$$\mathcal{H} : \mathcal{C}^\infty(\mathcal{C}\ell(\Omega\times\mathcal{O});\Lambda^{p,q}) \to \mathcal{C}^\infty(\mathcal{C}\ell(\Omega\times\mathcal{O});\Lambda^{p,q-1})$$

such that

$$f = (\bar{\partial}_z + d_t)\mathcal{H}f + \mathcal{H}(\bar{\partial}_z + d_t)f. \tag{VI.7.4}$$

Our construction is one among many that are possible. Actually we shall restrict ourselves to the case $p = 0$. For $p > 0$ we may rewrite the expression of f (in (VI.7.3)) as

$$f = \sum_{|I|=p} f_I\wedge dz_I,\ f_I \in \mathcal{C}^\infty(\mathcal{C}\ell(\Omega\times\mathcal{O});\Lambda^{0,q}),$$

and set

$$\mathcal{H}f = \sum_{|I|=p} (\mathcal{H}f_I)\wedge dz_I.$$

Below q will always stand for an integer ≥ 1. First suppose $m \geq q$. Consider the Bochner-Martinelli-Leray-Koppelman operator acting on smooth (m,q)-forms in the closure of the open ball $\Omega \subset \mathbb{C}^m$, as defined in sections VI.1 and VI.2. Call $\mathcal{T}$ the isomorphism $\varphi_0 \to \varphi_0\wedge dz$ of $\mathcal{C}^\infty(\mathcal{C}\ell\Omega;\Lambda^{0,q})$ onto $\mathcal{C}^\infty(\mathcal{C}\ell\Omega;\Lambda^{m,q})$, and set

$$G^{(q)}\varphi_0 = (-1)^{n-q}\mathcal{T}^{-1}(\mathcal{K}_\Omega + \mathcal{B}_{\partial\Omega})\mathcal{T}\varphi_0 \in \mathcal{C}^\infty(\mathcal{C}\ell\Omega;\Lambda^{0,q-1}).$$

Note that $d\mathcal{T}\varphi_0 = \mathcal{T}\bar\partial\varphi_0$. According to (VI.2.12), we have

$$\varphi_0 = \bar\partial G^{(q)}\varphi_0 + G^{(q+1)}\bar\partial\varphi_0. \tag{VI.7.5}$$

Below we write G rather than $G^{(q)}$.

Let $K^{(q)}$ be the homotopy operator for the exterior derivative d_t in the open ball $\mathbb{O}$ as defined in (II.6.4); the operators $K^{(q)}$ satisfy the identity (II.6.5). We shall write K rather than $K^{(q)}$.

We decompose the differential form under study,

$$f = \sum_{|I|+|J|=q} f_{I,J}(x,y,t)\, d\bar z_I\wedge dt_J, \tag{VI.7.6}$$

as follows:

$$f = \sum_{r=0}^{q} \sum_{|J|=r} f_J\wedge dt_J, \tag{VI.7.7}$$

with f_J a $\mathcal{C}^\infty$ differential form of bidegree $(0,q-r)$ in $\mathcal{C}\ell\Omega \subset \mathbb{C}^m$ depending smoothly on the parameter $t \in \mathcal{C}\ell\mathbb{O}$. We then define

$$Gf = \sum_{r=0}^{q-1} \sum_{|J|=r} (Gf_J)\wedge dt_J. \tag{VI.7.8}$$

Notice that

$$(\bar\partial_z + d_t)Gf = \sum_{r=0}^{q-1} \sum_{|J|=r} (\bar\partial_z Gf_J)\wedge dt_J +$$

$$\sum_{r=0}^{q-1}(-1)^{q-r+1}\sum_{|J|=r}\sum_{\ell=1}^{n'} (G[(\partial/\partial t_\ell)f_J])\wedge dt_\ell\wedge dt_J.$$

On the other hand,

$$G[(\bar\partial_z + d_t)f] = \sum_{r=0}^{q} \sum_{|J|=r} G(\bar\partial_z f_J)\wedge dt_J +$$

$$\sum_{r=0}^{q-1}(-1)^{q-r}\sum_{|J|=r}\sum_{\ell=1}^{n'}(\mathrm{G}[(\partial/\partial t_\ell)f_J])\wedge \mathrm{d}t_\ell\wedge \mathrm{d}t_J.$$

By adding these two equations and availing ourselves of (VI.7.5) we get

$$f = (\bar{\partial}_z+\mathrm{d}_t)(\mathrm{G}f) + \mathrm{G}[(\bar{\partial}_z+\mathrm{d}_t)f] + g, \tag{VI.7.9}$$

where

$$g = \sum_{|J|=q}[f_J-\mathrm{G}(\bar{\partial}_z f_J)]\mathrm{d}t_J$$

(when $|J| = q, f_J$ and $\mathrm{G}\bar{\partial}_z f_J$ are zero-forms). By (VI.7.5) applied to $\varphi_0 = \bar{\partial}_z f_J$ we get

$$\bar{\partial}_z g \equiv 0, \tag{VI.7.10}$$

and therefore also

$$\bar{\partial}_z \mathrm{K}g \equiv 0. \tag{VI.7.11}$$

By virtue of (II.6.5), (VI.7.10), and (VI.7.11) we get

$$g = \mathrm{d}_t\mathrm{K}g - \mathrm{K}\mathrm{d}_t g = (\bar{\partial}_z+\mathrm{d}_t)\mathrm{K}g - \mathrm{K}(\bar{\partial}_z+\mathrm{d}_t)g.$$

By combining this property with (VI.7.9) and by setting

$$\mathscr{H}f = \mathrm{G}f + \mathrm{K}\left\{\sum_{|J|=q}[f_J-\mathrm{G}(\bar{\partial}_z f_J)]\mathrm{d}t_J\right\}, \tag{VI.7.12}$$

we see that $\mathscr{H}$ does indeed satisfy (VI.7.4).

Suppose now $n' = n - m \geq q$. We can decompose the form (VI.7.6) as

$$f = \sum_{r=0}^{q}\sum_{|I|=r}\tilde{f}_I\wedge \mathrm{d}\bar{z}_I,$$

where the $\tilde{f}_I$ are now forms of degree $q-|I|$ in $\mathbb{C}$ whose coefficients are $\mathscr{C}^\infty$ functions in $\mathscr{O}(\Omega\times\mathbb{C})$. In this case set

$$\mathrm{K}f = \sum_{r=0}^{q-1}\sum_{|I|=r}(\mathrm{K}\tilde{f}_I)\wedge \mathrm{d}\bar{z}_I.$$

The operator $\mathscr{H}$ defined by

$$\mathscr{H}f = \mathrm{K}f + \mathrm{G}\left\{\sum_{|I|=q}[\tilde{f}_I-\mathrm{K}(\mathrm{d}_t\tilde{f}_I)]\mathrm{d}\bar{z}_I\right\} \tag{VI.7.13}$$

also satisfies (VI.7.4).

REMARK VI.7.1. Obviously there are other ways of combining the action of

G with that of K, intermediate between the two extremes (VI.7.12) and (VI.7.13), which still yield a linear operator $\mathcal{H}$ verifying (VI.7.4). ∎

REMARK VI.7.2. Formula (VI.7.4) entails that, in an elliptic structure on $\mathcal{M}$, any basis $\{L_1,\ldots,L_n\}$ of $\mathcal{V}$ over an open set Ω of $\mathcal{M}$ is hypo-elliptic in Ω (Definition III.5.2). Indeed, suppose $f \in \mathcal{D}'(\Omega)$ and $L_j f \in \mathcal{C}^\infty(\Omega)$, $j = 1,\ldots,n$. Let $U \subset \Omega$ be the domain of local coordinates $z_1,\ldots,z_\nu$, $t_1,\ldots,t_{n-\nu}$ in which Formula (VI.7.4) is valid: by hypothesis, $(\bar{\partial}_z + d_t)f \in \mathcal{C}^\infty(U)$. Since $\mathcal{H}f \equiv 0$ and $\mathcal{H}$ maps $\mathcal{C}^\infty(U;\Lambda^{0,1})$ into $\mathcal{C}^\infty(U)$ (cf. Theorem VI.3.1) we conclude that $f \in \mathcal{C}^\infty(U)$. ∎

Levi Flat Structures

Elliptic structures are a subclass of the following more general class of involutive structures:

DEFINITION VI.7.1. *We say that the involutive structure of $\mathcal{M}$ is* Levi flat *if the Whitney sum $\mathcal{V} + \overline{\mathcal{V}}$ is a vector subbundle of $\mathbb{C}T\mathcal{M}$ and defines an involutive structure on $\mathcal{M}$.*

If $\mathcal{V}$ is Levi flat and if L_1, L_2 are two smooth sections of $\mathcal{V}$ over some open subset Ω of $\mathcal{M}$, then the commutator $[L_1,\overline{L}_2]$ is a section of $\mathcal{V} + \overline{\mathcal{V}}$ over Ω and therefore is orthogonal to the characteristic set T^0. It follows that the Levi form (Definition I.8.1) in the structure defined by $\mathcal{V}$ vanishes identically. This is the motivation for the terminology introduced in Definition VI.7.1.

PROPOSITION VI.7.1. *Every elliptic structure is Levi flat.*

This is obvious: for, if the structure of $\mathcal{M}$ is elliptic, $\mathcal{V} + \overline{\mathcal{V}} = \mathbb{C}T\mathcal{M}$.

Assume the involutive structure of $\mathcal{M}$ is Levi flat. The involutive structure defined by $\mathcal{V} + \overline{\mathcal{V}}$ on $\mathcal{M}$ is real (Definition I.2.3) and therefore, by the Frobenius theorem, it defines a foliation in $\mathcal{M}$ (see section I.10). Every section of $\mathcal{V} + \overline{\mathcal{V}}$ over a subset of a leaf $\mathcal{L}$ of that foliation is tangent to $\mathcal{L}$ and $\mathcal{V} + \overline{\mathcal{V}}$ defines an elliptic structure on $\mathcal{L}$.

The intersection $\mathcal{V}\cap\overline{\mathcal{V}}$ is a vector subbundle of $\mathbb{C}T\mathcal{M}$; and it clearly defines an involutive real structure on $\mathcal{M}$, whose associated foliation is "finer" than the foliation defined by $\mathcal{V} + \overline{\mathcal{V}}$: an arbitrary leaf of $\mathcal{V}\cap\overline{\mathcal{V}}$ is contained in one (and only one) leaf of $\mathcal{V} + \overline{\mathcal{V}}$. Call n' the rank of $\mathcal{V}\cap\overline{\mathcal{V}}$. Along an arbitrary leaf $\mathcal{L}$ in the foliation defined by $\mathcal{V} + \overline{\mathcal{V}}$ we have

$$\mathbb{C}T\mathcal{L} \cong (\mathcal{V}\cap\overline{\mathcal{V}})\oplus(\mathcal{V}/\mathcal{V}\cap\overline{\mathcal{V}})\oplus(\overline{\mathcal{V}}/\mathcal{V}\cap\overline{\mathcal{V}})$$

which shows that $\dim \mathcal{L} - n'$ is an even number, $2m'$.

It was first proved in Nirenberg [1] that every Levi flat structure is locally integrable. Here we shall content ourselves with proving a weaker result:

THEOREM VI.7.2. *Suppose that the involutive structure of $\mathcal{M}$ is Levi flat. Then, given any integer $\mu \geq 1$, each point $\mathfrak{p}$ of $\mathcal{M}$ has an open neighborhood in which* T′ *has a basis made up of the differentials of m $\mathcal{C}^\mu$ functions.*

PROOF. Every point $\mathfrak{p} \in \mathcal{M}$ has an open neighborhood U in which there are local coordinates $\xi_1,\ldots,\xi_{2m'}, \eta_1,\ldots,\eta_r, t_1,\ldots,t_{n'}$ ($2m' + n' + r = \dim \mathcal{M}$) such that the leaves of the foliation associated with $\mathcal{V} + \overline{\mathcal{V}}$ are defined in U by the equations $\eta = const.$, while those of the foliation associated with $\mathcal{V}\cap\overline{\mathcal{V}}$ are defined by the equations $\xi = const.$, $\eta = const.$ The vector fields $\partial/\partial t_1,\ldots, \partial/\partial t_{n'}$ form a smooth basis of $\mathcal{V}\cap\overline{\mathcal{V}}$ over U. To obtain a full basis of $\mathcal{V}$ over U (possibly contracted about 0) we adjoin to them m' additional $\mathcal{C}^\infty$ vector fields that, after a linear substitution and a relabeling of the coordinates ξ_i (cf. beginning of the proof of Theorem VI.7.1), can be taken to be of the kind

$$\mathrm{L}_i = \partial/\partial\xi_{m'+i} + \sum_{j=1}^{m'} a_{ij}\partial/\partial\xi_j,\ i = 1,\ldots,m'.$$

Because the vector fields L_i and $\partial/\partial t_k$ commute, the coefficients a_{ij} only depend (smoothly) on (ξ,η). By focusing on the vector fields L_i alone we see that we are dealing with a family of complex structures depending on the parameter η (which varies in some open neighborhood W of 0 in $\mathbb{R}^r$).

Inspection of the proof of Theorem VI.5.1 shows easily that if a complex structure on a manifold depends in $\mathcal{C}^\infty$ fashion on $\eta \in$ W, then whatever the integer $\mu \geq 1$, one can find, in a suitable neighborhood of an arbitrary point, first integrals that are of class $\mathcal{C}^\mu$ with respect to η. In our situation this means that, to each integer $\mu \geq 1$ there exist an open neighborhood V of 0 in (ξ,η)-space $\mathrm{R}^{2m'}\times\mathbb{R}^r$ and $\mathcal{C}^\mu$ functions in V, $Z_i(\xi,\eta)$ $(1 \leq i \leq m')$, satisfying the conditions

$$d_\xi Z_1\wedge\cdots\wedge d_\xi Z_{m'} \neq 0, \tag{VI.7.14}$$

$$\mathrm{L}_j Z_i = 0,\ i,j = 1,\ldots,m'. \tag{VI.7.15}$$

The ellipticity of the system $\{\mathrm{L}_1,\ldots,\mathrm{L}_{m'}\}$ entails that the functions Z_i are $\mathcal{C}^\infty$ with respect to ξ. Returning now to (ξ,η,t)-space it is clear that the differentials $dZ_1,\ldots,dZ_{m'}, d\eta_1,\ldots,d\eta_r$ make up a basis of T′ over a suitably small neighborhood of the central point $\mathfrak{p}$. ∎

In the notation of the proof of Theorem VI.7.2 we may take $x_i = \mathcal{R}e Z_i$, $y_j = \mathcal{I}m Z_j$ $(1 \leq i, j \leq m')$ as new ξ-coordinates. Note, however, that these new coordinates are only of class $\mathcal{C}^\mu$. The tangent structure bundle $\mathcal{V}$ is spanned by the vector fields $\partial/\partial\bar{z}_j$, $\partial/\partial t_k$ $(1 \leq j \leq m', 1 \leq k \leq n')$. We may then make use of the homotopy formula (VI.7.4); it suffices to regard η as a parameter.

VI.8. Partial Local Group Structures

Let $\mathcal{M}$ be a $\mathcal{C}^\infty$ manifold. In the present section, we suppose given a *finite dimensional* Lie subalgebra $\mathfrak{g}$ of $\mathcal{C}^\infty(\mathcal{M};\mathrm{T}\mathcal{M})$: $\mathfrak{g}$ is a finite dimensional real vector space, stable for the commutation bracket. If $\mathfrak{p}$ is a point of $\mathcal{M}$, we call $\mathfrak{g}_\mathfrak{p}$ the subspace of $\mathrm{T}_\mathfrak{p}\mathcal{M}$ obtained by freezing at $\mathfrak{p}$ every vector field belonging to $\mathfrak{g}$. We reason under the hypothesis that

$$\dim \mathfrak{g}_\mathfrak{p} = \dim \mathfrak{g},\ \forall\, \mathfrak{p} \in \mathcal{M}. \tag{VI.8.1}$$

We write $n' = \dim \mathfrak{g}$, $m' = \dim \mathcal{M} - n'$. Because of (VI.8.1) the union of the vector spaces $\mathfrak{g}_\mathfrak{p}$ is a vector subbundle $\mathcal{W}_0{}^\mathfrak{g}$ of $\mathrm{T}\mathcal{M}$, of rank n'; we shall denote by $\mathcal{W}^\mathfrak{g}$ its complexification, regarded as a subbundle of $\mathbb{C}\mathrm{T}\mathcal{M}$. Both these vector subbundles, $\mathcal{W}_0{}^\mathfrak{g}$ and $\mathcal{W}^\mathfrak{g}$, are involutive.

We are going to need the following classical result (see Spivak [1], I, chap. 10):

THEOREM VI.8.1. *Let $\mathfrak{g}$ be a Lie subalgebra of $\mathcal{C}^\infty(\mathcal{M};\mathrm{T}\mathcal{M})$ whose dimension is equal to $n' \leq \dim \mathcal{M}$ and which satisfies* (VI.8.1).

Each point $\mathfrak{p}$ of $\mathcal{M}$ has an open neighborhood in which there are coordinates $x_1,\ldots,x_{m'}$, $u_1,\ldots,u_{n'}$, vanishing at $\mathfrak{p}$, such that the vector fields belonging to $\mathfrak{g}$ all have expressions

$$\sum_{j=1}^{n'} c_j(u)\partial/\partial u_j \tag{VI.8.2}$$

with coefficients c_j that are real-valued analytic functions of $u = (u_1,\ldots,u_{n'})$ in some open neighborhood of 0 *in* $\mathbb{R}^{n'}$.

PROOF. First we apply the Frobenius theorem (Theorem I.10.1) and select local coordinates $y_1,\ldots,y_{m'}$, $v_1,\ldots,v_{n'}$ in some open neighborhood Ω of $\mathfrak{p} \in \mathcal{M}$, vanishing at $\mathfrak{p}$ and such that the leaves of the foliation associated with $\mathcal{W}^\mathfrak{g}$ (see first part of section I.1.10) are defined, in Ω, by the equations $y = const$. We shall make use of a basis of the Lie algebra $\mathfrak{g}$, made up of vector fields

$$\mathrm{M}_j = \sum_{k=1}^{n'} a_{jk}(y,v)\partial/\partial v_k,\quad j = 1,\ldots,n'. \tag{VI.8.3}$$

The coefficients a_{jk} are real-valued and $\mathcal{C}^\infty$ in Ω; the matrix $A = (a_{jk})_{1\leq j,k\leq n'}$ is nonsingular. We have

$$[\mathrm{M}_i,\mathrm{M}_j] = \sum_{\ell=1}^{n'} C_{ij,\ell}\mathrm{M}_\ell. \tag{VI.8.4}$$

The *constants of structure* $C_{ij,\ell}$ satisfy the identities:

$$C_{jk,\ell} = -C_{kj,\ell}, \tag{VI.8.5}$$

$$\sum_{p=1}^{n'} C_{ij,p}C_{pk,\ell} + C_{jk,p}C_{pi,\ell} + C_{ki,p}C_{pj,\ell} = 0. \tag{VI.8.6}$$

The basis (VI.8.3) enables us to introduce the forms

$$\beta_k = \sum_{\ell=1}^{n'} b_{k\ell}(y,v)\mathrm{d}v_\ell \ (k = 1,\ldots,n')$$

defined by the conditions

$$\langle\beta_k,\mathrm{M}_\ell\rangle = \delta_{k\ell},\ 1 \le k,\ \ell \le n'. \tag{VI.8.7}$$

For any $\mathscr{C}^1$ function f in Ω we have $\mathrm{d}_v f = \sum_{k=1}^{n'}(\mathrm{M}_k f)\beta_k$, and therefore

$$\sum_{\substack{j,k=1\\ j<k}}^{n'} ([\mathrm{M}_j,\mathrm{M}_k]f)\beta_j\wedge\beta_k + \sum_{\ell=1}^{n'}(\mathrm{M}_\ell f)\mathrm{d}_v\beta_\ell = 0,$$

whence, by (VI.8.4),

$$\mathrm{d}_v\beta_\ell = -\tfrac{1}{2}\sum_{j,k=1}^{n'} C_{jk,\ell}\beta_j\wedge\beta_k,\ \ell = 1,\ldots,n'. \tag{VI.8.8}$$

We consider the following initial value problem for a system of linear ODE, in which the unknowns are differential one-forms in u-space $\mathbb{R}^{n'}$:

$$\dot\alpha_i = \mathrm{d}u_i - \sum_{p,q=1}^{n'} C_{pq,i}u_p\alpha_q, \tag{VI.8.9}$$

$$\alpha_i\big|_{t=0} = 0. \tag{VI.8.10}$$

Its unique solutions $\{\alpha_i\}_{1\le i\le n'}$ are real forms and their coefficients (in the basis $\mathrm{d}u_1,\ldots,\mathrm{d}u_{n'}$) extend as entire functions of (u,t) in $\mathbb{C}^{n'}\times\mathbb{C}$. In the immediate sequel d stands for the exterior derivative in u-space. We derive from (VI.8.9):

$$\mathrm{d}\dot\alpha_i = -\sum_{p,q=1}^{n'} C_{pq,i}\mathrm{d}u_p\wedge\alpha_q - \sum_{p,q=1}^{n'} C_{pq,i}u_p\mathrm{d}\alpha_q =$$

$$-\sum_{p,q=1}^{n'} C_{pq,i}\dot\alpha_p\wedge\alpha_q - \sum_{p,q=1}^{n'} C_{pq,i}\sum_{j,k=1}^{n'} C_{jk,p}u_j\alpha_k\wedge\alpha_q - \sum_{p,q=1}^{n'} C_{pq,i}u_p\mathrm{d}\alpha_q.$$

By (VI.8.5) we have

$$(\partial/\partial t)\left\{\sum_{p,q=1}^{n'} C_{pq,i}\alpha_p\wedge\alpha_q\right\} = 2\sum_{p,q=1}^{n'} C_{pq,i}\dot{\alpha}_p\wedge\alpha_q,$$

and thus, if we set $\vartheta_i = \left\{d\alpha_i + \frac{1}{2}\sum_{p,q=1}^{n'} C_{pq,i}\alpha_p\wedge\alpha_q\right\}$,

$$\dot{\vartheta}_i + \sum_{p,q=1}^{n'} C_{pq,i}u_p\vartheta_q =$$

$$-\sum_{p,q=1}^{n'} C_{p\ell,i}\sum_{j,k=1}^{n'} C_{jk,p}u_j\alpha_k\wedge\alpha_\ell + \frac{1}{2}\sum_{j,p=1}^{n'} C_{pj,i}u_j\sum_{k,\ell=1}^{n'} C_{\ell k,p}\alpha_k\wedge\alpha_\ell =$$

$$-\sum_{j,k,\ell=1}^{n'}\left\{\sum_{p=1}^{n'} C_{jk,p}C_{p\ell,i} + \tfrac{1}{2}C_{k\ell,p}C_{pj,i}\right\}u_j\alpha_k\wedge\alpha_\ell.$$

By exchanging the summation with respect to k and to ℓ and taking (VI.8.5) into account, we get

$$\sum_{k,\ell=1}^{n'} C_{jk,p}C_{p\ell,i}\alpha_k\wedge\alpha_\ell = \sum_{k,\ell=1}^{n'} C_{\ell j,p}C_{pk,i}\alpha_k\wedge\alpha_\ell$$

and therefore

$$\sum_{j,k,\ell=1}^{n'}\left\{\sum_{p=1}^{n'} C_{jk,p}C_{p\ell,i} + \tfrac{1}{2}C_{k\ell,p}C_{pj,i}\right\}u_j\alpha_k\wedge\alpha_\ell =$$

$$\frac{1}{2}\sum_{j,k,\ell=1}^{n'}\left\{\sum_{p=1}^{n'} C_{jk,p}C_{p\ell,i} + C_{\ell j,p}C_{pk,i} + C_{k\ell,p}C_{pj,i}\right\}u_j\alpha_k\wedge\alpha_\ell = 0$$

by (VI.8.6). In summary, we have proved that

$$\dot{\vartheta}_i + \sum_{p,q=1}^{n'} C_{pq,i}u_p\vartheta_q = 0.$$

Since, by (VI.8.10) and the definition of ϑ_i, we have $\vartheta_i|_{t=0} = 0$, we conclude, by the uniqueness in the fundamental theorem on ODE, that $\vartheta_i = 0$ for all t. We have thus proved that the one-forms in u-space, $\alpha_i(t)$ $(1 \le i \le n')$, also satisfy the identity (VI.8.8) (with d_u substituted for d_v) whatever $t \in \mathbb{R}$.

It follows from (VI.8.9) that, when $u = 0$, then $\dot{\alpha}_i = du_i$ for all t and thus $\alpha_i(t)|_{u=0} = t\,du_i$. We reach the following conclusion:

there is an open neighborhood $\mathcal{A}$ of the origin in $\mathbb{R}^{n'}$ in which the one-forms $\alpha_1(1),\ldots,\alpha_{n'}(1)$ are linearly independent. (VI.8.11)

From now on we fix $t = 1$ and write α_i to mean $\alpha_i(1)$.

It is convenient to take the neighborhood Ω of p to be a product $\mathcal{U}\times\mathcal{B}$ with $\mathcal{U}$ (resp., $\mathcal{B}$) an open neighborhood of 0 in y-space $\mathbb{R}^{m'}$ (resp., in v-space $\mathbb{R}^{n'}$). Consider the following one-forms in the product $\mathcal{A}\times\mathcal{B}\subset\mathbb{R}^{n'}\times\mathbb{R}^{n'}$,

$$\varphi_k = \pi_1{}^*\alpha_k - \pi_2{}^*\beta_k. \tag{VI.8.12}$$

We have denoted by π_1 (resp., π_2) the projection $\mathcal{A}\times\mathcal{B}\to\mathcal{A}$ (resp.,$\to\mathcal{B}$); $\pi_1{}^*$, $\pi_2{}^*$ are the corresponding pullback maps. Let d now denote the exterior derivative in (u,v)-space; the identities (VI.8.8) for the forms α_i and β_i entail

$$-2d\varphi_k = \sum_{p,q=1}^{n'} C_{pq,k}(\pi_1{}^*\alpha_p\wedge\pi_1{}^*\alpha_q - \pi_2{}^*\beta_p\wedge\pi_2{}^*\beta_q) =$$

$$\sum_{p,q=1}^{n'} C_{pq,k}(\pi_1{}^*\alpha_p\wedge\varphi_q + \varphi_p\wedge\pi_2{}^*\beta_q).$$

This proves that the system of forms φ_k defines an involutive real structure on $\mathcal{A}\times\mathcal{B}$. By the Frobenius theorem $\mathcal{A}\times\mathcal{B}$ is foliated by the leaves of this structure; the leaves are smooth submanifolds of dimension n'. Let Λ denote the leaf that passes through the point $(0,0)\in\mathcal{A}\times\mathcal{B}$.

The one-forms $\pi_1{}^*\alpha_i$, $\pi_2{}^*\beta_j$ $(1\le i,j\le n')$ span $T^*(\mathcal{A}\times\mathcal{B})$; it follows that the one-forms $\pi_1{}^*\alpha_i$, φ_j also do, and so do the one-forms φ_i, $\pi_2{}^*\beta_j$. Since the leaves of the structure defined by $\{\varphi_1,\dots,\varphi_{n'}\}$ are precisely the manifolds on which the pullbacks of the φ_i all vanish, we see that the pullbacks to Λ of the $\pi_1{}^*\alpha_i$ span $T^*\Lambda$; and so do the pullbacks of the $\pi_2{}^*\beta_j$. Finally, since the α_i (resp., β_j) make up a basis of $T^*\mathcal{A}$ (resp., $T^*\mathcal{B}$) we conclude that the restriction of π_1 (resp., π_2) is a diffeomorphism of an open neighborhood Λ' of $(0,0)$ in Λ onto an open neighborhood $\mathcal{A}'$ (resp., $\mathcal{B}'$) of the origin in $\mathcal{A}$ (resp., $\mathcal{B}$). The compose

$$F = \pi_2\circ(\pi_2|_{\Lambda'})^{-1}$$

is a diffeomorphism of $\mathcal{A}'$ onto $\mathcal{B}'$, with the obvious property that

$$\alpha_i = F^*\beta_i,\ i = 1,\dots,n'.$$

We must now take into account the fact that β_i depends smoothly on $y\in\mathcal{U}$; it follows at once that the submanifold Λ will also depend smoothly on y, and so will the diffeomorphism F. This makes it possible to introduce the diffeomorphism from $\mathcal{U}\times\mathcal{A}'$ onto $\mathcal{U}\times\mathcal{B}'$,

$$y = x,\ v = F(x,u). \tag{VI.8.13}$$

If we now regard β_i as a form in $\Omega = \mathcal{U}\times\mathcal{B}$, its pullback under the map (VI.8.13) is a form

$$\tilde{\alpha}_i = \alpha_i + \sum_{j=1}^{n'} \sum_{k=1}^{m'} b_{ij}(x,F(x,u))\,(\partial F_j/\partial x_k)\mathrm{d}x_k.$$

The pullback under the map (VI.8.13) of the vector field M_j is a vector field

$$\tilde{M}_j = \sum_{k=1}^{n'} \tilde{a}_{jk}(x,u)\partial/\partial u_k,$$

and we have, of course, $\langle\tilde{\alpha}_i, \tilde{M}_j\rangle = \delta_{ij}$. But this is equivalent to

$$\langle\alpha_i, \tilde{M}_j\rangle = \delta_{ij}\ (1 \le i, j \le n'). \tag{VI.8.14}$$

This means that the matrix of the coefficients of the vector fields $\tilde{M}_j$ is the inverse of the matrix of the coefficients of the forms α_i; it entails that the coefficients of the $\tilde{M}_j$ are analytic functions of u, independent of x, in $\mathcal{A}' \times \mathcal{B}'$. ■

As in the proof of Theorem VI.8.1 we reason in an open neighborhood $\Omega = \mathcal{U} \times \mathcal{A}$ of the point $p \in \mathcal{M}$: $\mathcal{U}$ is an open neighborhood of 0 in x-space $\mathbb{R}^{m'}$, $\mathcal{A}$ one in u-space $\mathbb{R}^{n'}$, and we assume that an arbitrary vector field $M \in \mathfrak{g}$ has an expression (VI.8.2) in the local chart $(\Omega, x_1,\ldots,x_{m'}, u_1,\ldots,u_{n'})$. This allows us to regard $\mathfrak{g}$ as a Lie algebra of real-analytic (and real) vector fields in $\mathcal{A}$.

To each $M \in \mathfrak{g}$ we can associate its flow $\Phi_M(t)$. We recall its definition: given an arbitrary point $a \in \mathcal{A}$, the initial value problem $\dot{u} = M|_u$, $u|_{t=0} = a$ has a unique solution, defined in some interval $\mathcal{I}(M,a) \subset \mathbb{R}^1$ containing zero; one sets $u(t) = \Phi_M(t)a$ for $t \in \mathcal{I}(M,a)$. For each $t \in \mathcal{I}(M,a)$ $\Phi_M(t)$ is a $\mathcal{C}^\omega$ diffeomorphism of an open neighborhood of a onto one of $\Phi_M(t)a$. It is checked at once that $\Phi_M(0)$ is the identity map of $\mathcal{A}$, and that $\Phi_M(t)\Phi_M(t')a = \Phi_M(t+t')a$, when this makes sense, i.e., when $t' \in \mathcal{I}(M,a)$, $t \in \mathcal{I}(M,\Phi_M(t')a)$, $t+t' \in \mathcal{I}(M,a)$. Notice also that there is an open neighborhood $\mathbb{O}$ of the origin in $\mathfrak{g}$, an open neighborhood $U_{\mathbb{O}}$ of the origin in $\mathcal{A}$ and an open interval $\mathcal{I}(\mathbb{O})$ in $\mathbb{R}^1$ containing zero, such that the following is true:

$\Phi_M(t)$ is a diffeomorphism of $U_{\mathbb{O}}$ onto an open subset of $\mathcal{A}$, whatever $M \in \mathbb{O}$ and $t \in \mathcal{I}(\mathbb{O})$. (VI.8.15)

Let M belong to $\mathbb{O}$ and N be an arbitrary $\mathcal{C}^\infty$ vector field in $\mathcal{A}$. We may let N act on the pushforward, under the diffeomorphism $\Phi_M(t)$ ($t \in \mathcal{I}(\mathbb{O})$), of a function defined and smooth in a neighborhood of $a \in U_{\mathbb{O}}$, and pull back the result to a via the same diffeomorphism. This operation defines a tangent vector at a, and by letting a vary, a vector field in $U_{\mathbb{O}}$ denoted by $\Phi_M(t)_*N$. It is checked at once that

$$\frac{\mathrm{d}}{\mathrm{d}t}(\Phi_M(t)_*N) = -\,[M,\Phi_M(t)_*N]. \tag{VI.8.16}$$

An arbitrary element M of $\mathfrak{g}$ defines a homomorphism of the finite dimensional Lie algebra $\mathfrak{g}$, Ad M: (Ad M)(N) = [M,N], N $\in \mathfrak{g}$. Thus, restricted to vector fields N $\in \mathfrak{g}$ equation (VI.8.16) implies

$$\frac{d}{dt}\Phi_M(t)_*|_{t=0} = -\text{ Ad M}. \tag{VI.8.17}$$

Since $\Phi_M(t)_*\Phi_M(t')_* = \Phi_M(t+t')_*$ for admissible values of t, t', we reach the conclusion that the restriction of $\Phi_M(t)_*$ to $\mathfrak{g}$ is equal to $\exp(-t(\text{Ad M}))$.

Now, the function $t \to \exp(-t(\text{Ad M})) \in \text{Aut}(\mathfrak{g})$, the group of automorphisms of the Lie algebra $\mathfrak{g}$, can be extended to all real values of t. Call G the smallest closed subgroup of Aut($\mathfrak{g}$) containing all the automorphisms $\exp(-t(\text{Ad M}))$ when M ranges over $\mathfrak{g}$ and t over $\mathbb{R}$. It is a Lie group whose Lie algebra is canonically isomorphic to $\mathfrak{g}$ (the Lie algebra of G is alternatively defined as the tangent space to G at the identity, or as the Lie algebra of left-invariant vector fields on G).

In substance we have described above a $\mathscr{C}^\omega$ diffeomorphism of an open neighborhood of 0 in $\mathscr{A}$ onto an open neighborhood of the identity in G. Indeed, given any point $a \in \mathscr{A}$ sufficiently close to 0 there is a unique vector field $M_a \in \mathfrak{g}$ such that 1 belongs to the interval of definition $\mathscr{I}(M_a,0)$ and that $a = \Phi_{M_a}(1)0$. The sought diffeomorphism is the map $a \to \exp(-\text{Ad } M_a)$.

Returning to the manifold $\mathcal{M}$, we may state:

THEOREM VI.8.2. *Suppose there is a finite dimensional Lie subalgebra* $\mathfrak{g}$ *of* $\mathscr{C}^\infty(\mathcal{M};T\mathcal{M})$ *satisfying* (VI.8.1).

Then an arbitrary point $\mathfrak{p}$ *of* $\mathcal{M}$ *has an open neighborhood* Ω *such that there is a diffeomorphism* $\Omega \cong \mathscr{U} \times \mathscr{A}$, *with* $\mathscr{U}$ *an open neighborhood of* 0 *in* $\mathbb{R}^{m'}$ ($m' = \dim \mathcal{M} - \dim \mathfrak{g}$) *and* $\mathscr{A}$ *an open neighborhood of the identity in the Lie group* G, *which transforms each vector field* M $\in \mathfrak{g}$ *into a left-invariant vector field on* G (regarded as a vector field in $\mathscr{U} \times \mathscr{A}$ independent of the point $x \in \mathscr{U}$).

When the situation described in the above statement obtains we can regard the set of transformations of $\mathscr{U} \times \mathscr{A}$, $(x,g) \to (x,hg)$, with $h \in$ G suitably close to the unit element, as a *group of local transformations* (also called a *local dynamical system*) of $\mathcal{M}$. This particular kind of dynamical system is associated to the (finite dimensional) Lie group G.

For reference in the next section we mention the particular case of an *abelian* Lie algebra $\mathfrak{g} \subset \mathscr{C}^\infty(\mathcal{M};T\mathcal{M})$, i.e., the case where the vector fields belonging to $\mathfrak{g}$ commute.

THEOREM VI.8.3. *Let* $\mathfrak{g}$ *be an abelian Lie subalgebra of* $\mathscr{C}^\infty(\mathcal{M};T\mathcal{M})$ *whose dimension is equal to* $n' \leq \dim \mathcal{M}$ *and which satisfies* (VI.8.1).

Each point $\mathfrak{p}$ of $\mathcal{M}$ has an open neighborhood in which there are coordinates $x_1,\ldots,x_{m'}$, $u_1,\ldots,u_{n'}$, vanishing at $\mathfrak{p}$, such that $\mathfrak{g}$ is spanned by $\partial/\partial u_1,\ldots,\partial/\partial u_{n'}$.

PROOF. We take advantage of Theorem VI.8.1 and introduce a basis $\mathrm{M}_1,\ldots,\mathrm{M}_{n'}$ as in (VI.8.3), except that now the coefficients are independent of y. Define the one-forms β_k by (VI.8.7). Since now the constants of structure (see (VI.8.4)) are all equal to zero we derive from (VI.8.8) that every form β_k is exact in some open neighborhood of $\mathfrak{p}$. Set there $\beta_k = du_k$; it follows from (VI.8.7) that $\mathrm{M}_k = \partial/\partial u_k$. ■

VI.9. Involutive Structures with Transverse Group Action. Rigid Structures. Tube Structures

Throughout the present section $\mathcal{M}$ will denote a $\mathcal{C}^\infty$ manifold equipped with an involutive structure; as usual, the tangent and cotangent structure bundles will be denoted by $\mathcal{V}$ and T', and their rank by n and m respectively.

THEOREM VI.9.1. *Suppose that an arbitrary point $\mathfrak{p}$ of $\mathcal{M}$ has an open neighborhood* U *in which there is a finite dimensional Lie subalgebra $\mathfrak{g}$ of* $\mathcal{C}^\infty(\mathrm{U};\mathrm{T}\mathcal{M})$ *satisfying* (VI.8.1) *such that the following is true, over* U:

$$\mathcal{V}\cap\mathcal{W}^{\mathfrak{g}} \text{ is a vector bundle;} \qquad \text{(VI.9.1)}$$

$$\mathbb{C}\mathrm{T}\mathcal{M} = \mathcal{V} + \overline{\mathcal{V}} + \mathcal{W}^{\mathfrak{g}}; \qquad \text{(VI.9.2)}$$

$$[\mathfrak{g},\mathcal{V}] \subset \mathcal{V}. \qquad \text{(VI.9.3)}$$

Then the involutive structure of $\mathcal{M}$ is locally integrable.

We recall that $\mathcal{W}^{\mathfrak{g}}$ is the subbundle of $\mathbb{C}\mathrm{T}\mathcal{M}$ spanned by $\mathfrak{g}$; Condition (VI.9.2) is a kind of transversality property between $\mathfrak{g}$ and $\mathcal{V} + \overline{\mathcal{V}}$ (+ is the Whitney sum, not necessarily direct). The meaning of (VI.9.3) is clear: the commutation bracket of any vector field belonging to $\mathfrak{g}$ with any $\mathcal{C}^\infty$ section of $\mathcal{V}$ over an open subset of U is a section of $\mathcal{V}$ over that open set.

PROOF. Property (VI.9.1) is equivalent to the fact that $\mathcal{V} + \mathcal{W}^{\mathfrak{g}}$ is a vector bundle over U; it then follows from (VI.9.2) that $\mathcal{V} + \mathcal{W}^{\mathfrak{g}}$ is elliptic. Denote by $n + d$ its fibre dimension. Theorem VI.6.1 states that, possibly after contracting U about $\mathfrak{p}$, there are local coordinates ξ_i, η_j, τ_k in U such that the vector fields $\partial/\partial\bar{\zeta}_j$, $\partial/\partial\tau_k$ ($\zeta_j = \xi_j + i\eta_j$, $1 \le i, j \le \nu$, $1 \le k \le n+d-\nu$) form a basis of $\mathcal{V} + \mathcal{W}^{\mathfrak{g}}$ over U. Every element of $\mathfrak{g}$ is a real vector field and thus it must belong to the span of the $\partial/\partial\tau_k$; in particular we must have $n' = \dim \mathfrak{g} \le n+d-\nu$. Each leaf of the foliation associated with $\mathcal{W}^{\mathfrak{g}}$ is contained in a submanifold $\zeta = const$. We may change the coordinates τ_k in such a way that each one of those leaves is defined by equations $\zeta = const.$, $t = const.$, where we now write $t_\ell = \tau_{n+d-\nu-n'+\ell}$ ($\ell = 1,\dots,m'-2\nu$). In the coordinates $x_1,\dots,x_{m'}$, $u_1,\dots,u_{n'}$ of Theorem VI.8.1 those same leaves are defined by $x = const$. It follows that $\xi = F(x)$, $\eta = G(x)$, $t = H(x)$, and that the Jacobian matrix of (F,G,H) with respect to x is invertible. We therefore have the right to use the local coordinates ξ_i, η_j, u_k, t_ℓ ($1 \leq i, j \leq \nu$, $1 \leq k \leq n'$, $1 \leq \ell \leq m'-2\nu$). In this coordinate system the elements of $\mathfrak{g}$ still have the expressions (VI.8.2).

We select a basis $M_1,\dots,M_{n'}$ of $\mathfrak{g}$ such that $M_1,\dots,M_p$ span a supplementary of $\mathcal{V}$ in $\mathcal{V} + \mathcal{W}^{\mathfrak{g}}$ ($0 \leq p \leq n'$; the case $p = 0$ is not precluded: it means that $\mathcal{W}^{\mathfrak{g}} \subset \mathcal{V}$ and therefore, that $\mathcal{V}$ is elliptic). We may select a smooth basis of $\mathcal{V}$ in U (possibly contracted further about $\mathfrak{p}$) consisting of vector fields

$$L_i = \partial/\partial\bar{\zeta}_i + \sum_{k=1}^{p} \lambda_{ik}(\xi,\eta,u,t)M_k,\ i = 1,\dots,\nu; \tag{VI.9.4}$$

$$L_{\nu+j} = M_{p+j} + \sum_{k=1}^{p} \lambda'_{jk}(\xi,\eta,u,t)M_k,\ j = 1,\dots,n'-p; \tag{VI.9.5}$$

$$L_{\nu+n'-p+\ell} = \partial/\partial t_\ell + \sum_{k=1}^{p} \lambda''_{\ell k}(\xi,\eta,u,t)M_k,\ \ell = 1,\dots,m'-2\nu. \tag{VI.9.6}$$

We now avail ourselves of Condition (VI.9.3) by writing that the commutators $[M_h,L_i]$ are sections of $\mathcal{V}$, i.e., they belong to the span of the $L_{\nu+j}$ ($1 \leq j \leq n'-p$). Take first $i \leq \nu$. We get, in the notation of (VI.8.4),

$$\sum_{k=1}^{p}\left\{(M_h\lambda_{ik})M_k + \lambda_{ik}\sum_{\ell=1}^{n'} C_{hk,\ell}M_\ell\right\} = \sum_{j=1}^{n'-p} \gamma_{hi,j}L_{\nu+j}$$

which implies at once

$$\gamma_{hi,j} = \sum_{q=1}^{p} C_{hq,p+j}\lambda_{iq},\ j = 1,\dots,n'-p; \tag{VI.9.7}$$

and then, after taking (VI.9.7) into account,

$$M_h\lambda_{ik} + \sum_{q=1}^{p} C_{hq,k}\lambda_{iq} = \sum_{j=1}^{n'-p}\sum_{q=1}^{p} C_{hq,p+j}\lambda_{iq}\lambda'_{jk},$$
$$1 \leq h \leq n',\ 1 \leq i \leq \nu,\ 1 \leq k \leq p. \tag{VI.9.8}$$

Exactly the same reasoning, using (VI.9.6) instead of (VI.9.4), shows that

$$M_h\lambda''_{ik} + \sum_{q=1}^{p} C_{hq,k}\lambda''_{iq} = \sum_{j=1}^{n'-p}\sum_{q=1}^{p} C_{hq,p+j}\lambda''_{iq}\lambda'_{jk},$$
$$1 \leq h \leq n',\ \nu+n'-p+1 \leq i \leq n,\ 1 \leq k \leq p. \tag{VI.9.9}$$

On the other hand, if $\nu+1 \le i \le \nu+n'-p$,

$$[\mathrm{M}_h,\mathrm{L}_i] = \sum_{q=1}^{n'} C_{h(n-p+i),q}\mathrm{M}_q + \sum_{k=1}^{p}\left\{(\mathrm{M}_h\lambda'_{ik})\mathrm{M}_k + \lambda'_{ik}\sum_{q=1}^{n'} C_{hk,q}\mathrm{M}_q\right\} =$$

$$\sum_{j=1}^{n'-p} \gamma'_{hi,j}\mathrm{L}_{\nu+j},$$

whence (cf. (VI.9.7))

$$\gamma'_{hi,j} = C_{h(n-p+i),p+j} + \sum_{\ell=1}^{p} C_{h\ell,p+j}\lambda'_{i\ell},\ j = 1,\dots,n'-p, \tag{VI.9.10}$$

and then (cf. (VI.9.8))

$$\mathrm{M}_h\lambda'_{ik} + C_{h(n-p+i),k} + \sum_{\ell=1}^{p} C_{h\ell,k}\lambda'_{i\ell} =$$

$$\sum_{j=1}^{n'-p}\left\{C_{h(n-p+i),p+j}\lambda'_{jk} + \sum_{\ell=1}^{p} C_{h\ell,p+j}\lambda'_{j\ell}\lambda'_{ik}\right\},$$

$$1 \le h \le n',\ \nu+1 \le i \le \nu+n'-p,\ 1 \le k \le p. \tag{VI.9.11}$$

Since $\mathrm{M}_1,\dots,\mathrm{M}_{n'}$ span the tangent space in u-space and since their coefficients are analytic functions of u alone we see that the system of equations (VI.9.8), (VI.9.9), and (VI.9.11) is equivalent to a system of equations

$$\mathrm{d}_u\Lambda = B(u,\Lambda). \tag{VI.9.12}$$

In (VI.9.12) $\Lambda = (\lambda,\lambda',\lambda'')$ is a $\mathscr{C}^\infty$ map of an open ball $\mathscr{A} = \{\, u \in \mathbb{R}^{n'};\ |u| < \delta \,\}$ into $\mathbb{C}^{np}$, which depends in $\mathscr{C}^\infty$ fashion on the parameters ξ, η, t; $B(u,\Lambda)$ is a function of (u,Λ) valued in $\mathbb{C}^{nn'p}$, analytic with respect to u and a polynomial of degree two with respect to Λ. It is an easy exercise to prove that (VI.9.12) entails that Λ is analytic with respect to u.

We extend the coefficients of the vector fields L_i ($1 \le i \le n$) as holomorphic functions of u in an open neighborhood of the origin in $\mathbb{C}^{n'}$; we extend each $\partial/\partial u_k$ as the anti-Cauchy-Riemann vector field. The complex conjugate of M_j, $\overline{\mathrm{M}}_j$, is a linear combination of the Cauchy-Riemann vector fields $\partial/\partial\bar{u}_k$ ($1 \le k \le n'$) with coefficients that are antiholomorphic functions of u, independent of (ξ,η,t). It follows that $[\mathrm{L}_i,\overline{\mathrm{M}}_j] = 0$, $1 \le i \le n$, $1 \le j \le n'$. On the other hand, the commutation brackets of the $\overline{\mathrm{M}}_j$ belong to the span of the $\overline{\mathrm{M}}_j$ by (VI.8.4). It follows that the vector fields

$$\mathrm{L}_1,\dots,\mathrm{L}_n,\ \overline{\mathrm{M}}_1,\dots,\overline{\mathrm{M}}_{n'} \tag{VI.9.13}$$

define an involutive structure on an open neighborhood of the origin in

$\mathbb{C}^{\nu}\times\mathbb{C}^{n'}\times\mathbb{R}^{m'-2\nu}$ (where the variable is (ζ,u,t)). It is clear that this structure is elliptic. By Theorem VI.7.1 it is locally integrable. Notice that $2\nu + 2n' + (m'-2\nu) = \dim \mathrm{M} + n' = m + n + n'$. There are exactly m $\mathscr{C}^\infty$ functions $Z_1,\dots,Z_m$ in an open neighborhood $\mathcal{O}$ of the origin in $\mathbb{C}^{\nu}\times\mathbb{C}^{n'}\times\mathbb{R}^{m'-2\nu}$ whose differentials span the vector subbundle of $\mathbb{C}T^*(\mathbb{C}^{\nu}\times\mathbb{C}^{n'}\times\mathbb{R}^{m'-2\nu})$ over $\mathcal{O}$ which is orthogonal to the vector fields (VI.9.13). These functions are holomorphic with respect to u. It follows that their restrictions to $(\mathbb{C}^{n}\times\mathbb{R}^{n'}\times\mathbb{R}^{m'-2\nu})\cap\mathcal{O}$ are solutions, there, in the original involutive structure of $\mathcal{M}$; $dZ_1,\dots,dZ_m$ form a smooth basis of T′ over that same set. ■

Inspection of the proof of Theorem VI.9.1 shows that, under certain hypotheses, some precision can be added to its conclusion. The first of those hypotheses is that the Lie algebra $\mathfrak{g}$ be abelian. We are going to see that this property is related to the following concept (cf. Definition IV.5.2):

DEFINITION VI.9.1. *A hypo-analytic structure on a $\mathscr{C}^\infty$ manifold $\mathcal{M}$ will be called* rigid *if there exist in $\mathcal{M}$ a set of $\mathscr{C}^\infty$ functions $t_1,\dots,t_{n-\nu}$ and a set of hypo-analytic functions, $z_1,\dots,z_\nu$, $w_1,\dots,w_p$ ($\nu \le n$, $\nu + p = m$), with the following properties*:

every point of $\mathcal{M}$ has an open neighborhood in which $\mathcal{R}e z_i$, $\mathcal{I}m z_j$ ($1 \le i, j \le \nu$), $\mathcal{R}e w_k$ ($1 \le k \le p$), t_ℓ ($1 \le \ell \le n-\nu$) are local coordinates; (VI.9.14)

$\mathcal{I}m w_k$ ($1 \le k \le p$) is independent of $\mathcal{R}e w = (\mathcal{R}e w_1,\dots,\mathcal{R}e w_p)$. (VI.9.15)

If (VI.9.14) holds we take $x_i = \mathcal{R}e z_i$, $y_j = \mathcal{I}m z_j$ ($1 \le i, j \le \nu$), $s_k = \mathcal{R}e w_k$ ($1 \le k \le p$), t_ℓ ($1 \le \ell \le n-\nu$) as local coordinates in some open neighborhood U of an arbitrary point $\mathfrak{p}$ of $\mathcal{M}$. Then if (VI.9.15) holds we will have

$$w_k = s_k + \iota\varphi_k(x,y,t),\ k = 1,\dots,p. \tag{VI.9.16}$$

The "rigidity" of the structure refers to the fact that the φ_k are independent of s (cf. (I.7.4)).

REMARK VI.9.1. It is important to note that there is no hypothesis here that the differentials $d\varphi_k$ vanish at any point of U, as prescribed in Condition (I.7.5). ■

THEOREM VI.9.2. *The following properties are equivalent*:

every point of $\mathcal{M}$ has an open neighborhood in which the involutive structure induced by $\mathcal{M}$ underlies a rigid hypo-analytic structure; (VI.9.17)

an arbitrary point $\mathfrak{p}$ of $\mathcal{M}$ has an open neighborhood U *in* (VI.9.18) *which there is an abelian finite dimensional Lie subalgebra* $\mathfrak{g}$ *of* $\mathscr{C}^\infty(U;T\mathcal{M})$ *satisfying* (VI.8.1), (VI.9.1), (VI.9.2), (VI.9.3) *over* U.

PROOF. (VI.9.17) $\Rightarrow$ (VI.9.18). Suppose that we have basic hypo-analytic functions in an open neighborhood U of $\mathfrak{p}$, $z_i = x_i + \iota y_i$ ($1 \leq i \leq \nu$) and w_k ($1 \leq k \leq p$) given by (VI.9.16). The following vector fields make up a basis of $\mathcal{V}$ over U:

$$L_j = \partial/\partial \bar{z}_j - \iota \sum_{k=1}^{p} (\partial \varphi_k/\partial \bar{z}_j)\partial/\partial s_k,\ j = 1,\ldots,\nu, \qquad \text{(VI.9.19)}$$

$$L_{\nu+\ell} = \partial/\partial t_\ell - \iota \sum_{k=1}^{p} (\partial \varphi_k/\partial t_\ell)\partial/\partial s_k,\ \ell = 1,\ldots,n-\nu. \qquad \text{(VI.9.20)}$$

Since φ_k does not depend on s we see that the vector fields $L_1,\ldots,L_n$ commute with $\partial/\partial s_1,\ldots,\partial/\partial s_d$ and therefore with their span over the real numbers, $\mathfrak{g}$.

(VI.9.18) $\Rightarrow$ (VI.9.17). We reason in a suitably small open neighborhood U of an arbitrary point $\mathfrak{p}$ of $\mathcal{M}$; the analysis takes place in the same setup as that of the proof of Theorem VI.9.1, and uses the same notation. We go back to equations (VI.9.8), (VI.9.9), and (VI.9.11). Thanks to Theorem VI.8.3 we may take $M_k = \partial/\partial u_k$, $k = 1,\ldots,n'$. Since $\mathfrak{g}$ is abelian, the constants of structure $C_{jk,\ell}$ (see (VI.8.4)) are all equal to zero. It follows that Equation (VI.9.12) reads, under the present hypotheses, $d_u\Lambda \equiv 0$, which means that the coefficients λ_{ik}, λ'_{jk}, and $\lambda''_{\ell k}$ in the vector fields (VI.9.4), (VI.9.5), and (VI.9.6) are all independent of u. Furthermore, in the present situation, the vector fields L_i ($1 \leq i \leq n$) must commute, otherwise their commutators, which belong to the span of $\partial/\partial u_1,\ldots,\partial/\partial u_p$, would not be sections of $\mathcal{V}$. This property is equivalent to the fact that the one-forms

$$\omega_k = \sum_{i=1}^{p} \lambda_{ik} d\bar{z}_i + \sum_{j=1}^{n'-p} \lambda'_{jk} du_{p+j} + \sum_{\ell=1}^{m'-2\nu} \lambda''_{\ell k} dt_\ell,\ k = 1,\ldots,p,$$

are closed in the sense of the differential complex $D = \bar{\partial}_z + d_u + d_t$. We apply formula (VI.7.4): $\omega_k = D\psi_k$ with $\psi_k(\xi,\eta,u'',t)$ independent of $u' = (u_1,\ldots,u_p)$ (using the notation $u'' = (u_{p+1},\ldots,u_{n'})$). The following are "first integrals" (in U, for the involutive structure of $\mathcal{M}$):

$$\zeta_j = \xi_j + \iota\eta_j\ (1 \leq j \leq \nu),\ w_k = u_k - \psi_k(\xi,\eta,u'',t)\ (1 \leq k \leq p).$$

Taking $u_k - \mathcal{R}e\,\psi_k(\xi,\eta,u'',t)$ as new coordinate s_k for $k = 1,\ldots,p$, shows that the functions w_k are indeed of the kind (VI.9.16). (The variable called here (u'',t) is called t in (VI.9.16).) ■

REMARK VI.9.2. We have succeeded in eliminating from $\mathcal{I}m\,w_k$ all the variables s_j, $j = 1,\ldots,p$. We recall that the vector fields $\partial/\partial s_k$ $(1 \le k \le p)$ span a direct summand of $\mathcal{V}$ in the Whitney sum $\mathcal{V} + \mathcal{W}^{\mathfrak{g}}$. ∎

A further specialization of the rigid structures turns out to be of interest. In the notation of Definition VI.9.1 these special structures correspond to the case $\nu = 0$: the cotangent structure bundle of $\mathcal{M}$ is spanned, over $\mathcal{M}$, by m hypo-analytic functions

$$w_j = s_j + \imath\varphi_j(t),\ j = 1,\ldots,m. \tag{VI.9.21}$$

When $\mathcal{M} = \mathbb{R}^m \times \Omega$, with Ω an open subset of $\mathbb{R}^n$, one recognizes the tube structures introduced in section I.12. The following is a (slight) generalization of Definition I.12.1:

DEFINITION VI.9.2. *A hypo-analytic structure on a* $\mathcal{C}^\infty$ *manifold* $\mathcal{M}$ *will be called a* tube structure, *or said to be* tubular, *if there exist in* $\mathcal{M}$ $\mathcal{C}^\infty$ *functions* $t_1,\ldots,t_n$ *and hypo-analytic functions* $w_1,\ldots,w_m$ *with the following properties*:

every point of $\mathcal{M}$ *has an open neighborhood in which* $\mathcal{R}e\, w_k$ $(1 \le k \le m)$, t_ℓ $(1 \le \ell \le n)$ *are local coordinates*; (VI.9.22)

for all $j = 1,\ldots,m$, $\mathcal{I}m\, w_j$ *is independent of* $\mathcal{R}e\, w = (\mathcal{R}e\, w_1,\ldots,\mathcal{R}e\, w_m)$. (VI.9.23)

The "tube" version of Theorem VI.9.2 can be stated as follows.

THEOREM VI.9.3. *The following properties are equivalent*:

every point of $\mathcal{M}$ *has an open neighborhood in which the involutive structure induced by* $\mathcal{M}$ *underlies a tubular hypo-analytic structure*; (VI.9.24)

an arbitrary point $\mathfrak{p}$ *of* $\mathcal{M}$ *has an open neighborhood* U *in which there is an abelian finite dimensional Lie subalgebra* $\mathfrak{g}$ *of* $\mathcal{C}^\infty(\mathrm{U};T\mathcal{M})$ *satisfying over* U *the conditions* (VI.8.1) *and* (VI.9.3), *as well as the following one*, (VI.9.25)

$$\mathbb{C}T\mathcal{M} = \mathcal{V} + \mathcal{W}^{\mathfrak{g}}. \tag{VI.9.26}$$

PROOF. It suffices to observe that (VI.9.26) is equivalent to the fact that there is (over U) a direct summand of $\mathcal{V}$ in $\mathcal{V} + \mathcal{W}^{\mathfrak{g}}$ with rank equal to m, and to take Remark VI.9.2 into account. ∎

REMARK VI.9.3. In both Theorems VI.9.2 and VI.9.3 we may as well assume

$$\mathcal{V} \cap \mathcal{W}^{\mathfrak{g}} = 0. \tag{VI.9.27}$$

Indeed, we may replace $\mathfrak{g}$ by any one of its subspaces, $\mathfrak{g}_0$, which spans a direct summand of $\mathcal{V}$ in $\mathcal{V} + \mathcal{W}^{\mathfrak{g}}$. ■

Rigid and tubular hypo-analytic structures that at the same time are CR structures are noteworthy. Let us first look at the equations (VI.9.16). The CR nature of the structure expresses itself in the fact that $n-\nu \leq p$ and that the rank of the Jacobian matrix

$$(\partial\varphi_k/\partial t_\ell)_{1\leq k\leq p, 1\leq \ell\leq n-\nu}$$

must be equal to $n-\nu$. This means that we may select $n-\nu$ functions φ_k as new variables t_ℓ. We end up with the basic hypo-analytic functions

$$\begin{gathered} z_j = x_j + \imath y_j \ (1 \leq j \leq \nu),\ w_k = s_k + \imath t_k \ (1 \leq k \leq n-\nu), \\ w_\ell = s_\ell + \imath\varphi_\ell(x,y,t) \ (n-\nu+1 \leq \ell \leq p), \end{gathered} \tag{VI.9.28}$$

where one recognizes the typical functions that define a CR structure with the added feature that the φ_ℓ only depend on x, y, t. It may of course happen that $\nu = n$, in which case one gets the standard presentation of a rigid CR structure:

$$z_j = x_j + \imath y_j \ (1 \leq j \leq n),\ w_k = s_k + \imath\varphi_k(x,y) \ (1 \leq k \leq p). \tag{VI.9.29}$$

Notice that the structure defined by (VI.9.28) would still be rigid if the φ_ℓ ($\ell > n-\nu$) were to depend on s_k, $k \leq n-\nu$.

The same reasoning applies to Equations (VI.9.21). In order for the tubular structure defined by the w_j to be CR it is necessary and sufficient that the map $\varphi = (\varphi_1,\dots,\varphi_m)$ (from an open subset of $\mathbb{R}^n$ to $\mathbb{R}^m$) have rank $n \leq m$. After selecting n functions φ_j as new coordinates t, we end up with basic hypo-analytic functions

$$w_j = s_j + \imath t_j \ (1 \leq j \leq n),\ w_k = s_k + \imath\varphi_k(t) \ (n+1 \leq k \leq m). \tag{VI.9.30}$$

Of course, in (VI.9.30) a further $\mathbb{C}$-linear substitution of the basic hypo-analytic functions achieves that the differentials $d\varphi_k$, $k > n$, vanish at the central point. The analogous remark is valid for (V.9.28) and (V.9.29).

We close this section with a word about Condition (V.9.3), in the "general case" (i.e., we do not assume $\mathfrak{g}$ to be abelian). We choose the neighborhood Ω of $\mathfrak{p} \in \mathcal{M}$ and local coordinates in Ω as indicated at the end of section VI.8: the coordinates define a diffeomorphism $\Omega \cong \mathcal{U} \times \mathcal{A}$ with $\mathcal{U}$ an open neighborhood of the origin in $\mathbb{R}^{m'}$ and $\mathcal{A}$ an open neighborhood of the identity in the Lie group G. To start with we may use the coordinates x_i ($1 \leq i \leq m'$) in $\mathcal{U}$ and the coordinates u_j ($1 \leq j \leq n'$) in $\mathcal{A}$, as in Theorem VI.8.1. But then, as we have done in the proof of Theorem VI.9.1, we may change the coordinates x_i in such a way that $\mathcal{V}$ be spanned over Ω by n vector fields L_i as in (VI.9.4), (VI.9.5), and (VI.9.6), whose coefficients are analytic with respect to u. Let us go back to Equation (VI.8.16), where we take $N = L_i$ for some i, $1 \leq i \leq n$. Because the coefficients of L_i are analytic with respect to u, it is seen at once that $\Phi_M(t)_*L_i$ depends analytically on t in some open interval in $\mathbb{R}^1$, centered at zero. We repeatedly differentiate both sides of (VI.8.16) with respect to t and take (VI.9.3) into account. Induction on the order of the derivatives shows that the value at $t = 0$ of every t-derivative of $\Phi_M(t)_*L_i$ is a section of $\mathcal{V}$ in Ω, thereby proving that $\Phi_M(t)_*L_i$ is a section of $\mathcal{V}$ in a suitable neighborhood of p contained in Ω for all t in a suitably small interval centered at 0, in $\mathbb{R}^1$. Recall that the local diffeomorphism $\Phi_M(t)$ is a map $(x,g) \to (x,h_M(t)g)$ with $t \to h_M(t)$ the one-parameter subgroup of G generated by $M \in \mathfrak{g}$, and that the elements $h_M(t)$, $M \in \mathfrak{g}$, $t \in \mathbb{R}^1$, generate G. We may rephrase the property just described by saying that the vector bundle $\mathcal{V}$ is invariant under the left-translations of G—provided, of course, these left-translations are sufficiently close to the identity.

When $\mathfrak{g}$ is abelian and since, here, the analysis is purely local, we may as well assume that G is the additive group $\mathbb{R}^{n'}$. We have seen that the vector fields L_i can be chosen so that their coefficients are independent of u: they are invariant under the translations in u-space and so is the vector bundle $\mathcal{V}$ they span (over Ω).

Notes

The study of the Cauchy-Riemann differential complex, which is often referred to as the *Dolbeault complex*, goes back at least to Dolbeault [1]. The Bochner-Martinelli formula (VI.1.17) was first established for functions in Martinelli [1,2] and in Bochner [1]. It was generalized to differential forms in Koppelman [1]. The Leray-Koppelman formula (VI.2.12) first appeared in Lieb [1] and Øvrelid [1]. Extensions of these formulas, also to the $\bar{\partial}_b$-complex, can be found in Harvey and Polking [1] and in Treves [11].

The Newlander-Nirenberg theorem in a complex structure (Theorem VI.5.1) was proved for the first time in Newlander and Nirenberg [1] (see also Nijenhuis and Woolf [1]). It is an easy consequence of the solution of the $\bar{\partial}$-Neumann problem, as first shown in Kohn [1]. A very different proof, based on the analyticity of solutions of a certain determined system of first-order nonlinear PDE with "analytic coefficients," can be found in Malgrange [1]. The proof given in section IV.5 is essentially taken from Webster [1]. The local integrability of all elliptic structures (Theorem VI.7.1) and that of all Levi flat structures (cf. Theorem VI.7.2) was first established in Nirenberg [1]. The results in sections VI.8 and VI.9 are taken from Baouendi, Rothschild, and Treves [1]. More general results can be found in Baouendi and Rothschild [2,3].

In recent years there have been numerous efforts to obtain local embeddings of CR structures of the hypersurface type (i.e., whose characteristic set is a line bundle). The first case to be settled was that of a CR structure on a manifold of dimension ≥ 9 whose Levi form is definite (Kuranishi [1]); Kuranishi's method was further refined in Akahori [1] to establish the local embeddability in dimension 7. The problem is still unsolved in dimension 5; in dimension 3 embedding is not possible, in general (see section VII.5). A shorter proof of the Kuranishi-Akahori theorem was published in Webster [2]; it is an elaboration of the method followed in this book to prove the Newlander-Nirenberg theorem. Local embeddability of almost complex manifolds with boundary (see chapter V) has been proved in the strongly pseudoconvex case, in Hanges and Jacobowitz [1]; and in the weakly pseudoconvex case, by D. Catlin.

VII

Examples of Nonintegrability and of Nonsolvability

As announced in the introduction to chapter VI, the contents of the present chapter consist of a number of examples of involutive structures that are not locally integrable or in which *local solvability* does not hold. Local solvability means that the inhomogeneous equations

$$L_j u = f_j,\ j = 1,\dots,n, \tag{1}$$

admit a local (distribution) solution whatever choice one makes of the right-hand sides, the *smooth* functions $f_1,\dots,f_n$, provided, of course, these satisfy the compatibility conditions

$$L_j f_k = L_k f_j,\ j,\ k = 1,\dots,n. \tag{2}$$

(We are assuming that the vector fields L_j commute.)

The simplest structure in which solvability does not hold is the structure on $\mathbb{R}^2$ (with coordinates x, t) defined by the *Mizohata vector field*

$$\partial/\partial t - it\partial/\partial x. \tag{3}$$

The Mizohata structure is evidently integrable (it is real-analytic!); it admits the first integral $Z = x + \imath t^2/2$. A general *Mizohata structure* on a manifold $\mathcal{M}$ is defined (Definition VII.1.1) as a nonelliptic involutive structure in which the cotangent structure bundle is a complex line bundle and the Levi form is nondegenerate at every point of the characteristic set $T°$. The requirement forces $T°$ to vanish except on a smooth one-dimensional submanifold γ over which it is a real line bundle. The *signature* of the Levi form, which is to say, the absolute value of the difference between the number of eigenvalues of one sign and the number of those of the opposite sign, is locally constant on the two-dimensional manifold $T°|_{\gamma}$.

Suppose $\mathcal{M}$ carries a Mizohata structure. In a suitably small open neighborhood U of a point $p \in \mathcal{M}$ such that $T_p^\circ \neq 0$ local coordinates $(x, t_1,\dots,t_n)$ (vanishing at p) can be chosen so as to ensure that the tangent structure bundle $\mathcal{V}$ be spanned, over U, by vector fields

$$L_j = \partial/\partial t_j - \imath(\varepsilon_j t_j + \rho_j(x,t))\partial/\partial x,\ j = 1,\ldots,n. \tag{4}$$

In (4) $\varepsilon_j = +1$ or -1 (the Levi matrix on T_p° is conjugate to the diagonal matrix with entries $\varepsilon_1,\ldots,\varepsilon_n$). The smooth functions ρ_j vanish to infinite order at $t = 0$. The Mizohata structure is integrable (in a neighborhood of $\mathfrak{p}$) if and only if the coordinates x, t_j can be chosen in such a manner that $\rho_j \equiv 0$ for all $j = 1,\ldots,n$ (Theorem VII.1.1). In the latter case there is an obvious first integral $Z = x + \imath\mathfrak{Q}(t)$, with $\mathfrak{Q}(t)$ a nondegenerate quadratic form in t-space $\mathbb{R}^n$. Note that the *fibres* of the mapping Z are connected, except when the signature of $\mathfrak{Q}$ is equal to $|n-2|$. The latter is precisely the case in which nonsolvability occurs (section VII.2; actually, when $n = 1$ the hypothesis that the structure is locally integrable is not needed).

If $\varepsilon_1 = +1$ and (when $n \geq 2$) $\varepsilon_j = -1$ for $j = 2,\ldots,n$, it is possible to select a priori the functions ρ_j in such a fashion that the Mizohata structure not be locally integrable at $\mathfrak{p}$, specifically, that any $\mathcal{C}^1$ solution h of the homogeneous equations

$$L_j h = 0,\ j = 1,\ldots,n, \tag{5}$$

in a neighborhood of $\mathfrak{p}$, be such that $h_x = 0$ (hence $dh = 0$) at $\mathfrak{p}$. One can go further (section VII.3) when the dimension of the base manifold $\mathcal{M}$ is equal to *two* (and therefore $n = 1$). In this case the local integrability of the structure is equivalent to the "triviality" of the Sjöstrand invariant (Definition VII.3.1).

The properties outlined above exemplify the intimate but not well understood link between solvability and integrability. The link is further confirmed by what happens in involutive structures in which the cotangent structure bundle T' is a complex line bundle (section VII.4). Suppose such a structure is locally integrable: near an arbitrary point $\mathfrak{p} \in \mathcal{M}$ the bundle T' is spanned by the differential of a $\mathcal{C}^\infty$ function Z. If the fibres of the map Z (in any neighborhood of $\mathfrak{p}$) are not connected, then it is possible to construct a set of $\mathcal{C}^\infty$ functions f_j satisfying condition (2) but such that no distribution satisfies (1) in any neighborhood of the central point. (When the fibres are connected it is always possible to solve Equations (1), but this result is not proved in the present book.) Furthermore, when the fibres are not connected, one can modify the basic vector fields in a manner that the new vector fields still satisfy the Frobenius condition but do not admit a first integral in any neighborhood of $\mathfrak{p}$ (the new vector fields agree with the original ones to infinite order at $\mathfrak{p}$).

It has been known since the early years of the solvability theory of a (single) linear PDE that the Mizohata and the Lewy equations share some important properties. Microlocally one can be transformed into the other. Reflecting this kinship, the results in sections VII.5 and VII.6 parallel those in section VII.2: it is shown that the Lewy equation is not solvable (in any neighborhood of any point) and that, if modified by an appropriate "flat perturbation," it gives rise to an equation that is not integrable. It is further shown that nonsolvability occurs in any hypersurface of $\mathbb{C}^{n+1}$ whose Levi form has the signature $|n-2|$

and that the natural CR structure of such a hypersurface can be modified (again by a flat perturbation) to transform it into a CR manifold that is not any more locally embeddable in $\mathbb{C}^{n+1}$.

At the end of section VII.5 we consider the smooth vector fields in $\mathbb{R}^3$ which have the peculiarity that the only functions (of class $\mathscr{C}^{1+\delta}$—but the result remains valid for $\mathscr{C}^1$ functions) they annihilate, in any open subset, are the constant functions. These vector fields we call *aberrant*. It is shown that the aberrant (single) vector fields are dense in the space of all smooth complex vector fields in $\mathbb{R}^3$, duly topologized (it is naturally isomorphic to the cubic power of $\mathscr{C}^\infty(\mathbb{R}^3)$). More than that: the nonaberrant vector fields make up a small set indeed: it is the union of a sequence of interiorless closed sets.

Finally, section VII.8 presents the example of a CR structure on $\mathbb{R}^5$ that is one-sided but not two-sided locally integrable: when equipped with it, no neighborhood of the origin in $\mathbb{R}^5$ is isomorphic (as a CR manifold) to a hypersurface in $\mathbb{C}^6$, yet such a neighborhood is isomorphic to the *the boundary of a complex manifold* (of complex dimension three).

VII.1. Mizohata Structures

In the present section $\mathcal{M}$ will be a $\mathscr{C}^\infty$ manifold equipped with an involutive structure. As usual the tangent and cotangent structure bundles are denoted by $\mathcal{V}$ and T′. We shall assume, throughout, that dim $\mathcal{M} \geq 2$ and that T′ *is a complex line bundle*, i.e., the fibre dimensions of $\mathcal{V}$ and T′ are respectively equal to $n = \dim \mathcal{M} - 1$ and to 1.

Definition VII.1.1. *The involutive structure of $\mathcal{M}$ will be called a* Mizohata structure *if* T′ *has rank one, if* $\mathrm{T}^\circ \neq 0$ *and if the Levi form* (Definition I.8.1) *is nondegenerate at every point of* $\mathrm{T}^\circ \backslash 0$.

When $\mathcal{M}$ is equipped with a Mizohata structure, we denote by Σ° the closure of $\mathrm{T}^\circ \backslash 0$ in $\mathrm{T}^*\mathcal{M}$. Clearly Σ° is a *conic* subset of $\mathrm{T}^*\mathcal{M}$, i.e., $(p,\omega) \in \Sigma^\circ \Rightarrow (p,c\omega) \in \Sigma^\circ$ for all $c > 0$. We shall denote by π the base projection $\mathrm{T}^*\mathcal{M} \to \mathcal{M}$. In the complement of $\pi\Sigma^\circ$ the involutive structure of $\mathcal{M}$ induces an elliptic structure.

The terminology in Definition VII.1.1 is motivated by the following:

Example VII.1.1. *The two-dimensional model case*

Equip $\mathcal{M} = \mathbb{R}^2$ with the hypo-analytic structure defined by $Z = x + iy^2/2$. The underlying tangent structure bundle is spanned by the Mizohata vector field (Example I.2.7) $\mathrm{L} = \partial/\partial y - iy\partial/\partial x$. The characteristic set T° is equal to zero when $y \neq 0$ and is spanned by dx over the real axis $y = 0$. The Levi form at a point $((x,0),\xi dx) \in \mathrm{T}^\circ$, $\xi \neq 0$, is given by

$$\langle \xi dx, (2\iota)^{-1}[cL, \bar{c}\bar{L}] \rangle = \xi |c|^2, \tag{VII.1.1}$$

and thus it is nondegenerate. ∎

EXAMPLE VII.1.2. *The multidimensional model case*

Let x, $t_1,\ldots,t_n$ denote the coordinates in $\mathbb{R}^{n+1}$. Equip $\mathbb{R}^{n+1}$ with the hypo-analytic structure defined by the function

$$Z = x + \iota\mathcal{Q}(t) \tag{VII.1.2}$$

with $\mathcal{Q}(t) = (t_1^2 + \cdots + t_\nu^2 - t_{\nu+1}^2 - \cdots - t_n^2)/2$. Then $\mathcal{V}$ is spanned by the vector fields

$$L_j = \partial/\partial t_j - \iota\varepsilon_j t_j \partial/\partial x, \; j = 1,\ldots,n, \tag{VII.1.3}$$

with $\varepsilon_j = +1$ if $j \leq \nu$ and $\varepsilon_j = -1$ if $j > \nu$. By using the basis $\{L_1,\ldots,L_n\}$ to identify the fibres of $\mathcal{V}$ to $\mathbb{R}^n$, one can identify the Levi form (at the point dx of the fibre of $\mathcal{V}$ under consideration) to the quadratic form $2\mathcal{Q}$. ∎

EXAMPLE VII.1.3. Let Θ be an n-dimensional $\mathcal{C}^\infty$ manifold, countable at infinity ($n \geq 1$). Denote by t the variable point in Θ. Let $\varphi(t)$ be a real-valued $\mathcal{C}^\infty$ functions in Θ. Equip $\mathbb{R}^1 \times \Theta$ with the hypo-analytic structure defined by the function $Z = x + \iota\varphi(t)$ (x is the variable in $\mathbb{R}^1$). In a local coordinate patch $(U, x, t_1,\ldots,t_n)$ in $\mathbb{R}^1 \times \Theta$ the tangent structure bundle $\mathcal{V}$ is spanned by the (commuting) vector fields

$$L_j = \partial/\partial t_j - \iota(\partial\varphi/\partial t_j)\partial/\partial x, \; j = 1,\ldots,n. \tag{VII.1.4}$$

The characteristic set T° is not zero only at those points (x,t) such that $d\varphi(t) = 0$, i.e., such that t is a critical point of φ. At those points T° is generated by dx. It is checked at once that, if in the coordinate patch $(U, x, t_1,\ldots,t_n)$ we identify the fibre of $\mathcal{V}$ to $\mathbb{R}^n$ by means of the basis $L_1,\ldots,L_n$, then we can identify the Levi form to the Hessian of φ with respect to the variables t_j. In order that the underlying involutive structure of $\mathbb{R}^1 \times \Theta$ be a Mizohata structure it is necessary and sufficient that every critical point of φ be nondegenerate, i.e., that φ be a *Morse function* on Θ. In particular, the critical points of φ are isolated. In this example $\pi\Sigma^\circ$ is the set of points (x,t) with $x \in \mathbb{R}^1$ and t a critical point of φ. ∎

We return to the general case. Let $\mathfrak{p} \in \pi\Sigma^\circ$. We refer the reader to the first part of section I.5. There are local coordinates x, $t_1,\ldots,t_n$ in an open neighborhood U of $\mathfrak{p}$, vanishing at $\mathfrak{p}$, and a $\mathcal{C}^\infty$ one-form spanning T' over U,

$$\varpi = dx - \sum_{j=1}^{n} \lambda_j(x,t)dt_j, \tag{VII.1.5}$$

with

$$\lambda_j(0,0) = 0,\ j = 1,\dots,n \tag{VI.1.6}$$

(cf. (I.5.22), (I.5.23)). The base projection of $\Sigma°$ is defined, in U, by the condition

$$\mathcal{I}m\lambda_j(x,t) = 0,\ j = 1,\dots,n. \tag{VII.1.7}$$

(I.8.10) yields the following expression for the Levi form at the point p:

$$\langle \varpi,(2\imath)^{-1}[\mathrm{L},\bar{\mathrm{L}}]\rangle = (2\imath)^{-1}[\mathrm{L},\bar{\mathrm{L}}]x|_p \tag{VII.1.8}$$

for any smooth section L of $\mathcal{V}$ in U.

Since the form (VII.1.5) spans T′ over U, the following vector fields span $\mathcal{V}$ over U:

$$\mathrm{L}_j = \partial/\partial t_j + \lambda_j(x,t)\partial/\partial x,\ j = 1,\dots,n. \tag{VII.1.9}$$

They commute. By virtue of (VII.1.6) this implies

$$\partial\lambda_j/\partial t_k = \partial\lambda_k/\partial t_j \ \mathit{at}\ p\ (1 \le j,\ k \le n). \tag{VII.1.10}$$

If we then write $\mathrm{L} = \zeta_1\mathrm{L}_1 + \cdots + \zeta_n\mathrm{L}_n$, it follows from (VII.1.7) and (VII.1.8) that the Levi form at p can be identified to the quadratic form in ζ-space $\mathbb{C}^n$,

$$\zeta \to (2\imath)^{-1}\sum_{j,k=1}^{n} \zeta_j\bar{\zeta}_k(\partial\bar{\lambda}_k/\partial t_j - \partial\lambda_j/\partial t_k)|_p.$$

By (VII.1.10) this is equal to the quadratic form

$$\zeta \to -\sum_{j,k=1}^{n} \zeta_j\bar{\zeta}_k\ \partial(\mathcal{I}m\lambda_j)/\partial t_k|_p. \tag{VII.1.11}$$

Our hypothesis is that the quadratic form (VII.1.11) is nondegenerate, i.e., that the $n \times n$ matrix with generic entry $\partial(\mathcal{I}m\lambda_j)/\partial t_k$, is nonsingular. This entails that $(x,t) \to (x,\mathcal{I}m\lambda)$ (where $\lambda = (\lambda_1,\dots,\lambda_n)$) is a diffeomorphism of an open neighborhood of 0 in $\mathbb{R}^{n+1}$ onto another such neighborhood. In particular, it shows that $\pi\Sigma°$ is a $\mathcal{C}^\infty$ curve.

We recall that the real cotangent bundle $\mathrm{T}^*\mathcal{M}$ carries a natural *symplectic structure*, defined by the symplectic one-form σ. The value of σ at an arbitrary point $(p,\omega) \in \mathrm{T}^*\mathcal{M}$ can be defined as follows: let $\mathbf{u}$ be a vector tangent to $\mathrm{T}^*\mathcal{M}$ at (p,ω) and let $\mathrm{D}\pi(\mathbf{u})$ denote its projection to the base, $\mathcal{M}$; $\mathrm{D}\pi(\mathbf{u})$ is a tangent vector to $\mathcal{M}$ at p. Then

$$\langle \sigma|_{(p,\omega)},\mathbf{u}\rangle = \langle \omega,(\mathrm{D}\pi)\mathbf{u}\rangle. \tag{VII.1.12}$$

(Sometimes σ is called "the tautological form.") A submanifold Σ of $\mathrm{T}^*\mathcal{M}$ is called *symplectic* if the pullback of $d\sigma$ to Σ defines a nondegenerate bilinear functional on each tangent space $\mathrm{T}_p\Sigma$, $p \in \Sigma$. The whole manifold $\mathrm{T}^*\mathcal{M}$ itself is a symplectic manifold.

PROPOSITION VII.1.1. *Suppose that* $\mathcal{M}$ *carries a Mizohata structure. Then the closure* Σ° *of* $T^\circ\backslash 0$ *in* $T^*\mathcal{M}$ *is a conic symplectic two-dimensional submanifold of* $T^*\mathcal{M}$ *whose base projection is a closed one-dimensional submanifold of* $\mathcal{M}$.

PROOF. We have just seen that $\pi\Sigma^\circ$ is a smooth one-dimensional submanifold of $\mathcal{M}$; it is clearly closed. We have $\dim \Sigma^\circ = 2$ since the fibres of T° at points of Σ° are lines. The hypothesis that the quadratic form (VII.1.11) is nondegenerate implies that the one-forms

$$dx,\ d(\mathcal{I}m\lambda_j),\ j = 1,\ldots,n,$$

make up a basis of $T^*\mathcal{M}$ in an open neighborhood of p in $\mathcal{M}$, which we take to be U. Denote by ξ, $\eta_1,\ldots,\eta_n$ the coordinates with respect to this basis, in the fibres of $T^*\mathcal{M}$ at points of U. Then we can write $\sigma = \xi dx + d(\mathcal{I}m\lambda\cdot\eta)$. But Σ° is defined, in $T^*\mathcal{M}|_U$, by the equations $\mathcal{I}m\lambda = \eta = 0$ and thus the pullback of σ to Σ° is equal to ξdx; the two-form $d\xi\wedge dx$ on $\Sigma^\circ\cap(T^*\mathcal{M}|_U)$ is nondegenerate. ■

Any symplectic manifold is orientable, and so is Σ°.

PROPOSITION VII.1.2. *Suppose that* $\mathcal{M}$ *carries a Mizohata structure. When* $\dim \mathcal{M}$ *is even the curve* $\pi\Sigma^\circ$ *has a natural orientation.*

PROOF. The sign of $\omega \in T^\circ_p$, $\omega \neq 0$, is defined to be the sign of the product of the eigenvalues of the Levi form at (p,ω) (none of which is equal to zero); replacing ω by $c\omega$ amounts to multiplying that product by c^n with $n = \dim \mathcal{M} - 1$ odd. This defines a preferred orientation on each fibre of Σ°. The preferred orientation on $\pi\Sigma^\circ$ can then be chosen in such a way that, if $\mathbf{u}$ and $\mathbf{v}$ are tangent vectors to $\pi\Sigma^\circ$ at p and to T°_p at $(p,0)$ respectively, which point forward, necessarily the value of the fundamental symplectic two-form on Σ° at the exterior product $\mathbf{u}\wedge\mathbf{v}$ is > 0. ■

A local change of variables can be used to bring the basis of $\mathcal{V}$ over a neighborhood of a point into a "standard form":

THEOREM VII.1.1. *Assume that* $\mathcal{M}$ *is equipped with a Mizohata structure.*

There exist coordinates $x, t_1,\ldots,t_n$ *in an open neighborhood* U *of an arbitrary point* $p \in \mathcal{M}$, *a nondegenerate quadratic form* $\mathfrak{Q}$ *in* $\mathbb{R}^n$ *and* n $\mathcal{C}^\infty$ *functions* ρ_j *in* U $(1 \leq j \leq n)$ *such that the following is true:*

$$\rho_j \textit{ vanishes to infinite order at } t = 0, \textit{ for every } j = 1,\ldots,n; \tag{VII.1.13}$$

there is a $\mathcal{C}^\infty$ *basis of* $\mathcal{V}$ *over* U *consisting of the vector fields* (VII.1.14)

$$L_j = \partial/\partial t_j - (\imath\partial\mathfrak{Q}/\partial t_j - \rho_j)\partial/\partial x,\ j = 1,\ldots,n. \tag{VII.1.15}$$

In order for the involutive structure of $\mathcal{M}$ to be integrable in some open neighborhood of $\mathfrak{p}$ it is necessary and sufficient that there exist coordinates x and t_j and a quadratic form $\mathfrak{Q}$, as above, such that (VII.1.14) *is valid with* $\rho_j \equiv 0$ *for all* $j = 1,\ldots,n$.

PROOF. We start by using coordinates x, t_j in some open neighborhood of $\mathfrak{p}$, in which the vector fields L_j have the expressions (VII.1.9). It is clear that the vector fields $\partial/\partial t_j$ ($j = 1,\ldots,n$) are transverse, near $\mathfrak{p}$, to the curve $\pi\Sigma^\circ$. As a consequence, there is a $\mathcal{C}^\infty$ map F of some interval $|x| < r$ in $\mathbb{R}^1$, into $\mathbb{R}^n$, such that $F(0) = 0$, and such that the curve $\pi\Sigma^\circ$ is defined, near the point $\mathfrak{p}$, by the equation $t = F(x)$. After the change of variable $t \to t - F(x)$ the curve $\pi\Sigma^\circ$ will be defined, near $\mathfrak{p}$, by the equation $t = 0$ and the cotangent structure bundle T$'$ will be spanned (over a neighborhood of $\mathfrak{p}$) by the form (VII.1.5) where now $\mathcal{I}m\lambda_j(x,0) \equiv 0$ for every $j = 1,\ldots,n$ (cf. (VII.1.7). Furthermore, if we carry out the change of variable

$$x \to x + \mathcal{R}e \sum_{j=1}^{n} \lambda_j(x,0)t_j$$

and multiply the form ϖ by a nonvanishing factor, we end up in a situation where (VII.1.6) and (VII.1.9) are valid, and so is the property

$$\lambda|_{t=0} = 0. \tag{VII.1.16}$$

It is convenient, below, to take $U = \mathcal{I} \times \Omega$ with $\mathcal{I}$ an interval in the x-line $\mathbb{R}^1$ and Ω an open neighborhood of the origin in t-space $\mathbb{R}^n$. We apply a variant of Lemma VI.5.1. The only difference in the proof is that there is only one variable x (the proof of Lemma VI.5.1 did not rely on the number of variables x, nor make use of the fact that the vector fields L_j defined a complex structure). According to this variant, there is a $\mathcal{C}^\infty$ function $u(x,t)$ in $U = \mathcal{I} \times \Omega$ such that

$$L_j u \textit{ vanishes to infinite order at } t = 0 \ (1 \le j \le n); \tag{VII.1.17}$$

$$u|_{t=0} = x. \tag{VII.1.18}$$

We derive from (VII.1.16) and (VII.1.17):

$$\partial^2 u/\partial t_i \partial t_j|_{t=0} = -(\partial\lambda_j/\partial t_i)(\partial u/\partial x)|_{t=0} = -\partial\lambda_j/\partial t_i|_{t=0}$$

since $u_x(x,0) = 1$ by (VII.1.18). It follows at once from this equality and from our hypothesis that the form (VII.1.11) is nondegenerate, that the Hessian of $\mathcal{I}m u$ with respect to t is nondegenerate in some open neighborhood of the origin. This allows us to apply Morse's lemma and write

$$\mathcal{I}m\, u(x,t) = \mathfrak{Q}(G(x,t)),$$

where $\mathfrak{Q}$ is a quadratic form in $\mathbb{R}^n$ and G a diffeomorphism of an open neighborhood of the origin in $\mathbb{R}^n$ into $\mathbb{R}^n$, depending smoothly on x and such that $G(x,0) \equiv 0$. Taking (VII.1.18) into account allows us to carry out the change of coordinates $(x,t) \to (x^\#,t^\#)$, with

$$x^\# = \mathcal{R}e\, u, \; t^\# = G(x,t)$$

in a neighborhood of $\mathfrak{p}$. In the new coordinates the expressions of the vector fields (VII.1.9) are given by

$$\mathrm{L}_j = \sum_{k=1}^{n} (\mathrm{L}_j G_k)\partial/\partial t_k^\# + [\mathrm{L}_j(\mathcal{R}e\, u)]\partial/\partial x^\#.$$

If we apply to the system $\{\mathrm{L}_1,\ldots,\mathrm{L}_n\}$ the inverse of the matrix $(\mathrm{L}_j G_k)_{1\le j,k\le n}$ (which is invertible in a suitably small open neighborhood of $\mathfrak{p}$), we end up with a system of vector fields like those in (VII.1.9) but in the new coordinates $x^\#$, $t_j^\#$. We delete the superscripts #. Now we have $u = x + \imath\mathfrak{Q}(t)$ and, as a consequence,

$$\mathrm{L}_j u = \imath\partial\mathfrak{Q}/\partial t_j + \lambda_j,$$

thereby getting the first part of the statement with $\rho_j = \mathrm{L}_j u$.

If the functions $\rho_j = \mathrm{L}_j u$ vanish identically, then, according to (VI.1.15), the vector fields L_j have the expressions

$$\mathrm{L}_j = \partial/\partial t_j - \imath(\partial\mathfrak{Q}/\partial t_j)\partial/\partial x, \; j = 1,\ldots,n. \tag{VII.1.19}$$

Conversely, if (VII.1.19) is true and if we take $Z = x + \imath\mathfrak{Q}(t)$, then clearly $\mathrm{L}_j Z = 0$ whatever $j = 1,\ldots,n$, and $\mathrm{d}Z$ spans T' over an open neighborhood of $\mathfrak{p}$. ■

VII.2. Nonsolvability and Nonintegrability when the Signature of the Levi Form is $|n - 2|$

Let $\mathcal{M}$ be equipped with an involutive structure such that $n \geq 1$ and $m = 1$. Until further notice we shall not assume that it is a Mizohata structure. We begin by establishing a link between local integrability and local solvability (under the hypothesis $m = 1$).

We reason in a neighborhood U of a point of $\mathcal{M}$ denoted by 0 and referred to as the origin. The open set U is the domain of local coordinates x, t_j ($1 \leq j \leq n$), all of which vanish at 0. The tangent structure bundle $\mathcal{V}$ is spanned, over U, by vector fields (VII.1.9); these commute.

Here we shall denote by $\mathscr{C}^\infty(\mathrm{U};\Lambda^{0,q})$ the space of differential forms

$$f = \sum_{|J|=q} f_J(x,t)\mathrm{d}t_J \tag{VII.2.1}$$

with coefficients $f_J \in \mathcal{C}^\infty(U)$. The differential operator L acts on such a form according to the formula

$$\mathrm{L}f = \sum_{|J|=q} \sum_{k=1}^{n} \mathrm{L}_k f_J(x,t)\, \mathrm{d}t_k \wedge \mathrm{d}t_J \tag{VII.2.2}$$

(cf. (I.7.30)). Consider the following one-forms in U,

$$\lambda = \sum_{j=1}^{n} \lambda_j \mathrm{d}t_j,\ \lambda_x = \sum_{j=1}^{n} (\partial\lambda_j/\partial x)\mathrm{d}t_j. \tag{VII.2.3}$$

Notice that, in this notation, the differential operator L can be written as

$$\mathrm{L} = \mathrm{d}_t + \lambda\wedge\partial_x \tag{VII.2.4}$$

where the partial derivative with respect to x, ∂_x, acts coefficientwise on forms (VII.2.1). The property that the vector fields (VII.1.9) commute is equivalent to the fact that $\mathrm{L}^2 = 0$ and to the property that

$$\mathrm{L}\lambda = 0. \tag{VII.2.5}$$

LEMMA VII.2.1. *The one-form* λ_x *is* L-*closed.*

PROOF. If we differentiate (VII.2.5) with respect to x, we get

$$\mathrm{L}_j\lambda_{kx} + \lambda_{jx}\lambda_{kx} = \mathrm{L}_k\lambda_{jx} + \lambda_{kx}\lambda_{jx},$$

which shows that λ_x is L-closed. ■

PROPOSITION VII.2.1. *The following two properties are equivalent:*

there is a $\mathcal{C}^\infty$ *function* Z *in some open neighborhood* U *of* 0 *such that* $\mathrm{L}Z = 0$ *and* $\mathrm{d}Z \neq 0$ *at every point of* U; (VII.2.6)

the equation $\mathrm{L}v = -\lambda_x$ *has a* $\mathcal{C}^\infty$ *solution in some open neighborhood of* 0. (VII.2.7)

PROOF. Suppose first that (VII.2.6) holds. Differentiate the equation $\mathrm{L}Z = 0$ with respect to x; one gets $\mathrm{L}Z_x = -\lambda_x Z_x$. We have $Z_x \neq 0$ at every point of U, otherwise $\mathrm{L}Z = \mathrm{d}_t Z + Z_x\lambda = 0$ would entail $\mathrm{d}Z = 0$ at some point of U. After contracting U about 0 we may set $v = \log Z_x$; clearly $\mathrm{L}v = -\lambda_x$.

Conversely, suppose there is $v \in \mathcal{C}^\infty(U)$ satisfying the latter equation; set

$$u(x,t) = \int_0^x e^{v(y,t)}\mathrm{d}y. \tag{VII.2.8}$$

We have

$$\mathrm{L}u(x,t) = \int_0^x e^{v(y,t)}\, \mathrm{d}_t v(y,t)\, \mathrm{d}y + e^{v(x,t)}\lambda(x,t) =$$

$$- \int_0^x e^{v(y,t)}[\lambda_y(y,t) + \lambda(y,t)v_y(y,t)]dy + e^{v(x,t)}\lambda(x,t) = e^{v(0,t)}\lambda(0,t).$$

It follows from this that $e^v\lambda|_{x=0}$ is L-closed; but this means that this form is d_t-closed. Let $w(t)$ be any $\mathscr{C}^\infty$ function in some open neighborhood of the origin in t-space $\mathbb{R}^n$ satisfying $d_t w = e^v\lambda|_{x=0}$; $Z = u - w$ satisfies the requirements in Proposition VII.2.1. ■

Throughout the remainder of the present section we assume that $\mathcal{M}$ is equipped with a Mizohata structure. By the *signature* of the Levi form at a point we shall mean the number $|n - 2\nu|$ where ν is the number of eigenvalues > 0.

We continue to reason in the neighborhood U of the origin and use the coordinates x, t_j as before. Let us first observe that when the structure of $\mathcal{M}$ is *not* locally integrable at the point 0, there is an L-closed one-form, namely λ_x, which is not L-exact in any neighborhood of 0. In principle this does not preclude that there might be a distribution u such that $Lu = \lambda_x$ in some neighborhood of 0 (it only excludes that there be a $\mathscr{C}^\infty$ function u with this property). But it is known, and will be shown in Volume 2 of the present book, that when the signature of the Levi form is $< n$, the differential operator L acting on zero-forms is hypo-elliptic (cf. Definition III.5.2), which implies that the distribution u is in fact a $\mathscr{C}^\infty$ function, and therefore cannot exist.

THEOREM VII.2.1. *Suppose that $\mathcal{M}$ is equipped with a locally integrable Mizohata structure and that the signature of the Levi form at 0 is equal to $|n - 2|$.*

Then there is an L*-closed one-form $f \in \mathscr{C}^\infty(U;\Lambda^{0,1})$ such that there is no distribution u in any open neighborhood* $U' \subset U$ *of* 0 *that satisfies* $Lu = f$ *in* U'.

PROOF. By Theorem VII.1.1 we may choose the coordinates x and t_j in such a way that the vector fields spanning $\mathcal{V}$ over U be given by (VII.1.19). Thus, in the present circumstances, $\lambda = -\iota d_t\mathfrak{Q}$. After a linear change of t variables we may even assume that the L_j have the expressions (VII.1.3). The hypothesis on the signature of the Levi form allows us to take $\varepsilon_1 = +1$, $\varepsilon_j = -1$ for $j = 2,\ldots,n$, i.e., $Z = x + \frac{1}{2}\iota(t_1^2 - |t'|^2)$ where $t' = (t_2,\ldots,t_n)$.

Then consider a sequence of closed disks K_ν ($\nu = 1,2,\ldots$) in the upper half-plane $\mathscr{I}m z > 0$ such that

$$0 < \sup_{K_{\nu+1}} \mathscr{R}e z < \inf_{K_\nu} \mathscr{R}e z,$$

and which converge to the set $\{0\}$ as $\nu \to +\infty$. Let $\tilde{F}$ be a $\mathscr{C}^\infty$ function in $\mathbb{R}^2$ such that $\tilde{F} \equiv 0$ off the union of the disks K_ν and $\tilde{F} \neq 0$ in the interior $\mathscr{I}nt K_\nu$

of K_ν, whatever the integer ν. We are going to reason under the hypothesis that there is a point $e^{i\theta}$ in the unit circle such that the following is true:

to every $\varepsilon >$ *there is an integer* $N(\varepsilon) \geq 0$ *such that, for all* $\nu \geq N(\varepsilon)$, (VII.2.9)

$$\sup_{\mathcal{I}nt\, K_\nu} \left| \tilde{F}/|\tilde{F}| - e^{i\theta} \right| \leq \varepsilon.$$

Call F^+ the function equal to the pullback $\tilde{F}\circ Z$ for $t_1 \geq |t'|$ and to zero otherwise. Note that $F^+ \in \mathscr{C}^\infty(U)$ since $\tilde{F}$ vanishes to infinite order as $t_1 - |t'| > 0$ tends to zero, and $\tilde{F}\circ Z \equiv 0$ in the region $|t'| > |t_1|$. In the region $t_1 > 0$,

$$LF^+ = L(\tilde{F}\circ Z) = [(\partial\tilde{F}/\partial\bar{z})\circ Z]L\bar{Z} = [(\partial\tilde{F}/\partial\bar{z})\circ Z]L(\bar{Z}-Z) =$$

$$- 2\imath[(\partial\tilde{F}/\partial\bar{z})\circ Z]d_t\mathcal{Q},$$

whence

$$LF^+\wedge d_t\mathcal{Q} \equiv 0, \tag{VII.2.10}$$

which is then obviously valid in the whole of U. Set then

$$f = -2\imath F^+ d\mathcal{Q}(t). \tag{VII.2.11}$$

It follows at once from (VII.2.10) that f is L-closed.

We are now going to suppose that there is a distribution u in some open neighborhood $U' \subset U$ of 0 satisfying $Lu = f$ and prove that this leads to a contradiction. By Remark VI.7.2 we know that u is a $\mathscr{C}^\infty$ function in the complement of $t = 0$. It is convenient to take $U' = \mathcal{I} \times \Omega$ where $\mathcal{I}$ is an open interval in $\mathbb{R}^1$ and Ω an open ball in $\mathbb{R}^n$, both centered at the origin. Define $\Omega^+ = \{ t \in \Omega;\ t_1 > 0 \}$, $\Omega^- = \{ t \in \Omega;\ t_1 < 0 \}$. In $\mathcal{I}\times\Omega^+$ (resp., $\mathcal{I}\times\Omega^-$) we may use the coordinates x, $y = \mathcal{Q}(t)$, $t_2,\ldots,t_n$. Over $\mathcal{I}\times(\Omega^+\cup\Omega^-)$ the hypo-analytic structure of $\mathcal{M}$ is defined by $Z = x + \imath y\ (= z)$ and $\mathcal{V}$ is spanned by the vector fields $L_0 = \partial/\partial y - \imath\partial/\partial x$ and $\partial/\partial t_j$ $(2 \leq j \leq n)$. On the other hand, in these coordinates, (VII.2.11) reads

$$f = - 2\imath\tilde{F}(x,y)dy \text{ in } \mathcal{I}\times\Omega^+,\ f \equiv 0 \text{ in } \mathcal{I}\times\Omega^-.$$

Therefore we must have:

$$\begin{aligned} L_0u &= - 2\imath\tilde{F}(x,y) \text{ in } \mathcal{I}\times\Omega^+,\ L_0u = 0 \text{ in } \mathcal{I}\times\Omega^-; \\ \partial u/\partial t_j &= 0 \text{ in } \mathcal{I}\times(\Omega^+\cup\Omega^-)\ (2 \leq j \leq n). \end{aligned} \tag{VII.2.12}$$

It follows that, in $\mathcal{I}\times(\Omega^+\cup\Omega^-)$, u is a $\mathscr{C}^\infty$ function of (x,y) independent of t'; moreover, in $\mathcal{I}\times\Omega^-$, it is holomorphic with respect to $z = x + \imath y$. It is also a holomorphic function of z in the complement of supp F^+ in $\mathcal{I}\times\Omega^+$. Define u^+ as the function of (x,y) equal to u in $\mathcal{I}\times\Omega^+$ and u^- the one equal to u in $\mathcal{I}\times\Omega^-$; we may regard both u^+ and u^- as defined in regions of the plane. We can in fact delimit those regions. Indeed, let R be the radius of the ball Ω and take $t_1^2 = \varepsilon|t'|^2$, $|t|^2 = (1-\varepsilon)R^2$ for $\varepsilon > 0$ arbitrarily small; then $\mathcal{Q}(t) = -(1-\varepsilon)^2R^2/2(1+\varepsilon)$. This shows that both u^+ and u^- are defined in the rectangle

$$x \in \mathcal{I},\ |y| < R^2/2. \tag{VII.2.13}$$

In some neighborhood of a point $(x_*,0,t'_*)$, $x_* \in \mathcal{I}$, $|t'_*| < R$, u is a solution, i.e., $\mathrm{L}u = 0$. By Theorem II.3.1, in such a neighborhood u is the uniform limit of a sequence of polynomials $P_\nu(x+\imath\mathcal{Q}(t))$ ($\nu = 1, 2,\ldots$). Since the symmetry $t_1 \to -t_1$ has no effect on $\mathcal{Q}(t)$ we conclude that u^+ and u^- are the limits of the same sequence of holomorphic polynomials $P_\nu(z)$ in one and the same open subset in the portion $y < 0$ of the rectangle (VII.2.13). They must therefore be equal in the entire complement in (VII.2.13) of the union of the disks K_ν.

Let K'_ν be a closed disk contained in the rectangle (VII.2.13) that does not intersect any K_μ for $\mu \neq \nu$ and whose interior contains K_ν. Since u^- is holomorphic everywhere in (VII.2.13), the integral over $\partial\mathrm{K}'_\nu$ of the one-form $u^-\,\mathrm{d}z$ vanishes, whence

$$0 = \int_{\partial\mathrm{K}'_\nu} u^+\mathrm{d}z = \int_{\mathrm{K}'_\nu} (\partial u^+/\partial\bar{z})\,\mathrm{d}\bar{z}\wedge\mathrm{d}z = \int_{\mathrm{K}_\nu} \tilde{F}\,\mathrm{d}z\wedge\mathrm{d}\bar{z}$$

by (VII.2.12). At this point we apply (VII.2.9): for ν sufficiently large we will have $\mathcal{R}e(\mathrm{e}^{-\imath\theta}\tilde{F}) > 0$ in the interior of K_ν, which implies

$$\mathcal{I}m\left\{\mathrm{e}^{-\imath\theta}\int_{\mathrm{K}_\nu}\tilde{F}dz\wedge\mathrm{d}\bar{z}\right\} < 0,$$

whence the announced contradiction. ■

The argument in the proof of Theorem VII.2.1 can be adapted to yield the proof of the following result:

THEOREM VII.2.2. *Same hypotheses as in Theorem* VII.2.1. *Let* $F^+ \in \mathcal{C}^\infty(\mathrm{U})$ *be the function defined in the proof of Theorem* VII.2.1; *let* L_j $(1 \leq j \leq n)$ *be the vector fields in* U *given by* (VII.1.19); *and set*

$$\mathrm{L}_j^{\#} = \mathrm{L}_j + 2\imath F^+(\partial\mathcal{Q}/\partial t_j)\partial/\partial x,\ j = 1,\ldots,n.$$

Then

$$[\mathrm{L}_j^{\#},\mathrm{L}_k^{\#}] = 0,\ \forall\, j, k = 1,\ldots,n. \tag{VII.2.14}$$

And if $h \in \mathcal{C}^\infty(\mathrm{U})$ *satisfies*

$$\mathrm{L}_j^{\#}h = 0,\ j = 1,\ldots,n, \tag{VII.2.15}$$

in some open neighborhood $\mathrm{U}' \subset \mathrm{U}$ *then perforce* $\mathrm{d}h|_0 = 0$.

PROOF. First of all, (VII.2.10) entails

$$[L_j^\#, L_k^\#]/2\imath = (\partial\mathfrak{Q}/\partial t_k)L_j F^+ - (\partial\mathfrak{Q}/\partial t_j)L_k F^+ = 0,$$

whence (VII.2.14).

Let now $h \in \mathscr{C}^\infty(U)$ satisfy (VII.2.15), i.e.,

$$Lh = h_x f \tag{VII.2.16}$$

in the notation of (VII.2.11). In each region $\mathcal{J} \times \Omega^+$ and $\mathcal{J} \times \Omega^-$ we may use the coordinates x, $y = \mathfrak{Q}(t)$, t_j ($j = 2,\ldots,n$) as in the proof of Theorem VII.2.1 and the basis $\{ L_0, \partial/\partial t_2, \ldots, \partial/\partial t_n \}$ of $\mathcal{V}$ ($L_0 = \partial/\partial y - \imath\partial/\partial x$). Then, for any $\mathscr{C}^\infty$ function g in U,

$$Lg = (L_0 g)dy + \sum_{j=2}^{n} (\partial g/\partial t_j)dt_j.$$

In particular, $L\mathfrak{Q} = dy$, and (VII.2.16) reads

$$L_0 h = -2\imath F^+ h_x, \quad \partial h/\partial t_j \equiv 0 \ (2 \le j \le n).$$

This allows us to push forward from $\mathcal{J} \times \Omega^+$ to the (x,y)-plane, under the map $(x,y,t') \to x + \imath y$, the function h, getting thus a function $\tilde{h}$.

Suppose we had $h_x(0) \neq 0$. Then the product $\tilde{h}_x \tilde{F}$ would have the same properties as $\tilde{F}$, in particular (VII.2.9). But according to the proof of Theorem VII.2.1 in which we replace the form f by the form $h_x f$, Equation (VII.2.16) could not be satisfied, whence a contradiction.

Thus $h_x(0) = 0$. But then Equation (VII.2.16), which reads

$$d_t h = h_x(f - \lambda),$$

entails $d_t h|_0 = 0$, hence $dh|_0 = 0$. ∎

The meaning of Theorem VII.2.2 is clear: one can modify in a neighborhood of 0 the locally integrable structure of $\mathcal{M}$ to get a new involutive structure that is not locally integrable in such a neighborhood. Furthermore, the new structure can be taken to differ from the original one by a perturbation that vanishes to infinite order at 0; in particular, it is a Mizohata structure.

VII.3. Mizohata Structures on Two-Dimensional Manifolds

In this section we assume dim $\mathcal{M} = 2$. Proposition VII.1.2 applies: in any Mizohata structure on $\mathcal{M}$ the curve $\pi\Sigma^\circ$ has a natural orientation. In such a structure the number ν of positive eigenvalues of the Levi form is either zero or one; in both cases $|n - 2\nu| = 1 = |n - 2|$ and thus Theorems VII.2.1 and VII.2.2 will also apply.

The reasoning will be purely local and we may as well restrict ourselves to dealing with pairs $(\mathrm{U},\mathcal{T})$ consisting of an open neighborhood U of 0 in $\mathbb{R}^2$ and of a Mizohata structure $\mathcal{T}$ on U such that $\mathrm{T}_0^\circ \neq 0$ (i.e., $0 \in \pi\Sigma^\circ$, a property we express by saying that the origin is *characteristic*). We say that two such pairs, $(\mathrm{U},\mathcal{T})$ and $(\mathrm{U}',\mathcal{T}')$, define the same *germ of Mizohata structure on the germ of space* $(\mathbb{R}^2,0)$ if $\mathcal{T} = \mathcal{T}'$ in an open neighborhood $\mathrm{U}'' \subset \mathrm{U} \cap \mathrm{U}'$ of the origin.

Let $\mathcal{T}$ be a Mizohata structure on the open neighborhood U of 0 in $\mathbb{R}^2$ (such that 0 is characteristic in $\mathcal{T}$). By a *morphism* of $(\mathrm{U},\mathcal{T})$ into another such pair, $(\mathrm{U}',\mathcal{T}')$, we mean a map $f\colon \mathrm{U} \to \mathrm{U}'$, which preserves the origin and such that the pushforward under f of $\mathcal{V}$, the tangent structure bundle in $\mathcal{T}$, is contained in $\mathcal{V}'$, the tangent structure bundle in $\mathcal{T}'$. The morphism will be an *isomorphism* if f is a diffeomorphism of U onto U′. In this case the pushforward map is a diffeomorphism of $\mathcal{V}$ onto $\mathcal{V}'$. Below we use the following terminology: we are going to say that two germs of Mizohata structures on $(\mathbb{R}^2,0)$, $\dot{\mathcal{T}}$ and $\dot{\mathcal{T}}'$, are *equivalent* if there is a representative $(\mathrm{U},\mathcal{T})$ of $\dot{\mathcal{T}}$ and one, $(\mathrm{U}',\mathcal{T}')$, of $\dot{\mathcal{T}}'$ that are isomorphic. We shall denote by $\mathcal{M}iz(2)$ the quotient for this equivalence relation of the set of all germs of Mizohata structures on $(\mathbb{R}^2,0)$.

Let $\mathcal{T}$ be a Mizohata structure on U (in which 0 is characteristic). Take $\mathrm{U} = \mathcal{I} \times]-T,T[$ (with $\mathcal{I}$ an open interval in $\mathbb{R}^1$ centered at zero and $T > 0$) such that the tangent structure bundle $\mathcal{V}$ in $\mathcal{T}$ is spanned over U by the vector field

$$\mathrm{L} = \partial/\partial t - [\imath t - \rho(x,t)]\partial/\partial x \qquad \text{(VII.3.1)}$$

(Theorem VII.1.1). Among other things, this means that the curve $\pi\Sigma^\circ \subset \mathrm{U}$ is defined by $t = 0$. It also means that the natural orientation of $\pi\Sigma^\circ$ (Proposition VII.1.2) is the standard orientation of the x-axis. Below we write $\mathrm{U}^+ = \{(x,t) \in \mathrm{U}; t > 0\}$, $\mathrm{U}^- = \{(x,t) \in \mathrm{U}; t < 0\}$.

Lemma VII.3.1. *Let $f = (f_1,f_2)$ be a continuous map of* U *into $\mathbb{R}^2$ that preserves the origin and has the following properties:*

$$f_2|_{t=0} \equiv 0; \qquad \text{(VII.3.2)}$$

in each open set U^+, U^- the functions f_1, f_2 are $\mathcal{C}^\infty$ and each one of their derivatives extends as a continuous function to $\mathcal{C}\ell\,\mathrm{U}^+$ and to $\mathcal{C}\ell\,\mathrm{U}^-$; (VII.3.3)

in $\mathrm{U}^+ \cup \mathrm{U}^-$ the pushforward of L *under f is collinear to a vector field* (VII.3.4)

$$L^\# = \partial/\partial t - [\imath t - \rho^\#(x,t)]\partial/\partial x,$$

with $\rho^\#$ a $\mathcal{C}^\infty$ function vanishing to infinite order at $t = 0$.

Then f_1 and f_2^2 belong to $\mathcal{C}^\infty(\mathrm{U};\mathbb{R})$ and

$$\partial f_1/\partial x = (\partial f_2/\partial t)^2 \text{ at } t = 0. \qquad \text{(VII.3.5)}$$

PROOF. Condition (VII.3.4) is equivalent to

$$\mathrm{L}f_1 + (\iota f_2 - \rho^{\#}\circ f)\mathrm{L}f_2 \equiv 0.$$

If we disregard all quantities that vanish to infinite order as $t \to 0$, the preceding equation is equivalent to the pair of relations

$$\partial_t f_1 = -tf_2\partial_x f_2, \; t\partial_x f_1 = f_2\partial_t f_2. \qquad \text{(VII.3.6)}$$

Because of (VII.3.2) the second equation (VII.3.6) implies (VII.3.5). We also derive, by differentiating $k-1$ times ($k \geq 1$) both sides in both equations (VII.3.6) and putting $t = 0$:

$$\begin{aligned} \partial_t f_1 &= \partial_t f_2^2 = 0, \\ \partial_t^k f_1 &= -(k-1)\, \partial_t^{k-2}\partial_x f_2^2/2, \qquad \text{(VIII.3.7)} \\ \partial_t^k f_2^2/2 &= (k-1)\partial_t^{k-2}\partial_x f_1 \; \textit{for}\; k \geq 2. \end{aligned}$$

By induction on k we derive from (VII.3.7) that every derivative of f_1 and f_2^2 at $t = 0$ is uniquely determined by the traces $f_1(x,0)$, $f_2(x,0)$, and therefore the restrictions of f_1 and f_2^2 to U^+ and U^- agree to infinite order at $t = 0$. ■

LEMMA VII.3.2. *Same hypotheses as in Lemma* VII.3.1. *Assume moreover that* $(\partial f_1/\partial x)(x,0) \neq 0$, $\forall\, x \in \mathcal{I}$. *Then* f *is a* $\mathcal{C}^\infty$ *diffeomorphism in a neighborhood of the set* $\mathcal{I} \times \{0\} \subset \mathrm{U}$ *and*

$$\partial f_1/\partial x > 0,\; \partial f_2/\partial t \neq 0 \;\textit{at every point}\; (x,0),\; x \in \mathcal{I}. \qquad \text{(VII.3.8)}$$

PROOF. From Equation (VII.3.5) we derive that if $\partial_x f_1(x,0) \neq 0$, then the limits of $\partial_t f_2(x,t)$ as $t \to 0$ from above and from below are $\neq 0$, and then the square of both these limits must be equal to $\partial_x f_1(x,0)$. On the x-axis, $\partial_t f_2 = (\partial_x f_1)^{1/2}$ or $\partial_t f_2 = -(\partial_x f_1)^{1/2}$. These cases are mutually exclusive since $\partial_x f_1(x,0) > 0$. But the choice is determined by the values of f when $t \neq 0$. We can then exploit (VII.3.7) to determine the traces $\partial_t^k f(x,0)$ for all values of $k \in \mathbb{Z}_+$; they do not depend on whether $t \to 0$ from above or from below. This proves that $f \in \mathcal{C}^\infty(\mathrm{U};\mathbb{R}^2)$. Also thanks to (VII.3.2) we see that, when $t = 0$, the Jacobian determinant $\mathrm{D}(f_1,f_2)/\mathrm{D}(x,t)$ is equal to $(\partial_x f_1)(\partial_t f_2)$. Then select an open neighborhood $\mathbb{O}$ of $\mathcal{I} \times \{0\}$ in U in which $\partial_x f_1 > 0$; clearly the map $(x,t) \to (f_1(x,t),t)$ is injective in $\mathbb{O}$. Let $\mathbb{O}'$ denote the image of $\mathbb{O}$ under that map, and call $g_1(x',t)$ the solution of $f_1(g_1,t) = x'$ such that $g_1(0,0) = 0$. If the open set $\mathbb{O}$ is suitably chosen, the map $(x',t) \to (x', f_2(g_1(x',t),t))$ is injective in $\mathbb{O}'$. ■

Property (VII.3.8) shows that f preserves the orientation of the x-axis. On the other hand, none of the two possible cases, $tf_2 \geq 0$ or $tf_2 \leq 0$, is precluded, and thus f does not necessarily preserve the orientation of U.

View $\mathcal{R} = \mathcal{I} \times]-T^2/2, T^2/2[$ as a subset of the (x,y)-plane. Define the following two functions in $\mathcal{R}$:

$$\tilde{\rho}^+(x,y) = \rho(x,\sqrt{2y})/\sqrt{2y},\ \tilde{\rho}^-(x,y) = \rho(x,-\sqrt{2y})/\sqrt{2y} \text{ for } y \geq 0;$$
$$\tilde{\rho}^+ \equiv \tilde{\rho}^- \equiv 0 \text{ for } y < 0.$$

That both $\tilde{\rho}^+$ and $\tilde{\rho}^-$ are $\mathcal{C}^\infty$ functions in $\mathcal{R}$ follows from the fact that $\rho(x,t)$ vanishes to infinite order at $t = 0$. The vector fields in $\mathcal{R}$,

$$\mathcal{L}^+ = \partial/\partial y - \iota(1-\tilde{\rho}^+)\partial/\partial x,\ \mathcal{L}^- = \partial/\partial y - \iota(1+\tilde{\rho}^-)\partial/\partial x, \quad \text{(VII.3.9)}$$

define two complex structures on $\mathcal{R}$, in general different. The map $(x,t) \to (x,t^2/2)$ is a diffeomorphism of U^+ onto $\mathcal{R}^+ = \{(x,y) \in \mathcal{R}; y > 0\}$, transforming L into $\sqrt{2y}\mathcal{L}^+$. The map $(x,t) \to (x,-t^2/2)$ is a diffeomorphism of U^- onto $\mathcal{R}^+$ transforming L into $-\sqrt{2y}\mathcal{L}^-$. Note that both diffeomorphisms preserve the orientation.

After decreasing the size of the rectangle $\mathcal{R}$ if necessary, we may assume, thanks to the Newlander-Nirenberg theorem, that there are two $\mathcal{C}^\infty$ functions ζ^+ and ζ^- such that

$$\mathcal{L}^+\zeta^+ = \mathcal{L}^-\zeta^- = 0 \text{ in } \mathcal{R},\ \zeta^+(0) = \zeta^-(0) = 0,$$

which can be viewed as diffeomorphisms of $\mathcal{R}$ onto some open neighborhoods of 0 in $\mathbb{C}$. These diffeomorphisms preserves the orientation of $\mathcal{R}$ since

$$d\zeta^+ \wedge d\bar{\zeta}^+ = g^+ dz \wedge d\bar{z},\ d\zeta^- \wedge d\bar{\zeta} = g^- dz \wedge d\bar{z},$$

with $g^+ > 0$, $g^- > 0$ $(z = x + \iota y)$. In particular, both ζ^+ and ζ^- map $\mathcal{R}^+$ onto open sets Ω^+ and Ω^- respectively. If $\mathcal{R}$ is small enough we may assume that both Ω^+ and Ω^- are simply connected (they are of course connected). Note that the boundary $\partial\Omega^+$ contains the arc of curve $\zeta^+(\mathcal{I} \times \{0\})$ (on which lies the origin), and $\partial\Omega^-$ contains $\zeta^-(\mathcal{I} \times \{0\})$. By the Riemann mapping theorem there is a biholomorphism H^+ of Ω^+ onto the upper half unit disk $D^+ = \{z \in \mathbb{C}; |z| < 1, \mathcal{I}m\, z > 0\}$; H^+ extends as a $\mathcal{C}^\infty$ function up to any piece of the boundary of Ω^+ that is of class $\mathcal{C}^\infty$, in particular to the arc of curve $\zeta^+(\mathcal{I} \times \{0\})$. We require $H^+(0) = 0$ and $[(\partial/\partial z)H^+](0) > 0$. Thus $H^+ \circ (\zeta|_{\mathcal{R}^+})$ preserves the origin and the orientation. Likewise, there is a biholomorphism H^- of Ω^- onto D^+ that extends as a $\mathcal{C}^\infty$ function up to the arc of curve $\zeta^-(\mathcal{I} \times \{0\})$ and such that $H^-(0) = 0$, $[(\partial/\partial z)H^-](0) > 0$. Then define

$$\begin{aligned} Z^+(x,t) &= H^+(\zeta^+(x,t^2/2)) \text{ if } x \in \mathcal{I},\ t \geq 0; \\ Z^-(x,t) &= H^-(\zeta^-(x,t^2/2)) \text{ if } x \in \mathcal{I},\ t \leq 0. \end{aligned} \quad \text{(VII.3.10)}$$

Observe that Z^+ and Z^- are solutions of the homogeneous equation $\mathrm{L}h = 0$ in U^+ and U^- respectively.

The function Z^+ defines a diffeomorphism of U^+ onto D^+; it extends in $\mathscr{C}^\infty$ fashion to the interval $\mathscr{I}\times\{0\}$, and thus it induces a diffeomorphism ξ^+ of $\mathscr{I}$ onto the open interval $]-1,+1[$. Likewise Z^- defines a diffeomorphism of U^- onto D^+ and induces a diffeomorphism ξ^- of $\mathscr{I}$ onto $]-1,+1[$. Both ξ^+ and ξ^- preserve the orientation. It follows that $\xi^+\circ(\xi^-)^{-1}$ is a diffeomorphism of the interval $]-1,+1[$ onto itself that preserves 0 as well as the orientation. We shall denote by $\chi(\mathcal{T})$ the germ at zero of the diffeomorphism $\xi^+\circ(\xi^-)^{-1}$.

Let $\mathscr{C}^\infty \mathit{Diff}_0^+$ be the group of germs at zero of monotone increasing $\mathscr{C}^\infty$ diffeomorphisms of the real line that preserve the origin. Denote by $\mathscr{C}^\omega \mathit{Diff}_0^+$ the analogue in the analytic category. Let us say that two elements f and g of $\mathscr{C}^\infty \mathit{Diff}_0^+$ are *equivalent* if there exist germs $h_1, h_2 \in \mathscr{C}^\omega \mathit{Diff}_0^+$ such that $g = h_1 f h_2$. We denote by

$$\mathscr{D}_0^+ = \mathscr{C}^\omega \mathit{Diff}_0^+ \backslash \mathscr{C}^\infty \mathit{Diff}_0^+ / \mathscr{C}^\omega \mathit{Diff}_0^+ \tag{VII.3.11}$$

the set of cosets for this equivalence relation.

THEOREM VII.3.1. *The map $\mathcal{T} \to \chi(\mathcal{T})$ induces a bijection of $\mathscr{M}iz\,(2)$ onto $\mathscr{D}_0^+$.*

PROOF. I. *The map $\mathcal{T} \to \chi(\mathcal{T})$ induces a map $\dot{\chi}$: $\mathscr{M}iz\,(2) \to \mathscr{D}_0^+$.*

Consider two Mizohata structures $\mathcal{T}$ and $\mathcal{T}'$ on open neighborhoods of the origin in $\mathbb{R}^2$ in which 0 is characteristic. For convenience we assume that the two structures $\mathcal{T}$ and $\mathcal{T}'$ are defined in one and the same rectangle U of the (x,t)-plane by two vector fields L and L$'$ of the kind (VII.3.1). Let Z^+, Z^- be the maps defined in (VII.3.10) and their analogues Z'^+, Z'^- for the structure $\mathcal{T}'$. Suppose there is a diffeomorphism f that transforms $\mathcal{T}$ into the restriction of $\mathcal{T}'$ over some open neighborhood of 0 contained in U; it has perforce the properties hypothesized in Lemma VII.3.1. From (VII.3.8) we derive that $x \to f_1(x,0)$ is a monotone increasing diffeomorphism of $\mathscr{I}$ onto an open interval $\mathscr{I}' \subset \mathscr{I}$.

The pullback of Z'^+ under the diffeomorphism f, $Z'^+\circ f$, is also a solution of the homogeneous equation $\mathrm{L}h = 0$ in U^+. Near every point of U^+ it is holomorphic with respect to Z^+, which means that there is a holomorphic function φ^+ in D^+ such that $Z'^+\circ f = \varphi^+(Z^+)$; moreover, φ^+ is $\mathscr{C}^\omega$ up to the part $|\mathscr{R}e\, z| < 1$, $\mathscr{I}m\, z = 0$ of the boundary of D^+, by Schwarz reflection. Similarly we have $Z'^-\circ f = \varphi^-(Z^-)$ for some holomorphic function φ^- in D^+, $\mathscr{C}^\omega$ up to the part $|\mathscr{R}e\, z| < 1$, $\mathscr{I}m\, z = 0$ of the boundary. We get

$$\xi'^+\circ(f_1|_{t=0}) = \varphi^+\circ\xi^+, \; \xi'^-\circ(f_1|_{t=0}) = \varphi^-\circ\xi^-,$$

whence

$$\xi'^+\circ(\xi'^-)^{-1} = \varphi^+\circ\xi^+\circ(\xi^-)^{-1}\circ(\varphi^-)^{-1}, \tag{VII.3.12}$$

which shows that $\chi(\mathcal{T}') = \varphi^+\circ\chi(\mathcal{T})\circ(\varphi^-)^{-1}$ (in the sense of germs).

II. *The map* $\dot{\chi}$ *is injective.*

We start from (VII.3.12) where φ^+ and φ^- are analytic transformations of an interval in $\mathbb{R}^1$ centered at zero, such that $\varphi^+(0) = \varphi^-(0) = 0$. We extend each of them as a biholomorphism (also denoted by φ^+ and φ^- respectively) of some open neighborhood of 0 in $\mathbb{C}$ onto another such neighborhood. Let $U_0 = \mathcal{I}_0 \times]-T_0,T_0[$ be an open rectangle in which the composed maps

$$f^+ = (Z'^+)^{-1} \circ \varphi^+ \circ Z^+, f^- = (Z'^-)^{-1} \circ \varphi^- \circ Z^-$$

are defined (we assume that Z'^+ and Z'^- are defined in some rectangle U containing U_0; cf. (VII.3.10)). We take the interval $\mathcal{I}_0$ small enough that the restrictions $f^+|_{t=0}, f^-|_{t=0}$ will be diffeomorphisms of $\mathcal{I}_0$ onto some open interval containing zero. But note that (VII.3.12) entails $f^+|_{t=0} = f^-|_{t=0}$. We define f in U_0 to be equal to f^+ for $t > 0$ and to f^- for $t < 0$. Let us write $f = (f_1,f_2)$; since f preserves the x-axis, condition (VII.3.2) is also satisfied. Condition (VII.3.3) is satisfied trivially. Since the pullback $Z'^+ \circ f$ is a holomorphic function of Z^+ the pushforward of L under f is collinear to L′, and thus Property (VII.3.4) is satisfied with $L^\# = L'$. By Lemma VII.3.1 $f \in \mathcal{C}^\infty(U)$, and by Lemma VII.3.2 f is a diffeomorphism of an open neighborhood of 0 in U_0 onto some other open neighborhood of 0 in $\mathbb{R}^2$.

III. *The map* $\dot{\chi}$ *is surjective.*

Let f_0 be a monotone-increasing diffeomorphism of a compact interval (centered at 0) onto another such interval; assume $f_0(0) = 0$. We extend f_0 as a $\mathcal{C}^\infty$ function in the whole real line. According to Lemma VI.5.1 there is a $\mathcal{C}^\infty$ function h in $\mathbb{R}^2$ such that

$$h|_{y=0} = f_0, \tag{VII.3.13}$$

$$\partial h/\partial \bar{z} \text{ vanishes to infinite order at } y = 0. \tag{VII.3.14}$$

Set $Z(x,t) = h(x,t^2/2)$ for $(x,t) \in \mathbb{R}^2$. We have

$$dZ = Z_x[dx + (\imath t - \lambda)dt],$$

where $\lambda = 2\imath t h_{\bar{z}}/h_x$ vanishes to infinite order at $t = 0$. Then define the following function in $\mathbb{R}^2$: $\rho(x,t) = \lambda(x,t)$ if $t \geq 0$, $\rho \equiv 0$ if $t < 0$. If the rectangle U $= \mathcal{I} \times]-T,T[$ is sufficiently small, we can equip it with the Mizohata structure defined by means of the vector field (VII.3.1). Going now to (VII.3.9), we see that

$$\tilde{\rho}^+ = 2\imath h_x^{-1}\partial h/\partial \bar{z} \textit{ for } y \geq 0,\ \tilde{\rho}^- \equiv 0.$$

As a consequence, we may take $\zeta^- = x + \imath y$. Let ζ^+, H^+, and Z^+ be as in (VII.3.10); since $LZ = 0$ we have, in U^+, $Z^+ = \varphi^+ \circ Z$ with φ^+ a holomorphic function in $Z(U^+)$, $\mathcal{C}^\infty$ up to the arc of curve $Z(\mathcal{I} \times \{0\})$. But we know that $Z(x,0) = h(x,0) = f_0(x)$ and thus $Z(\mathcal{I} \times \{0\})$ is an interval in the real line (containing 0). By construction this is also true of $Z^+(\mathcal{I} \times \{0\})$.

This means that φ^+ maps an interval in the real line, containing zero, onto another such interval. By the Schwarz reflection principle it implies that φ^+ is analytic. We have $\xi^+ = \varphi^+ \circ f_0$ and since ξ^- is the identity map, we conclude that $\xi^+ \circ (\xi^-)^{-1}$ belongs to the class of f_0 in $\mathfrak{D}_0^+$. ∎

DEFINITION VII.3.1. *Let $\mathcal{T}$ be a Mizohata structure on an open neighborhood of* 0 *in* $\mathbb{R}^2$ *and* $\dot{\mathcal{T}}$ *its germ at* 0; $\dot{\chi}(\mathcal{T})$ ($\in \mathfrak{D}_0^+$) *will be called the* Sjöstrand invariant *of $\mathcal{T}$ at the origin.*

Since $\dot{\chi}(\mathcal{T})$ is truly invariant, i.e., coordinate free, we are allowed to extend Definition VII.3.1 to any two-dimensional manifold $\mathcal{M}$ equipped with a Mizohata structure and talk of the Sjöstrand invariant of $\mathcal{M}$ at any one of its points.

COROLLARY VII.3.1. *In order for two germs of Mizohata structures at the origin in* $\mathbb{R}^2$ *to be equivalent at* 0 (see p. 325) *it is necessary and sufficient that their Sjöstrand invariant be equal.*

Consider the model Mizohata structure on $\mathbb{R}^2$ (Example VII.1.1). In this case the vector fields $\mathcal{L}^+$ and $\mathcal{L}^-$ are both equal to $\partial/\partial y - \imath\partial/\partial x$ and we may take $\zeta^+ = \zeta^- =$ the identity map. Then $Z^+ = Z^- = x + \imath t^2/2$ and $\xi^+ = \xi^- =$ identity map of $]-1,+1[$: *the Sjöstrand invariant at the origin of the model Mizohata structure on* $\mathbb{R}^2$ *is the class of the identity.*

COROLLARY VII.3.2. *In order for the Mizohata structure $\mathcal{T}$ on $\mathcal{M}$ to be locally integrable it is necessary and sufficient that its Sjöstrand invariant be the class of the identity.*

PROOF. The condition is sufficient, by Corollary VII.3.1. It is necessary by Theorem VII.1.1 and Corollary VII.3.1. ∎

VII.4. Nonintegrability and Nonsolvability when the Cotangent Structure Bundle has Rank One

We propose now to generalize the results of section VII.2 to a class of involutive structures in which *the cotangent structure bundle is a complex line bundle*. The results in this case will "explain" why the signature $|n - 2|$ is critical when dealing with Mizohata structures.

We consider a $\mathcal{C}^\infty$ manifold $\mathcal{M}$ equipped with an involutive structure such that $m = 1$, $n \geq 1$ (thus $\dim \mathcal{M} \geq 2$). We go back to the concepts and notation of the beginning of section VII.2: the analysis is carried out in the local chart $(U, x, t_1, \ldots, t_n)$; we assume that the tangent structure bundle $\mathcal{V}$ is spanned by the vector fields (VII.1.9); λ and λ_x are the one-forms (VII.2.3). We introduce a modification of the vector fields L_j:

$$L_j^{\#} = L_j - g\lambda_j\partial/\partial x,\ j = 1,\dots,n, \tag{VII.4.1}$$

with $g \in \mathscr{C}^\infty(U)$.

LEMMA VII.4.1. *Suppose that*

$$[Lg - g(1-g)\lambda_x]\wedge\lambda \equiv 0. \tag{VII.4.2}$$

Then

$$[L_j^{\#}, L_k^{\#}] = 0,\ j,\ k = 1,\dots,n. \tag{VII.4.3}$$

PROOF. Since the vector fields L_j commute and therefore $L\lambda \equiv 0$, we have

$$[L_j^{\#}, L_k^{\#}] = \left\{\lambda_j[L_k g - g(1-g)\lambda_{kx}] - \lambda_k[L_j g - g(1-g)\lambda_{jx}]\right\}\partial/\partial x,$$

whence the result. ■

We shall now reason under the assumption that there is a solution in U,

$$Z = x + \imath\Phi(x,t), \tag{VII.4.4}$$

with $\Phi \in \mathscr{C}^\infty(U)$ real-valued and $\Phi(0,0) = 0$. Note that Z_x is nowhere equal to zero in U, and that differentiating the homogeneous equation $LZ = 0$ with respect to x leads to

$$LZ_x^{-1} = Z_x^{-1}\lambda_x. \tag{VII.4.5}$$

LEMMA VII.4.2. *Let F be a function in* U *with the property that to each point $\mathfrak{p} \in$ U there is a $\mathscr{C}^\infty$ function $\tilde{F}_{\mathfrak{p}}$ in the plane, such that $F = \tilde{F}_{\mathfrak{p}}\circ Z$ in some neighborhood of $\mathfrak{p}$. If $|Z_x^{-1}F| < 1$ in* U, *then the function*

$$g = -Z_x^{-1}F/(1 - Z_x^{-1}F) \tag{VII.4.6}$$

satisfies (VII.4.2) *in* U.

PROOF. Near $\mathfrak{p} \in$ U we have, in the notation of the statement,

$$LF = [(\partial\tilde{F}_{\mathfrak{p}}/\partial\bar{z})\circ Z]L\bar{Z}$$

and since $L\bar{Z} = L(Z+\bar{Z}) = 2\lambda$, by patching up together these local representations, one concludes that there is a function $F_1 \in \mathscr{C}^\infty(U)$ such that

$$LF = F_1\lambda.$$

Combining this with (VII.4.5) yields

$$L(Z_x^{-1}F) = (Z_x^{-1}F)\lambda_x + Z_x^{-1}F_1\lambda.$$

It follows that letting L act on the function (VII.4.6) yields

$$Lg = g(1-g)\lambda_x - (1-Z_x^{-1}F)^{-2}Z_x^{-1}F_1\lambda,$$

whence the sought conclusion. ■

We are now going to make the hypothesis that $\mathcal{M}$ and its involutive structure are *analytic*. We then use analytic coordinates x, t_j in U. It is convenient to take U to be a product $\mathcal{I} \times \Omega$ with $\mathcal{I}$ an open interval in the x-line $\mathbb{R}^1$ and Ω an open ball in t-space $\mathbb{R}^n$, both centered at the origin.

The analyticity hypothesis allows one to choose the solution Z to be analytic and such that

$$Z|_{t=0} = x \text{ (i.e., } \Phi(x,0) \equiv 0). \tag{VII.4.7}$$

Suppose that we had $d_t\Phi|_0 \neq 0$; then the structure would be elliptic in an open neighborhood of 0; there would not be much that could be said beyond the results of chapter VI. We shall therefore make the opposite hypothesis: at the origin,

$$d_t\Phi = 0. \tag{VII.4.8}$$

An important ingredient in the analysis will be the *fibres* of the structure near 0 (cf. Proposition I.10 and Corollary II.3.1). These are the germs of sets at 0 whose representatives in U are defined by equations

$$x = x_0, \; \Phi(x_0,t) = y_0. \tag{VII.4.9}$$

The set defined by (VII.4.9) is a fibre of the map Z in U. It is an *analytic variety* that we denote by $\mathcal{F}(z_0)$ (setting $z_0 = x_0 + \iota y_0$). Such a variety is said to be *singular* if there is $t \in \Omega$ satisfying (VII.4.9) and also $d_t\Phi(x_0,t) = 0$, in which case y_0 is a *critical value* of $\Phi(x_0,t)$. We shall refer to z_0 as a critical value of Z. The fibre of the origin is singular and 0 is a critical value of Z.

Actually Equation (VII.4.8) defines an analytic variety $\mathcal{N}$ in U (or a complex-analytic variety $\tilde{\mathcal{N}}$ in a complexification of U). Recall that an analytic variety is a disjoint union of connected analytic submanifolds. In the neighborhood of a given point the number of such submanifolds is finite. By contracting U we may assume that $\mathcal{N}$ itself is connected and also that, for $x_0 \in \mathcal{I}$, the number of critical values y_0 of $\Phi(x_0,t)$ is bounded independently of x_0. Thus, in any slice $\{x_0\} \times \Omega$, all fibres, except at most finitely many, are regular. The regular fibres can be identified to analytic hypersurfaces in Ω (indeed, they have dimension $n - 1$). This does not preclude the possibility that a whole slice $\{x_0\} \times \Omega$ be a fibre, i.e., that $\Phi(x_0,t) = y_0$ for all $t \in \Omega$. But in this case, evidently, the fibre is singular. This phenomenon can only happen at isolated points $x_0 \in \mathcal{I}$, unless $\Phi \equiv 0$, a case of no interest to us. By contracting the interval $\mathcal{I}$ we may assume that, if it happens at all, it only happens when $x_0 = 0$.

As (VII.4.8) holds identically in $\mathcal{N}$, the function $t \to \Phi(x,t)$ must be locally constant on each slice $\mathcal{N} \cap (\{x_0\} \times \Omega)$ (such slices need not be connected). After some further contracting of $\mathcal{I}$ and Ω about 0, one may assume that the slice $\mathcal{N} \cap (\{0\} \times \Omega)$ is connected; $\Phi(0,t)$ must be constant on it, which means that it must vanish identically on it, i.e., $\mathcal{N} \cap (\{0\} \times \Omega)$ is contained in the fibre of the origin. All other fibres $\mathcal{F}(z)$ such that $x = \mathcal{R}e\, z = 0$ are analytic submanifolds of U of codimension two.

Since we have taken $\mathcal{N}$ to be connected, the same is true of $Z(\mathcal{N})$, the set of all critical values of Z; $Z(\mathcal{N})$ is also the union of finitely many analytic submanifolds, perforce of dimension ≤ 1. Thus $Z(\mathcal{N})$ is a union of finitely many open arcs of analytic curves and finitely many points. By further contracting the interval $\mathcal{I}$ about zero we can cut off all those points, except 0 if it is one of them. Then in the region $x > 0$ the set $Z(\mathcal{N})$ will be the disjoint union of finitely many analytic curves $y = \psi_j^+(x)$ $(j = 1,\ldots,r^+)$ with $\psi_j^+(x) \to 0$ as $x \to +0$; likewise, in the region $x < 0$, it will be the disjoint union of the analytic curves $y = \psi_j^-(x)$ $(j = 1,\ldots,r^-;\ \psi_j^-(x) \to 0$ as $x \to -0)$.

We now introduce the crucial hypothesis of this section:

There are points $z_\nu \in \mathbb{C}$, $z_\nu \to 0$, *and a basis of neighborhoods of* 0, $V_\nu \subset U$ $(\nu = 1,2,\ldots)$, *such that, for each* ν, V_ν *intersects two distinct connected components of the fibre* $\mathcal{F}(z_\nu)$. (VII.4.10)

EXAMPLE VII.4.1. Suppose that the involutive structure of $\mathcal{M}$ is a Mizohata structure (Definition VII.1.1). According to Theorem VII.1.1, we may assume that $Z = x + \imath\mathcal{Q}(t)$ with $\mathcal{Q}(t)$ a nondegenerate quadratic form. Property (VII.4.10) means that there is a sequence of points $y_\nu \to 0$ in $\mathbb{R}^n$ such that the hypersurfaces $\mathcal{Q}(t) = y_\nu$ are not connected. In order for this to be the case it is necessary and sufficient that the signature of $\mathcal{Q}$ be equal to $|n - 2|$. ■

THEOREM VII.4.1. *Suppose* (VII.4.10) *holds. Given any open neighborhood of* 0, U′, *whose closure is compact and contained in* U, *there is a* $\mathcal{C}^\infty$ *function* F *in* U′ *that vanishes to infinite order at* 0 *and has the following properties:*

$$|Z_x^{-1}F| < 1 \text{ and the one-form } F\lambda \text{ is L-closed in } U'; \qquad \text{(VII.4.11)}$$

whatever the open neighborhood $U'' \subset U'$ *of* 0:

there is no distribution u *satisfying* $Lu = F\lambda$ *in* U″; (VII.4.12)

if g *is defined by* (VII.4.6), *then any* $\mathcal{C}^1$ *function* h *satisfying* (VII.4.13)

$$L_j h - g\lambda_j \partial h/\partial x = 0,\ j = 1,\dots,n,\ in\ U'', \tag{VII.4.14}$$

is such that $dh|_0 = 0$.

PROOF. We take $U' = \mathcal{I}' \times \Omega'$, $U'' = \mathcal{I}'' \times \Omega''$ where $\mathcal{I}'' \subset\subset \mathcal{I}' \subset\subset \mathcal{I}$ are open intervals and $\Omega'' \subset\subset \Omega' \subset\subset \Omega$ are open balls, all centered at the origin. Suppose that a fibre $\mathcal{F}(z_0)$ has a connected component $\mathcal{A}$ such that U'' has a nonempty intersection both with $\mathcal{A}$ and with $\mathcal{F}(z_0)\backslash\mathcal{A}$ ($z_0 = x_0 + \imath y_0$). Now select two points $t_0, t_1 \in \Omega''$ with the following property: if ℓ denotes the closed straight-line segment joining t_0 to t_1 ($\ell \subset\subset \Omega''$), then the segment $\{x_0\} \times \ell$ intersects $\mathcal{F}(z_0)$ only at its end points, and $(x_0,t_0) \in \mathcal{A}$, $(x_0,t_1) \in \mathcal{F}(z_0)\backslash\mathcal{A}$. The image of ℓ under the map $t \to \Phi(x_0,t)$ is a segment of the kind $[y_0, y_0 + c]$ or of the kind $[y_0 - c, y_0]$ ($c > 0$). It follows that there are points in ℓ, $t_\nu \to t_0$ and $\tilde{t}_\nu \to \tilde{t}_0$ such that $y_\nu = \Phi(x_0,t_\nu) = \Phi(x_0,\tilde{t}_\nu)$ ($\to y_0$) and y_ν is not a critical value of the map $t \to \Phi(x_0,t)$.

We claim that, for ν large enough, U'' intersects two distinct components of $\mathcal{F}(x_0 + \imath y_\nu) \cap (\mathcal{C}\ell U')$. Suppose it were not so. There would be a smooth curve γ_ν on the hypersurface $\Phi(x_0,t) = y_\nu$ in Ω joining t_ν to $\tilde{t}_\nu$. Let $\mathcal{O}$ be an open set containing $(\mathcal{C}\ell U'') \cap \mathcal{A}$ and whose closure $\mathcal{C}\ell\mathcal{O}$ is a compact subset of U' that does not intersect $\mathcal{F}(z_0)\backslash\mathcal{A}$. For each ν the curve γ_ν would intersect $\partial\mathcal{O}$ and there would be a subsequence of points $\tau_{\nu_k} \in \gamma_{\nu_k} \cap \partial\mathcal{O}$ converging as $k \to +\infty$ to a point $\tau \in \partial\mathcal{O}$. But since $\Phi(x_0,t) = y_\nu$ for $t \in \gamma_\nu \cap \partial\mathcal{O}$ we would necessarily have $\Phi(x_0,\tau) = y_0$, an impossibility.

The upshot of this is that, possibly after replacing U by a relatively compact open subneighborhood of 0, we may assume that none of the points z_ν in (VII.4.10) is a critical value of the map Z and that none is a real number or a purely imaginary one. After eliminating a subsequence it is possible to assume that $0 < |\mathcal{R}e\, z_{\nu+1}| < |\mathcal{R}e\, z_\nu|$ ($\nu \in \mathbb{Z}_+$) and that the numbers $\mathcal{R}e\, z_\nu$ all have the same sign. In the argument below we assume that they are all > 0.

Once again select $U' = \mathcal{I}' \times \Omega' \subset\subset U$ and a basis of open neighborhoods of 0, $\{V_\nu\}_{\nu=0,1,\dots}$, such that, for every ν, $V_{\nu+1} \subset\subset V_\nu \subset\subset U'$ and V_ν intersects two distinct components C_ν and C'_ν of $\mathcal{F}(z_\nu)$. We then select two disks centered at z_ν, $K_\nu \subset D_\nu$, D_ν open, K_ν compact, whose radii will be chosen suitably small. We begin by requiring that $Z(\mathcal{N}) \cap D_\nu = \emptyset$. We also require that

$$\sup_{D_\nu} (\mathcal{R}e\, z) < \inf_{D_{\nu-1}} (\mathcal{R}e\, z). \tag{VII.4.15}$$

Last, we require that there be two connected components Γ_ν and Γ'_ν of $\overset{-1}{Z}(D_\nu)$ endowed with the following properties:

$$(\mathcal{C}\ell\Gamma_\nu) \cap (\mathcal{C}\ell\Gamma'_\nu) = \emptyset,\ C_\nu \subset \Gamma_\nu,\ C'_\nu \subset \Gamma'_\nu; \tag{VII.4.16}$$

any connected component of $\mathcal{F}(z) \cap U'$ ($z \in D_\nu$) *that intersects* Γ_ν *or* Γ'_ν *is entirely contained in that set;* (VII.4.17) *no two distinct connected components of the same fibre in* U', $\mathcal{F}(z) \cap U'$, *intersect* Γ_ν, *nor do they intersect* Γ'_ν; (VII.4.18)

there are two-dimensional analytic submanifolds of Γ_ν *and* Γ'_ν, Σ_ν *and* Σ'_ν *respectively, each mapped diffeomorphically by* Z *onto* D_ν. (VII.4.19)

Now let the function $\tilde{F} \in \mathcal{C}^\infty(\mathbb{R}^2)$ be chosen exactly as in the proof of Theorem VII.2.1. In particular $\tilde{F}$ will have property (VII.2.9). Call F the function in U' equal to $\tilde{F} \circ Z$ in the union of the sets Γ_ν and to zero everywhere else.

We have

$$LF = [(\partial \tilde{F}/\partial \bar{z}) \circ Z] L\bar{Z} = [(\partial \tilde{F}/\partial \bar{z}) \circ Z] L(Z + \bar{Z}) = 2[(\partial \tilde{F}/\partial \bar{z}) \circ Z]\lambda,$$

whence (cf. (VII.2.10))

$$\lambda \wedge LF \equiv 0. \tag{VII.4.20}$$

It follows from (VII.2.5) and (VII.4.20) that $F\lambda$ is L-closed.

Now suppose there were a distribution u in $U'' = \mathcal{J}'' \times \Omega'' \subset\subset U'$ satisfying

$$Lu = F\lambda \tag{VII.4.21}$$

in U''. If ν_0 is large enough, the closures of Σ_ν and Σ'_ν will be contained in U'' for all $\nu > \nu_0$. Since L is elliptic in $\Gamma_\nu \cup \Gamma'_\nu$, u is a $\mathcal{C}^\infty$ function there (Remark VI.7.2). By pushing forward, under the diffeomorphism Z, the restrictions of u to Σ_ν and to Σ'_ν respectively, we get two $\mathcal{C}^\infty$ functions $\tilde{u}^+$ and $\tilde{u}^-$ in D_ν. But notice that $Lu = 0$ in $\Gamma_\nu \backslash [\Gamma_\nu \cap Z^{-1}(K_\nu)]$ and in Γ'_ν. It implies at once that $\tilde{u}^+$ is a holomorphic function in $D_\nu \backslash K_\nu$ and $\tilde{u}^-$ is one in D_ν. We are going to show that, in $D_\nu \backslash K_\nu$,

$$\tilde{u}^+ = \tilde{u}^-. \tag{VII.4.22}$$

Let $\mathcal{J}_\nu$ be an interval contained in the projection of $D_\nu \backslash K_\nu$ on the real axis. Because of (VII.4.15) the vertical slab $\mathcal{J}_\nu \times \mathbb{R}$ does not intersect any K_μ ($\mu \in \mathbb{Z}_+$). It follows that $Lu = 0$ in $\mathcal{J}_\nu \times \Omega''$. To any point $\tilde{z}_\nu = \tilde{x}_\nu + \imath \tilde{y}_\nu \in (D_\nu \backslash K_\nu) \cap (\mathcal{J}_\nu \times \mathbb{R})$ there are unique points $(\tilde{x}_\nu, \tilde{t}_\nu) \in \Sigma_\nu$, $(\tilde{x}_\nu, \tilde{t}'_\nu) \in \Sigma'_\nu$ such that

$$\tilde{x}_\nu \in \mathcal{J}_\nu,\ \Phi(\tilde{x}_\nu, \tilde{t}) = \Phi(\tilde{x}_\nu, \tilde{t}'_\nu) = \tilde{y}_\nu. \tag{VII.4.23}$$

Write $t(s) = s\tilde{t}'_\nu + (1-s)\tilde{t}_\nu, -\varepsilon < s < 1 + \varepsilon$, with $\varepsilon > 0$ small enough that the image ℓ_ν of the map $s \longrightarrow t(s)$ will be contained in Ω'' and $Z(\tilde{x}_\nu, t(s))$ will not be a critical value of the map Z for any s such that $|s| < \varepsilon$ or $|1-s| < \varepsilon$. In particular, ℓ_ν cannot be contained in a singular fibre of the map $t \longrightarrow Z(\tilde{x}_\nu, t)$. It follows that ℓ_ν intersects the singular fibres of $Z(\tilde{x}_\nu, \cdot)$ only at a finite number of points, $s_0 = -\varepsilon < s_1 < \cdots < s_r < s_{r+1} = 1 + \varepsilon$. Let s belong to any one of the open intervals $\mathcal{J}_j =]s_j, s_{j+1}[$ $(0 \leq j \leq r)$. The point $(\tilde{x}_\nu, t(s))$ has an open neighborhood in $\mathcal{J}_\nu \times \Omega''$ in which the map Z is open, and u is $\mathcal{C}^\infty$ and constant on the fibres of the map Z. As a consequence u can be pushed forward via Z, to a holomorphic function in some open neighborhood of $Z(\tilde{x}_\nu,t(s))$. As long as s remains in the same interval $\mathcal{J}_j$ these holomorphic functions agree on overlaps. For each $j = 0, 1,\dots,r$, we thus get a holomorphic function $\tilde{u}_j$ in an open neighborhood $\mathcal{U}_j$, in the plane, of the image $\tilde{\mathcal{J}}_j$ of the interval $\mathcal{J}_j$ under the map $s \rightarrow Z(\tilde{x}_\nu,t(s))$. Note that this image is a vertical segment made up of the points $\tilde{x}_\nu + \iota y$, $\Phi(\tilde{x}_\nu,t(s_{j-1})) < y < \Phi(\tilde{x}_\nu,t(s_j))$. We may therefore take $\mathcal{U}_j$ to be the union of horizontal segments centered at the points $Z(\tilde{x}_\nu,t(s))$ with $s \in \mathcal{J}_j$.

Suppose now that $s = s_j$ for some j such that $1 \leq j \leq r$. In some open neighborhood $\mathcal{O}_j$ of $(\tilde{x}_\nu,t(s_j))$ u is the distribution limit of a sequence of "holomorphic polynomials" $P_k \circ Z$ (Corollary II.3.3 and Remark II.3.2). As a consequence, in the open set $Z(\mathcal{O}_j) \backslash Z(\mathcal{N})$, u defines a holomorphic function $\tilde{v}_j$. Keep in mind that the image of $\mathcal{N} \cap (\mathcal{J}_\nu \times \Omega'')$ under the map Z is contained in a finite union of disjoint analytic curves $y = \psi_i^+(x)$. It follows that, if $\mathcal{O}_j$ is small enough, $Z(\mathcal{O}_j) \backslash Z(\mathcal{N})$ will either consist of a single connected component $\mathcal{A}_j$ lying on one side of the relevant arc of curve $y = \psi_i^+(x)$, or else it will consist of two connected components $\mathcal{A}_j$ and $\mathcal{B}_j$ lying on opposite sides of that arc of curve. We can take both $\mathcal{A}$ and $\mathcal{B}$ to have simple shapes, close to that of a half-disk. In the former case, $\tilde{v}_j$ will agree with $\tilde{u}_j$ in $\mathcal{A}_j \cap \mathcal{U}_j$ and with $\tilde{u}_{j+1}$ in $\mathcal{A}_j \cap \mathcal{U}_{j+1}$ and, consequently, $\tilde{u}_j = \tilde{u}_{j+1}$ in the whole of $\mathcal{U}_j \cap \mathcal{U}_{j+1}$ (because of the shape of $\mathcal{U}_j$). In the latter case, by unique continuation across the arc of curve $y = \psi_i^+(x)$, $\tilde{v}_j$ extends as a holomorphic function in a disk Δ_j centered at $Z(\tilde{x}_\nu,t(s_j))$ and agrees with $\tilde{u}_j$ in $\Delta_j \cap \mathcal{U}_j$ and with $\tilde{u}_{j+1}$ in $\Delta_j \cap \mathcal{U}_{j+1}$. If we repeat this observation for $s = 1,\dots,r$, we reach the conclusion that there is a unique holomorphic function $\tilde{u}$ defined in an open set containing all the vertical segments $\tilde{\mathcal{J}}_j$ $(0 \leq j \leq r+1)$ such that $u = \tilde{u} \circ Z$ in some open set containing all the straight-line segments $\{ (\tilde{x}_\nu,t(s)); s \in \mathcal{J}_j \}$. But $\tilde{\mathcal{J}}_0 \cap \tilde{\mathcal{J}}_{r+1}$ contains $\tilde{z}_\nu$ and we conclude that (VII.4.22) holds in an open neighborhood of that point, and therefore in the whole of the open set $D_\nu \backslash K_\nu$ since it is an annulus.

Let D'_ν be an open disk centered at $\tilde{z}_\nu$, with $K_\nu \subset D'_\nu \subset\subset D_\nu$; let c_ν be its circumference, oriented counterclockwise. By (VII.4.22) we must have

$$\int_{c_\nu} \tilde{u}^+ dz = 0. \tag{VII.4.24}$$

On the other hand, the pullback to Σ_ν of $d(\tilde{u}^+ dz)$ under the map Z is equal to the pullback of $d(u \wedge dZ) = F\lambda \wedge dZ$ by (VII.4.21). Note that $2\lambda = L(Z + \overline{Z}) = L\overline{Z}$ and thus $2\lambda \wedge dZ = d\overline{Z} \wedge dZ$. This means that the pushforward to D_ν of $2F\lambda \wedge dZ$ (as a two-form on Σ_ν) under the map $Z|_{\Sigma_\nu}$ is equal to $\tilde{F} d\bar{z} \wedge dz$. We derive from (VII.4.24):

$$\int_{K_\nu} \tilde{F}\, d\bar{z} \wedge dz = 0,$$

an impossibility for large enough ν, due to (VII.2.9).

This completes the proof of (VII.4.12).

Now let g be given by (VII.4.6) and $h \in \mathscr{C}^1(\mathrm{U}'')$ satisfy (VII.4.14). This means that h is a solution of the equation (with unknown u)

$$\mathrm{L}u = gh_x\lambda. \tag{VII.4.25}$$

Suppose $h_x(0) \neq 0$. We may assume that $Z_x - 1$ and F are as close to zero as we wish, provided the neighborhood U″ is small enough. If this is so, then Condition (VII.2.9) will be satisfied by gh_x in the place of F. But we have just seen that Equation (VII.4.25) could not have a solution in U″, whence a contradiction. Thus we must have $h_x(0) = 0$. But then (VII.4.14) entails $\mathrm{d}_t h|_0 = 0$, whence the desired conclusion, $\mathrm{d}h|_0 = 0$. ■

Suppose that the function F is as described in the proof of Theorem VII.4.1, and let g be given by (VII.4.6) and the vector fields $\mathrm{L}_j^\#$ by (VII.4.1). Lemmas VII.4.1 and VII.4.2 state that these vector fields define an involutive structure in some open neighborhood of 0. Theorem VII.4.1 tells us that this structure is not locally integrable in any subneighborhood of 0. The structure defined by the $\mathrm{L}_j^\#$ coincides to infinite order at the origin with the involutive structure originally given on $\mathcal{M}$.

VII.5. Nonintegrability and Nonsolvability in Lewy Structures. The Three-Dimensional Case

DEFINITION VII.5.1. *The involutive structure of* $\mathcal{M}$ *will be called a* Lewy structure *if it is a CR structure, if the characteristic set* T° *is a line bundle, and if the Levi form is nondegenerate at every point of* $\mathrm{T}^\circ \backslash 0$. *When equipped with a Lewy structure* $\mathcal{M}$ *will be called a* Lewy manifold.

We continue to denote by m (resp., n) the rank of the cotangent (resp., tangent) structure bundle T′ (resp., $\mathcal{V}$) on $\mathcal{M}$. If $\mathcal{M}$ is a Lewy manifold, necessarily $m = n+1$. CR structures with this property are often said to be "of hypersurface type." The reason is that when they are locally integrable, they can be locally embedded in $\mathbb{C}^{n+1}$ as hypersurfaces (indeed, $\dim \mathcal{M} = 2n+1$). Thus we shall refer to a Lewy manifold whose CR structure is locally integrable as a *locally embeddable* Lewy manifold.

The terminology of Definition VII.5.1 is motivated by Example I.7.1, which can be thought of as the *three-dimensional model case*. In the present section we are going to look at general three-dimensional Lewy manifolds. In this case $m = 2$, $n = 1$.

Until otherwise specified we deal with a locally embeddable three-dimensional Lewy manifold $\mathcal{M}$. The analysis takes place in an open neighborhood U of a point $0 \in \mathcal{M}$ (called "the origin") in which are defined local coordinates x, y, s, and a $\mathscr{C}^\infty$ function $w = s + \iota\varphi(z,s)$ ($z = x+\iota y$), all vanishing at 0, such that $T'|_U$ is spanned by dz and dw. We also assume that $d\varphi|_0 = 0$. We take U to be a product

$$U = \Delta \times \mathcal{I}, \qquad \text{(VII.5.1)}$$

where Δ is a polydisk in $\mathbb{C}$, $\mathcal{I}$ an open interval in $\mathbb{R}$, both centered at the origin. Throughout the remainder of the section we assume that the Levi form at 0 does not vanish. There is no loss of generality, then, in hypothesizing that

$$\varphi(z,s) = |z|^2 + O(|s||z| + s^2 + |z|^3) \qquad \text{(VII.5.2)}$$

(cf. Example I.9.3). We introduce the vector fields in U, L, M defined by the relations

$$Lz = Lw = 0,\ L\bar{z} = 1,\ Mz = 1,\ M\bar{z} = Mw = 0.$$

We have

$$L = \partial/\partial\bar{z} + \lambda\partial/\partial s,\ M = \partial/\partial z + \mu\partial/\partial s, \qquad \text{(VII.5.3)}$$

with

$$\lambda = -\iota w_s^{-1}\partial\varphi/\partial\bar{z},\ \mu = -\iota w_s^{-1}\partial\varphi/\partial z. \qquad \text{(VII.5.4)}$$

We also introduce

$$M_0 = w_s^{-1}\partial/\partial s.$$

Note that the vector fields L, M, and M_0 commute.

We avail ourselves of (VII.5.2). Possibly after contracting Δ and $\mathcal{I}$ about the origin we see that there is a $\mathcal{C}^\infty$ function $z(s)$ of s in $\mathcal{I}$ such that $z(0) = 0$ and such that, for some constant $C > 0$ and all $z \in \Delta$,

$$|z - z(s)|^2/C \le |\varphi(z,s) - \varphi_0(s)| \le C|z - z(s)|^2, \qquad \text{(VII.5.5)}$$

where $\varphi_0(s) = \varphi(z(s),s)$. Note that

$$\varphi_0(0) = d\varphi_0/ds|_0 = 0. \qquad \text{(VII.5.6)}$$

We can therefore find a number $\kappa > 0$ small enough that the set $\{ (s,t) \in \mathbb{R}^2;\ |s| < \kappa t \}$ lies above the curve $t = \varphi_0(s)$. It is a consequence of (VII.5.5) that there is $\delta > 0$ such that, given any point $(s,t) \in \mathbb{R}^2$ in the set

$$s^2 + t^2 < \delta^2,\ |s| < \kappa t,\ t > \varphi_0(s), \qquad \text{(VII.5.7)}$$

the equation

$$\varphi(z,s) = t \qquad \text{(VII.5.8)}$$

defines a smooth closed curve $\gamma(s,t)$ in the z-plane, contained in the disk Δ and winding around $z(s)$.

We select two sequences $\{K_\nu\}$, $\{K'_\nu\}$ ($\nu = 1,2\ldots$) of closed disks in the region (VII.5.7) of the plane, submitted to the requirement that

$$\sup_{K_\nu} s < \inf_{K'_\nu} s < \sup_{K'_\nu} s < \inf_{K_{\nu-1}} s. \tag{VII.5.9}$$

Then we select two functions f, $g \in \mathscr{C}^\infty(\mathbb{R}^2)$ such that $f \equiv 0$ off the union of the disks K_ν, $g \equiv 0$ off the union of the K'_ν, and moreover such that, for a suitable choice of the angles θ_0 and θ'_0,

to every $\varepsilon > 0$ *there is an integer* $N(\varepsilon) \geq 0$ *such that, for all* $\nu \geq N(\varepsilon)$, (VII.5.10)

$$\sup_{\mathcal{I}nt\, K_\nu} |f/|f| - e^{\iota\theta_0}| + \sup_{\mathcal{I}nt\, K'_\nu} |g/|g| - e^{\iota\theta'_0}| \leq \varepsilon.$$

We shall use the notation $f \circ w$ to signify the function $f(s,\varphi(z,s))$ in U. Observe that there is a basis of open neighborhoods of 0 in $\mathbb{R}^3$, $\{\Omega_\nu\}$ $(\nu = 0,1,\ldots)$, such that the restrictions of $f \circ w$ and $g \circ w$ to each Ω_ν have compact support in Ω_ν.

In view of this remark the next statement shows that whatever the open neighborhood Ω of 0 in $\mathbb{R}^3$, there exist functions $v \in \mathscr{C}_c^\infty(\Omega)$ such that the equation $Lu = v$ has no $\mathscr{C}^1$ solution in Ω:

THEOREM VII.5.1. *Let Ω be an open neighborhood of* 0 *in* $\Delta \times \mathcal{I}$ *and let P, Q be two continuous functions in Ω. If there is a $\mathscr{C}^\infty$ function $u \in \mathscr{C}^1(\Omega)$ satisfying*

$$Lu = (f \circ w)P + (g \circ w)Q \tag{VII.5.11}$$

in Ω, then necessarily $P = Q = 0$ at the origin.

PROOF. We may take $\Omega = \Delta_0 \times \mathcal{I}_0$ and select $\delta > 0$ small enough that whenever (s,t) lies in the region (VII.5.7), then the closed curve $\gamma(s,t)$ is entirely contained in Δ_0 and $s \in \mathcal{I}_0$. Call $\mathscr{A}$ the complement in (VII.5.7) of the union of all sets K_ν and K'_ν. We have

$$Lu = 0 \text{ if } (z,s) \in \Omega \text{ and } s + \iota\varphi(z,s) \in \mathscr{A}. \tag{VII.5.12}$$

Consider the function of $(s,t) \in \mathscr{A}$,

$$I(s,t) = \oint_{\gamma(s,t)} u(z,s)dz.$$

We contend that $I \equiv 0$ in $\mathscr{A}$. It suffices to show that $\bar{\partial} I \equiv 0$ in $\mathscr{A}$. For then I is a holomorphic function of $s + \iota t$ in the open set $\mathscr{A}$. But when $t \to \varphi_0(s)$, $I(s,t) \to 0$ since then the curve $\gamma(s,t)$ contracts to the single point $\{z(s)\}$. Taking (VII.5.6) and (VII.5.9) into account shows that $I(s,t)$ must vanish identically in $\mathscr{A}$.

For any $e^{\iota\theta} \in S^1$ there is a unique point $z = \zeta(s,t,\theta)$ on the curve $\gamma(s,t)$ such that $[z - z(s)]/[\bar{z} - \bar{z}(s)] = e^{\iota\theta}$. This defines a $\mathscr{C}^\infty$ map

$$\mathscr{A} \times S^1 \ni (s,t,\theta) \to (\zeta(s,t,\theta),s) \in \Omega.$$

Because of (VII.5.12) the pullback to $\mathcal{A}\times S^1$ of du is equal to that of (Mu)dz + (M_0u)dw, which demands that

$$\partial u/\partial\bar{w} = (\mathrm{M}u)\partial\zeta/\partial\bar{w},\ \partial u/\partial\theta = (\mathrm{M}u)\partial\zeta/\partial\theta. \qquad \text{(VII.5.13)}$$

Here ζ stands for $\zeta(s,t,\theta)$ and w for $s+\imath t$. For each fixed $(s,t)\in\mathcal{A}$, as $e^{\imath\theta}$ winds around S^1 the point ζ winds around $z(s)$ along the curve $\gamma(s,t)$. Thus

$$I = \int_0^{2\pi} u(\zeta,s)(\partial\zeta/\partial\theta)\mathrm{d}\theta,$$

and by (VII.5.13)

$$\begin{aligned}\partial I/\partial\bar{w} &= \int_0^{2\pi}\left\{(\mathrm{M}u)(\partial\zeta/\partial\bar{w})(\partial\zeta/\partial\theta)+u(\partial^2\zeta/\partial\bar{w}\partial\theta)\right\}\mathrm{d}\theta\\ &= \int_0^{2\pi}(\partial/\partial\theta)[u(\partial\zeta/\partial\bar{w})]\mathrm{d}\theta = 0,\end{aligned}$$

which is what we wanted to prove.

Availing ourselves once again of (VII.5.9), we select a closed disk $\mathcal{O}_\nu$ (resp., $\mathcal{O}'_\nu$) containing K_ν (resp., K'_ν), whose closure does not intersect K'_ν (resp., K_ν) nor any set K_μ or K'_μ for $\mu\neq\nu$. Since $I\equiv 0$ in $\mathcal{A}$ we have trivially

$$\int_{\partial\mathcal{O}_\nu}\oint_{\gamma(s,t)} u(z,s)\mathrm{d}z\mathrm{d}w = 0, \qquad \text{(VII.5.14)}$$

and likewise with $\mathcal{O}'_\nu$. For each ν sufficiently large, the mapping

$$(s,t,\theta)\rightarrow(\zeta(s,t,\theta),s)$$

is a diffeomorphism of $\partial\mathcal{O}_\nu\times S^1$ (resp., $\partial\mathcal{O}'_\nu\times S^1$) onto a two-dimensional torus $\mathcal{T}_\nu$ (resp., $\mathcal{T}'_\nu$) contained in Ω. We call $\hat{\mathcal{T}}_\nu$ (resp., $\hat{\mathcal{T}}'_\nu$) the solid torus having $\mathcal{T}_\nu$ (resp., $\mathcal{T}'_\nu$) as its boundary. The meaning of (VII.5.14) is that the integral over $\mathcal{T}_\nu$ of the two-form $u\mathrm{d}z\wedge\mathrm{d}w$ vanishes. By Stokes' theorem the integral over $\hat{\mathcal{T}}_\nu$ of the three-form $\mathrm{d}(u\mathrm{d}z\wedge\mathrm{d}w)$ must vanish. The same is true over $\hat{\mathcal{T}}'_\nu$. By (VII.5.11) and the fact that $f\circ w$ (resp., $g\circ w$) vanishes identically in $\hat{\mathcal{T}}'_\nu$ (resp., $\hat{\mathcal{T}}_\nu$) we conclude that

$$\int_{\hat{\mathcal{T}}_\nu} f(w)P(z,s)w_s\mathrm{d}z\wedge\mathrm{d}\bar{z}\wedge\mathrm{d}s = 0, \qquad \text{(VII.5.15)}$$

$$\int_{\hat{\mathcal{T}}_\nu} g(w)Q(z,s)w_s\mathrm{d}z\wedge\mathrm{d}\bar{z}\wedge\mathrm{d}s = 0, \qquad \text{(VII.5.16)}$$

The intersection of supp($f\circ w$) with $\hat{\mathcal{T}}_\nu$ is determined by the fact that $w = s + \imath\varphi(z,s)\in K_\nu$; similarly, the intersection of supp($g\circ w$) with $\hat{\mathcal{T}}'_\nu$ is determined by the fact that $w \in K'_\nu$. If we had $P(0,0) \neq 0$ (resp., $Q(0,0) \neq 0$) then, for ν suitably large, (VII.5.10) would contradict (VII.5.15) as well as (VII.5.16)). ∎

Next we propose to show that a suitable perturbation of the vector field L defines a Lewy structure on a neighborhood of the origin in $\mathbb{R}^3$ that is not locally integrable there. We shall make use of the following:

LEMMA VII.5.1. *Let $f \in \mathscr{C}^\infty(\mathbb{R}^2)$ have its support in the sector $|s| \leq t$. If the neighborhood of* 0 *in* $\mathcal{M}$, U, *is small enough then, in* U, *whatever the integer* N, *$(f \circ w)/z^N$ vanishes to infinite order at $z = 0$.*

PROOF. It follows from our choice of f and from (VII.5.2) that, if U is small enough, then $|s| \leq 2|z|^2$ in U$\cap$[supp($f \circ w$)]. Clearly f vanishes to infinite order at the origin (in $\mathbb{R}^2$) and, thus, to every integer $N \geq 1$ there are constants C_N, $C'_N > 0$ such that, in U,

$$|f(s,\varphi(z,s))| \leq C_N(|s| + |\varphi(z,s)|)^N \leq C'_N|z|^{2N},$$

whence the lemma. ∎

Let f and g be the functions in $\mathbb{R}^2$ described above. According to Lemma VII.5.1, the functions $(f \circ w)/z$ and $(g \circ w)/z$ are smooth in U.

THEOREM VII.5.2. *Consider the vector field in* U.

$$\Lambda = \mathrm{L} - (\lambda/z)[(f \circ w)\mathrm{M}_0 + (g \circ w)\mathrm{M}], \qquad \text{(VII.5.17)}$$

and let Ω be any open neighborhood of 0 *in* $\mathcal{M}$ *contained in* U. *If a function h $\in \mathscr{C}^1(\Omega)$ satisfies $\Lambda h = 0$ in Ω, then $dh|_0 = 0$.*

PROOF. Since $[1 + \imath\varphi_s(z,s)]\lambda = -\imath\partial\varphi/\partial\bar{z}$, we derive from (VII.5.2)

$$\lambda = -\imath z + \mathrm{O}(|s| + |z|^2)$$

and therefore (cf. proof of Lemma VII.5.1), on supp($f \circ w$)

$$|\lambda + \imath z| \leq \mathit{const.}\ |z|^2.$$

It follows that

$$(\lambda/z)\mathrm{M}_0 h = P_1 h_s, \ (\lambda/z)\mathrm{M}h = Q_1(h_z + \mu h_s),$$

with P_1, Q_1 continuous functions in $\Delta \times \mathcal{J}$ such that $P_1 Q_1|_0 \neq 0$. Since $\Lambda h = 0$ we have, in Ω,

$$\mathrm{L}h = (f \circ w)P_1 h_s + (g \circ w)Q_1(h_z + \mu h_s). \qquad \text{(VII.5.18)}$$

We derive from (VII.5.18) and from Theorem VII.5.1 that, necessarily,

$$h_s|_0 = h_z|_0 = 0. \qquad \text{(VII.5.19)}$$

Recalling the expression of the vector field L, one sees that (VII.5.19) entail $\bar{\partial}_z h|_0 = 0$, hence $dh|_0 = 0$. ■

We remark that L $-$ Λ is a vector field in U whose coefficients vanish to infinite order at the origin. It follows that, in a suitably small open neighborhood Ω of 0 in $\mathcal{M}$, the Levi form of the vector field Λ cannot vanish, and thus Λ defines indeed a Lewy structure on Ω. According to Theorem VII.5.2, this Lewy structure is not locally integrable at 0.

Let us now consider the space $\mathbb{D}^\infty(\mathbb{R}^3)$ of all smooth vector fields in $\mathbb{R}^3$. We denote by x, y, s the real coordinates in $\mathbb{R}^3$ and write $z = x + \imath y$. An element of $\mathbb{D}^\infty(\mathbb{R}^3)$ is a linear combination

$$\mathrm{L} = A(x,y,s)\partial/\partial z + B(x,y,s)\partial/\partial\bar{z} + C(x,y,s)\partial/\partial s$$

with A, B, $C \in \mathscr{C}^\infty(\mathbb{R}^3)$. Thus $\mathbb{D}^\infty(\mathbb{R}^3) \cong (\mathscr{C}^\infty(\mathbb{R}^3))^3 \cong \mathscr{C}^\infty(\mathbb{R}^3;\mathbb{C}^3)$. We equip $\mathbb{D}^\infty(\mathbb{R}^3)$ with the Fréchet space structure transferred from $(\mathscr{C}^\infty(\mathbb{R}^3))^3$.

Definition VII.5.2. *A vector field* L $\in \mathbb{D}^\infty(\mathbb{R}^3)$ *will be called* aberrant *if given any open subset* Ω *of* $\mathbb{R}^3$ *and any number* $\delta > 0$, *every solution* $h \in \mathscr{C}^{1+\delta}(\Omega)$ *of the homogeneous equation* L$h = 0$ *is constant in* Ω.

Theorem VII.5.3. *The set of vector fields* L $\in \mathbb{D}^\infty(\mathbb{R}^3)$ *that are not aberrant is the countable union of a sequence of closed sets whose interior is empty.*

Proof. We shall make use of a basis of the topology of $\mathbb{R}^3$ consisting of closed balls B_ν ($\nu = 1,2,\ldots$). For any number $r \geq 1$ call $\mathscr{F}^r_\nu$ the set of vector fields L $\in \mathbb{D}^\infty(\mathbb{R}^3)$ such that the following is true:

$$\text{\textit{There is a solution } } h \in \mathscr{C}^r(\mathrm{B}_\nu) \text{\textit{ of the equation} } \mathrm{L}h = 0 \text{\textit{ whose norm in} } \mathscr{C}^r(\mathrm{B}_\nu) \text{\textit{ is}} \leq \nu \text{\textit{ and such that} } \min_{\mathrm{B}_\nu} |dh| \geq 1/\nu. \tag{VII.5.20}$$

Suppose $r > 1$ and let $\{\mathrm{L}_j\}_{j=1,2,\ldots}$ be a sequence in $\mathscr{F}^r_\nu$ converging to some L $\in \mathbb{D}^\infty(\mathbb{R}^3)$. For each $j = 1,2,\ldots$, let $h_j \in \mathscr{C}^r(\mathrm{B}_\nu)$ satisfy $\mathrm{L}_j h_j = 0$ in B_ν and have the properties listed in (VII.5.20). By compactness of the embedding $\mathscr{C}^r(\mathrm{B}_\nu) \to \mathscr{C}^1(\mathrm{B}_\nu)$ the sequence $\{h_j\}$ contains a subsequence that converges in $\mathscr{C}^1(\mathrm{B}_\nu)$ to a limit h. Perforce L$h = 0$ in B_ν, the norm of h in $\mathscr{C}^1(\mathrm{B}_\nu)$ does not exceed ν and $|dh| \geq 1/\nu$ everywhere in B_ν. This proves that the closure of $\mathscr{F}^r_\nu$ is contained in $\mathscr{F}^1_\nu$.

We claim that the interior of $\mathscr{F}^1_\nu$ is empty. An arbitrary element L $\in \mathscr{F}^1_\nu$ is the limit in $\mathbb{D}^\infty(\mathbb{R}^3)$ of a sequence of vector fields $\mathrm{L}_j \in \mathbb{D}^\infty(\mathbb{R}^3)$ such that L_j, $\bar{\mathrm{L}}_j$, and $[\mathrm{L}_j,\bar{\mathrm{L}}_j]$ are linearly independent at the center p_ν of the ball B_ν. We may even assume that the coefficients of the vector fields L_j are real analytic. It follows that p_ν has an open neighborhood in which there exist two $\mathscr{C}^\omega$ solutions

f_j and g_j of the equation $L_j h = 0$ such that $df_j \wedge d\bar{f}_j \wedge dg_j \neq 0$ at $\mathfrak{p}_\nu$. We apply Theorem VII.5.2 to each vector field L_j: there is a vector field $R_j \in \mathbb{D}^\infty(\mathbb{R}^3)$ that vanishes to infinite order at $\mathfrak{p}_\nu$ and is such that, if Ω is any open neighborhood of $\mathfrak{p}_\nu$, any solution $h \in \mathscr{C}^1(\Omega)$ of the equation

$$(L_j + R_j)h = 0 \tag{VII.5.21}$$

satisfies $dh(\mathfrak{p}_\nu) = 0$. Then select $\chi_j \in \mathscr{C}_c^\infty(B_\nu)$, $\chi_j \equiv 1$ in some open neighborhood of $\mathfrak{p}_\nu$ and such that $\chi_j R_j \to 0$ in $\mathbb{D}^\infty(\mathbb{R}^3)$. It is clear that $L_j + \chi_j R_j$ converges to L in $\mathbb{D}^\infty(\mathbb{R}^3)$ and that any solution $h \in \mathscr{C}^1(B_\nu)$ of (VII.5.21) must be such that $dh(\mathfrak{p}_\nu) = 0$. It follows that $L_j + \chi_j R_j \notin \mathscr{F}_\nu^1$.

Let $L \in \mathbb{D}^\infty(\mathbb{R}^3)$ be a vector field that does not belong to $\mathscr{C}\ell\mathscr{F}_\nu^{1+1/\nu}$ whatever $\nu = 1,2,\ldots$; and let h be a $\mathscr{C}^{1+\delta}$ solution in some open set $\Omega \subset \mathbb{R}^3$ of the equation $Lh = 0$. Denote by K the compact closure of an open subset Ω' of Ω. There is a sequence of integers $\nu \nearrow +\infty$ such that $B_\nu \subset \Omega'$; as soon as such an integer ν exceeds $1/\delta$ there is a point $\mathfrak{p}'_\nu \in B_\nu$ such that $|dh(\mathfrak{p}'_\nu)| \leq 1/\nu$. We conclude that $dh(\mathfrak{p}) = 0$ for some $\mathfrak{p} \in K$. Because of the arbitrariness of $\Omega' \subset\subset \Omega$ such points $\mathfrak{p}$ are dense in Ω and therefore $dh \equiv 0$ in Ω. ■

VII.6. Nonintegrability in Lewy Structures. The Higher-Dimensional Case

In the present section we deal with a Lewy manifold (Definition VII.5.1) of dimension $2n + 1 \geq 5$.

EXAMPLE VII.6.1. *The multidimensional model case*

Let x_i, y_j $(1 \leq i, j \leq n)$ and s denote the coordinates in $\mathbb{R}^{2n+1}$ and equip $\mathbb{R}^{2n+1}$ with the hypo-analytic structure defined by the $n+1$ functions $z_j = x_j + \imath y_j$ $(1 \leq j \leq n)$, $w = s + \imath\mathfrak{Q}(z)$ with $\mathfrak{Q}$ a nondegenerate hermitian form on $\mathbb{C}^n$:

$$\mathfrak{Q}(z) = \sum_{j,k=1}^{n} c_{jk} z_j \bar{z}_k, \tag{VII.6.1}$$

$$c_{jk} = \bar{c}_{kj}, \ \det(c_{jk})_{1 \leq j,k \leq n} \neq 0. \tag{VII.6.2}$$

The underlying tangent structure bundle is spanned by the Lewy vector fields

$$L_j = \partial/\partial\bar{z}_j - \imath(\partial\mathfrak{Q}/\partial\bar{z}_j)\partial/\partial s \ (j = 1,\ldots,n). \tag{VII.6.3}$$

The characteristic set T° is spanned by the real one-form

$$\varpi = ds + \imath(\bar{\partial}_z\mathfrak{Q} - \partial_z\mathfrak{Q}) \tag{VII.6.4}$$

(cf. (I.8.12)). Since

$$(2\imath)^{-1}[L_j, \bar{L}_k] = (\partial\mathfrak{Q}/\partial\bar{z}_j\partial z_k)\partial/\partial s, \tag{VII.6.5}$$

it is customary to identify the Levi form (Definition I.8.1) to the quadratic form $\mathcal{Q}$. ■

Until otherwise specified we assume the Lewy manifold $\mathcal{M}$ to be locally embeddable. The analysis takes place in an open neighborhood U of a point 0 $\in \mathcal{M}$ (called "the origin") in which are defined local coordinates x_i, y_j ($1 \leq i, j \leq n$) and s, and a function $w = s + \imath\varphi(z,s)$, all vanishing at 0, such that $T'|_U$ is spanned by the dz_j ($z_j = x_j + \imath y_j$) and dw. We also assume that $d\varphi|_0 = 0$. We take U to be a product

$$U = \Delta \times \mathcal{I}, \tag{VII.6.6}$$

where now Δ is a polydisk in $\mathbb{C}^n$, $\mathcal{I}$ an open interval in $\mathbb{R}^1$, both centered at the origin.

Throughout the remainder of the section we assume that the signature of the Levi form at 0 is equal to $|n - 2|$ (cf. section VII.2.). There is no loss of generality, then, in hypothesizing that

$$\varphi(z,s) = |z_1|^2 - |z'|^2 + O(|s||z| + s^2 + |z|^3), \tag{VII.6.7}$$

where $z' = (z_2,\dots,z_n)$ (Example I.9.3). We introduce the vector fields in U, L_j, M_j ($1 \leq j \leq n$) defined by the relations

$$L_j z_i = L_j w = 0,\ L_j \bar{z}_i = \delta_{ij},\ M_j z_i = \delta_{ij},\ M_j \bar{z}_i = M_j w = 0$$
$$(1 \leq i, j \leq n).$$

We have

$$L_j = \partial/\partial\bar{z}_j + \lambda_j \partial/\partial s,\ M_j = \partial/\partial z_j + \mu_j \partial/\partial s, \tag{VII.6.8}$$

$$\lambda_j = -\imath w_s^{-1} \partial\varphi/\partial\bar{z}_j,\ \mu_j = -\imath w_s^{-1}\partial\varphi/\partial z_j. \tag{VII.6.9}$$

We recall that $w_s = 1 + \imath\varphi_s$. Also, set $M_0 = w_s^{-1}\partial/\partial s$. Note that the vector fields L_i, M_j, and M_0 commute, which entails

$$L_j\lambda_k = L_k\lambda_j,\ M_0\lambda_j = L_j(w_s^{-1}). \tag{VII.6.10}$$

We shall also define

$$M = \sum_{k=1}^{n} z_k M_k = \sum_{k=1}^{n} z_k \partial/\partial z_k + \mu\partial/\partial s, \tag{VII.6.11}$$

with $\mu = \sum_{k=1}^{n} z_k\mu_k$. Obviously, M and each L_j commute, i.e.,

$$L_j\mu = M\lambda_j,\ j = 1,\dots,n. \tag{VII.6.12}$$

Also note, for use below, that

$$L_j\bar{w} = 2\lambda_j,\ M\bar{w} = 2\mu. \tag{VII.6.13}$$

The analogue of Lemma VII.5.1 is also valid when $n > 1$.

LEMMA VII.6.1. *Let $f \in \mathscr{C}^\infty(\mathbb{R}^2)$ have its support in the sector $|s| \leq t$. If the neighborhood of* 0 *in* $\mathcal{M}$, U, *is small enough then, in* U, *whatever the integer* N, $(f\circ w)/z_1^N$ *vanishes to infinite order at* $z_1 = 0$.

PROOF. It is the same as the proof of Lemma VII.5.1 if we note that if U is small enough, then $|s| + |z|^2 \leq 3|z_1|^2$ in $\mathrm{U}\cap[\mathrm{supp}(f\circ w)]$. ∎

Let both f, $g \in \mathscr{C}^\infty(\mathbb{R}^2)$ have their supports in the sector $|s| \leq t$ and define the functions in U,

$$F = (f\circ w)/[z_1 + w_s^{-1}(f\circ w)],\ G = (g\circ w)/[z_1^2 - \mu(g\circ w)]. \qquad \text{(VII.6.14)}$$

Here μ is the same function as in (VII.6.11). Throughout the remainder of this section we shall reason under the hypothesis that *in* U, *both functions* $w_s^{-1}(f\circ w)/z_1$ *and* $\mu(g\circ w)/z_1^2$ *are very small compared to* 1 *and vanish to infinite order at* $z_1 = 0$. Then, by Lemma V.6.1, both F and G are smooth in U.

LEMMA VII.6.2. *The vector fields in* U,

$$\tilde{\mathrm{L}}_j = \mathrm{L}_j - \lambda_j F\mathrm{M}_0,\ j = 1,\ldots,n, \qquad \text{(VII.6.15)}$$

commute pairwise. So do the vector fields

$$\mathrm{L}_j^\# = \mathrm{L}_j + \lambda_j G\mathrm{M},\ j = 1,\ldots,n. \qquad \text{(VII.6.16)}$$

PROOF. By (VII.6.10) we have $[\tilde{\mathrm{L}}_j,\tilde{\mathrm{L}}_k] = \tilde{\psi}_{jk}\mathrm{M}_0$, with

$$\tilde{\psi}_{jk} = \mathrm{L}_j(\lambda_k F) - \mathrm{L}_k(\lambda_j F) - F^2(\lambda_j\mathrm{M}_0\lambda_k - \lambda_k\mathrm{M}_0\lambda_j) =$$

$$\lambda_k[\mathrm{L}_j F + F^2\mathrm{L}_j(w_s^{-1})] - \lambda_j[\mathrm{L}_k F + F^2\mathrm{L}_k(w_s^{-1})].$$

If we make use of (VII.6.10) and (VII.6.12), we get $[\mathrm{L}_j^\#,\mathrm{L}_k^\#] = \psi_{jk}^\#\mathrm{M}$ with

$$\psi_{jk}^\# = \mathrm{L}_j(\lambda_k G) - \mathrm{L}_k(\lambda_j G) + G^2(\lambda_j\mathrm{M}\lambda_k - \lambda_k\mathrm{M}\lambda_j) =$$

$$\lambda_k(\mathrm{L}_j G - G^2\mathrm{L}_j\mu) - \lambda_j(\mathrm{L}_k G - G^2\mathrm{L}_k\mu).$$

Straightforward computation shows that

$$\mathrm{L}_j F + F^2\mathrm{L}_j(w_s^{-1}) = [1 + (f\circ w)/w_s z_1]^{-2}\mathrm{L}_j[(f\circ w)/z_1], \qquad \text{(VII.6.17)}$$

$$\mathrm{L}_j G - G^2\mathrm{L}_j\mu = [1 - \mu(g\circ w)/z_1^2]^{-2}\mathrm{L}_j[g\circ w)/z_1^2]. \qquad \text{(VII.6.18)}$$

By (VII.6.13) we have

$$\mathrm{L}_j[(f\circ w)/z_1] = 2\lambda_j(\bar{\partial}f\circ w)/z_1,\ \mathrm{L}_j[(g\circ w)/z_1^2] = 2\lambda_j(\bar{\partial}g\circ w)/z_1^2,$$

where $\bar{\partial}f = \frac{1}{2}(\partial f/\partial s + \imath\partial f/\partial t)$, and likewise with g. Combining this with

(VII.6.17) and (VII.6.18) shows that there are $\mathscr{C}^\infty$ functions in U, F_1 and G_1, such that, for all $j = 1,\ldots,n$,

$$L_jF + F^2L_j(w_s^{-1}) = F_1\lambda_j,\ L_jG - G^2L_j\mu = G_1\lambda_j.$$

An immediate consequence of this is that $\tilde{\psi}_{jk} \equiv \psi^{\#}_{jk} \equiv 0$. ■

COROLLARY VII.6.1. *If $fg \equiv 0$ then the n vector fields in* U,

$$\Lambda_j = L_j - \lambda_j(FM_0 - GM),\ j = 1,\ldots,n, \tag{VII.6.19}$$

commute pairwise.

PROOF. Indeed, at each point of U and for every j, either $\Lambda_j - \tilde{L}_j$ or else $\Lambda_j - L^{\#}_j$ vanish to infinite order. ■

REMARK VII.6.1. For each $j = 1,\ldots,n$, $\Lambda_j - L_j$ vanishes to infinite order at the origin. ■

We now introduce a parameter $a = (a_1,\ldots, a_n)$ (we write $(a' = (a_2,\ldots, a_n))$, which will vary in the piece of hyperboloid

$$\mathscr{H} = \{ a \in \mathbb{R}^n;\ a_1^2 - |a'|^2 = 1,\ |a| < 2 \}. \tag{VII.6.20}$$

Below $\zeta = \xi + \imath\eta$ will denote a complex variable. There is an open disk D $\subset \mathbb{C}^1$ centered at 0, whose image D^a under the map $\zeta \to \zeta a$ is contained in the polydisk Δ for all points $a \in \mathscr{H}$; D^a is a relatively open subset of the complex line $z = \zeta a$ in $\mathbb{C}^n$, and $U^a = D^a \times \mathscr{I}$ is a relatively open subset of the zero-set of the linear functions $a_jz_k - a_kz_j$ $(1 \le j, k \le n)$. We shall use the notation

$$\varphi^a(\zeta,s) = \varphi(\zeta a,s),\ w^a = s + \imath\varphi^a(\zeta,s).$$

According to (VII.6.7), we have

$$\varphi^a(\zeta,s) = |\zeta|^2 + O(|s||\zeta| + s^2 + |\zeta|^3). \tag{VII.6.21}$$

The functions ζ and w^a define a CR structure on D$\times\mathscr{I}$; the tangent structure bundle in this structure is spanned by the vector field

$$L^a = \partial/\partial\bar{\zeta} + \lambda^a\partial/\partial s, \tag{VII.6.22}$$

where $\lambda^a = -\imath(w_s^a)^{-1}\partial\varphi^a/\partial\bar{\zeta}$. Note that λ^a is equal to the pullback to D$\times\mathscr{I}$ under the map $(\zeta,s)\to(\zeta a,s)$ of the function $a\cdot\lambda = a_1\lambda_1 + \cdots + a_n\lambda_n$. The pushforward to U^a of L^a under the map $(\zeta,s)\to(\zeta a,s)$ is the vector field $\sum_{j=1}^n a_jL_j$. We also introduce the vector fields in D$\times\mathscr{I}$,

$$M^a = \partial/\partial\zeta + \mu^a\partial/\partial s,\ M_0^a = (w_s^a)^{-1}\partial/\partial s, \tag{VII.6.23}$$

where $\mu^a = -\imath(w_s^a)^{-1}\partial\varphi^a/\partial\zeta$. Note, in passing, that the vector field M preserves the linear functions of z, $a_j z_k - a_k z_j$ $(1 \le j, k \le n)$, and thus it is tangent to U^a. The pushforward of the vector field ζM^a under the map $(\zeta, s) \to (\zeta a, s)$ is equal to M (keep in mind that $M = 0$ when $z = 0$); the pushforward of M_0^a is equal to M_0 (which is trivially tangent to U^a).

We may pull back to $D \times \mathcal{J}$ the restrictions to U^a of the functions F and G defined in (VII.6.14). We get two smooth functions in $D \times \mathcal{J}$, which we denote by F^a and G^a respectively. We may then form the analogue of the vector fields (VII.6.19),

$$\Lambda^a = L^a - \lambda^a(F^a M_0^a - G^a \zeta M^a).$$

The pushforward to U^a of Λ^a under the map $(\zeta, s) \to (\zeta a, s)$ is the vector field $\sum_{j=1}^n a_j \Lambda_j$.

We may now apply the results when $n = 1$: possibly after contracting D and $\mathcal{J}$ about the origin, there is a $\mathscr{C}^\infty$ function $\zeta^a(s)$ of (a, s) in the product manifold $\mathscr{H} \times \mathcal{J}$ (cf. (VII.6.20)) such that $\zeta^a(0) = 0$ and that, for some $C > 0$ and all $\zeta \in D$,

$$|\zeta - \zeta^a(s)|^2/C \le |\varphi^a(\zeta, s) - \varphi_0^a(s)| \le C|\zeta - \zeta^a(s)|^2, \quad \text{(VII.6.24)}$$

where $\varphi_0^a(s) = \varphi^a(\zeta^a(s), s)$. Notice that

$$\varphi_0^a(0) = (d\varphi_0^a/ds)(0) = 0. \quad \text{(VII.6.25)}$$

We can therefore find a number $\kappa > 0$ small enough that the set $\{(s,t) \in \mathbb{R}^2; |s| < \kappa t\}$ lies above the curve $t = \varphi_0^a(s)$ whatever $a \in \mathscr{H}$. It is a consequence of (VII.6.24) that there is $\delta > 0$ such that, given any point $(s,t) \in \mathbb{R}^2$ in the set

$$s^2 + t^2 < \delta^2, \ |s| < \kappa t, \ t > \varphi_0^a(s), \quad \text{(VII.6.26)}$$

and any $a \in \mathscr{H}$, the equation

$$\varphi^a(\zeta, s) = t \quad \text{(VII.6.27)}$$

defines a smooth closed curve $\gamma^a(s,t)$ in the ζ-plane, contained in the disk D and winding around $\zeta^a(s)$.

We select two sequences $\{K_\nu\}$, $\{K'_\nu\}$ $(\nu = 1,2,\ldots)$ of closed disks in the region and the functions f and g exactly as in section VII.5: $f \equiv 0$ off the union of the K_ν, $g \equiv 0$ off the union of the K'_ν; (VII.5.9) and (VII.5.10) hold. We may then apply Theorem VII.5.1 with L^a and w^a substituted for L and w.

From the analogue of Theorem VII.5.1 we derive the analogue of Theorem VII.5.2. We recall that the signature of the Levi form in the structure under study is assumed to be equal to $|n - 2|$.

THEOREM VII.6.1. *Let Λ_j $(1 \le j \le n)$ be the vector fields* (VII.6.19) *and let* U′ *be any open neighborhood of* 0 *in* $\mathcal{M}$ *contained in* U. *If a function* $h \in \mathscr{C}^1(U')$ *satisfies*

$$\Lambda_j h = 0, j = 1,\ldots,n \tag{VII.6.28}$$

in U′, *then* $\mathrm{d}h|_0 = 0$.

PROOF. Let Ω be an open neighborhood of 0 in $\mathbb{C}\times\mathbb{R}$ mapped into U′ by $(\zeta,s) \to (\zeta\,a,s)$ and let h^a denote the pullback to $\mathrm{D}\times\mathcal{J}$ of h under that same map. We have

$$\mathrm{L}^a h^a = \sum_{j=1}^{n} a_j(\mathrm{L}_j h)^a = \sum_{j=1}^{n} a_j \lambda_j^a [F^a(\mathrm{M}_0 h)^a - G^a(\mathrm{M}h)^a],$$

hence

$$\mathrm{L}^a h^a = \lambda^a (F^a \mathrm{M}_0^a h^a - G^a \zeta \mathrm{M}^a h^a). \tag{VII.6.29}$$

According to (VII.6.14), we have

$$F^a = (f \circ w^a)P_1 / a_1\zeta, \; G^a\zeta = (g\circ w^a)Q_1/a_1\zeta,$$

where P_1 and Q_1 are smooth functions in $\mathrm{D}\times\mathcal{J}$, both equal to 1 at the origin. Starting from this we reason exactly as in the proof of Theorem VII.5.2, to derive from Lemma VII.5.1 and from (VII.6.29) that, necessarily, $h_s^a|_0 = h_\zeta^a|_0 = 0$. We get at once

$$h_s|_0 = 0, \tag{VII.6.30}$$

and $\sum_{j=1}^{n} a_j(\partial h/\partial z_j)|_0 = 0$. But the piece of hyperboloid $\mathcal{H}$ spans the whole space $\mathbb{R}^n$ and therefore we must have

$$\partial_z h|_0 = 0. \tag{VII.6.31}$$

Recalling the expressions of the vector fields Λ_j (see (VII.6.3) and (VII.6.19)), one checks that (VII.6.30) and (VII.6.31) entail $\bar{\partial}_z h|_0 = 0$, hence $\mathrm{d}h|_0 = 0$. ■

VII.7. Example of a CR Structure that is not Locally Integrable but is Locally Integrable on One Side

Denote by $z = x + \imath y$ and s the coordinates in $\mathbb{C}\times\mathbb{R}$. Let $f \in \mathcal{C}_c^\infty(\mathbb{R}^2)$ be the same function as in Theorem VII.5.1. Consider the following functions of z and $w = s + \imath t$:

$$f_+(z,w) = (2\pi)^{-1}\int_{\sigma>0}\int_{\mathbb{R}} e^{\imath\sigma(w-s')} f(s',|z|^2)\mathrm{d}s'\mathrm{d}\sigma,$$

$$g_+(z,w) = \tfrac{1}{2}\pi^{-2}\int_{\sigma>0}\iiint_{\mathbb{R}^3} e^{\imath\sigma(w+\imath|z|^2 - s' + \imath|z'|^2 - 2\imath z\bar{z}')}(z-z')^{-1}.$$

$$f(s',|z'|^2)dx'dy'ds'd\sigma.$$

Both functions f_+ and g_+ are defined in the region $\mathcal{I}m\, w \geq 0$ of $\mathbb{C}^2$, since

$$\mathcal{I}m(w + \imath|z|^2 - s' + \imath|z'|^2 - 2\imath z\bar{z}') = \mathcal{I}m\, w + |z - z'|^2.$$

One can easily prove, by integration by parts, that both f_+ and g_+ are smooth up to the boundary, in the half-space $\mathcal{I}m\, w \geq 0$. Clearly they are both holomorphic with respect to w in the open half-space $\mathcal{I}m\, w > 0$.

Lemma VII.8.1. *We have, in the half-space* $\mathcal{I}m\, w \geq 0$:

$$(\partial/\partial\bar{z} - \imath z\partial/\partial s)g_+ = f_+. \tag{VII.8.1}$$

Proof. It suffices to take into account the fact that $(\partial/\partial\bar{z})[(z-z')^{-1}] = \pi\delta(z-z')$, where δ is the Dirac distribution. ∎

Lemma VII.8.2. *The function* $f - f_+|_{t=0}$ *extends as a function* f_- *of* (z,w) *in the half-space* $\mathcal{I}m\, w \leq 0$, $\mathcal{C}^\infty$ *up to the boundary* $\mathcal{I}m\, w = 0$, *and holomorphic with respect to w in the open half-space* $\mathcal{I}m\, w < 0$.

Proof. It suffices to observe that

$$f(s,|z|^2) = (2\pi)^{-1}\int\int_{\mathbb{R}^2} e^{\imath\sigma(s-s')}f(s',|z|^2)ds'd\sigma, \tag{VII.8.2}$$

and to define

$$f_-(z,w) = (2\pi)^{-1}\int_{\sigma<0}\int_{\mathbb{R}} e^{\imath\sigma(w-s')}f(s',|z|^2)ds'd\sigma. \blacksquare$$

Now let x_i, y_j $(i, j = 1,2)$ and s denote the coordinates in $\mathbb{R}^5$; and set, as usual, $z_j = x_j + \imath y_j$. We shall write below

$$L_0 = \partial/\partial\bar{z}_1 - \imath z_1\partial/\partial s$$

and make use of the pair of vector fields in $\mathbb{R}^5$,

$$L = L_0 - f(s,|z_1|^2)\partial/\partial z_2,\ \partial/\partial\bar{z}_2. \tag{VII.8.3}$$

It is clear that L, $\partial/\partial\bar{z}_2$, $\bar{L}$, and $\partial/\partial z_2$ are linearly independent, and that L and $\partial/\partial\bar{z}_2$ commute: the vector fields (VII.8.3) define a CR structure $\mathcal{S}_0$ on $\mathbb{R}^5$; they annihilate the functions z_1 and $w = s + \imath|z_1|^2$.

Proposition VII.8.1. *Let* U *be an arbitrary open neighborhood of* 0 *in* $\mathbb{R}^5$. *There is no function* $h \in \mathcal{C}^1(U)$ *that satisfies, in* U,

$$Lh = \partial h/\partial\bar{z}_2 = 0, \tag{VII.8.4}$$

$$dz_1 \wedge dw \wedge dh \neq 0. \tag{VII.8.5}$$

PROOF. By (VII.8.4) we have

$$L_0 h = (f \circ w)(\partial h/\partial z_2).$$

Theorem VII.5.1 demands that $\partial h/\partial z_2$ vanish when $z_1 = s = 0$. On the other hand, h must be holomorphic with respect to z_2. The equation $Lh = 0$ would imply, *at the origin*, $\partial h/\partial \bar{z}_1 = 0$ and therefore $dh = A dz_1 + B dw$, which contradicts (VII.8.5). ■

Thus the CR structure $\mathcal{S}_0$ on $\mathbb{R}^5$ is *not* locally integrable at the origin.

Next, we embed $\mathbb{R}^5$ into $\mathbb{C}^3$ (in which the coordinates are z_1, z_2 and $w = s + \imath t$) as the hyperplane $t = 0$. We regard the half-space

$$\mathcal{M}_- = \{ (z,w) \in \mathbb{C}^3;\ t \leq 0 \}$$

as a manifold with boundary. We regard f and f_- as functions in $\mathbb{R}^5$, independent of z_2. We extend f_- (as we have the right to do) as a $\mathcal{C}^\infty$ function in $\mathcal{M}_-$ such that $\partial f_-/\partial \bar{w} \equiv 0$. We call G the extension of the function $g_+(z_1,s)$ to $\mathbb{C}^3$ as a function independent of t (and of z_2). We regard L_0 as a vector field in the whole of $\mathbb{C}^3$; it does not involve any partial differentiation with respect to t.

Since L_0 and $\partial/\partial \bar{w}$ commute, we see that the three vector fields in $\mathcal{M}_-$,

$$\Lambda_1 = L_0 - (f_- + L_0 G)\partial/\partial z_2,\ \Lambda_2 = \partial/\partial \bar{w} - (\partial G/\partial \bar{w})\partial/\partial z_2,\ \partial/\partial \bar{z}_2, \tag{VII.8.6}$$

commute, and thus they define a complex structure $\mathcal{S}_-$ on $\mathcal{M}_-$.

PROPOSITION VII.8.2. *The complex structure $\mathcal{S}_-$ of $\mathcal{M}_-$ induces on $\partial\mathcal{M}_-$ the CR structure $\mathcal{S}_0$.*

PROOF. The vector fields Λ_1 and $\partial/\partial \bar{z}_2$ are tangent to the boundary of $\mathcal{M}_-$, $\partial\mathcal{M}_-$ (defined by $t = 0$). When $t = 0$ we have $L_0 G = f_+$, by (VII.8.1). ■

Thus the CR structure $\mathcal{S}_0$ on $\mathbb{R}^5$ is locally integrable on one side, at the origin (Definition V.1.3).

Notes

The whole question of local solvability of a linear PDE was given impetus by H. Lewy's discovery of his celebrated "counter-example" (see Lewy [2]). A first necessary condition for the local solvability of a general first-order equation (later extended to higher order PDE) was proved in Hörmander [1]; it "explained" the Lewy example. It also showed that the equation now referred to as Mizohata's (see Mizohata [1]) provided the simplest example of an analytic vector field L in the plane such that, in any open neighborhood of the origin, the inhomogeneous equation $Lu = f$ has no distribution solution for "most" $\mathscr{C}^\infty$ right-hand sides f. Essential features of the Lewy equation, or of the Mizohata equation, can be generalized to systems of vector fields, as shown in the present chapter; and it has seemed appropriate to attach to these more general hypo-analytic structures the names of Lewy and of Mizohata.

Thus, the scalar version (i.e., $n = 1$) of Theorem VII.2.1 is a particular case of Theorem 3.1 in Hörmander [1]; the kind of constructive proof given here goes back to Grushin [1]. That the involutive structure defined by a suitable perturbation of the Mizohata equation is not locally integrable (see Theorem VII.2.2) was first shown in Nirenberg [2]. Soon afterward (in Nirenberg [3]) the analogous question for the Lewy equation was settled (in the present text, this corresponds to Theorem VII.5.2). The Sjöstrand invariant (Definition VII.3.1) was introduced in Sjöstrand [1]. The density of aberrant vector fields in three-dimensional space (Theorem VII.5.3) was first observed in Jacobowitz and Treves [1].

Going now to systems of more than one equation, as pointed out in the text, Theorems VII.2.1 and VII.2.2 are particular cases of Theorem VII.4.1. The latter was proved in Treves [8]. The analogous result in Lewy structures (Theorem VII.6.1) was first proved in Jacobowitz and Treves [2].

The example, in section VII.7, of a CR structure that can be "realized" as a boundary of a complex manifold without being locally integrable is taken from Hill [1].

It ought to be mentioned that the range (in $\mathscr{C}^\infty$ functions, in distributions or in hyperfunctions) of the Lewy operator, as well as that of the Mizohata operator, can be described by means of a special "projector." On this topic, to which we shall return in Volume 2 from the microlocal standpoint, we refer the reader to section 2.3, chapter 3 in Sato, Kawai, and Kashiwara [1] and to Greiner, Kohn, and Stein [1].

VIII

Necessary Conditions for the Vanishing of the Cohomology. Local Solvability of a Single Vector Field

The examples of nonsolvability (as well as those of nonintegrability) in chapter VII were "constructed." In the search for conditions necessary for solvability or, more generally, necessary for exactness in the differential complex associated with the given involutive structure (see section I.6), another approach is by way of *a priori estimates*. These are shown to follow from the hypothesis of exactness, by a Functional Analysis argument that is a routine adaptation of a classical lemma of Hörmander [1]. The claim here is that if exactness holds at the (p,q) level, then the absolute value of the integral of $f \wedge v$, where f is a d'-*closed* $\mathscr{C}^\infty$ section of $\Lambda^{p,q}$ and v a compactly supported $\mathscr{C}^\infty$ section of $\Lambda^{m-p,n-q}$ (see (I.6.16), (I.6.19)), is bounded by the product $N(f)N(\mathrm{d}'v)$ of norms (of the $\mathscr{C}^r$ kind) of f and of $\mathrm{d}'v$. Regardless of whether $\mathrm{d}'f$ vanishes, the exterior product $f \wedge v$ may be identified to a true differential form of degree $m + n$ (= dimension of the base manifold $\mathcal{M}$). When $\mathcal{M}$ is orientable, the integral of this form does not vanish for all v unless $f \equiv 0$; its natural extension establishes a separating duality between the distribution sections of $\Lambda^{p,q}$ and the smooth sections of $\Lambda^{m-p,n-q}$ (with the proviso that the support of one or the other be compact; see section VIII.1).

The next step is to construct, under appropriate circumstances, two sequences of sections f_ν and v_ν such that $\int f_\nu \wedge v_\nu = +1$ while $N(f_\nu)N(\mathrm{d}'v_\nu) \to 0$, thus contradicting the a priori estimate and, as a consequence, the exactness hypothesis. We take $f_\nu = fe^{\nu h}$, $v_\nu = ve^{-\nu h}$ with $\int f \wedge v = +1$, $\mathrm{d}'f = 0$ and h such that $\mathcal{R}e\, h \leq 0$ on the support of f and $\mathcal{R}e\, h > 0$ on that of $\mathrm{d}'v$. At this juncture we encounter a difficulty not present in the traditional situation where one deals with a single vector field (or a single differential operator): the requirement that f_ν be d′-closed. When the critical degree q is highest, i.e., when $q = n$ (which is the only case to consider when the tangent structure bundle $\mathcal{V}$ is a line bundle), every section of $\Lambda^{p,q}$ is trivially d′-closed and the requirement imposes no limitation on h (see section VIII.2). When $q < n$, h must be a solution, i.e., must itself satisfy $d'h = 0$ (Theorem VIII.1.1). This demands that the involutive structure of $\mathcal{M}$ admit "enough" nontrivial solutions. There lies the reason for limiting our attention to locally integrable structures. In the absence of local integrability, when $q < n$ we do not know how to begin constructing the cocycle f, not to speak of how to find solutions h with the desired properties.

From section VIII.3 onward the analysis proceeds within the framework of a locally integrable structure. Under the additional hypothesis that the Levi form (see section I.9) at some characteristic point lying above the point $\mathfrak{p} \in \mathcal{M}$ is nondegenerate and has q eigenvalues of one sign and $n-q$ ones of the opposite sign, it is shown, in section VIII.3, that local exactness at $\mathfrak{p}$ does not hold in degree q. This generalizes the classical result in realizable CR structures.

The remainder of the chapter is devoted to two classes of locally integrable structures that, in a sense, represent extreme cases: the first class is characterized by the fact that the characteristic set at the central point is a real line; the second one, by the fact that the tangent structure bundle $\mathcal{V}$ is a line bundle. Examples of the first class are, on one hand, the CR structures of real hypersurfaces in complex space; on the other hand, the structures in which $m = 1$, i.e., the cotangent structure bundle T′ is a line bundle, locally spanned by the differential dZ of a single first integral. The latter stand in contraposition to the structures in the second class, where one is dealing, at the local level, with a single vector field L.

The last four sections of chapter VIII are devoted to characterizing the local solvability of the single scalar equation (in $m+1$ variables)

$$Lu = f. \tag{1}$$

When Equation (1) is locally solvable (for all choices of smooth right-hand sides f), then L possesses m first integrals; this follows from a solvability result in $\mathcal{C}^\infty$ in Hörmander [3] and is not proved here. Actually, we establish, under the solvability condition $(\mathcal{P})$, the existence of an L^2 solution u of (1) without invoking the hypothesis of local integrability. But local integrability allows us to apply the Approximate Poincaré Lemma (see section II.6) and the Mittag-Leffler procedure to derive, from the existence of L^2 solutions, that of $\mathcal{C}^\infty$ solutions. It also enables us to give a streamlined transcription of Moyer's proof of the necessity of condition $(\mathcal{P})$.

Sections VIII.4, VIII.5, and VIII.6 are devoted to establishing a necessary condition for local exactness of the differential operator d′ in degree q, under the hypothesis that the characteristic set at the central point has dimension one.

Section VIII.4 presents the statement and the proof of a technical result (Theorem VIII.4.1), which is then applied, in section VIII.5, to the case $m = 1$, and in section VIII.5 to cases $m \geq 1$. When $m = 1$, the necessary condition for local exactness in degree q derived from Theorem VIII.4.1 is that the singular holology of the fibres of the structure (see Corollary II.3.1) be trivial in dimension $q-1$; the condition is invariant and it is known to be also sufficient when $q = 1$ and when $q = n$. Actually, in all cases, whether $m = 1$ or $m > 1$, the condition in Theorem VIII.4.1 lends itself to a topological interpretation. Lack of space has prevented us from describing it here (see Notes).

VIII.1 Preliminary Necessary Conditions for Exactness

Let the $\mathscr{C}^\infty$ manifold $\mathcal{M}$ be endowed with an involutive structure. The cotangent and tangent structure bundles are denoted by T′ and $\mathcal{V}$, as usual; the rank of T′ is m, that of $\mathcal{V}$ is n; dim $\mathcal{M} = m + n = N$. Below $\Omega \neq \emptyset$ denotes an open subset of $\mathcal{M}$.

We select a *locally finite open covering* of $\mathcal{M}$, $\{U_\iota\}_{\iota\in I}$, consisting of the domains U_ι of local coordinates $\{x_{\iota,j}\}_{j=1,\ldots,N}$. Furthermore, we require that, in each open set U_ι, there be a smooth basis $\{\psi_{\iota,1},\ldots,\psi_{\iota,m},\varpi_{\iota,1},\ldots,\varpi_{\iota,n}\}$ of $\mathbb{C}T^*\mathcal{M}$ such that $\psi_{\iota,1},\ldots,\psi_{\iota,m}$ span T′ over U_ι. Necessarily $\varpi_{\iota,1},\ldots,\varpi_{\iota,n}$ span a supplementary of T′ in $\mathbb{C}T^*\mathcal{M}$, again over U_ι. For each pair of multi-indices $J = \{j_1,\ldots,j_p\}$, $K = \{k_1,\ldots,k_q\}$ $(1 \leq j_1 <\cdots< j_p \leq m,\ 1 \leq k_1 <\cdots< k_q \leq n)$, let $\mathbf{e}_{\iota;J,K}$ denote the equivalence class, in the vector bundle $\Lambda^{p,q}$, of

$$\psi_{\iota,j_1}\wedge\cdots\wedge\psi_{\iota,j_p}\wedge\varpi_{\iota,k_1}\wedge\cdots\wedge\varpi_{\iota,k_q}. \tag{VIII.1.1}$$

In U_ι any section $f \in \mathscr{C}^\infty(\Omega;\Lambda^{p,q})$ has an expression

$$f = \sum_{|J|=p}\sum_{|K|=q} f_{\iota;J,K}\mathbf{e}_{\iota;J,K}, \tag{VIII.1.2}$$

with $f_{\iota;J,K} \in \mathscr{C}^\infty(U_\iota;\Lambda^{p,q})$. We also select a $\mathscr{C}^\infty$ partition of unity $\{\chi_\iota\}_{\iota\in I}$ subordinate to the covering $\{U_\iota\}_{\iota\in I}$. If A is then a compact subset of Ω, we shall write

$$\|f\|_{A,\ell} = \sup_{A}\sum_{\iota\in I}\sum_{|J|=p}\sum_{|K|=q}\sum_{|\alpha|\leq\ell}|(\partial/\partial x_\iota)^\alpha(\chi_\iota f_{\iota;J,K})|, \tag{VIII.1.3}$$

with $(\partial/\partial x_\iota)^\alpha = (\partial/\partial x_{\iota,1})^{\alpha_1}\cdots(\partial/\partial x_{\iota,N})^{\alpha_N}$. (The intersection of supp$(\chi_\iota f_{J,K})$ with A is a compact subset of U_ι.) The topology of $\mathscr{C}^\infty(\Omega;\Lambda^{p,q})$ can be defined by means of the seminorms (VIII.1.3) as A ranges over the family of all compact subsets of Ω and ℓ over $\mathbb{Z}_+$.

We denote by $\mathscr{C}_c^\infty(A;\Lambda^{p,q})$ the subspace of $\mathscr{C}^\infty(\Omega;\Lambda^{p,q})$ consisting of the sections whose support is contained in the compact subset A of Ω; $\mathscr{C}_c^\infty(A;\Lambda^{p,q})$ will be endowed with the topology induced by that of $\mathscr{C}^\infty(\Omega;\Lambda^{p,q})$; it is a Fréchet space. The space $\mathscr{C}_c^\infty(\Omega;\Lambda^{p,q})$ of the compactly supported $\mathscr{C}^\infty$ sections of $\Lambda^{p,q}$ in Ω, i.e., the union of the spaces $\mathscr{C}_c^\infty(A;\Lambda^{p,q})$ as A ranges over the family of all compact subsets of Ω, is equipped with the inductive limit topology. It suffices, here, to recall two properties of this topology: a sequence $\{f_\nu\}_{\nu=1,2,\ldots}$ in $\mathscr{C}_c^\infty(\Omega;\Lambda^{p,q})$ converges if the support of f_ν is contained in some compact subset A of Ω independent of ν, and if the sections f_ν converge in $\mathscr{C}^\infty(\Omega;\Lambda^{p,q})$; a linear functional λ in $\mathscr{C}_c^\infty(\Omega;\Lambda^{p,q})$ is continuous if and only if its restriction to each subspace $\mathscr{C}_c^\infty(A;\Lambda^{p,q})$ is continuous, or, which amounts to the same, if λ is *sequentially continuous*.

Throughout the remainder of the chapter we shall assume the manifold $\mathcal{M}$ to be orientable. In dealing with a nonorientable manifold one must introduce *twisted* forms and currents, by going to the oriented covering of the base manifold. For the sake of simplicity we shall not do so here and content ourselves with referring the reader to Schwartz [1], chap. 10.

Proposition VIII.1.1. *Suppose the manifold $\mathcal{M}$ is orientable. There is then an isomorphism of the vector bundle $\Lambda^{m-p,n-q}$ onto the dual bundle of $\Lambda^{p,q}$.*

Proof. To say that $\mathcal{M}$ is orientable is to say that there is a $\mathscr{C}^\infty$ section λ of the line bundle Λ^N ($= \Lambda^N \mathbb{C}T^*\mathcal{M}$, $N = \dim \mathcal{M}$) such that $\lambda \neq 0$ at every point of $\mathcal{M}$. Once and for all we make a choice of such a section λ. Given an arbitrary point p of $\mathcal{M}$ let $\theta \in T_p'^{p,q}$ be any representative of a class $\dot\theta \in \Lambda_p^{p,q}$ and $\theta' \in T_p'^{m-p,n-q}$ any representative of a class $\dot\theta' \in \Lambda_p^{m-p,n-q}$. The exterior product $\theta\wedge\theta'$ is an element of Λ_p^N. Since $\theta\wedge\theta' = 0$ if $\theta \in T_p'^{p+1,q-1}$ or $\theta' \in T_p'^{m-p+1,n-q-1}$ (or both), $\theta\wedge\theta'$ is independent of the choice of the representatives θ and θ'. We may therefore define

$$\theta\wedge\theta' = \langle\dot\theta,\dot\theta'\rangle\lambda. \tag{VIII.1.4}$$

By using smooth bases of T′ and of a supplementary of T′ in $\mathbb{C}T^*\mathcal{M}$ over an open neighborhood U of p the following claims are easy to ascertain: given any nonzero element $\theta \in T_p'^{p,q}$, there is $\theta' \in T_p'^{m-p,n-q}$ such that $\langle\dot\theta,\dot\theta'\rangle \neq 0$; if $\dot\theta$ and $\dot\theta'$ are the values at p of smooth sections, of $\Lambda^{p,q}$ and $\Lambda^{m-p,n-q}$ respectively, the complex-valued function $\langle\dot\theta,\dot\theta'\rangle$ varies smoothly. Thus the map $\dot\theta' \to (\dot\theta \to \langle\dot\theta,\dot\theta'\rangle)$ defines a $\mathscr{C}^\infty$ vector bundle isomorphism of $\Lambda^{m-p,n-q}$ onto the bundle whose fibre at any given point is the dual of that of $\Lambda^{p,q}$. ∎

Let us return to the open subset Ω of $\mathcal{M}$. Closely related to Proposition VIII.1.1 is the fact that the bilinear functional

$$(\dot f,\dot v) \to \int \dot f\wedge\dot v \left(= \int\langle\dot f,\dot v\rangle\lambda\right) \tag{VIII.1.5}$$

on $\mathscr{C}_c^\infty(\Omega;\Lambda^{p,q})\times\mathscr{C}_c^\infty(\Omega;\Lambda^{m-p,n-q})$ is *separating*. We can extend it by continuity to $\mathscr{C}_c^\infty(\Omega;\Lambda^{p,q})\times\mathscr{D}'(\Omega;\Lambda^{m-p,n-q})$ and to $\mathscr{C}^\infty(\Omega;\Lambda^{p,q})\times\mathscr{E}'(\Omega;\Lambda^{m-p,n-q})$. The proof of the next statement is left to the reader.

PROPOSITION VIII.1.2. *Suppose $\mathcal{M}$ orientable. The bracket* (or "integral")

$$\int \dot{f} \wedge \dot{v} \qquad \text{(VIII.1.6)}$$

turns the spaces

$$\mathscr{C}_c^\infty(\Omega;\Lambda^{p,q}) \text{ and } \mathscr{D}'(\Omega;\Lambda^{m-p,n-q}) \text{ (resp., } \mathscr{C}^\infty(\Omega;\Lambda^{p,q}) \text{ and } \mathscr{E}'(\Omega;\Lambda^{m-p,n-q}))$$

into the dual of one another.

Both spaces $\mathscr{C}^\infty(\Omega;\Lambda^{p,q})$ and $\mathscr{C}_c^\infty(\Omega;\Lambda^{p,q})$ are *reflexive*: they are naturally isomorphic to the (strong) duals of $\mathscr{E}'(\Omega;\Lambda^{n-p,m-q})$ and $\mathscr{D}'(\Omega;\Lambda^{n-p,m-q})$ respectively (the latter carry their strong dual topologies; cf. Schwartz [1], chap. 3).

By taking the support of the sections $\dot{f}$ and $\dot{v}$ in (VIII.1.5) suitably small and using representatives f and v, one checks at once that

$$\int \mathrm{d}'\dot{f} \wedge \dot{v} = (-1)^{p+q-1}\int \dot{f} \wedge \mathrm{d}'\dot{v}. \qquad \text{(VIII.1.7)}$$

Indeed, $\mathrm{d}f \wedge v$ is a representative of $\mathrm{d}'\dot{f} \wedge \dot{v}, f \wedge \mathrm{d}v$ is one of $\dot{f} \wedge \mathrm{d}'\dot{v}$; we have

$$\mathrm{d}(f \wedge v) = \mathrm{d}f \wedge v + (-1)^{p+q} f \wedge \mathrm{d}v$$

and it suffices to apply Stokes' theorem. Formula (VIII.1.7) can be restated as follows:

PROPOSITION VIII1.3. *Suppose $\mathcal{M}$ orientable. In the sense of the duality bracket* (VIII.1.6), *the operator* d' *acting on distribution sections of $\Lambda^{m-p,n-q}$ is the transpose of* $(-1)^{p+q-1}\mathrm{d}'$ *acting on $\mathscr{C}^\infty$ sections of $\Lambda^{p,q}$.*

COROLLARY VIII.1.1. *Suppose $\mathcal{M}$ orientable. In order for a distribution section of $\Lambda^{m-p,n-q}$* (i.e., a cochain) *to be* d'*-closed* (i.e., to be a cocycle) *it is necessary and sufficient that it be orthogonal, in the sense of the bracket* (VIII.1.6), *to all $\mathscr{C}^\infty$ sections of $\Lambda^{p,q}$ that are* d'*-exact* (i.e., to all smooth coboundaries). *For a distribution section of $\Lambda^{m-p,n-q}$ to be* d'*-exact it is necessary that it be orthogonal to all $\mathscr{C}^\infty$ sections of $\Lambda^{p,q}$ that are* d'*-closed.*

The topology on the spaces $\mathscr{C}^\infty(\Omega;\Lambda^{p,q})$ introduced above enables us to prove a necessary condition in order that the cohomology space $H'^{p,q}(\Omega)$ of the differential complex (I.6.16) (or (I.6.18)) ($0 \le p \le m$, $1 \le q \le n$) vanish. Actually, we shall consider a slightly more general property, which involves an open subset $\Omega' \neq \emptyset$ of Ω:

> *Given any cocycle $f \in \mathscr{C}^\infty(\Omega;\Lambda^{p,q})$ there is a distribution section $u \in \mathscr{D}'(\Omega';\Lambda^{p,q-1})$ such that $\mathrm{d}'u = f$ in Ω'.* $\qquad$ (VIII.1.8)$_{p,q}$

(Henceforth we omit the dots on the top that signal that we are dealing with sections of $\Lambda^{p,q}$ which, we recall, are equivalence classes of forms or currents; see section I.6.)

LEMMA VIII.1.1. *If* $(\text{VIII}.1.8)_{p,q}$ *holds, to every compact subset* K' *of* Ω' *there exist a compact subset* K *of* Ω *and numbers* $C > 0$, $\ell \in \mathbb{Z}_+$ *such that, whatever the cocycle* $f \in \mathscr{C}^\infty(\Omega;\Lambda^{p,q})$ *and the cochain* $v \in \mathscr{C}_c^\infty(\Omega';\Lambda^{m-p,n-q})$ *with* $\operatorname{supp} v \subset K'$, *we have*

$$\left|\int f\wedge v\right| \leq C\, \|f\|_{K,\ell}\|d'v\|_{K',\ell}. \qquad (\text{VIII}.1.9)_{p,q}$$

PROOF. Let $\mathbf{E}$ denote the kernel of the differential operator d' in $\mathscr{C}^\infty(\Omega;\Lambda^{p,q})$ equipped with the topology inherited from $\mathscr{C}^\infty(\Omega;\Lambda^{p,q})$; $\mathbf{E}$ is a Fréchet space. Let $\mathbf{F}$ denote the subspace of $\mathscr{C}^\infty(\Omega';\Lambda^{m-p,n-q})$ consisting of the sections whose support is contained in the compact set K'. We equip $\mathbf{F}$ with the topology defined by the seminorms $v \rightarrow \|d'v\|_{K';\ell}$ ($\ell = 0,1,\ldots,$). Unless $q = n$ $\mathbf{F}$ is not Haudorff; the closure of $\{0\}$ in $\mathbf{F}$ is the subspace $\mathfrak{Z}$ made up of the cocycles that belong to $\mathbf{F}$; the associated Hausdorff space is the quotient space $\mathbf{F}_0 = \mathbf{F}/\mathfrak{Z}$ equipped with the quotient topology; $\mathbf{F}_0$ is a metrizable space.

Restrict to the product $\mathbf{E}\times\mathbf{F}$ the bilinear functional

$$(f,v) \rightarrow \int f\wedge v \text{ (see (VIII.1.5))}. \qquad (\text{VIII}.1.10)$$

Fix $v \in \mathbf{F}$; the continuity of the linear functional $f \rightarrow \int f\wedge v$ on $\mathbf{E}$ is obvious. Now assume that $(\text{VIII}.1.8)_{p,q}$ holds and fix $f \in \mathbf{E}$. Select $u \in \mathscr{D}'(\Omega';\Lambda^{p,q-1})$ such that $d'u = f$. Since $v \in \mathscr{C}_c^\infty(\Omega';\Lambda^{m-p,n-q})$,

$$d'u\wedge v = d'(u\wedge v) + (-1)^{p+q}\, u\wedge d'v,$$

and since $u\wedge v$ has compact support (contained in Ω'),

$$\int d'u\wedge v = (-1)^{p+q}\int u\wedge d'v,$$

whence the continuity in $\mathbf{F}$ of the linear functional $v \rightarrow \int f\wedge v$. It follows that this functional induces a continuous linear functional on $\mathbf{F}_0$. Thus (VIII.1.10) induces a bilinear functional on $\mathbf{E}\times\mathbf{F}_0$ that is separately continuous. It is a consequence of the Banach-Steinhaus theorem that any separately continuous bilinear functional on the product of a Fréchet space with a metrizable space is continuous, whence the continuity of the lift back to $\mathbf{E}\times\mathbf{F}$, which is what $(\text{VIII}.1.9)_{p,q}$ expresses. ∎

We recall that, in this book, any function or any distribution h such that $d'h \equiv 0$ is called a *solution* (in other words, a solution is a *zero cocycle*).

THEOREM VIII.1.1. *Let* p, q *be integers such that* $0 \leq p \leq m$, $1 \leq q \leq n$. *Assume that the following condition holds:*

there exist a $\mathscr{C}^\infty$ solution h in Ω, a cocycle $f \in \mathscr{C}^\infty(\Omega;\Lambda^{p,q})$, and a cochain $v \in \mathscr{C}_c^\infty(\Omega';\Lambda^{m-p,n-q})$ such that (VIII.1.11)$_{p,q}$

$$\mathcal{R}e\, h \leq 0 \text{ on supp } f,\ \mathcal{R}e\, h > 0 \text{ on supp d}'v,\ \int f \wedge v \neq 0.$$

Then Property (VIII.1.8)$_{p,q}$ *does not hold.*

PROOF. We reason under Hypothesis (VIII.1.11)$_{p,q}$. There is a number $c > 0$ such that $\mathcal{R}e\, h \geq c$ on supp $\mathrm{d}'v$. As a consequence, to each integer $\ell > 0$ there is a constant $C_\ell > 0$ such that, if $\mathrm{K}' = \text{supp } v$ ($\subset\subset \Omega'$), we have, for all numbers $\rho > 0$,

$$\|\mathrm{d}'(\mathrm{e}^{-\rho h}v)\|_{\mathrm{K}',\ell} \leq C_\ell \rho^\ell \mathrm{e}^{-c\rho}. \tag{VIII.1.12}$$

On the other hand, whatever the compact subset K of Ω, there is $C_{\mathrm{K},\ell} > 0$ such that

$$\|\mathrm{e}^{\rho h} f\|_{\mathrm{K},\ell} \leq C_{\mathrm{K},\ell}\rho^\ell. \tag{VIII.1.13}$$

We apply (VIII.1.9)$_{p,q}$ after substituting $\mathrm{e}^{\rho h} f$ for f and $\mathrm{e}^{-\rho h} v$ for v, and we take (VIII.1.12) and (VIII.1.13) into account. If (VIII.1.8)$_{p,q}$ were true we would get, for a suitable choice of K,

$$\left|\int f \wedge v\right| \leq CC_\ell C_{\mathrm{K},\ell}\rho^{2\ell}\mathrm{e}^{-c\rho}.$$

Letting ρ go to $+\infty$ would entail $\int f \wedge v = 0$, contradicting (VIII.1.11)$_{p,q}$. ∎

VIII.2. Exactness of Top-Degree Forms

When $p = m$, $q = n$, the hypothesis in Theorem VIII.1.1 can be modified:

THEOREM VIII.2.1. *Suppose there is a $\mathscr{C}^\infty$ function h in an open neighborhood $\mathcal{O}$ of a compact subset $\mathrm{K}_0 \neq \emptyset$ of $\mathcal{M}$ endowed with the following properties:*

$$\mathcal{R}e\, h \equiv 0 \text{ in } \mathrm{K}_0 \text{ and } \mathcal{R}e\, h > 0 \text{ in } \mathcal{O}\backslash\mathrm{K}_0; \tag{VIII.2.1}$$

to each pair $(k,\ell) \in \mathbb{Z}_+^2$ there is $C_{k,\ell} > 0$ such that (VIII.2.2)

$$\|\mathrm{d}'h\|_{\{p\},\ell} \leq C_{k,\ell}[\mathcal{R}e\, h(p)]^k,\ \forall\, p \in \mathcal{O}.$$

Then, whatever the open subsets Ω and Ω' of $\mathcal{M}$ such that $\mathcal{O} \subset \Omega' \subset \Omega$, Property (VIII.1.8)$_{m,n}$ *does not hold.*

PROOF. There is a number $c > 0$ such that $\mathrm{K}_0 \subset \mathrm{K}_1 = \{p \in \mathcal{O};\ \mathcal{R}e\, h(p) \leq c\} \subset\subset \mathcal{O}$. We call K a compact subset of $\mathcal{O}$ whose interior contains K_1. There exists (cf. Lemma VIII.8.A1) a sequence of functions $\chi_\nu \in \mathscr{C}_c^\infty(\mathcal{O})$ ($\nu \in \mathbb{Z}_+$, $\nu > c^{-1}$) with the following properties:

$$\chi_\nu = 1 \text{ if } \mathscr{R}\!e h \leq 1/2\nu,\ \chi_\nu = 0 \text{ if } \mathscr{R}\!e h > 1/\nu; \qquad \text{(VIII.2.3)}$$

to every $\ell \in \mathbb{Z}_+$ *there is a constant* $C_\ell > 0$ *such that, for all* $\nu \in \mathbb{Z}_+$, $\nu > c^{-1}$, (VIII.2.4)

$$\|\chi_\nu\|_{K,\ell} \leq C_\ell \nu^\ell.$$

We also select a function $\chi \in \mathscr{C}_c^\infty(K)$ equal to one in K_1.

Since $\mathcal{O}$ is orientable, we can find a real, $\mathscr{C}^\infty$ differential form ϖ in $\mathcal{O}$ of degree $m+n$, nowhere vanishing. We are going to apply Lemma VIII.1.1 with

$$f = e^{\nu h}\chi_\nu\varpi,\ v = e^{-\nu h}\chi;$$

$\nu \in \mathbb{Z}_+$ will be chosen below (but $\nu > c^{-1}$). We have, for a suitable constant $C > 0$,

$$\|f\|_{K,\ell} \leq C\nu^\ell. \qquad \text{(VIII.2.5)}$$

On the other hand, $d'v = e^{-\nu h}(d'\chi - \nu\chi d'h)$. Provided C is large enough,

$$\|e^{-\nu h}d'\chi\|_{K,\ell} \leq Ce^{-c\nu},$$

and thanks to (VIII.2.2),

$$\|e^{-\nu h}\chi d'h\|_{K,\ell} \leq C'_{k,\ell}\nu^\ell \operatorname*{Max}_{\mathrm{supp}\chi}\left\{(\mathscr{R}\!e h)^k e^{-\nu\mathscr{R}\!e h}\right\} \leq C''_{k\ell}\nu^{\ell-k}.$$

Thus

$$\|d'v\|_{K,\ell} \leq Ce^{-c\nu} + C_k\nu^{\ell-k}. \qquad \text{(VIII.2.6)}$$

Combining (VIII.2.5) with (VIII.2.6), and taking $k = 2\ell + k_0$ ($k_0 \in \mathbb{Z}_+$), shows that, for a suitably large $C_0 > 0$,

$$\|f\|_{K,\ell}\|d'v\|_{K,\ell} \leq C_0\nu^{-k_0}. \qquad \text{(VIII.2.7)}$$

The left-hand side in $(\text{VIII.1.9})_{m,n}$ is equal to $\left|\int \chi_\nu\varpi\right|$. Let $\mathfrak{p}$ be an arbitrary point of K_0 and let $x_1,\ldots,x_N$ ($N = \dim \mathcal{M} = m + n$) be coordinates in an open neighborhood $U \subset \mathcal{O}$ of $\mathfrak{p}$, all vanishing at $\mathfrak{p}$. Possibly after changing the sign of x_1 we may assume that, in U,

$$\varpi = g dx \ (dx = dx_1\wedge\cdots\wedge dx_N,\ g > 0).$$

Since $\mathscr{R}\!e h \geq 0$ in $\mathcal{O}$, and possibly after contracting U about $\mathfrak{p}$, we see that there is a constant $C > 0$ such that

$$\mathscr{R}\!e h(x) \leq C|x|^2,\ \forall\, x \in U.$$

As soon as ν is large enough, the ball $\mathcal{B}_\nu = \{ x \in U; |x|^2 \leq 1/2C\nu \}$ will be contained in U. On the other hand, $\chi_\nu = 1$ in $\mathcal{B}_\nu$, and thus

$$\int \chi_\nu \varpi \geq \int_{\mathcal{B}_\nu} g dx \geq c_0 \nu^{-N/2}.$$

According to (VIII.1.9)$_{m,n}$ and to (VIII.2.7), we should have $c_0\nu^{-N/2} \leq C\nu^{-k_0}$ for all sufficiently large integers ν, which is absurd if $k_0 > N/2$. ∎

COROLLARY VIII.2.1. *Suppose that*

there is a $\mathcal{C}^\infty$ solution h in an open neighborhood $\mathcal{O}$ of a compact subset $K_0 \neq \emptyset$ *of $\mathcal{M}$ satisfying* (VIII.2.1). (VIII.2.8)

Then, whatever the open subsets Ω and Ω' of $\mathcal{M}$ such that $\mathcal{O} \subset \Omega' \subset \Omega$, Property (VIII.1.8)$_{m,n}$ *does not hold.*

Indeed, if h is a solution, (VIII.2.2) is automatically satisfied.

COROLLARY VIII.2.2. *Suppose there is a point* $(0,\theta) \in T^\circ$ *at which the Levi form of the involutive structure of $\mathcal{M}$ is definite. Then, whatever the open subsets Ω and Ω' of $\mathcal{M}$ such that $0 \in \Omega' \subset \Omega$, Property* (VIII.1.8)$_{m,n}$ *does not hold.*

PROOF. We refer the reader to the proof of Theorem VIII.3.1. The difference between the situation in section VIII.3 and the present one is that, here, we are not assuming local integrability, and thus there do not necessarily exist solutions (see chapter VII). But the same argument as in the beginning of the proof of Theorem VIII.3.1 can be developed in the ring of formal power series (see Lemma IV.1.1). It follows (cf. (VIII.3.7)) that we may select coordinates $x_1,\ldots,x_N$ in an open neighborhood U of 0, vanishing at 0 and such that there is a formal solution

$$w = x_m + \imath \left[|x|^2 + \sum_{|\alpha| \geq 3} c_\alpha x^\alpha \right] \quad (c_\alpha \in \mathbb{R}).$$

We introduce a cutoff function $\chi \in \mathcal{C}^\infty(\mathbb{R}^N)$, $\chi \geq 0$ everywhere, $\chi(x) = 1$ if $|x| < \frac{1}{2}$, $\chi(x) = 0$ if $|x| \geq 1$, and select numbers $R_k \nearrow +\infty$ such that the series

$$\sum_{k=3}^{+\infty} \chi(x/R_k) \sum_{|\alpha|=k} c_\alpha x^\alpha$$

converges in $\mathcal{C}^\infty(U)$, to a function $\mathfrak{R}(x)$. We then set

$$h = |x|^2 + \mathfrak{R}(x) - \imath x_N.$$

Since the Taylor expansion of h at 0 is equal to $-\imath w$, we see that $\mathrm{d}'h$ vanishes to infinite order at 0. Then if Ω' is any open neighborhood of 0, there is an open neighborhood $\mathcal{O} \subset \mathrm{U} \cap \Omega'$ of 0 in which

$$\tfrac{1}{2}|x|^2 \leq \mathcal{R}e h \leq 2|x|^2.$$

It is then readily checked that Conditions (VIII.2.1) and (VIII.2.2) are satisfied with $\mathrm{K}_0 = \{0\}$. ∎

It can be shown, under certain additional hypotheses, that the negation of the hypothesis in Corollary VIII.2.1 entails $(\mathrm{VIII.1.8})_{m,n}$, and thus also, by Theorem VIII.2.1, the negation of $(\mathrm{VIII.1.11})_{m,n}$. In those cases, therefore, the hypothesis in Corollary VIII.2.1 is seen to be equivalent to $(\mathrm{VIII.1.11})_{m,n}$. But, as we now show, this equivalence can be established directly in a number of special cases.

We shall limit ourselves to locally integrable structures, and actually to open sets Ω in which there exist coordinates x_j $(1 \leq j \leq m)$, t_k $(1 \leq k \leq n)$ and first integrals Z_i $(1 \leq i \leq m)$ as in (I.7.23). We shall make use of the vector fields (I.7.27) in Ω. An arbitrary section f of $\Lambda^{p,q}$ over Ω will be identified to a differential form (I.7.29) and the coboundary $\mathrm{d}'f$ to the form (I.7.30). Note also that, over Ω, the map $f_0 \to \mathrm{d}Z \wedge f_0$ defines an isomorphism of $\Lambda^{0,q}$ onto $\Lambda^{m,q}$, and $\Lambda^{p,q} \cong \Lambda^p \otimes \Lambda^q$.

LEMMA VIII.2.1. *Property* $(\mathrm{VIII.1.11})_{p,n}$ *is equivalent to the following property:*

there are functions $h \in \mathscr{C}^\infty(\Omega)$, $\chi \in \mathscr{C}_c^\infty(\Omega')$ *such that* $\mathrm{L}h \equiv 0$ *in* Ω *and* (VIII.2.9)

$$\mathcal{R}e h > 0 \text{ on } \operatorname{supp} \mathrm{L}\chi; \qquad \text{(VIII.2.10)}$$

$$\exists\, \mathfrak{p} \in \Omega',\ \mathcal{R}e h(\mathfrak{p}) \leq 0,\ \chi(\mathfrak{p}) \neq 0. \qquad \text{(VIII.2.11)}$$

PROOF. Suppose $(\mathrm{VIII.1.11})_{p,n}$ holds and write

$$f = \sum_{|J|=p} f_J \mathrm{d}Z_J \wedge \mathrm{d}t_1 \wedge \cdots \wedge \mathrm{d}t_n,\ v = \sum_{|J|=p} v_J \mathrm{d}Z_{J^*},$$

where J^* is the ordered complement of J with respect to $[1,\ldots,m]$. If $\int f \wedge v \neq 0$ we must have, for some multi-index J, $|J| = p$,

$$\int f_J v_J \mathrm{d}Z \wedge \mathrm{d}t \neq 0.$$

We take $\chi = v_J$. Since $\mathcal{R}e h \leq 0$ on $\operatorname{supp} f$, we must have $\mathcal{R}e h \leq 0$ at some point of $\operatorname{supp} \chi$. On the other hand,

$$\mathrm{L}v = \sum_{|I|=p} \mathrm{L}v_I \wedge \mathrm{d}Z_{I^*}$$

and thus the support of $\mathrm{L}v$ is equal to the union of the supports of the forms $\mathrm{L}v_I$. It follows that we have $\mathcal{R}e\, h > 0$ on supp $\mathrm{L}\chi$.

Assume now that (VIII.2.9) holds. We have $\mathcal{R}e\, h > \delta > 0$ on supp $\mathrm{L}\chi$. We select a function $f_0 \in \mathcal{C}_c^\infty(\Omega)$ such that

$$f_0(\mathfrak{p}) \neq 0 \textit{ and } \mathcal{R}e\, h \leq \delta/2 \textit{ on } \mathrm{supp}\, f_0; \tag{VIII.2.12}$$

$$\chi f_0 \det(\partial Z/\partial x) \geq 0 \quad \textit{everywhere}. \tag{VIII.2.13}$$

This is possible thanks to (VIII.2.11). Define then

$$f = f_0 \mathrm{d}Z_1 \wedge \cdots \wedge \mathrm{d}Z_p \wedge \mathrm{d}t_1 \wedge \cdots \wedge \mathrm{d}t_n,\ v = \chi \mathrm{d}Z_{p+1} \wedge \cdots \wedge \mathrm{d}Z_m.$$

We have

$$\int f \wedge v = \int \chi f_0 \det(\partial Z/\partial x)\mathrm{d}x\mathrm{d}t \neq 0.$$

Of course $\mathrm{L}f \equiv 0$, $\mathcal{R}e(h - \delta/2) \leq 0$ on supp f, $\mathcal{R}e(h - \delta/2) > 0$ on supp $\mathrm{L}v$, which shows that $(\mathrm{VIII.1.11})_{p,n}$ holds. ∎

Let Σ be an arbitrary orbit of $\mathcal{V}$ in $\mathcal{O}$ (Definition I.11.1). Since $\mathcal{V}|_\Sigma$ is contained in $\mathbb{C}T\Sigma$ it can be regarded as the tangent structure bundle of an involutive structure on Σ, the structure inherited from $\mathcal{M}$. We recall that it is locally integrable as soon as the structure of $\mathcal{M}$ is locally integrable.

THEOREM VIII.2.2. *Assume that the following conditions are satisfied:*

no orbit of $\mathcal{V}$ in Ω' is contained in a compact subset of Ω'; (VIII.2.14)

given any orbit Σ of $\mathcal{V}$ in Ω' and any open set $\mathcal{O} \subset \Omega'$, every nonempty connected component of $\Sigma \cap \mathcal{O}$ is an orbit of $\mathcal{V}$ in $\mathcal{O}$. (VIII.2.15)

Suppose moreover that every orbit of $\mathcal{V}$ in Ω' is either an open subset of Ω' or else inherits an elliptic structure from $\mathcal{M}$.

Then Property (VIII.2.9) *entails Property* (VIII.2.8).

PROOF. Suppose (VIII.2.9) holds. There is $\delta > 0$ such that $\mathcal{R}e\, h \geq \delta$ on supp $\mathrm{L}\chi$. Call μ the minimum of $\mathcal{R}e\, h$ on supp $\chi \subset\subset \Omega'$. We know by (VIII.2.11) that $\mu \leq 0$. We shall prove that

$$\mathcal{R}e\, h > \mu \textit{ on } \partial(\mathrm{supp}\ \chi). \tag{VIII.2.16}$$

This will imply (VIII.2.8) simply by taking K_0 to be the set of points in supp χ at which $\mathcal{R}e\, h = \mu$ and Ω'' to be the interior of supp χ.

Suppose (VIII.2.16) were false. We would then have $\mathcal{R}e\, h(\mathfrak{p}) = \mu$ at some point $p \in \partial(\mathrm{supp}\,\chi)$, whence $L\chi \equiv 0$ in some open neighborhood of p, $U \subset \Omega$. If the orbit of $\mathcal{V}$ in Ω' through p, Σ_p, were an open subset of $\mathcal{M}$, it would follow from (VIII.2.15) that the orbit of $\mathcal{V}$ in U through p, which is a connected component of $\Sigma_p \cap U$, is an open set. By Theorem II.3.3 we would perforce have $\chi \equiv 0$ in a full neighborhood of p, a contradiction.

Let us therefore assume that the structure inherited by Σ_p from $\mathcal{M}$ is elliptic. By Proposition III.5.1 (see also Theorem VI.5.1) this structure is hypocomplex. Theorem III.5.3 states that the (local) maximum principle, for the real parts of solutions, holds in Σ_p. On the other hand, by virtue of the fact that the restriction of χ to Σ_p is a solution in $U \cap \Sigma_p$, and once again by Theorem II.3.3, we see that $\Sigma_p \cap \mathrm{supp}\,\chi$ is a neighborhood $\mathcal{N}$ of p in Σ_p. We have $\mathcal{R}e\,h \geq \mu = \mathcal{R}e\,h(p)$ in $\mathcal{N}$, which implies $\mathcal{R}e\,h = \mu$ everywhere in Σ_p. But then $L\chi \equiv 0$ in a full neighborhood $\mathcal{O}$ of Σ_p. On the other hand, if $p_1 \in \Sigma_p$ there are points p_2 arbitrarily close to p_1 such that $\chi \neq 0$ at some point (located near p) of the orbit of $\mathcal{V}$ in $\mathcal{O}$ through p_2, Σ_{p_2}. Since the involutive structure inherited by Σ_{p_2} from $\mathcal{M}$ is elliptic or else Σ_{p_2} is an open set, the support of the restriction of χ to Σ_{p_2} is perforce equal to the whole of Σ_{p_2}. We conclude that $p_1 \in \mathrm{supp}\,\chi$, i.e., $\Sigma_p \subset \mathrm{supp}\,\chi$, which would demand that the closure of Σ_p be a compact subset of Ω', contrary to hypothesis (VIII.2.14). ■

Corollary VIII.2.3. *Suppose $\mathcal{V}$ is of finite type in Ω'* (Definition I.11.2). *Then* (VIII.2.9) *entails* (VIII.2.8).

Proof. If $\mathcal{V}$ is of finite type in Ω', the only orbit of $\mathcal{V}$ in any connected open subset $\mathcal{O}$ of Ω' is the set $\mathcal{O}$ itself. From this (VIII.2.14) and (VIII.2.15) follow at once. As shown, (VIII.2.9) entails (VIII.2.16). ■

When $\mathcal{M}$ and $\mathcal{V}$ are of class $\mathcal{C}^\omega$, Theorem I.11.3 applies.

Corollary VIII.2.4. *Assume that the manifold $\mathcal{M}$ and the vector bundle $\mathcal{V}$ are both real-analytic. Suppose, moreover, that in addition to* (VIII.2.14), *the following condition is satisfied:*

the dimension of the characteristic set, at any point of Ω', is ≤ 1. (VIII.2.17)

Then (VIII.2.9) *entails* (VIII.2.8).

Proof. That (VIII.2.15) holds is an immediate consequence of the fact that the orbits of $\mathcal{V}$ in $\mathcal{O}$ are the *integral manifolds* of the Lie algebra $\mathfrak{g}(\mathcal{V},\mathcal{O})$ generated by the vector fields $\mathcal{R}e\,L$, $L \in \mathcal{C}^\infty(\mathcal{O};\mathcal{V})$ (see section I.11). Let then Σ be an orbit of $\mathcal{V}$ in Ω'. Its conormal bundle is orthogonal to all the vector fields $\mathcal{R}e\,L$ with L a section of $\mathcal{V}$ over Σ, and thus the conormal bundle is contained in the characteristic set T°. As a consequence, $\mathrm{codim}\,\Sigma \leq 1$. If $\mathrm{codim}\,\Sigma = 1$

the characteristic set along Σ is entirely contained in the conormal bundle of Σ, whence it follows that the involutive structure inherited by Σ from $\mathcal{M}$ is elliptic. ■

VIII.3. A Necessary Condition for Local Exactness Based on the Levi Form

We define a locally integrable structure on an open neighborhood $\mathcal{O}$ of the origin in $\mathbb{C}^\nu \times \mathbb{R}^d \times \mathbb{R}^{n'}$ by means of the functions z_i $(1 \leq i \leq \nu)$ and w_k $(1 \leq k \leq d)$ given by (I.7.4) with (I.7.5) holding. We modify slightly the notation and write $\varphi_k(z,s,t)$ rather than $\varphi_k(x,y,s,t)$. As pointed out following (I.5.14), the characteristic set at the origin T_0° is spanned by $\mathrm{d}s_1,\ldots,\mathrm{d}s_d$; any characteristic covector at the origin will have an expression $\sigma \cdot \mathrm{d}s$ with $\sigma \in \mathbb{R}^d$. We recall that the Levi form at a characteristic point $(0,\sigma\cdot\mathrm{d}s)$ is a quadratic form on the fibre at the origin, $\mathcal{V}_0$, of the tangent structure bundle $\mathcal{V}$.

Theorem VIII.3.1. *Suppose $q \geq 1$ and that there is $\sigma \in \mathbb{R}^d$ such that*

the Levi form at $(0,\sigma\cdot\mathrm{d}s)$ has q eigenvalues > 0 and $n-q$ eigenvalues < 0, and its restriction to $\mathcal{V}_0 \cap \overline{\mathcal{V}}_0$ is nondegenerate. (VIII.3.1)

Then, given any two sufficiently small open neighborhoods $\mathrm{V} \subset \mathrm{U} \subset\subset \mathcal{O}$ *of the origin, there is a cocycle* $f \in \mathcal{C}^\infty(\mathrm{U};\Lambda^{0,q})$ *such that no section* $u \in \mathcal{C}^\infty(\mathrm{V};\Lambda^{0,q-1})$ *satisfies* $\mathrm{d}'u = f$ *in* V.

Proof. Our starting point will be formula (I.9.19); thanks to Proposition I.9.1 (cf. Remark I.9.2) all the coefficients a_{jk} vanish. Because of the different meaning of q in the present context we shall modify the notation: we shall write

$$z' = (z_1,\ldots,z_\mu),\ z'' = (z_{\mu+1},\ldots,z_\nu),$$

$$t' = (t_1,\ldots,t_{\mu'}),\ t'' = (t_{\mu'+1},\ldots,t_{n'}).$$

We shall then rewrite formula (I.9.19) as follows:

$$\sigma_1\varphi_1 + \cdots + \sigma_d\varphi_d =$$

$$|z'|^2 - |z''|^2 + |t'|^2 - |t''|^2 + \mathrm{O}(|z|^3 + |t|^3 + |s|(|z| + |s| + |t|)). \quad \text{(VIII.3.2)}$$

In accordance with the considerations in section I.9 and with our hypothesis (VIII.3.1), the number of eignevalues of the levi form that are > 0 is equal to

$$q = \mu + \mu'. \quad \text{(VIII.3.3)}$$

It is convenient to effect right away a linear change of the variables s_k, and a concomitant change of the functions w_k, which transforms σ into the vector $(0,\ldots,0,1)$. Let ε then be a small number > 0. We carry out the change of scale

$$z_i = \varepsilon\tilde{z}_i\,(1 \le i \le \nu),\ s_j = \varepsilon^2\tilde{s}_j\,(1 \le j \le d),\ t_k = \varepsilon\tilde{t}_k\,(1 \le k \le n-\nu),$$

and we set $\tilde{w}_k = w_k/\varepsilon^2$. After deleting the tildas we may write

$$w_k = s_k + \imath[\mathcal{Q}_k(z,t) + \varepsilon\mathfrak{R}_k(z,t,s,\varepsilon)], \tag{VIII.3.4}$$

where $\mathcal{Q}_k$ is a quadratic form and

$$|\mathfrak{R}_k(z,t,s,\varepsilon)\,| \le const.(|z|^3 + |t|^3 + |s|(|z| + |s| + |t|)) \tag{VIII.3.5}$$

$(1 \le k \le d$; the constant is independent of $\varepsilon)$. According to (VIII.3.2), we have

$$\mathcal{Q}_d(z,t) = |z'|^2 - |z''|^2 + |t'|^2 - |t''|^2. \tag{VIII.3.6}$$

Now let ε' be another positive number. We define

$$h = w_d + \imath\varepsilon'\sum_{k=1}^{d} w_k^2. \tag{VIII.3.7}$$

We have

$$\mathcal{I}m\,h = \mathcal{Q}_d + \varepsilon'|s|^2 + \varepsilon\mathfrak{R}_d - \varepsilon'\sum_{k=1}^{d}(\mathcal{Q}_k + \varepsilon\mathfrak{R}_k)^2.$$

This implies straightaway

$$\left|\mathcal{I}m\,h - \mathcal{Q}_d - \varepsilon'|s|^2\right| \le C\left\{\varepsilon(|z|^2 + |s|^2 + |t|^2) + \varepsilon'(|z|^4 + |s|^4 + |t|^4)\right\},$$

where C is a suitably large positive constant. We require

$$|z|^2 + |t|^2 < 1,\ C|s|^2 < 1/4 \text{ in } \mathrm{U}. \tag{VIII.3.8}$$

Then

$$\left|\mathcal{I}m\,h - \mathcal{Q}_d - \varepsilon'|s|^2\right| \le (C\varepsilon + \varepsilon'/4)|s|^2 + C(\varepsilon + \varepsilon')(|z|^2 + |t|^2).$$

We choose

$$C\varepsilon < \varepsilon'/4,\ C(\varepsilon + \varepsilon') < 1/2. \tag{VIII.3.9}$$

We obtain, in U,

$$\tfrac{1}{2}(|z'|^2 + \varepsilon'|s|^2 + |t'|^2) - \tfrac{3}{2}(|z''|^2 + |t''|^2) \le$$

$$\mathcal{I}m\,h \le \tfrac{3}{2}(|z'|^2 + \varepsilon'|s|^2 + |t'|^2) - \tfrac{1}{2}(|z''|^2 + |t''|^2). \tag{VIII.3.10}$$

Next we introduce the following functions:

$$h_1 = -\imath h - 4(|z'|^2+|t'|^2) - 4\sum_{k=1}^{d} w_k^2,$$

$$h_2 = \imath h - 4(|z''|^2+|t''|^2).$$

We have (cf. (I.7.15), (I.7.16)):

$$L_j h_1 = 0 \text{ if } \mu+1 \le j \le \nu \text{ or if } \nu+\mu'+1 \le j \le n; \quad \text{(VIII.3.11)}$$

$$L_j h_2 = 0 \text{ if } 1 \le j \le \mu \text{ or if } \nu+1 \le j \le \mu'. \quad \text{(VIII.3.12)}$$

Moreover, according to (VIII.3.10) and provided U and the constant $a > 0$ are sufficiently small, we have

$$\mathcal{R}e\, h_i(z,s,t) \le -a(|z|^2+|s|^2+|t|^2),$$
$$\forall\,(z,s,t) \in U \ (i = 1,2). \quad \text{(VIII.3.13)}$$

Let $V \subset\subset U$ then be an open neighborhood of the origin; toward the end of the proof we shall determine how small the number $\delta = \operatorname*{Max}_V (|z|^2+|s|^2+|t|^2)$ must be. We introduce a cutoff function $\chi \in \mathcal{C}_c^\infty(V)$, $\chi \equiv 1$ in some neighborhood of 0. We set, for each $\rho > 0$,

$$f_\rho = e^{\rho h_1} d\bar{z}_1\wedge\cdots\wedge d\bar{z}_\mu\wedge dt_1\wedge\cdots\wedge dt_{\mu'},$$

$$v_\rho = \rho^{\frac{1}{2}(m+n)} e^{\rho h_2}\chi d\bar{z}_{\mu+1}\wedge\cdots\wedge d\bar{z}_\nu\wedge dt_{\mu'+1}\wedge\cdots\wedge dt_{n'}\wedge dz\wedge dw$$

($dz = dz_1\wedge\cdots\wedge dz_\nu$, $dw = dw_1\wedge\cdots\wedge dw_d$); f_ρ defines an equivalence class in $\mathcal{C}^\infty(U;\Lambda^{0,q})$, v_ρ one in $\mathcal{C}_c^\infty(V;\Lambda^{m,n-q})$ (we shall denote the equivalence classes by f_ρ and v_ρ like their representatives; recall that $q = \mu+\mu'$). It follows at once from (VIII.3.11) that $d'f_\rho = 0$; and from (VIII.3.12), that $d'v_\rho$ is defined by the form

$$\rho^{\frac{1}{2}(m+n)} e^{\rho h_2} d'\chi\wedge d\bar{z}_{\mu+1}\wedge\cdots\wedge d\bar{z}_\nu\wedge dt_{\mu'+1}\wedge\cdots\wedge dt_{n'}\wedge dz\wedge dw.$$

Thus, on supp $d'v_\rho$ we have $|z|^2+|s|^2+|t|^2 \ge a_0$ and therefore, by (VIII.3.13), $\mathcal{R}e\, h_2 \le -b$ (a_0, $b > 0$). We further derive from (VIII.3.13) that given any compact subset K of U, with $V \subset K$, and any integer $\ell \ge 0$, we have

$$\|f_\rho\|_{K,\ell}\|d'v_\rho\|_{K,\ell} \le const.\ \rho^{\frac{1}{2}(m+n)+2\ell} e^{-b\rho} \quad \text{(VIII.3.14)}$$

(the norms are those introduced in section VIII.1).

On the other hand,

$$\int f_\rho\wedge v_\rho = c\rho^{\frac{1}{2}(m+n)} \int e^{\rho(h_1+h_2)}\chi\Delta\ dxdydsdt, \quad \text{(VIII.3.15)}$$

where $\Delta = \det(I_d+\imath\varphi_s)$ and $0 \ne c \in \mathbb{R}$ (I_d : $d\times d$ identity matrix). We make the change of variables $(x,y,s,t) \to \rho^{-\frac{1}{2}}(x,y,s,t)$. Observe that

$$H(x,y,s,t,\rho) = \rho(h_1+h_2)(\rho^{-\frac{1}{2}}x,\rho^{-\frac{1}{2}}y,\rho^{-\frac{1}{2}}s,\rho^{-\frac{1}{2}}t) =$$
$$-\ 4(|z|^2+|s|^2+|t|^2) + 4\rho|\varphi(\rho^{-\frac{1}{2}}x,\rho^{-\frac{1}{2}}y,\rho^{-\frac{1}{2}}s,\rho^{-\frac{1}{2}}t)|^2 -$$
$$8\,\iota\rho^{\frac{1}{2}}s\cdot\varphi(\rho^{-\frac{1}{2}}x,\rho^{-\frac{1}{2}}y,\rho^{-\frac{1}{2}}s,\rho^{-\frac{1}{2}}t).$$

If we take into account the fact that φ and $d\varphi$ vanish at the origin, we conclude that, as $\rho \to +\infty$, $H(x,y,s,t,\rho)$ converges pointwise to $-4(|z|^2+|s|^2+|t|^2)$. On the other hand, since φ and $d\varphi$ vanish at 0, we have

$$\rho|\varphi(\rho^{-\frac{1}{2}}x,\rho^{-\frac{1}{2}}y,\rho^{-\frac{1}{2}}s,\rho^{-\frac{1}{2}}t)|^2 \leq C(|z|^2+|s|^2+|t|^2)^2/\rho$$

with a constant $C > 0$ that depends only on a bound for the second partial derivatives of φ. We make use of the fact that $(|z|^2+|s|^2+|t|^2)/\rho \leq \delta$ on the support of $\chi(\rho^{-\frac{1}{2}}x,\rho^{-\frac{1}{2}}y,\rho^{-\frac{1}{2}}s,\rho^{-\frac{1}{2}}t)$, to obtain

$$\rho|\varphi(\rho^{-\frac{1}{2}}x,\rho^{-\frac{1}{2}}y,\rho^{-\frac{1}{2}}s,\rho^{-\frac{1}{2}}t)|^2 \leq C\delta(|z|^2+|s|^2+|t|^2).$$

We require $C\delta \leq 1/2$; we reach the following conclusion:

$$\mathcal{R}e H(x,y,s,t,\rho) \leq -\ 2(|z|^2+|s|^2+|t|^2) \qquad \text{(VIII.3.16)}$$

on the support of $\chi(\rho^{-\frac{1}{2}}x,\rho^{-\frac{1}{2}}y,\rho^{-\frac{1}{2}}s,\rho^{-\frac{1}{2}}t)$.

We are in a position to apply Lebesgue's dominated convergence theorem to the integral (VIII.3.15): the integrand converges pointwise to $e^{-4(|z|^2+|s|^2+|t|^2)}$ since $\chi(0)\Delta(0) = 1$; the absolute value of the integrand does not exceed

$$const.\ e^{-2(|z|^2+|s|^2+|t|^2)}.$$

We conclude that

$$\int f_\rho \wedge v_\rho \to c\int e^{-4(|z|^2+|s|^2+|t|^2)}dxdydsdt \neq 0.$$

Combining this fact with (VIII.3.14) yields a contradiction of $(\text{VIII.1.9})_{m,q}$; therefore, by Lemma VIII.1.1, Property $(\text{VIII.1.8})_{m,q}$ cannot hold with $\Omega = U$, $\Omega' = V$. ■

VIII.4. A Result about Structures whose Characteristic Set has Rank at Most Equal to One

Let 0 be a point of $\mathcal{M}$ (referred to in the sequel as the origin). In the remainder of the chapter we shall be concerned with the validity of the property now defined:

DEFINITION VIII.4.1. *We say that the differential equation*

$$d'u = f \qquad \text{(VIII.4.1)}$$

is locally solvable at the point 0 in degree *q if, given any open neighborhood of* 0 *in* $\mathcal{M}$, Ω, *there is an open neighborhood of* 0, $\Omega' \subset \Omega$, *such that* (VIII.1.8)$_{0,q}$ *holds.*

Since $\Lambda^{p,q} \cong \Lambda^{p,0} \otimes \Lambda^{0,q}$ the local solvability at 0 in degree q implies the same "in bidegree" (p,q) for any $p \geq 1$.

In the next three sections we limit our attention to locally integrable structures in which the following hypothesis is satisfied:

$$\textit{The rank of the characteristic set at the origin, } T_0^\circ, \textit{ is } \leq 1. \qquad \text{(VIII.4.2)}$$

When the dimension of T_0° is zero, the structure is elliptic in a full neighborhood of the origin; in this case the results presented here are all trivial. We shall therefore assume

$$\dim T_0^\circ = 1. \qquad \text{(VIII.4.3)}$$

Let w then be *any* $\mathscr{C}^\infty$ solution in some open neighborhood Ω of 0 in $\mathcal{M}$ whose differential spans the characteristic set at the origin, T_0° (Ω will not be changed in what follows). Select $m-1 = \nu$ other $\mathscr{C}^\infty$ solutions $z_1,\ldots,z_\nu$ in some open neighborhood of 0 such that $dz_1 \wedge \cdots \wedge dz_\nu \wedge dw \neq 0$ at 0 (we write $z = (z_1,\ldots,z_\nu)$). These hypotheses are equivalent to the validity, *at the origin,* of the following property:

$$dz_1 \wedge \cdots \wedge dz_\nu \wedge d\bar{z}_1 \wedge \cdots \wedge d\bar{z}_\nu \wedge d(\mathcal{R}e\, w) \neq 0,\ d(\mathcal{I}m\, w) = 0. \qquad \text{(VIII.4.4)}$$

We take $x_i = \mathcal{R}e\, z_i$, $y_j = \mathcal{I}m\, z_j$ $(1 \leq i, j \leq \nu)$ and $s = \mathcal{R}e\, w$ as coordinates, and adjoin to them coordinates $t_1,\ldots,t_n$, to obtain a complete system of coordinates in Ω (possibly contracted about 0). By adding constants we may assume that all the coordinates, as well as $\mathcal{I}m\, w$, vanish at 0. The differentials of the functions

$$z_j = x_j + \imath y_j \ (1 \leq j \leq \nu),\ w = s + \imath\varphi(z,s,t) \qquad \text{(VIII.4.5)}$$

span T′ over Ω; we have

$$\varphi|_0 = 0,\ d\varphi|_0 = 0. \qquad \text{(VIII.4.6)}$$

The characteristic set at the origin is spanned by $ds|_0$. From this we can construct a commuting system of vector fields that span $\mathcal{V}$ over Ω (cf. section I.7):

$$\begin{aligned} L_j &= \partial/\partial\bar{z}_j - \imath\varphi_{\bar{z}_j} N,\ j = 1,\ldots,\nu; \\ L_{\nu+\ell} &= \partial/\partial t_\ell - \imath\varphi_{t_\ell} N,\ \ell = 1,\ldots,n' = n-\nu, \end{aligned} \qquad \text{(VIII.4.7)}$$

where $N = (1 + \imath\varphi_s)^{-1}\partial/\partial s$.

A section f of $\Lambda^{0,q}$ (over Ω) is represented by a unique differential form or current, depending on whether the coefficients $f_{J,K}$ are functions or distributions,

$$f = \sum_{|J|+|K|=q} f_{J,K} \mathrm{d}\bar{z}_J \wedge \mathrm{d}t_K, \tag{VIII.4.8}$$

and the action of the operator d′ on the class $\dot{f}$ of f is represented by

$$\mathrm{L}f = \sum_{|J|+|K|=q} \left\{ \sum_{j=1}^{\nu} \mathrm{L}_j f_{J,K} \mathrm{d}\bar{z}_j + \sum_{k=1}^{n-\nu} \mathrm{L}_{\nu+k} f_{J,K} \mathrm{d}t_k \right\} \wedge \mathrm{d}\bar{z}_J \wedge \mathrm{d}t_K \tag{VIII.4.9}$$

(cf. (I.7.34), (I.7.35)). We recall that, since $\Lambda^{m,q} = \mathrm{T}'^{m,q}$, a section of $\Lambda^{m,q}$ is truly a form or a current, $F = f \wedge \mathrm{d}z \wedge \mathrm{d}w$ (we shall always write $\mathrm{d}z = \mathrm{d}z_1 \wedge \cdots \wedge \mathrm{d}z_\nu$) with f as in (VIII.4.8); and $\mathrm{d}'F = \mathrm{d}F = \mathrm{L}F = \mathrm{L}f \wedge \mathrm{d}z \wedge \mathrm{d}w$ with $\mathrm{L}f$ given by (VIII.4.9).

In the remainder of this section we regard the open set $\Omega \subset \mathcal{M}$ as a $\mathscr{C}^\infty$ manifold equipped with the hypo-analytic structure defined by the first integrals (VIII.4.5). A hypo-analytic function in a neighborhood of a point $\mathfrak{p} \in \Omega$ can be equated, in a possibly smaller neighborhood of $\mathfrak{p}$, to a holomorphic function of (z,w).

We denote by $\mathscr{HA}$ the *sheaf of germs of hypo-analytic functions* in Ω: $\mathscr{HA}$ is the pullback under the map (z,w): $\Omega \ni (x,y,s,t) \to (x+\imath y, s+\imath\varphi(z,s,t)) \in \mathbb{C}^m$ of the sheaf of germs of holomorphic functions in $\mathbb{C}^m$. And we shall denote by $\mathscr{HA}^m$ the sheaf of germs of differential forms $\mathrm{d}f_1 \wedge \cdots \wedge \mathrm{d}f_m$ with $f_1,\ldots,f_m$ hypo-analytic functions; $\mathscr{HA}^m$ is the pullback under the map (z,w) of the sheaf of germs of closed differential forms of type $(m,0)$, i.e., forms $f\ \mathrm{d}z \wedge \mathrm{d}w$ with f holomorphic.

We shall look at the *level sets* of w, i.e., sets such as

$$\mathcal{S} = \{ (z,s,t) \in \Omega;\ s = \mathscr{R}e\, w_0,\ \varphi(z,s,t) = \mathscr{I}m\, w_0 \}\ (w_0 \in \mathbb{C}). \tag{VIII.4.10}$$

It will be convenient, in the sequel, to assume that $\mathcal{S}$ is a *noncritical level set* of w, which will mean that the two-form $\mathrm{d}w \wedge \mathrm{d}\bar{w}$ does not vanish at any point of $\mathcal{S}$ or, equivalently, that $\mathscr{I}m\, w_0$ is a noncritical value of the function $(z,t) \to \varphi(z, \mathscr{R}e\, w_0, t)$ (by Sard's lemma the noncritical values are dense among all the values).

Let $\mathcal{N}_{\mathcal{S}}$ then denote the sheaf of germs of hypo-analytic functions in $\mathcal{M}$ that vanish on $\mathcal{S}$. We shall call $\mathscr{HA}_{\mathcal{S}}$ the restriction to $\mathcal{S}$ of the quotient sheaf $\mathscr{HA}/\mathcal{N}_{\mathcal{S}}$. Likewise let $\mathcal{N}^m_{\mathcal{S}}$ denote the sheaf of germs of differential forms $\mathrm{d}f_1 \wedge \cdots \wedge \mathrm{d}f_m$ that vanish on $\mathcal{S}$, with $f_1,\ldots,f_m$ hypo-analytic functions. Note that every section of $\mathscr{HA}^m|_\Omega$ is divisible by $\mathrm{d}w$. We shall call $\mathscr{HA}^\nu_{\mathcal{S}}$ the pull-back to $\mathcal{S}$ of the quotient sheaf $(\mathscr{HA}^m|_\Omega/\mathrm{d}w)/(\mathcal{N}^m_{\mathcal{S}}/\mathrm{d}w)$ (recall that $\nu = m - 1$).

Below we consider a polydisk $\Delta(z_0,r) = \{ z \in \mathbb{C}^\nu;\ |z_i - z_{0_i}| < r,\ i = 1, \ldots, \nu \}$ and an open ball $\mathscr{B} = \{ t \in \mathbb{R}^{n'};\ |t| < r' \}$ (with $r, r' > 0$).

LEMMA VIII.4.1. *Let $\mathcal{S}$ be a noncritical level set of w and let $(z_0,s_0,t_0) \in \mathcal{S}$. Suppose that $\Delta(z_0,r) \times \{s_0\} \times \mathscr{B} \subset \Omega$. If a holomorphic function of z in the polydisk $\Delta(z_0,r)$, $\tilde{h}$, vanishes identically in the set $\{ z \in \Delta(z_0,r);\ \exists\ t \in \mathscr{B},\ \varphi(z,s_0,t) = \mathscr{I}m\, w_0 \}$, then $\tilde{h} \equiv 0$ in $\Delta(z_0,r)$.*

PROOF. We regard $\tilde{h}$ as a smooth function in (z,t)-space $\Delta(z_0,r)\times\mathcal{B}$, which satisfies $\bar{\partial}_z\tilde{h} + d_t\tilde{h} = 0$. Since $\mathcal{S}$ is noncritical the equation $\varphi(z,s_0,t) = \mathcal{I}m\, w_0$ defines a smooth hypersurface in $\Delta(z_0,r)\times\mathcal{B}$. The claim then follows by uniqueness in the Cauchy problem (Corollary II.3.7). ∎

Denote by $\mathcal{A}$ the sheaf of germs of functions in Ω which are holomorphic with respect to z and are independent of (s,t); and by $\mathcal{A}^\nu$ the sheaf of germs of ν-forms $f\, dz_1\wedge\cdots\wedge dz_\nu$ with f a holomorphic function of z (regarded as a function in Ω).

PROPOSITION VIII.4.1. *If $\mathcal{S}$ is a noncritical level set of w we have natural sheaf isomorphisms*

$$\mathcal{HA}_{\mathcal{S}} \cong \mathcal{A}|_{\mathcal{S}},\ \mathcal{HA}^\nu_{\mathcal{S}} \cong \mathcal{A}^\nu|_{\mathcal{S}}. \tag{VIII.4.11}$$

PROOF. By Lemma VIII.4.1 we have an injection $\mathcal{A}|_{\mathcal{S}} \to \mathcal{HA}_{\mathcal{S}}$. Let $\tilde{h}(z,w)$ be a hypo-analytic function in a neighborhood V of a point $p \in \mathcal{S}$; then $\tilde{h}(z,w) - \tilde{h}(z,w_0)$ defines a continuous secton of $\mathcal{N}_{\mathcal{S}}$ over $V\cap\mathcal{S}$ while $\tilde{h}(z,w_0)$ defines one of $\mathcal{A}$. The same argument applies with superscripts ν. ∎

In the sequel, unless otherwise specified, it is assumed that the level set $\mathcal{S}$ in Ω is noncritical. We may define the following *hypo-analytic homology and cohomology* spaces on $\mathcal{S}$, for $q = 0,1,\ldots,n-1$:

$${}_a\mathrm{H}_q(\mathcal{S}) = \mathrm{H}_c^{n-1-q}(\mathcal{S},\mathcal{HA}^\nu_{\mathcal{S}});\ {}_a\mathrm{H}^q(\mathcal{S}) = \mathrm{H}^q(\mathcal{S},\mathcal{HA}_{\mathcal{S}}) \tag{VIII.4.12}$$

(the subscript c indicates that we are considering the cohomology with compact support; see H. Cartan [1], Godement [1]). It follows from Proposition VIII.4.1 that

$${}_a\mathrm{H}_q(\mathcal{S}) \cong \mathrm{H}_c^{n-1-q}(\mathcal{S},\mathcal{A}^\nu|_{\mathcal{S}});\ {}_a\mathrm{H}^q(\mathcal{S}) \cong \mathrm{H}^q(\mathcal{S},\mathcal{A}|_{\mathcal{S}}). \tag{VIII.4.13}$$

We can of course embed $\mathcal{S}$ into (z,t)-space. As before write $s_0 = \mathcal{R}e\, w_0$ and set $\Omega_0 = \{(z,t) \in \mathbb{C}^\nu\times\mathbb{R}^{n'};\ (z,s_0,t) \in \Omega\}$; then $\mathcal{S} \cong \mathcal{S}_0 = \{(z,t) \in \Omega_0;\ \varphi(z,s_0,t) = \mathcal{I}m\, w_0\}$. This embedding of $\mathcal{S}$ allows us to make use of a special cohomological resolution of $\mathcal{A}$, to take into account the dependence of φ on t. We introduce the vector bundle $\tilde{\Lambda}^{0,q}$ over Ω_0 whose sections are differential forms (VIII.4.8) (their coefficients depend only on z, t); and the vector bundle $\tilde{\Lambda}^{\nu,q}$ whose sections are differential forms $f\wedge dz$ with f a section of $\tilde{\Lambda}^{0,q}$ $(0 \leq q \leq n)$. Throughout the sequel $p = 0$ or ν. Let $\mathcal{C}^\infty\tilde{\Lambda}^{p,q}$ denote the sheaf of $\mathcal{C}^\infty$ sections of $\tilde{\Lambda}^{p,q}$; the differential operator $\mathrm{D} = \bar{\partial}_z + d_t$ defines a sheaf homomorphism $\mathcal{C}^\infty\tilde{\Lambda}^{p,q} \to \mathcal{C}^\infty\tilde{\Lambda}^{p,q+1}$. Regarding $\mathcal{A}$ and $\mathcal{A}^\nu$ as sheaves over Ω_0 we have the *De Rham-Dolbeault* resolution:

$$0 \to \mathcal{A}^p \xrightarrow{i} \mathcal{C}^\infty\tilde{\Lambda}^{p,0} \xrightarrow{\mathrm{D}} \mathcal{C}^\infty\tilde{\Lambda}^{p,1} \xrightarrow{\mathrm{D}} \cdots \xrightarrow{\mathrm{D}} \mathcal{C}^\infty\tilde{\Lambda}^{p,n} \to 0 \tag{VIII.4.14}$$

(i: natural injection; $\mathscr{A}^\circ = \mathscr{A}$; of course every sheaf $\mathscr{C}^\infty\tilde{\Lambda}^{p,q}$ is fine).

One can pullback to $\mathscr{S}_0$ the sheaves $\mathscr{C}^\infty\tilde{\Lambda}^{p,q}$; (VIII.4.12) and (VIII.4.13) yield isomorphisms

$$ {}_a\mathrm{H}_q(\mathscr{S}) \cong $$
$$ \{ \sigma \in \Gamma_c(\mathscr{S}_0;\mathscr{C}^\infty\tilde{\Lambda}^{\nu,n-q-1});\ \mathrm{D}\sigma = 0 \}/\{ \mathrm{D}\tau;\ \tau \in \Gamma_c(\mathscr{S}_0;\mathscr{C}^\infty\tilde{\Lambda}^{\nu,n-q-2}) \}; \tag{VIII.4.15} $$

$$ {}_a\mathrm{H}^q(\mathscr{S}) \cong $$
$$ \{ \sigma \in \Gamma(\mathscr{S}_0;\mathscr{C}^\infty\tilde{\Lambda}^{0,q});\ \mathrm{D}\sigma = 0 \}/\{ \mathrm{D}\tau;\ \tau \in \Gamma(\mathscr{S}_0;\mathscr{C}^\infty\tilde{\Lambda}^{0,q-1}) \}, \tag{VIII.4.16} $$

where $\Gamma(\mathscr{S}_0;\mathscr{F})$ denotes the space of sections over $\mathscr{S}_0$ of a sheaf $\mathscr{F}$ and $\Gamma_c(\mathscr{S}_0;\mathscr{F})$ the subspace of compactly supported such sections. The denominator at the right in (VIII.4.15) vanishes when $q = n-1$, whereas it vanishes in (VIII.4.16) when $q = 0$.

We may also introduce the reduced hypo-analytic homology of $\mathscr{S}$ in dimension zero. We notice in (VIII.4.15) that a representative of a class in ${}_a\mathrm{H}_0(\mathscr{S})$ can be identified to a compactly supported, smooth, D-closed form

$$ u = \sum_{|J|+|K|=n-1} u_{J,K}\mathrm{d}z\wedge \mathrm{d}\bar{z}_J\wedge \mathrm{d}t_K $$

whose degree, $\nu+n-1 = 2\nu+n'-1$, is precisely equal to dim $\mathscr{S}$. We have therefore the right to integrate over $\mathscr{S}_0$ any product hu with $h \in \mathscr{C}^\infty(\Omega_0)$.

Definition VIII.4.2. *By the* reduced hypo-analytic homology in dimension zero *of the noncritical level set* $\mathscr{S}$ *we shall mean the quotient modulo* $\{ \mathrm{D}\tau;\ \tau \in \Gamma_c(\mathscr{S}_0;\mathscr{C}^\infty\tilde{\Lambda}^{\nu,n-2}) \}$ *of the linear space of sections* $\sigma \in \Gamma_c(\mathscr{S}_0;\mathscr{C}^\infty\tilde{\Lambda}^{\nu,n-1})$ *that satisfy* $\mathrm{D}\sigma = 0$ *as well as*

$$ \int_{\mathscr{S}_0} i^*(h\sigma) = 0 \text{ if } h \in \Gamma(\Omega_0;\mathscr{A}). $$

We shall denote by ${}_a\tilde{\mathrm{H}}_0(\mathscr{S},\Omega)$ *the quotient space thus defined.*

(We have denoted by i^* the pullback map to $\mathscr{S}_0$.)

The above remark allows us also to integrate over $\mathscr{S}_0$ any cup product $u\wedge v$ with $u \in \Gamma(\mathscr{S}_0;\ \mathscr{C}^\infty\tilde{\Lambda}^{0,q})$ & $v \in \Gamma_c(\mathscr{S}_0;\ \mathscr{C}^\infty\tilde{\Lambda}^{\nu,n-q-1})$ (see Godement [1], chap.2, 6.6). This leads to the following definition:

Suppose $q \geq 1$ and let $[\beta] \in {}_a\mathrm{H}^q(\mathscr{S})$, $[\gamma] \in {}_a\mathrm{H}_q(\mathscr{S})$ be two classes with representatives $\beta \in \Gamma(\mathscr{S}_0;\ \mathscr{C}^\infty\tilde{\Lambda}^{0,q})$, $\gamma \in \Gamma_c(\mathscr{S}_0;\mathscr{C}^\infty\tilde{\Lambda}^{\nu,n-q-1})$ ($\mathrm{D}\beta = 0$, $\mathrm{D}\gamma = 0$). We set

$$ I^q_{\mathscr{S}}([\beta],[\gamma]) = \int_{\mathscr{S}_0} i^*(\beta\wedge\gamma). \tag{VIII.4.17} $$

When $q = 0$ we adopt the same definition (VIII.4.17) except that $[\gamma] \in {}_a\tilde{H}_0(\mathcal{S},\Omega)$ and the representative γ is "orthogonal" to all holomorphic functions of z (regarded as functions in Ω_0). We shall refer to the bilinear functional (VIII.4.17) as the *intersection number of $\mathcal{S}$ in dimension q*.

Actually we must "relativize" the preceding definition. We introduce two open subsets $V \subset U$ of Ω. We write $U_0 = \{ (z,t) \in \Omega_0;\ (z,s_0,t) \in U \}$, $V_0 = \{(z,t) \in \Omega_0;\ (z,s_0,t) \in V \}$. Sections $\beta \in \Gamma(\mathcal{S}_0 \cap U_0; \mathcal{C}^\infty\tilde{\Lambda}^{0,q})$, $\gamma \in \Gamma_c(\mathcal{S}_0 \cap V_0; \mathcal{C}^\infty\tilde{\Lambda}^{\nu,n-q-1})$ such that $D\beta = 0$, $D\gamma = 0$ (with γ "orthogonal" to all holomorphic functions of z in V_0 when $q = 0$) define respectively two classes $[\beta] \in {}_aH^q(\mathcal{S}\cap U)$, $[\gamma] \in {}_aH_q(\mathcal{S}\cap V)$ ($[\gamma] \in {}_a\tilde{H}_0(\mathcal{S}\cap V,V)$ when $q = 0$). We set

$$I^q_{U,V,\mathcal{S}}([\beta],[\gamma]) = \int_{\mathcal{S}_0} i^*(\beta\wedge\gamma). \qquad \text{(VIII.4.18)}$$

Since the integration in (VIII.4.18) is restricted to a compact subset of $\mathcal{S}_0 \cap V_0$, we might as well only require that the level set $\mathcal{S}$ be noncritical in V; this has the advantage of freeing the definition (VIII.4.18) from any dependence on Ω.

DEFINITION VIII.4.3. *Let* $V \subset U$ *be two open subsets of* Ω, $\mathcal{S}$ *a level set of w in* Ω *which is noncritical in* V, *q an integer*, $0 \le q \le n$. *We shall refer to the bilinear functional*

$$I^q_{U,V,\mathcal{S}}:\ {}_aH^q(\mathcal{S}\cap U) \times {}_aH_q(\mathcal{S}\cap V) \to \mathbb{C} \text{ if } 1 \le q \le n-1$$

$$\left[\text{resp., } I^0_{U,V,\mathcal{S}}:\ {}_aH^0(\mathcal{S}\cap U) \times {}_a\tilde{H}_0(\mathcal{S}\cap V,V) \to \mathbb{C} \text{ if } q = 0\right],$$

as the intersection number of $\mathcal{S}$ in dimension q relative to the pair (U,V).

The intersection number $I^q_{U,V,\mathcal{S}}$ is a hypo-analytic invariant attached to the triplet $(U,V,\mathcal{S})$.

DEFINITION VIII.4.4. *Let q be an integer*, $0 \le q \le n$. *We say that the hypo-analytic function w is* acyclic at 0 in dimension *q if to each open neighborhood of* 0, $U \subset \Omega$, *there is another open neighborhood of* 0, $V \subset U$, *such that the intersection number in dimension q of* $\mathcal{S}$ *relative to the pair* (U,V) *vanishes identically, whatever the level set* $\mathcal{S}$ *of w in* Ω *which is noncritical in* V.

Obviously, the acyclicity of w at the origin is a hypo-analytically invariant property of the germ at 0 of w and we shall apply the adjective "acyclic" to germs of hypo-analytic functions. Thus, if we stick to the hypo-analytic structure in Ω defined by the first integrals (VIII.4.5), we may state:

THEOREM VIII.4.1. *If the differential equation* $d'u = f$ *is locally solvable at the point* 0 *in degree* q, *then every germ at* 0 *of a hypo-analytic function whose differential spans the characteristic set at* 0 *is acyclic in dimension* q.

VIII.5. Proof of Theorem VIII.4.1

We begin by equating the nonvanishing of the intersection number of a level set of the hypo-analytic function w with a more directly usable property. Let $\mathrm{V} \subset \mathrm{U}$ be open subsets of Ω and let $\mathcal{S}$ be the level set $w = w_0$ in Ω, assumed to be noncritical in V. We conform to the notation in section VIII.4; thus $\mathcal{S}_0 = \{(z,t); (z,s_0,t) \in \mathcal{S}\}$ $(s_0 = \mathcal{R}e\, w_0)$. We write $r_0 = \mathcal{I}m\, w_0$ and

$$\mathrm{U}_0^+ = \{(z,t) \in \mathrm{U}_0;\ \varphi(z,s_0,t) > r_0\},\ \mathrm{U}_0^- = \{(z,t) \in \mathrm{U}_0;\ \varphi(z,s_0,t) < r_0\};$$

by hypothesis $d[\varphi(z,s_0,t)]$ does not vanish at any point of $\mathcal{S}_0 \cap \mathrm{V}_0$.

Let $q \in \mathbb{Z}_+$, $0 \le q \le n-1$; we shall be interested in the following property:

$$\exists\, f \in \mathcal{C}^\infty(\mathrm{U}_0;\tilde{\Lambda}^{0,q}),\ u \in \mathcal{C}_c^\infty(\mathrm{V}_0;\tilde{\Lambda}^{\nu,n-q-1}) \text{ such that } \quad (\mathrm{VIII.5.1})_q^+$$

$$\operatorname{supp} \mathrm{D}f \subset \mathrm{U}_0^-,\ \operatorname{supp} \mathrm{D}u \subset \mathrm{U}_0^+ \cap \mathrm{V}_0,\ \int f \wedge \mathrm{D}u \neq 0.$$

We recall that $\mathrm{D} = \bar{\partial}_z + \mathrm{d}_t$. By $(\mathrm{VIII.5.1})_q^-$ we shall mean the analogous property, with U_0^+ and U_0^- exchanged.

PROPOSITION VIII.5.1. *If either* $(\mathrm{VIII.5.1})_q^+$ *or* $(\mathrm{VIII.5.1})_q^-$ *holds, the intersection number of* $\mathcal{S}$ *in dimension* q *relative to the pair* (U,V) *does not vanish identically.*

PROOF. Suppose for instance that $(\mathrm{VIII.5.1})_q^+$ holds; then $\mathrm{D}f \equiv 0$ and $\mathrm{D}u \equiv 0$ in a full neighborhood of $\mathcal{S}_0 \cap \mathrm{U}_0$. The germs of f and of u at points of $\mathcal{S}_0$ define sections $\beta \in \Gamma(\mathcal{S}_0 \cap \mathrm{U}_0;\ \mathcal{C}^\infty\tilde{\Lambda}^{0,q})$, $\gamma \in \Gamma_c(\mathcal{S}_0 \cap \mathrm{V}_0;\ \mathcal{C}^\infty\tilde{\Lambda}^{\nu,n-q-1})$, which are D-closed. In turn β and γ define classes $[\beta] \in {}_a\mathrm{H}^q(\mathcal{S} \cap \mathrm{U})$, $[\gamma] \in {}_a\mathrm{H}_q(\mathcal{S} \cap \mathrm{V})$. By Stokes' theorem we have

$$\int_{\mathcal{S}_0} i^*(\beta \wedge \gamma) = \int_{\mathcal{S}_0} f \wedge u = \pm \int_{\mathrm{U}_0^+} f \wedge \mathrm{D}u \neq 0.$$

(The choice of the sign in front of the last integral depends on the choice of the orientation.) This proves the result when $q \ge 1$.

Suppose $q = 0$. In addition to what precedes we must show that u defines a reduced homology class in $\mathcal{S}_0 \cap \mathrm{V}_0$ (Definition VIII.4.2). Let h be an arbitrary holomorphic function of z in V_0. Keeping in mind that $\operatorname{supp} u \subset\subset \mathrm{V}_0$ we get, once again by Stokes' theorem,

$$\int_{\mathcal{S}_0} hu = \pm \int_{\mathrm{U}_0^-} h\mathrm{D}u = 0. \blacksquare$$

In order to prove a converse to Proposition VIII.5.1 it is convenient to assume that the geometries of U and V are very simple. We shall take

$$\mathrm{U} = \Delta \times \mathcal{I} \times \mathcal{B} \qquad \text{(VIII.5.2)}$$

with Δ (resp., $\mathcal{I}$, resp. $\mathcal{B}$) an open polydisk (resp., an open interval, resp., an open ball) in z-space $\mathbb{C}^\nu$ (resp., in the s-line $\mathbb{R}$, resp., in t-space $\mathbb{R}^{n'}$) centered at the origin; and likewise $\mathrm{V} = \Delta_1 \times \mathcal{I}_1 \times \mathcal{B}_1$.

PROPOSITION VIII.5.2. *Suppose* $\mathrm{U} = \Delta \times \mathcal{I} \times \mathcal{B}$, $\mathrm{V} = \Delta_1 \times \mathcal{I}_1 \times \mathcal{B}_1$. *If* $I^q_{\mathrm{U,V},\mathcal{S}}$ *does not vanish identically, then at least one of the two properties* (VIII.5.1)^+_q *or* (VIII.5.1)^-_q *holds.*

PROOF. Suppose there are D-closed sections $\beta \in \Gamma(\mathcal{S}_0 \cap \mathrm{U}_0; \mathcal{C}^\infty\tilde{\Lambda}^{0,q})$, $\gamma \in \Gamma_c(\mathcal{S}_0 \cap \mathrm{V}_0; \mathcal{C}^\infty\tilde{\Lambda}^{\nu,n-q-1})$ such that $\int_{\mathcal{S}_0} i^*(\beta \wedge \gamma) \neq 0$. To β and γ there correspond D-closed $\mathcal{C}^\infty$ sections f_0 and u_0 of the vector bundles $\tilde{\Lambda}^{0,q}$ and $\tilde{\Lambda}^{\nu,n-q-1}$ respectively, over one and the same open neighborhood $\mathcal{O}$ of $\mathcal{S}_0 \cap \mathrm{U}_0$. We can arrange matters in such a way that there will be $\chi \in \mathcal{C}^\infty(\mathrm{U}_0)$ with supp $\chi \subset \mathcal{O}$, $\chi \equiv 1$ in a neighborhood of $\mathcal{S}_0 \cap \mathrm{U}_0$ and (supp χ) $\cap$(supp u_0) $\subset\subset \mathrm{V}_0$. Define

$$f_1 = \chi f_0 \in \mathcal{C}^\infty(\mathrm{U}_0;\tilde{\Lambda}^{0,q}),\ u_1 = \chi u_0 \in \mathcal{C}^\infty_c(\mathrm{V}_0;\tilde{\Lambda}^{\nu,n-q-1}).$$

Then $\mathrm{D}f_1 \equiv 0$, $\mathrm{D}u_1 \equiv 0$ in a neighborhood of $\mathcal{S}_0 \cap \mathrm{U}_0$; moreover,

$$\int_{\mathcal{S}_0} f_1 \wedge u_1 \neq 0. \qquad \text{(VIII.5.3)}$$

Define $g = \mathrm{D}f_1$ in U_0^+, $g = -\mathrm{D}f_1$ in U_0^-. We have $g \in \mathcal{C}^\infty(\mathrm{U}_0;\tilde{\Lambda}^{0,q+1})$, $\mathrm{D}g \equiv 0$ in U_0; $g \equiv 0$ in a neighborhood of $\mathcal{S}_0 \cap \mathrm{U}_0$. But since $\mathrm{U}_0 = \Delta \times \mathcal{B}$, there is $h \in \mathcal{C}^\infty(\mathrm{U}_0;\tilde{\Lambda}^{0,q})$ such that $\mathrm{D}h = g$ in U_0. Define

$$f_2^+ = (f_1 - h)/2,\ f_2^- = (f_1 + h)/2;$$

thus $f_1 = f_2^+ + f_2^-$. Moreover $\mathrm{D}f_2^+ = (\mathrm{D}f_1 - g)/2$ vanishes identically in a neighborhood of $\mathcal{C}\ell\mathrm{U}_0^+$; likewise, $\mathrm{D}f_2^- \equiv 0$ in a neighborhood of $\mathcal{C}\ell\mathrm{U}_0^-$. In view of (VIII.5.3) we cannot have $\int_{\mathcal{S}_0} f_2^+ \wedge u_1 = \int_{\mathcal{S}_0} f_2^- \wedge u_1 = 0$. Suppose, for instance, that

$$0 \neq \int_{\mathcal{S}_0} f_2^+ \wedge u_1 = \pm \int_{\mathrm{U}_0^+} f_2^+ \wedge \mathrm{D}u_1 \qquad \text{(VIII.5.4)}$$

(again, by Stokes' theorem).

We now take advantage of the hypothesis that $\mathrm{V}_0 = \Delta_1 \times \mathcal{B}_1$: if $q \geq 1$ there is $u \in \mathcal{C}^\infty_c(\mathrm{V}_0;\tilde{\Lambda}^{\nu,n-q-1})$ such that $\mathrm{D}u = \mathrm{D}u_1$ in $\mathrm{V}_0 \cap \mathrm{U}_0^+$, $\mathrm{D}u = 0$ in

$V_0\backslash(V_0\cap U_0^+)$. Selecting $f = f_2^+$ completes the proof of $(VIII.5.1)_q^+$ in this case.

When $q = 0$ the section $\gamma \in \Gamma_c(\mathcal{S}_0\cap V_0;\mathscr{C}^\infty\tilde{\Lambda}^{\nu,n-1})$ is such that, for all holomorphic functions of z in V_0, h,

$$\int_{\mathcal{S}_0} hu_0 = 0. \tag{VIII.5.5}$$

We define u_1 and reason as before; in particular we reach the conclusion (VIII.5.4) (for instance). Since $u_0 = u_1$ on $\mathcal{S}$ we derive from (VIII.5.4), by Stokes' theorem,

$$\int_{\mathcal{S}_0\cap U_0^+} hDu_1 = 0. \tag{VIII.5.6}$$

Let v^+ denote the section of $\mathscr{C}_c^\infty(V_0;\tilde{\Lambda}^{\nu,n})$ equal to Du_1 in U_0^+ and to 0 in $U_0\backslash U_0^+$; we have $\int_{V_0} hv^+ = 0$ for all holomorphic functions of z in V_0, h. Since $V_0 = \Delta_1\times\mathcal{B}_1$ here also there is $u \in \mathscr{C}_c^\infty(V_0;\tilde{\Lambda}^{\nu,n-1})$ such that $Du = v^+$ in V_0; selecting $f = f_2^+$ enables to complete the proof of $(VIII.5.1)_0^+$.■

We now proceed with the proof of Theorem VIII.4.1. We shall hypothesize that

the hypo-analytic function w is not acyclic at 0 *in dimension q;* (VIII.5.7)

i.e., that

there is an open neighborhood $U \subset \Omega$ *of* 0 *such that, given any open neighborhood of* 0, $V \subset U$, *there is a level set* $\mathcal{S}$ *of w in* Ω, *noncritical in* V, *whose intersection number in dimension q relative to the pair* (U,V) *does not vanish identically.* (VIII.5.8)

Since there is a natural "restriction" map ${}_a\mathrm{H}^q(\mathcal{S}\cap U) \to {}_a\mathrm{H}^q(\mathcal{S}\cap U')$ if $U' \subset U$ and a natural "injection" map ${}_a\mathrm{H}_q(\mathcal{S}\cap V') \to {}_a\mathrm{H}_q(\mathcal{S}\cap V)$ if $V' \subset V$, the nonvanishing of the intersection number $I^q_{U,V,\mathcal{S}}$ remains valid if we decrease U and increase V. We may therefore assume that, while $U = \Delta\times\mathcal{J}\times\mathcal{B}$ as in (VIII.5.2), V ranges over a family of product sets $V^{(\iota)} = \Delta^{(\iota)}\times\mathcal{J}^{(\iota)}\times\mathcal{B}^{(\iota)}$ ($\iota = 1,2,\ldots$) with diam $V^{(\iota)} \to 0$. We hypothesize that there is a sequence of values $w^{(\iota)} \to 0$ such that:

whatever $\iota = 1,2,\ldots$, *when* $w_0 = w^{(\iota)}$ *and* $V = V^{(\iota)}$ *at least one of the two properties* $(VIII.5.1)_q^+$ *or* $(VIII.5.1)_q^-$ *holds.* (VIII.5.9)

In the remainder of the proof we take U as in (VIII.5.2) and we let $V = \Delta_1\times\mathcal{J}_1\times\mathcal{B}_1$ stand for one of the product sets $V^{(\iota)}$, $w_0 = s_0+\iota r_0$ for the corresponding value $w^{(\iota)}$. We assume that $(VIII.5.1)_q^+$ holds; f and u will be the forms in $(\mathrm{VIII}.5.1)_q^+$. We shall conclude that $(\mathrm{VIII}.1.8)_{0,q}$ is not valid if Ω is replaced by U and Ω' by V. A similar argument applies if $(\mathrm{VIII}.5.1)_q^-$ holds.

There is a real number $\rho > 0$ such that

$$\mathcal{I}m\, w > r_0 + \rho \quad on \text{ supp } Du. \tag{VIII.5.10}$$

Since there is a constant $A > 0$ such that

$$|\varphi(z,s,t) - \varphi(z,s_0,t)| \leq A|s - s_0|, \ \forall\ (z,s,t) \in \mathrm{U},$$

it is easily derived from $(\mathrm{VIII}.5.1)_q^+$ that there exists a number $\eta > 0$ such that $[s_0 - \eta,\ s_0 + \eta] \subset \mathcal{J}_1$ and such that if $|s - s_0| < \eta$, then

$$\varphi(z,s,t) > r_0 + 3\rho/4 \ \ if\ (z,t) \in \text{supp } \mathrm{D}u; \tag{VIII.5.11}$$

$$\varphi(z,s_0,t) < r_0 + \rho/4 \ \ if\ (z,t) \in \text{supp } \mathrm{D}f. \tag{VIII.5.12}$$

Next select a number η', $0 < \eta' < \eta$, a function $\chi \in \mathcal{C}_c^\infty(\mathcal{J}_1)$ such that $\chi(s) = 1$ if $|s - s_0| < \frac{1}{2}\eta'$, $\chi(s') = 0$ if $|s - s_0| > \eta'$, and a function $g \in \mathcal{C}^\infty(\mathbb{R}^2)$ such that supp $g \subset \mathfrak{R} = \{\, (s,r) \in \mathbb{R}^2;\ |s - s_0| \leq \delta, |r - r_0 - \frac{1}{2}\rho| \leq \varepsilon \,\}$, $g > 0$ in the interior of $\mathfrak{R}$, with δ and ε positive numbers, chosen below. We require right away

$$\delta < \tfrac{1}{2}\eta',\ \varepsilon < \tfrac{1}{4}\rho. \tag{VIII.5.13}$$

We regard χ as a function in U and we denote by g_* the restriction to U of the function $g \circ w$. Call π the coordinate projection $\mathrm{U} \ni (z,s,t) \to (z,t) \in \Delta \times \mathcal{B}$ and π^* the corresponding pullback map. The differential form $F = g_* d\overline{w} \wedge \pi^* f$ defines a section belonging to $\mathcal{C}^\infty(\mathrm{U};\Lambda^{0,q+1})$ (also denoted by F). Since $\overset{-1}{\pi}(\text{supp } u) \cap \text{supp } \chi \subset\subset \mathrm{V}$ the form $v = \chi \mathrm{d}w \wedge \pi^* u$ defines a section belonging to $\mathcal{C}_c^\infty(\mathrm{V};\Lambda^{m,n-q-1})$ (also denoted by v).

We claim that

$$\mathrm{d}'F = 0. \tag{VIII.5.14}$$

Indeed,

$$\mathrm{d}'F = [\mathrm{d}'(g_*\mathrm{d}\overline{w})] \wedge \pi^* f \ - \ g_* \mathrm{d}\overline{w} \wedge \pi^*(\mathrm{D}f).$$

The section $\mathrm{d}'(g_* d\overline{w})$ of $\Lambda^{0,2}$ over U has the representative

$$\mathrm{d}(g_*\mathrm{d}\overline{w}) = [(\partial g/\partial w) \circ w]\mathrm{d}w \wedge \mathrm{d}\overline{w},$$

which is a section of $\mathrm{T}'^{1,1}$. This means that $\mathrm{d}'(g_*\mathrm{d}\overline{w}) \equiv 0$. On the other hand, (VIII.5.12) and (VIII.5.13) entail $\pi^*(\mathrm{D}f) \equiv 0$ in a neighborhood of supp g_*, which proves (VIII.5.14).

Since $\chi = 1$ on supp g_*, we derive

$$\int F \wedge v = \pm \int g_* \mathrm{d}\bar{w} \wedge \mathrm{d}w \wedge \pi^*(f \wedge u). \qquad \text{(VIII.5.15)}$$

Define

$$G(s,r) = \int_{-\infty}^{r} g(s,\rho)\mathrm{d}\rho.$$

We note that

$$\begin{aligned} r < r_0 + \tfrac{1}{4}\rho &\Rightarrow G(s,r) = 0, \\ r \geq r_0 + \tfrac{1}{2}\rho + \varepsilon &\Rightarrow G(s,r) = C(s), \end{aligned} \qquad \text{(VIII.5.16)}$$

where $C(s)$ is a $\mathscr{C}^\infty$ function of s, $C(s) = 0$ if $|s - s_0| > \delta$ and $C(s) > 0$ if $|s - s_0| < \delta$. In particular, by (VIII.5.11) and (VIII.5.16),

$$(z,t) \in \text{supp } \mathrm{D}u,\ |s - s_0| < \delta \Rightarrow G(w(z,s,t)) = C(s). \qquad \text{(VIII.5.17)}$$

We claim that

$$\int F \wedge v \neq 0. \qquad \text{(VIII.5.18)}$$

We have, in U, $g_* d\bar{w} \wedge \mathrm{d}w = -2\imath \mathrm{d}[(G \circ w)\mathrm{d}s]$ and thus, by (VIII.5.15),

$$\int F \wedge v = \pm 2\imath \int \mathrm{d}[(G \circ w)\mathrm{d}s] \wedge \pi^*(f \wedge u)$$

$$= \pm 2\imath \left\{ \int (G \circ w)\mathrm{d}s \wedge \pi^*(\mathrm{D}f \wedge u) \pm \int (G \circ w)\mathrm{d}s \wedge \pi^*(f \wedge \mathrm{D}u) \right\}$$

by Stokes' theorem. But according to (VIII.5.12) and (VIII.5.16), we have $(G \circ w)\pi^*(\mathrm{D}f) \equiv 0$. On the other hand, by virtue of (VIII.5.17),

$$\int (G \circ w)\mathrm{d}s \wedge \pi^*(f \wedge \mathrm{D}u) = \left[\int C(s)\mathrm{d}s \right] \int f \wedge \mathrm{D}u \neq 0,$$

which proves (VIII.5.18).

Let us introduce the three quantities

$$M = \text{Max}\ \{ \varphi(z,s,t);\ |s - s_0| \leq \eta',\ (z,t) \in \text{supp } \mathrm{D}u \},$$

$$M_+ = \text{Max}\ \{ \varphi(z,s,t);\ \tfrac{1}{2}\eta' \leq |s - s_0| \leq \eta',\ (z,t) \in \text{supp } u \},$$

$$M_- = \text{Min}\ \{ \varphi(z,s,t);\ \tfrac{1}{2}\eta' \leq |s - s_0| \leq \eta',\ (z,t) \in \text{supp } u \},$$

and the two subsets of the w-plane

$$\mathscr{A} = \{ s + \imath r \in \mathbb{C};\ |s - s_0| \leq \eta',\ r_0 + 3\rho/4 \leq r \leq M \},$$

$$\mathscr{B} = \{ s + \imath r \in \mathbb{C};\ \tfrac{1}{2}\eta' \leq |s - s_0| \leq \eta',\ M_- \leq r \leq M_+ \}.$$

Since

$$d'v = d'\chi \wedge dw \wedge \pi^* u - \chi \wedge dw \wedge \pi^* Du,$$

we conclude that $w(\mathrm{supp}\, d'v) \subset \mathcal{A} \cup \mathcal{B}$.

On the other hand, $w(\mathrm{supp}\, F)$ is contained in the rectangle $\mathfrak{R}$. We want to show that if the numbers ε and δ are chosen suitably small, then there is an entire holomorphic function H in the plane such that $\mathcal{R}e\, H > 0$ in $\mathcal{A} \cup \mathcal{B}$ whereas $\mathcal{R}e\, H < 0$ on $\mathfrak{R}$. The simple geometric situation makes it clear that there is an open, bounded, connected, and simply connected subset $\mathbb{O}$ of the plane whose closure contains the compact set $\mathcal{A} \cup \mathcal{B} \cup \mathfrak{R}$ and whose boundary $\partial \mathbb{O}$ intersects $\mathcal{A} \cup \mathcal{B} \cup \mathfrak{R}$ at the single point $s_0 + \imath(r_0 + \frac{1}{4}\rho)$. By the Riemann mapping theorem there is a biholomorphism H_0 of $\mathbb{O}$ onto the disk

$$\Delta_1 = \{\, s + \imath r \in \mathbb{C};\ |s + \imath r - 1| < 1 \,\},$$

which maps the point $s_0 + \imath(r_0 + \frac{1}{4}\rho)$ onto the origin. There is a number $c > 0$ such that $\mathcal{R}e\, H_0 > 2c$ on $\mathcal{A} \cup \mathcal{B}$. Possibly after further decreasing δ and ε we may achieve that $\mathcal{R}e\, H_0 < c/4$ on $\mathfrak{R}$. Since the closure of $\mathbb{O}$ is a Runge compact subset of $\mathbb{C}$, we can approximate H_0 by an entire function H satisfying:

$$\mathcal{R}e\, H > c \text{ on } \mathcal{A},\ \mathcal{R}e\, H < 3c/4 \text{ on } \mathfrak{R}.$$

The function $h = H \circ w - c$ is a hypo-analytic function in Ω that satisfies

$$\mathcal{R}e\, h > 0 \text{ on } \mathrm{supp}\, d'v, \tag{VIII.5.19}$$

$$\mathcal{R}e\, h < 0 \text{ on } \mathrm{supp}\, F. \tag{VIII.5.20}$$

All conditions in Theorem VIII.1.1 are satisfied, whence the desired conclusion. ■

VIII.6. Applications of Theorem VIII.4.1

Acyclicity when $m = 1$

When $m = 1$, i.e., $\nu = 0$, $n = n'$, there are no first integrals z_i. In Ω the hypo-analytic structure is entirely defined by the single function $w = s + \imath\varphi(s,t)$. The tangent structure bundle $\mathcal{V}$ is spanned over Ω by the vector fields

$$L_j = \partial/\partial t_j - \imath\varphi_{t_j}(1 + \imath\varphi_s)^{-1}\partial/\partial s,\ j = 1,\ldots,n. \tag{VIII.6.1}$$

As before let $\mathcal{S}$ denote the level set $w = w_0$ in Ω. Both sheaves $\mathcal{HA}_{\mathcal{S}}$ and $\mathcal{HA}'_{\mathcal{S}}$ are naturally isomorphic to the constant sheaf on $\mathcal{S}$ and thus, by (VIII.4.12),

$$_a\mathrm{H}_*(\mathcal{S}) = \mathrm{H}_*(\mathcal{S}),\ _a\mathrm{H}^*(\mathcal{S}) = \mathrm{H}^*(\mathcal{S}), \tag{VIII.6.2}$$

with the standard homology and cohomology theories at the right: $H_*(\mathcal{S})$ = singular homology, $H^*(\mathcal{S})$ = singular cohomology with complex coefficients. When $\mathcal{S}$ is smooth they may be interpreted according to De Rham's theory. If $\mathcal{S}$ is identified to the hypersurface $\mathcal{S}_0$ in t-space $\mathbb{R}^n$ and if Ω_0 is connected, Definition VIII.4.2 shows that

$$ {}_a\tilde{H}_0(\mathcal{S},\Omega) \cong $$

$$ \{u \in \mathcal{C}_c^\infty(\mathcal{S};\Lambda^{n-1}\mathbb{C}T^*\mathcal{S});\ \int_{\mathcal{S}} u = 0\}/\{\, dv;\ v \in \mathcal{C}_c^\infty(\mathcal{S};\Lambda^{n-2}\mathbb{C}T^*\mathcal{S})\,\}, $$

(Λ^q: q^{th} exterior power), whence

$$ {}_a\tilde{H}_0(\mathcal{S},\Omega) \cong \tilde{H}_0(\mathcal{S}), \qquad \text{(VIII.6.3)} $$

the reduced homology in dimension zero in De Rham's sense. In particular, $\tilde{H}_0(\mathcal{S}) = 0$ if and only if $\mathcal{S}$ is connected.

The isomorphisms (VIII.6.2) and (VIII.6.3) equate $I^q_{\mathcal{S}}$ to the standard pairing between De Rham cohomology classes and singular homology classes of the same dimension (reduced homology in dimension zero), the "intersection number" of $\mathcal{S}$. It turns those spaces (equipped with their natural topologies) into the dual of one another.

Proposition VIII.6.1. *Suppose $m = 1$ and the level set $\mathcal{S}$ noncritical. In $\mathcal{S}$ a q-cycle γ is homologous to zero if and only if $I^q_{\mathcal{S}}([\beta],[\gamma]) = 0$ for all q-cocycles β; a q-cocycle β is cohomologous to zero if and only if $I^q_{\mathcal{S}}([\beta],[\gamma]) = 0$ for all q-cycles γ.*

Follows at once from Theorem 17′ in De Rham [1]. We say that $\mathcal{S}$ is *homologically trivial in dimension q* if $H_q(\mathcal{S}) = 0$ when $q \geq 1$ or if $\tilde{H}_0(\mathcal{S}) = 0$ when $q = 0$.

Corollary VIII.6.1. *Let $m = 1$, $0 \leq q \leq n-1$. The intersection number $I^q_{\mathcal{S}}$ vanishes identically if and only if $\mathcal{S}$ is homologically trivial in dimension q.*

Let $V \subset U$ be two open subsets of Ω. All that precedes can be relativized in standard fashion. We at once obtain the following:

Proposition VIII.6.2. *Suppose $m = 1$, $0 \leq q \leq n-1$, and that the level set $\mathcal{S}$ is noncritical in* V. *The intersection number $I^q_{U,V,\mathcal{S}}$ of $\mathcal{S}$ relative to the pair* (U,V) *vanishes identically if and only if*

$$ \textit{every q-cycle in } \mathcal{S}\cap V \textit{ bounds in } \mathcal{S}\cap U. \qquad \text{(VIII.6.4)} $$

We recall that all cycles are compactly supported. Property (VIII.6.4) asserts the vanishing of the natural map

$$ H_q(\mathcal{S}\cap V) \to H_q(\mathcal{S}\cap U) \ \textit{if } q \geq 1, \qquad \text{(VIII.6.5)} $$

$$\tilde{H}_0(\mathcal{S}\cap V) \to \tilde{H}_0(\mathcal{S}\cap U) \text{ if } q = 0. \tag{VIII.6.6}$$

PROPOSITION VIII.6.3. *When* $m = 1, 0 \leq q \leq n-1$, *the hypo-analytic function* w *is acyclic at* 0 *in dimension* q *if and only if every open neighborhood* U $\subset \Omega$ *of* 0 *contains another open neighborhood,* V, *such that all level sets of* w *in* Ω *satisfy condition* (VIII.6.4).

Sard's lemma allows us to dispense with the requirement that the level sets of w be noncritical: since the noncritical values of w are dense, if the map (VIII.6.5) (resp., (VIII.6.6)) vanishes for all noncritical level sets it also vanishes for all level sets.

It is a consequence of Corollary II.3.1 that the germs at 0 of the level sets of w are invariants of the locally integrable structure that underlies the hypo-analytic structure of Ω. Thus the acyclicity of w at 0 is really a property of the fibres of the locally integrable structure of $\mathcal{M}$ at 0.

DEFINITION VIII.6.1. *When* $m = 1, 0 \leq q \leq n-1$, *we say that the integrable structure of the manifold* $\mathcal{M}$, *defined by the vector bundles* $\mathcal{V}$ *and* T′, *is* acyclic at 0 in dimension q *if* w *is acyclic at* 0 *in dimension* q.

Theorem VIII.4.1 entails

THEOREM VIII.6.1. *Suppose* $m = 1$. *If the differential equation* $d'u = f$ *is locally solvable at the point* 0 *in degree* q *then the locally integrable structure of* $\mathcal{M}$ *is acyclic at* 0 *in dimension* q.

When $m = 1$ the acyclicity of the locally integrable structure at 0 in dimension *zero* is equivalent to the following property:

there is a basis of neighborhoods of 0 *in* Ω *consisting of sets in which every fibre of the structure is connected.* (VIII.6.7)

Indeed, let U be an arbitrary open neighborhood of 0 contained in Ω; let V $\subset$ U be an open neighborhood of 0 such that (VIII.6.4) holds with $q = 0$, whatever the level set $\mathcal{S}$ of w in Ω. This means that there is a unique connected component of $\mathcal{S}\cap$U that contains $\mathcal{S}\cap$V; let $\tilde{V}$ denote the union of all the connected components of $\mathcal{S}\cap$U that intersect V (as $\mathcal{S}$ varies). We have V $\subset \tilde{V} \subset$ U and if $\mathcal{S}$ is any level set of w in Ω, $\mathcal{S}\cap\tilde{V}$ is connected. Conversely, suppose there is a neighborhood $\tilde{V} \subset$ U of 0 with the latter property; its interior V satisfies (VIII.6.4). ■

Continue to assume $m = 1$. In top degree $q = n-1$ (with $n > 1$, otherwise $q = 0$) the acyclicity of the locally integrable structure is equivalent to the following property:

there is a basis of neighborhoods of 0 *in* Ω *consisting* (VIII.6.8) *of open sets in which no fibre of the structure has a compact connected component.*

Indeed, the fact that no level set $\mathcal{S}$ of w in Ω has a compact connected component contained in an open neighborhood of 0, $U \subset \Omega$, is equivalent to $H_{n-1}(\mathcal{S} \cap U) = 0$. The latter property can be equated to the property that $\mathcal{I}m\, w$ does not have any local extremum in a neighborhood of 0 (such a formulation covers also the case $n = 1$; cf. Corollary VIII.2.1).

Acyclicity in Top-Degree

In the present subsection we allow m to be ≥ 1 but we focus on the case $q = n-1$.

PROPOSITION VIII.6.4. *In order that* ${}_aH_{n-1}(\mathcal{S}) = 0$ *it is necessary and sufficient that* $\mathcal{S}$ *not have any compact connected component.*

PROOF. If $\mathcal{S}$ has a connected component K the section equal to the germ of dz at points of K and to zero everywhere else belongs to $\Gamma_c(\mathcal{S};\mathcal{A}^\nu|_{\mathcal{S}})$.

Conversely suppose there is a section $\sigma \in \Gamma_c(\mathcal{S};\mathcal{A}^\nu|_{\mathcal{S}})$ that is not identically zero. The set K of points $\mathfrak{p} \in \mathcal{S}$ such that $\sigma(\mathfrak{p}) \neq 0$ is closed. If we show that K is open in $\mathcal{S}$, K will be a finite union of compact connected components of $\mathcal{S}$. In some open and connected neighborhood $\mathcal{O}$ of $\mathfrak{p}$ we may equate $\sigma(\mathfrak{p})$ to a ν-form $f dz$ with f a holomorphic function of z (independent of t). Since f cannot vanish identically in any open subset of $\mathcal{O}$ we must have $\sigma(\mathfrak{p}') \neq 0$, $\forall\, \mathfrak{p}' \in \mathcal{O} \cap \mathcal{S}$. ∎

PROPOSITION VIII.6.5. *Consider two open neighborhoods of the origin,* $V = \Delta_1 \times \mathcal{J}_1 \times \mathcal{B}_1 \subset U = \Delta \times \mathcal{J} \times \mathcal{B} \subset \Omega$ (cf. (VIII.5.2)). *If the level set* $\mathcal{S}$ *is noncritical in* V *the following properties are equivalent:*

there is a connected component $\mathcal{O}$ *of* $V_0 \backslash \mathcal{S}_0$ *whose closure is a compact subset of* V_0; (VIII.6.9)

${}_a\tilde{H}_0(\mathcal{S} \cap V, V) \neq 0$ *if* $n = 1$; ${}_aH_{n-1}(\mathcal{S} \cap V) \neq 0$ *if* $n \geq 2$; (VIII.6.10)

$I^{n-1}_{U,V,\mathcal{S}}$ *does not vanish identically.* (VIII.6.11)

We recall that $U_0 = \Delta \times \mathcal{B}$, $V_0 = \Delta_1 \times \mathcal{B}_1 \subset \mathbb{C}^\nu \times \mathbb{R}^{n'}$.

PROOF. When $2\nu + n' = 1$ necessarily $\nu = 0$, $m = n = 1$. In this case (see case $m = 1$ above) (VIII.6.10) is equivalent to (VIII.6.11) and to the fact that $\mathcal{S} \cap V$ ($\cong \mathcal{S}_0 \cap V_0$) is a finite set of points consisting of more than a single point, thus not connected, which is here a restatement of (VIII.6.9).

When $2\nu + n' \geq 2$ (VIII.6.9) entails that $\mathcal{S}_0 \cap V_0$ *has a compact connected component* (the boundary $\partial\mathcal{O}$ of $\mathcal{O}$). Conversely, due to the simple geometry of V_0, any compact connected component of $\mathcal{S}_0 \cap V_0$ must bound an open set $\mathcal{O} \subset\subset V_0$. Then (VIII.6.10) is equivalent to (VIII.6.9) by Proposition VIII.6.4; (VIII.6.11) $\Rightarrow$ (VIII.6.10) trivially. We are going to prove that, when $2\nu+n' \geq 2$, (VIII.6.9) $\Rightarrow$ (VIII.6.11). After a translation we may assume that $0 \in \mathcal{O}$.

We make use of the current $E_{\nu,n'} \in \mathcal{D}'(\mathbb{C}^\nu \times \mathbb{R}^{n'};\Lambda^{0,\nu+n'-1})$ defined in the Appendix to the present section. Let $\psi \in \mathcal{C}_c^\infty(V_0)$ be equal to 1 in an open neighborhood of $\mathcal{C}\ell\mathcal{O}$ with $(\mathrm{supp}\,\psi) \cap \mathcal{S}_0 = \partial\mathcal{O}$. By Lemma VIII.6.A1 $DE_{\nu,n'} \equiv 0$ in a neighborhood of $\mathcal{S}_0$. Clearly $D(\psi dz) \equiv 0$ in a neighborhood of $\mathcal{C}\ell\mathcal{O}$. Thus $E_{\nu,n'}$ defines a continuous section over $\mathcal{S}_0$ and therefore, by restriction, one over $\mathcal{S}_0 \cap U_0$, β, of the sheaf $\mathcal{C}^\infty\Lambda^{0,n-1}$, while ψdz defines one, γ, of $\mathcal{C}^\infty\Lambda^{\nu,0}$; $D\beta = 0$, $D\gamma = 0$; $\mathrm{supp}\,\gamma \subset\subset V$. By Stokes' theorem and by Lemma VIII.6.A1,

$$I^{n-1}_{U,V,\mathcal{S}}([\beta],[\gamma]) = \int_{\mathcal{S}} E_{\nu,n'} \wedge \psi dz = \pm \int_{\mathcal{O}} dE_{\nu,n'} \wedge dz \neq 0.$$

When $n = 1$ one must show, in addition to what precedes, that the form ψdz defines an element of the *reduced* homology ${}_a\tilde{H}_0(\mathcal{S},\Omega)$. But $n = \nu + n' = 1$ and $2\nu + n' \geq 2$ demand $\nu = 1$, $n' = 0$: U_0 is a disk in the complex plane and $\partial\mathcal{O}$ is a closed $\mathcal{C}^\infty$ curve bounding a compact subset of U_0. Given any holomorphic function h in U_0 we have $\int_{\mathcal{S}_0} h\psi dz = \int_{\partial\mathcal{O}} h dz = 0$. ∎

Proposition VIII.6.5 makes it clear that when dealing with the acyclic property in top dimension, $n-1$, there is no need to "relativize": properties (VIII.6.9) and (VIII.6.10) contain no reference to the larger neighborhood U. We leave the proof of the next statement to the reader:

PROPOSITION VIII.6.6. *In order for w to be acyclic at* 0 *in dimension* $n-1$ *it is necessary and sufficient that there be an open neighborhood* $U = \Delta \times \mathcal{J} \times \mathcal{B}$ *of* 0 *in* Ω *that has the following property:*

for no $s_0 + \imath r_0 \in \mathbb{C}$ *does the set* $\{(z,t) \in \Delta \times \mathcal{B};\ \varphi(z,s_0,t) \neq r_0\}$ *have a connected component with compact closure contained in* $\Delta \times \mathcal{B}$. (VIII.6.12)

REMARK VIII.6.1. It is possible to prove that if w is not acyclic at 0 in dimension $n-1$ then, for every open neighborhood $U \subset \Omega$ of the origin, there is a $\mathcal{C}^\infty$ solution h in Ω whose real part has a local extremum in U. This relates the acyclicity hypothesis to the solvability condition in Corollary VIII.2.1. ∎

Local solvability at 0 in degree $n-1$ entails the following condition:

there is an open neighborhood $U \subset \Omega$ *of the origin such that, given any point* $\mathfrak{p} \in U$, *every hypo-analytic function in* Ω *whose differential spans the characteristic set at* $\mathfrak{p}$ *is acyclic at* $\mathfrak{p}$ *in dimension* $n-1$. (VIII.6.13)

In the CR case we can equate Property (VIII.6.13) with a property that does not involve all the hypo-analytic functions whose differentials span the characteristic set. There are no coordinates t_k and the map $(x,y,s) \to (z,w)$ effects an embedding of Ω into $\mathbb{C}^{n+1}$; it is a diffeomorphism of Ω onto the hypersurface Σ in $\mathbb{C}^{n+1}$ defined by the equation $\mathcal{I}m\, w = \varphi(z, \mathcal{R}e\, w)$. The tangent structure bundle $\mathcal{V}$ is spanned, over Ω, by the vector fields

$$L_j = \partial/\partial\bar{z}_j - \imath\varphi_{\bar{z}_j}(1+\imath\varphi_s)^{-1}\partial/\partial s,\ j = 1,\ldots,n. \qquad \text{(VIII.6.14)}$$

We have $(2\imath)^{-1}[L_j,\overline{L}_k] = \lambda_{jk}\partial/\partial s$ with

$$\lambda_{jk} = \varphi_{z_k\bar{z}_j}/(1+\varphi_s^2) + O(|\varphi_z|). \qquad \text{(VIII.6.15)}$$

As is customary we identify the matrix $\mathcal{L}(z,s) = (\lambda_{jk}(z,s))_{1\le j,k\le n}$ to the Levi form at the point $(z,s) \in \Omega$ (cf. section I.9).

Proposition VIII.6.7. *Suppose* $\mathcal{M}$ *is a CR manifold. If there is a neighborhood of* 0 *in* $\mathcal{M}$ *at no point of which the Levi form is definite, then every hypo-analytic function in* Ω *whose differential spans the characteristic set at* 0 *is acyclic at* 0 *in dimension* $n-1$.

Proof. We take the closure of $U = \Delta \times \mathcal{J}$ to be a compact subset of Ω (as before, Δ is an open polydisk in z-space, $\mathcal{J}$ an open interval in the s-line, both centered at 0). By (VIII.6.15) there is a constant $C > 0$ such that

$$|(1+\varphi_s^2)\lambda_{jk} - \varphi_{z_k\bar{z}_j}| \le C|\varphi_z|. \qquad \text{(VIII.6.16)}$$

Suppose there is an open subset A of the polydisk Δ and a number $s_0 \in \mathcal{J}$ such that the following is true:

$$\operatorname{diam} A = \delta < (2C)^{-1}, \qquad \text{(VIII.6.17)}$$

$$\mu_1 = \min_{z\in\partial A} \varphi(z,s_0) > \min_{z\in A} \varphi(z,s_0) = \mu_2. \qquad \text{(VIII.6.18)}$$

Let $z_0 \in A$ be such that $\varphi(z_0,s_0) = \mu_2$. Define

$$\psi(z) = \varphi(z,s_0) - (\mu_1-\mu_2)|z-z_0|^2/2\delta^2.$$

We have $\psi(z_0) = \mu_2$ and if $z \in \partial A$, $\psi(z) \ge \mu_1 - \frac{1}{2}(\mu_1-\mu_2) > \mu_2$. Thus there is a point $z_* \in A$ at which ψ reaches its minimum (over $\overline{A}$). We have

$$\psi_z(z_*) = 0,\ \psi_{z\bar{z}}(z_*) \ge 0 \text{ (positive semidefinite)}.$$

From the first condition we derive $|\varphi_z(z_*,s_0)| \leq (\mu_1 - \mu_2)/\delta\sqrt{2}$; from the second condition,

$$\varphi_{z\bar{z}}(z^*,s_0) \geq [(\mu_1 - \mu_2)/2\delta^2]I \ (I\text{: identity } n\times n \text{ matrix}).$$

Let χ be an eigenvalue of $\mathscr{L}(z_*,s_0)$ with eigenvector $\mathbf{v}$, $|\mathbf{v}| = 1$. We have, at the point (z_*,s_0),

$$(1+\varphi_s^2)\chi = \mathbf{v}\cdot[(1+\varphi_s^2)\mathscr{L} - \varphi_{z\bar{z}}]\mathbf{v} + \mathbf{v}\cdot\varphi_{z\bar{z}}\mathbf{v} \geq$$

$$[(\mu_1 - \mu_2)/2\delta^2](1-\sqrt{2}C\delta) > 0.$$

If instead of (VIII.6.18) we had assumed that

$$\operatorname*{Max}_{z\in\partial A} \varphi(z,s_0) < \operatorname*{Max}_{z\in A} \varphi(z,s_0),$$

we would have concluded that there is a point $z_* \in A$ such that $\mathscr{L}(z_*,s_0)$ is definite negative. ■

Proposition VIII.6.8. *Suppose $\mathcal{M}$ is a CR manifold. Property* (VIII.6.13) *is equivalent to the property that there is an open neighborhood of* 0 *at no point of which the Levi form is definite.*

Proof. When the Levi form is definite at $\mathfrak{p} \in \Omega$ there is a hypo-analytic function h in some open neighborhood of 0, e.g. the function (VIII.3.7), whose differential spans the characteristic set at $\mathfrak{p}$ and whose real part has a local extremum at $\mathfrak{p}$. The latter precludes h being acyclic at 0 in dimension $n-1$. It suffices to combine this remark with Proposition VIII.6.7. ■

Thus for CR manifolds of hypersurface type and $q = n-1$, Theorem VIII.4.1 coincides with Theorem VIII.3.1.

Applications when $m > 1$

Consider as before two open neighborhoods of the origin, $\mathrm{U} = \Delta\times\mathscr{I}\times\mathscr{B} \subset\subset \Omega$ and $\mathrm{V} = \Delta_1\times\mathscr{I}_2\times\mathscr{B}_1 \subset\subset \mathrm{U}$; s_0, U_0, V_0 have the meaning given to them earlier. We partition each set of variables z_i and t_j into two subsets, as in section VIII.3:

$$z' = (z_1,\ldots,z_\mu),\ z'' = (z_{\mu+1},\ldots,z_{m-1}),$$

$$t' = (t_1,\ldots,t_{\mu'}),\ t'' = (t_{\mu'+1},\ldots,t_{n'}).$$

We reason under the following basic hypothesis:

There is a number $a_0 > 0$ such that the following two conditions are satisfied: (VIII.6.19)

$$\forall\ (z,t) \in \overline{\mathrm{U}}_0,\ \varphi(z,s_0,t) = a_0 \Rightarrow |z'|+|t'| \neq 0; \tag{VIII.6.20}$$

$\{ (z,t) \in \mathrm{U}_0;\ z'' = 0,\ t'' = 0,\ \varphi(z,s_0,t) \leq a_0 \}$ *is a compact subset of* V_0. (VIII.6.21)$^+$

LEMMA VIII.6.1. *Suppose* $q = \mu+\mu'-1 \geq 0$ *and that* (VIII.6.19) *holds. Then Property* (VIII.5.1)$_q^+$ *holds.*

PROOF. By compactness we derive from (VIII.6.20) that there is a positive number $a < a_0$ such that

$$\forall (z,t) \in \mathrm{U}_0,\ \varphi(z,s_0,t) = a \Rightarrow |z'|+|t'| \neq 0. \tag{VIII.6.22}$$

Let b be any number $> a$. Suppose there were a point (z_0,t_0) such that

$$\varphi(z_0,s_0,t_0) = b \text{ and } z_0' = 0,\ t_0' = 0.$$

The intersection of U_0 with the subspace $\mathcal{H}'' = \{ (z,t);\ z' = 0,\ t' = 0 \}$ is convex, hence it contains a straight-line segment ℓ joining 0 to (z_0,t_0). On ℓ the function φ must take the value a, in contradiction to (VIII.6.22). We reach the conclusion that

$$\forall\ (z,t) \in \mathrm{U}_0,\ \varphi(z,s_0,t) \geq a \Rightarrow |z'|+|t'| \neq 0. \tag{VIII.6.23}$$

Below we choose once for all a number r_0, $a < r_0 < a_0$, which is a noncritical value of the function $(z,t) \to \varphi(z,s_0,t)$, and a function $\chi \in \mathscr{C}^\infty(\mathbb{R})$ such that

$$\chi(\tau) = 0 \text{ if } \tau < a,\ \chi(\tau) = 1 \text{ if } \tau > \tfrac{1}{2}(a+r_0). \tag{VIII.6.24}$$

This allows us to consider the differential form in U_0,

$$f = \chi(\varphi(z,s_0,t))E_{\mu,\mu'}$$

where $E_{\mu,\mu'}$ is the current defined in the Appendix to the present section (here $E_{\mu,\mu'}$ is regarded as a current in (z,t)-space). The differential form f defines a smooth section of $\tilde{\Lambda}^{0,\mu+\mu'-1}$ in U_0; we have, by Lemma VIII.6.A1,

$$\mathrm{D}f = \mathrm{D}[\chi(\varphi(z,s_0,t))]\wedge E_{\mu,\mu'}\ (\mathrm{D} = \bar{\partial}_z+\mathrm{d}_t)$$

and thus, by (VIII.6.24), $\varphi < r_0$ on supp $\mathrm{D}f$ (in U_0).

On the other hand, let $\theta(z',t')$ be a $\mathscr{C}^\infty$ function in $\mathbb{C}^\mu \times \mathbb{R}^{\mu'}$ that vanishes identically if $\varphi(z',0,s_0,t',0) > a_0$ and equals one if $\varphi(z',0,s_0,t',0) < r$, with $a < r_0 < r < a_0$. We also make use of a cutoff function $\chi_1 \in \mathscr{C}_c^\infty(\mathbb{C}^{m-\mu-1} \times \mathbb{R}^{n'-\mu'})$, $\chi_1 \geq 0$ everywhere, supp $\chi_1 = \mathscr{B}_\delta''$, the open ball with radius $\delta > 0$ and center the origin. When $\mathrm{d}\theta \neq 0$ and $\chi_1 \neq 0$ we must have, for some positive constant C, $\varphi(z,s_0,t) \geq r - C\delta$. Therefore, if δ is sufficiently small, there will be a number $r_1 > r_0$ such that

$$\varphi(z,s_0,t) > r_1 \text{ if } \chi_1(z'',t'')\mathrm{d}\theta(z',t') \neq 0. \tag{VIII.6.25}$$

We then define

$$u = \theta(z',t')\chi_1(z'',t'')d\bar{z}_{\mu+1}\wedge\cdots\wedge d\bar{z}_{m-1}\wedge dt_{\mu'+1}\wedge\cdots\wedge dt_{n'}\wedge dz,$$

where as before $dz = dz_1\wedge\cdots\wedge dz_{m-1}$. It follows at once from (VIII.6.25) that $\varphi > r_0$ on supp $\mathrm{D}u$. Thanks to (VIII.6.21)$^+$ and provided δ is small enough, supp $u \subset V_0$; thus u defines a $\mathscr{C}^\infty$ section of $\tilde{\Lambda}^{m-1,n-\mu-\mu'}$ with compact support contained in V_0.

Finally we look at the integral in (z,t)-space, $\int f\wedge \mathrm{D}u$. Recall that $\chi(\varphi(z,s_0,t)) = 1$ if $\varphi(z,s_0,t) > r_0$, which is true on supp $\mathrm{D}u$. It follows that

$$\int f\wedge \mathrm{D}u =$$

$$\int E_{\mu,\mu'}(z',t')\wedge d\theta(z',t')\wedge\chi_1(z'',t'')d\bar{z}_{\mu+1}\wedge\cdots\wedge d\bar{z}_{m-1}\wedge dt_{\mu'+1}\wedge\cdots\wedge dt_{n'}\wedge dz =$$

$$\pm\int \theta(z',t)\chi_1(z'',t'')dE_{\mu,\mu'}(z',t')\wedge d\bar{z}_{\mu+1}\wedge\cdots\wedge d\bar{z}_{m-1}\wedge dt_{\mu'+1}\wedge\cdots\wedge dt_{n'}\wedge dz =$$

$$\pm\int \chi_1(z'',t'')dz_{\mu+1}\wedge\cdots\wedge dz_{m-1}\wedge d\bar{z}_{\mu+1}\wedge\cdots\wedge d\bar{z}_{m-1}\wedge dt_{\mu'+1}\wedge\cdots\wedge dt_{n'} \neq 0$$

by Lemma VIII.6.A1 and the fact that $\chi_1 \geq 0$ is not identically zero. ■

REMARK VIII.6.2. If instead of (VIII.6.21)$^+$ we had made the hypothesis

$$\{ (z,t) \in U_0;\ z'' = 0,\ t'' = 0,\ \varphi(z,s_0,t) \geq -a_0 \} \subset\subset V_0, \qquad \text{(VIII.6.21)}^-$$

we would have reached the conclusion that Property (VIII.5.1)$_q^-$ holds. ■

REMARK VIII.6.3. Suppose $\varphi(0,z'',s_0,0,t'') \leq 0$ for all (z'',t''); and also that, for some number $a_0 > 0$, $\{ (z,t) \in U_0;\ \varphi(z,s_0,t) \leq a_0,\ z'' = 0,\ t'' = 0\} \subset\subset V_0$. Then $\varphi(z,s_0,t) \leq \mathit{const.}(|z'|+|t'|)$ and therefore (VIII.6.20) holds. ■

EXAMPLE VIII.6.1. Partition each set of variables z_i and t_j as in (VIII.6.19) and suppose that

$$\varphi(z,s_0,t) = \varphi_1(z',t') - \varphi_2(z'',t''), \qquad \text{(VIII.6.26)}$$

$$\varphi_1 \geq 0,\ \varphi_2 \geq 0 \ \textit{everywhere in}\ \Omega,\ \varphi_1|_0 = \varphi_2|_0 = 0. \qquad \text{(VIII.6.27)}$$

Assume furthermore that, for some number $a_0 > 0$,

$$\{ (z,t) \in U_0;\ z'' = 0,\ t'' = 0,\ \varphi_1(z',t') \leq a_0 \} \subset\subset V_0. \qquad \text{(VIII.6.28)}$$

Then Condition (VIII.6.19) is satisfied. Indeed, (VIII.6.28) is obviously equivalent to (VIII.6.21)$^+$; and Condition (VIII.6.20) holds, for if $\varphi_1(z',t') = a_0 + \varphi_2(z'',t'') \geq a_0 > 0$, we could not possibly have $z' = 0$ and $t' = 0$ since $\varphi_1|_0 = 0$. ■

EXAMPLE VIII.6.2. Suppose the structure of $\mathcal{M}$ is CR, i.e., there are no variables t_k, and that Condition (VIII.6.19) is satisfied. If necessary, and possibly after contracting U about 0, we may avail ourselves of Sard's lemma and slightly decrease a_0 to ensure that a_0 is a noncritical value of $\varphi(z,s_0)$. Then the equation $\varphi(z,s_0) = a_0$ defines a $\mathscr{C}^\infty$ hypersurface that does not intersect the z''=coordinate space ($z' = 0$) and whose intersection with the z'-coordinate space ($z'' = 0$) is a kind of compact singular cycle of codimension one. By further decreasing a_0, if need be, we can ensure that the equations $\varphi(z',0,s_0) = a_0$ defines a compact $\mathscr{C}^\infty$ hypersurface in z'-space $\mathbb{C}^\mu$. One of the bounded components of its complement contains the origin. ∎

Combining Proposition VIII.5.1 and Lemma VIII.6.1 enables us to derive from Theorem VIII.4.1 the following:

THEOREM VIII.6.2. *Suppose* $q = \mu+\mu'-1 \geq 0$ *and that there are an open neighborhood of* 0 *in* Ω, $\mathrm{U} = \Delta\times\mathcal{S}\times\mathcal{B}$, *a basis of open neighborhoods of* 0, $\mathrm{V}^{(\iota)} = \Delta^{(\iota)}\times\mathcal{S}^{(\iota)}\times\mathcal{B}^{(\iota)} \subset \mathrm{U}$, *and sequences of numbers* $s^{(\iota)} \in \mathcal{S}^{(\iota)}$, $a^{(\iota)} > 0$ ($\iota = 1,2,\ldots$) *such that Property* (VIII.6.19) *holds if* $\mathrm{V} = \mathrm{V}^{(\iota)}$, $s_0 = s^{(\iota)}$, $a_0 = a^{(\iota)}$. *Then the differential equation* $\mathrm{d}'u = f$ *is not locally solvable at* 0 *in degree* q.

In the preceding statement the numbers $s^{(\iota)}$, $a^{(\iota)}$ necessarily converge to zero.

EXAMPLE VIII.6.3. Suppose the hypo-analytic structure of $\mathcal{M}$ is CR; partition the set of variables z as in (VIII.6.19). Suppose that when $s = 0$ we have $\varphi = |z'|^2+\psi(z)$ with $|\psi(z',0)| \leq const.|z'|^3$ and $\psi(0,z'') \leq 0$. Then (Remark VIII.6.2) there is no local solvability at the origin in degree $q = \mu-1$ (≥ 0). This for instance is the situation if the Levi form at 0 has μ eignvalues of one sign and $n-\mu$ of the opposite sign (cf. Theorem VIII.3.1). ∎

EXAMPLE VIII.6.4. Take $\mathcal{M} = \mathbb{C}^2\times\mathbb{R}$ and

$$\varphi(z,s) = |z_1|^2 + \lambda|z_2|^2 \tag{VIII.6.29}$$

with λ a real number. The tangential Cauchy-Riemann equations are not locally solvable at 0 in degree *zero* if $\lambda \leq 0$ (Example VIII.6.3). On the other hand, it is well known that they are locally solvable at 0 in degree zero if $\lambda > 0$. We have here, for $\lambda = 0$, the example of a CR structure of the hypersurface type that is the limit of a sequence of CR structures in which local solvability at 0 holds, and of a sequence in which it does not. Note also that, when $\lambda = 0$, the Levi form of the structure is everywhere positive semidefinite. ∎

EXAMPLE VIII.6.5. Here also take $\mathcal{M} = \mathbb{C}^2\times\mathbb{R}$ and $\varphi(z,s) = |z_1|^2+|z_2|^2$. There is local solvability at 0 in degree zero. But now take

$U_0 = \{ z \in \mathbb{C}^2; |z_1| < r_1, |z_2| < r_2 \}$, $V_0 = \{ z \in \mathbb{C}^2; |z_1| < \rho_1, |z_2| < \rho_2 \}$ with $\rho_2 < r_2 < \rho_1 < r_1$. Then, given any two intervals $\mathcal{I}_1 \subset \mathcal{I}$ centered at 0, there is a $\bar{\partial}_b$-closed section $f \in \mathcal{C}^\infty(U_0 \times \mathcal{I}; \Lambda^{0,1})$ such that no distribution u in $V_0 \times \mathcal{I}_1$ satisfies there the equation $\bar{\partial}_b u = f$. If $r_2^2 < a_0 < \rho_1^2$ both Conditions (VIII.6.20) and (VIII.6.21)$^+$ are satisfied (with $s_0 = 0$). The nonsolvability result follows from Lemma VIII.6.1 by the arguments of section VIII.5. ■

Appendix to Section VIII.6: The Current $E_{\mu,\mu'}$

In accordance with the notation in section VIII.3 we write $z' = (z_1,\dots,z_\mu)$, $t' = (t_1,\dots,t_{\mu'})$. If $\mu \geq 1$ and $\mu' \geq 1$ we define the following form in $\mathbb{C}^\mu \times \mathbb{R}^{\mu'}$:

$$\Omega = 2\sum_{i=1}^{\mu} (-1)^{i-1}\bar{z}_i d\bar{z}_1\wedge\cdots\wedge\widehat{d\bar{z}_i}\wedge\cdots\wedge d\bar{z}_\mu\wedge dt_1\wedge\cdots\wedge dt_{\mu'} +$$

$$(-1)^\mu d\bar{z}_1\wedge\cdots\wedge d\bar{z}_\mu\wedge \sum_{j=1}^{\mu'} (-1)^{j-1} t_j dt_1\wedge\cdots\wedge\widehat{dt_j}\wedge\cdots\wedge dt_{\mu'}, \qquad \text{(VIII.6.A1)}$$

where the hatted factors must be omitted. When $\mu \geq 1$ and $\mu' = 0$, we set

$$\Omega = 2\sum_{i=1}^{\mu} (-1)^{i-1}\bar{z}_i d\bar{z}_1\wedge\cdots\wedge\widehat{d\bar{z}_i}\wedge\cdots\wedge d\bar{z}_\mu, \qquad \text{(VIII.6.A2)}$$

whereas when $\mu = 0$ and $\mu' = 1$, we set

$$\Omega = \sum_{j=1}^{\mu'} (-1)^{j-1} t_j dt_1\wedge\cdots\wedge\widehat{dt_j}\wedge\cdots\wedge dt_{\mu'}. \qquad \text{(VIII.6.A3)}$$

We have

$$d\Omega = (2\mu+\mu')d\bar{z}_1\wedge\cdots\wedge d\bar{z}_\mu\wedge dt_1\wedge\cdots\wedge dt_{\mu'}. \qquad \text{(VIII.6.A4)}$$

Let us now set

$$D = |z'|^2 + |t'|^2. \qquad \text{(VIII.6.A5)}$$

It is an immediate consequence of the definitions of Ω and of D that

$$dD\wedge\Omega = 2D d\bar{z}_1\wedge\cdots\wedge d\bar{z}_\mu\wedge dt_1\wedge\cdots\wedge dt_{\mu'} \bmod (dz_1,\dots,dz_\mu). \qquad \text{(VIII.6.A6)}$$

Notice that the coefficients of the differential form

$$E_{\mu,\mu'} = \Gamma(\mu+\mu'/2)(\pi D)^{-\mu-\frac{1}{2}\mu'}\,\Omega$$

are locally L^1-functions in $\mathbb{C}^\mu \times \mathbb{R}^{\mu'}$.

LEMMA VIII.6.A1. *Let* $\delta(z',t')$ *denote the Dirac distribution in* (z',t')*-space* $\mathbb{C}^{\mu}\times\mathbb{R}^{\mu'}$; *then*

$$dE_{\mu,\mu'} \equiv \delta(z',t')d\bar{z}_1\wedge\cdots\wedge d\bar{z}_\mu\wedge dt_1\wedge\cdots\wedge dt_{\mu'} \; mod \; (dz_1,\dots,dz_\mu). \tag{VIII.6.A7}$$

PROOF. Suppose first $2\mu+\mu' \geq 3$. Then Formula (VIII.6.A7) can be rewritten as

$$\sum_{i=1}^{\mu}(\partial/\partial\bar{z}_i)(\bar{z}_i/D^{\mu+\frac{1}{2}\mu'}) + \tfrac{1}{2}\sum_{j=1}^{\mu'}(\partial/\partial t_j)(t_j/D^{\mu+\frac{1}{2}\mu'}) =$$

$$[\pi^{\mu+\frac{1}{2}\mu'}/\Gamma(\mu+\mu'/2)]\delta(z',t'),$$

which, in turn, is equivalent to

$$\left\{4\sum_{i=1}^{\mu}\partial^2/\partial z_i\partial\bar{z}_i + \sum_{j=1}^{u'}\partial^2/\partial t_j^2\right\}(1/D^{\mu+\frac{1}{2}\mu'-1}) =$$

$$-[2(2\mu+\mu'-2)\pi^{\mu+\frac{1}{2}\mu'}/\Gamma(\mu+\mu'/2)]\delta(z',t'),$$

a restatement of the fact that $-\Gamma(\mu+\mu'/2)/[2(2\mu+\mu'-2)\pi^{\mu+\frac{1}{2}\mu'}D^{\mu+\frac{1}{2}\mu'-1}]$ is a fundamental solution of the Laplacian in $\mathbb{R}^{2\mu+\mu'}$.

When $\mu = 1$, $\mu' = 0$, Formula (VIII.6.A7) reads $(\partial/\partial\bar{z})(1/z) = \pi\delta(z)$. When $\mu = 0$ and $\mu' = 2$ it reads

$$(\partial/\partial t_1)(t_1/|t|^2) + (\partial/\partial t_2)(t_2/|t|^2) =$$

$$[(\partial/\partial t_1)^2+(\partial/\partial t_2)^2](\log|t|) = 2\pi\delta(t).$$

Finally, when $\mu = 0$ and $\mu' = 1$, it reads $(\partial/\partial t)(t/|t|) = 2\delta(t)$. ■

VIII.7. The Case of a Single Vector Field: Property $(\mathcal{P})$

The remaining sections of the present chapter will be devoted to the study of solvability in the case $n = 1$. Thus $\mathcal{M}$ shall be a $\mathcal{C}^\infty$ manifold, countable at infinity, carrying an involutive structure in which the tangent structure bundle $\mathcal{V}$ is a line bundle. As usual the cotagent structure bundle is called T' and its rank is $m \geq 1$.

If Γ is any one-dimensional orbit of the structure in an open subset $\mathcal{O}$ of $\mathcal{M}$ (Definition I.11.1) $\mathcal{V}|_\Gamma = \overline{\mathcal{V}}|_\Gamma$ is spanned by any nonvanishing vector field tangent to Γ. If Σ is a two-dimensional orbit in $\mathcal{O}$, $\mathcal{V}|_\Sigma \subset \mathbb{C}T\Sigma$ is the tangent structure bundle in the involutive structure that Σ inherits from $\mathcal{M}$; the cotangent structure bundle in this structure, T'_Σ, i.e., the pullback of T' to Σ, is a line bundle (since $\dim \Sigma = 2$ and rank $\mathcal{V} = 1$). We cannot have $T'_\Sigma = \overline{T}'_\Sigma$ throughout Σ; otherwise Σ would be foliated by integral curves of $\mathcal{V}$ and could not be an orbit. Keep in mind, however, that $T'_\Sigma = \overline{T}'_\Sigma$ can occur in nonempty open subsets of Σ.

DEFINITION VIII.7.1. *We shall say that the involutive structure of* $\mathcal{M}$ *has* Property ($\mathcal{P}$) in the open set $\mathcal{O}$ *if the following two requirements are satisfied:*

$$\textit{the orbits in } \mathcal{O} \textit{ have dimension} \leq 2; \tag{VIII.7.1}$$

$$\textit{every two-dimensional orbit } \Sigma \textit{ in } \mathcal{O} \textit{ is orientable and } T'_\Sigma \wedge \overline{T}'_\Sigma/2\imath \textit{ does not change sign on } \Sigma \text{ (Definition III.6.1).} \tag{VIII.7.2}$$

We say that Property ($\mathcal{P}$) holds at a point $\mathfrak{p} \in \mathcal{M}$ *if there is an open neighborhood of* $\mathfrak{p}$ *in which Property* ($\mathcal{P}$) *holds.*

If Condition ($\mathcal{P}$) holds in $\mathcal{O}$ then it will hold in every open subset of $\mathcal{O}$. In particular, ($\mathcal{P}$) is an "open" property.

Let U be an open neighborhood of $\mathfrak{p} \in \mathcal{M}$ in which Condition ($\mathcal{P}$) holds. Let $x_1,\ldots,x_m,t$ denote local coordinates in U; we shall assume that they all vanish at $\mathfrak{p}$. We take U small enough that $\mathcal{V}$ is spanned over U by a vector field

$$L = \partial/\partial t + \sum_{k=1}^{m} \lambda_k(x,t)\partial/\partial x_k,$$

with $\lambda_k \in \mathcal{C}^\infty(U)$. After contracting U about $\mathfrak{p}$, we can (for instance, by applying the Frobenius theorem, Theorem I.10.1) choose the x_j $(1 \leq j \leq m)$ in such a way that

$$\mathcal{R}e L = \partial/\partial t + \sum_{k=1}^{m} a_k(x,t)\partial/\partial x_k \ (a_k = \mathcal{R}e\lambda)$$

be proportional to $\partial/\partial t$, and thus, after division by a nonvanishing factor, assume

$$L = \partial/\partial t + \imath\sum_{k=1}^{m} b_k(x,t)\partial/\partial x_k, \tag{VIII.7.3}$$

with $b_k \in \mathcal{C}^\infty(\Omega)$ real-valued for all $k = 1,\ldots,m$. We see that the following one-forms make up a basis of T' over U:

$$\varpi_k = dx_k - \imath b_k(x,t)dt, \ k = 1,\ldots,m. \tag{VIII.7.4}$$

It is convenient to take U in the product form $U = \Omega \times \mathcal{J}$, with Ω an open set in x-space $\mathbb{R}^m$ and $\mathcal{J}$ an open interval in the real t-line, both containing the origin. We define the vector-valued function in U,

$$b(x,t) = (b_1(x,t),\ldots,b_m(x,t)) \ (\in \mathbb{R}^m).$$

Let Γ be a one-dimensional orbit of $\mathcal{V}$ in U; Γ must perforce be a line segment $\{x_0\} \times \mathcal{J}$ $(x_0 \in \Omega)$ since $\partial/\partial t$ is tangent to Γ. In this case we must have

$$b(x_0,t) \equiv 0, \forall\, t \in \mathcal{J}. \qquad \text{(VIII.7.5)}$$

Let Σ be a two-dimensional orbit in U. The vector field $\partial/\partial t$ is tangent to Σ, which must therefore be a union of segments $\{x_0\}\times\mathcal{J}$ $(x_0 \in \Omega)$. The intersection of Σ with the hyperplane $t = 0$ is a smooth curve γ; let

$$v = \sum_{k=1}^{m} v_k(x)\partial/\partial x_k$$

be a $\mathcal{C}^\infty$ vector field tangent to γ and nowhere zero. Below we also use the notation

$$\boldsymbol{v}(x) = (v_1(x),\ldots,v_m(x)) \ (\in \mathbb{R}^m);$$

we may as well assume $|\boldsymbol{v}| = 1$. We may extend $\boldsymbol{v}$ as a vector field tangent to Σ by "parallel transport" along the t-lines $\{x\}\times\mathcal{J}$ $(x \in \gamma)$. Since $\mathcal{I}m\mathrm{L}$ is tangent to Σ we have, necessarily,

$$\boldsymbol{b}(x,t) = \pm\, |\boldsymbol{b}(x,t)|\boldsymbol{v}(x),\ \forall\ (x,t) \in \Sigma.$$

Along Σ we have

$$\mathrm{L} = \partial/\partial t \pm \imath|\boldsymbol{b}(x,t)| \sum_{k=1}^{m} v_k(x)\partial/\partial x_k.$$

And along γ we can find a smooth function $\chi(x)$ satisfying

$$\sum_{k=1}^{m} v_k\partial\chi/\partial x_k = 1. \qquad \text{(VIII.7.6)}$$

We may use (χ,t) as coordinates on Σ and thus write, in Σ,

$$\mathrm{L} = \partial/\partial t \pm \imath|\boldsymbol{b}(x,t)|\partial/\partial\chi.$$

The two-form $\mathrm{d}\chi\wedge\mathrm{d}t$ orients Σ. The line bundle T'_Σ is spanned by the one-form

$$\varpi = -\,\mathrm{d}\chi \pm \imath|\boldsymbol{b}(x,t)|\mathrm{d}t.$$

We have

$$\varpi\wedge\bar{\varpi}/2\imath = \pm\, |\boldsymbol{b}(x,t)|\mathrm{d}\chi\wedge\mathrm{d}t. \qquad \text{(VIII.7.7)}$$

Notice that, for each $x_0 \in \gamma$, there is $t_0 \in \mathcal{J}$ such that $\boldsymbol{b}(x_0,t_0) \neq 0$, otherwise $\{x_0\}\times\mathcal{J}$ would be an orbit in U, and Σ would not be one. It follows then from (VIII.7.2) that the sign in the right-hand side of (VIII.7.7) is constant throughout Σ. After redefining $\boldsymbol{v}$ if need be we may assume it is the *plus* sign. It is also clear that $\boldsymbol{v}$ is then determined in a full neighborhood $\mathrm{V}_0 \subset \Omega$ of x_0 (in which $\boldsymbol{b}(x,t_0) \neq 0$) by the formula

$$\boldsymbol{v}(x) = \boldsymbol{b}(x,t_0)/|\boldsymbol{b}(x,t_0)|;$$

$\boldsymbol{v}$ is smooth in V_0. Note that also the function $|\boldsymbol{b}(x,t)|$ is smooth in V_0. Indeed, at each point $x \in V_0$ there must be an index k, $1 \leq k \leq m$, such that $|v_k(x)| \geq 1/\sqrt{m}$, and $|\boldsymbol{b}(x,t)| = b_k(\chi,t)/v_k(x)$. Thus, Properties (VIII.7.1) and (VIII.7.2) entail

$$\boldsymbol{b}(x,t) = |\boldsymbol{b}(x,t)|\boldsymbol{v}(x), \ \forall\ (x,t) \in \mathrm{U}, \tag{VIII.7.8}$$

$$\mathrm{L} = \partial/\partial t + \imath|\boldsymbol{b}(x,t)| \sum_{k=1}^{m} v_k(x)\partial/\partial x_k \ \textit{in}\ \mathrm{U}, \tag{VIII.7.9}$$

$$\mathrm{L} = \partial/\partial t + \imath|\boldsymbol{b}(x,t)|\partial/\partial\chi \ \textit{in}\ \Sigma. \tag{VIII.7.10}$$

It is then natural to introduce the following subset of Ω:

$$\mathcal{N}_0 = \{\, x \in \Omega;\ \forall\ t \in \mathcal{I},\ \boldsymbol{b}(x,t) = 0\,\}. \tag{VIII.7.11}$$

The product $\mathcal{N} = \mathcal{N}_0 \times \mathcal{I}$ is the union of all one-dimensional orbits in U; $\mathrm{U}\backslash\mathcal{N}$ is the union of all two-dimensional orbits in U; $\boldsymbol{v}(x)$ is a $\mathcal{C}^\infty$ map $\Omega\backslash\mathcal{N}_0 \to S^{m-1}$. Every orbit in U is orientable; if Σ is a two-dimensional orbit, $\mathrm{T}'_\Sigma \wedge \overline{\mathrm{T}}'_\Sigma/2\imath$ does not change sign on Σ (if $\dim \Sigma = 1$, $\mathrm{T}'_\Sigma \wedge \overline{\mathrm{T}}'_\Sigma \equiv 0$). The relations (VIII.7.8), (VIII.7.9), and (VIII.7.10) hold. Conversely, if this is so, Property ($\mathcal{P}$) is valid in U. In view of (VIII.7.8) these facts can be restated as follows:

Proposition VIII.7.1. *The validity of Property ($\mathcal{P}$) in* U *is equivalent to each one of the following properties:*

for every $x \in \Omega$, the vector-valued function $\mathcal{I} \ni t \to \boldsymbol{b}(x,t) \in \mathbb{R}^m$ never changes direction; (VIII.7.12)

for every $x \in \Omega$ and every $\xi \in \mathbb{R}^m$, the real-valued function (VIII.7.13)

$$\mathcal{I} \ni t \to \xi\cdot\boldsymbol{b}(x,t) = \sum_{k=1}^{m} b_k(x,t)\xi_k$$

never changes sign.

Evident. The reader will notice that $\xi\cdot\boldsymbol{b}(x,t)$ is the *symbol* of $\mathcal{I}m\mathrm{L}$ and that, for x fixed, $\{x\}\times\mathcal{I}$ is an integral curve of $\mathcal{R}e\mathrm{L}$. If we denote by τ the dual variable to t, we can view (x,t,ξ,τ)-space as the real cotangent bundle of $\mathbb{R}^{m+1}$, $\mathrm{T}^*\mathbb{R}^{m+1}$. The symbol of $\mathcal{R}e\mathrm{L}$ is equal to τ and the Hamiltonian field of that symbol is $\partial/\partial t$, regarded as a vector field in $\mathrm{T}^*\mathbb{R}^{m+1}$. Then, for (x,ξ) fixed, the line segment $\{(x,\xi,\tau)\}\times\mathcal{I}$ is an integral curve of the Hamiltonian field of the symbol of $\mathcal{R}e\mathrm{L}$, i.e., what is called a *bicharacteristic* of $\mathcal{R}e\mathrm{L}$. One may as well limit one's attention to those bicharacteristics of $\mathcal{R}e\mathrm{L}$ on which the symbol of $\mathcal{R}e\mathrm{L}$ vanishes identically, i.e., on which $\tau = 0$; they are called *null bicharacteristics* of $\mathcal{R}e\mathrm{L}$. This leads us to one of the standard formulations of Property ($\mathcal{P}$):

the symbol of $\mathcal{I}m$L does not change sign along any null bicharacteristic of $\mathcal{R}e$L. (VIII.7.14)

So stated, Condition ($\mathcal{P}$) has been shown to be necessary and sufficient in order that L, a linear partial differential operator of any order ≥ 1, of principal type, be locally solvable (see Hörmander [4], chap. 26). However, even in the case of a vector field L, the invariance of (VIII.7.14) under replacement of L by a nonvanishing multiple gL is by no means evident, whereas Properties (VIII.7.1) and (VIII.7.2) clearly are invariant.

We go back to an involutive structure on the manifold $\mathcal{M}$, with $n = 1$.

EXAMPLE VIII.7.1. In $\mathbb{R}^2$ (coordinates x,t) consider the vector field

$$\mathrm{L} = \partial/\partial t - \imath t^p \partial/\partial x. \qquad \text{(VIII.7.15)}$$

Condition ($\mathcal{P}$) is satisfied at the points $t = 0$ if and only if the nonnegative integer p is even. ■

EXAMPLE VIII.7.2. Let $\mathcal{M}$ be a three-dimensional CR manifold. The tangent structure bundle $\mathcal{V}$ in such a structure is one-dimensional. Let L be any nowhere vanishing $\mathcal{C}^\infty$ section of $\mathcal{V}$ over an open subset Ω of $\mathcal{M}$. In Ω L and $\overline{\mathrm{L}}$ are linearly independent and, as a consequence, the dimension of every orbit must be ≥ 2.

Suppose that the CR structure of $\mathcal{M}$ satisfies condition ($\mathcal{P}$) at every point of $\mathcal{M}$. As the dimension of every orbit is equal to two, then, at every point of Ω, the bracket $[\mathrm{L},\overline{\mathrm{L}}]$ must lie in the span of L and $\overline{\mathrm{L}}$, i.e., the structure must be Levi flat (Definition VI.7.1). This is only possible if

$$(2\imath)^{-1}[\mathrm{L},\overline{\mathrm{L}}] = [\mathcal{R}e\mathrm{L},\mathcal{I}m\mathrm{L}] = 0 \qquad \text{(VIII.7.16)}$$

everywhere in Ω; (VIII.7.16) means that $\mathcal{R}e$L and $\mathcal{I}m$L (which are linearly independent) define a real structure on Ω. Thus the image of $\mathcal{V}$ under the map $\mathcal{R}e$:$\mathbb{C}T\mathcal{M} \to T\mathcal{M}$ is a real plane bundle. By the Frobenius theorem this structure is locally integrable. In particular, in some open neighborhood of an arbitrary point of Ω, there are local coordinates (x',y',s') such that, after division of L by a nonvanishing factor, we have

$$\mathrm{L} = \tfrac{1}{2}(\partial/\partial x' + \imath\partial/\partial y').$$

The manifold $\mathcal{M}$ is foliated by the orbits of L. These orbits are two-dimensional (immersed) submanifolds of $\mathcal{M}$ which inherit a complex structure from $\mathcal{M}$. ■

The validity of Property ($\mathcal{P}$) at every point of an open set $\Omega \subset \mathcal{M}$ does not entail its validity in Ω:

EXAMPLE VIII.7.3. Let $b_k \in \mathscr{C}^\infty(\mathbb{R})$, $k = 1,2$, $b_1(t) > 0$ for $t > 1$, $b_1(t) = 0$ for $t < 1$, $b_2(t) > 0$ for $t < -1$, $b_2(t) = 0$ for $t > -1$. In the tube structure on $\mathbb{R}^3$ defined by the vector field

$$L = \partial/\partial t + \imath b_1(t)\partial/\partial x_1 + \imath b_2(t)\partial/\partial x_2,$$

the only orbit is $\mathbb{R}^3$ itself. Yet Property $(\mathscr{P})$ holds in each half-space $t < 1$, $t > -1$. ■

When the manifold $\mathcal{M}$ and its involutive structure are of class $\mathscr{C}^\omega$ the validity of $(\mathscr{P})$ at every point implies its validity in $\mathcal{M}$, for in this case the orbits are the integral manifolds of the Lie algebra $\mathfrak{g}(\mathscr{V})$ (see section I.11). One can then state:

PROPOSITION VIII.7.2. *Let $\mathcal{M}$ be a real-analytic manifold equipped with an analytic involutive structure in which the tangent structure bundle $\mathscr{V}$ has rank one.*

If Property $(\mathscr{P})$ holds at every point of $\mathcal{M}$, then it holds in $\mathcal{M}$ and every two-dimensional orbit inherits a hypocomplex structure (Definition III.5.1) *from $\mathcal{M}$.*

PROOF. If Σ is any two-dimensional orbit in $\mathcal{M}$, p any point of Σ, U any open neighborhood of p, then the connected components of $\Sigma \cap U$ are orbits of the structure in U. Take U as before: U is the domain of coordinates x_i $(1 \leq i \leq m)$ and t; (VIII.7.8), (VIII.7.9), and (VIII.7.10) are valid. If we had $T' = 0$ on some nonempty arc of curve $\gamma \subset \Sigma \cap U$, this arc would perforce lie in some segment $\{x_0\} \times \mathcal{J}$. But since $\boldsymbol{b}(x,t)$ is real-analytic, (VIII.7.5) would be true, implying $\gamma \subset \mathcal{N}$ and contradicting the fact that Σ is a two-dimensional orbit. It suffices then to apply Theorem III.6.2. ■

VIII.8. Sufficiency of Condition $(\mathscr{P})$: Existence of L^2 Solutions

Same setup as in section VIII.7. We shall reason under the hypothesis that the involutive structure of the manifold $\mathcal{M}$ satisfies Condition $(\mathscr{P})$. The analysis will be carried out in an open neighborhood U of a point of $\mathcal{M}$, referred to as "the origin", in which there are local coordinates x_i $(1 \leq i \leq m)$ and t; $U = \Omega \times \mathcal{J}$ as in section VIII.7. We assume that there is a $\mathscr{C}^\infty$ vector field L given by (VIII.7.3) which spans $\mathscr{V}$ over U. For the sake of simplicity the coefficients b_k will be assumed to be (real-valued) $\mathscr{C}^\infty$ functions in an open neighborhood $\mathscr{O}$ of the closure of U; (VIII.7.8) will hold in $\mathscr{O}$. Below $\|\cdot\|_0$ stands for the L^2 norm in (x,t)-space $\mathbb{R}^{m+1}$.

THEOREM VIII.8.1. *Let Ω' be any relatively compact open subset of Ω. There* *is a constant $C > 0$ such that, given any sufficiently small number $\varepsilon > 0$, we have*

$$\|u\|_0 \leq C\varepsilon\|Lu\|_0,\ \forall\ u \in \mathscr{C}_c^\infty(\Omega' \times]-\varepsilon,\varepsilon[). \qquad \text{(VIII.8.1)}$$

PROOF. Let $\mathcal{N}_0$ be the set (VIII.7.11) and $\mathcal{N} = \mathcal{N}_0 \times \mathcal{I}$. Let χ be the characteristic function of the set $U\backslash\mathcal{N}$. We have, for any pair of functions $u, f \in \mathscr{C}_c^\infty(U)$,

$$\langle L[(1-\chi)u], f\rangle = \int u(1-\chi)\ {}^t\!Lf\ dxdt,$$

where at the left is the bracket of the duality between test-functions and distributions, and where ${}^t\!L$ is the transpose of the differential operator L:

$${}^t\!L = -L - \imath c(x,t),\ c = \sum_{k=1}^{m} \partial b_k/\partial x_k. \qquad \text{(VIII.8.2)}$$

If $x_0 \in \mathcal{N}_0$ and $c(x_0,t_0) \neq 0$ for some $t_0 \in \mathcal{I}$ there is an open neighborhood $\mathcal{O}_0$ of x_0 in Ω such that $\mathcal{N}_0 \cap \mathcal{O}_0$ is contained in a submanifold $\{ x \in \mathcal{O}_0;\ b_k(x,t_0) = 0,\ db_k(x,t_0) \neq 0 \}$, and thus $\mathcal{N}_0 \cap \mathcal{O}_0$ has measure zero. It follows that $(1-\chi)c = 0$ almost everywhere. We conclude that, in the distribution sense,

$$L[(1-\chi)u] = (1-\chi)u_t = (1-\chi)Lu. \qquad \text{(VIII.8.3)}$$

Now, if $u \equiv 0$ for $|t| > \varepsilon$, we have

$$\int |u|^2 dt \leq 2\varepsilon^2 \int |u_t|^2 dt,$$

whence

$$\|(1-\chi)u\|_0 \leq \sqrt{2}\varepsilon\|(1-\chi)Lu\|_0,\ \forall\ u \in \mathscr{C}_c^\infty(\Omega \times [-\varepsilon,\varepsilon]). \qquad \text{(VIII.8.4)}$$

It will therefore suffice to prove that, for a suitable constant $C > 0$,

$$\|\chi u\|_0 \leq C\varepsilon\|\chi Lu\|_0,\ \forall\ u \in \mathscr{C}_c^\infty(\Omega' \times]-\varepsilon,\varepsilon[). \qquad \text{(VIII.8.5)}$$

We define the following function in Ω,

$$\rho(x) = \sup_{t \in \mathcal{I}} |\boldsymbol{b}(x,t)|. \qquad \text{(VIII.8.6)}$$

The zero set of ρ is precisely the set $\mathcal{N}_0$. Also, for some constant $C_1 > 0$,

$$|\rho(x) - \rho(x')| \leq C_1|x - x'|,\ \forall\ x, x' \in \Omega. \qquad \text{(VIII.8.7)}$$

Proof of (VIII.8.7): For any $x \in \Omega$ let t_x be a point in $\mathcal{C}\ell\,\mathcal{I}$ at which the function $t \to |\boldsymbol{b}(x,t)|$ reaches its maximum, $\rho(x)$. If C_1 is large enough, we have

$$\rho(x) = |\boldsymbol{b}(x,t_x)| \leq |\boldsymbol{b}(x',t_x)| + C_1|x-x'| \leq \rho(x') + C_1|x-x'|,$$

whence (VIII.8.7) after exchange of x and x'. ■

Given $x_0 \in \Omega\backslash\mathcal{N}_0$ select a point $t_0 \in \mathcal{C}\ell\mathcal{S}$ such that $|\boldsymbol{b}(x_0,t_0)| = \rho(x_0)$. There is an open neighborhood $\mathcal{O}_0$ of x_0 in Ω in which $|\boldsymbol{b}(x,t_0)| > 0$. As a consequence, using the elementary fact that $|\partial_x|\boldsymbol{b}(x,t_0)|| \leq |\partial_x \boldsymbol{b}(x,t_0)|$, we obtain, in $\mathcal{O}_0$,

$$|\partial_x \boldsymbol{v}(x)| \leq 2|\partial_x \boldsymbol{b}(x,t_0)|/|\boldsymbol{b}(x,t_0)|.$$

Putting $x = x_0$ in this inequality leads to the conclusion that there is a constant $A > 0$ such that

$$|\partial_x \boldsymbol{v}(x)| \leq A\rho(x)^{-1}, \ \forall\, x \in \Omega\backslash\mathcal{N}_0. \tag{VIII.8.8}$$

We introduce a number Δ_0 such that

$$\rho(x_0) < \Delta_0 < 2\rho(x_0), \tag{VIII.8.9}$$

and a small number $\theta > 0$, which will eventually be chosen (independently of x_0) and kept fixed. Next we introduce a number of cubes in $\mathbb{R}^m$, all centered at x_0 and with edges parallel to the coordinate axes; and first of all the cube $\mathcal{Q}_1$ with diameter $\theta\Delta_0$. By (VIII.8.7) and (VIII.8.9),

$$x \in \mathcal{Q}_1 \Rightarrow |\rho(x) - \rho(x_0)| \leq 2C_1\rho(x_0)\theta,$$

and therefore, if $\theta < 1/4C_1$,

$$\tfrac{1}{2}\rho(x_0) < \rho(x) < 2\rho(x_0), \ \forall\, x \in \mathcal{Q}_1. \tag{VIII.8.10}$$

This implies $\mathcal{Q}_1 \subset\subset \Omega\backslash\mathcal{N}_0$.

Since $|\boldsymbol{v}| \equiv 1$ we must have $|v_j(x_0)| \geq 1/\sqrt{m}$ for some j, $1 \leq j \leq m$. For simplicity suppose this is true for $j = 1$. From (VIII.8.8) and from the definition of $\mathcal{Q}_1$ we get

$$|v_1(x) - v_1(x_0)| \leq A\theta, \ \forall\, x \in \mathcal{Q}_1,$$

and thus, provided $2A\theta \leq 1/\sqrt{m}$,

$$|v_1(x)| \geq 1/2\sqrt{m}, \ \forall\, x \in \mathcal{Q}_1. \tag{VIII.8.11}$$

This shows that, in $\mathcal{Q}_1$,

$$|\partial_x(v_j/v_1)| \leq 2\sqrt{m}|\partial_x v_j| + 4m|\partial_x v_1|,$$

whence, by (VIII.8.8)

$$|\partial_x(v_j/v_1)| \leq 6Am/\rho \ \textit{in}\ \mathcal{Q}_1\ (j = 2,\ldots,m). \tag{VIII.8.12}$$

We define a mapping $y \to x(y)$ from the cube $\mathcal{Q}_2$ centered at x_0, with edges parallel to the coordinate axes and diameter $\theta^2\Delta_0$, into $\mathcal{Q}_1$, as follows: $x(y)$ is the solution of the system of integral equations:

$$x_1 = y_1,\ x_j(y) = y_j + \int_{x_{01}}^{y_1} (v_j/v_1)(s,x'(s,y'))ds,\ j = 2,\ldots,m, \tag{VIII.8.13}$$

($x' = (x_2,\ldots,x_m)$, $y' = (y_2,\ldots,y_m)$). We have, for each $j = 2,\ldots,m$,

$$\partial x_j/\partial y_1 = (v_j/v_1)(x),\ x_j|_{y_1=x_{0_1}} = y_j \qquad \text{(VIII.8.14)}$$

Suppose y belongs to $\mathfrak{Q}_2$. We define a sequence of functions $x'_n(y)$ ($n = 0,1,\ldots$) by induction as follows:

$$x'_0 = y',\ x'_n(y) = y' + \int_{x_{0_1}}^{y_1} (v'/v_1)(s,x'_{n-1}(s,y'))ds,$$

where $v' = (v_2,\ldots,v_m)$. Suppose (x_1,x'_{n-1}) stays inside $\mathfrak{Q}_1$; then, by (VIII.8.11),

$$|x'_n - y'| \le 2\sqrt{m}|y_1 - x_{0_1}|,$$

whence

$$|x_n - x_0| \le (1+4m)^{\frac{1}{2}}|y_1 - x_{0_1}| \le (1+m)^{\frac{1}{2}}\theta^2\Delta_0,$$

and thus, if $(1+m)^{\frac{1}{2}}\theta < 1$, (x_1,x'_n) will also stay inside $\mathfrak{Q}_1$. This remark allows us to apply the Picard method and to conclude, in standard fashion, that the sequence $\{x'_n\}_{n=0,1\ldots}$ converges uniformly to a map $x'(y)$: $\mathfrak{Q}_2 \to \mathbb{R}^{m-1}$. Also, that $y \to x(y) = (y_1,x'(y))$ is a diffeomorphism of $\mathfrak{Q}_2$ onto an open neighborhood of x_0 in $\mathfrak{Q}_1$.

Let us then introduce two more cubes, $\mathfrak{Q}_3$ and $\mathfrak{Q}_4$, defined just like $\mathfrak{Q}_1$ except that their diameters are equal to $\theta^3\Delta_0$ and $\theta^4\Delta_0$ respectively. From (VIII.8.11) and (VIII.8.13) we derive, for all $y \in \mathfrak{Q}_2$,

$$|x(y) - y| \le \sqrt{m}|x_1 - x_0|. \qquad \text{(VIII.8.15)}$$

Suppose then that $x(y) \in \mathfrak{Q}_4$. This implies $|y - x_0| \le (1+m)^{\frac{1}{2}}\theta^4\Delta_0$, i.e., $y \in \mathfrak{Q}_3$. This means that $\mathfrak{Q}_2\backslash\mathfrak{Q}_3$ is mapped by $y \to x(y)$ into $\mathfrak{Q}_1\backslash\mathfrak{Q}_4$ and therefore the image of $\mathfrak{Q}_3$ must necessarily contain $\mathfrak{Q}_4$.

By (VIII.8.14) the pushforward, under the diffeomorphism $y \to x(y)$, of the vector field in $\mathfrak{Q}_2 \times \mathcal{J}$,

$$\tilde{L} = \partial/\partial t + \imath\tilde{b}(y,t)\partial/\partial y_1,$$

is equal to L; here $\tilde{b}(y,t) = b_1(x(y),t)$. Note that if u is any function belonging to $\mathscr{C}_c^\infty(\mathfrak{Q}_4 \times \mathcal{J})$, its pullback $u(x(y),t)$ is compactly supported in $\mathfrak{Q}_3 \times \mathcal{J}$.

The intersection $\mathscr{R}(a')$ of $\mathfrak{Q}_2 \times \mathcal{J}$ with the affine subspace $y' = a'$ is mapped into the intersection of a two-dimensional orbit Σ with $\mathfrak{Q}_0 \times \mathcal{J}$. By the hypothesis ($\mathscr{P}$) $\tilde{b}(y,t)$ never changes sign in $\mathscr{R}(a')$ and a fortiori in $\mathscr{R}(a') \cap \mathfrak{Q}_3$. The latter intersection can be identified to the open rectangle

$$\mathscr{R}_1 = \{\ (y_1,t) \in \mathbb{R}^2;\ |y_1 - x_{0_1}| < \theta^3\Delta_0,\ t \in \mathcal{J}\ \}.$$

We are going to apply Lemma III.6.3 in $\mathscr{R}_1$ with $\tilde{L}$ substituted for ^{t}L. The bounds on the constant C in (III.6.13) are of outmost importance to us. Inspection of the proof of Lemma III.6.3 shows that such bounds derive from bounds on the norm of the pseudodifferential operator (of order zero)

$$(\mathbf{D}_{y_1}^{\pm})^{\frac{1}{2}}[(\mathbf{D}_{y_1}^{\pm})^{\frac{1}{2}}, \tilde{b}(y,t)] \tag{VIII.8.16}$$

as a bounded linear operator on $L^2(\mathcal{R}_1)$. Keep in mind that the role of x in section III.6 is played here by y_1; y' plays the role of a parameter. Let us also note that

$$\partial\tilde{b}/\partial y_1 = \sum_{k=1}^{m} v_k/v_1(\partial b_1/\partial x_k),$$

whence

$$|\partial\tilde{b}/\partial y_1| \leq \sqrt{m}|\partial_x b_1|, \tag{VIII.8.17}$$

$$|\partial^2\tilde{b}/\partial y_1^2| \leq B/\tilde{\rho}(y), \tag{VIII.8.18}$$

where $\tilde{\rho}(y) = \rho(x(y))$ and B is a positive constant. Actually we are going to replace $\tilde{b}(y,t)$ by $\varphi(y_1)\tilde{b}(y,t)$, $\varphi \in \mathscr{C}_c^\infty(\mathbb{R})$ satisfying the following conditions: $0 \leq \varphi \leq 1$ everywhere, $\varphi \equiv 1$ if $|y_1 - x_{0_1}| < \theta^3\Delta_0$, $\varphi \equiv 0$ if $|y_1 - x_{0_1}| > \theta^2\Delta_0$. We shall further require that to each integer $N \geq 1$ there be a constant $C_N > 0$ such that

$$|\varphi^{(N)}| \leq C_N(\theta\Delta_0)^{-N}. \tag{VIII.8.19}$$

Combining (VIII.8.17), (VIII.8.18), and (VIII.8.19) yields:

$$|(\partial/\partial y_1)[\varphi(y_1)\tilde{b}(y,t)]| \leq C\theta^{-2}, \tag{VIII.8.20}$$

as well as

$$|(\partial/\partial y_1)^2[\varphi(y_1)\tilde{b}(y,t)]| \leq C\theta^{-4}\Delta_0^{-1}.$$

(We have made use of the fact that $|\tilde{b}_1| \leq \rho(x_0)$, and also of (VIII.8.9).) From the last inequality we derive

$$\int |(\partial/\partial y_1)^2[\varphi(y_1)\tilde{b}(y,t)]| dy_1 \leq C\theta^{-2}. \tag{VIII.8.21}$$

In the sequel we shall not modify any more the constant θ.

We shall provisionally forget the variables y_j, $j \geq 2$, as well as t; until specified otherwise, we write x instead of y_1. We shall be dealing with a $\mathscr{C}^\infty$ function β in $\mathbb{R}$ endowed with properties analogous to those of $y_1 \to \varphi(y_1)\tilde{b}(y,t)$. In particular, the x-projection of the support of β is contained in a compact subset of the real line. Below we use the notation

$$N(\beta) = \text{Max}\ |\beta| + \text{Max}\ |\beta'| + \|\beta''\|_{L^1}.$$

We are going to show that the commutator

$$(\mathbf{D}^+)^{\frac{1}{2}}[(\mathbf{D}^+)^{\frac{1}{2}},\beta] = \mathbf{D}^+\beta - (\mathbf{D}^+)^{\frac{1}{2}}\beta(\mathbf{D}^+)^{\frac{1}{2}}$$

defines a bounded linear operator on $L^2(\mathbb{R})$.

LEMMA VIII.8.1. *There is a constant* $C > 0$ *such that*

$$\left|\int \beta\{u\mathbf{D}^+\bar{v} - [(\mathbf{D}^+)^{\frac{1}{2}}u][(\mathbf{D}^+)^{\frac{1}{2}}\bar{v}]\}dx\right| \leq CN(\beta), \qquad \text{(VIII.8.22)}$$

for all $u, v \in \mathscr{C}_c^\infty(\mathbb{R})$ *such that* $\|u\|_0 \leq 1$, $\|v\|_0 \leq 1$.

Proof of Lemma VIII.8.1: We have

$$4\pi^2 \int \beta\left[u\mathbf{D}^+\bar{v} - [(\mathbf{D}^+)^{\frac{1}{2}}u](\mathbf{D}^+)^{\frac{1}{2}}\bar{v}\right]dx =$$
$$\int_{\xi>0}\int_{\eta>0} \hat{\beta}(\xi-\eta)(\xi^{\frac{1}{2}}-\eta^{\frac{1}{2}})\xi^{\frac{1}{2}}\hat{u}(\eta)\overline{\hat{v}(\xi)}d\xi d\eta \;+$$
$$\int_{\xi>0}\int_{\eta<0} \hat{\beta}(\xi-\eta)\hat{u}(\eta)\xi\overline{\hat{v}(\xi)}d\xi d\eta. \qquad \text{(VIII.8.23)}$$

Up to the factor 2 the first term in the right-hand side in (VIII.8.23) is equal to

$$2\int_{\xi>0}\int_{\eta>0} (\xi-\eta)\hat{\beta}(\xi-\eta)(\xi^{\frac{1}{2}}+\eta^{\frac{1}{2}})^{-1}\xi^{\frac{1}{2}}\hat{u}(\eta)\overline{\hat{v}(\xi)}d\xi d\eta =$$
$$\int_{\xi>0}\int_{\eta>0} (\xi-\eta)\hat{\beta}(\xi-\eta)(\xi^{\frac{1}{2}}+\eta^{\frac{1}{2}})^{-1}(\xi^{\frac{1}{2}}-\eta^{\frac{1}{2}})\hat{u}(\eta)\overline{\hat{v}(\xi)}d\xi d\eta \;+$$
$$\int_{\xi>0}\int_{\eta>0} (\xi-\eta)\hat{\beta}(\xi-\eta)\hat{u}(\eta)\overline{\hat{v}(\xi)}d\xi d\eta. \qquad \text{(VIII.8.24)}$$

The absolute value of the first term in the right-hand side of (VIII.8.24) is equal to

$$\left|\int_{\xi>0}\int_{\eta>0} (\xi-\eta)\hat{\beta}(\xi-\eta)(\xi^{\frac{1}{2}}+\eta^{\frac{1}{2}})^{-1}(\xi^{\frac{1}{2}}-\eta^{\frac{1}{2}})\hat{u}(\eta)\overline{\hat{v}(\xi)}d\xi d\eta\right| =$$

$$\left|\int_{\xi>0}\int_{\eta>0} (\xi-\eta)^2\hat{\beta}(\xi-\eta)(\xi^{\frac{1}{2}}+\eta^{\frac{1}{2}})^{-2}\hat{u}(\eta)\overline{\hat{v}(\xi)}d\xi d\eta\right| \leq$$

$$\left[\int|\beta''(x)|dx\right]\int_{\xi>0}\int_{\eta>0} (\xi+\eta)^{-1}|\hat{u}(\eta)\hat{v}(\xi)|d\xi d\eta \leq \mathit{const.}\int |\beta''(x)|dx.$$

As for the second term at the right in (VIII.8.24), it is equal to

$$\int_{\xi>0}\int_{\eta>0}(\xi-\eta)\hat{\beta}(\xi-\eta)\hat{u}(\eta)\overline{\hat{v}(\xi)}d\xi d\eta =$$

$$-\ 4\iota\pi^2\int \beta'(x)\mathbf{P}^+u(x)\overline{\mathbf{P}^+v(x)}\ dx,$$

where $\mathbf{P}^+$ is the operator so denoted in the proof of Lemma III.6.3. Thus

$$\left|\int_{\xi>0}\int_{\eta>0}(\xi-\eta)\hat{\beta}(\xi-\eta)\hat{u}(\eta)\overline{\hat{v}(\xi)}d\xi d\eta\right| \leq 4\pi^2\text{Max}\ |\beta'|.$$

The absolute value of the second term, in the right-hand side of (VIII.8.23), does not exceed

$$\int_{\xi>0}\int_{\eta<0}|(\xi-\eta)^2\hat{\beta}(\xi-\eta)|(\xi-\eta)^{-1}|\hat{u}(\eta)|\ |\hat{v}(\xi)|d\xi d\eta$$

$$\leq \textit{const.}\int|\beta''(x)|dx. \blacksquare$$

By Lemma III.6.3 we reach the conclusion that there is a constant $C > 0$, independent of $x_0 \in \Omega\backslash\mathcal{N}_0$ and of $\varepsilon > 0$ such that $\|u\|_0 \leq C\varepsilon\|\tilde{\mathrm{L}}u\|_0$ for all $u \in \mathscr{C}_c^\infty(\mathbb{R}^{m+1})$, supp $u \subset \{(y,t);\ y \in \mathfrak{Q}_3,\ t \in \mathcal{I}\}$. Reverting to the variables x we obtain

$$\|u\|_0 \leq C\varepsilon\|\mathrm{L}u\|_0,\ \forall\ u \in \mathscr{C}_c^\infty(\mathfrak{Q}_4\times]-\varepsilon,\varepsilon[). \qquad \text{(VIII.8.25)}$$

At this stage we apply the Whitney lemma stated and proved in the Appendix to the present section. Let $\Omega' \subset\subset \Omega$ be open. We select a function $g \in \mathscr{C}_c^\infty(\Omega)$, $g \equiv 1$ in some open neighborhood Ω'' of $\mathscr{Cl}\,\Omega'$, and define $f = g\rho$. We make use of the covering of the open set $\{x \in \Omega; f(x) \neq 0\}$ by cubes $\mathfrak{Q}^{(j)}$ ($j \in \mathbb{Z}_+$) constructed in the Appendix. In applying (VIII.8.25) we take $\mathfrak{Q}_4$ to be one of these cubes. The number A in the Appendix will be taken to be $\frac{1}{4}\theta^{-4}$ (θ will be chosen in accordance with the requirements specified earlier). That Condition (VIII.8.A1) is satisfied follows at once from the definition of $\mathfrak{Q}_4$ and from (VIII.8.9), (VIII.8.10). We use the partition of unity $\{\psi_j\}$, subordinate to the covering $\{\mathfrak{Q}^{(j)}\}$, of Lemma VIII.8.A1. Substituting $g(x)\boldsymbol{b}(x,t)$ for $\boldsymbol{b}(x,t)$ only modifies L outside $\Omega''\times\mathcal{I}$, and thus (VIII.8.25) implies

$$\|\psi_j u\|_0 \leq C\varepsilon\|\mathrm{L}(\psi_j u)\|_0,\ \forall\ u \in \mathscr{C}_c^\infty(\Omega'\times]-\varepsilon,\varepsilon[). \qquad \text{(VIII.8.26)}$$

We have

$$\sum_{j=0}^{+\infty}\|\mathrm{L}(\psi_j u)\|_0^2 \leq 2\|\Psi\mathrm{L}u\|_0^2 + 2\sum_{j=0}^{+\infty}\|(\boldsymbol{b}\cdot\partial_x\psi_j)u\|_0^2,$$

where $\Psi = (\Sigma_{j=1}^{+\infty}|\psi_j^2|)^{\frac{1}{2}}$. We apply (VIII.8.A3):

$$\sum_{j=0}^{+\infty} |\boldsymbol{b}\cdot\partial_x\psi_j|^2 \leq C'\chi,$$

whence, by adding up the estimates (VIII.8.26) for $j = 0,1,\ldots,$

$$\|\Psi u\|_0^2 \leq 2C\varepsilon\left\{\|\Psi \mathrm{L} u\|_0 + C'\|\chi u\|_0\right\}, \ \forall\, u \in \mathscr{C}_c^\infty(\Omega' \times]-\varepsilon,\varepsilon[).$$

Since $\{\psi_j\}$ is a partition of unity we have $\Psi \leq \chi$ in $\Omega'\backslash(\Omega'\cap\mathscr{N}_0)$ (we recall that $\chi = 1$ in $\Omega\backslash\mathscr{N}_0$, $\chi = 0$ in $\mathscr{N}_0$). On the other hand, by virtue of Property (VIII.8.A2) we have

$$(\Sigma_{j=1}^{+\infty}\psi_j)^2 \leq C(\nu)\Psi^2,$$

and thus, for all $u \in \mathscr{C}_c^\infty(\Omega' \times]-\varepsilon,\varepsilon[)$,

$$\|\chi u\|_0^2 \leq C''\varepsilon\left\{\|\chi \mathrm{L} u\|_0^2 + \|\chi u\|_0^2\right\},$$

whence (VIII.8.5) if we require $C''\varepsilon < \frac{1}{2}$.

The proof of Theorem VIII.8.1 is complete. ∎

Corollary VIII.8.1. *There is a constant $C > 0$ such that, given any sufficiently small number $\varepsilon > 0$ and any function $u \in \mathscr{C}_c^\infty(\Omega' \times]-\varepsilon,\varepsilon[)$,*

$$\|u\|_0 \leq C\varepsilon\|{}^t\mathrm{L} u\|_0. \qquad \text{(VIII.8.27)}$$

Proof. By (VIII.8.2) we have $\|\mathrm{L}u\|_0 \leq \|{}^t\mathrm{L}u\|_0 + (\underset{\mathrm{U}}{\mathrm{Max}}|c|)\|u\|_0$. Putting this into (VIII.8.1) and requiring $C\varepsilon\underset{\mathrm{U}}{\mathrm{Max}}|c| < 1/2$ yields what we want. ∎

Corollary VIII.8.2. *Let C and ε be as in Corollary* VIII.8.1. *There is a bounded linear operator* $\mathbf{G}$ *on* $L^2(\Omega' \times]-\varepsilon,\varepsilon[)$, *with norm* $\leq C\varepsilon$, *such that, for all* $f \in L^2(\Omega' \times]-\varepsilon,\varepsilon[)$,

$$\mathrm{L}\mathbf{G}f = f. \qquad \text{(VIII.8.28)}$$

Corollary VIII.8.2 follows immediately from Lemma III.6.A3 and Corollary VIII.8.1.

If we apply Corollary VIII.8.1 in conjunction with Lemmas III.6.A2, III.6.A3, we can state:

Corollary VIII.8.3. *Suppose* (VIII.7.9) *holds. To each real number s there is an open neighborhood $\mathrm{U}_s \subset \mathrm{U}$ of the origin and a bounded linear operator $\mathbf{G}_s$ on the Sobolev space $\mathrm{H}^s(\mathbb{R}^{m+1})$ such that, for all $f \in \mathrm{H}^s(\mathbb{R}^{m+1})$,*

$$\mathrm{L}\mathbf{G}_s f = f \text{ in } \mathrm{U}_s. \qquad \text{(VIII.8.29)}$$

Appendix to Section VIII.8: A Whitney Lemma

The variable in $\mathbb{R}^m$ is denoted $x = (x_1,\dots,x_m)$. Let f be a nonnegative function in $\mathbb{R}^m$ such that, for some constant $K > 0$ and all $x, x' \in \mathbb{R}^m$,

$$|f(x)-f(x')| \leq K|x-x'|.$$

We write

$$\overline{\mathcal{N}} = \{ x \in \mathbb{R}^m; f(x) = 0 \}.$$

Throughout the Appendix by a *cube* we shall always mean a cube whose edges are parallel to the coordinate axes.

LEMMA VIII.8.A1. *There is an integer $\nu \geq 1$ and a constant $C > 0$ such that the following is true, whatever the number $A > 1$. There is a covering of $\mathbb{R}^m\backslash\overline{\mathcal{N}}$ by open cubes $\mathfrak{Q}^{(j)}$ ($j \in \mathbb{Z}_+$) endowed with the following properties:*

$$A \text{ diam } \mathfrak{Q}^{(j)} \leq \inf_{\mathfrak{Q}^{(j)}} f(x); \tag{VIII.8.A1}$$

each cube $\mathfrak{Q}^{(j)}$ intersects at most ν cubes $\mathfrak{Q}^{(j')}$, $j' \neq j$; (VIII.8.A2)

there is a partition of unity $\{\psi_j\}$ in $\mathcal{C}_c^\infty(\mathbb{R}^m\backslash\overline{\mathcal{N}})$, subordinate to the covering $\{\mathfrak{Q}^{(j)}\}$ of $\mathbb{R}^m\backslash\overline{\mathcal{N}}$, and to each $\alpha \in \mathbb{Z}_+^m$, a constant $C_\alpha > 0$ such that (VIII.8.A3)

$$\sum_{j=0}^{+\infty} |\partial_x^\alpha \psi_j| \leq C_\alpha[1+A/f(x)]^{|\alpha|}.$$

PROOF. Denote by $\mathcal{F}$ the family of *closed* cubes with vertices at points of $\mathbb{Z}^m$ (the lattice of points in $\mathbb{R}^m$ with integral coordinates) and with edge-length 1. Define $\mathcal{F}_0$ to be the set of cubes $\mathfrak{Q} \in \mathcal{F}$ such that

$$\inf_{\mathfrak{Q}} f \geq 2B \text{ diam } \mathfrak{Q},\ B = A+K. \tag{VIII.8.A4}$$

Here, of course, diam $\mathfrak{Q} = \sqrt{m}$. Subdivide each cube in $\mathcal{F}\backslash\mathcal{F}_0$ into 2^m closed cubes with vertices at points of $\frac{1}{2}\mathbb{Z}^m$, with edge-length 1/2. Call $\mathcal{F}_1$ the set of the latter cubes in which (VIII.8.A4) holds. Repeat the operation: for each $p \in \mathbb{Z}_+$, $\mathcal{F}_p$ consists of closed cubes with vertices at points in $2^{-p}\mathbb{Z}^m$ with edge-length 2^{-p}, in each of which (VIII.8.A4) holds true.

We claim that the union $\mathcal{G}$ of the families $\mathcal{F}_p$ ($p = 0,1,\dots$) constitutes a covering of $\mathbb{R}^m\backslash\overline{\mathcal{N}}$. Suppose x does not belong to any of the cubes $\mathfrak{Q} \in \mathcal{G}$. This means that there is a sequence of closed cubes $\mathfrak{Q}_p$ ($p = 0,1,\dots$) each of which contains x and has the following properties: (i) the edge-length of $\mathfrak{Q}_p$ is equal to 2^{-p}; (ii) the minimum of f on $\mathcal{Q}_p$, reached at a point x_p, is less than $B\sqrt{m}2^{-p+1}$. We have

$$f(x) \leq f(x_p) + \sqrt{m}2^{-p} K \leq \sqrt{m}2^{-p}(2B+K).$$

By letting p go to $+\infty$ we conclude that $x \in \overline{\mathcal{N}}$.

Another claim is that if $\mathcal{Q}$ is any cube belonging to $\mathcal{G}$ but not to $\mathcal{F}_0$, then

$$\forall\, x \in \mathcal{Q}, f(x) \leq 6B \text{ diam } \mathcal{Q}. \qquad \text{(VIII.8.A5)}$$

Indeed, suppose $f(x) > 6B$ diam $\mathcal{Q}$ for some $x \in \mathcal{Q}$ and let x' be any point such that $|x - x'| \leq$ diam $\mathcal{Q}$. We would have

$$f(x') \geq 6B \text{ diam } \mathcal{Q} - 2K \text{ diam } \mathcal{Q} \geq 4B \text{ diam } \mathcal{Q}.$$

But this would entail that the cube whose subdivision should have yielded $\mathcal{Q}$ satisfies (VIII.8.A4), and therefore would not have been subdivided.

Call $\mathcal{H}$ the family of *open* cubes $\mathcal{Q}$ obtained as follows: if x_0 is the center of $\mathcal{Q}$ the contraction $x \to \frac{1}{2}(x - x_0)$ transforms $\mathcal{Q}$ into a cube whose closure belongs to $\mathcal{G}$. Let x then be an arbitrary point in $\mathcal{Q}$; we have $f(x) \geq f(x_0) - K$ diam $\mathcal{Q}$, whence (VIII.8.A1), thanks to (VIII.8.A4). It follows from this that every cube belonging to $\mathcal{H}$ is contained in $\mathbb{R}^m \backslash \overline{\mathcal{N}}$; and since $\mathcal{G}$ is a closed covering of $\mathbb{R}^m \backslash \overline{\mathcal{N}}$, $\mathcal{H}$ is an open covering of $\mathbb{R}^m \backslash \overline{\mathcal{N}}$.

On the other hand, (VIII.8.A5) implies, for all $\mathcal{Q} \in \mathcal{H}$ whose edge-length is ≤ 1,

$$\exists\, x \in \mathcal{Q}, f(x) \leq 3B \text{ diam } \mathcal{Q}. \qquad \text{(VIII.8.A6)}$$

Let another cube $\mathcal{Q}' \in \mathcal{H}$ intersect $\mathcal{Q}$; let the edge-length of $\mathcal{Q}$ (resp., $\mathcal{Q}'$) be equal to 2^{-p} (resp., $2^{-p'}$); suppose both integers p and p' to be ≥ 0. For any $x \in \mathcal{Q} \cap \mathcal{Q}', f(x) \geq A\sqrt{m}2^{-p}$. But we may avail ourselves of (VIII.8.A6) and use the fact that there is $x' \in \mathcal{Q}'$ such that $f(x') \leq 3B\sqrt{m}2^{-p'}$, whence

$$A2^{-p} - 3B2^{-p'} \leq K2^{-p'}, \text{ i.e., } 2^{-p} \leq (3 + 4K/A)2^{-p'},$$

and after exchange of p and p',

$$|p - p'| \leq \log_2(3 + 4K). \qquad \text{(VIII.8.A7)}$$

This immediately implies that there is an integer ν, depending only on the Lipschitz constant K, which bounds the number of cubes $\mathcal{Q}' \in \mathcal{H}$ that intersect a given cube $\mathcal{Q} \in \mathcal{H}$ and are distinct from $\mathcal{Q}$, whence (VIII.8.A2).

Another consequence of (VIII.8.A6) is that, if $\mathcal{Q} \in \mathcal{H}$ and if the edge-length of $\mathcal{Q}$ does not exceed 1, then

$$\forall\, x \in \mathcal{Q}, f(x) \leq (3 + 4K)A\text{diam } \mathcal{Q}. \qquad \text{(VIII.8.A8)}$$

Let us then order the cubes in the family $\mathcal{H}$ into a sequence $\{\mathcal{Q}_j\}_{j=0,1,\ldots}$; denote by c_j the center of the cube $\mathcal{Q}_j$ and by ℓ_j *half* its edge-length. Let φ be a $\mathcal{C}^\infty$ function in $\mathbb{R}^m$, $\varphi \geq 0$ everywhere, $\varphi \equiv 1$ in an open neighborhood of the closed cube centered at the origin and with edge-length one, $\varphi \equiv 0$ outside some compact subset of the open cube centered at the origin and with edge-length two. The support of the function $\varphi((x-c_j)/\ell_j)$ is obviously contained in $\mathcal{Q}_j$ and $\varphi((x-c_j)/\ell_j) \equiv 1$ in some cube centered at c_j with edge-length $> \ell_j$. It follows from the fact that the family $\mathcal{G}$ is a closed covering of $\mathbb{R}^m \backslash \overline{\mathcal{N}}$ that the functions φ_j below make up a partition of unity in $\mathbb{R}^m \backslash \overline{\mathcal{N}}$:

$$\varphi_j(x) = D(x)^{-1}\varphi((x-c_j)/\ell_j)$$

where

$$D(x) = \sum_{k=0}^{+\infty} \varphi((x-c_k)/\ell_k).$$

We have, for any $\alpha \in \mathbb{Z}_+^m$,

$$|\partial_x^\alpha[\varphi((x-c_j)/\ell_j)]| \leq \ell_j^{-1}\,\mathrm{Max}|\partial_x^\alpha\varphi|,$$

$$|\partial_x^\alpha D(x)| \leq \mathrm{Max}|\partial_x^\alpha\varphi| \sum_{x\in\mathcal{Q}_k} \ell_k^{-|\alpha|} \leq (\nu+1)\mathrm{Max}|\partial_x^\alpha\varphi| \sup_{x\in\mathcal{Q}_k} \ell_k^{-|\alpha|}.$$

But it follows from (VIII.8.A8) that, if $x \in \mathcal{Q}_k$,

$$\ell_k^{-1} \leq 2(3+4K)Af(x)^{-1},$$

whence easily (VIII.8.A3). ■

VIII.9. Applications of the Approximate Poincaré Lemma to the Existence of Smooth Solutions

We return provisionally to a manifold $\mathcal{M}$ equipped with an involutive structure in which the tangent structure bundle is not necessarily a line bundle. But we shall now assume that the structure of $\mathcal{M}$ is locally integrable.

THEOREM VIII.9.1. *Let the manifold $\mathcal{M}$ be equipped with a locally integrable structure. Let p, q be integers, $0 \leq p \leq m$, $1 \leq q \leq n$. Suppose an open neighborhood* U *of a point* $\mathfrak{p} \in \mathcal{M}$ *has the following property, for some real number s:*

$$\forall f \in \mathcal{C}^\infty(\mathcal{C}\ell\,\mathrm{U};\Lambda^{p,q}) \text{ such that } \mathrm{d}'f = 0,\ \exists\, u \in \mathrm{H}^s_{loc}(\mathrm{U};\Lambda^{p,q-1}) \text{ such that } \mathrm{d}'u = f \text{ in } \mathrm{U}. \tag{VIII.9.1}$$

Then there is an open neighborhood $\mathrm{U}_0 \subset \mathrm{U}$ *of* $\mathfrak{p}$ *endowed with the following property:*

$$\forall f \in \mathscr{C}^\infty(\mathscr{C}\ell U;\Lambda^{p,q}) \text{ such that } d'f = 0, \exists\, u \in \mathscr{C}^\infty(U_0;\Lambda^{p,q-1}) \text{ such that } d'u = f \text{ in } U_0. \qquad \text{(VIII.9.2)}$$

PROOF. After contracting U about p we can choose coordinates, $x_1,\ldots,x_m$, t, all vanishing at p, and $\mathscr{C}^\infty$ solutions in U, $Z_j = x_j + \iota\Phi_j(x,t)$ $(j = 1,\ldots,m)$, such that (I.7.23) and (I.7.24) hold. Further contracting of U about p ensures that the second-order differential operator $\Delta_{L,\kappa M}$ (see section II.4) is elliptic in U. In U we identify any section f of $\Lambda^{p,q}$ to its standard representative (I.7.29), and the differential operator d′ to the operator L defined by (I.7.30). The differential operator $\Delta_{L,\kappa M}$ will act on differential forms (I.7.29) coefficientwise.

Let $f \in \mathscr{C}^\infty(\mathscr{C}\ell U;\Lambda^{p,q})$ be such that $Lf = 0$. For each integer $\nu \geq 0$ let $w_\nu \in H^s_{loc}(U;\Lambda^{p,q-1})$ be a solution of $Lw_\nu = \Delta^\nu_{L,\kappa M} f$. Let $v_\nu \in H^{s+2\nu}_{loc}(U;\Lambda^{p,q-1})$ satisfy $\Delta^\nu_{L,\kappa M} v_\nu = w_\nu$. We have $Lv_\nu - f = g_\nu$, with $g_\nu \in \mathscr{C}^\infty(\mathscr{C}\ell U;\Lambda^{p,q})$ and $\Delta^\nu_{L,\kappa M} g_\nu = 0$. We apply Lemma II.4.2 and Proposition II.4.2: there is a differential form

$$\tilde{g}_\nu(z,t) = \sum_{|J|=p} \sum_{|K|=q} \tilde{g}_{\nu,J,K}(z,t)\, dZ_J \wedge dt_K,$$

whose coefficients are holomorphic functions of (z,t), in an open neighborhood, in $\mathbb{C}^{m+1}$, of the image of U under the map $(x,t) \to (Z(x,t),t)$, and whose pullback under the latter map, $\pi^*\tilde{g}_\nu$, is equal to g_ν. Since we obviously have $Lg_\nu = 0$ we must have $d_t\tilde{g}_\nu = 0$. Let $K^{(q)}$ denote the homotopy operator (II.6.4). We set

$$u_\nu = v_\nu - \pi^*(K^{(q)}\tilde{g}_\nu).$$

We have $u_\nu \in H^{s+2\nu}_{loc}(U;\Lambda^{p,q-1})$ and $Lu_\nu = f$ in U.

Actually, after taking ν sufficiently large and relabeling the differential forms u_ν, we may assume that $u_\nu \in \mathscr{C}^\nu(U;\Lambda^{p,q-1})$ for every $\nu \in \mathbb{Z}_+$. We shall apply Theorem II.2.1 if $q = 1$ and Theorem II.6.1 if $q \geq 2$. If the neighborhood $U_0 \subset U$ of p is appropriately chosen, the following obtains: there is an increasing sequence of compact subsets K_ν of U_0 whose union is equal to U_0 and, for each $\nu \geq 1$, a differential form $h_\nu \in \mathscr{C}^\infty(U_0;\Lambda^{p,q-1})$ such that the following is true:

$$\text{if } q = 1,\ Lh_\nu \equiv 0;$$

$$\text{if } q \geq 2,\ h_\nu = L\psi_\nu \text{ for some } \psi_\nu \in \mathscr{C}^\infty(U_0;\Lambda^{p,q-2});$$

$$\text{whatever } q,\ 1 \leq q \leq n,\ \|u_\nu - u_{\nu-1} - h_\nu\|_{\mathscr{C}^\nu(K_\nu)} \leq 2^{-\nu}. \qquad \text{(VIII.9.3)}$$

The series

$$u_0 + \sum_{\nu=1}^{+\infty} (u_\nu - u_{\nu-1} - h_\nu)$$

converges in $\mathscr{C}^\infty(U_0;\Lambda^{p,q-1})$, to a solution of $Lu = f$. ■

COROLLARY VIII.9.1. *Suppose the manifold* $\mathcal{M}$ *and its involutive structure are real-analytic. Then* (VIII.9.1) *entails that there be a subneighborhood* $U_0 \subset U$ *such that* (VIII.9.2) *holds.*

COROLLARY VIII.9.2. *Suppose the manifold* $\mathcal{M}$ *is* $\mathscr{C}^\infty$ *and its involutive structure is locally integrable. Suppose furthermore that the tangent structure bundle* $\mathcal{V}$ *is a line bundle and that condition* ($\mathcal{P}$) *is satisfied at the point* $\mathfrak{p}$. *Then there is an open neighborhood* U_0 *of* $\mathfrak{p}$ *such that the following holds:*

$$\forall f \in \mathscr{C}^\infty(\mathscr{C}\ell U_0),\ \exists\, u \in \mathscr{C}^\infty(U_0) \text{ such that } Lu = f. \quad \text{(VIII.9.4)}$$

PROOF. Indeed, according to Theorem VIII.8.1 there is an open neighborhood U of $\mathfrak{p}$ in which condition (VIII.9.1), with $p = 0$, $q = 1$ and $s = 0$, is satisfied. ■

Microlocal analysis makes it possible to prove the conclusion in Corollary VIII.9.2 without the hypothesis of local integrability (see Hörmander [4], sect. 26.11). The next result shows the link between local solvability and local integrability.

THEOREM VIII.9.2. *Let the manifold* $\mathcal{M}$ *be equipped with an involutive structure. Suppose* $\dim \mathcal{M} = m+1$ *and that the tangent structure bundle* $\mathcal{V}$ *is a line bundle. Suppose moreover that there is an open neighborhood* U *of a point* $\mathfrak{p} \in \mathcal{M}$ *that has the following property:*

$$\forall f \in \mathscr{C}^\infty(\mathscr{C}\ell U;\Lambda^{0,1}),\ \exists\, u \in \mathscr{C}^\infty(U) \text{ such that } d'u = f \text{ in } U. \quad \text{(VIII.9.5)}$$

Then there is an open neighborhood $U_0 \subset U$ *of* $\mathfrak{p}$ *in which there exist* m $\mathscr{C}^\infty$ *solutions* $Z_1,\ldots,Z_m$ *such that* $dZ_1\wedge\cdots\wedge dZ_m \neq 0$ *at every point of* U_0.

PROOF. Possibly after contracting U about $\mathfrak{p}$ we select coordinates $x_1,\ldots,x_m,t$ in U such that the tangent structure bundle $\mathcal{V}$ is spanned by the vector field

$$L = \partial/\partial t + \sum_{k=1}^{m} \lambda_k(x,t)\partial/\partial x_k.$$

As before it is convenient to take U to be a product $\Omega \times \mathcal{I}$. We apply an obvious extension of Lemma VI.5.1 to nonelliptic structures: there are $\mathscr{C}^\infty$ functions in U, Z_{0_j} ($j = 1,\ldots,m$), such that, for each j,

$$R_j = LZ_{0_j} \in \mathscr{C}^\infty(U) \text{ vanishes to infinite order at } t = 0; \quad \text{(VIII.9.6)}$$

$$Z_{0_j}|_{t=0} = x_j. \quad \text{(VIII.9.7)}$$

We introduce a cutoff function $\chi \in \mathcal{C}^\infty(\mathbb{R}^{m+1})$, $\chi \equiv 0$ off the ball centered at 0 with radius one, $\chi \equiv 1$ in the ball centered at 0 with radius 1/2. Set $\chi_r(x,t) = \chi(x/r,t/r)$ for any $r > 0$. Given any integer $N \geq 0$ there are positive constants A_N, $B_N > 0$ such that, for all $\alpha \in \mathbb{Z}_+^m$, $\beta \in \mathbb{Z}_+$, $|\alpha|+\beta \leq N$, and for all $r > 0$ small enough that $\overline{\mathcal{B}}_r = \{ (x,t) \in \mathbb{R}^{m+1}; |x|^2+t^2 \leq r^2 \} \subset \mathrm{U}$,

$$|\partial_x^\alpha \partial_t^\beta \chi_r| \leq A_N r^{-|\alpha|-\beta}, \ |\partial_x^\alpha \partial_t^\beta R_j| \leq B_N r^{N+1} \ in \ \mathcal{B}_r.$$

We derive from this

$$|\partial_x^\alpha \partial_t^\beta (\chi_r R_j)| \leq C_N r, \ |\alpha|+|\beta| \leq N. \tag{VIII.9.8}$$

This implies that if r ranges over a sequence of positive numbers that converges to zero, the functions $\chi_r R_j$ converge to zero in $\mathcal{C}_c^\infty(\mathrm{U})$. But since, by (VIII.9.5), L is a continuous linear map of the Fréchet space $\mathcal{C}^\infty(\mathrm{U})$ onto itself, it follows easily from the open mapping theorem that there is a sequence of functions $\psi_{j,r} \in \mathcal{C}^\infty(\mathrm{U})$ that converges to zero and such that $\mathrm{L}\psi_{j,r} = \chi_r R_j$ for each pair (j,r). Thus, for r sufficiently small, we shall have Max $|\mathrm{d}\psi_{j,r}|$ very small in comparison to one. It will ensue that the Jacobian determinant, with respect to $x_1,\ldots,x_m$, of the functions $Z_j = Z_{0_j} - \psi_{j,r}$ $(j = 1,\ldots,m)$, is very close to 1 in some neighborhood of the origin. But $\mathrm{L}Z_j = (1-\chi_r)R_j$, hence $\mathrm{L}Z_j = 0$ in the ball $\mathcal{B}_{r/2}$. ∎

VIII.10. Necessity of Condition ($\mathcal{P}$)

We consider a vector field L in an open neighborhood U of the origin in $\mathbb{R}^{m+1}$ (where the coordinates are x_i, $i = 1,\ldots,m$, t); L will be given by (VIII.7.3). We take $\mathrm{U} = \Omega \times \mathcal{J}$ as before; Ω will be an open ball in $\mathbb{R}^m$, $\mathcal{J}$ an open interval in $\mathbb{R}$, both centered at the origin. We shall reason under the hypothesis:

To each $f \in \mathcal{C}^\infty(\mathcal{C}\ell\,\mathrm{U};\Lambda^{m,1})$ *there is* $u \in \mathcal{D}'(\mathrm{U};\Lambda^{m,0})$ *satisfying* $\mathrm{L}u = f$ *in* U. (VIII.10.1)

Let $\mathcal{B} \subset \mathrm{U}$ be an open ball. Any element f_0 of $\mathcal{C}^\infty(\mathcal{C}\ell\,\mathcal{B})$ can be extended as a $\mathcal{C}^\infty$ function in $\mathbb{R}^{m+1}$ and thus any differential form $f = f_0 \mathrm{d}x_1 \wedge \cdots \wedge \mathrm{d}x_m \wedge \mathrm{d}t \in \mathcal{C}^\infty(\mathcal{C}\ell\,\mathcal{B};\Lambda^{m,1})$ can be extended as an element of $\mathcal{C}^\infty(\mathcal{C}\ell\,\mathrm{U};\Lambda^{m,1})$. Restricting to $\mathcal{B}$ the solutions $u \in \mathcal{D}'(\mathrm{U};\Lambda^{m,0})$ of $\mathrm{L}u = f$ shows that (VIII.10.1) implies the same property with $\mathcal{B}$ substituted for U.

Proposition VIII.10.1. *Suppose* (VIII.10.1) *hold. Then*

$$\forall\ (x,t) \in \mathrm{U},\ \xi \in \mathbb{R}^m,\ \xi \cdot \boldsymbol{b}(x,t) = 0 \Rightarrow \xi \cdot \boldsymbol{b}_t(x,t) = 0.$$

Here, as in the sequel, subscripts indicate partial differentiations.

PROOF. The symbol of $[L,\bar{L}]/2\imath$ is equal to $\xi\cdot \boldsymbol{b}_t(x,t)$. Thus, if we had $\xi\cdot\boldsymbol{b}(x,t) = 0$ and $\xi\cdot\boldsymbol{b}_t(x,t) \neq 0$ it would mean that the Levi form of L at the characteristic point $(x,t,\xi,0) \in U\times\mathbb{R}^{m+1}$ does not vanish. Since here $n = 1$ this is equivalent to saying that the Levi form is definite, contradicting (VIII.10.1) by Corollary VIII.2.2. ■

In the present section we shall reason under the hypothesis that the involutive structure of $\mathcal{M}$ is locally integrable. We call the attention of the reader to the fact that, by virtue of Theorem VIII.9.2, this is a consequence of (VIII.9.5). After a contraction of U about 0 we assume that there exist "first integrals" $Z_j \in \mathcal{C}^\infty(\mathcal{C}\ell U)$, $j = 1,\dots,m$. We take $Z(0) = 0$ and, after substituting $Z_x(0)^{-1}Z$ for Z and further contracting U about 0, we also assume that the Jacobian matrix of the functions $\mathcal{R}e Z_j$ with respect to the x_i is equal to the identity $m\times m$ matrix at the origin and is nonsingular at every point of U, and that the differential of every function $\mathcal{I}m Z_j$ with respect to the x_i vanishes at the origin. Note also that, when local integrability holds, Property (VIII.10.1) is equivalent to the following one:

To each $f \in \mathcal{C}^\infty(\mathcal{C}\ell U;\Lambda^{0,1})$ there is $u \in \mathcal{D}'(U)$ satisfying $Lu = f$ *in* U, (VIII.10.2)

as seen by making use of the isomorphism $f \to f\wedge dZ_1\wedge\cdots\wedge dZ_m$ of $\mathcal{C}^\infty(\mathcal{C}\ell U;\Lambda^{0,1})$ (resp. $\mathcal{D}'(U)$) onto $\mathcal{C}^\infty(\mathcal{C}\ell U;\Lambda^{m,1})$ (resp., $\mathcal{D}'(U;\Lambda^{m,0})$).

Next we look at a $\mathcal{C}^\infty$ function $h = \varphi + \imath\psi$ in U, not necessarily a solution, but endowed with the following property:

the Hessian φ_{xx} of φ with respect to x is positive-definite at every point of U. (VIII.10.3)

We associate with h the following real vector field in (x,t,ξ)-space $U\times\mathbb{R}^m$:

$$\vartheta = \partial/\partial t - A\cdot\partial/\partial x + B\cdot\partial/\partial\xi,$$

with

$$A = \varphi_{xx}^{-1}\varphi_{xt},\ B = \psi_{xt} - \psi_{xx}A.$$

Note that φ_{xt} and ψ_{xt} are real m-vectors. For later reference we note that, since $\vartheta\varphi_x = \varphi_{xt} - \varphi_{xx}A = 0$, $\vartheta(\xi-\psi_x) = B - \psi_{xt} + \psi_{xx}A = 0$,

on each integral curve of ϑ, φ_x *and* $\xi-\psi_x$ *are constant.* (VIII.10.4)

Let $y \in \Omega$ and $\eta \in \mathbb{R}^m$ be such that

$$\varphi_x(y,0) = 0,\ \psi_x(y,0) = \eta. \qquad \text{(VIII.10.5)}$$

Notice that hypothesis (VIII.10.3) entails the uniqueness of y, and therefore also that of η. Denote by $\Phi_\vartheta(t)$ the flow of the vector field $-(A\cdot\partial/\partial x -$

$B\cdot\partial/\partial\xi$) at time t and set $(x(y,t,\eta),\xi(y,t,\eta)) = \Phi_\vartheta(t)(y,\eta)$. By (VIII.10.4) and (VIII.10.5), if $x = x(y,t,\eta)$, $\xi = \xi(y,t,\eta)$, then necessarily

$$\varphi_x(x,t) = 0,\ \xi = \psi_x(x,t). \tag{VIII.10.6}$$

Let then h depend in $\mathscr{C}^\infty$ fashion on the parameter (y,η) (allowed to range over $\Omega\times\mathbb{R}^m$) in such a way that (VIII.10.5) holds. The next statement expresses an obvious property of the flow Φ_ϑ.

PROPOSITION VIII.10.2. *Let Ω' be an arbitrary relatively compact open subset of Ω and Ξ be one in $\mathbb{R}^m$. There is an open interval $\mathscr{I}(\Omega'\times\Xi)\subset\mathscr{I}$, containing zero, such that the map*

$$(y,t,\eta)\to(x(y,t,\eta),t,\xi(y,t,\eta)) \tag{VIII.10.7}$$

defines a diffeomorphism of $\Omega'\times\mathscr{I}(\Omega'\times\Xi)\times\Xi$ onto an open subset of $\mathrm{U}\times\mathbb{R}^m$.

At this stage we take h to be a solution in U (in (x,t)-space).

PROPOSITION VIII.10.3. *If* U *is sufficiently small there are solutions in* $\mathscr{C}^\infty(\mathrm{U}\times\Omega\times\mathbb{R}^m)$ *that satisfy* (VIII.10.3) *and* (VIII.10.5).

PROOF. Take the function

$$h(x,t;y,\eta) = \iota\eta\cdot Z_x(y,0)^{-1}[Z(x,t)-Z(y,0)] + \tfrac{1}{2}\kappa(1+|\eta|^2)\langle Z(x,t)-Z(y,0)\rangle^2. \tag{VIII.10.8}$$

We have

$$h_x(x,t;y,\eta) = {}^tZ_x(x,t)\left[\iota{}^tZ_x(y,0)^{-1}\eta + \kappa(1+|\eta|^2)[Z(x,t)-Z(y,0)]\right],$$

whence $h_x(y,0;y,0) = \iota\eta$. Also,

$$h_{xx}(x,t;y,\eta) = {}^tZ_{xx}(x,t)\left[\iota{}^tZ_x(y,0)^{-1}\eta + \kappa(1+|\eta|^2)[Z(x,t)-Z(y,0)]\right] + \kappa(1+|\eta|^2){}^tZ_x(x,t)Z_x(x,t).$$

Since

$$\|h_{xx}-\kappa(1+|\eta|^2)I\| \le C\left[|\eta| + \kappa(1+|\eta|^2)(|Z| + \|Z_x-I\|)\right]$$

and since $Z|_0 = 0$ and $Z_x|_0 = I$, the identity $m\times m$ matrix, $\mathscr{R}e\,h_{xx}$ will be positive-definite as soon as the neighborhood U is small enough and the constant κ is large enough. ∎

Note that any $\mathscr{C}^\infty$ solution $h = \varphi + \imath\psi$ satisfies

$$\varphi_t = \boldsymbol{b}\cdot\psi_x, \ \psi_t = -\boldsymbol{b}\cdot\varphi_x. \qquad \text{(VIII.10.9)}$$

We introduce the function in $\mathrm{U}\times\mathbb{R}^m$,

$$F(x,t,\xi) = \boldsymbol{b}(x,t)\cdot\xi. \qquad \text{(VIII.10.10)}$$

PROPOSITION VIII.10.4. *At every point of* $\mathrm{U}\times\mathbb{R}^m$ *at which* (VIII.10.6) *holds,*

$$\mathrm{d}F = 0 \Rightarrow A = B = 0.$$

PROOF. We derive from (VIII.10.9):

$$\varphi_{xt} = \boldsymbol{b}_x\psi_x + \psi_{xx}\boldsymbol{b}, \ \psi_{xt} = -\boldsymbol{b}_x\varphi_x - \varphi_{xx}\boldsymbol{b}$$

($\boldsymbol{b}_x$ is an $m\times m$ real matrix). If we substitute in the expressions of A and B we get

$$A = \varphi_{xx}^{-1}(\boldsymbol{b}_x\psi_x + \psi_{xx}\boldsymbol{b}), \ -B = \boldsymbol{b}_x\varphi_x + \varphi_{xx}\boldsymbol{b} + \psi_{xx}A.$$

If (VIII.10.6) is valid then

$$A = \varphi_{xx}^{-1}(\boldsymbol{b}_x\xi + \psi_{xx}\boldsymbol{b}), \ -B = \varphi_{xx}\boldsymbol{b} + \psi_{xx}A. \qquad \text{(VIII.10.11)}$$

On the other hand $\mathrm{d}F = 0 \Rightarrow \boldsymbol{b} = 0$, $\boldsymbol{b}_x\xi = 0$, whence $A = B = 0$. ■

We continue to deal with a $\mathscr{C}^\infty$ solution $h = \varphi + \imath\psi$ in U that depends, in $\mathscr{C}^\infty$ fashion, on $(y,\eta) \in \Omega\times\mathbb{R}^m$ and satisfies (VIII.10.5). We regard both h and ϑ as defined in the submanifold Σ of (x,t,ξ,y,η)-space $\mathrm{U}\times\mathbb{R}^m\times\Omega\times\mathbb{R}^m$ defined by the $2m$ equations (VIII.10.6). Thus $\dim\Sigma = 2m+1$; Σ is coordinatized by means of y_i, t, η_j $(1 \le i, j \le m)$.

In the following statement the integral curves of a (nonvanishing) vector field are oriented in the direction of the vector field; this gives meaning to "later points."

PROPOSITION VIII.10.5. *Suppose* (VIII.10.2) *holds, as well as the following condition:*

> *if* $\vartheta\varphi < 0$ *at some point of any integral curve of the vector field* ϑ *in* Σ *then* $\vartheta\varphi \le 0$ *at every later point on the same curve.* (VIII.10.12)

Then there is an open neighborhood $\mathrm{U}' \subset \mathrm{U}$ *of the origin in which the following property holds:*

> *if, for some* $(x,t,\xi) \in \mathrm{U}'\times\mathbb{R}^m$, $\boldsymbol{b}(x,t)\cdot\xi < 0$ *then* $\boldsymbol{b}(x,t')\cdot\xi \le 0$ *for every* $t' > t$ *such that* $(x,t') \in \mathrm{U}'$. (VIII.10.13)

PROOF. Select $\mathrm{U}' = \Omega'\times\mathscr{I}'$ and an open ball $\Xi \subset \mathbb{R}^m$ centered at 0 such that the coordinate projection $(x,t,\xi,y,\eta) \to (x,t,\xi)$ is a diffeomorphism of an open subset Σ' of Σ onto $U' \times \Xi$. This is possible, by Proposition VIII.10.2. Let ϑ_* denote the pushforward of the vector field ϑ from Σ' to $U' \times \Xi$ under this coordinate projection.

In Σ, $\xi = \psi_x$ and thus, by the first equation (VIII.10.9), $\varphi_t = \boldsymbol{b}(x,t)\cdot\xi$; also $\vartheta\varphi = \varphi_t - A\varphi_x = \varphi_t$. This implies that, in Σ', $\vartheta\varphi$ is equal to the pull-back, via the coordinate projection $(x,t,\xi,y,\eta) \to (x,t,\xi)$, of the function F defined in (VIII.10.10). As a consequence of this, and of Hypothesis (VIII.10.12), we get

if $F < 0$ at some point of an integral curve of ϑ_* in $U' \times \Xi$ then $F \le 0$ at any later point on the same curve. (VIII.10.14)

We apply Lemma VIII.10.A2 with $\mathbf{v} = \partial/\partial t$, $\mathbf{v}_0 = \vartheta_*$. Proposition VIII.10.4 shows that $dF = 0 \Rightarrow \vartheta_* = \partial/\partial t$ and Proposition VIII.10.1 shows that $F = 0 \Rightarrow F_t = 0$. It follows that (VIII.10.14) implies indeed (VIII.10.13). ■

THEOREM VIII.10.1. *Assume the involutive structure of $\mathcal{M}$ to be locally integrable. If* (VIII.10.2) *holds then Condition* ($\mathcal{P}$) *is satisfied at the origin.*

PROOF. Let $U' = \Omega' \times \mathcal{I}'$ and Ξ be the sets so denoted in the proof of Proposition VIII.10.5. If ($\mathcal{P}$) is not satisfied at the origin then

$$\exists\ (x,\xi) \in \Omega' \times \Xi,\ s,\ t \in \mathcal{I}',\ s < t \text{ such that } \boldsymbol{b}(x,s)\cdot\xi < 0,\ \boldsymbol{b}(x,t)\cdot\xi > 0.$$

It follows from Proposition VIII.10.5 that Condition (VIII.10.12) is violated and thus there is an integral curve γ of ϑ such that $\vartheta\varphi < 0$ at some point $p_1 \in \gamma$, $\vartheta\varphi > 0$ at a later point $p_2 \in \gamma$. On γ y and η are constant: $y = x_0$, $\eta = \xi_0$. On the open arc of curve γ between p_1 and p_2 the function $\mathcal{R}e\,h(x,t;x_0,\xi_0)$ reaches its minimum α_0 on a compact set K_0 (perforce away from the end points p_1 and p_2). By virtue of (VIII.10.3) there is an open neighborhood $\mathcal{O}$ of the arc γ such that $\mathcal{R}e\,h(x,t;x_0,\xi_0) > \alpha_0$ in $\mathcal{O}\backslash\gamma$, whence (VIII.2.8). ■

COROLLARY VIII.10.1. *If* (VIII.9.5) *holds, then Condition* ($\mathcal{P}$) *is satisfied at the origin.*

Appendix to Section VIII.10: Lemmas about Real Vector Fields

Let $\mathbf{v}$ be a Lipschitz-continuous real vector field in an open subset Ω of $\mathbb{R}^N$ and $\Phi_{\mathbf{v}}(t)$ denote the flow of $\mathbf{v}$. Given any $x \in \Omega$, $\Phi_{\mathbf{v}}(t)x$ is the solution, in an interval $0 \le t < T(x)$, of the equation

$$\dot{y} = \mathbf{v}(y),\ y(0) = x. \tag{VIII.10.A1}$$

Let A be a positive constant such that $|\mathbf{v}(x)-\mathbf{v}(x')| \leq A|x-x'|$, $\forall\, x, x' \in \Omega$. It is then elementary to show that, for all sufficiently small $t \geq 0$,

$$|\Phi_{\mathbf{v}}(t)x-\Phi_{\mathbf{v}}(t)x'| \leq e^{At}|x-x'|. \qquad \text{(VIII.10.A2)}$$

LEMMA VIII.10.A1. *Let $\mathcal{S}$ be a closed subset of Ω. The following conditions are equivalent:*

$$\forall\, x \in \mathcal{S},\ 0 < t < T(x),\ \Phi_{\mathbf{v}}(t)x \in \mathcal{S}; \qquad \text{(VIII.10.A3)}$$

$$\forall\, x \in \mathcal{S},\ \lim_{\varepsilon\to+0} \tfrac{1}{\varepsilon}\operatorname{dist}(x+\varepsilon\mathbf{v}(x),\mathcal{S}) = 0. \qquad \text{(VIII.10.A4)}$$

PROOF. Suppose (VIII.10.A3) holds. Then

$$\tfrac{1}{\varepsilon}\operatorname{dist}(x+\varepsilon\mathbf{v}(x),\mathcal{S}) \leq |\tfrac{1}{\varepsilon}[\Phi_{\mathbf{v}}(\varepsilon)x-x]-\mathbf{v}(x)| \to 0 \text{ as } \varepsilon\to+0,$$

whence (VIII.10.A4).

Suppose now that (VIII.10.A4) holds. For $x \in \mathcal{S}$ and $0 \leq t < T(x)$ define $g(t) = \operatorname{dist}(\Phi_{\mathbf{v}}(t)x,\mathcal{S})$. For each $t \in [0,T(x)[$ there is a point $z_t \in \mathcal{S}$ such that $g(t) = |\Phi_{\mathbf{v}}(t)x-z_t|$. Henceforth let x and t be fixed. If $0 < \varepsilon < \operatorname{Min}[T(x)-t,T(z_t)]$ we have $\Phi_{\mathbf{v}}(t+\varepsilon)x = \Phi_{\mathbf{v}}(\varepsilon)\Phi_{\mathbf{v}}(t)x$ and

$$g(t+\varepsilon) \leq$$

$$|\Phi_{\mathbf{v}}(t+\varepsilon)x-\Phi_{\mathbf{v}}(\varepsilon)z_t| + |\Phi_{\mathbf{v}}(\varepsilon)z_t-z_t-\varepsilon\mathbf{v}(z_t)| + \operatorname{dist}(z_t+\varepsilon\mathbf{v}(z_t),\mathcal{S}) \leq$$

$$e^{A\varepsilon}g(t) + |\Phi_{\mathbf{v}}(\varepsilon)z_t-z_t-\varepsilon\mathbf{v}(z_t)| + \operatorname{dist}(z_t+\varepsilon\mathbf{v}(z_t),\mathcal{S}),$$

by (VIII.10.A2), whence

$$\tfrac{1}{\varepsilon}[g(t+\varepsilon)-g(t)] \leq$$

$$\tfrac{1}{\varepsilon}(e^{A\varepsilon}-1)g(t) + |\tfrac{1}{\varepsilon}[\Phi_{\mathbf{v}}(\varepsilon)z_t-z_t]-\mathbf{v}(z_t)| + \tfrac{1}{\varepsilon}\operatorname{dist}(z_t+\varepsilon\mathbf{v}(z_t),\mathcal{S}),$$

and therefore

$$\limsup_{\varepsilon\to+0} \tfrac{1}{\varepsilon}[g(t+\varepsilon)-g(t)] \leq Ag(t),\ \forall\, t \in [0,T(x)].$$

A standard argument shows that the function $e^{-At}g(t)$ is monotone decreasing in the interval $[0,T(x)]$, whence the result since $g(0) = 0$. ■

LEMMA VIII.10.A2. *Let $\mathbf{v}_0$, $\mathbf{v}$ be two Lipschitz-continuous real vector fields and F a real-valued $\mathcal{C}^1$ function in the open set Ω. Suppose the following is true:*

$$F(x) = 0,\ dF(x) = 0 \Rightarrow \mathbf{v}(x) = \mathbf{v}_0(x); \qquad \text{(VIII.10.A5)}$$

$$F(x) = 0 \Rightarrow \mathbf{v}F(x) \le 0. \qquad \text{(VIII.10.A6)}$$

Suppose that every integral curve γ of $\mathbf{v}_0$ has the following property:

$$\text{if } F < 0 \text{ at some point of } \gamma \text{ then } F \le 0 \text{ at any later point of } \gamma. \qquad \text{(VIII.10.A7)}$$

Then every integral curve of $\mathbf{v}$ *has also property* (VIII.10.A7).

In (VIII.10.A7) the meaning of "later point" is determined by the natural orientation of the integral curves of a vector field.

PROOF. Let $\mathcal{S}$ denote the closure of the set of points that are later, on the integral curve of $\mathbf{v}_0$ on which they lie, than some point x such that $F(x) < 0$. By (VIII.10.A7) we have $F \le 0$ in $\mathcal{S}$. We are going to show that (VIII.10.A4) holds.

When $F(x) < 0$, x belongs to $\mathcal{I}nt\mathcal{S}$ and (VIII.10.A4) is trivially true. If $F(x) = 0$ and $\mathrm{d}F(x) = 0$ then $\mathbf{v} = \mathbf{v}_0$ by (VIII.10.A5) and $x + \varepsilon\mathbf{v}(x) = x + \varepsilon\mathbf{v}_0(x)$. Since (VIII.10.A3) holds for $\mathbf{v}_0$ we have, here also,

$$\lim \varepsilon^{-1}\mathrm{dist}(x+\varepsilon\mathbf{v}(x),\mathcal{S}) = 0.$$

Finally, suppose $F(x) = 0$, $\mathrm{d}F(x) \ne 0$; we have, for some $K > 0$ and all y in a suitably small neighborhood $\mathrm{U} \subset \Omega$ of x, $F(y) \le -\nabla F(x)\cdot(x-y) + K|x-y|^2$, and thus $\mathcal{N} = \{ y \in \mathrm{U};\ -\nabla F(x)\cdot(x-y) + K|x-y|^2 < 0 \} \subset \mathcal{I}nt\mathcal{S}$. If $y = x+\varepsilon\mathbf{v}(x) - \varepsilon^{3/2}\alpha$, with $\alpha = \nabla F(x)/|\nabla F(x)|^2$, then, by virtue of (VIII.10.A6),

$$-\nabla F(x)\cdot(x-y) + K|x-y|^2 = \varepsilon\mathbf{v}F(x) - \varepsilon^{3/2} + \mathrm{O}(\varepsilon^2)$$

will be < 0 as soon as ε is small enough. On the other hand,

$$\tfrac{1}{\varepsilon}\mathrm{dist}(x+\varepsilon\mathbf{v}(x),\mathcal{S}) \le \tfrac{1}{\varepsilon}|x+\varepsilon\mathbf{v}(x)-y| = \varepsilon^{\frac{1}{2}}|\alpha| \to 0 \text{ as } \varepsilon \to +0. \ \blacksquare$$

Notes

Lemma VIII.1.1 is the version of the classical a priori estimate in Hörmander [1], which serves in the study of the differential complex associated with a locally integrable structure, with which chapter VIII is concerned. When the "right-hand sides" are top-degreee differential forms the relevant statement is Theorem VIII.2.1; its consequence, Corollary VIII.2.2, generalizes the similar construction in Hörmander [1], as well as that in Cordaro and Hounie [1]. The last article provided much of the inspiration, and the material, for section VIII.2. The absence of local exactness when the Levi form is nondegenerate on the intersection $\mathcal{V}\cap\overline{\mathcal{V}}$ (Theorem VIII.3.1) is the straighrforward generalization of the classical result in Andreotti and Hill [2] for hypersurfaces, and in Andreotti, Fredericks, and Nacinovich [1] for generic submanifolds of higher codimension in complex space. Sections VIII.4, VIII.5, and VIII.6 are directly extracted from Cordaro and Treves [1]. In the situation of section VIII.5, i.e., when the cotangent structure bundle T′ is a line bundle, the sufficiency for local solvability of the homological condition on the fibres derived from Theorem VIII.5.3 is known to be true in two cases: if $q = 1$ (lowest degree forms), according to Mendoza and Treves [1] when the coefficients are $\mathscr{C}^\infty$; if $q = n$ (top-degree forms), in the analytic category, according to Cordaro and Hounie [1]. At the time of this writing the problem is open when the right-hand side is a "differential form" of degree q with $1 < q < n$. In 1962 Nirenberg and Treves observed that the differential operators $\partial/\partial t - it^k\partial/\partial x$ (in $\mathbb{R}^2$) are locally solvable at the origin if and only if the integer k is even. This led them to the solvability condition $(\mathscr{P})$ introduced in Nirenberg and Treves [1]; the proof of its sufficiency for L^2 solvability, in section VIII.8, is the one in Treves [3]. The proof of its necessity for solvability in distrubutions is an adaptaton of the argument in Moyer [1] and is taken from Treves [10]. Statements and proofs in the Appendix to section VIII.10 are exactly those in Brézis [1].

IX

FBI Transform in a Hypo-Analytic Manifold

To its discoverer the Fourier transform was a tool to analyze the solutions of a particular linear partial differential equation. Since then its range of application has not ceased to expand, and now encompasses the most diverse fields of mathematical analysis, pure and applied. Today linear PDE theory (as well as certain types of nonlinear differential equations) is but one of its areas of application. In the context of differential equations it shows up under a variety of guises, such, for instance, as the Laplace transform and the Fourier-Borel transform. Circa 1968 Brós and Iagolnitzer introduced a version particularly well suited to the study of (real) analyticity. Subsequently it was also shown to be the right concept in the study of *microlocal analyticity*. It is only natural that the Fourier-Brós-Iagolnitzer transform, the *FBI transform* as it is now customarily called, turns up as the tool of choice in the theory of hypo-analytic structures, which are a natural generalization of the analytic structures. This aspect will be more deeply explored, down to the microlocal level, in Volume 2 of the present book.

In Volume 1 we limit ourselves to giving the simplest local definition (section IX.1) and to showing (in sections IX.3 and IX.4 respectively) how the decay, either exponential or "rapid," of the FBI transform of a distribution u characterizes the local (as opposed to microlocal) regularity, respectively hypo-analytic or $\mathscr{C}^\infty$, of u. In section IX.7 we describe one instance of the effectiveness of the FBI transform: how it enables one to prove the propagation of hypo-analyticity along elliptic submanifolds.

It is only in section IX.5 that the FBI transform in a hypo-analytic manifold is defined. The first four sections of the chapter concern themselves solely with the FBI transform on a totally real submanifold $\mathscr{X}$, of maximum real dimension, m, of complex space $\mathbb{C}^m$ (Definition IX.1.1). The "classical" situation, which is the situation considered by Brós and Iagolnitzer, corresponds to $\mathscr{X} = \mathbb{R}^m$. The basic idea is that of the "resolution of the identity":

$$u(x) = (2\pi)^{-m}\int\int_{\mathbb{R}^m\times\mathbb{R}^m} e^{\imath\xi\cdot(x-x')-|\xi||x-x'|^2}u(x')\Delta(x-x',\xi)dx'd\xi, \tag{1}$$

where $u \in \mathscr{E}'(\mathbb{R}^m)$ and $\Delta(x,\xi)$ is the Jacobian determinant of the map

$$\xi \to \xi + \imath|\xi|x. \tag{2}$$

Formula (1) follows at once from the classical Fourier inversion formula by deforming the domain of ξ-integration from $\mathbb{R}^m$ to the image of $\mathbb{R}^m$ under the map (2).

Our task is to adapt the identity (1) to compactly supported distributions in the submanifold $\mathscr{X}$: x and x' must denote variable points in $\mathscr{X}$, ξ will range over an $\mathbb{R}$-linear subspace Ξ of $\mathbb{C}^m$ (of real dimension m, possibly depending on x or on x'); the absolute values $|\xi|$ and $|x-x'|^2$ are to be replaced by their *holomorphic* extensions (denoted by $\langle\xi\rangle$ and $\langle x-x'\rangle^2$); those extensions are well defined provided ξ, x and x' are nearly real. Moreover, in order that the integral at the right in (1) make sense, even as a duality bracket between currents and differential forms in $\mathscr{X}$, the absolute value of the exponential in the integrand needs to be bounded as $|\xi| \to +\infty$. Actually the advantages of the FBI transform accrue when the submanifold $\mathscr{X}$ and the linear subspace Ξ satisfy the strong positivity condition

$$\mathscr{I}m\{\xi\cdot(x-x') + \imath\langle\xi\rangle\langle x-x'\rangle^2\} \geq c|\xi|\,|x-x'|^2,$$
$$\forall\, x, x' \in \mathscr{X}, \xi \in \Xi\ (c>0). \tag{3}$$

This forces on us the choice of the subspace Ξ: it has to be the fibre, either at the point x or at the point x', of what we call the *real structure bundle* of $\mathscr{X}$, namely the preimage of the real cotangent bundle of $\mathscr{X}$, $T^*\mathscr{X}$, under the isomorphism $T'^{1,0}|_{\mathscr{X}} \cong \mathbb{C}T^*\mathscr{X}$ defined by pullback to $\mathscr{X}$.

It remains to extract the expression of the FBI transform out of the identity (1) in which the domain of integration is $\mathscr{X}\times\Xi$ instead of $\mathbb{R}^m\times\mathbb{R}^m$. Natural choices are

$$\int_{\mathscr{X}} e^{\imath\zeta\cdot(z-x')-\langle\zeta\rangle\langle z-x'\rangle^2}u(x')dx' \tag{4}$$

or

$$\int_{\mathscr{X}} e^{\imath\zeta\cdot(z-x')-\langle\zeta\rangle\langle z-x'\rangle^2}u(x')\Delta(z-x',\zeta)dx'. \tag{5}$$

(Here z is allowed to range over $\mathbb{C}^m$ and ζ over a suitable cone in complex space.) The simpler integral (4) has the drawback that the inversion formula (essentially formula (IX.4.1) in the text; see Lemma IX.4.1) is fairly complicated. As made obvious by (1) the inversion of the transform (5) is as simple as one could possibly wish: it suffices to integrate in ζ space (with respect to the measure $(2\pi)^{-m}d\zeta$). In the present book we have made the choice (5). In Volume 2, the theory of hypo-analytic pseudodifferential operators (and Fourier integral operators) will provide a unified approach to the integrals (4), (5) and to similar ones. The choice (5) suggests that the FBI transform of a function (or a distribution) u be defined as a differential form of degree m in Ξ depending holomorphically on $z \in \mathbb{C}^m$, specifically

$$\int_{\mathscr{X}} e^{\imath\theta\cdot(z-x')}u(x')dx' \wedge d\theta,$$

where $\theta = \zeta + \imath\langle\zeta\rangle(z-x')$ and the integration is carried out in x'-space. We do not pursue this approach further here.

In a hypo-analytic manifold $\mathcal{M}$ the hypo-analytic charts used most often in the book consist of a domain U, of first integrals Z_i, of coordinates x_j and t_k such that $Z_i = x_i + \sqrt{-1}\Phi_i(x,t)$ $(1 \le i, j \le m,\ 1 \le k \le n)$. The maximally real submanifolds of $\mathcal{M}$ over which the FBI integration is carried out are the submanifolds $\mathscr{X}_t$ of U defined by $t = const.$; $\mathscr{X}_t$ can be identified to its image $\tilde{\mathscr{X}}_t$ under the map $(x,t) \to Z(x,t)$; $\tilde{\mathscr{X}}_t$ is a totally real submanifold of $\mathbb{C}^m$, of real dimension m. We deal solely with distributions u in U that are $\mathscr{C}^\infty$ functions of t valued in the space of distributions with respect to x; all "solutions" are of that type, according to Proposition I.4.3. The trace of such a distribution u on the submanifolds $\mathscr{X}_t$ is well defined and we can transfer it to $\tilde{\mathscr{X}}_t$ by means of the diffeomorphism $(x,t) \to Z(x,t)$. Once this is done we can then make use of the integral (5) to define the FBI transform of u "at time" t, $\mathscr{F}u(t;z,\zeta)$ (section IX.5).

One of the most interesting aspects of the FBI transform is its effect on the action of the basic vector fields in the hypo-analytic chart (U,Z). We refer to the vector fields M_i and L_j given by (I.7.26) and (I.7.27). Reasoning, as one must, modulo functions that decay exponentially as $|\zeta| \to +\infty$, one readily sees that

$$\mathscr{F}M_i u \approx (\partial/\partial z_i)\mathscr{F}u,\ \mathscr{F}L_j u \approx (\partial/\partial t_j)\mathscr{F}u.$$

Thus $\mathscr{F}$ transforms the differential complex associated with the hypo-analytic structure, and represented by the differential operator L (see (I.7.30)), into the De Rham complex in t space. This has profound consequences, further explored in Volume 2.

In what was said so far only the "coarse embedding" (see section II.1), defined by the above first integrals Z_i, was considered. But one can equally well make use of the fine embedding (see section II.7) and define the *FBI minitransform*. In the latter the integration is performed not over a maximally real submanifold; instead it is performed over a *minimal noncharacteristic* (hypo-analytic) submanifold. This has the advantage of bringing down the dimension of the domain of integration to that of the characteristic set (at the central point) and, by the same token, of reducing to what is essentially the characteristic set the integration with respect to the dual variable in the inversion formula. The gain is especially clear when dealing with a hypo-analytic structure in which the characteristic set has dimension ≤ 1, for instance with a CR manifold of the hypersurface type (section IX.6).

IX.1. FBI Transform in a Maximally Real Submanifold of Complex Space

The variable point in $\mathbb{C}^m$ will be denoted by z or z'; the complex coordinates of z will be denoted by z_i. "Dual" coordinates will be ζ_j $(1 \leq j \leq m)$. For any number $\kappa > 0$ we shall write

$$\mathscr{C}_\kappa = \{ \zeta \in \mathbb{C}^m;\ |\mathscr{I}m\zeta| < \kappa|\mathscr{R}e\zeta| \}.$$

For any $z \in \mathbb{C}^m$ we write $\langle z\rangle^2 = z\cdot z = z_1^2 + \cdots + z_m^2$; and for any $\zeta \in \mathscr{C}_1$ we write $\langle\zeta\rangle = (\zeta\cdot\zeta)^{\frac{1}{2}}$ (main branch of the square root). We shall also use systematically the notation

$$\Delta(z,\zeta) = \det\{I + \imath(z\odot\zeta)/\langle\zeta\rangle\} \tag{IX.1.1}$$

where $z\odot\zeta$ denotes the matrix $(z_i\zeta_j)_{1\leq i,j\leq m}$; $\Delta(z,\zeta)$ is the Jacobian determinant of the map $\zeta \to \zeta + \imath\langle\zeta\rangle z$ (here $z \in \mathbb{C}^m$, $\zeta \in \mathscr{C}_1$).

Let $\mathscr{X}$ now be a totally real submanifold of $\mathbb{C}^m$, with $\dim_{\mathbb{R}} \mathscr{X} = m$; i.e., $\mathscr{X}$ is a maximally real submanifold of $\mathbb{C}^m$.

Definition IX.1.1. *Let u be a compactly supported distribution in the manifold $\mathscr{X}$. The duality bracket*

$$\int_{\mathscr{X}} e^{\imath\zeta\cdot(z-z') - \langle\zeta\rangle\langle z-z'\rangle^2} u(z')\Delta(z-z',\zeta)dz',$$

in which $z \in \mathbb{C}^m$ and $\zeta \in \mathscr{C}_1$, will be called the Fourier-Brós-Iagolnitzer (in short, FBI) transform *of u and denoted by $\mathscr{F}u(z,\zeta)$.*

In Definition IX.1.1 the distribution u is regarded as a zero-current on $\mathscr{X}$ (see Schwartz [1], chap. 9); it is evaluated on the $\mathscr{C}^\infty$ m-form $\mathscr{E}dz'$ where

$$\mathscr{E} = e^{\imath\zeta\cdot(z-z') - \langle\zeta\rangle\langle z-z'\rangle^2}\Delta(z-z',\zeta)$$

and $dz' = dz_1'\wedge\cdots\wedge dz_m'$; z and ζ enter as parameters. Notice that the m-form in ζ-space, $\mathscr{E}d\zeta$, is the pullback under the map

$$\mathscr{C}_1 \ni \zeta \to \zeta + \imath\langle\zeta\rangle(z-z') \in \mathbb{C}^m \tag{IX.1.2}$$

of the m-form $e^{\imath\zeta\cdot(z-z')}d\zeta$.

Call M_i (cf. section I.7) the vector fields on $\mathscr{X}$ defined by the relations

$$M_i(z_j|_{\mathscr{X}}) = \delta_{ij}$$

(δ_{ij}: Kroenecker index; $i, j = 1,\ldots,m$). The vector fields $M_1,\ldots,M_m$ form a smooth basis of $\mathbb{C}T\mathscr{X}$. The structure theorem for compactly supported distributions (cf. (II.1.21)) tells us that any such distribution, u, can be represented as a sum

$$u = \sum_{|\alpha|\leq r} M^\alpha u_\alpha, \tag{IX.1.3}$$

where for each $\alpha = (\alpha_1,\ldots,\alpha_m) \in \mathbb{Z}_+^m$, u_α is a continuous function in $\mathscr{X}$ whose support is compact and contained in an arbitrary neighborhood of supp u ($r \in \mathbb{Z}_+$, $M^\alpha = M_1^{\alpha_1}\cdots M_m^{\alpha_m}$). By linearity we have

$$\mathscr{F}u(z,\zeta) = \sum_{|\alpha|\leq r} \mathscr{F}M^\alpha u_\alpha(z,\zeta). \tag{IX.1.4}$$

Now, integration by parts (cf. Lemma II.1.1) shows that

$$\mathscr{F}M^\alpha u(z,\zeta) = \int_{\mathscr{X}} e^{\imath\zeta\cdot(z-z')-\langle\zeta\rangle\langle z-z'\rangle^2}\mathscr{P}_\alpha(z-z',\zeta)u(z')dz', \tag{IX.1.5}$$

where

$$\mathscr{P}_\alpha(z,\zeta) = e^{-\imath\zeta\cdot z+\langle\zeta\rangle\langle z\rangle^2}M^\alpha\left\{\Delta(z,\zeta)e^{\imath\zeta\cdot z-\langle\zeta\rangle\langle z\rangle^2}\right\} =$$

$$e^{-\imath\zeta\cdot z+\langle\zeta\rangle\langle z\rangle^2}(\partial/\partial z)^\alpha\left\{\Delta(z,\zeta)e^{\imath\zeta\cdot z-\langle\zeta\rangle\langle z\rangle^2}\right\}.$$

It is clear that $\mathscr{P}_\alpha(z,\zeta)$ is a polynomial with respect to z, $\zeta/\langle\zeta\rangle$, $\langle\zeta\rangle$. To every compact subset K of $\mathbb{C}^m$ there is a constant $C_K > 0$ such that

$$|\mathscr{P}_\alpha(z,\zeta)| \leq C_K(1+|\zeta|)^{|\alpha|},\ \forall\ z \in K,\ \zeta \in \mathscr{C}_1. \tag{IX.1.6}$$

We also see that

$$\mathscr{F}M^\alpha u = (\partial/\partial z)^\alpha \mathscr{F}u. \tag{IX.1.7}$$

Combining (IX.1.3), (IX.1.4), and (IX.1.7) yields

$$\mathscr{F}u(z,\zeta) = \sum_{|\alpha|\leq r}(\partial/\partial z)^\alpha\mathscr{F}u_\alpha(z,\zeta) =$$

$$\sum_{|\alpha|\leq r}\int_{\mathscr{X}} e^{\imath\zeta\cdot(z-z')-\langle\zeta\rangle\langle z-z'\rangle^2}\mathscr{P}_\alpha(z-z',\zeta)u_\alpha(z')dz'. \tag{IX.1.8}$$

We note that

$$\int_{\mathscr{X}} e^{\imath\zeta\cdot(z-z')-\langle\zeta\rangle\langle z-z'\rangle^2}\mathscr{P}_\alpha(z-z',\zeta)u_\alpha(z')dz'$$

is the integral over $\mathcal{X}$ of a compactly supported *continuous function* in $\mathcal{X}$; differentiation under the integral sign shows that it defines a holomorphic function of (z,ζ) in $\mathbb{C}^m \times \mathcal{C}_1$.

As a consequence of what precedes, in particular of the estimate (IX.1.6), we may state:

PROPOSITION IX.1.1. *The FBI transform of a compactly supported distribution u in $\mathcal{X}$, $\mathcal{F}u$, is a holomorphic function of (z,ζ) in $\mathbb{C}^m \times \mathcal{C}_1$. To each compact subset* K *of* $\mathbb{C}^m$ *and to each number* κ, $0 < \kappa < 1$, *there are constants C, $R > 0$ such that*

$$|\mathcal{F}u(z,\zeta)| \le Ce^{R|\zeta|},\ \forall\ z \in \mathrm{K},\ \zeta \in \mathcal{C}_\kappa. \tag{IX.1.9}$$

In order to refine this estimate and to invert the FBI transform we must introduce some new concepts.

IX.2. The Real Structure Bundle of a Maximally Real Submanifold. Well-Positioned Maximally Real Submanifolds of $\mathbb{C}^m$. Inverting the FBI Transform of a Compactly Supported Distribution

We revert provisionally to a general involutive structure on a $\mathcal{C}^\infty$ manifold $\mathcal{M}$; as usual, the tangent and cotangent structure bundles are denoted by $\mathcal{V}$ and T' and their fibre dimensions by m and n, respectively. Let $\mathcal{X}$ be a maximally real (embedded) submanifold of $\mathcal{M}$. We recall that this is equivalent to the property that the pullback map $\mathbb{C}\mathrm{T}^*\mathcal{M}|_{\mathcal{X}} \to \mathbb{C}\mathrm{T}^*\mathcal{X}$ induces an isomorphism $\mathrm{T}'|_{\mathcal{X}} \cong \mathbb{C}\mathrm{T}^*\mathcal{X}$.

DEFINITION IX.2.1. *We shall refer to the image of the real cotangent bundle of $\mathcal{X}$, $\mathrm{T}^*\mathcal{X}$, under the natural isomorphism $\mathbb{C}\mathrm{T}^*\mathcal{X} \cong \mathrm{T}'|_{\mathcal{X}}$ as the* real structure bundle *of $\mathcal{X}$ and we shall denote it by $\mathbb{R}\mathrm{T}'_{\mathcal{X}}$.*

Of course, $\mathbb{R}\mathrm{T}'_{\mathcal{X}}$ is a real vector bundle over $\mathcal{X}$ of fibre dimension equal to m. Next we introduce the characteristic set T^0 of $\mathcal{M}$ (Definition I.2.2). We regard it as a subset of T'.

PROPOSITION IX.2.1. *We have* $\mathrm{T}^0|_{\mathcal{X}} \subset \mathbb{R}\mathrm{T}'_{\mathcal{X}}$.

PROOF. Follows at once from the fact that the diagram

$$\begin{array}{ccc} \mathrm{T}^0|_{\mathcal{X}} & \to & \mathrm{T}^*\mathcal{X} \\ \downarrow & & \downarrow \\ \mathrm{T}'|_{\mathcal{X}} & \cong & \mathbb{C}\mathrm{T}^*\mathcal{X} \end{array} \tag{IX.2.1}$$

is commutative (the vertical arrows are the natural injections, the horizontal maps are those defined by pullback to $\mathcal{X}$). ■

Now let $\mathcal{M}$ be a hypo-analytic manifold. We shall make use of a local parametrization of the maximally real submanifold $\mathcal{X}$: we assume that there is a $\mathcal{C}^\infty$ diffeomorphism $s \to f(s)$ of an open neighborhood Ω_0 of 0 in $\mathbb{R}^m$ onto an open neighborhood Ω in $\mathcal{X}$, of $f(0)$. The diffeomorphism $s \to f(s)$ defines an isomorphism of $\mathrm{T}^*\mathbb{R}^m|_{\Omega_0}$ onto $\mathrm{T}^*\mathcal{X}|_\Omega$, namely the map

$$(s,\sigma) \to (f(s), {}^t\mathrm{D}f^{-1}(s)\sigma),$$

where $\mathrm{D}f(s)$ is the differential of f at the point s and ${}^t\mathrm{D}f^{-1}$ is the transpose of the inverse, i.e., the *contragredient*, of $\mathrm{D}f$.

We assume furthermore that Ω is contained in the domain U of a hypo-analytic chart (U,Z). The variable point in any fibre of T′ at a point of U can be represented by a one-form

$$\zeta\cdot\mathrm{dZ} = \sum_{j=1}^{m} \zeta_j \mathrm{dZ}_j.$$

Its pullback to $\mathcal{X}$ at $\mathfrak{p} = f(s)$, $\zeta\cdot\mathrm{dZ}|_{\mathcal{X},\mathfrak{p}}$, will belong to $\mathbb{R}\mathrm{T}'_{\mathcal{X}}|_\mathfrak{p}$ if the pullback to $\mathbb{R}^m$, under the map f, of the complex cotangent vector $\zeta\cdot\mathrm{dZ}|_{\mathcal{X},\mathfrak{p}}$ is equal to $\sigma\cdot\mathrm{d}s$ for some $\sigma \in \mathbb{R}^m$. The latter pullback is equal to

$$\sum_{j,k=1}^{m} \zeta_j(\partial/\partial s_k)[\mathrm{Z}_j(f(s))]\mathrm{d}s_k = {}^t[(\mathrm{Z}|_{\mathcal{X}})\circ f]_s\zeta\cdot\mathrm{d}s.$$

By $[(\mathrm{Z}|_{\mathcal{X}})\circ f]_s$ we have denoted the Jacobian, with respect to the variables s_j, of the functions in Ω_0, $(\mathrm{Z}_i|_{\mathcal{X}})\circ f$ $(1 \le i, j \le m)$. We reach the conclusion that we must have

$$\zeta = {}^t[(\mathrm{Z}|_{\mathcal{X}})\circ f]_s^{-1}\sigma,\ \sigma \in \mathbb{R}^m. \tag{IX.2.2}$$

One can devise a closer fitting of the hypo-analytic functions Z_i and of the local parametric representation of $\mathcal{X}$. We may select (cf. beginning of section II.1) a hypo-analytic local chart in $\mathcal{M}$, (U,Z), centered at an arbitrary point 0 of $\mathcal{X}$, and local coordinates x_i, t_j $(1 \le i \le m,\ 1 \le j \le n)$ in U, all vanishing at 0, such that

$$\mathcal{X}\cap\mathrm{U} = \{ (x,t) \in \mathrm{U};\ t = 0 \}; \tag{IX.2.3}$$

$$Z(x,t) = x + \imath\Phi(x,t), \tag{IX.2.4}$$

with Φ: $\mathrm{U} \to \mathbb{R}^m$, $\Phi(0,0) = 0$ (Φ is $\mathcal{C}^\infty$ since Z is hypo-analytic). We shall assume that $\mathrm{U} = \mathrm{V}\times\mathrm{W}$ with V (resp., W) an open neighborhood of 0 in x-space $\mathbb{R}^m$ (resp., t-space $\mathbb{R}^n$). We take V to be sufficiently small that $x \to Z(x,0)$ is a diffeomorphism of V onto an open neighborhood of 0 in $\mathcal{X}$. One may then use the local parametrization of $\mathcal{X}$ defined by $f(x) = Z(x,0) = x + \imath\Phi(x,0)$.

From (IX.2.2) we deduce that the pullback to $\mathcal{X}$ of a complex covector $\zeta\cdot dZ$ at a point $(x,0) \in \mathcal{X}$ will belong to $\mathbb{R}T'_{\mathcal{X}}$ if and only if there is a vector $\xi \in \mathbb{R}^m$ such that

$$\zeta\cdot dZ = \xi\cdot dx, \tag{IX.2.5}$$

which is equivalent to

$$\zeta = {}^tZ_x^{-1}(x,0)\xi. \tag{IX.2.6}$$

Let $M_1,\ldots,M_m$ be the vector fields in U defined by the orthonormality relations (I.7.25) and whose expressions are to be found in (I.7.26). Thanks to (IX.2.3) we see that the vector fields M_i are all tangent to $\mathcal{X}$. The symbol (Definition I.2.1) of M_i at the point $\xi \in T^*_p\mathcal{X}$ is given by

$$\sigma(M_i)(x,0,\xi) = \langle \xi\cdot dx, M_i\rangle|_p, \tag{IX.2.7}$$

assuming that $p = (x,0)$. Thus, when (IX.2.5) holds, we have

$$\zeta_i = \sigma(M_i)(x,0,\xi),\ i = 1,\ldots,m. \tag{IX.2.8}$$

In the remainder of the present section we limit ourselves to the case $\mathcal{M} = \mathbb{C}^m$ and to a totally real submanifold $\mathcal{X}$ of $\mathbb{C}^m$, $\dim_{\mathbb{R}} \mathcal{X} = m$. We use systematically the notation of section IX.1. We identify the complex cotangent space to $\mathbb{C}^m$ at any one of its points to the space $\mathbb{C}^{2m}$ by means of the isomorphism

$$(\zeta,\zeta') \to \zeta\cdot dz + \zeta'\cdot d\bar{z} = \sum_{j=1}^{m} \zeta_j dz_j + \zeta'_j d\bar{z}_j. \tag{IX.2.9}$$

The map (IX.2.9) allows us to identify the fibres of the complex structure bundle of $\mathbb{C}^m$ (often denoted by $T'^{1,0}$) to $\mathbb{C}^m \times \{0\}$, i.e., to the set of points $(\zeta,0)$ in $\mathbb{C}^{2m}$. Thus the whole bundle $T'^{1,0}$ can be identified to $\mathbb{C}^m \times \mathbb{C}^m$. We may then regard each fibre of $\mathbb{R}T'_{\mathcal{X}}$ as a vector subspace of $\mathbb{C}^m$.

DEFINITION IX.2.2. *We shall say that the maximally real submanifold $\mathcal{X}$ of $\mathbb{C}^m$ is* well positioned *at one of its points*, z_0, *if there is a number* κ, $0 < \kappa < 1$, *and an open neighborhood* Ω *of* z_0 *in* $\mathcal{X}$ *such that the following is true*:

$$\textit{Whatever } z, z' \in \Omega \textit{ and } \zeta \in (\mathbb{R}T'_{\mathcal{X}}|_z)\cup(\mathbb{R}T'_{\mathcal{X}}|_{z'}), \tag{IX.2.10}$$

$$|\mathcal{I}m\,\zeta| < \kappa|\mathcal{R}e\,\zeta|; \tag{IX.2.11}$$

$$\mathcal{I}m[\zeta\cdot(z-z') + i\langle\zeta\rangle\langle z-z'\rangle^2] \geq (1-\kappa)|\zeta|\,|z-z'|^2. \tag{IX.2.12}$$

We shall say that $\mathcal{X}$ *is* very well positioned *at* z_0 *if, given* any *number* κ, $0 \leq \kappa < 1$, *there is an open neighborhood* Ω *of* z_0 *in* $\mathcal{X}$ *such that* (IX.2.10) *is true*.

REMARK IX.2.1. Notice that the conditions (IX.2.11) and (IX.2.12) are unchanged if we exchange z and z' and replace ζ by $-\zeta$. It would therefore have sufficed to require that ζ belong to $\mathbb{R}T'_{\mathscr{X}}|_z$. ■

PROPOSITION IX.2.2. *Given any maximally real $\mathscr{C}^\infty$ submanifold $\mathscr{X}$ of $\mathbb{C}^m$ and any point z_0 of $\mathscr{X}$ there is a biholomorphism H of an open neighborhood $\mathscr{O}$ of z_0 in $\mathbb{C}^m$ onto an open neighborhood of the origin, with $H(z_0) = 0$, such that $H(\mathscr{X}\cap\mathscr{O})$ is very well positioned at* 0.

PROOF. A translation and a $\mathbb{C}$-linear transformation in $\mathbb{C}^m$ brings us right away to the situation in which $z_0 = 0$ and the tangent space to $\mathscr{X}$ at the origin is the real space $\mathscr{I}m z = 0$. This means that near 0 $\mathscr{X}$ is equal to the image of some open neighborhood U of 0 in $\mathbb{R}^m$ under a map $x \to Z(x)$ with

$$Z(x) = x + \imath\Phi(x), \tag{IX.2.13}$$

with Φ a $\mathscr{C}^\infty$ map $U \to \mathbb{R}^m$, $\Phi(0) = 0$, $D\Phi(0) = 0$.

Let k be any integer ≥ 2 and $\mathscr{P}$ be a polynomial map $\mathbb{R}^m \to \mathbb{R}^m$ such that

$$\Phi(x) - \mathscr{P}(x) = 0(|x|^{k+1}).$$

The map $z \to z + \imath\mathscr{P}(z)$ has a holomorphic inverse H in an open neighborhood $\mathscr{O}$ of 0 in $\mathbb{C}^m$ and for any $s \in \mathbb{R}^m$ sufficiently close to 0,

$$H(s+\imath\Phi(s)) - s = 0(|s|^{k+1}).$$

We may change the parametrization of $H(\mathscr{X}\cap\mathscr{O})$ near 0, from s to $\mathscr{R}e H(Z(s))$, which shows that $H(\mathscr{X}\cap\mathscr{O})$ is equal, near 0, to the image of a function (IX.2.13) with the added property that

$$|\Phi(x)| \leq const.|x|^{k+1}. \tag{IX.2.14}$$

In the remainder of the proof we assume that $\mathscr{X}$ itself is defined as the image of (IX.2.13) and that (IX.2.14) holds. We prove then that $\mathscr{X}$ is very well positioned at the origin. Given any number κ, $0 < \kappa < 1$, we are going to select an open ball $\mathscr{B}_0$ in $\mathbb{R}^m$, centered at 0, whose radius is so small that (IX.2.10) will be valid when $\Omega = Z(\mathscr{B}_0)$.

A first consequence of (IX.2.14) is that, given any $\varepsilon > 0$, if the radius of $\mathscr{B}_0$ is small enough, then, for all $x, x' \in \mathscr{B}_0$,

$$\|Z_x(x) - Z_x(x')\| \leq \varepsilon|x-x'|. \tag{IX.2.15}$$

In particular, since $Z_x(0) = I$, the $m \times m$ identity matrix, we have, for all $x \in \mathscr{B}_0$,

$$\|Z_x(x) - I\| \leq \varepsilon|x|. \tag{IX.2.16}$$

Likewise we may assume that we have, for all $x, x' \in \mathscr{B}_0$,

$$|x-x'| \leq (1+\varepsilon)|Z(x)-Z(x')| \leq (1+\varepsilon)^2|x-x'|, \tag{IX.2.17}$$

$$|Z_x(x)^{-1}[Z(x)-Z(x')] - (x-x')| \leq \varepsilon|x-x'|^2, \tag{IX.2.18}$$

and by combining (IX.2.17) and (IX.2.18), and redefining ε,

$$|Z_x(x)^{-1}[Z(x)-Z(x')] - (x-x')| \leq \varepsilon|Z(x)-Z(x')|^2. \tag{IX.2.19}$$

To describe $\mathbb{R}T'_{\mathcal{X}}$ (near 0) we apply the relation (IX.2.6). Keep in mind that, in the present context, there are no variables t. Thus $(z,\zeta) \in \mathbb{R}T'_{\mathcal{X}}$, with $z \in Z(U)$, means that there is $x \in U$ and $\xi \in \mathbb{R}^m$ such that

$$\zeta = {}^tZ_x(x)^{-1}\xi. \tag{IX.2.20}$$

Thanks to (IX.2.16) we can choose the radius of $\mathcal{B}_0$ small enough that

$$1-\varepsilon \leq |\mathcal{R}e\langle\zeta\rangle|/|\xi| \leq |\zeta|/|\xi| \leq 1+\varepsilon \tag{IX.2.21}$$

over $Z(\mathcal{B}_0)$. This easily implies (IX.2.11) provided ε is appropriately small.

Also, (IX.2.20) enables us to derive from (IX.2.19), if $z = Z(x)$, $z' = Z(x')$, $x, x' \in \mathcal{B}_0$ (and if the radius of $\mathcal{B}_0$ is sufficiently small),

$$|\mathcal{I}m[\zeta\cdot(z-z')]| \leq \varepsilon|\zeta|\,|z-z'|^2. \tag{IX.2.22}$$

We can also ensure, by decreasing the radius of $\mathcal{B}_0$ if necessary, that

$$\varepsilon^{-1}|\mathcal{I}m\langle z-z'\rangle^2| + (1-\varepsilon)|z-z'|^2 \leq \mathcal{R}e\langle z-z'\rangle^2. \tag{IX.2.23}$$

The conjunction of (IX.2.20), (IX.2.22) and (IX.2.23) implies at once (IX.2.12) for all $z, z' \in Z(\mathcal{B}_0)$, $\zeta \in \mathbb{R}T'_{\mathcal{X}}|_z$. Remark IX.2.1 allows us to prove (IX.2.10) in full. ∎

To close the present section we discuss two consequences of the hypothesis that $\mathcal{X}$ is well positioned at a point z_0, namely the slow growth of the FBI transform of a compactly supported distribution when the transform is regarded as a function in the real structure bundle $\mathbb{R}T'_{\mathcal{X}}$; and the inversion formula.

THEOREM IX.2.1. *Suppose the maximally real submanifold $\mathcal{X}$ of $\mathbb{C}^m$ is well positioned at one of its points, z_0. Then there is an open neighborhood Ω of z_0 in $\mathcal{X}$ with the following property:*

To each distribution u in $\mathcal{X}$ whose support is compact and contained in Ω there is an integer $k > 0$ and a constant $C > 0$ such that

$$|\mathcal{F}u(z,\zeta)| \leq C(1+|\zeta|)^k, \ \forall\, (z,\zeta) \in \mathbb{R}T'_{\mathcal{X}}|_{\Omega}. \tag{IX.2.24}$$

PROOF. Suppose (IX.2.10) holds. Represent $u \in \mathcal{E}'(\Omega)$ by a formula (IX.1.3) with supp $u_\alpha \subset\subset \Omega$ for all α, $|\alpha| \leq r$. Then the estimate (IX.2.24) follows at once from (IX.1.8) and (IX.2.12). ∎

Below we only use the property that, if u is any compactly supported distribution in $\mathcal{X}$ then its FBI transform $\mathcal{F}u(z,\zeta)$ is a holomorphic function in $\mathbb{C}^m \times \mathcal{C}_1$ satisfying an estimate (IX.1.9). Define, for any $\varepsilon > 0$ and $z \in \mathbb{C}^m$,

$$u^{\varepsilon}(z) = (2\pi)^{-m}\int_{\mathbb{R}^m} e^{-\varepsilon\langle\zeta\rangle^2}\mathcal{F}u(z,\zeta)d\zeta \qquad \text{(IX.2.25)}$$

(of course, here $\langle\zeta\rangle = |\zeta|$). In other words, by Definition IX.1.1,

$$u^{\varepsilon}(z) =$$

$$(2\pi)^{-m}\iint e^{\imath\zeta\cdot(z-z')-\langle\zeta\rangle\langle z-z'\rangle^2-\varepsilon\langle\zeta\rangle^2}u(z')\Delta(z-z',\zeta)dz'd\zeta, \qquad \text{(IX.2.26)}$$

with the "integration" with respect to (z',ζ) carried out over $\mathcal{X}\times\mathbb{R}^m$. For every $\varepsilon > 0$ u^{ε} is an entire holomorphic function in $\mathbb{C}^m$.

THEOREM IX.2.2. *Suppose that* $0 \in \mathcal{X}$ *and that* $\mathcal{X}$ *is well positioned at the origin* (Definition IX.2.2). *There is an open neighborhood* Ω *of* 0 *in* $\mathcal{X}$ *such that, whatever* $u \in \mathcal{E}'(\Omega)$, *as* $\varepsilon \to +0$, u^{ε} *converges to* u *in* $\mathcal{D}'(\Omega)$.

PROOF. In the integral at the right, in (IX.2.26), we restrict the variation of z to Ω. If Ω is small enough it will have property (IX.2.10) for some number κ, $0 < \kappa < 1$. This enables one to deform within the cone $\mathcal{C}_\kappa$ (thanks to (IX.2.11)) the domain of ζ-integration from $\mathbb{R}^m$ to $\mathbb{R}T'_{\mathcal{X}}|_z$. If $z' \in \Omega$ and $(z,\zeta) \in \mathbb{R}T'_{\mathcal{X}}|_\Omega$ we will have, by (IX.2.12),

$$\mathcal{I}m[\zeta\cdot(z-z') + \imath\langle\zeta\rangle\langle z-z'\rangle^2] \geq (1-\kappa)|\zeta|\,|z-z'|^2.$$

Set $\theta = \zeta + \imath\langle\zeta\rangle(z-z')$. By contracting Ω about 0 we can also achieve that $|\mathcal{I}m\theta| \leq \frac{1}{2}(1+\kappa)|\mathcal{R}e\theta|$, which enables us to introduce the number $\langle\theta\rangle$. Then the integral at the right, in (IX.2.26), has the same limit, when $\varepsilon \to +0$, as the integral

$$\mathcal{J}^{\varepsilon}u(z) = (2\pi)^{-m}\iint e^{\imath\theta\cdot(z-z')-\varepsilon\langle\theta\rangle^2}u(z')dz'd\theta,$$

in which we deform the contour of θ-integration to $\mathbb{R}^m + \imath(z-z')/2\varepsilon$. By carrying out the θ-integration we get

$$\mathcal{J}^{\varepsilon}u(z) = (4\varepsilon\pi)^{-\frac{1}{2}m}\int_{\Omega} e^{-\langle z-z'\rangle^2/4\varepsilon}u(z')dz'$$

which converges to u in $\mathcal{D}'(\Omega)$ as $\varepsilon \to +0$ (cf. Lemma II.3.1). ∎

REMARK IX.2.2. Suppose the maximally real submanifold $\mathcal{X}$ of $\mathbb{C}^m$ is well positioned at z_0. Thanks to (IX.2.11) we can, for each z, $z' \in \Omega$, deform the domain of ζ-integration in the integral at the right in (IX.2.26) from $\mathbb{R}^m$ to $\mathbb{R}T'_{\mathscr{X}}|_{z'}$ within the cone $\mathscr{C}_\kappa$. We conclude that the integration with respect to (z',ζ) in that same integral can be carried out over $\mathbb{R}T'_{\mathscr{X}}$. ∎

IX.3. Holomorphic Extendability of a Distribution Characterized by the Rate of Decay of its FBI Transform.

We shall use the "conic" terminology: by a *cone* in a complex vector space E we mean a subset of E\{0} invariant under all dilations $\mathbf{v} \to \rho\mathbf{v}$, $\rho > 0$. Thus all cones will have their vertices at the origin. A subset of a complex vector bundle will be called *conic* if its intersection with each fibre is a (possibly empty) cone. It will be called *conically compact* if it is conic and closed, and if its base projection is compact. This terminology will also be used in $\mathbb{C}^m \times \mathbb{C}^p$ which we view as a vector bundle over $\mathbb{C}^m$ $(1 \le p < +\infty)$.

THEOREM IX.3.1. *Let $\mathscr{X}$ be a maximally real submanifold of $\mathbb{C}^m$ passing through, and well positioned at, the origin. If the open neighborhood of* 0 *in $\mathscr{X}$, Ω, is sufficiently small, then the following properties of a compactly supported distribution in Ω, u, are equivalent*:

There is an open neighborhood $\mathbb{O}$ of the origin in $\mathbb{C}^m$ and a holomorphic function $\tilde{u}$ in $\mathbb{O}$ whose restriction to $\mathbb{O}\cap\Omega$ is equal to u. (IX.3.1)

There is a conic and open neighborhood $\mathscr{C}$ of $(\mathbb{R}T'_{\mathscr{X}}\backslash 0)|_0$ *in* $T'\backslash 0$ *and constants C, R* > 0 *such that* (IX.3.2)

$$|\mathscr{F}u(z,\zeta)| \le Ce^{-|\zeta|/R},\ \forall\ (z,\zeta) \in \mathscr{C}. \qquad \text{(IX.3.3)}$$

PROOF. We take Ω so small that (IX.2.10) holds for some κ, $0 < \kappa < 1$.

I. (IX.3.1) ⇒ (IX.3.2)

Suppose first that $u \equiv 0$ in some open neighborhood of 0 in Ω. We use the representation (IX.1.3), assuming that all u_α vanish identically in one and the same open neighborhood $\Omega' \subset \Omega$ of 0. By virtue of (IX.2.12) there is a constant $c > 0$ such that

$$\mathscr{I}m[\zeta\cdot(z-z') + \imath\langle\zeta\rangle\langle z-z'\rangle^2] \ge c|\zeta|$$

for $z = 0$ and all $z' \in \operatorname{supp} u \subset\subset \Omega\backslash\Omega'$, $\zeta \in \mathbb{R}T'_{\mathscr{X}}|_0$. With a smaller $c > 0$ this will remain true if (z,ζ) varies in an appropriately "thin" conic neighborhood $\mathscr{C}$ of $\mathbb{R}T'_{\mathscr{X}}\backslash 0|_0$ in $T'/0$. We shall therefore have, in the integrals at the right in (IX.1.8),

$$|e^{\imath\zeta\cdot(z-z')-\langle\zeta\rangle\langle z-z'\rangle^2}\mathcal{P}_\alpha(z-z',\zeta)| \le const.(1+|\zeta|)^{|\alpha|}e^{-c|\zeta|},$$

whence (IX.3.3) in this particular case.

Next we deal with the general case: $u = \tilde{u}$ in $\Omega\cap\mathcal{O}$. Let a function $v \in \mathcal{C}_c^\infty(\Omega\cap\mathcal{O})$ be equal to u in a neighborhood of 0. By the first part of the proof we know that $u - v$ has property (IX.3.2). It suffices to show that v also has it. In other words we may assume that u itself is a $\mathcal{C}^\infty$ function whose support is contained in a neighborhood of 0 in Ω as small as we wish.

Introduce $\chi \in \mathcal{C}_c^\infty(\Omega\cap\mathcal{O})$, $0 \le \chi \le 1$ everywhere, $\chi \equiv 1$ in some neighborhood of 0 in Ω. There is a number $\lambda_0 > 0$ small enough that the image of supp u via the map

$$z' \to \tilde{z} = z' - \imath\lambda\chi(z')\bar{\zeta}/|\zeta| \tag{IX.3.4}$$

remains in a compact subset of $\Omega\cup\mathcal{O}$, whatever $\zeta \in \mathbb{C}^m\backslash\{0\}$ and λ, $0 \le \lambda \le \lambda_0$. We carry out the deformation of the domain of z'-integration under the map (IX.3.4) in the integral defining $\mathcal{F}u(z,\zeta)$ (see Definition IX.1.1) and look at the real part of the exponent of the exponential,

$$-\mathcal{I}m[\zeta\cdot(z-\tilde{z}) + \imath\langle\zeta\rangle\langle z-\tilde{z}\rangle^2] =$$
$$-\mathcal{I}m[\zeta\cdot(z-z') + \imath\langle\zeta\rangle\langle z-z'\rangle^2] - \lambda\chi(z')|\zeta| + \mathcal{R}(z,z',\zeta),$$

where, for some $C > 0$ and any $\varepsilon > 0$,

$$|\mathcal{R}(z,z',\zeta)|/|\zeta| \le C\left\{\lambda\chi(z')[|z-z'| + \lambda\chi(z')]\right\}$$
$$\le \varepsilon|z-z'|^2 + C\varepsilon^{-1}\lambda^2\chi^2(z').$$

By taking $z = 0$, $\zeta \in \mathbb{R}T'_{\mathfrak{X}}|_0$ and by choosing ε sufficiently small, we get, thanks to (IX.2.12), for all $z' \in$ supp u,

$$\mathcal{I}m[\zeta\cdot(z-\tilde{z}) + \imath\langle\zeta\rangle\langle z-\tilde{z}\rangle^2]/|\zeta| \ge$$
$$(1-\kappa-\varepsilon)|z-z'|^2 + \lambda\chi(z')(1-C\varepsilon^{-1}\lambda). \tag{IX.3.5}$$

We fix $\varepsilon < 1-\kappa$ and $\lambda < \varepsilon/C$. It is clear that the right-hand side in (IX.3.5) is bounded away from zero on supp u; this remains true even when z varies in a small neighborhood of 0 in $\mathbb{C}^m$ and ζ in a cone in $T'\backslash 0|_0$ containing $\mathbb{R}T'_{\mathfrak{X}}\backslash 0|_0$, whence (IX.3.2).

II. (IX.3.2) $\Rightarrow$ (IX.3.1)

We take Ω small enough to satisfy Condition (IX.2.10) as well as the requirements in Theorem IX.2.2. In the integral at the right in (IX.2.26) we can deform within the cone $\mathcal{C}_1$ the domain of ζ-integration from $\mathbb{R}^m$ to $\mathbb{R}T'_{\mathfrak{X}}|_0$. The inequality (IX.3.3) implies that, as $\varepsilon \to +0$, the functions u^ε converge uniformly in a small neighborhood $\mathcal{O}$ of the origin in $\mathbb{C}^m$, to a holomorphic function f. On the other hand, by Theorem IX.2.2, their restrictions converge in $\mathcal{D}'(\Omega \cap \mathcal{O})$ to u, whence (IX.3.1). ■

Theorem IX.3.1 allows us to define the FBI transform *at the origin* of an arbitrary distribution u in Ω—provided we interpret the transform not as a true function of (z,ζ) but as an equivalence class of holomorphic functions, in a set such as $\mathcal{C}$ in (IX.3.2), modulo functions that decay exponentially as $|\zeta| \to +\infty$. Indeed, we may define $\mathcal{F}u$ as the class of $\mathcal{F}v$ with v any compactly supported distribution in Ω which equals u in some open neighborhood of 0 in $\mathcal{X}$. By Theorem IX.3.1 if v_1 is any other such distribution $\mathcal{F}(v - v_1)$ decays exponentially.

Of course, what this means is that we are not dealing with true distributions, but rather with the germs at 0 of distributions in $\mathcal{X}$. Actually the relevant congruence among distributions is not that two equivalent distributions coincide in some open neighborhood of 0: it is that they differ, in such a neighborhood, by a hypo-analytic function, i.e., the restriction of a holomorphic function in some open neighborhood of the origin in $\mathbb{C}^m$. Let $\mathcal{D}'_0(\mathcal{X})$ denote the space of germs at the point 0 of distributions in $\mathcal{X}$ and let $\mathcal{A}_0(\mathcal{X})$ denote the space of the restrictions to $\mathcal{X}$ of the germs at 0 of holomorphic functions in $\mathbb{C}^m$. The FBI transform $\mathcal{F}\dot{u}$ of an element $\dot{u}$ of the quotient vector space $\mathcal{D}'_0(\mathcal{X})/\mathcal{A}_0(\mathcal{X})$ shall be an equivalence class of holomorphic functions in a conic and open neighborhood $\mathcal{C}$ of $\mathbb{R}T'_{\mathcal{X}}\backslash 0|_0$ in $T'\backslash 0$ modulo holomorphic functions in $\mathcal{C}$ that decay exponentially. Specifically, $\mathcal{F}\dot{u}$ will be the class of the FBI transform of any compactly supported representative of $\dot{u}$.

In this terminology a restatement of Theorem IX.3.1 is that *the FBI transform in $\mathcal{D}'_0(\mathcal{X})/\mathcal{A}_0(\mathcal{X})$ is injective*. According to Proposition IX.1.1 and Theorem IX.2.1, the FBI transforms of elements of $\mathcal{D}'_0(\mathcal{X})/\mathcal{A}_0(\mathcal{X})$ are classes of holomorphic functions in some open and conic subset of $T'\backslash 0$ containing $\mathbb{R}T'_{\mathcal{X}}\backslash 0|_0$, which grow at most exponentially and whose restrictions to $\mathbb{R}T'_{\mathcal{X}}$ grow at most polynomially.

IX.4. Smoothness of a Distribution Characterized by the Rate of Decay of its FBI Transform

As in the preceding section $\mathcal{X}$ denotes a maximally real submanifold of $\mathbb{C}^m$ passing through, and well positioned at, the origin, Ω an open neighborhood of 0 in $\mathcal{X}$ satisfying condition (IX.2.10). The next theorem will characterize the smoothness near 0 of a distribution $u \in \mathcal{E}'(\Omega)$ in terms of the "rapid decay" of its FBI transform. We are going to need a modification of the inversion formula (IX.2.25).

Let then K be a compact subset of 0 in Ω and define, for any $\varepsilon > 0$,

$$u_K^\varepsilon(z) =$$

$$(2\pi^3)^{-m/2}\int\int e^{\iota\zeta\cdot(z-z')-\langle\zeta\rangle(z-z')^2-\varepsilon\langle\zeta\rangle^2}\mathcal{F}u(z',\zeta)\langle\zeta\rangle^{\frac{1}{2}m}dz'd\zeta, \quad \text{(IX.4.1)}$$

where the integration with respect to (z',ζ) is carried out over $\mathbb{R}T'_{\mathcal{X}}|_K$. Property (IX.2.11) ensures that u_K^ε is an entire holomorphic function in $\mathbb{C}^m$.

Lemma IX.4.1. *Let $\mathcal{X}$ be a maximally real submanifold of $\mathbb{C}^m$ passing through, and well positioned at, the origin. If the open neighborhood Ω of 0 in $\mathcal{X}$ is sufficiently small, then given any compact neighborhood of the origin in Ω, K, the following is true.*

There are two open neighborhoods of the origin, $\mathcal{O}$ and Ω_0, in $\mathbb{C}^m$ and in Ω respectively, such that if u is any distribution with compact support in Ω_0, then, as $\varepsilon \to +0$, the restriction to $\mathcal{O}\cap\Omega_0$ of the holomorphic function u_K^ε converges in the distribution sense to that of u + h, where h is a holomorphic function in $\mathcal{O}$.

Proof. We shall assume that Ω has Property (IX.2.10) for some κ, $0 < \kappa < 1$. Also take Ω small enough that Theorem IX.2.1 and IX.2.2 apply. Introduce a diffeomorphism Z of an open neighborhood U of 0 in $\mathbb{R}^m$ onto Ω, such that $Z(0) = 0$. Property (IX.2.11) entails, for all $x \in U$ and $\xi \in \mathbb{R}^m$,

$$|\mathcal{I}m\,{}^tZ_x(x)^{-1}\xi| \leq \kappa|\mathcal{R}e\,{}^tZ_x(x)^{-1}\xi|. \quad \text{(IX.4.2)}$$

If $\mathcal{R}e\,{}^tZ_x(0)\eta = 0$, $\eta \in \mathbb{R}^m$, putting $x = 0$ and $\xi = \iota\,{}^tZ_x(0)\eta$ in (IX.4.2) shows that $\eta = 0$. This means that after contracting Ω about 0, we can take $\mathcal{R}eZ(x)$ as new variable x and redefine Φ so as to have $Z(x) = x + \iota\Phi(x)$. Then substituting $(I + {}^t\Phi_x(x)^2)\xi$ for ξ in (IX.4.2) shows that, for all $x \in U$, $\|\Phi_x(x)\| \leq \kappa$. Taking U to be a convex set enables us to conclude that, for all $x, x' \in U$,

$$|\Phi(x) - \Phi(x')| \leq \kappa|x - x'|. \quad \text{(IX.4.3)}$$

Now, Properties (IX.2.12) and (IX.2.24) imply

$$\left|e^{\iota\zeta\cdot(z-z')-\langle\zeta\rangle(z-z')^2}\langle\zeta\rangle^{\frac{1}{2}m}\mathcal{F}u(z',\zeta)\right| \leq$$

$$C(1+|\zeta|)^{k+\frac{1}{2}m}e^{-(1-\kappa)|\zeta||z-z'|^2}, \quad \text{(IX.4.4)}$$

for some $C > 0$ and all $z \in \Omega$ and $(z',\zeta) \in \mathbb{R}T'_{\mathcal{X}}|_\Omega$. Let K′ then be another compact neighborhood of 0 in Ω. Call K″ the closure of $[K\cap(\Omega\backslash K')]\cup[K'\cap(\Omega\backslash K)]$. By virtue of (IX.4.4) there are constants $C, R > 0$ such that

$$\left|e^{\iota\zeta\cdot(z-z')-\langle\zeta\rangle(z-z')^2}\langle\zeta\rangle^{\frac{1}{2}m}\mathcal{F}u(z',\zeta)\right| \leq Ce^{-|\zeta|/R}$$

for all $(z',\zeta) \in \mathbb{R}T'_{\mathcal{X}}|_{K''}$ and all z in a sufficiently small open neighborhood $\mathcal{O}$ of 0 in $\mathbb{C}^m$. It follows at once from this that $u_K^\varepsilon - u_{K'}^\varepsilon$ converges uniformly in $\mathcal{O}$, to a holomorphic function. In other words, it will suffice to prove the conclusion in Lemma IX.4.1 for one particular compact neighborhood K of 0 in Ω. We shall take K to be the image of a closed ball $\mathcal{B}$ centered at 0 (in $U \subset \mathbb{R}^m$) under the map $x \to Z(x)$.

We are dealing with the "integral"

$$(2\pi^3)^{-m/2} \iiint e^{i\zeta\cdot(z-z'') - \langle\zeta\rangle[\langle z-z'\rangle^2 + \langle z'-z''\rangle^2] - \varepsilon\langle\zeta\rangle^2} u(z'')\cdot$$

$$\langle\zeta\rangle^{\frac{1}{2}m} \Delta(z'-z'',\zeta)\mathrm{d}z'\mathrm{d}z''\mathrm{d}\zeta, \tag{IX.4.5}$$

where $z \in \Omega_0$, $z' \in K$, $z'' \in \operatorname{supp} u$ and $\zeta \in \mathbb{R}T'_{\mathcal{X}}|_{z'}$. Actually, the integral with respect to z'' is a duality bracket, between $u \in \mathcal{E}'(\Omega)$ and a $\mathcal{C}^\infty$ m-form. We deform the domain of ζ-integration from the fibre of $\mathbb{R}T'_{\mathcal{X}}$ at z' to the fibre at z'', i.e., we move the ζ-integration from ${}^tZ_x(x')^{-1}\mathbb{R}^m$ to ${}^tZ_x(x'')^{-1}\mathbb{R}^m$ ($x' = \mathcal{R}e z'$, $x'' = \mathcal{R}e z''$).

Next we apply Stokes' theorem and deform the domain of integration with respect to z' in (IX.4.5) from K to $\mathcal{B}\cup\Sigma$ where Σ is the piece of vertical cylinder,

$$\Sigma = \{ (x,y) \in \mathbb{R}^{2m};\, x \in \partial\mathcal{B},\, y = t\Phi(x),\, 0 < t < 1 \}$$

($\mathcal{B}\cup\Sigma$ is equipped with the appropriate orientation). If the open neighborhood of 0 in $\mathbb{C}^m$, $\mathcal{O}$, is sufficiently small, there are constants $C, R > 0$ such that

$$\left| e^{-\langle\zeta\rangle[\langle z-z'\rangle^2 + \langle z'-z''\rangle^2]} \right| \le Ce^{-(1+|z'|^2)|\zeta|/R}$$

for all $z \in \mathcal{O}$, $z' \in \Sigma\cup(\mathbb{R}^m\backslash\mathcal{B})$, $z'' \in \Omega_0 = \Omega\cap\mathcal{O}$ and all $\zeta \in \mathbb{R}T'_{\mathcal{X}}|_{z''}$. We conclude that the integral (IX.4.5), in which the z'-integration is carried out either over Σ or over $\mathbb{R}^m\backslash\mathcal{B}$, will converge uniformly, as $\varepsilon \to 0$, to a holomorphic function in $\mathcal{O}$. It will therefore suffice to consider the integral (IX.4.5) in which the integration with respect to z' is carried out over $\mathbb{R}^m$.

We shall apply an elementary identity: first of all, if A, B, λ, μ are complex numbers and if $\mathcal{R}e\mu > 0$,

$$(\mu/\pi)^{\frac{1}{2}} \int_{-\infty}^{+\infty} e^{-\mu y^2}[A + B(\lambda - y)]\mathrm{d}y = A + B\lambda. \tag{IX.4.6}$$

Next we observe that $\Delta(z,\zeta)$ is a polynomial of degree m with respect to z (see (IX.1.1)). But viewed as a function of z_j for each j, depending on the parameters z_k, $k \neq j$, it is a polynomial of degree 1. We have thus the right to apply (IX.4.6) in each variable separately. We get

$$(\mu/\pi)^{\frac{1}{2}m} \int_{\mathbb{R}^m} e^{-\mu\langle y\rangle^2} \Delta(z-y,\zeta)\mathrm{d}y = \Delta(z,\zeta). \tag{IX.4.7}$$

We derive from (IX.4.7):

$$(2\langle\zeta\rangle/\pi)^{m/2}\int_{\mathbb{R}^m} e^{-\langle\zeta\rangle[\langle z'-z\rangle^2+\langle z'-z''\rangle^2]}\Delta(z'-z'',\zeta)dz' =$$

$$(2\langle\zeta\rangle/\pi)^{m/2}e^{-\frac{1}{2}\langle\zeta\rangle\langle z-z''\rangle^2}\int_{\mathbb{R}^m} e^{-2\langle\zeta\rangle\langle z'-\frac{1}{2}(z+z'')\rangle^2}\Delta(z'-z'',\zeta)dz' =$$

$$(2\langle\zeta\rangle/\pi)^{m/2}e^{-\frac{1}{2}\langle\zeta\rangle\langle z-z''\rangle^2}\int_{\mathbb{R}^m} e^{-2\langle\zeta\rangle|y|^2}\Delta(\tfrac{1}{2}(z-z'')-y,\zeta)dy =$$

$$e^{-\frac{1}{2}\langle\zeta\rangle\langle z-z''\rangle^2}\Delta(\tfrac{1}{2}(z-z''),\zeta).$$

We conclude that the integral (IX.4.5), in which the integration with respect to z' is carried out over $\mathbb{R}^m$, is equal to

$$(2\pi)^{-m}\int\int e^{\imath\zeta\cdot(z-z'')-\frac{1}{2}\langle\zeta\rangle\langle z-z''\rangle^2-\varepsilon\langle\zeta\rangle^2}u(z'')\Delta(\tfrac{1}{2}(z-z''),\zeta)dz''d\zeta.$$

Set $z = 2\tilde{z}$, $z'' = 2\tilde{z}''$, $\zeta = \tilde{\zeta}/2$, $\tilde{u}(\tilde{z}) = u(2z)$. The preceding integral is seen to be equal to

$$(2\pi)^{-m}\int\int e^{\imath\tilde{\zeta}\cdot(\tilde{z}-\tilde{z}'')-\langle\tilde{\zeta}\rangle\langle\tilde{z}-\tilde{z}''\rangle^2-\frac{1}{4}\varepsilon\langle\tilde{\zeta}\rangle^2}u(2\tilde{z}'')\Delta(\tilde{z}-\tilde{z}'',\tilde{\zeta})d\tilde{z}''d\tilde{\zeta} =$$

$$(2\pi)^{-m}\int e^{-\frac{1}{4}\varepsilon\langle\zeta\rangle^2}\mathcal{F}\tilde{u}(\tilde{z},\zeta)d\zeta,$$

which converges to u in $\mathcal{D}'(\Omega)$ as $\varepsilon \to +0$, by Theorem IX.2.2. ■

THEOREM IX.4.1. *Let $\mathcal{X}$ and Ω be as in Lemma* IX.4.1. *In order for $u \in \mathcal{E}'(\Omega)$ to be a $\mathcal{C}^\infty$ function in some open neighborhood of the origin in Ω it is necessary and sufficient that there be a compact neighborhood* K *of the origin in Ω such that the following is true*:

to every integer $k \geq 0$ there is a constant $C_k \geq 0$ such that, for all $(z,\zeta) \in \mathbb{R}T'_{\mathcal{X}}|_K$, (IX.4.8)

$$|\mathcal{F}u(z,\zeta)| \leq C_k(1+|\zeta|)^{-k}.$$

PROOF. I. *Necessity*. Assume that (IX.2.10) holds in Ω. Thanks to Theorem IX.3.1 we have the right to replace u by any other distribution equal to u in an open neighborhood of 0 in Ω and whose support is as small as we wish. Thus, if we assume that u is $\mathcal{C}^\infty$ in some open neighborhood of 0 we may as well take $u \in \mathcal{C}_c^\infty(\Omega)$. Integration by parts shows that

$$(1+\langle\zeta\rangle^2)^k\mathcal{F}u(z,\zeta) =$$

$$\int_{\mathcal{X}} e^{\imath\zeta\cdot(z-z')}(1+\Delta'_M)^k\left\{e^{-\langle\zeta\rangle\langle z-z'\rangle^2}u(z')\Delta(z-z',\zeta)\right\}dz'. \quad \text{(IX.4.9)}$$

Here $\Delta'_M = M_1'^2 + \cdots + M_m'^2$ and M'_j is the vector field on $\mathcal{X}$ denoted by M_j in section IX.1 but now acting in the variables z'. There is a constant $A_k > 0$ such that

$$\left| e^{\langle\zeta\rangle(z-z')^2}(1+\Delta'_M)^k\left\{ e^{-\langle\zeta\rangle(z-z')^2}u(z')\Delta(z-z',\zeta)\right\}\right| \le$$
$$A_k[(1+|\zeta|)(1+|\zeta||z-z'|^2)]^k \sum_{|\alpha|\le 2k}|M^\alpha u(z')|.$$

By (IX.2.12) we get, for a suitable constant $B_k > 0$,

$$\left|\int_{\mathscr{X}} e^{i\zeta\cdot(z-z')}(1+\Delta'_M)^k\left\{ e^{-\langle\zeta\rangle(z-z')^2}u(z')\Delta(z-z',\zeta)\right\}dz'\right| \le$$
$$B_k(1+|\zeta|)^k\int_{\operatorname{supp} u}(1+|\zeta||z-z'|^2)^k e^{-(1-\kappa)|\zeta||z-z'|^2}dz'. \qquad \text{(IX.4.10)}$$

Since $|1+\langle\zeta\rangle^2|^k \ge c_k(1+|\zeta|)^{2k}$ $(c_k > 0)$ by virtue of (IX.2.11), we derive at once the estimate in (IX.4.8) from (IX.4.9) and (IX.4.10).

II. *Sufficiency*. Take K in (IX.4.1) to be the same as in the statement of Lemma IX.4.1. Once again, thanks to Theorem IX.3.1, we may assume that $\operatorname{supp} u \subset\subset \Omega_0$ (Ω_0 is the neighborhood of 0 in Lemma IX.4.1). Let k be any integer $\ge 2m$ and let $(\partial/\partial z)^\alpha$, $|\alpha| \le k-2m$, act under the integral sign on the right-hand side of (IX.4.1). The hypotheses (IX.2.10) and (IX.4.8) entail that the absolute value of the differentiated integrand does not exceed $const.(1+|\zeta|)^{-3m/2}$. It follows that the restriction to $\mathscr{X}$ of $\partial^\alpha u_K^\varepsilon/\partial z^\alpha$ converges uniformly as $\varepsilon \to +0$ in some open neighborhood of 0 independent of α, whence the sought conclusion, by Theorem IX.2.2. ∎

We may remove the requirement that supp u be compact and consider the following property:

If $v \in \mathscr{E}'(\mathscr{X})$ is equal to $u \in \mathscr{D}'(\mathscr{X})$ in some open neighborhood of 0 *in $\mathscr{X}$ there is a conic neighborhood $\mathscr{C}$ of* $\mathbb{R}T'_{\mathscr{X}}\backslash 0|_0$ *in* $\mathbb{R}T'_{\mathscr{X}}\backslash 0$ *such that to every integer $k \ge 0$ there is a constant $C_k > 0$ such that* (IX.4.11)

$$|\mathscr{F}v(z,\zeta)| \le C_k(1+|\zeta|)^{-k},\ \forall\,(z,\zeta) \in \mathscr{C}.$$

We define, as we must, the FBI transform of u at the origin as the equivalence class of the FBI transforms of the distributions $v \in \mathscr{E}'(\mathscr{X})$ equal to u in some neighborhood of 0 in $\mathscr{X}$. Since the FBI transforms of these distributions v only differ by functions that decay exponentially as $|\zeta| \to +\infty$, one can rephrase Property (IX.4.11) by saying that

as $|\zeta| \to +\infty$ the FBI transform of u decays, in a conic neighborhood of $\mathbb{R}T'_{\mathscr{X}}|_0$ *in* $\mathbb{R}T'_{\mathscr{X}}$, *faster than any power of* $1/|\zeta|$. (IX.4.12)

As a matter of fact it would be more rigorous to make use of the terminology of germs: one is dealing here with an element u of $\mathcal{D}'_0(\mathcal{X})/\mathcal{A}_0(\mathcal{X})$ and the word "neighborhood" in (IX.4.12) must be interpreted as the germ of a conic and open subset of $\mathbb{R}T'_{\mathcal{X}}$, at $\mathbb{R}T'_{\mathcal{X}}\backslash 0|_0$. Better still would be to switch to the sphere bundle associated with $\mathbb{R}T'_{\mathcal{X}}$, i.e., to the quotient of $\mathbb{R}T'_{\mathcal{X}}$ for the equivalence relation $(z,\zeta) \approx (z',\zeta')$ meaning $z = z'$, $\zeta = \rho\zeta'$ for some $\rho > 0$. We leave the details to the reader.

We may state:

COROLLARY IX.4.1. *Let $\mathcal{X}$ pass through and be well positioned at the origin. In order that $u \in \mathcal{D}'(\mathcal{X})$ be equal to a $\mathcal{C}^\infty$ function in some open neighborhood of the origin it is necessary and sufficient that* (IX.4.12) *hold.*

IX.5. FBI Transform in a Hypo-Analytic Manifold

We now consider a $\mathcal{C}^\infty$ manifold $\mathcal{M}$ equipped with a hypo-analytic structure. As usual, the underlying cotangent structure bundle T′ will have rank m, the tangent structure bundle, $\mathcal{V}$, rank n. In the present section we are going to make use of the *coarse local embedding* (section II.1). The analysis will take place in a hypo-analytic chart (U,Z) centered at a point $0 \in \mathcal{M}$ (point to which we refer as the origin); U is the domain of local coordinates x_i, t_j $(1 \leq i \leq m$, $1 \leq j \leq n)$ all vanishing at 0 and such that $Z = x + \imath\Phi(x,t)$ with $\Phi\colon \mathrm{U} \to \mathbb{R}^m$ a $\mathcal{C}^\infty$ map. We take $\mathrm{U} = \mathrm{V} \times \mathrm{W}$ with V (resp., W) an open neighborhood of 0 in x-space $\mathbb{R}^m$ (resp., t-space $\mathbb{R}^n$). Composing Z with a biholomorphism of an open neighborhood of 0 in $\mathbb{C}^m$ onto another such neighborhood yields a new map Z with the property that

$$|\Phi(x,t)| \leq const.(|x|^{k+1} + |t|), \tag{IX.5.1}$$

with k an integer ≥ 2 (cf. (IX.2.14)); this entails $\Phi|_0 = 0$, $\Phi_x|_0 = 0$. And thanks to it, by contracting both V and W about the origin one can indeed obtain a local embedding, i.e., *the map*

$$(x,t) \to (Z(x,t),t) \tag{IX.5.2}$$

is a diffeomorphism of U *onto a $\mathcal{C}^\infty$ submanifold Σ of* $\mathbb{C}^m \times \mathbb{R}^n$. Throughout the present section we shall always reason under these hypotheses.

For each $t \in \mathrm{W}$ we call $\mathcal{X}_t$ the submanifold $\mathrm{V} \times \{t\}$ of U; it is a maximally real submanifold of $\mathcal{M}$. For each $t \in \mathrm{W}$,

the map $x \to Z(x,t)$ is a diffeomorphism of $\mathcal{X}_t$ onto the submanifold of $\mathbb{C}^m$, (IX.5.3)

$$\Sigma_t = \{ z \in \mathbb{C}^m;\ (z,t) \in \Sigma \};$$

Σ_t is a maximally real submanifold of $\mathbb{C}^m$.

We refer the reader to the beginning of section IX.2 for a description of the real structure bundle $\mathbb{R}T'_{\Sigma_t}$ of $\Sigma_t \subset \mathbb{C}^m$ (Definition IX.2.1). We only recall that if we use the parametric equations $z = Z(x,t)$ $(x \in V)$ to define Σ_t then the fibre of $\mathbb{R}T'_{\Sigma_t}$ at a point $z = Z(x,t)$ consists of those covectors $\zeta \in \mathbb{C}^m$ such that $\zeta = {}^tZ_x(x,t)^{-1}\xi$ for some $\xi \in \mathbb{R}^m$ (cf. (IX.2.2)).

PROPOSITION IX.5.1. *Suppose that* (IX.5.1) *holds. Then* Σ_0 *is very well positioned at the origin and if* $t \in W$ *is sufficiently close to* 0, Σ_t *is well positioned at the point* $Z(0,t)$ (see Definition IX.2.2).

PROOF. Take $x, x' \in V$, $\zeta = {}^tZ_x(x,t)^{-1}\xi$. Then

$$\zeta\cdot[Z(x,t)-Z(x',t)] = \xi\cdot Z_x(x,t)^{-1}[Z(x,t)-Z(x',t)] =$$

$$\xi\cdot[x-x' + R(x,x',t)(x-x')\cdot(x-x')],$$

where R is an $m \times m$ matrix whose entries are smooth functions in $V \times V \times W$ with values in $\mathbb{R}^m$. By (VII.5.1) we have

$$\|R(x,x',t)\| \leq const.(|x|+|x'|+|t|).$$

It follows that

$$|\mathcal{Im}\{\zeta\cdot[Z(x,t)-Z(x',t)]\}| \leq const.(|x|+|x'|+|t|)|x-x'|^2|\xi|. \tag{IX.5.4}$$

On the other hand, again by (VII.5.1),

$$|\langle\zeta\rangle/|\xi|-1| \leq const.(|x|+|x'|+|t|); \tag{IX.5.5}$$

$$|\Phi(x,t)-\Phi(x',t)| \leq const.(|x|+|x'|+|t|)|x-x'|. \tag{IX.5.6}$$

If we combine (IX.5.4), (IX.5.5), and (IX.5.6) we get, for some constant $C > 0$,

$$\mathcal{Im}\left\{\zeta\cdot[Z(x,t)-Z(x',t)] + \imath\langle\zeta\rangle\langle Z(x,t)-Z(x',t)\rangle^2\right\} \geq$$

$$[1-C(|x|+|x'|+|t|)]|x-x'|^2|\xi|. \tag{IX.5.7}$$

This is the same inequality as (IX.2.12) where $z = Z(x,t)$, $z' = Z(x',t)$, if we require $|x|+|x'|+|t|$ to be sufficiently small to ensure

$$[1-C(|x|+|x'|+|t|)]|\xi| \geq (1-\kappa)|\zeta|. \tag{IX.5.8}$$

Since $\zeta = {}^tZ_x(x,t)^{-1}\xi$, and thanks to (IX.5.1), $|\xi|/|\zeta|$ can be made to be as close to 1 as one wishes provided $|x|+|t|$ is correspondingly small. It is also clear that if $|x|+|x'|+|t|$ is sufficiently small, (IX.5.5) will imply (IX.2.11). From these observations the claims in Proposition IX.5.1 follow at once. ■

Henceforth we assume that W is small enough that for each $t \in W$, Σ_t is well positioned at 0. Actually we shall strengthen this requirement. We shall reason under the hypothesis that there is a number c_0, $0 < c_0 < 1$, such that, for every $t \in \mathrm{W}$,

$$\mathscr{Im}\{\zeta\cdot(z-z') + \imath\langle\zeta\rangle\langle z-z'\rangle^2\} \geq c_0|\zeta|\,|z-z'|^2 \qquad \text{(IX.5.9)}$$

for all $z, z' \in \Sigma_t$, $\zeta \in (\mathbb{R}\mathrm{T}'_{\Sigma_t}|_z)\cup(\mathbb{R}\mathrm{T}'_{\Sigma_t}|_{z'})$. We shall also assume that those same covectors ζ belong to $\mathscr{C}_\kappa$ for some κ, $0 < \kappa < 1$. By (IX.5.5) and (IX.5.7) it is clear that these conditions can be satisfied by selecting V and W sufficiently small.

We are going to deal with distributions $u \in \mathscr{C}^\infty(\mathrm{W};\mathscr{D}'(\mathrm{V}))$: they are $\mathscr{C}^\infty$ functions of t in W valued in the space of distributions of x in V; they have traces on each submanifold $\mathscr{X}_t$; the trace can be identified to an element of $\mathscr{D}'(\mathrm{V})$. We denote by $\tilde{u}(z,t)$ the push-forward to Σ of the distribution $u(x,t)$ under the map (IX.5.2). At first we shall also require that, for each $t \in \mathrm{W}$, the support of $u(x,t)$ (as a distribution with respect to x) be a compact subset of V, a property we express by writing $u \in \mathscr{C}^\infty(\mathrm{W};\mathscr{E}'(\mathrm{V}))$. In such a case, for each $t \in \mathrm{W}$, the support of $\tilde{u}(z,t)$ is a compact subset of the submanifold Σ_t.

We can define the *FBI transform* of $u \in \mathscr{C}^\infty(\mathrm{W};\mathscr{E}'(\mathrm{V}))$ as the "integral" (which, in reality, is a duality bracket)

$$\mathscr{F}u(t;z,\zeta) = \int_{\Sigma_t} e^{\imath\zeta\cdot(z-z')-\langle\zeta\rangle\langle z-z'\rangle^2}\tilde{u}(z',t)\Delta(z-z',\zeta)dz' \qquad \text{(IX.5.10)}$$

(cf. Definition IX.1.1). Of course,

$$\mathscr{F}u(t;z,\zeta) =$$

$$\int_{\mathscr{X}_t} e^{\imath\zeta\cdot(z-Z(x',t))-\langle\zeta\rangle\langle z-Z(x',t)\rangle^2}u(x',t)\Delta(z-Z(x',t),\zeta)dZ(x',t). \qquad \text{(IX.5.11)}$$

The right-hand side in (IX.5.11) is the duality bracket between the zero-current $u(x,t)$ and the $\mathscr{C}^\infty$ m-form in $\mathscr{X}_t$,

$$e^{\imath\zeta\cdot(z-Z(x',t))-\langle\zeta\rangle\langle z-Z(x',t)\rangle^2}\Delta(z-Z(x',t),\zeta)dZ(x',t).$$

The FBI transform $\mathscr{F}u(t;z,\zeta)$ is a $\mathscr{C}^\infty$ function of (t,z,ζ) in $\mathrm{W}\times\mathbb{C}^m\times\mathscr{C}_1$ (see beginning of section IX.1), holomorphic with respect to (z,ζ).

We introduce the standard vector fields L_j and M_i ($1 \leq j \leq n$, $1 \leq i \leq m$) in U (see (I.7.25), (I.7.26), (I.7.27)). Their pushforward to Σ under the map (IX.5.2) are the vector fields $\tilde{\mathrm{L}}_j$, $\tilde{\mathrm{M}}_i$ of section II.1. In particular, they act on $\mathscr{C}^\infty$ functions $\tilde{f}(z,t)$ that are holomorphic with respect to z, according to (II.1.9). On Σ_t, $\tilde{\mathrm{M}}_i$ is the analogue of the vector field called M_i in section IX.1. We deduce from (IX.1.7):

$$\mathscr{F}(\mathrm{M}_i u) = (\partial/\partial z_i)\mathscr{F}u. \qquad \text{(IX.5.12)}$$

The novelty here is the action of the vector fields L_j:

PROPOSITION IX.5.2. *Whatever* $u \in \mathcal{C}^\infty(\mathrm{W};\mathcal{E}'(\mathrm{V}))$ *and* $j = 1,\ldots,n$,

$$\mathcal{F}(\mathrm{L}_j u) = (\partial/\partial t_j)\mathcal{F}u. \qquad \text{(IX.5.13)}$$

If one makes use of the expression (IX.5.11) the proof of (IX.5.13) (like that of (IX.5.12)) is a direct application of Lemma II.1.1; it is essentially the same as the proof of (II.2.9), and we shall not repeat it here.

The meaning of (IX.5.13) is obvious: when acting on $\mathcal{C}^\infty(\mathrm{W};\mathcal{E}'(\mathrm{V}))$, the FBI integral transforms the differential operator L into the exterior derivative in t-space. Likewise, the meaning of (IX.5.12) is that the FBI transformation transforms the "vertical" differential operator M (see (I.7.37)) into the anti-Cauchy-Riemann operator $\partial_{\bar{z}}$.

The same remarks apply when we remove the constraint that u have compact support as a distribution of x, provided we define the FBI transform $\mathcal{F}u$ as the equivalence class, modulo functions that decay exponentially as $|\zeta| \to +\infty$, of the transforms $\mathcal{F}v$ with $v \in \mathcal{C}^\infty(\mathrm{W};\mathcal{E}'(\mathrm{V}))$ equal to u in a set $\mathrm{W}\times\mathrm{V}'$, where $\mathrm{V}' \subset \mathrm{V}$ is an open neighborhood of 0 in $\mathbb{R}^m$. Often, in the sequel, we shall take the representative v of the kind χu, with $\chi \in \mathcal{C}_c^\infty(\mathrm{V})$ equal to 1 in some subneighborhood $\mathrm{V}_0 \subset \mathrm{V}$ of the origin. Somewhat abusively we are going to write χu for the distribution $\chi(x)u(x,t)$. Thus $\mathcal{F}u$ will be the class of $\mathcal{F}(\chi u)$.

There are situations in which a careful choice of the size of the sets supp χ and V_0 can be useful.

PROPOSITION IX.5.3. *There are open balls* $\mathrm{V}_0 \subset \mathrm{V}_1 \subset \mathrm{V}$ *in* $\mathbb{R}^m$ *and* $\mathrm{W}_0 \subset \mathrm{W}$ *in* $\mathbb{R}^n$, *all centered at the origin, and constants* r, κ, $R > 0$ *such that, if* $\chi \in \mathcal{C}_c^\infty(\mathrm{V}_1)$ *is equal to one in* V_0, *then, to every solution* h *in* U *there is a constant* $C > 0$ *such that*

$$|\mathcal{F}(\chi h)(t;z,\zeta) - \mathcal{F}(\chi h)(0;z,\zeta)| \leq C|t|e^{-|\zeta|/R} \qquad \text{(IX.5.14)}$$

in the region

$$t \in \mathrm{W}_0,\ z \in \mathbb{C}^m,\ |z| < r,\ \zeta \in \mathcal{C}_\kappa \text{ (see (IX.1.1))}. \qquad \text{(IX.5.15)}$$

We recall that any solution in U is an element of $\mathcal{C}^\infty(\mathrm{W};\mathcal{D}'(\mathrm{V}))$, according to Proposition I.4.3.

PROOF. An integration along rays issued from 0 in t-space shows that, in the region (IX.5.15),

$$|\mathcal{F}(\chi h)(t;z,\zeta) - \mathcal{F}(\chi h)(0;z,\zeta)| \leq |t| \operatorname*{Max}_{\mathrm{W}_0} |\mathrm{d}_s\mathcal{F}(\chi h)(s;z,\zeta)|. \qquad \text{(IX.5.16)}$$

We avail ourselves of (IX.5.13):

$$\mathrm{d}_t\mathcal{F}(\chi h) = \mathcal{F}(\mathrm{L}(\chi h)) = \mathcal{F}((\mathrm{L}\chi)h).$$

But $L\chi \equiv 0$ for $x \in V_0$, i.e., $d\chi \neq 0$ implies $|x| \geq \delta$ ($\delta > 0$ is the radius of the ball V_0). Then, according to (IX.5.9),

$$\mathscr{I}m\{\zeta\cdot(z - Z(x,t)) + \imath\langle\zeta\rangle\langle z - Z(x,t)\rangle^2\} \geq c_0\delta^2|\zeta| \qquad \text{(IX.5.17)}$$

for $t \in W$, $x \in \text{supp}(d\chi)$, $z = 0$ and $\zeta = {}^tZ_x(x,t)^{-1}\xi$, $\xi \in \mathbb{R}^m\backslash\{0\}$. We take the radius of V_1 to be equal to 2δ. Then, by virtue of (IX.5.1), we get, in V_1,

$$\|{}^tZ_x(x,t)^{-1} - I\| \leq const.(\delta^2 + |t|),$$

whence, by (IX.5.1) again, and by (IX.5.17),

$$\mathscr{I}m\{\xi\cdot(z - Z(x,t)) + \imath|\xi|\langle z - Z(x,t)\rangle^2\} \geq [\delta^2(c_0 - C\delta) - |t|]|\xi|$$

for $t \in W$, $x \in \text{supp}(d\chi)$, $z = 0$, $\xi \in \mathbb{R}^m\backslash\{0\}$. We require $\delta < c_0/4C$. We can then select the radius of W and the numbers $r, \kappa > 0$ small enough that, in the region (IX.5.15),

$$\mathscr{I}m\{\xi\cdot(z - Z(x,t)) + \imath|\xi|\langle z - Z(x,t)\rangle^2\} \geq \tfrac{1}{2}c_0\delta^2|\xi|. \qquad \text{(IX.5.18)}$$

When $h \in \mathscr{C}^0(U)$ we derive easily from (IX.5.18), for some $c > 0$,

$$\operatorname*{Max}_{W_0} |d_s\mathscr{F}(\chi h)(s;z,\zeta)| \leq const.e^{-c|\zeta|}. \qquad \text{(IX.5.19)}$$

When the solution h is arbitrary we use a representation (II.5.1) in conjunction with the formulas (IX.5.12), and the Cauchy estimates in z-space. We leave the details to the reader. ∎

Proposition IX.5.3 adds precision to the statement that, when h is a solution in U, and provided t is sufficiently close to 0, then

$$\mathscr{F}(\chi h)(t;z,\zeta) \approx \mathscr{F}(\chi h)(0;z,\zeta) \qquad \text{(IX.5.20)}$$

modulo exponentially decaying functions.

The equivalence relation (IX.5.20) is consistent with the hypo-analyticity result one derives from Theorem IX.3.1: applied to $\mathscr{X} = \mathscr{X}_0$ Theorem IX.3.1 states that the trace of h on $\mathscr{X}_0$ is hypo-analytic if and only if, in the notation of (IX.5.20),

$$\mathscr{F}(\chi h)(t;z,\zeta) \approx 0 \qquad \text{(IX.5.21)}$$

for $t = 0$. By (IX.5.20) this will be true for all t close enough to the origin. On the other hand, if $h|_{\mathscr{X}_0} = \tilde{h}\circ(Z|_{t=0})$ with $\tilde{h}$ holomorphic in some open neighborhood of 0 in $\mathbb{C}^m$, standard uniqueness (see, e.g., Corollary II.3.5) demands that $h \equiv \tilde{h}\circ Z$ in a full neighborhood of 0 which, again by Theorem IX.5.1, means that (IX.5.21) holds for small $|t|$.

In summary, the validity of (IX.5.21) (for $t = 0$ or, equivalently, for all t in a neighborhood of 0) is a necessary and sufficient condition in order that the solution $h \in \mathscr{D}'(\Omega)$ be hypo-analytic at 0.

IX.6. The FBI Minitransform

In section IX.5 we have used the coarse local embedding (see section II.1) to define the FBI transform (at the point 0). In the present section we are going to make use of the *fine local embedding* to define a transform of the same kind. We shall use the concepts and notation of section II.7 to which we refer the reader. In particular, the map (II.7.1) will be a diffeomorphism of the neighborhood of $0 \in \mathcal{M}$, U, onto a $\mathcal{C}^\infty$ submanifold Σ of $\mathbb{C}^\nu \times \mathbb{C}^d \times \mathbb{C}^{n-\nu}$ and U will be the product (II.7.2). Actually, it is convenient to make the hypothesis

$$|\varphi(z,s,t)| \leq const.(|z| + |s|^{k+1} + |t|) \qquad \text{(IX.6.1)}$$

with k an integer ≥ 2 (cf. (IX.5.1)).

For each $(z,t) \in \Delta \times W$ we call $\mathcal{X}_{z,t}$ the submanifold $\{z\} \times V \times \{t\}$ (see (II.7.2)). Its image $\Sigma_{z,t}$ under the map $s \rightarrow s + \imath\varphi(z,s,t)$ is a maximally real submanifold of $\mathbb{C}^d$. The fibre of the real structure bundle of $\Sigma_{z,t}$, $\mathbb{R}T'_{\Sigma_{z,t}}$, at a point $w = s + \imath\varphi(z,s,t)$, consists of the covectors $\sigma = {}^t w_s(z,s,t)^{-1}\rho$, $\rho \in \mathbb{R}^d$. The submanifold $\Sigma_{0,0}$ of $\mathbb{C}^d$ is very well positioned at 0, and if Δ and W are small enough, then $\Sigma_{z,t}$ is well positioned at $w(z,0,t)$ (cf. Proposition IX.5.1). In what follows we reason under the hypothesis that there is c_0, $0 < c_0 < 1$, such that

$$\mathcal{I}m\{\sigma\cdot(w-w') + \imath\langle\sigma\rangle\langle w-w'\rangle^2\} \geq c_0|\sigma|\,|w-w'|^2 \qquad \text{(IX.6.2)}$$

for all $w, w' \in \Sigma_{z,t}$, $\sigma \in (\mathbb{R}T'_{\Sigma_{z,t}}|_w) \cup (\mathbb{R}T'_{\Sigma_{z,t}}|_{w'})$ whatever $(z,t) \in \Delta \times W$. We shall also assume that those same covectors σ belong to the cone $\mathcal{C}_\kappa = \{\sigma \in \mathbb{C}^d;\ |\mathcal{I}m\sigma| < \kappa|\mathcal{R}e\sigma|\}$ with $\kappa < 1 - c_0$.

Here we deal with distributions $u \in \mathcal{C}^\infty(\Delta \times W; \mathcal{D}'(V))$: they are $\mathcal{C}^\infty$ functions of (z,t) in $\Delta \times W$ valued in the space of distributions with respect to s, in V. Every solution in U is of this kind, by Proposition I.4.3. We denote by $\mathcal{C}^\infty(\Delta \times W; \mathcal{E}'(V))$ the space of those distributions $u \in \mathcal{C}^\infty(\Delta \times W; \mathcal{D}'(V))$ such that the support of $u(z,s,t)$, viewed as a distribution in V, is compact for each $(z,t) \in \Delta \times W$. We call $\tilde{u}(z,w,t)$ the pushforward to Σ, under the map $(z,s,t) \rightarrow s + \imath\varphi(z,s,t)$, of $u \in \mathcal{C}^\infty(\Delta \times W; \mathcal{D}'(V))$. If $u \in \mathcal{C}^\infty(\Delta \times W; \mathcal{E}'(V))$ then, for each $(z,t) \in \Delta \times W$, $\tilde{u}(z,w,t) \in \mathcal{E}'(\Sigma_{z,t})$. We use the notation (IX.1.1) in the present context, with w substituted for z and σ for ζ.

Definition IX.6.1. *By the* FBI minitransform *in* U *of the distribution* $u \in \mathcal{C}^\infty(\Delta \times W; \mathcal{E}'(V))$ *we shall mean the duality bracket*

$$\mathcal{F}u(z,t;w,\sigma) =$$

$$\int_{\Sigma_{z,t}} e^{\imath\sigma\cdot(w-w') - \langle\sigma\rangle\langle w-w'\rangle^2} \tilde{u}(z,w',t)\Delta(w-w',\sigma)dw'. \qquad \text{(IX.6.3)}$$

(We are making use of the same letter $\mathcal{F}$ to denote the FBI minitransform as we did to denote the FBI "maxitransform." The nature of the argument

$(z,t;w,\sigma)$ on which the minitransform depends should suffice to avoid a possible confusion.)

We have (cf. (IX.5.11))

$$\mathcal{F}u(z,t;w,\sigma) =$$

$$\int_{\mathcal{X}_{z,t}} e^{\iota\sigma\cdot(w-s'-\iota\varphi(z,s',t))-\langle\sigma\rangle\langle w-s'-\iota\varphi(z,s',t)\rangle^2}u(z,s',t)\cdot$$

$$\Delta(w-s'-\iota\varphi(z,s',t),\sigma)dw(z,s',t). \tag{IX.6.4}$$

The inversion formula for the FBI minitransform can be expressed as follows (cf. Theorem IX.2.2):

$$u(z,s,t) = (2\pi)^{-d}\int \mathcal{F}u(z,t;w,\sigma)d\sigma, \tag{IX.6.5}$$

with the integration performed over the real structure bundle $\mathbb{R}T'_{\Sigma_{z,t}}|_w$ where $w = s + \iota\varphi(z,s,t)$. The validity of (IX.6.5) is of course dependent on the hypothesis that the submanifolds $\Sigma_{z,t}$ of $\mathbb{C}^d$ are well positioned at $w(z,0,t)$.

We make use of the vector fields L_j given by (I.7.15) and (I.7.16); also of the vector fields M_i given by (I.7.18) and N_k given by (I.7.17). If $u \in \mathcal{C}^\infty(\Delta\times W;\mathcal{E}'(V))$, we have (cf. Lemma II.7.1)

$$\mathcal{F}(M_iu) = (\partial/\partial z_i)\mathcal{F}u,\ i = 1,\ldots,\nu; \tag{IX.6.6}$$

$$\mathcal{F}(N_ku) = (\partial/\partial w_k)\mathcal{F}u,\ k = 1,\ldots,d; \tag{IX.6.7}$$

$$\begin{aligned}\mathcal{F}(L_ju) &= (\partial/\partial \bar z_j)\mathcal{F}u,\quad j = 1,\ldots,\nu;\\ \mathcal{F}(L_{\nu+\ell}u) &= (\partial/\partial t_\ell)\mathcal{F}u,\quad \ell = 1,\ldots,n-\nu.\end{aligned} \tag{IX.6.8}$$

Thus the FBI minitransformation transforms the differential operator L, acting on $\mathcal{C}^\infty(\Delta\times W;\mathcal{E}'(V))$, into the operator $\bar\partial_z + d_t$. As for the "vertical" differential operator M (see (I.7.39)) it is transformed into $\partial_z + \partial_w$.

REMARK IX.6.1. The CR case is worthy of special attention. In this case there are no variables t_ℓ; $n = \nu$. Then the FBI minitransform transforms the differential operator L acting on $\mathcal{C}^\infty(\Delta;\mathcal{E}'(V))$ into the Cauchy-Riemann operator in z-space, $\bar\partial_z$. ■

All the preceding assertions remain valid when we remove the constraint that u have compact support with respect to s. Then, of course, we must define the FBI minitransform of u as an equivalence class of functions modulo functions that decay exponentially as $|\sigma| \to +\infty$.

The analogue of Proposition IX.5.3 can be stated here as follows:

PROPOSITION IX.6.1. *There exist an open polydisk $\Delta_0 \subset \Delta$ in $\mathbb{C}^\nu$, open balls $V_0 \subset V_1 \subset V$ in $\mathbb{R}^d$, $W_0 \subset W \subset \mathbb{R}^{n-\nu}$, $\mathcal{O} \subset \mathbb{C}^d$, all centered at the origin, and* *constants* κ, $R > 0$ *such that, if* $\chi \in \mathscr{C}_c^\infty(\mathrm{V}_1)$ *is equal to* 1 *in* V_0, *then to each solution h in* U *there is a holomorphic function* $\mathscr{G}h$ *in* $\Delta \times \mathbb{O} \times \mathscr{C}_\kappa$ *and a constant* $C > 0$ *such that, in* $\Delta_0 \times \mathrm{W}_0 \times \mathbb{O} \times \mathscr{C}_\kappa$,

$$|\mathscr{F}(\chi h)(z,t;w,\sigma) - \mathscr{G}h(z,w,\sigma)| \leq C\mathrm{e}^{-|\sigma|/R}. \qquad \text{(IX.6.9)}$$

PROOF. We may apply Proposition IX.5.3 in (t,w,σ)-space, regarding z as a parameter. By suitable choices of Δ_0, V_0, V_1, etc., we obtain that

$$|\mathscr{F}(\chi h)(z,t;w,\sigma) - \mathscr{F}(\chi h)(z,0;w,\sigma)| \leq C|t|\mathrm{e}^{-|\sigma|/R} \qquad \text{(IX.6.10)}$$

in the region $\Delta_0 \times \mathrm{W}_0 \times \mathbb{O} \times \mathscr{C}_\kappa$. It will therefore suffice to show that there is a holomorphic function $\mathscr{G}h(z,w,\sigma)$ in $\Delta_0 \times \mathbb{O} \times \mathscr{C}_\kappa$ such that, in that same set,

$$|\mathscr{F}(\chi h)(z,0;w,\sigma) - \mathscr{G}h(z,w,\sigma)| \leq C\mathrm{e}^{-|\sigma|/R}. \qquad \text{(IX.6.11)}$$

In other words, we may now reason as if there were no variables t. Applying (IX.6.8) to $u = \chi h$ yields $\bar{\partial}_z \mathscr{F}(\chi h) = \mathscr{F}((\mathrm{L}\chi)h)$. Let $\delta > 0$ be the radius of V_0; we have $\mathrm{L}\chi \equiv 0$ in V_0, whence, by (IX.6.2),

$$\mathscr{I}m\{\sigma\cdot[w - s - \imath\varphi(z,s,0)] + \imath\langle\sigma\rangle\langle w - s - \imath\varphi(z,s,0)\rangle^2\} \geq c_0\delta^2|\sigma| \qquad \text{(IX.6.12)}$$

for all $z \in \Delta$, $s \in \mathrm{supp}(\mathrm{L}\chi)$, $w = 0$ and $\sigma = {}^t w_s(z,s,0)^{-1}\sigma_0$, $\sigma_0 \in \mathbb{R}^d$. We adapt the argument in the proof of Proposition IX.5.3: take the radius of V_1 to be equal to 2δ; then, by (IX.6.1),

$$\|{}^t w_s^{-1}(z,s,0) - I\| \leq const.(\delta^2 + |z|),\ \forall\, z \in \Delta,\ s \in \mathrm{V}_1,$$

and therefore, by (IX.6.1) and (IX.6.12),

$$\mathscr{I}m\{\sigma\cdot[w - s - \imath\varphi(z,s,0)] + \imath\langle\sigma\rangle\langle w - s - \imath\varphi(z,s,0)\rangle^2\} \geq [\delta^2(c_0 - C\delta) - |z|]|\sigma|$$

for $z \in \Delta$, $s \in \mathrm{supp}(\mathrm{d}\chi)$, $w = 0$, $\sigma \in \mathbb{R}^d$. We require $\delta < c_0/4C$. If then Δ_0, $\mathbb{O}$ and κ are sufficiently small we shall have

$$\mathscr{I}m\{\sigma\cdot[w - s - \imath\varphi(z,s,0)] + \imath\langle\sigma\rangle\langle w - s - \imath\varphi(z,s,0)\rangle^2\} \geq \tfrac{1}{2}c_0\delta^2|\sigma|,$$

$$\forall\, z \in \Delta_0,\ s \in \mathrm{supp}(\mathrm{d}\chi),\ w \in \mathbb{O},\ \sigma \in \mathscr{C}_\kappa. \qquad \text{(IX.6.13)}$$

When $h \in \mathscr{C}^0(\mathrm{U})$, (IX.6.13) entails that, for a suitable constant $C > 0$,

$$|\bar{\partial}_z \mathscr{F}(\chi h)(z,0;w,\sigma)| \leq C\mathrm{e}^{-|\sigma|/R} \qquad \text{(IX.6.14)}$$

in $\Delta_0 \times \mathbb{O} \times \mathscr{C}_\kappa$. We may take $R = 4/\delta^2$. When h is an arbitrary distribution solution we avail ourselves of a representation of the kind (II.7.26). In this manner we obtain an estimate (IX.6.14) with any $R > 4/\delta^2$. The point here is that $1/R$ may be selected independently of the particular solution h under consideration.

Then, by making use of one of the inverses of the Cauchy-Riemann operator (the inhomogeneous Cauchy formula provides the inverse best adapted to polydisks), we can find a solution g of the equation

$$\bar{\partial}_z g(z,w,\sigma) = \bar{\partial}_z \mathcal{F}(\chi h)(z,0;w,\sigma),$$

which is a $\mathcal{C}^\infty$ function in $\Delta_0 \times \mathcal{O} \times \mathcal{C}_\kappa$, holomorphic with respect to (w,σ), and which satisfies

$$|g(z,w,\sigma)| \leq C e^{-|\sigma|/R}$$

for some $C > 0$ possibly larger than the constant in (IX.6.14) but with the same number R. The function

$$\mathcal{G}h(z,w,\sigma) = \mathcal{F}(\chi h)(z,0;w,\sigma) - g(z,w,\sigma)$$

satisfies (IX.6.11). ∎

Proposition IX.6.1 adds precision to the statement that, when h is a solution in U, and provided (z,t) is sufficiently close to the origin, then $\mathcal{F}h(z,t;w,\sigma)$ is equivalent, modulo functions which decay exponentially as $|\sigma| \to +\infty$, to a function that is holomorphic with respect to (z,w,σ) and is independent of t, i.e.,

$$(\bar{\partial}_z + \mathrm{d}_t)\mathcal{F}h \approx 0. \tag{IX.6.15}$$

We close this section by briefly indicating how the FBI minitransform can be used to characterize the hypo-analyticity at a point. Let U be the neighborhood of 0 in $\mathcal{M}$ introduced at the beginning of the present section.

THEOREM IX.6.1. *In order that a distribution solution h in* U *be hypo-analytic at* 0 *it is necessary and sufficient that* $\mathcal{F}h \approx 0$.

PROOF. Theorem IX.3.1 applied to the manifold $\Sigma_{z,t}$ shows directly that if a distribution h in U is hypo-analytic at 0, then $\mathcal{F}(\chi h)(z,t;w,\sigma)$ decays exponentially as $|\sigma| \to +\infty$ (χ is a cutoff function in s-space, of the kind introduced earlier).

Conversely, let h be a distribution solution in U. Suppose that $\mathcal{F}(\chi h)$ decays exponentially. Then we conclude, again by Theorem IX.3.1, that there is a $\mathcal{C}^\infty$ function $\tilde{h}(z,w,t)$ in some open neighborhood of 0 in $\mathbb{C}^\nu \times \mathbb{C}^d \times \mathbb{R}^{n-\nu}$, holomorphic with respect to w, such that

$$h(z,s,t) = \tilde{h}(z,s+\iota\varphi(z,s,t),t)$$

in some open neighborhood $\mathrm{U}' \subset \mathrm{U}$ of 0. But note that

$$\mathrm{L}[\tilde{h}(z,s+\iota\varphi(z,s,t))] = (\bar{\partial}_z + \mathrm{d}_t)\tilde{h}(z,w,t)|_{w=s+\iota\varphi(z,s,t)}.$$

Since $\mathrm{L}h \equiv 0$ in U the pullback to $\Sigma \subset \mathbb{C}^\nu \times \mathbb{C}^d \times \mathbb{R}^{n-\nu}$ of $(\bar{\partial}_z + \mathrm{d}_t)\tilde{h}$ vanishes identically in some neighborhood of the origin. Recall that $\Sigma_{z,t}$ is maximally real in $\mathbb{C}^d$. It follows that $(\bar{\partial}_z + \mathrm{d}_t)\tilde{h}$ vanishes in some neighborhood of 0 in w-space $\mathbb{C}^d$ for each $(z,t) \in \Delta \times \mathrm{W}$ and therefore that it vanishes in an open neighborhood of 0 in $\mathbb{C}^\nu \times \mathbb{C}^d \times \mathbb{R}^{n-\nu}$. We conclude that, in that neighborhood,

$\tilde{h}$ is holomorphic with respect to z and independent of t. Thus h is hypo-analytic at 0. ∎

Among the elements u of $\mathcal{C}^\infty(\Delta \times \mathrm{V};\mathcal{D}'(\mathrm{V}))$ that satisfy $\mathcal{F}u \approx 0$ the solutions that are hypo-analytic at the origin are characterized by Property (IX.6.15).

IX.7. Propagation of Hypo-Analyticity

Let $\mathcal{M}$ be a $\mathcal{C}^\infty$ manifold equipped with a hypo-analytic structure (with cotangent structure bundle T' of rank m, tangent structure bundle $\mathcal{V}$ of rank n). In the present section, by a submanifold we shall always mean an embedded $\mathcal{C}^\infty$ submanifold. We are going to prove that certain connected submanifolds of $\mathcal{M}$ propagate hypo-analyticity. This means the following: let $\mathcal{N}$ denote the submanifold, and let h be any distribution solution in some open set containing $\mathcal{N}$. Suppose h is hypo-analytic at some point of $\mathcal{N}$ (Definition III.1.2). Then necessarily h is hypo-analytic at every point of $\mathcal{N}$.

We have seen (Theorem II.3.3) that orbits propagate the zero-set of solutions. But they do not, in general, propagate hypo-analyticity, nor for that matter $\mathcal{C}^\infty$ regularity, of solutions, as shown in the following:

EXAMPLE IX.7.1. In the plane $\mathbb{R}^2$ equipped with the Mizohata structure defined by $Z = x + \imath y^2$ there is a single orbit, $\mathbb{R}^2$ itself, yet there are solutions that are nonsmooth at a single point, e.g., any square root of Z. ∎

We introduce the class of submanifolds that will serve as propagators of hypo-analyticity. In the next two definitions $\mathcal{M}$ need not be equipped with a hypo-analytic structure; all it needs is an involutive structure.

DEFINITION IX.7.1. *By an* elliptic submanifold *of $\mathcal{M}$ we shall mean a submanifold $\mathcal{N}$ of $\mathcal{M}$ that inherits from $\mathcal{M}$ an elliptic structure. We say that $\mathcal{N}$ is a* complex submanifold *of $\mathcal{M}$ if $\mathcal{N}$ inherits a complex structure* (see Definition I.2.3).

Any elliptic submanifold of $\mathcal{M}$ is compatible with the involutive structure of $\mathcal{M}$ (Definition I.3.1).

EXAMPLE IX.7.2. Let (x,y,s,t) be the coordinates in $\mathbb{C}\times\mathbb{R}^2$. Equip $\mathbb{C}\times\mathbb{R}^2$ with the hypo-analytic structure defined by the functions $z = x+\imath y$ and s. Then the subspace $s = 0$ is an elliptic submanifold of $\mathbb{C}\times\mathbb{R}^2$. ∎

PROPOSITION IX.7.1. *Any elliptic submanifold $\mathcal{N}$ of a CR manifold $\mathcal{M}$ is a complex submanifold.*

PROOF. Since $\mathbb{C}T^*\mathcal{M} = T' + \overline{T}'$ the structure induced on $\mathcal{N}$ is CR. It must therefore be complex if it is to be elliptic. ∎

EXAMPLE IX.7.3. Let (x,y,s) be the coordinates in $\mathbb{C}\times\mathbb{R}$. Equip $\mathbb{C}\times\mathbb{R}$ with the hypo-analytic CR structure defined by z. Then the plane $s = 0$ is a complex submanifold of $\mathbb{C}\times\mathbb{R}$. ∎

Proposition IX.7.1 makes it clear that, in general, $\mathcal{M}$ will not contain any elliptic submanifold (in general a CR manifold does not contain any complex submanifold). But it can happen that, through some point of $\mathcal{M}$, there pass many different elliptic submanifolds:

EXAMPLE IX.7.4. Let x_j, y_k $(1 \le j, k \le 2)$ and s denote the coordinates in $\mathbb{C}^2\times\mathbb{R}$. Equip $\mathbb{C}^2\times\mathbb{R}$ with the CR structure defined by the functions $z_j = x_j + \iota y_j$ $(j = 1, 2)$ and $w = s + \iota|z_1 z_2|^2$. There are two complex curves passing through the origin, defined by $z_j = s = 0$ $(j = 1, 2)$. ∎

EXAMPLE IX.7.5. Equip $\mathbb{C}^2\times\mathbb{R}$ with the CR structure defined by the z_j $(j = 1, 2)$ and $w = s + \iota(|z_1|^2 - |z_2|^2)$. There are infinitely many complex curves passing through the origin: they are defined by the equations $z_1 = e^{\iota\theta}z_2$ $(\theta \in \mathbb{R})$, $s = 0$. ∎

We describe another class of elliptic manifolds:

DEFINITION IX.7.2. *We shall say that the involutive structure of* $\mathcal{M}$ vanishes *on the submanifold* $\mathcal{N}$ *if the pullback of* $T'|_{\mathcal{N}}$ *to* $\mathcal{N}$, $T'_{\mathcal{N}}$, *is identically zero.*

The condition $T'_{\mathcal{N}} = 0$ is equivalent to each one of the following two properties:

$$T'|_{\mathcal{N}} \subset \mathbb{C}N^*\mathcal{N}; \tag{IX.7.1}$$

$$\mathbb{C}T\mathcal{N} \subset \mathcal{V}|_{\mathcal{N}}. \tag{IX.7.2}$$

If they are satisfied $\mathcal{N}$ is (obviously) elliptic.

EXAMPLE IX.7.6. Equip $\mathbb{R}^2$ with the hypo-analytic structure defined by $Z = xe^{\iota y}$. This hypo-analytic structure vanishes on the y-axis $x = 0$ (cf. Examples I.4.2, I.11.1). ∎

Next we give a local description of an elliptic submanifold.

PROPOSITION IX.7.2. *Let* $\mathcal{N}$ *be an elliptic submanifold of* $\mathcal{M}$. *An arbitrary point* $0 \in \mathcal{N}$ *is the center of a local chart* $(U, x_1,\dots,x_\nu, y_1,\dots,y_\nu, s_1,\dots,s_d,$

$t_1,\ldots,t_{n-\nu}$) such that in U *the hypo-analytic structure of $\mathcal{M}$ is defined by the functions $z_j = x_j + \imath y_j$ ($1 \le j \le \nu$) and $w_k = s_k + \imath\varphi_k(z,s,t)$ ($1 \le k \le d$), that* (I.7.5) *holds and that the following is true.*

In U *the submanifold $\mathcal{N} \cap$ U is defined by equations*

$$z_j = 0, \quad j = \nu'+1,\ldots,\nu; \tag{IX.7.3}$$

$$s = 0; \tag{IX.7.4}$$

$$t_\ell = 0, \; \ell = r+1,\ldots,n-\nu, \tag{IX.7.5}$$

where ν' and r are integers such that $0 \le \nu' \le \nu$, $0 \le r \le n-\nu$. Furthermore, throughout $\mathcal{N} \cap$ U,

$$\varphi_k(z,s,t) \equiv 0, \; 1 \le k \le d. \tag{IX.7.6}$$

That $0 \in$ U is the center of the local chart in the preceding statement means that all the coordinates x_i, y_j, s_k, and t_ℓ vanish at 0; by (I.7.6) so do the functions φ_k. It should also be understood that when $\nu' = \nu$, there are no equations (IX.7.3), whereas when $r = n-\nu$ there are no equations (IX.7.5).

PROOF. Let U be a local chart like the one in the statement but in which $\mathcal{N} \cap$ U is not necessarily defined in U by Equations (IX.7.3), (IX.7.4), (IX.7.5) (and (IX.7.6) does not necessarily hold). The pullbacks to $\mathcal{N} \cap$ U of the differentials dz_j and dw_k must span a vector subbundle of $\mathbb{C}T^*\mathcal{N}$ whose intersection with the span of the $d\bar{z}_j$ and $d\overline{w}_k$ is equal to 0. By (I.7.5) this requires that the pullbacks to $\mathcal{N}$ of the dw_k vanish at 0. After contracting U about 0 and effecting a $\mathbb{C}$-linear substitution of the z_j we may assume that the cotangent structure bundle of the elliptic structure induced on $\mathcal{N}$, $T'_{\mathcal{N}}$, is spanned by $dz_1,\ldots,dz_{\nu'}$ ($0 \le \nu' \le \nu$; $\nu' = 0 \Rightarrow T'_{\mathcal{N}} = 0$). Below we write $z' = (z_1,\ldots,z_{\nu'})$, $z'' = (z_{\nu'+1},\ldots,z_\nu)$. The structure of $\mathcal{N}$ being elliptic is hypocomplex (Proposition III.5.1) and thus there are holomorphic functions f_j ($j = \nu'+1,\ldots,\nu$) and g_k ($k = 1,\ldots,d$) in some open neighborhood of the origin in $\mathbb{C}^{\nu'}$ such that, when restricted to $\mathcal{N} \cap$ U,

$$z_j = f_j(z'), \; j = \nu'+1,\ldots,\nu; \; w_k = g_k(z'), \; k = 1,\ldots,d.$$

Note, in passing, that the differential of every function g_k must vanish at the origin since the pullback of dw_k to $\mathcal{N}$ does. A consequence of this is that if we take the functions $z_j - f_j(z')$ as new coordinates z_j (for $\nu' < j \le \nu$), (IX.7.3) will be valid on $\mathcal{N} \cap$ U. If we take the functions $w_k - g_k(z')$ as new functions w_k and $s_k - \mathcal{R}e\, g_k(z')$ as new coordinates s_k, (IX.7.4) and (IX.7.6) will hold on $\mathcal{N} \cap$ U.

Of course, the pullbacks to $\mathcal{N}$ of the differentials dz_i and $d\bar{z}_j$ ($1 \le i, j \le \nu'$) must be linearly independent. Note also that codim $\mathcal{N} \ge 2(\nu-\nu') + d$, and therefore, since $m = \nu + d$, $2\nu' \le \dim \mathcal{N} \le n - \nu + 2\nu'$. Let therefore r be an integer, $0 \le r \le n - \nu$, such that

$$\dim \mathcal{N} = 2\nu' + r. \tag{IX.7.7}$$

We may assume that $\mathcal{N}$ is defined in U by Equations (IX.7.3) and (IX.7.4) and by additional real equations

$$h_\ell(x',y',t) = 0,\ \ell = r+1,\dots,n-\nu,$$

such that $dh_{r+1}\wedge\cdots\wedge dh_{n-\nu} \neq 0$. The Jacobian matrix of $(h_{r+1},\dots,h_{n-\nu})$ with respect to t must have rank $n-\nu-r$ on $\mathcal{N}\cap\mathrm{U}$ otherwise there would be a linear relation between the dz_i and the $d\bar{z}_j$. After relabeling the coordinates t_ℓ we may assume that the Jacobian determinant of $(h_{r+1},\dots,h_{n-\nu})$ with respect to $(t_{r+1},\dots,t_{n-\nu})$ is $\neq 0$. A further contraction of U about 0 allows us then to take h_ℓ as new coordinate t_ℓ for $\ell = r+1,\dots,n-\nu$. ■

Below we reason under the hypothesis

$$\varphi_k = 2\mathcal{R}e(A_k\cdot z'') + B_k\cdot s + C_k\cdot t''\ (1 \leq k \leq d), \tag{IX.7.8}$$

where A_k, B_k, C_k are smooth functions in U with values in $\mathbb{C}^{\nu-\nu'}$, $\mathbb{R}^d$, $\mathbb{R}^{n-\nu-r}$ respectively, all vanishing at the origin. Actually, it is convenient to carry out one more holomorphic substitution of the functions w_k, namely replace w_k by

$$w_k - \tfrac{1}{2}\imath \sum_{\ell,\ell'=1}^{d} (\partial^2\varphi_k/\partial s_\ell \partial s_{\ell'}|_0) w_\ell w_{\ell'},$$

and one more change of the variables s_k, namely replace s_k by

$$s_k + \tfrac{1}{2} \sum_{\ell,\ell'=1}^{d} (\partial^2\varphi_k/\partial s_\ell \partial s_{\ell'}|_0) \mathcal{I}m(w_\ell w_{\ell'}).$$

As the reader can readily check this allows us to assume that, in addition to (IX.7.8), we also have

$$|\varphi(z,t,s)| \leq const.(|z| + |t| + |s|^3). \tag{IX.7.9}$$

We shall take U to be a product:

$$\mathrm{U} = \Delta' \times \Delta'' \times \mathrm{V} \times \mathrm{W}' \times \mathrm{W}'', \tag{IX.7.10}$$

with Δ' (resp., Δ'') an open polydisk in z'-space $\mathbb{C}^{\nu'}$ (resp., z''-space $\mathbb{C}^{\nu-\nu'}$), V an open ball in s-space $\mathbb{R}^d$, W′ (resp., W″) an open ball in t'-space $\mathbb{R}^r$ (resp., t''-space $\mathbb{R}^{n-\nu-r}$). Thus $\Delta = \Delta' \times \Delta''$, $\mathrm{W} = \mathrm{W}' \times \mathrm{W}''$.

Notice that if $z = (z',0)$, $z' \in \Delta'$, $t = (t',0)$, $t' \in \mathrm{W}'$, the origin lies on the maximally real submanifold $\Sigma_{z,t}$ of $\mathbb{C}^d$. By availing ourselves of (IX.7.9) (cf. Proposition IX.5.1), after possibly contracting Δ, V, W about 0, we can ensure not only that $\Sigma_{z,t}$ be well positioned at $w(z,0,t)$ $(= i\varphi(z,0,t))$ but also that the open subset of $\Sigma_{z,t}$ defined by $s \in \mathrm{V}$ have property (IX.2.10) for a number κ, $0 < \kappa < 1$, *independent of* $(z,t) \in \Delta \times \mathrm{W}$. Without our recalling it this will be our hypothesis throughout the sequel: for all $z \in \Delta$, $t \in \mathrm{W}$, and all $w \in \Sigma_{z,t}$, $(w',\sigma) \in \mathbb{R}\mathrm{T}'_{\Sigma_{z,t}}$,

$$|\mathscr{Im}\,\sigma| < \kappa|\mathscr{Re}\,\sigma|, \qquad \text{(IX.7.11)}$$

$$\mathscr{Im}\{\sigma\cdot(w-w') + \imath\langle\sigma\rangle\langle w-w'\rangle^2\} \geq (1-\kappa)|\sigma|\,|w-w'|^2. \qquad \text{(IX.7.12)}$$

THEOREM IX.7.1. *Let $\mathcal{N}$ be a connected elliptic submanifold of $\mathcal{M}$ and let h be a distribution solution in some open set Ω containing $\mathcal{N}$.*

If $\mathcal{N}$ intersects the hypo-analytic singular support of h (Definition III.1.3), *then $\mathcal{N}$ is entirely contained in it.*

PROOF. Let 0 be an arbitrary point of $\mathcal{N}$ and let U be an open neighborhood of $0 \in \mathcal{N}$ exactly like the one described above. We deal with a distribution solution h in U which is hypo-analytic at $0 \in \mathcal{N}\cap$U. In what follows $\chi \in \mathscr{C}_c^\infty(\mathrm{V})$ will be a cutoff function like that in Proposition IX.6.1.

First we draw some consequences from (IX.7.8) and (IX.7.12). If we let w vary in some open (and bounded) neighborhood $\mathcal{O}$ of the origin in $\mathbb{C}^d$ and recall that $\imath\varphi(z,0,t) \in \Sigma_{z,t}$ we get, for some constant $C > 0$, independent of $(z,s,t) \in \mathrm{U}$, $w \in \mathcal{O}$, $\sigma \in {}^t w_s(z,s,t)^{-1}\mathbb{R}^d$,

$$\mathscr{Im}\{\sigma\cdot(w-s-\imath\varphi(z,s,t)) + \langle\sigma\rangle\langle w-s-\imath\varphi(z,s,t)\rangle^2\}/|\sigma| \geq$$
$$(1-\kappa)|s|^2 - C(|w|+|z''|+|t''|). \qquad \text{(IX.7.13)}$$

We return to the proofs of Propositions IX.5.3 and IX.6.1, which we adapt to the present context. We are now dealing with the FBI minitransform and reason under the present more precise hypotheses. We observe that $|s| \geq \delta > 0$ on the support of $\mathrm{d}\chi$ and thus, by (IX.7.13), if the radii of Δ'', W$''$, and $\mathcal{O}$, as well as $\kappa > 0$, are sufficiently small, we shall have

$$\mathscr{Im}\{\sigma\cdot[w-s-\imath\varphi(z,s,t)] + \imath\langle\sigma\rangle\langle w-s-\imath\varphi(z,s,t)\rangle^2\}/|\sigma| \geq \tfrac{1}{4}\delta^2,$$

$$\forall\, z \in \Delta,\ s \in \operatorname{supp}(\mathrm{d}\chi),\ t \in \mathrm{W},\ w \in \mathcal{O},\ \sigma \in \mathscr{C}_\kappa. \qquad \text{(IX.7.14)}$$

The crucial difference between (IX.7.14) and the analogous inequalities in the proofs of Propositions IX.5.3 and IX.6.1 is that (IX.7.14) has been derived without contracing Δ' or W$'$. It follows from (IX.6.8) and (IX.7.14) that here one gets, for suitably large constants $C, R > 0$,

$$|(\bar\partial_z + \mathrm{d}_t)\mathscr{F}(\chi h)(t,z;w,\sigma)| \leq Ce^{-|\sigma|/R},$$

$$\forall\, z \in \Delta,\ t \in \mathrm{W},\ w \in \mathcal{O},\ \sigma \in \mathscr{C}_\kappa. \qquad \text{(IX.7.15)}$$

When h is a $\mathscr{C}^0$ solution, we may take $R = 4/\delta^2$ (see (IX.7.11)). When h is an arbitrary distribution solution, we avail ourselves of a representation of the kind (II.7.26). In this manner we obtain an estimate (IX.7.15) with any $R > 4/\delta^2$. The point here is that $1/R$ is independent of the particular solution h we are dealing with.

First, by integration along rays in t-space we derive (cf. (IX.5.14)) that

$$|\mathscr{F}(\chi h)(t,z;w,\sigma) - \mathscr{F}(\chi h)(0,z;w,\sigma)| \leq C|t|e^{-|\sigma|/R} \qquad \text{(IX.7.16)}$$

is also valid in $\Delta \times W \times \mathcal{O} \times \mathcal{C}_\kappa$.

Then (cf. proof of Proposition IX.6.1) we introduce a solution g of the equation

$$\bar{\partial}_z g(z,w,\sigma) = \bar{\partial}_z \mathcal{F}(\chi h)(z,0;w,\sigma),$$

which is a $\mathcal{C}^\infty$ function in $\Delta \times \mathcal{O} \times \mathcal{C}_\kappa$, holomorphic with respect to (w,σ), and which satisfies

$$|g(z,w,\sigma)| \le Ce^{-|\sigma|/R}$$

for some $C > 0$ possibly larger than the constant in (IX.7.15) but with the same number R. The difference with the proof of Proposition IX.6.1 is that we were forced to contract the whole polydisk Δ, whereas now we only contract Δ''. The function

$$\mathcal{G}h(z,w,\sigma) = \mathcal{F}(\chi h)(z,0;w,\sigma) - g(z,w,\sigma)$$

is holomorphic in the open set $\Delta \times \mathcal{O} \times \mathcal{C}_\kappa$ and satisfies there

$$|\mathcal{F}(\chi h)(z,0;w,\sigma) - \mathcal{G}h(z,w,\sigma)| \le Ce^{-|\sigma|/R}. \tag{IX.7.17}$$

By combining (IX.7.16) and (IX.7.17) we get, possibly with an increased constant $C > 0$,

$$|\mathcal{F}(\chi h)(z,t;w,\sigma) - \mathcal{G}h(z,w,\sigma)| \le Ce^{-|\sigma|/R},$$
$$\forall\ (z,t,w,\sigma) \in \Delta \times W \times \mathcal{O} \times \mathcal{C}_\kappa. \tag{IX.7.18}$$

Another consequence of (IX.7.13) is that, given any $\varepsilon > 0$, we can choose Δ'', W'', $\mathcal{O}$, and κ sufficiently small that

$$\mathcal{Im}\{\sigma\cdot[w-s-\imath\varphi(z,s,t)] + \imath\langle\sigma\rangle\langle w-s-\imath\varphi(z,s,t)\rangle^2\} \ge -\varepsilon|\sigma|/2,$$
$$\forall\ (z,s,t) \in \Delta \times V \times W,\ (w,\sigma) \in \mathcal{O} \times \mathcal{C}_\kappa. \tag{IX.7.19}$$

Here again the significance lies in the fact that in order for (IX.7.19) to be true, we only need to modify the sets Δ'', V, W'', $\mathcal{O}$ and possibly decrease the positive number κ. The set $\Delta' \times W'$ can remain as initially given. We derive easily from (IX.7.19) that, in the set $\Delta \times W \times \mathcal{O} \times \mathcal{C}_\kappa$,

$$|\mathcal{F}(\chi h)(z,t;w,\sigma)| \le Ce^{\varepsilon|\sigma|}. \tag{IX.7.20}$$

In the set $\Delta \times \mathcal{O} \times \mathcal{C}_\kappa$ we have, by (IX.7.18) and (IX.7.20),

$$|\mathcal{G}h(z,w,\sigma)| \le Ce^{\varepsilon|\sigma|}. \tag{IX.7.21}$$

Our hypothesis is that h is hypo-analytic at 0. By Theorem IX.6.1 we can select a polydisk $\Delta_* \subset \Delta$, a ball $W_* \subset W$, both centered at 0, the ball $\mathcal{O}$ and the numbers κ, C, $c > 0$ in such a way that

$$|\mathcal{F}(\chi h)(z,t;w,\sigma)| \le Ce^{-c|\sigma|} \tag{IX.7.22}$$

in the set $\Delta_* \times W_* \times \mathcal{O} \times \mathcal{C}_\kappa$. By (IX.7.18) and (IX.7.22) we get, in the set $\Delta_* \times \mathcal{O} \times \mathcal{C}_\kappa$ and for some number $c > 0$,

$$|\mathcal{G}h(z,w,\sigma)| \le Ce^{-c|\sigma|}. \tag{IX.7.23}$$

Below we denote by r (resp., r_*) the polyradius of the polydisk Δ' (resp., Δ'_*). Let r' be a number such that $r_* \le r' \le r$. Let z'' be a point in Δ''_* and (w,σ) one in $\mathcal{O} \times \mathcal{C}_\kappa$. Then denote by $M(r';z'',w,\sigma)$ the supremum of the function $z' \to \mathcal{G}h(z,w,\sigma)$ in the polydisk $\{ z' \in \mathbb{C}^{\nu'};\ |z_j| \le r', j = 1,\dots,\nu' \}$. Let θ, $0 < \theta < 1$, be such that $r' = r_*^{1-\theta} r^\theta$. As $\mathcal{G}h$ is holomorphic we may apply Hadamard's three-circle theorem:

$$M(r';z'',w,\sigma) \le M(r_*;z'',w,\sigma)^{1-\theta} M(r,z'',w,\sigma)^\theta. \tag{IX.7.24}$$

If we then take (IX.7.21) and (IX.7.23) into account, we get

$$|\mathcal{G}h(z,w,\sigma)| \le Ce^{-[(1-\theta)c - \theta\varepsilon]|\sigma|}. \tag{IX.7.25}$$

If we choose $\varepsilon < (1-\theta)c/\theta$ and combine with (IX.7.18), we conclude that h is hypo-analytic at any point $(z',0,0,t',0)$, $z' \in \Delta'$, $t' \in W'$.

Every point of $\mathcal{N}$ is the center of an open set of the kind $\Delta' \times W'$ above. It is clear that the radii of the polydisks Δ' and those of the balls W' can be taken to vary continuously as the central point moves over $\mathcal{N}$, and in particular, to be bounded from below, away from zero, on compact subsets of $\mathcal{N}$. This fact implies that the intersection of $\mathcal{N}$ with sing supp$_{ha}$ h is open in $\mathcal{N}$. Since that intersection is obviously closed in $\mathcal{N}$ it must either be empty or else be equal to the whole of $\mathcal{N}$, whence the theorem. ∎

Corollary IX.7.1. *Let $\hat{\mathcal{M}}$ be a complex manifold, $\mathcal{M}$ a $\mathcal{C}^\infty$ submanifold of $\hat{\mathcal{M}}$ inheriting from $\hat{\mathcal{M}}$ a CR structure, $\mathcal{N}$ a connected complex submanifold of $\hat{\mathcal{M}}$ contained in $\mathcal{M}$. Let h be a CR distribution on $\mathcal{M}$.*

If a point of $\mathcal{N}$ has an open neighborhood in $\hat{\mathcal{M}}$ to which h extends as a holomorphic function, the same is true of every point of $\mathcal{N}$.

Proof. Indeed, the complex submanifold $\mathcal{N} \subset \mathcal{M}$ is an elliptic submanifold of $\mathcal{M}$ (cf. Proposition IX.7.1) and the hypo-analyticity of h at a point is equivalent to its holomorphic extendability to a full neighborhood of that point in $\hat{\mathcal{M}}$. ∎

Corollary IX.7.2. *Let $\mathcal{M}$ be a hypo-analytic manifold, $\mathcal{N}$ a connected submanifold of $\mathcal{M}$ on which the involutive structure of $\mathcal{M}$ vanishes. If h is a distribution solution in an open set $\Omega \supset \mathcal{N}$, then either $\mathcal{N} \subset$ sing supp$_{ha}$ h or else $\mathcal{N} \cap$ sing supp$_{ha}$ $h = \emptyset$.*

One can for instance apply Corollary IX.7.2 to the structure on $\mathbb{R}^2$ in Example IX.7.6: the y-axis propagates hypo-analyticity.

We close this section by showing that elliptic submanifolds, even submanifolds on which the involutive structure of $\mathcal{M}$ vanishes, do not necessarily propagate smoothness of solutions.

EXAMPLE IX.7.7. Equip $\mathbb{R}^2$ (coordinates: x, t) with the hypo-analytic structure defined by $Z = x(1+\imath t^2)$. The axis $x = 0$ is an elliptic submanifold.

Let us call $\mathscr{E}$ the space of *continuous solutions* in the whole plane that are $\mathscr{C}^\infty$ in the complement of the axis $t = 0$. Denote by $\mathscr{B}_{j,k,\ell}$ (j, k, ℓ: integers ≥ 1) the subset of $\mathscr{E}$ of those solutions h whose restrictions to the disk $\Delta_j = \{ (x,t) \in \mathbb{R}^2; x^2+t^2 \leq 1/j \}$ belong to $\mathscr{C}^k(\Delta_j)$ and whose derivatives of order $\leq k$ have their absolute value not exceeding ℓ in Δ_j.

Suppose that the t-axis propagates $\mathscr{C}^\infty$ regularity. Then, perforce, every function belonging to $\mathscr{E}$ must be smooth in some open neighborhood of 0, and thus, whatever k_0, $\mathscr{E}$ must be the union of the sets $\mathscr{B}_{j,k,\ell}$ such that $k > k_0$. Equip $\mathscr{E}$ with its natural topology: the topology of uniform convergence of the functions on every compact subset of the plane, and of each one of their derivatives on every compact subset of the region $t \neq 0$. It makes a Fréchet space out of $\mathscr{E}$. Baire's category theorem then requires that the closure of one of the sets $\mathscr{B}_{j,k,\ell}$ have an interior point. Let $\{h_\nu\}_{\nu\in\mathbb{Z}_+}$ be a sequence in $\mathscr{B}_{j,k,\ell}$ which converges to h in $\mathscr{E}$. Since $\mathscr{B}_{j,k,\ell}$ is bounded in $\mathscr{C}^k(\Delta_j)$ the sequence $\{h_\nu\}_{\nu\in\mathbb{Z}_+}$ contains a subsequence that converges in $\mathscr{C}^{k-1}(\Delta_j)$, necessarily to h. We conclude that

$$\mathscr{C}\ell\mathscr{B}_{j,k,\ell} \subset \mathscr{B}_{j,k-1,\ell}.$$

It follows that $\mathscr{B}_{j,k-1,\ell}$ has an interior point. Since $\mathscr{B}_{j,k-1,\ell}$ is convex and stable under the symmetry $h \to -h$ its interior must contain the zero function and, as a consequence, its dilations must fill the whole space $\mathscr{E}$. In other words, to each integer $k \geq 0$ there is an integer $j \geq 1$ such that the restriction to Δ_j of any solution $h \in \mathscr{E}$ belongs to $\mathscr{C}^k(\Delta_j)$.

For any $\varepsilon > 0$ the function $Z-\varepsilon = x-\varepsilon+\imath xt^2$ vanishes only at the point $x = \varepsilon$, $t = 0$. When $\mathscr{R}e(Z-\varepsilon) = 0$ we have $\mathscr{I}m(Z-\varepsilon) \geq 0$. This allows us to define the square-root $(Z-\varepsilon)^{\frac{1}{2}}$ (say, the branch that is > 0 for $x > \varepsilon$). This is a continuous solution, $\mathscr{C}^\infty$ in the complement of the point $(\varepsilon,0)$. But given any disk Δ centered at the origin, we can find $\varepsilon > 0$ such that $(Z-\varepsilon)^{\frac{1}{2}}$ is not $\mathscr{C}^1$ in Δ, whence a contradiction. ∎

Notes

What we call the FBI transform was first introduced in Iagolnitzer and Stapp [1] and elaborated in Brós and Iagolnitzer [1]. A summary of the work of Brós and Iagolnitzer on the subject can be found in Brós and Iagolnitzer [2]. Many versions of the same transform have been studied, by mathematicians as well as physicists. Some versions, such as that in Cordoba and Fefferman [1], are geared towards the study of $\mathscr{C}^\infty$ pseudodifferential operators. A very general approach, aimed at the study of the analytic singularities of solutions of linear PDE (with analytic coefficients) and close to the theory of Fourier integral operators with complex phase, is presented in Sjöstrand [2]. An early application to hypo-analytic structures, of the tube type, is given in Baouendi and Treves [3]. The extension in full generality to hypo-analytic manifolds was achieved in Baouendi, Chang, and Treves [1]. How to apply the general theory of Sjöstrand [2] to embedded CR manifolds is shown in the preprint Sjöstrand [3]. All the works cited above, beginning with those of Brós and Iagolnitzer, focus on the microlocal applications of the FBI transform, to which we shall return in Volume 2.

A first application of the FBI *minitransform*, to rigid CR structures, was presented in Baouendi, Rothschild, Treves [1]. Further applications, to semirigid structures, can be found in Baouendi and Rothschild [2]. The result on propagation of hypo-analyticity in section IX.7 was first proved, using the FBI "maxitransform," in Hanges and Treves [1].

X

Involutive Systems of Nonlinear First-Order Differential Equations

How much of the theory of locally integrable structures can be carried over to systems of first-order *nonlinear* partial differential equations? The last chapter of the present volume gives the beginning of an answer to this question. In essence our approach is "microlocal": an involutive system of first-order nonlinear PDE is a $\mathcal{C}^\infty$ submanifold Σ, submitted to a number of conditions, of the complexified one-jet bundle $\mathbb{C}\mathcal{J}^1\mathcal{M}$ over the manifold $\mathcal{M}$ (the equations are set in $\mathcal{M}$). The base projection must map Σ *onto* $\mathcal{M}$. Unless the PDE under consideration are linear, the intersection of Σ with a fibre of $\mathbb{C}\mathcal{J}^1\mathcal{M}$ is not a linear subspace; but it is a holomorphic submanifold. In other words, the equations that locally define Σ are holomorphic with respect to the fibre variables; this makes it possible to input the one-jet of complex-valued solutions. The submanifold Σ is submitted to a *rank condition*: the codimension of Σ is equal to $2n$ and the complex codimension of the intersections of Σ with each complex cotangent space to $\mathcal{M}$ is equal to n; and to a requirement of *involution*: the ideal of the $\mathcal{C}^\infty$ functions, defined in an open subset of $\mathbb{C}\mathcal{J}^1\mathcal{M}$ and holomorphic with respect to the fibre variables, which vanish on Σ, is stable under the Poisson bracket in $\mathbb{C}\mathcal{J}^1\mathcal{M}$. These conditions generalize those in the linear case: that the tangent structure bundle $\mathcal{V}$ be a vector bundle of rank n, and that it satisfy the Frobenius formal integrability condition. The leading parts of the Hamiltonians of those functions define an involutive structure on Σ (with the proviso that all functions and vector fields be holomorphic along the fibres; the Hamiltonians are first-order linear differential operators). Actually, there is a natural symplectic structure on $\mathbb{C}\mathcal{J}^1\mathcal{M}$, "horizontally" real and "vertically" complex, in which Σ may be regarded as an involutive (or co-isotropic) submanifold; the Hamiltonians in question span the symplectic orthogonal of what plays the role here of the tangent space to Σ, namely the space of the first-order differential operators (frozen at the point) whose principal parts are tangent to Σ and have "vertical" components of type (1,0) (section X.1).

Any $\mathscr{C}^2$ solution u in an open set $\Omega \subset \mathcal{M}$ determines a $\mathscr{C}^1$ diffeomorphism, $p \to (p,u(p),du(p))$, of Ω onto a $\mathscr{C}^1$ submanifold Λ_u of Σ. The principal parts of the Hamiltonians are tangent to Λ_u; their restrictions to Λ_u can be regarded as complex vector fields in the base, of course depending on the solution u. The behavior of these vector fields is crucial to the investigations of the nonlinear equations. The rank condition allows us to locally select "space" coordinates x_i $(1 \leq i \leq m)$ and "time" coordinates t_j $(1 \leq j \leq n)$ in which to write the given system of equations in the classical form

$$u_{t_j} = f_j(x,t,u,u_x),\ j = 1,\ldots,n. \tag{1}$$

This means that, if ζ_i, τ_j are the complex coordinates in the complex cotangent spaces dual of the coordinates x_i, t_j, then, in the region of $\mathbb{C}\mathcal{J}^1\mathcal{M}$ under scrutiny, the submanifold Σ may be defined by the equations

$$\tau_j - f_j(x,t,\zeta_0,\zeta) = 0,\ j = 1,\ldots,n. \tag{2}$$

This permits one to use the coordinates x_i, t_j, ζ_0, ζ_i in Σ; along Λ_u the coordinates x_i, t_j suffice. The Hamiltonian operators that matter here are those of the left-hand sides in (2), H_{F_j} (see (X.2.5)). The "canonical" involutive structure on Σ is then defined by the principal parts of H_{F_j}, $\mathrm{H}^0_{F_j}$, together with the vertical Cauchy-Riemann vector fields $\partial/\partial\zeta_0$, $\partial/\partial\bar{\zeta}_i$ (throwing in the latter insure that all solutions will be holomorphic with respect to the fibre variables). Along the N-dimensional submanifold Λ_u the vector field $\mathrm{H}^0_{F_j}$ is equal to

$$\mathscr{L}_j^0 = \partial/\partial t_j - \sum_{i=1}^{m}(\partial f_j/\partial\zeta_i)\partial/\partial x_i \tag{3}$$

(section X.2).

Section X.3 introduces the basic hypothesis under which all subsequent results are arrived at: that the canonical involutive structure on Σ is *locally integrable*. This is always the case in the analytic category (one can then take the functions f_j to be real-analytic) or when the canonical involutive structure on Σ is *elliptic* (this is equivalent to the hypothesis that the vector fields (3) form an elliptic system; local integrability is then a consequence of the Newlander-Nirenberg theorem). The role played by the "horizontal" first integrals Z_i, respectively equal to x_i up to second order with respect to (x, ζ), is different from that played by the "vertical" first integrals Ξ_h, respectively equal to ζ_h up to second-order with respect to (x, ζ) $(1 \leq i \leq m, 0 \leq h \leq m)$. What count are the restrictions of these first integrals to the submanifold Λ_u, Z_i^u and Ξ_h^u. Under the hypothesis that the functions f_j, as well as the initial value of the solution, $u|_{t=0}$, are real-analytic, the restrictions Ξ_h^u can be expressed as holomorphic functions of the Z_i^u. This is also true (by hypocomplexity, see chapter III) when the system of vector fields (3) is elliptic (with no hypothesis on the Cauchy data). In both these cases the implicit function theorem allows us to determine uniquely the solution from its initial value, and to conclude in the former case that u is analytic, and in the latter that u is smooth (section X.3).

The last three sections of the chapter extend part of these results to the general case, always under the local (or rather microlocal) integrability hypothesis. The treatment of the fully nonlinear case can be reduced to that of a related system of *quasilinear* equations provided we allow the latter to apply to vector-valued functions. However, the principal part of this matrix system is a scalar multiple of the identity matrix; only the zero-order terms are truly matricial. Section X.4 redoes for such systems what was done in the previous sections for scalar systems. In section X.5 it is shown how the implicit function theorem can be combined with a Gaussian approximation very similar to the one in the linear case (see section II.2) to yield the approximation of $\mathscr{C}^1$ solutions of the quasilinear systems and, as a consequence, of $\mathscr{C}^2$ solutions of the fully nonlinear systems, by solutions that are determined by the Cauchy data. The trick is to apply the implicit function theorem, not to the restrictions (to Λ_u) of the vertical integrals Ξ_h^u $(0 \leq h \leq m)$, but rather to their Gaussian approximations (Theorem X.5.1). Section X.6 describes some of the consequences of the approximation: uniqueness in the Cauchy problem, stability of the domains of uniqueness under the orbital flow, approximation of $\mathscr{C}^2$ solutions by solutions that are as smooth ($\mathscr{C}^\infty$ or real-analytic) as the equations.

X.1. Involutive Systems of First-Order Nonlinear PDE

Let U be an open subset of the manifold $\mathcal{M}$ ($\dim \mathcal{M} = N$). We suppose that U is the domain of local coordinates $x_1,\dots,x_N$; we denote by $\zeta_1,\dots,\zeta_N$ the associated complex coordinates in the complex tangent spaces $\mathbb{C}T_x^*\mathcal{M}$ at points x of U. We identify $\mathbb{C}T^*\mathcal{M}|_U$ to $U \times \mathbb{C}^N$ by means of those coordinates. Actually we shall reason in an open subset $\mathcal{O}$ of $\mathbb{C} \times (\mathbb{C}T^*\mathcal{M}|_U) \cong U \times \mathbb{C}^{N+1}$; the coordinate in the first factor $\mathbb{C}$ will be denoted by ζ_0. The product $\mathbb{C} \times (\mathbb{C}T^*\mathcal{M}|_U)$ must be thought of as the portion of the complexified *one-jet bundle* of $\mathcal{M}$ (see below) that lies over U.

Suppose then given n functions $F_1,\dots,F_n$ in $\mathcal{O}$, of class $\mathscr{C}^\infty$ and holomorphic with respect to (ζ_0,ζ). In the present chapter we take a look at the differential equations:

$$F_j(x,u,u_x) = 0, \; j = 1,\dots,n. \tag{X.1.1}$$

We propose to generalize to systems (X.1.1) some of the properties established in chapters I and II for linear equations. Recall, however, that those properties were secured under special hypotheses and, first of all, hypotheses of linear independence and of involution (Frobenius condition). We must begin by extending these.

The linear independence of the vector fields generalizes as follows:

$$\mathrm{d}_\zeta F_1 \wedge \cdots \wedge \mathrm{d}_\zeta F_n \neq 0 \textit{ at every point of } \mathcal{O}. \tag{X.1.2}$$

It ought to be made clear that d_ζ is the exterior derivative with respect to

$(\zeta_1,\ldots,\zeta_N)$ and does not involve ζ_0. Condition (X.1.2) demands $n \leq N$. Moreover, suppose that all the functions $F_1,\ldots,F_n$ vanish at some point $\mathfrak{p}$ of $\mathcal{O}$; a consequence of (X.1.2) is that the set

$$(x,\zeta_0,\zeta) \in \mathcal{O};\ F_j(x,\zeta_0,\zeta) = 0,\ j = 1,\ldots,n, \tag{X.1.3}$$

is a $\mathscr{C}^\infty$ submanifold of $\mathcal{O}$ whose intersection with each fibre $\mathbb{C}^N$ ($\cong$ ζ-space) is a holomorphic submanifold of complex codimension n.

The involution condition bears on the *holomorphic Poisson brackets* of the functions F_j. The expression of the Poisson brackets of functions defined in a subset of the one-jet bundle is somewhat more complicated than that of functions defined in a subset of the cotangent bundle. But there is an easy trick that transforms the former into the latter. Introduce an additional real variable x_0, varying in the open interval $]-1,1[$, i.e., substitute $]-1,1[\times\mathcal{M}$ for $\mathcal{M}$, and associate to an arbitrary $\mathscr{C}^1$ function F in $]-1,1[\times\mathcal{O}$, holomorphic with respect to (ζ_0,ζ), the function

$$F^\#(x_0,x,\zeta_0,\zeta) = (1+x_0)F(x,\zeta_0,(1+x_0)^{-1}\zeta).$$

In passing note that with such notation and if we set, in accordance with the same rule, $v(x) = (1+x_0)u(x)$, (X.1.1) reads

$$F_j^\#(x_0,x,v_{x_0},v_x) = 0,\ j = 1,\ldots,n. \tag{X.1.4}$$

Let G and $G^\#$ be another similar pair of functions in $]-1,1[\times\mathcal{O}$. Regarding $F^\#$ and $G^\#$ as functions in an open subset (still $]-1,1[\times\mathcal{O}$) of $\mathbb{C}T^*(]-1,1[\times\mathcal{M})$ one can form their holomorphic Poisson bracket:

$$\{F^\#,G^\#\} = \sum_{i=0}^{N}(\partial F^\#/\partial\zeta_i)\partial G^\#/\partial x_i - (\partial F^\#/\partial x_i)\partial G^\#/\partial\zeta_i.$$

Expressing $\{F^\#,G^\#\}$ in terms of F and G and putting $x_0 = 0$ defines the holomorphic Poisson bracket of F, G:

$$\{F,G\} = \sum_{i=1}^{N}(\partial F/\partial\zeta_i)\partial G/\partial x_i - (\partial F/\partial x_i)\partial G/\partial\zeta_i +$$

$$(\partial F/\partial\zeta_0)\left[G - \sum_{i=1}^{N}\zeta_i\partial G/\partial\zeta_i\right] - (\partial G/\partial\zeta_0)\left[F - \sum_{i=1}^{N}\zeta_i\partial F/\partial\zeta_i\right]. \tag{X.1.5}$$

This allows us to define the *holomorphic Hamiltonian* of F:

$$\mathrm{H}_F = \sum_{i=1}^{N}(\partial F/\partial\zeta_i)\partial/\partial x_i - (\partial F/\partial x_i + \zeta_i\partial F/\partial\zeta_0)\partial/\partial\zeta_i +$$

$$\left[\sum_{i=1}^{N}\zeta_i\partial F/\partial\zeta_i - F\right]\partial/\partial\zeta_0 + \partial F/\partial\zeta_0. \tag{X.1.6}$$

Thus

$$\{F,G\} = \mathrm{H}_F G = -\mathrm{H}_G F. \tag{X.1.7}$$

The Jacobi identity is valid:

$$\{F,\{G,H\}\} + \{G,\{H,F\}\} + \{H,\{F,G\}\} = 0, \tag{X.1.8}$$

equivalent to

$$[\mathrm{H}_F,\mathrm{H}_G] = \mathrm{H}_{\{F,G\}}. \tag{X.1.9}$$

It is also clear that when F and G are both independent of ζ_0 $\{F,G\}$ is their ordinary Poisson bracket.

Returning to Equations (X.1.1), the involution condition can be stated as follows:

$$\forall\, k, \ell = 1,\ldots,n,\ \{F_k,F_\ell\} \equiv 0 \textit{ on the set } (\mathrm{X}.1.3). \tag{X.1.10}$$

PROPOSITION X.1.1. *Properties* (X.1.2) *and* (X.1.10) *entail that the holomorphic Hamiltonians* $\mathrm{H}_{F_1},\ldots,\mathrm{H}_{F_n}$ *are holomorphic tangent to the set* (X.1.3) *and satisfy, in that set, the formal integrability condition*:

$$[\mathrm{H}_{F_j},\mathrm{H}_{F_k}] = \sum_{\ell=1}^{n} c_{jk\ell}\mathrm{H}_{F_\ell},\ j,\ k = 1,\ldots,n. \tag{X.1.11}$$

PROOF. That H_{F_j} is *holomorphic tangent* to the set (X.1.3) means that given any $\mathscr{C}^\infty$ function G in a neighborhood of (X.1.3) in $\mathbb{O}$ that is holomorphic with respect to (ζ_0,ζ) and vanishes on (X.1.3), then $\mathrm{H}_{F_j}G \equiv 0$ in the set (X.1.3). Thanks to hypothesis (X.1.2) we must have, in a suitable neighborhood $\mathbb{O}'$ of (X.1.3) in $\mathbb{O}$,

$$G = \sum_{\ell=1}^{n} a_\ell F_\ell,$$

with $a_1,\ldots,a_n$ $\mathscr{C}^\infty$ functions in $\mathbb{O}'$. But then, on the set (X.1.3),

$$\mathrm{H}_G = \sum_{\ell=1}^{n} a_\ell \mathrm{H}_{F_\ell}. \tag{X.1.12}$$

Applying (X.1.12) to $G = \{F_j,F_k\}$ and taking (X.1.9) into account yields (X.1.11). ∎

To globalize the concept of a nonlinear system of the kind (X.1.1) one must introduce the complex one-jet bundle $\mathbb{C}\mathscr{J}^1\mathscr{M}$. We recall briefly its definition. Denote by $\mathscr{F}_x$ the vector space of complex-valued functions defined and $\mathscr{C}^\infty$ in some open neighborhood of $x \in \mathscr{M}$ (the neighborhood may vary with the function: what we are dealing with here are *germs* of $\mathscr{C}^\infty$ functions at x). Denote by $\mathcal{N}_x^1$ the subspace of those functions that vanish to second order at x: the functions, as well as their differentials, vanish at x. Then, by definition,

$$\mathbb{C}\mathcal{J}_x^1\mathcal{M} = \mathcal{F}_x/\mathcal{N}_x^1. \tag{X.1.13}$$

The image of $f \in \mathcal{F}_x$ under the quotient map $\mathcal{F}_x \to \mathbb{C}\mathcal{J}_x^1\mathcal{M}$ may be denoted by

$$(f(x), \mathrm{d}f(x)),$$

and we have a natural isomorphism $\mathbb{C}\mathcal{J}_x^1\mathcal{M} \cong \mathbb{C}\times\mathbb{C}\mathrm{T}_x^*\mathcal{M}$. As x ranges over $\mathcal{M}$ the complex vector spaces $\mathbb{C}\mathcal{J}_x^1\mathcal{M}$ make up the one-jet bundle of $\mathcal{M}$,

$$\mathbb{C}\mathcal{J}^1\mathcal{M} \cong \mathbb{C}\times\mathbb{C}\mathrm{T}^*\mathcal{M}. \tag{X.1.14}$$

An element of $\mathbb{C}\mathcal{J}^1\mathcal{M}$ will be a triple (x,a,ω) consisting of a point $x \in \mathcal{M}$, a complex number a, and a complex cotangent vector ω at x. A continuous section $x \to (a(x),\omega(x))$ of $\mathbb{C}\mathcal{J}^1\mathcal{M}$ over an open subset Ω of $\mathcal{M}$ will be the Taylor expansion of order one at points of Ω, of a $\mathscr{C}^1$ function in Ω, if and only if $\omega = \mathrm{d}a$ in Ω. Below we denote by π: $(x,a,\omega) \to x$ the base projection of $\mathbb{C}\mathcal{J}^1\mathcal{M}$ onto $\mathcal{M}$.

There is a natural interpretation of the dual of $\mathbb{C}\mathcal{J}^1\mathcal{M}$, $\mathbb{C}\mathcal{D}^1\mathcal{M}$, the vector bundle over $\mathcal{M}$ whose fibre $\mathbb{C}\mathcal{D}_x^1\mathcal{M}$ at each point $x \in \mathcal{M}$ is the dual (over $\mathbb{C}$) of the fibre $\mathbb{C}\mathcal{J}_x^1\mathcal{M}$. A point in $\mathbb{C}\mathcal{D}^1\mathcal{M}$ is a triple $(x,b,\mathbf{v})$ consisting of a point $x \in \mathcal{M}$, of a complex number b and of a complex tangent vector $\mathbf{v}$ to $\mathcal{M}$ at x. The duality between $\mathbb{C}\mathcal{J}_x^1\mathcal{M}$ and $\mathbb{C}\mathcal{D}_x^1\mathcal{M}$ is expressed by the bracket

$$\langle(a,\omega),(b,\mathbf{v})\rangle = ab + \langle\omega,\mathbf{v}\rangle,\ (a,\omega) \in \mathbb{C}\mathcal{J}_x^1\mathcal{M},\ (b,\mathbf{v}) \in \mathbb{C}\mathcal{D}_x^1\mathcal{M}. \tag{X.1.15}$$

We shall always write $\mathbf{v}+b$ (or $b+\mathbf{v}$) rather than $(b,\mathbf{v})$. The reason for this is that a section of $\mathbb{C}\mathcal{D}^1\mathcal{M}$ over an open set Ω can be identified to a first-order differential operator with complex coefficients in Ω, i.e., to the sum $\mathrm{L}+c$ of a complex vector field L and of a complex-valued function c in Ω.

Note that the complex fibre dimension of $\mathbb{C}\mathcal{J}^1\mathcal{M}$—that is, its rank—is equal to $N+1$ and its total real dimension is equal to $3N+2$. Let $\mathcal{O}$ be an open subset of $\mathbb{C}\mathcal{J}^1\mathcal{M}$ such that $\pi(\mathcal{O}) \subset \mathrm{U}$, the domain of local coordinates $x_1,\ldots,x_N$. If F,G are two functions defined in $\mathcal{O}$, of class $\mathscr{C}^1$ and whose restriction to each fibre $\mathcal{O}\cap\mathbb{C}\mathcal{J}_x^1\mathcal{M}$ is holomorphic, their Poisson bracket (X.1.5) is a continuous function in $\mathcal{O}$ whose definition is independent of the choice of local coordinates x_i. Indeed, this is true of each one of the three vector fields

$$\sum_{i=1}^{N}(\partial F/\partial\zeta_i)\partial/\partial x_i - (\partial F/\partial x_i)\partial/\partial\zeta_i,\ \sum_{i=1}^{N}\zeta_i\partial/\partial\zeta_i,\ \partial/\partial\zeta_0.$$

Definition X.1.1. *An* involutive system of first-order differential equations *of rank n on $\mathcal{M}$ will be the datum of a closed $\mathscr{C}^\infty$ submanifold Σ of $\mathbb{C}\mathcal{J}^1\mathcal{M}$ endowed with the following properties*:

the base projection $\mathbb{C}\mathcal{J}^1\mathcal{M} \to \mathcal{M}$ maps Σ onto $\mathcal{M}$; (X.1.16) *each point* of Σ *has an open neighborhood* $\mathcal{O}$ *in which there are* n $\mathcal{C}^\infty$ *functions* $F_j(x,\zeta_0,\zeta)$ $(j = 1,\ldots,n)$, *holomorphic with respect to* (ζ_0,ζ), *such that* $\Sigma \cap \mathcal{O}$ *is the set* (X.1.3), *and that* (X.1.2) *and* (X.1.10) *hold true*. (X.1.17)

In the sequel we often abbreviate "differential equations" to DE.

Let $\Sigma \subset \mathbb{C}\mathcal{J}^1\mathcal{M}$ define an involutive system of first-order DE on $\mathcal{M}$. If $N = \dim \mathcal{M}$, the complex dimension of Σ_x is equal to $N+1-n$ and

$$\dim_{\mathbb{R}} \Sigma = 3N+2-2n. \tag{X.1.18}$$

The next statement confirms that Definition X.1.1 generalizes the concept of an involutive structure over $\mathcal{M}$ (see section I.1).

PROPOSITION X.1.2. *Every involutive structure on* $\mathcal{M}$ *defines an involutive system of first-order DE on* $\mathcal{M}$.

PROOF. Let T′ be the cotangent structure bundle in an involutive structure on $\mathcal{M}$. Call Σ the preimage of T′ under the natural projection $\mathbb{C}\mathcal{J}^1\mathcal{M} \to \mathbb{C}T^*\mathcal{M}$ (cf. (X.1.14)). Let $L_1,\ldots,L_n$ be n linearly independent, pairwise commuting, $\mathcal{C}^\infty$ sections of $\mathcal{V} = T'^{\perp}$ over some open neighborhood U. Then, over U, Σ is defined by the vanishing of the symbols $F_j(x,\zeta)$ of the vector fields L_j regarded as functions in $\mathbb{C}\mathcal{J}^1\mathcal{M}|_U$. Property (X.1.2) follows at once from the fact that the vector fields L_j are linearly independent, and Property (X.1.10) from the fact that they commute (in the present situation the holomorphic Poisson bracket {,} is the holomorphic extension of the standard Poisson bracket). ∎

If Σ is defined by an involutive structure on $\mathcal{M}$ whose structure cotangent bundle is T′ then the fibre Σ_x is the linear subspace $\mathbb{C}\times T'_x$ of $\mathbb{C}\mathcal{J}^1_x\mathcal{M} \cong \mathbb{C}\times\mathbb{C}T^*_x\mathcal{M}$. This means that the functions F_j in (X.1.17) can be chosen to be linear functions of ζ independent of ζ_0. More generally we shall say that the system Σ is *linear homogeneous* if one can choose the functions F_j in (X.1.17) to be linear with respect to (ζ_0,ζ), that the system Σ is *linear inhomogeneous* if one can choose the F_j to be affine functions, i.e., polynomials of degree one, of (ζ_0,ζ). If we use complex coordinates $\zeta_1,\ldots,\zeta_N$ in the complex cotangent spaces at points near some given point $0 \in \mathcal{M}$ we shall have, in $\mathcal{O}$:

$$F_j = \sum_{k=1}^{N} A_{jk}(x)\zeta_k + A_{j0}(x)\zeta_0 + B_j(x). \tag{X.1.19}$$

To say that the system is linear homogeneous is to say that $B_j \equiv 0$ for all $j = 1,\ldots,n$; to say that it is associated with an involutive structure is to say that not only $B_j \equiv 0$ but also $A_{j0} \equiv 0$ for all $j = 1,\ldots,n$.

We introduce two standard classes of nonlinear DE. The system Σ is said to be *quasilinear* if the functions F_j in (X.1.17) can be taken to be linear with respect to ζ but not necessarily with respect to ζ_0. In complex coordinates this means that

$$F_j = \sum_{k=1}^{N} A_{jk}(x,\zeta_0)\zeta_k + B_j(x,\zeta_0). \tag{X.1.20}$$

Note that condition (X.1.2) demands that

$$\mathrm{rank}(A_{jk})_{1\le j\le n, 1\le k\le N} = n. \tag{X.1.21}$$

Finally the system is called *semilinear* if the F_j can be chosen to have the form (X.1.20) with coefficients A_{jk} independent of ζ_0. When the system is not quasilinear it is sometimes said to be *fully nonlinear*.

There is a symplectic (and global) interpretation of Condition (X.1.10). Let $\mathcal{J}^{1,0}(\mathbb{C}\mathcal{J}^1\mathcal{M})$ denote the vector bundle over $\mathbb{C}\mathcal{J}^1\mathcal{M}$ consisting of the triples (p,A,μ) with $p \in \mathbb{C}\mathcal{J}^1\mathcal{M}$, $A \in \mathbb{C}$, μ a cotangent vector to $\mathbb{C}\mathcal{J}^1\mathcal{M}$ that is of type (1,0) with respect to (ζ_0,ζ). If $\pi(p)$ lies in the domain U of local coordinates x_i, we have

$$\mu = \sum_{i=1}^{N}(a_i dx_i + b_i d\zeta_i) + b_0 d\zeta_0;$$

μ does not involve any differential $d\bar{\zeta}_j$ nor $d\bar{\zeta}_0$. The dual of $\mathcal{J}^{1,0}(\mathbb{C}\mathcal{J}^1\mathcal{M})$, which we shall denote by $\mathcal{D}^{1,0}(\mathbb{C}\mathcal{J}^1\mathcal{M})$, can be interpreted as the bundle consisting of the elements $(p,\mathcal{L})$ with $p \in \mathbb{C}\mathcal{J}^1\mathcal{M}$ and, again if $\pi(p) \in$ U,

$$\mathcal{L} = \sum_{i=1}^{N}(\alpha_i\partial/\partial x_i + \beta_i\partial/\partial\zeta_i) + \beta_0\partial/\partial\zeta_0 + \gamma.$$

If F is a $\mathscr{C}^1$ function in $\mathcal{O} \subset \mathbb{C}\mathcal{J}^1\mathcal{M}$, holomorphic with respect to (ζ_0,ζ), the one-jet of F defines a section of $\mathcal{J}^{1,0}(\mathbb{C}\mathcal{J}^1\mathcal{M})$ over $\mathcal{O}$, $\mathcal{J}^1F$. Conversely, every element of the fibre $\mathcal{J}^{1,0}_p(\mathbb{C}\mathcal{J}^1\mathcal{M})$ is equal to the one-jet at $p \in \mathbb{C}\mathcal{J}^1\mathcal{M}$ of some $\mathscr{C}^1$ function F in a neighborhood of p. It follows that the Poisson bracket (X.1.5) defines a (smoothly varying) complex symplectic structure on each fibre $\mathcal{J}^{1,0}_p(\mathbb{C}\mathcal{J}^1\mathcal{M})$. As a consequence it defines a vector bundle isomorphism κ of $\mathcal{D}^{1,0}(\mathbb{C}\mathcal{J}^1\mathcal{M})$ onto $\mathcal{J}^{1,0}(\mathbb{C}\mathcal{J}^1\mathcal{M})$: let $\varpi \in \mathcal{D}^{1,0}_p(\mathbb{C}\mathcal{J}^1\mathcal{M})$, $\mathcal{L} \in \mathcal{D}^{1,0}_p(\mathbb{C}\mathcal{J}^1\mathcal{M})$ be arbitrary. Select two germs at p of $\mathscr{C}^1$ functions F and G, holomorphic with respect to (ζ_0,ζ) and such that $\mathcal{J}^1F|_p = \varpi$, $\mathcal{J}^1G|_p = \kappa\mathcal{L}$. Then, by definition of κ, we have:

$$\{F,G\}(p) = \langle\varpi,\mathcal{L}\rangle = \mathcal{L}F(p).$$

But this means that $\mathcal{L} = -\mathrm{H}_G|_p$; in other words, the isomorphism $\overset{-1}{\kappa}$ can be identified to the map $\mathcal{J}^1G|_p \to -\mathrm{H}_G|_p$. We can use this isomorphism to transfer the symplectic form to the fibres of $\mathcal{D}^{1,0}(\mathbb{C}\mathcal{J}^1\mathcal{M})$. Let $\mathbf{\Omega}$ denote the transferred complex symplectic form.

Let now ϑ be a $\mathscr{C}^1$ section of $\mathfrak{D}^{1,0}(\mathbb{C}\mathscr{J}^1\mathcal{M})$ over some open subset $\mathcal{O}$ of $\mathbb{C}\mathscr{J}^1\mathcal{M}$, and let F be a $\mathscr{C}^1$ function in $\mathcal{O}$ whose restriction to $\mathcal{O}\cap\mathbb{C}\mathscr{J}^1_x\mathcal{M}$ is holomorphic, whatever $x \in \mathcal{M}$. According to what we have just found we have

$$\boldsymbol{\Omega}(\vartheta,\mathrm{H}_F) = -\vartheta F. \tag{X.1.22}$$

Let Σ be a submanifold as in Definition X.1.1. Denote by $\mathfrak{D}^{1,0}(\Sigma)$ the vector subbundle of $\mathfrak{D}^{1,0}(\mathbb{C}\mathscr{J}^1\mathcal{M})|_\Sigma$ consisting of the pairs $(\mathfrak{p},\mathscr{L})$ with $\mathfrak{p} \in \Sigma$ and $\mathscr{L}$ *holomorphic tangent* to Σ at $\mathfrak{p}$ (recall that this means that if F is the germ at $\mathfrak{p}$ of a $\mathscr{C}^1$ function that is holomorphic with respect to (ζ_0,ζ) and that vanishes on the germ of Σ then $\mathscr{L}F(\mathfrak{p}) = 0$). Denote by $\mathfrak{D}^{1,0}(\Sigma)^\perp$ the vector subbundle of $\mathfrak{D}^{1,0}(\mathbb{C}\mathscr{J}^1\mathcal{M})|_\Sigma$ whose fibre at $\mathfrak{p} \in \Sigma$ is the orthogonal of $\mathfrak{D}^{1,0}_\mathfrak{p}(\Sigma)$ for the symplectic form $\boldsymbol{\Omega}$.

PROPOSITION X.1.3. *Let Σ be an involutive system of first-order DE on $\mathcal{M}$. Then*

$$\mathfrak{D}^{1,0}(\Sigma)^\perp \subset \mathfrak{D}^{1,0}(\Sigma). \tag{X.1.23}$$

Let $\mathcal{O}$, $F_1,\dots,F_n$ be as in (X.1.17); *then the holomorphic Hamiltonians $\mathrm{H}_{F_1},\dots,\mathrm{H}_{F_n}$ form a basis of $\mathfrak{D}^{1,0}(\Sigma)^\perp$ over $\Sigma\cap\mathcal{O}$.*

PROOF. Let x be an arbitrary point of $\mathcal{M}$. Recalling that $\Sigma_x \neq \emptyset$, we derive from (X.1.2) that $\operatorname{codim}_\mathbb{C}\Sigma_x = n$, hence the complex codimension of $\mathfrak{D}^{1,0}_\mathfrak{p}(\Sigma)$ as a vector subspace of $\mathfrak{D}^{1,0}_\mathfrak{p}(\mathbb{C}\mathscr{J}^1\mathcal{M})$ is equal to n ($\mathfrak{p} \in \Sigma$). The fact that the bilinear form $\boldsymbol{\Omega}$ is nondegenerate entails $\dim_\mathbb{C}\mathfrak{D}^{1,0}_\mathfrak{p}(\Sigma)^\perp = n$. By virtue of Proposition X.1.1 (X.1.23) will follow if we prove that $\mathrm{H}_{F_1},\dots,\mathrm{H}_{F_n}$ span $\mathfrak{D}^{1,0}(\Sigma)^\perp$ over $\Sigma\cap\mathcal{O}$. Since the latter are obviously linearly independent it suffices to show that they are sections of $\mathfrak{D}^{1,0}(\Sigma)^\perp$ over $\Sigma\cap\mathcal{O}$. We apply (X.1.22) with $F = F_j$. But $\vartheta F_j \equiv 0$ if ϑ is holomorphic tangent to Σ, whence the claim. ■

If we adapt the terminology of symplectic geometry to the present context (where we deal with first-order differential operators instead of vector fields and where holomorphy in the fibre variables must be preserved) we could say that Σ is *co-isotropic* (or involutive).

Let $\mathrm{H}^0_{F_j}$ denote the *principal part* of H_{F_j}, i.e., the difference $\mathrm{H}_{F_j} - \mathrm{H}^0_{F_j}$ is a differential operator of order zero, that is to say, it acts as multiplication by a smooth function; $\mathrm{H}^0_{F_j}$ is a complex vector field. Actually we restrict it to $\Sigma\cap\mathcal{O}$. By (X.1.11) we have

$$[\mathrm{H}^0_{F_j},\mathrm{H}^0_{F_k}] = \sum_{\ell=1}^{n} c_{jk\ell}\mathrm{H}^0_{F_\ell},\ j,\ k = 1,\dots,n. \tag{X.1.24}$$

Recall then the identity (X.1.12) valid for any $\mathscr{C}^\infty$ function G in a neighborhood of $\Sigma \cap \mathcal{O}$ that is holomorphic with respect to (ζ_0,ζ) and vanishes on $\Sigma \cap \mathcal{O}$; it implies

$$\mathrm{H}_G^0 = \sum_{\ell=1}^{n} a_\ell \mathrm{H}_{F_\ell}^0, \tag{X.1.25}$$

which in turn implies that if we modify our choice of the functions F_j it does not change the span of $\mathrm{H}_{F_1}^0,\ldots,\mathrm{H}_{F_n}^0$. We may state:

PROPOSITION X.1.4. *Let Σ be an involutive system of first-order DE on $\mathcal{M}$. There is an involutive structure $\mathcal{V}_\Sigma \subset \mathbb{C}T\Sigma$ on Σ defined by the following property: if $\mathcal{O}$, $F_1,\ldots,F_n$ are as in* (X.1.17) *then the principal parts of the holomorphic Hamiltonians $\mathrm{H}_{F_1}^0,\ldots,\mathrm{H}_{F_n}^0$ form a basis of $\mathcal{V}_\Sigma$ over $\Sigma \cap \mathcal{O}$.*

One can say that $\mathcal{V}_\Sigma$ is the "principal part" of $\mathfrak{D}^{1,0}(\Sigma)^\perp$.

All the definitions that have been introduced so far in the present section have their analogues in the *analytic* category: one may assume $\mathcal{M}$ to be a real-analytic manifold, Σ a real-analytic submanifold of $\mathbb{C}\mathcal{J}^1\mathcal{M}$, the defining functions F_i to be real-analytic (but in general complex-, not real-, valued), etc.

X.2. Local Representations

Given the involutive system of DE, Σ, in Definition X.1.1, one can try to select the defining functions F_i in a manner that facilitates the analysis of Equations (X.1.1). Let $\mathcal{O}$ denote an open neighborhood of a point $\mathfrak{p} \in \Sigma$, which will be contracted about $\mathfrak{p}$ as many times as needed. We avail ourselves of (X.1.2) by relabeling the coordinates x_i so as to have

$$\det\,(\partial F_j/\partial \zeta_i)_{1\le i,j\le n} \neq 0. \tag{X.2.1}$$

We apply the implicit function theorem and replace the functions F_j by new functions of the form $\zeta_j - f_j(x,\zeta_0,\zeta_{n+1},\ldots,\zeta_N)$, endowed with the same properties as the F_j. At this point we adopt a notation similar to that used for involutive structures (cf. (I.7.25)): we write t_j instead of x_j and τ_j instead of ζ_j if $1 \le j \le n$; x_i instead of x_{n+i} and ζ_i instead of ζ_{n+i} if $n < i \le m = N-n$. With this notation we see that Σ is defined in $\mathcal{O}$ by the equations

$$\tau_j - f_j(x,t,\zeta_0,\zeta) = 0,\ j = 1,\ldots,n, \tag{X.2.2}$$

where now $\zeta = (\zeta_1,\ldots,\zeta_m)$. On $\Sigma \cap \mathcal{O}$ we may use the following coordinates:

$$x_i\ (1 \le i \le m),\ t_j\ (1 \le j \le n),\ \mathcal{R}e\,\zeta_0,\ \mathcal{I}m\,\zeta_0,\ \mathcal{R}e\,\zeta_k,\ \mathcal{I}m\,\zeta_\ell\ (1 \le k, \ell \le m). \tag{X.2.3}$$

We shall denote by 0 the base projection of the point $\mathfrak{p} \in \Sigma$; we assume that the coordinates x_i, t_j are defined in an open neighborhood U of 0 and vanish at 0, and that the base projection of $\mathcal{O}$ is equal to U. Let us write $\mathfrak{p} = (0,a,\omega)$ with $a \in \mathbb{C}$ and $\omega \in \mathbb{C}T^*_0\mathcal{M}$.

Equations (X.1.1) can now be rewritten as

$$u_{t_j} = f_j(x,t,u,u_x),\ j = 1,\ldots,n. \tag{X.2.4}$$

Call F_j the left-hand sides in the equations (X.2.2). By virtue of Proposition X.1.1 we may regard the holomorphic Hamiltonian H_{F_j} as a differential operator on the manifold Σ (in $\Sigma\cap\mathcal{O}$). In the local coordinates (X.2.3) its expression is given by

$$\mathrm{H}_{F_j} = \partial/\partial t_j - \sum_{\ell=1}^{m}(f_{j\zeta_\ell}\partial/\partial x_\ell - f_{jx_\ell}\partial/\partial\zeta_\ell) + \\ f_{j\zeta_0}\rho - (\rho f_j)\partial/\partial\zeta_0 + f_j\partial/\partial\zeta_0 + f_{j\zeta_0}, \tag{X.2.5}$$

where $\rho = \sum_{i=1}^{m}\zeta_i\partial/\partial\zeta_i$. The proof of (X.2.5) is by direct computation and is left to the reader. In the sequel we use the following notation:

$$\mathrm{H}^0_{F_j} = \partial/\partial t_j - \sum_{\ell=1}^{m}(f_{j\zeta_\ell}\partial/\partial x_\ell - f_{jx_\ell}\partial/\partial\zeta_\ell) + \\ f_{j\zeta_0}\rho - (\rho f_j)\partial/\partial\zeta_0 + f_j\partial/\partial\zeta_0; \tag{X.2.6}$$

$$\mathcal{L}_{F_j} = \partial/\partial t_j - \sum_{\ell=1}^{m} f_{j\zeta_\ell}\partial/\partial x_\ell + f_{j\zeta_0}; \tag{X.2.7}$$

$$\mathcal{L}^0_{F_j} = \partial/\partial t_j - \sum_{\ell=1}^{m} f_{j\zeta_\ell}\partial/\partial x_\ell; \tag{X.2.8}$$

$\mathrm{H}^0_{F_j}$ is the *principal part* of H_{F_j}, $\mathcal{L}^0_{F_j}$ that of $\mathcal{L}_{F_j}$ and the latter is the pushdownward of H_{F_j} via the base projection.

In view of the expressions (X.2.5) the Frobenius relations (X.1.11) read here:

$$[\mathrm{H}_{F_j},\mathrm{H}_{F_k}] = 0,\ j,\ k = 1,\ldots,n. \tag{X.2.9}$$

Select a function $u \in \mathcal{C}^1(\mathrm{U})$ such that

$$(x,t) \in \mathrm{U} \Rightarrow (x,t,u(x,t),u_x(x,t),u_t(x,t)) \in \mathcal{O}. \tag{X.2.10}$$

We shall use the coordinates (X.2.3) on $\Sigma\cap\mathcal{O}$. Given any function $G(x,t,\zeta_0,\zeta)$ on $\Sigma\cap\mathcal{O}$ that is holomorphic with respect to (ζ_0,ζ), we define the function in U,

$$G^u(x,t) = G(x,t,u(x,t),u_x(x,t)).$$

In particular Equations (X.2.4) in U can be rewritten as $F^u_j \equiv 0$, $j = 1,\ldots,n$. We shall also use the notation:

$$\mathcal{L}^u_{F_j} = \partial/\partial t_j - \sum_{\ell=1}^{m} f^u_{j\zeta_\ell}\partial/\partial x_\ell + f^u_{j\zeta_0}; \tag{X.2.11}$$

$$\mathcal{L}^{0u}_{F_j} = \partial/\partial t_j - \sum_{\ell=1}^{m} f^u_{j\zeta_\ell}\partial/\partial x_\ell. \tag{X.2.12}$$

LEMMA X.2.1. *Let u be a $\mathcal{C}^2$ function in* U *such that* (X.2.10) *holds, $G(x,t,\zeta_0,\zeta)$ a $\mathcal{C}^1$ function in $\Sigma\cap\mathcal{O}$, holomorphic with respect to (ζ_0,ζ). Then, in* U, *for all $j = 1,\dots,n$,*

$$\mathcal{L}^{0u}_{F_j}G^u = (\mathrm{H}^0_{F_j}G)^u + (\partial G/\partial\zeta_0)^u F^u_j + \sum_{k=1}^{m}(\partial G/\partial\zeta_k)^u(\partial/\partial x_k)F^u_j. \tag{X.2.13}$$

PROOF. Let D be any of the partial derivatives $\partial/\partial x_i$ ($1 \le i \le m$) or $\partial/\partial t_j$ ($1 \le j \le n$). By the chain rule,

$$\mathrm{D}G^u = (\mathrm{D}G)^u + (\partial G/\partial\zeta_0)^u\mathrm{D}u + \sum_{k=1}^{m}(\partial G/\partial\zeta_k)^u\mathrm{D}(\partial u/\partial x_k), \tag{X.2.14}$$

whence

$$\mathcal{L}^{0u}_{F_j}G^u = (\mathcal{L}^0_{F_j}G)^u + G^u_{\zeta_0}\left[\partial u/\partial t_j - \sum_{k=1}^{m} f^u_{j\zeta_k}\partial u/\partial x_k\right] +$$

$$\sum_{k=1}^{m} G^u_{\zeta_k}\left[\partial/\partial t_j - \sum_{\ell=1}^{m} f^u_{j\zeta_\ell}\partial/\partial x_\ell\right]\partial u/\partial x_k.$$

If we apply (X.2.14) to f_j and substitute

$$(\partial/\partial x_k)f^u_j - (\partial f_j/\partial x_k)^u - (\partial f_j/\partial\zeta_0)^u\partial u/\partial x_k \text{ for } \sum_{\ell=1}^{m}(\partial f_j/\partial\zeta_\ell)^u\partial^2 u/\partial x_k\partial x_\ell,$$

we get

$$\mathcal{L}^{0u}_{F_j}G^u = (\mathcal{L}^0_{F_j}G)^u + G^u_{\zeta_0}\left[\partial u/\partial t_j - \sum_{k=1}^{m} f^u_{j\zeta_k}\partial u/\partial x_k\right] +$$

$$\sum_{k=1}^{m} G^u_{\zeta_k}(\partial/\partial x_k)F^u_j + \sum_{k=1}^{m} G^u_{\zeta_k}[(\partial f_j/\partial x_k)^u + (\partial f_j/\partial\zeta_0)^u\partial u/\partial x_k].$$

But we note that $u_{x_k}G^u_{\zeta_k} = (\zeta_k G_{\zeta_k})^u$ and likewise with f_j in the place of G. We reach the conclusion that

$$\mathcal{L}^{0u}_{F_j}G^u = (\mathcal{L}^0_{F_j}G)^u + G^u_{\zeta_0}F^u_j + G^u_{\zeta_0}f^u_j - G^u_{\zeta_0}\sum_{k=1}^{m}(\zeta_k f_{j\zeta_k})^u +$$

$$\sum_{k=1}^{m} G^u_{\zeta_k}(\partial/\partial x_k)F^u_j + \sum_{k=1}^{m} G^u_{\zeta_k}[(\partial f_j/\partial x_k)^u + (\zeta_k \partial f_j/\partial \zeta_0)^u] =$$

$$\left[\mathscr{L}^0_{F_j}G + f_x \cdot \partial_\zeta G + f_{j\zeta_0}(\zeta \cdot \partial_\zeta G) + f_j G_{\zeta_0} - (\zeta \cdot f_{j\zeta})G_{\zeta_0}\right]^u +$$

$$G^u_{\zeta_0}F^u_j + (G^u_\zeta \cdot \partial_x)F^u_j,$$

which, by (X.2.6) and (X.2.8), is what we wanted. ■

COROLLARY X.2.1. *If* $u \in \mathscr{C}^2(\mathrm{U})$ *satisfies* (X.2.10) *and is a solution of* (X.2.4) *in* U *then*

$$\mathscr{L}^{0u}_{F_j}G^u = (\mathrm{H}^0_{F_j}G)^u. \qquad \text{(X.2.15)}$$

By letting ∂_x act on both sides of Equations (X.2.4), we obtain

PROPOSITION X.2.1. *Suppose* $u \in \mathscr{C}^2(\mathrm{U})$ *satisfies* (X.2.10) *and is a solution of* (X.2.4). *Then its gradient* u_x *with respect to the variables* x_i $(1 \le i \le m)$ *is a solution of the following system of differential equations*:

$$\mathscr{L}^u_{F_j}u_x = (f_{jx})^u,\ j = 1,\ldots,n. \qquad \text{(X.2.16)}$$

Note that Equations (X.2.16) are relations between functions valued in $\mathbb{C}^m$.

We continue to suppose that the solution u belongs to $\mathscr{C}^2(\mathrm{U})$ and satisfies (X.2.10). The map

$$J^u\colon (x,t) \to (x,t,u(x,t),u_x(x,t),u_t(x,t)) \in \mathbb{C}\mathscr{J}^1\mathscr{M}$$

is a $\mathscr{C}^1$ diffeomorphism of U onto a $\mathscr{C}^1$ submanifold Λ_u of $\Sigma \cap \mathbb{O}$, obviously defined, in the coordinate system (X.2.3), by the equations

$$\zeta_0 - u(x,t) = 0,\ \zeta - u_x(x,t) = 0. \qquad \text{(X.2.17)}$$

At the point $(x,t,u(x,t),u_x(x,t)) \in \Lambda_u$ (again using the coordinates (X.2.3)) the principal part of the Hamiltonian H_{F_j} is equal to the vector field

$$\partial/\partial t_j - f^u_{j\zeta} \cdot \partial_x + (f^u_{jx} + f^u_{j\zeta_0}u_x) \cdot \partial_\zeta + (f^u_j - u_x \cdot f^u_{j\zeta})\partial_{\zeta_0}. \qquad \text{(X.2.18)}$$

It obviously annihilates $\zeta_0 - u(x,t)$. That it also annihilates $\zeta_k - u_{x_k}(x,t)$ is the meaning of the equations (X.2.16). We may state:

PROPOSITION X.2.2. *Let* $u \in \mathscr{C}^2(\mathrm{U})$ *satisfy* (X.2.10) *and be a solution of* (X.2.4). *Whatever* $j = 1,\ldots,n$, *the holomorphic Hamiltonian* H_{F_j} *is holomorphic tangent to the submanifold* Λ_u.

The restriction of the base projection to the submanifold Λ_u transforms H_{F_j} at the point $(x,t,u(x,t),u_x(x,t)) \in \Lambda_u$ into $\mathscr{L}^u_{F_j}$. From this observation and from (X.2.10) we derive

PROPOSITION X.2.3. *Suppose* $u \in \mathscr{C}^2(\mathrm{U})$ *satisfies* (X.2.10) *and is a solution of* (X.2.4). *Then, whatever* $j, k = 1,\dots,n$, *we have, in* U

$$[\mathscr{L}^u_{F_j},\mathscr{L}^u_{F_k}] = 0. \tag{X.2.19}$$

COROLLARY X.2.2. *We have, in* U, *for all* $j, k = 1,\dots,n$,

$$\mathscr{L}^u_{F_j}(f_{kx})^u = \mathscr{L}^u_{F_k}(f_{jx})^u. \tag{X.2.20}$$

The relations (X.2.20) are the *compatibility conditions* for the Equations (X.2.16).

There is another interpretation of the differential operators $\mathscr{L}^u_{F_j}$: as the *Fréchet derivative* of the map $u \to F^u_j$. Indeed, given any function $v \in \mathscr{C}^1(\mathrm{U})$, we have

$$\mathscr{L}^u_{F_j}v = v_{t_j} - \lim_{\lambda\to 0} \lambda^{-1}(f_j^{u+\lambda v} - f_j^u), \tag{X.2.21}$$

where, we recall, $f_j^{u+\lambda v} = f_j(x,t,u+\lambda v,u_x+\lambda v_x)$.

X.3. Microlocal Integrability. First Results on Uniqueness in the Cauchy Problem

As before $\mathfrak{p} = (0,a,\omega)$ denotes a point in the submanifold Σ of $\mathbb{C}\mathscr{J}^1\mathcal{M}$. We shall deal with the involutive structure $\mathscr{V}_\Sigma$ on Σ (see Proposition X.1.4). Actually, we shall "augment" $\mathscr{V}_\Sigma$ by adjoining to it the tangential Cauchy-Riemann vector fields in Σ; the resulting vector subbundle of $\mathbb{C}T\Sigma$, $\hat{\mathscr{V}}_\Sigma$, defines an involutive structure on Σ. In the coordinates system (X.2.3) on $\Sigma\cap\mathcal{O}$ the vector bundle $\hat{\mathscr{V}}_\Sigma$ is spanned by the vector fields

$$\mathrm{H}^0_{F_1},\dots,\mathrm{H}^0_{F_n},\ \partial/\partial\bar{\zeta}_0,\ \partial/\partial\bar{\zeta}_1,\dots,\partial/\partial\bar{\zeta}_m. \tag{X.3.1}$$

Observe that rank $\hat{\mathscr{V}}_\Sigma = m+n+1 = 1 + \dim \mathcal{M} = \dim \Sigma - 2m - 1$.

DEFINITION X.3.1. *We say that* Σ *is* microlocally integrable *at* $\mathfrak{p}$ *if the involutive structure* $\hat{\mathscr{V}}_\Sigma$ *is locally integrable over an open neighborhood of* $\mathfrak{p}$ *in* Σ.

The local integrability of $\hat{\mathscr{V}}_\Sigma$ means that there exist $\mathscr{C}^\infty$ functions Z_i ($i = 1,\dots,2m+1$), *holomorphic with respect to* (ζ_0,ζ), in an open neighborhood $\mathcal{O}$ of $\mathfrak{p}$ in which there exist $\mathscr{C}^\infty$ functions $F_1,\dots,F_n$ endowed with the properties listed in (X.1.17), such that the following holds, at every point of $\mathcal{O}$:

the pullbacks to Σ *of the differentials* $\mathrm{d}Z_i$ *are linearly independent*; (X.3.2)

$$\mathrm{H}^0_{F_j}Z_i = 0,\ j = 1,\dots,n. \tag{X.3.3}$$

It is convenient to make use of the coordinate system (X.2.3) in $\Sigma\cap\mathcal{O}$; $Z_1,\ldots,Z_{2m+1}$ make up a system of first integrals of the vector fields (X.3.1).

EXAMPLE X.3.1. Suppose the manifold $\mathcal{M}$ and the system Σ are analytic and choose the coordinates x_i, t_j to be real-analytic. Thanks to the expressions (X.2.6) we may solve, in the class of real-analytic functions of (x,t,ζ_0,ζ,τ) that are holomorphic with respect to the fibre variables (ζ_0,ζ,τ), the local Cauchy problems in $\mathcal{O}$ (perhaps contracted about the central point $\mathfrak{p}$):

$$\mathrm{H}^0_{F_j}h = 0,\ j = 1,\ldots,n,\ h|_{t=0} = h_0(x,\zeta_0,\zeta,\tau). \tag{X.3.4}$$

For $i = 1,\ldots,m$, call X_i (resp., Ξ_i) the analytic solution corresponding to the initial condition $h_0 = x_i$ (resp., ζ_i), Ξ_0 the solution corresponding to $h_0 = \zeta_0$. The functions $X_1,\ldots,X_m$, Ξ_0, $\Xi_1,\ldots,\Xi_m$ make up a complete system of first integrals of the vector fields (X.3.1). ■

EXAMPLE X.3.2. Suppose the system of vector fields (X.3.1) is *elliptic* on Σ at $\mathfrak{p}$, hence in a neighborhood of $\mathfrak{p}$ that we take to be the open set $\Sigma\cap\mathcal{O}$ (ellipticity of the system (X.3.1) demands $2(N+1) \geq \dim_{\mathbb{R}}\Sigma = 3N+2-2n$, i.e., $2n \geq N$). Theorem VI.7.1 states that the involutive structure $\mathcal{E}$ defined on $\Sigma\cap\mathcal{O}$ by the vector fields (X.3.1) is locally integrable and thus we can find first integrals $Z_1,\ldots,Z_{2m+1}$. ■

Let U be an open neighborhood of 0 in $\mathcal{M}$ and $u \in \mathcal{C}^2(\mathrm{U})$ a solution of Equations (X.2.4) such that

$$(x,t) \in \mathrm{U} \Rightarrow (x,t,u(x,t),du(x,t)) \in \mathcal{O};\ u(0) = a,\ du(0) = \omega. \tag{X.3.5}$$

By Proposition X.2.2 the vector fields $\mathrm{H}^0_{F_j}$ are holomorphic tangent to the submanifold Λ_u of Σ. The pullbacks to Λ_u of the differentials dZ_i span the orthogonal in $\mathbb{C}T^*\Lambda_u$ of the span (in $\mathbb{C}T\Lambda_u$) of $\mathrm{H}^0_{F_1},\ldots,\mathrm{H}^0_{F_n}$. Recall that $\dim\Lambda_u = \dim\mathcal{M} = m+n$; it follows that in a suitably small neighborhood of $\mathfrak{p}$, the pullbacks to Λ_u of exactly m differentials dZ_i must be linearly independent, which is equivalent to the following property:

the differentials dZ_i^u $(1 \leq i \leq 2m+1)$ *span a vector subbundle* T'^u *of rank m of* $\mathbb{C}T^*\mathcal{M}$ *over a neighborhood of* 0. (X.3.6)

It follows from Corollary X.2.1 that the vector bundle T'^u is the orthogonal of the vector subbundle $\mathcal{V}^u$ of $\mathbb{C}T\mathcal{M}$ spanned by the vector fields $\mathcal{L}^{0u}_{F_1},\ldots,\mathcal{L}^{0u}_{F_n}$.

REMARK X.3.1. Since u is of class $\mathcal{C}^2$ the functions $Z_i(x,t,u(x,t),u_x(x,t))$ are of class $\mathcal{C}^1$ and thus the vector bundle T'^u appears to be only of class $\mathcal{C}^0$. But we note that the coefficients of the vector fields $\mathcal{L}^{0u}_{F_j}$ are of class $\mathcal{C}^1$ and therefore span a vector bundle, $\mathcal{V}^u$, of class $\mathcal{C}^1$. Its orthogonal, T'^u, must also be of class $\mathcal{C}^1$. ■

Returning for a moment to the manifold Σ, one sees that the principal parts of the Hamiltonians H_{F_j}, $\mathrm{H}^0_{F_j}$, are (holomorphically) transverse to the vertical fibres $\Sigma_x = \Sigma \cap \mathbb{C}\mathcal{J}^1_x\mathcal{M}$ $(x \in \mathrm{U})$. The pullbacks to Σ_x of the differentials $\mathrm{d}Z_i$ span the holomorphic cotangent spaces to Σ_x over $\Sigma \cap \mathcal{O}$, and thus exactly $m + 1$ of them must be linearly independent. We may assume that the pullbacks to Λ_u of the remaining m differentials $\mathrm{d}Z_i$ are linearly independent. After relabeling we shall assume that the differentials of $Z_1,\ldots,Z_m$ span a "horizontal" supplementary of $\mathbb{C}T^*\Sigma_x$ in $\mathbb{C}T^*\Sigma$, and that those of $Z_{m+1},\ldots,Z_{2m+1}$ span $\mathbb{C}T^*\Sigma_x$. After a $\mathbb{C}$-linear substitution we can further arrange matters in such a manner that

the functions $Z_i - x_i$, $Z_{m+i} - \zeta_i$ $(1 \le i \le m)$, $Z_{2m+1} - \zeta_0$ *and their differentials with respect to* (x,ζ_0,ζ) *vanish at* $\mathfrak{p}$. (X.3.7)

Having done this we change notation: we write $\Xi_i = Z_{m+i}$ $(1 \le i \le m)$, $\Xi_0 = Z_{2m+1}$.

We are now in a position to further exploit Lemma X.2.1.

Proposition X.3.1. *Assume there exist first integrals of the systems of vector fields* (X.3.1), Z_i, Ξ_i $(1 \le i \le m)$ *and* Ξ_0 *as just described. Let* $u \in \mathscr{C}^2(\mathrm{U})$ *satisfy Condition* (X.3.5). *In order for* u *to be a solution of* (X.2.4) *in some open neighborhood* $\mathrm{U}' \subset \mathrm{U}$ *of* 0 *in* $\mathcal{M}$ *it is necessary and sufficient that the following equations hold in such a neighborhood*:

$$\mathscr{L}^{0u}_{F_j}\Xi^u_i = 0, \; i = 0,1,\ldots,m, \; j = 1,\ldots,n. \tag{X.3.8}$$

Proof. The condition is necessary, by Corollary X.2.1. By the definition of the functions Ξ_i and by (X.2.13), (X.3.8) entails

$$(\partial\Xi_i/\partial\zeta_0)^u F^u_j + \sum_{k=1}^{m}(\partial\Xi_i/\partial\zeta_k)^u(\partial/\partial x_k)F^u_j \equiv 0, \; i = 0,1,\ldots,m. \tag{X.3.9}$$

Thanks to hypotheses (X.3.5) and (X.3.7) we see that the matrix $(\partial\Xi_i/\partial\zeta_k)^u_{0\le i,k\le m}$ is equal to the $(m+1)\times(m+1)$ identity matrix at 0. It follows at once that we must have $F^u_j \equiv 0$ in some neighborhood of 0. ∎

We now discuss some applications of Proposition X.3.1 to uniqueness in the Cauchy problem. It is convenient to take $\mathrm{U} = \mathrm{V}\times\mathrm{W}$ with V (resp., W) an open neighborhood of the origin in x-space $\mathbb{R}^m$ (resp., t-space $\mathbb{R}^n$). Decompose $\omega \in \mathbb{C}T^*_0\mathcal{M}$ into an x-component, ξ_0, and a t-component, τ_0. We continue to reason under the microlocal integrability hypothesis of Proposition X.3.1. We write $Z = (Z_1,\ldots,Z_m)$, $\Xi = (\Xi_0,\ldots,\Xi_m)$.

We may avail ourselves of the implicit function theorem in the holomorphic category (version with parameters x_i, t_j) and solve the equations

$$\Xi_i(x,t,\zeta_0,\zeta) = w_i, \; i = 0,1,\ldots,m, \tag{X.3.10}$$

with respect to (ζ_0,ζ) in such a way that $\zeta_0 = a$, $\zeta = \xi$ when $x = 0$, $t = 0$, $w_i = \Xi_i(0,0,a,\xi)$. The unique solution will be denoted by $\Phi_i(x,t,w)$; it is holomorphic with respect to w. We see that if u satisfies Condition (X.3.5), then

$$u = \Phi_0(x,t,\Xi_0^u,\Xi^u),\ u_{x_i} = \Phi_i(x,t,\Xi_0^u,\Xi^u)\ (1 \le i \le m). \qquad \text{(X.3.11)}$$

We shall now be concerned with $\mathscr{C}^2$ solutions u of (X.2.4) in U that satisfy an initial value condition:

$$u|_{t=0} = u_0(x) \textit{ in } \mathrm{V}, \qquad \text{(X.3.12)}$$

with $u_0 \in \mathscr{C}^2(\mathrm{V})$. In view of (X.3.5) Condition (X.3.12) demands $u_0(0) = a$, $d_x u_0(0) = \xi$ and $\tau = d_t u(0,0) = (f_1(0,0,a,\xi),\ldots,f_n(0,0,a,\xi))$.

REMARK X.3.2. Proposition X.3.1 immediately yields a uniqueness result about systems of *semilinear* DE. In this case the coefficients of $\mathscr{L}^0_{F_1},\ldots,\mathscr{L}^0_{F_n}$ are independent of (ζ_0,ζ), and by hypothesis this system of vector fields is locally integrable. It follows that the solutions Ξ_i^u are completely determined by the values at $t = 0$. By virtue of (X.3.11) the same is true of u. Moreover, in the semilinear case, the regularity requirement on the solution u can be weakened to $u \in \mathscr{C}^1(\mathrm{U})$. ■

Back to fully nonlinear DE we introduce a new hypothesis:

There exist $m+1$ holomorphic functions $G_0,G_1,\ldots,G_m$ in an open neighborhood of the origin in $\mathbb{C}^m$ such that (X.3.13)

$$\Xi_i^u = G_i(Z^u),\ i = 0,1,\ldots,m. \qquad \text{(X.3.14)}$$

A priori the functions G_i could depend on the function u. Let us show that, in actuality, they only depend on the initial value $u_0(x) = u(x,0)$. Put $t = 0$ in both sides of (X.3.14):

$$\Xi_i(x,0,u_0(x),u_{0x}(x)) = G_i(Z(x,0,u_0(x),u_{0x}(x))).$$

Thanks to (X.3.7) we see that the map $x \to Z(x,0,u_0(x),u_{0x}(x))$ transforms V into a maximally real submanifold of $\mathbb{C}^m$, of class $\mathscr{C}^1$. The trace on it of the holomorphic function G_i completely determines G_i in a neighborhood of 0, which proves our claim.

THEOREM X.3.1. *Suppose* (X.3.13) *holds. There is a $\mathscr{C}^\infty$ function v in an open neighborhood* $\mathrm{U}' \subset \mathrm{U}$ *of* 0 *in* $\mathcal{M}$ *such that any $\mathscr{C}^2$ solution u of* (X.2.4) *in* U *which satisfies* (X.3.12) *must be equal to v in* U′.

Notice that the initial datum u_0 is only assumed to be of class $\mathscr{C}^2$ (in V).

PROOF. By Proposition X.3.1, (X.3.11), and (X.3.14) we have

$$u = \Phi_0(x,t,G_0(Z^u),G(Z^u)),\ u_{x_i} = \Phi_i(x,t,G_0(Z^u),G(Z^u))\ (1 \leq i \leq m). \tag{X.3.15}$$

Once again we avail ourselves of (X.3.7): applying the holomorphic implicit function theorem to the system of equations (in which $Z = Z(x,t,\zeta_0,\zeta)$)

$$\zeta_0 = \Phi_0(x,t,G_0(Z),G(Z)),\ \zeta_i = \Phi_i(x,t,G_0(Z),G(Z))\ (1 \leq i \leq m), \tag{X.3.16}$$

yields, in a suitable neighborhood of 0 in U, a unique set of $\mathscr{C}^\infty$ solutions

$$\zeta_0 = v_0(x,t),\ \zeta_i = v_i(x,t)\ (1 \leq i \leq m)$$

such that $v_0(0,0) = u_0(0)$, $v(0,0) = u_x(0)$. By (X.3.15) $u = v_0$, $u_{x_i} = v_i$ for all i. ■

COROLLARY X.3.1. *If there is a $\mathscr{C}^2$ solution u of* (X.2.4) *in* U *that satisfies* (X.3.12) *and if* (X.3.13) *holds, then u is a $\mathscr{C}^\infty$ function in some open neighborhood of* 0 *in* U, *and as a consequence u_0 is a $\mathscr{C}^\infty$ function in some open neighborhood of* 0 *in* V.

COROLLARY X.3.2. *Suppose the system* (X.3.1) *is elliptic in the open neighborhood $\mathcal{O}$ of $\mathfrak{p}$. Then there is a $\mathscr{C}^\infty$ function v in an open neighborhood* $\mathrm{U}' \subset \mathrm{U}$ *of* 0 *in $\mathcal{M}$ such that any $\mathscr{C}^2$ solution u of* (X.2.4) *satisfying* (X.3.12) *must be equal to v in* U′.

PROOF. The functions $Z_1^u,\ldots,Z_m^u$ form a set of first integrals of the system of vector fields $\mathscr{L}_{F_j}^{0u}$. The latter system is elliptic, by our hypothesis: indeed, if the system (X.3.1) is elliptic at $\mathfrak{p}$, the same is true of its base projection, which is made up of the vector fields $\mathscr{L}_{F_j}^{0}$. It follows that any solution in a neighborhood of 0 in $\mathcal{M}$ is equal to a holomorphic function of Z^u (also in a neighborhood of 0 in $\mathbb{C}^m$). In particular (X.3.13) holds. ■

REMARK X.3.3. Property (X.3.13) is a consequence of the hypothesis, more general than ellipticity, that the involutive structure defined by $\mathscr{L}_{F_1}^{0u},\ldots,\mathscr{L}_{F_n}^{0u}$ is hypocomplex at 0 (Definition III.5.1). ■

THEOREM X.3.2. *Suppose the manifold $\mathcal{M}$, the involutive system of DE Σ, and the function u_0 are of class $\mathscr{C}^\omega$. Then every $\mathscr{C}^2$ solution u of* (X.2.4) *in* U *that satisfies* (X.3.12) *is an analytic function in some open neighborhood of the origin.*

PROOF. In the present situation we may select the first integrals to be analytic and, in fact, to be such that

$$Z_i|_{t=0} = x_i,\ \Xi_i|_{t=0} = \zeta_i\ (1 \leq i \leq m),\ \Xi_0|_{t=0} = \zeta_0. \tag{X.3.17}$$

It follows from this that $\Xi^u|_{t=0} = u_{0x}(x)$, $\Xi_0^u|_{t=0} = u_0(x)$. Recall that (X.3.8) holds. On the other hand, $Z_1^u,\ldots,Z_m^u$ are first integrals of the system of vector fields $\mathcal{L}_{F_1}^{0u},\ldots,\mathcal{L}_{F_n}^{0u}$. It follows that $u_{0x_i}(Z^u)$, $u_0(Z^u)$ are annihilated by the latter vector fields, and clearly when $t = 0$ they are equal to Ξ_i^u, Ξ^u respectively. By the $\mathcal{C}^1$ version of the approximation formula (section II.3) we must have

$$\Xi_i^u \equiv u_{0x_i}(Z^u),\ \Xi^u \equiv u_0(Z^u) \tag{X.3.18}$$

in some neighborhood of 0, which shows that Condition (X.3.13) is satisfied.

In the present situation the solutions Φ_i $(0 \leq i \leq m)$ of (X.3.10) are analytic with respect to (x,t) and so are the solutions v_i of (X.3.16), whence the result. ■

Remark X.3.4. Under the hypotheses of Theorem X.3.2 all $\mathcal{C}^2$ solutions u of (X.2.4) that satisfy the initial condition (X.3.12) must coincide in an open neighborhood of $\mathrm{V}\times\{0\}$ in U, by the uniqueness part of the Cauchy-Kovalevska theorem. ■

In the analytic category we can strengthen the conclusion of Theorem X.3.1 without assuming that the initial datum u_0 is analytic.

Theorem X.3.3. *Suppose that the manifold $\mathcal{M}$ and the involutive system of DE Σ are analytic, and that* (X.3.13) *holds. There is an analytic function v in an open neighborhood* $\mathrm{U}' \subset \mathrm{U}$ *of* 0 *in $\mathcal{M}$ such that any $\mathcal{C}^2$ solution u of* (X.2.4) *in* U *which satisfies* (X.3.12) *must be equal to v in* U$'$.

Proof. It is the same as that of Theorem X.3.1 with the added feature that the functions $Z_i(x,t,\zeta_0,\zeta)$, $\Xi_i(x,t,\zeta_0,\zeta)$ $(1 \leq i \leq m)$, and $\Xi_0(x,t,\zeta_0,\zeta)$ are taken to be analytic, and so are the functions Φ_j $(0 \leq j \leq m)$ (all these functions are holomorphic with respect to (ζ_0,ζ) in a neighborhood of (a,ω)). The solutions $v_k(x,t)$ $(0 \leq k \leq m)$ of Equations (X.3.16) will therefore be analytic in a neighborhood of 0 in $\mathcal{M}$. ■

Corollary X.3.3. *Suppose that the manifold $\mathcal{M}$ and the involutive system of DE Σ are analytic, and that the system of vector fields* (X.3.1) *is elliptic in the open neighborhood $\mathcal{O}$ of $\mathfrak{p}$. Then every $\mathcal{C}^2$ solution u of* (X.2.4) *in* U *must be analytic in an open neighborhood* $\mathrm{U}' \subset \mathrm{U}$ *of* 0 *in $\mathcal{M}$.*

X.4. Quasilinear Systems of Differential Equations with Vector-Valued Unknown

In the present section we enlarge the class of overdetermined systems under consideration, to include a special kind of quasilinear systems in which the unknown function is vector-valued. We reason in an open neighborhood U of a point 0 of $\mathcal{M}$; U is the domain of local coordinates x_i $(1 \le i \le m)$, t_j $(1 \le j \le n = N-m)$ that vanish at 0. We look at a system of differential equations:

$$\partial \mathbf{u}/\partial t_j - \sum_{k=1}^{m} a_{jk}(x,t,\mathbf{u})\partial \mathbf{u}/\partial x_k = \mathbf{g}_j(x,t,\mathbf{u}),\ j = 1,\ldots,n. \qquad \text{(X.4.1)}$$

where $\mathbf{u}$ is a $\mathcal{C}^1$ map $\mathrm{U} \to \mathbb{C}^M$ $(M \ge 1)$. The coefficients a_{jk} are complex-valued $\mathcal{C}^\infty$ functions in $\mathrm{U} \times \Omega$ (Ω is an open subset of $\mathbb{C}^M$) and the right-hand sides $\mathbf{g}_j$ are $\mathcal{C}^\infty$ maps $\mathrm{U} \times \Omega \to \mathbb{C}^M$; a_{jk} and $\mathbf{g}_j$ are holomorphic with respect to the variable in Ω $(1 \le j \le n,\ 1 \le k \le m)$.

Systems such as (X.4.1) show up in our analysis in association with an involutive system of first-order differential equations of rank n on the manifold $\mathcal{M}$ (Definition X.1.1). To a fully nonlinear system such as (X.2.4) one can associate m quasilinear systems of rank n, provided the solution u is of class $\mathcal{C}^2$. If $u \in \mathcal{C}^2(\mathrm{U})$ we are allowed to differentiate (X.2.4) with respect to x_i. Setting $v_i = u_{x_i}$ we get

$$\partial v_i/\partial t_j = \sum_{k=1}^{m} (\partial f_j/\partial \zeta_k)(x,t,u,v)\partial v_i/\partial x_k +$$

$$(\partial f_j/\partial x_i)(x,t,u,v) + (\partial f_j/\partial \zeta_0)(x,t,u,v)v_i,\ j = 1,\ldots,n. \qquad \text{(X.4.2)}$$

Note also that the system (X.2.4) itself can be rewritten as follows:

$$\partial u/\partial t_j = \sum_{k=1}^{m} (\partial f_j/\partial \zeta_k)(x,t,u,v)\partial u/\partial x_k -$$

$$\sum_{k=1}^{m} (\partial f_j/\partial \zeta_k)(x,t,u,v)v_k + f_j(x,t,u,v),\ j = 1,\ldots,n. \qquad \text{(X.4.3)}$$

Combining Equations (X.4.2) and (X.4.3) for all values of $i = 1,\ldots,m$, leads to a single system of n quasilinear differential equations of the kind (X.4.1) if we take

$$M = m+1,\ \mathbf{u} = (u,v_1,\ldots,v_m),\ a_{jk} = (\partial f_j/\partial \zeta_k)(x,t,u,v),$$

$$\mathbf{g}_{j0} = -\sum_{k=1}^{m} (\partial f_j/\partial \zeta_k)(x,t,u,v)v_k + f_j(x,t,u,v),$$

$$\mathbf{g}_{jk} = (\partial f_j/\partial x_k)(x,t,u,v) + (\partial f_j/\partial \zeta_0)(x,t,u,v)v_k\ (1 \le k \le m).$$

(We are writing $\mathbf{g}_j = (\mathbf{g}_{j0},\ldots,\mathbf{g}_{jm})$.)

When they originate from (X.2.4) Equations (X.4.1) satisfy Frobenius-like conditions. We shall require that they do so in the general case as well. The first thing to do is to identify the analogues of the vector fields $\mathrm{H}^0_{F_j}$ (see (X.2.6)) and to prove the analogue of Proposition X.3.1. Denote by θ the variable in $\mathbb{C}^M$; let $\mathcal{O}$ be an open neighborhood of a point $p = (0,\theta_0)$ in $\mathcal{M}\times\mathbb{C}^M$ in which the functions a_{jk} and $\mathbf{g}_j$ are defined. We write, for $j = 1,\dots,n$,

$$\mathcal{L}_j^0 = \partial/\partial t_j - \sum_{k=1}^{m} a_{jk}(x,t,\theta)\partial/\partial x_k. \tag{X.4.4}$$

Throughout the sequel $\mathbf{u}$: $\mathrm{U} \to \mathbb{C}^M$ will be a $\mathscr{C}^1$ map such that

$$(x,t) \in \mathrm{U} \Rightarrow (x,t,\mathbf{u}(x,t)) \in \mathcal{O};\ \mathbf{u}(0) = \theta_0. \tag{X.4.5}$$

Let $\Psi(x,t,\theta) \in \mathscr{C}^\infty(\mathcal{O})$ be holomorphic with respect to θ. We set

$$\Psi^{\mathbf{u}}(x,t) = \Psi(x,t,\mathbf{u}),\ \mathcal{L}_j^{0\mathbf{u}} = \partial/\partial t_j - \sum_{k=1}^{m} a_{jk}^{\mathbf{u}}\partial/\partial x_k.$$

In this notation Equations (X.4.1) read

$$\mathcal{L}_j^{0\mathbf{u}}u_i = g_{ji}^{\mathbf{u}}\ (1 \le i \le M,\ 1 \le j \le n). \tag{X.4.6}$$

We introduce the notation

$$\mathrm{H}_j^0 = \mathcal{L}_j^0 + \sum_{i=1}^{M} g_{ji}\partial/\partial\theta_i. \tag{X.4.7}$$

Suppose that the quasilinear system under study is the system (X.4.2)–(X.4.3) associated with the fully nonlinear system of equations (X.2.4). In the notation of section X.3, with $(\zeta_0,\zeta_1,\dots,\zeta_m)$ substituted for θ, we have

$$\sum_{i=1}^{M} g_{ji}\partial/\partial\theta_i = \sum_{i=1}^{m}(\partial f_j/\partial x_i)\partial/\partial\zeta_i + (\partial f_j/\partial\zeta_0)\sum_{i=1}^{m}\zeta_i\partial/\partial\zeta_i -$$

$$\sum_{k=1}^{m}\zeta_k(\partial f_j/\partial\zeta_k)\partial/\partial\zeta_0 + f_j\partial/\partial\zeta_0.$$

Comparing with (X.2.6) shows that $\mathrm{H}_j^0 = \mathrm{H}_{F_j}^0$.

We return to general systems (X.4.1). The remainder of the present section parallels, and to a large extent duplicates, the argument in section X.3.

LEMMA X.4.1. *Let* $\mathbf{u}$ *be a* $\mathscr{C}^1$ *map* $\mathrm{U} \to \mathbb{C}^M$, Ψ *a* $\mathscr{C}^\infty$ *function in* $\mathcal{O}$, *holomorphic with respect to* θ. *Then, in some open neighborhood* $\mathrm{U}' \subset \mathrm{U}$ *of the origin,*

$$\mathcal{L}_j^{0\mathbf{u}}\Psi^{\mathbf{u}} = (\mathrm{H}_j^0\Psi)^{\mathbf{u}} + \sum_{i=1}^{M}(\partial\Psi/\partial\theta_i)^{\mathbf{u}}(\mathcal{L}_j^{0\mathbf{u}}u_i - g_{ji}^{\mathbf{u}}). \tag{X.4.8}$$

If $\mathbf{u} \in \mathscr{C}^1(\mathrm{U})$ satisfies (X.4.5), then $\Psi^{\mathbf{u}}$, $g_{ji}^{\mathbf{u}}$, and the coefficients of $\mathcal{L}_j^{0\mathbf{u}}$ are of class $\mathscr{C}^1$.

PROOF. The identity (X.4.8) follows at once from

$$\mathscr{L}_j^{0\mathbf{u}}\Psi^{\mathbf{u}} = (\mathscr{L}_j^{0}\Psi)^{\mathbf{u}} + \sum_{i=1}^{M}(\partial\Psi/\partial\theta_i)^{\mathbf{u}}\mathscr{L}_j^{0\mathbf{u}}u_i =$$

$$\left[\mathscr{L}_j^{0}\Psi + \sum_{i=1}^{M} g_{ji}\partial\Psi/\partial\theta_i\right]^{\mathbf{u}} + \sum_{i=1}^{M}(\partial\Psi/\partial\theta_i)^{\mathbf{u}}(\mathscr{L}_j^{0\mathbf{u}}u_i - g_{ji}^{\mathbf{u}}). \blacksquare$$

COROLLARY X.4.1. *Suppose* (X.4.1) *holds. Then*

$$\mathscr{L}_j^{0\mathbf{u}}\Psi^{\mathbf{u}} = (\mathrm{H}_j^{0}\Psi)^{\mathbf{u}}. \tag{X.4.9}$$

We are going to require

$$[\mathrm{H}_j^0,\mathrm{H}_{j'}^0] = 0,\ j,\ j' = 1,\dots,n. \tag{X.4.10}$$

DEFINITION X.4.1. *We shall say that the system* (X.4.1) *is* microlocally integrable at $\mathfrak{p}$ *if there is an open neighborhood* $\mathbb{O}$ *of* $\mathfrak{p}$ *in* $\mathrm{U}\times\mathbb{C}^M$ *in which the system of vector fields*

$$\mathrm{H}_j^0\ (1 \le j \le n),\ \partial/\partial\bar{\theta}_h\ (1 \le h \le M) \tag{X.4.11}$$

induces a locally integrable structure.

By reasoning as in section X.3 (and possibly after shrinking $\mathbb{O}$ about $\mathfrak{p}$), we can select smooth functions in $\mathbb{O}$, Z_k, Ξ_i holomorphic with respect to θ such that, for all $i = 1,\dots,M$, $k = 1,\dots,m$,

$$\mathrm{H}_j^0 Z_k = \mathrm{H}_j^0\Xi_i = 0,\ j = 1,\dots,n, \tag{X.4.12}$$

$Z_k - x_k$, $\Xi_i - \theta_i$ *and their differentials with respect to* (x,θ) *vanish at* $\mathfrak{p}$. (X.4.13)

PROPOSITION X.4.1. *Let* Z_k, Ξ_i $(1 \le i \le M,\ 1 \le k \le m)$ *be first integrals of the system of vector fields* (X.4.11) *satisfying* (X.4.13).

In order that a $\mathscr{C}^1$ *map* $\mathbf{u}\colon \mathrm{U} \to \mathbb{C}^M$ *such that* (X.4.5) *holds satisfy* (X.4.1) *in an open neighborhood* $\mathrm{U}' \subset \mathrm{U}$ *it is necessary and sufficient that the following equations hold in such a neighborhood*:

$$\mathscr{L}_j^{0\mathbf{u}}\Xi_i^{\mathbf{u}} = 0\ (1 \le i \le M,\ 1 \le j \le n). \tag{X.4.14}$$

PROOF. The condition is necessary by Corollary X.4.1. Suppose it is satisfied. Then, by (X.4.8), if Ψ is any of the functions Ξ_i $(1 \le i \le M)$,

$$\sum_{i=1}^{M}(\partial\Psi/\partial\theta_i)^{\mathbf{u}}(\mathscr{L}_j^{0\mathbf{u}}u_i - g_{ji}^{\mathbf{u}}) = 0,$$

for any $j = 1,\ldots,n$. According to (X.4.13), at the origin the matrix

$$(\partial \Xi_{i'}/\partial \theta_i)^{\mathbf{u}}_{1 \le i,i' \le M}$$

is equal to the $M \times M$ identity matrix. The claim follows at once from this fact. ■

Applying the implicit function theorem in the holomorphic category, with respect to θ (and with parameters x_k, t_j), to the equations

$$\Xi_i(x,t,\theta) = w_i,\ i = 1,\ldots,M \tag{X.4.15}$$

yields a unique solution $\theta = \Phi(x,t,w)$ such that $\theta = \theta_0$ when $x = 0$, $t = 0$, $w = \Xi(0,0,\theta_0)$; it is holomorphic with respect to w. Thus, if $\mathbf{u}$ satisfies (X.4.5), we shall have

$$\mathbf{u} = \Phi(x,t,\Xi^{\mathbf{u}}). \tag{X.4.16}$$

Let us now take $\mathrm{U} = \mathrm{V} \times \mathrm{W}$ as in section X.3. We deal with $\mathscr{C}^1$ solutions $\mathbf{u}$ of (X.4.1) submitted to the initial condition

$$\mathbf{u}|_{t=0} = \mathbf{u}_0(x) \text{ in } \mathrm{V}. \tag{X.4.17}$$

In view of (X.4.5) we must require $\mathbf{u}_0(0) = \theta_0$ $(1 \le i \le M)$. We introduce the analogue of Condition (X.3.13):

There exist holomorphic functions G_i in an open neighborhood of the origin in $\mathbb{C}^m$ such that (X.4.18)

$$\Xi_i^{\mathbf{u}} = G_i(Z^{\mathbf{u}})\ (1 \le i \le M,\ Z = (Z_1,\ldots,Z_m)).$$

Although a priori the functions G_h appear to depend on $\mathbf{u}$, actually they only depend on the initial value $\mathbf{u}_0$ (cf. remark following (X.3.14)).

THEOREM X.4.1. *Suppose* (X.4.18) *holds. There exist an open neighborhood* $\mathrm{U}' \subset \mathrm{U}$ *of* 0 *and a* $\mathscr{C}^\infty$ *map* $\mathbf{v}$: $\mathrm{U}' \to \mathbb{C}^M$ *such that any* $\mathscr{C}^1$ *solution* $\mathbf{u}$ *of* (X.4.1) *in* U *which satisfies* (X.4.17) *must be equal to* $\mathbf{v}$ *in* U'.

PROOF. By Proposition X.4.1, (X.4.16) and (X.4.18) we have $\mathbf{u} = \Phi(x,t,G(Z^{\mathbf{u}}))$. We make use of the fact that $Z - x$ and its differential with respect to (x,θ) vanish at $\mathfrak{p}$. The implicit function theorem applied to the equation $\theta = \Phi(x,t,G[Z(x,t,\theta)])$ yields, in a suitable neighborhood of 0, a solution $\theta = \mathbf{v}(x,t)$ such that $\mathbf{v}(0,0) = \theta_0$. Uniqueness then demands that $\mathbf{u} \equiv \mathbf{v}$. ■

COROLLARY X.4.2. *Under the hypotheses of Theorem* X.4.1 *the initial datum* $\mathbf{u}_0$ *must be a* $\mathscr{C}^\infty$ *function in a neighborhood of* 0 *in* V.

COROLLARY X.4.3. *Suppose the system of vector fields*

$$\partial/\partial t_j - \sum_{k=1}^{m} a_{jk}(0,0,\theta_0)\partial/\partial x_k \ (1 \le j \le n) \tag{X.4.19}$$

is elliptic. Then there are an open neighborhood $\mathrm{U}' \subset \mathrm{U}$ *of* 0 *and a* $\mathscr{C}^\infty$ *map* $\mathbf{v}: \mathrm{U}' \to \mathbb{C}^M$ *such that any* $\mathscr{C}^1$ *solution* $\mathbf{u}$ *of* (X.4.1) *that satisfies* (X.4.17) *must be equal to* $\mathbf{v}$.

PROOF. In view of (X.4.17), to say that the system (X.4.19) is elliptic is equivalent to saying that the system (X.4.11) is elliptic at $\mathfrak{p}$. Let Z_k, Ξ_i be first integrals of (X.4.11) in $\mathcal{O}$ satisfying (X.4.13). It follows at once from Corollary X.4.1 that the functions $\Xi_i^{\mathbf{u}}$ are annihilated by the vector fields $\mathscr{L}_j^{0\mathbf{u}}$ and therefore must be holomorphic functions of $Z^{\mathbf{u}}$. ∎

When the "coefficients" a_{jk} and $\mathbf{g}_j$ are of class $\mathscr{C}^\omega$ the analogues of Theorems X.3.2, X.3.3, and of Corollary X.3.3 are valid (the proofs are practically identical to those in section X.3):

THEOREM X.4.2. *Suppose* a_{jk} *and* g_j $(1 \le k \le m, 1 \le j \le n)$ *are real-analytic. If* $\mathbf{u}_0$ *is real analytic, then every* $\mathscr{C}^1$ *solution of* (X.4.1) *in* U *that satisfies* (X.4.17) *is an analytic function in some open neighborhood of* 0 *in* U.

In the next statements the initial datum $\mathbf{u}_0$ is not assumed to be real-analytic.

THEOREM X.4.3. *Suppose* a_{jk} *and* g_j $(1 \le k \le m, 1 \le j \le n)$ *are real-analytic and that* (X.4.18) *holds. There is an open neighborhood* $\mathrm{U}' \subset \mathrm{U}$ *of* 0 *and a* $\mathscr{C}^\omega$ *map* $\mathbf{v}: \mathrm{U}' \to \mathbb{C}^M$ *such that any* $\mathscr{C}^1$ *solution* $\mathbf{u}$ *of* (X.4.1) *in* U *that satisfies* (X.4.17) *must be equal to* $\mathbf{v}$ *in* U'.

COROLLARY X.4.4. *Suppose* a_{jk} *and* g_j $(1 \le k \le m, 1 \le j \le n)$ *are real-analytic and that the system of vector fields* (X.4.19) *is elliptic. Then every* $\mathscr{C}^1$ *solution* $\mathbf{u}$ *of* (X.4.1) *must be analytic in an open neighborhood of* 0 *in* U.

X.5. The Approximation Formula

We consider the system of quasilinear DE (X.4.1). Throughout the remainder of this last chapter we reason under the hypothesis of microlocal integrability (Definition X.4.1). We select $m+M$ $\mathscr{C}^\infty$ functions in the open neighborhood $\mathcal{O}$ of $\mathfrak{p} = (0,\theta_0)$, Z_k, Ξ_i $(1 \le k \le m, 1 \le i \le M)$, holomorphic with respect to θ, verifying (X.4.12) and (X.4.13). We deal with a solution $\mathbf{u} \in \mathscr{C}^1(\mathrm{U})$ of (X.4.1) that satisfies (X.4.5) and (X.4.17). We assume $\mathrm{U} = \mathrm{V} \times \mathrm{W}$.

Let $g \in \mathscr{C}_c^1(\mathrm{U})$. We define, for $z \in \mathbb{C}^m$, $t \in \mathrm{W}$,

$$\mathscr{E}_\nu^{\mathbf{u}} g(z,t) = (\nu/\pi)^{\frac{1}{2}m} \int_{\mathrm{V}\times\{t\}} e^{-\nu\langle z - Z^{\mathbf{u}}\rangle^2} g \, dZ^{\mathbf{u}}. \tag{X.5.1}$$

We have used the notation $\langle\zeta\rangle^2 = \sum_k \zeta_k^2$, $dZ^{\mathbf{u}} = dZ_1^{\mathbf{u}}\wedge\cdots\wedge dZ_m^{\mathbf{u}}$; let us also write

$$\Delta^{\mathbf{u}} = \det \partial Z^{\mathbf{u}}/\partial x.$$

The meaning of (X.5.1) is that

$$\mathscr{E}_\nu^{\mathbf{u}} g(z,t) = (\nu/\pi)^{\frac{1}{2}m}\int_V e^{-\nu\langle z - Z^{\mathbf{u}}(y,t)\rangle^2} g(y,t)\Delta^{\mathbf{u}}(y,t)dy. \tag{X.5.2}$$

LEMMA X.5.1. *There is an open neighborhood* $U' = V'\times W' \subset U$ *of the origin and a constant* $C_0 > 0$ *such that for all* $g \in \mathscr{C}^1(U)$ *with* supp $g \subset V'\times W$, *all* $\nu = 1,2,\ldots$ *and all* $(x,t)\in U'$,

$$|\mathscr{E}_\nu^{\mathbf{u}} g(Z^{\mathbf{u}}(x,t),t) - g(x,t)| \leq C_0\nu^{-\frac{1}{2}}. \tag{X.5.3}$$

PROOF. Thanks to (X.4.13) and to the fact that $\mathbf{u}$ is of class $\mathscr{C}^1$, we have

$$\mathscr{R}e\langle Z^{\mathbf{u}}(x,t) - Z^{\mathbf{u}}(y,t)\rangle^2 \geq c_0|x-y|^2 \ (c_0 > 0).$$

Defining $Z^{\mathbf{u}}(y,t) = y$ away from V we may write:

$$g(x,t) - \mathscr{E}_\nu^{\mathbf{u}} g(Z^{\mathbf{u}}(x,t),t) = (\nu/\pi)^{\frac{1}{2}m}\int_{\mathbb{R}^m\setminus V} e^{-\nu\langle Z^{\mathbf{u}}(x,t) - Z^{\mathbf{u}}(y,t)\rangle^2} g(y,t)\Delta^{\mathbf{u}}(y,t)dy +$$

$$(\nu/\pi)^{\frac{1}{2}m}\int_{\mathbb{R}^m} e^{-\nu\langle Z^{\mathbf{u}}(x,t) - Z^{\mathbf{u}}(y,t)\rangle^2}[g(x,t) - g(y,t)]\Delta^{\mathbf{u}}(y,t)dy$$

If $|x|$ is sufficiently small the absolute value of the first integral in the right-hand side is $\leq$ *const.*$e^{-c\nu}$. The change of variable $y \to x + y/\sqrt{\nu}$ and the hypothesis that g is of class $\mathscr{C}^1$ imply that the absolute value of the last integral is $\leq$ *const.*$\nu^{-\frac{1}{2}}$. ■

Let us now introduce the vector fields

$$M_h^{\mathbf{u}} = \sum_{k=1}^m \mu_{hk}^{\mathbf{u}}(x,t)\partial/\partial x_k,\ h = 1,\ldots,m,$$

where $(\mu_{hk}^{\mathbf{u}})_{1\leq h,k\leq m}$ is the inverse of the Jacobian matrix $\partial Z^{\mathbf{u}}/\partial x$. Integration by parts implies (cf. Lemma II.2.2):

$$(\partial/\partial z_h)\mathscr{E}_\nu^{\mathbf{u}} g(z,t) = \mathscr{E}_\nu^{\mathbf{u}} M_h^{\mathbf{u}} g(z,t),\ h = 1,\ldots,m, \tag{X.5.4}$$

$$(\partial/\partial t_j)\mathscr{E}_\nu^{\mathbf{u}} g(z,t) = \mathscr{E}_\nu^{\mathbf{u}} \mathscr{L}_j^{0\mathbf{u}} g(z,t),\ j = 1,\ldots,n. \tag{X.5.5}$$

LEMMA X.5.2. *If the neighborhoods* V′ *and* W′ *are sufficiently small, to every constant* $C_1 > 0$, *there is a constant* $C_2 > 0$ *such that for all* $g \in \mathscr{C}^1(\mathscr{C}\ell U)$ *with* supp $g \subset V'\times W$, *all* $\nu = 1,2,\ldots,$

$$|((\partial/\partial z_h)\mathscr{E}_\nu^{\mathbf{u}} g)(Z(x,t,\theta),t)| \leq C_2 \sup_U |M_h^{\mathbf{u}} g|,\ h = 1,\ldots,m, \tag{X.5.6}$$

in the region

$$(x,t) \in \mathrm{U}', \ \theta \in \mathbb{C}^M; \ |\theta - \mathbf{u}(x,t)| \leq C_1\nu^{-\frac{1}{2}}. \tag{X.5.7}$$

PROOF. We apply (X.5.4):

$$(\pi/\nu)^{\frac{1}{2}m}((\partial/\partial z_h)\mathscr{E}_\nu^{\mathbf{u}}g)(Z(x,t,\theta),t) =$$

$$\int_{\mathrm{V}} \mathrm{e}^{-\nu(Z^{\mathbf{u}}(x,t)-Z^{\mathbf{u}}(y,t)+[Z(x,t,\theta)-Z^{\mathbf{u}}(x,t)])^2}(\mathrm{M}_h^{\mathbf{u}}g)(y,t)\Delta^{\mathbf{u}}(y,t)\mathrm{d}y.$$

By the hypothesis (X.5.7) we have, for a suitable $C > 0$,

$$|Z(x,t,\theta) - Z^{\mathbf{u}}(x,t)| \leq CC_1\nu^{-\frac{1}{2}}.$$

Once again we effect the change of variables $y \to x + y/\sqrt{\nu}$. The assertion follows at once. ■

We now consider a *continuous* function ψ in the closure of U, $\mathscr{C}\ell\,\mathrm{U}$, which satisfies in U, in the distribution sense,

$$\mathscr{L}_j^{0\mathbf{u}}\psi = 0, \ j = 1,\ldots,n. \tag{X.5.8}$$

We select once for all a function $\chi \in \mathscr{C}_c^\infty(\mathrm{V}')$, $\chi(y) = 1$ if $|y| < \delta$ $(\delta > 0)$, and define

$$\mathscr{E}_\nu^{\mathbf{u}}(\chi\psi)(z,t) = (\nu/\pi)^{\frac{1}{2}m}\int_{\mathrm{V}} \mathrm{e}^{-\nu(z-Z^{\mathbf{u}}(y,t))^2}\chi(y)\psi(y,t)\Delta^{\mathbf{u}}(y,t)\mathrm{d}y. \tag{X.5.9}$$

Henceforth we assume that the neighborhood W′ is an open ball in $\mathbb{R}^n$ centered at the origin, and we denote by $\ell(t)$ the straight-line segment joining 0 to t. Observe that (X.5.5) makes sense when $g(x,t) = \chi(x)\psi(x,t)$. Indeed, the coefficients of $\mathscr{L}_j^{0\mathbf{u}}$ are of class $\mathscr{C}^1$ and therefore they act, as multipliers, on the partial derivatives of a continuous function. Actually, the Leibniz formula applies:

$$\mathscr{E}_\nu^{\mathbf{u}}(\chi\psi)(z,t) - \mathscr{E}_\nu^{\mathbf{u}}(\chi\psi)(z,0) = \int_{\ell(t)} \mathrm{d}_t\mathscr{E}_\nu^{\mathbf{u}}(\chi\psi)(z,t) =$$

$$(\nu/\pi)^{\frac{1}{2}m}\int_{\ell(t)}\int_{\mathrm{V}} \mathrm{e}^{-\nu(z-Z^{\mathbf{u}}(y,t))^2}[\mathscr{L}^{0\mathbf{u}}\chi(y)]\ \psi(y,t)\ \Delta^{\mathbf{u}}(y,t)\mathrm{d}y,$$

where

$$\mathscr{L}^{0\mathbf{u}}\chi(y) = \sum_{j=1}^n \mathscr{L}_j^{0\mathbf{u}}\chi(y)\ \mathrm{d}t_j$$

vanishes identically if $|y| \leq \delta$. Provided the neighborhood U′ is small enough we shall have for all $(y,t) \in (\operatorname{supp}\chi) \times \mathrm{W}'$,

$$|Z^{\mathbf{u}}(y,t) - y| \leq \delta/4,$$

hence $|\mathcal{R}e Z^{\mathbf{u}}(y,t)| \geq 3\delta/4$, $|\mathcal{I}m Z^{\mathbf{u}}(y,t)| \leq \delta/4$ if $|y| > \delta$. We may state:

LEMMA X.5.3. *Provided* V′ *and* W′ *are small enough, there are constants* C, $c_0 > 0$, *such that, for all functions* $\psi \in \mathscr{C}^0(\mathscr{C}\ell\, \mathrm{U})$ *satisfying* (X.5.8) *and all* $\nu = 1,2,\ldots,$

$$|\mathscr{E}_\nu^{\mathbf{u}}(\chi\psi)(z,t) - \mathscr{E}_\nu^{\mathbf{u}}(\chi\psi)(z,0)| \leq Ce^{-c_0\nu} \operatorname*{Max}_{\mathrm{U}} |\psi| \tag{X.5.10}$$

for all $z \in \mathbb{C}^m$, $|z| < \frac{1}{4}\delta$, *and all* $t \in \mathrm{W}'$.

We need a combination of Lemmas X.5.2 and X.5.3:

LEMMA X.5.4. *If the neighborhoods* V′ *and* W′ *are sufficiently small, there is an open neighborhood of the origin,* U″ ⊂ U′ (⊂ U), *such that the following is true: to every constant* $C_1 > 0$ *there is* $C_2 > 0$ *such that for all* $\psi \in \mathscr{C}^1(\mathscr{C}\ell\, \mathrm{U})$ *satisfying* (X.5.8) *and all* $\nu = 1,2,\ldots,$

$$|[(\partial/\partial z_h)\mathscr{E}_\nu^{\mathbf{u}}(\chi\psi)](Z(x,t,\theta),0)| \leq C_2 \sup_{\mathrm{U}}(|\psi| + |\mathrm{M}_h^{\mathbf{u}}\psi|),\ h = 1,\ldots,m, \tag{X.5.11}$$

in the region

$$(x,t) \in \mathrm{U}'',\ \theta \in \mathbb{C}^M;\ |\theta - \mathbf{u}(x,t)| \leq C_1\nu^{-\frac{1}{2}}. \tag{X.5.12}$$

PROOF. By (X.5.10) we have

$$|\mathscr{E}_\nu^{\mathbf{u}}(\chi \mathrm{M}_h^{\mathbf{u}}\psi)(z,t) - \mathscr{E}_\nu^{\mathbf{u}}(\chi \mathrm{M}_h^{\mathbf{u}}\psi)(z,0)| \leq Ce^{-c_0\nu} \operatorname*{Max}_{\mathrm{U}} |\mathrm{M}_h^{\mathbf{u}}\psi|$$

if $z \in \mathbb{C}^m$, $|z| < \frac{1}{4}\delta$. The argument in the proof of Lemma X.5.3 shows at once that

$$|\mathscr{E}_\nu^{\mathbf{u}}(\psi \mathrm{M}_h^{\mathbf{u}}\chi)(z,t) - \mathscr{E}_\nu^{\mathbf{u}}(\psi \mathrm{M}_h^{\mathbf{u}}\chi)(z,0)| \leq C'e^{-c_0\nu} \operatorname*{Max}_{\mathrm{U}} |\psi|,$$

whence, thanks to (X.5.4),

$$|(\partial/\partial z_h)\mathscr{E}_\nu^{\mathbf{u}}(\chi\psi)(z,t) - (\partial/\partial z_h)\mathscr{E}_\nu^{\mathbf{u}}(\chi\psi)(z,0)| \leq$$
$$Ce^{-c_0\nu} \operatorname*{Max}_{\mathrm{U}}(|\psi| + |\mathrm{M}_h^{\mathbf{u}}\psi|).$$

We select U″ small enough that if (X.5.12) is satisfied then $|Z(x,t,\theta)| < \frac{1}{4}\delta$. We may then combine the preceding estimate with (X.5.6) where we put $g = \chi\psi$; this yields (X.5.11). ■

We are going to apply what precedes with $\psi = \Xi_i^{\mathbf{u}}$ ($i = 1,\ldots,M$). Then consider the system of equations

$$\Xi_i(x,t,\theta) = \mathscr{E}_\nu^{\mathbf{u}}(\chi\Xi_i^{\mathbf{u}})(Z(x,t,\theta),0),\ i = 1,\ldots,M. \tag{X.5.13}$$

We shall make use of the unique $\mathscr{C}^\infty$ solution $\theta = \Phi(x,t,w)$ of the system of equations (X.4.15) such that $\theta = \theta_0$ when $x = 0$, $t = 0$, $w = \Xi(0,0,\theta_0)$; Φ is holomorphic with respect to w. We write $\Phi = (\Phi_1,\dots,\Phi_M)$, $\mathscr{E}_\nu^{\mathbf{u}}(\chi\Xi^{\mathbf{u}}) = (\mathscr{E}_\nu^{\mathbf{u}}(\chi\Xi_1^{\mathbf{u}}),\dots,\mathscr{E}_\nu^{\mathbf{u}}(\chi\Xi_M^{\mathbf{u}}))$. Thus Equations (X.5.13) are equivalent to

$$\theta = \Phi(x,t,\mathscr{E}_\nu^{\mathbf{u}}(\chi\Xi^{\mathbf{u}})(Z(x,t,\theta),0)). \tag{X.5.14}$$

THEOREM X.5.1. *There is an open neighborhood* U″ ⊂ U *of the origin in which the equations* (X.5.13) *admit unique* $\mathscr{C}^1$ *solutions* $\theta_i = v_{\nu,i}(x,t)$ *such that*

$$v_{\nu,i}(0,0) = \Phi_i(0,0,\mathscr{E}_\nu^{\mathbf{u}}(\chi\Xi^{\mathbf{u}})(Z(0,0,\theta_0),0)),\ i = 1,\dots,M. \tag{X.5.15}$$

Furthermore, for some constant $C_3 > 0$ *and all* $\nu = 1,2,\dots$, *we have, in* U″,

$$\sup_{1\le i\le M} |v_{\nu,i} - u_i| \le C_3\nu^{-\frac{1}{2}},\quad \sup_{1\le i\le M} |dv_{\nu,i}| \le C_3. \tag{X.5.16}$$

PROOF. We define a sequence of $\mathscr{C}^\infty$ functions (valued in $\mathbb{C}^M$) $\mathbf{v}_\nu^{(k)}(x,t)$ $(k \in \mathbb{Z}_+)$ by the induction rule

$$\mathbf{v}_\nu^{(0)} = \mathbf{u},$$

$$\mathbf{v}_\nu^{(k)}(x,t) = \Phi(x,t,\mathscr{E}_\nu^{\mathbf{u}}(\chi\Xi^{\mathbf{u}})(Z(x,t,\mathbf{v}_\nu^{(k-1)}(x,t)),0))\ \textit{if}\ k \ge 1.$$

We take $k \ge 1$ and hypothesize that there is a sufficiently small open neighborhood of the origin, U′ ⊂ U, and a constant $C_3 > 0$ such that

$$|\mathbf{v}_\nu^{(\ell)}(x,t) - \mathbf{u}(x,t)| \le C_3\nu^{-\frac{1}{2}},\ \forall\ (x,t) \in \mathrm{U}', \tag{X.5.17}$$

for all ℓ, $0 \le \ell < k$. It is obviously true for $k = 1$.

We have

$$\begin{aligned}
&|\mathbf{v}_\nu^{(k)}(x,t) - \mathbf{u}(x,t)| = \\
&|\Phi(x,t,\mathscr{E}_\nu^{\mathbf{u}}(\chi\Xi^{\mathbf{u}})(Z(x,t,\mathbf{v}_\nu^{(k-1)}(x,t)),0)) - \Phi(x,t,\Xi^{\mathbf{u}}(x,t))| \le \\
&C|\mathscr{E}_\nu^{\mathbf{u}}(\chi\Xi^{\mathbf{u}})(Z(x,t,\mathbf{v}_\nu^{(k-1)}(x,t)),0)) - \Xi^{\mathbf{u}}(x,t))| \le \\
&C|\mathscr{E}_\nu^{\mathbf{u}}(\chi\Xi^{\mathbf{u}})(Z(x,t,\mathbf{v}_\nu^{(k-1)}(x,t)),t) - \mathscr{E}_\nu^{\mathbf{u}}(\chi\Xi^{\mathbf{u}})(Z(x,t,\mathbf{v}_\nu^{(k-1)}(x,t)),0)| + \\
&C|\mathscr{E}_\nu^{\mathbf{u}}(\chi\Xi^{\mathbf{u}})(Z(x,t,\mathbf{v}_\nu^{(k-1)}(x,t)),t) - \mathscr{E}_\nu^{\mathbf{u}}(\chi\Xi^{\mathbf{u}})(Z^{\mathbf{u}}(x,t),t)| + \\
&C|\mathscr{E}_\nu^{\mathbf{u}}(\chi\Xi^{\mathbf{u}})(Z^{\mathbf{u}}(x,t),t) - \Xi^{\mathbf{u}}(x,t)| \le C'\mathrm{e}^{-c_0\nu} + \\
&C'|Z(x,t,\mathbf{v}_\nu^{(k-1)}(x,t)) - Z(x,t,\mathbf{u}(x,t))| + C|\mathscr{E}_\nu^{\mathbf{u}}(\chi\Xi^{\mathbf{u}})(Z^{\mathbf{u}}(x,t),t) - \Xi^{\mathbf{u}}(x,t)|
\end{aligned}$$

by (X.5.6) and (X.5.10). We avail ourselves once more of (X.4.13): to every $\varepsilon > 0$ there is $\delta > 0$ such that, if diam U′ $\le \delta$ and $|\theta - \theta_0| + |\theta' - \theta_0| \le \delta$, then

$$|Z(x,t,\theta) - Z(x,t,\theta')| \le \varepsilon|\theta - \theta'|.$$

We apply this with $\theta = \mathbf{v}_\nu^{(k-1)}(x,t)$, $\theta' = \mathbf{u}(x,t)$. We assume that U′ is small enough and ν large enough that (X.5.17) (for $\ell = k-1$) entail

$$|\mathbf{v}_\nu^{(k-1)}(x,t) - \theta_0| + |\mathbf{u}(x,t) - \theta_0| \leq \delta.$$

We obtain, for any $k \geq 2$,

$$|\mathbf{v}_\nu^{(k)}(x,t) - \mathbf{u}(x,t)| \leq C'\mathrm{e}^{-c_0\nu} + C'\varepsilon|\mathbf{v}_\nu^{(k-1)}(x,t) - \mathbf{u}(x,t)| + C|\mathscr{E}_\nu^{\mathbf{u}}(\chi\Xi^{\mathbf{u}})(Z^{\mathbf{u}}(x,t),t) - \Xi^{\mathbf{u}}(x,t)|.$$

According to Lemma X.5.1, we have, if $\chi \equiv 1$ in U′,

$$|\mathscr{E}_\nu^{\mathbf{u}}(\chi\Xi^{\mathbf{u}})(Z^{\mathbf{u}}(x,t),t) - \Xi^{\mathbf{u}}(x,t)| \leq C_0\nu^{-\frac{1}{2}},\ \forall\ (x,t) \in \mathrm{U}',$$

whence

$$|\mathbf{v}_\nu^{(k)} - \mathbf{u}| \leq CC_0\nu^{-\frac{1}{2}} + C'\mathrm{e}^{-c_0\nu} + C'\varepsilon|\mathbf{v}_\nu^{(k-1)} - \mathbf{u}| \leq C_3(CC_0/C_3 + C'\varepsilon)\nu^{-\frac{1}{2}} + C'\mathrm{e}^{-c_0\nu}.$$

To obtain (X.5.17) it suffices to select C_3 and $1/\varepsilon$ large enough.

Property (X.5.17), now valid for all $\ell \in \mathbb{Z}_+$, enables us to take advantage of Lemma X.5.4 and to apply the implicit function theorem. One concludes easily that the functions $\mathbf{v}_\nu^{(k)}$ converge to a function $\mathbf{v}_\nu = (v_{\nu,1},\ldots,v_{\nu,M}) \in \mathscr{C}^1(\mathrm{U}'';\mathbb{C}^M)$ whose components $v_{\nu,i}$ are the sought solutions of (X.5.13). The first estimate (X.5.16) is a direct consequence of (X.5.17); the second estimate (X.5.16) follows from (X.5.11) and from Equation (X.5.13). ∎

The definition of $\mathscr{E}_\nu^{\mathbf{u}}$, (X.5.2), shows that $\mathscr{E}_\nu^{\mathbf{u}}g(z,0)$ depends only on the "initial value" $\mathbf{u}_0(x)$. As a consequence the same is true of the solution $\mathbf{v}_\nu$ of (X.5.13). The definition of $\mathbf{v}_\nu$ can be reformulated as follows:

$$\mathbf{v}_\nu = \Phi(x,t,\Xi^{\mathbf{v}_\nu}),\ \Xi^{\mathbf{v}_\nu} = G_\nu^{\mathbf{u}}(Z^{\mathbf{v}_\nu}), \tag{X.5.18}$$

where $G_\nu^{\mathbf{u}}(z) = \mathscr{E}_\nu^{\mathbf{u}}(\chi\Xi^{\mathbf{u}})(z,0)$ (cf. (X.4.18)). From this and from Proposition X.4.1 applied with $\Xi - G_\nu^{\mathbf{u}}(Z)$ in the place of Ξ we derive:

PROPOSITION X.5.1. *Whatever* $\nu = 1,2,\ldots,$ *the function* $\mathbf{v}_\nu \in \mathscr{C}^1(\mathrm{U}'';\mathbb{C}^M)$ *is a solution of the system of quasilinear differential equations* (X.4.1) *in* U″.

X.6. Uniqueness in the Cauchy Problem

We pursue further the argument started in section X.5. Since, by the first inequality (X.5.16), $\mathbf{v}_\nu \to \mathbf{u}$ uniformly in U″, and since, as was pointed out at the end of section X.5, the solutions $\mathbf{v}_\nu$ depend only on the initial value $\mathbf{u}_0$ (such that $\mathbf{u}_0(0) = \theta_0$), we reach the following conclusion about uniqueness in the Cauchy problem for the equations (X.4.1):

THEOREM X.6.1. *Suppose the system of quasilinear differential equations* (X.4.1) *is microlocally integrable at* $\mathfrak{p} = (0,\theta_0)$ (Definition X.4.1). *If two* $\mathscr{C}^1$ *solutions* $\mathbf{u}_1$ *and* $\mathbf{u}_2$ *of* (X.4.1) *satisfy the same initial condition* (X.4.17), *then* $\mathbf{u}_1 = \mathbf{u}_2$ *in some open neighborhood of the origin.*

An immediate consequence of Theorem X.6.1 concerns the fully nonlinear system (X.2.4) via the associated quasilinear system (X.4.2)–(X.4.3). Here we must assume that $\mathfrak{p} = (0,a,\omega)$ and that the initial datum u_0 satisfies $u_0(0) = a$, $\mathrm{d}_x u_0(0) = \xi$ ($\omega = \xi\cdot\mathrm{d}x + \tau\cdot\mathrm{d}t$).

COROLLARY X.6.1. *Suppose the involutive system of differential equations* (X.2.4) *is microlocally integrable at* $\mathfrak{p}$ (Definition X.3.1). *If two* $\mathscr{C}^2$ *solutions* u_1 *and* u_2 *of* (X.2.4) *in* U *are both equal, in* V, *when* $t = 0$, *to* u_0, *then* $u_1 = u_2$ *in some open neighborhood of the origin.*

Theorem X.6.1 enables us to generalize Theorem II.3.3 to quasilinear systems. Let Ω be an open subset of $\mathcal{M}$ and $\mathbf{u} \in \mathscr{C}^1(\Omega)$ a solution of the system of equations (X.4.1) in Ω. In the present context the concept of orbit is relative to the vector subbundle $\mathcal{V}^{\mathbf{u}}$ of $\mathbb{C}T\mathcal{M}|_\Omega$ spanned by the vector fields $\mathcal{L}_j^{\mathbf{u}}$ (see section X.4). Although the subbundle $\mathcal{V}^{\mathbf{u}}$ is merely of class $\mathscr{C}^1$, Definition I.11.1 stands as it is: two points in Ω belong to the same orbit of $\mathcal{V}^{\mathbf{u}}$ if they can be joined by a curve γ that is the union of finitely many integral curves of some real, nonvanishing, vector field $\mathcal{R}e\mathrm{L}$, with L a $\mathscr{C}^1$ section of $\mathcal{V}^{\mathbf{u}}$.

THEOREM X.6.2. *Let* Ω *be an open subset of* $\mathcal{M}$ *and* $\mathbf{u} \in \mathscr{C}^1(\Omega)$ *a solution of the system of equations* (X.4.1) *in* Ω. *Suppose the system* (X.4.1) *is microlocally integrable at every point* $(y,\mathbf{u}(y))$, $y \in \Omega$. *Let* $\mathbf{v} \in \mathscr{C}^1(\Omega)$ *be another solution of* (X.4.1) *in* Ω *and let* Ω' *denote the interior of the subset of* Ω *in which* $\mathbf{u} = \mathbf{v}$. *Any orbit of* $\mathcal{V}^{\mathbf{u}}$ *that intersects* Ω' *is entirely contained in* Ω'.

PROOF. It parallels the proof of Theorem II.3.3: it exploits Theorem X.6.1 in the same manner as the proof of Theorem X.3.3 exploited Corollary II.3.6. It uses the fact that if L is a $\mathscr{C}^1$ section of $\mathcal{V}^{\mathbf{u}}$ over an open subset U of Ω and if $\mathcal{R}e\mathrm{L}$ does not vanish at any point of U then, possibly after contracting U, we may choose the coordinates x_i, t_j so that $\mathcal{R}e\mathrm{L} = \partial/\partial t$ (the changes of variables are of class $\mathscr{C}^1$). We leave the details to the reader. ■

Theorem X.6.2 entails a similar result about fully nonlinear equations provided we limit ourselves to a $\mathscr{C}^2$ solution u. The involutive structure that is relevant here is the one defined by the vector fields $\mathcal{L}_{F_j}^{\Omega u}$ (see section X.2); the vector subbundle of $\mathbb{C}T\mathcal{M}$ spanned by these vector fields is denoted, below, by $\mathcal{V}^u$; it is of class $\mathscr{C}^1$.

COROLLARY X.6.2. *Let Ω be an open subset of $\mathcal{M}$ and $u \in \mathscr{C}^2(\Omega)$ a solution of the system of equations* (X.2.4) *in Ω. Suppose the system* (X.2.4) *is microlocally integrable at every point $(y,u(y),du(y))$, $y \in \Omega$. Let $v \in \mathscr{C}^2(\Omega)$ be another solution of* (X.2.4) *in Ω and let Ω' denote the interior of the subset of Ω in which $u = v$. Any orbit of $\mathcal{V}^u$ that intersects Ω' is entirely contained in Ω'.*

Theorem X.6.1 has also the following consequence:

PROPOSITION X.6.1. *Suppose the quasilinear system* (X.4.2)–(X.4.3) *is microlocally integrable at $(0,a,\omega) \in \mathcal{M}\times\mathbb{C}^{m+1}$. Let $(u,v_1,\dots,v_m) \in \mathscr{C}^1(U;\mathbb{C}^{m+1})$ be a solution of* (X.4.2)–(X.4.3) *such that*

$$v_i|_{t=0} = (\partial u/\partial x_i)(x,0),\ i = 1,\dots,m, \tag{X.6.1}$$

with $u(0,0) = a$, $u_x(0,0) = \omega$. If $u \in \mathscr{C}^2(U)$ then, in some open neighborhood $U' \subset U$ of the origin, $v_i = \partial u/\partial x_i$ for every $i = 1,\dots,m$, and u is a solution of the fully nonlinear system (X.2.4).

PROOF. If u is of class $\mathscr{C}^2$, we have the right to differentiate with respect to x both sides in Equations (X.4.3); setting $w = u_x$ we get

$$\partial w/\partial t_j = \sum_{k=1}^{m}(\partial f_j/\partial\zeta_k)(x,t,u,v)\partial w/\partial x_k +$$

$$\sum_{k=1}^{m}(w_k - v_k)(\partial/\partial x)[(\partial f_j/\partial\zeta_k)(x,t,u,v)] + (\partial f_j/\partial x)(x,t,u,v) +$$

$$(\partial f_j/\partial\zeta_0)(x,t,u,v)w. \tag{X.6.2}$$

If we compare the systems of equations (X.4.2) with (X.6.2) we see that they are identical provided $v = w$ $(= u_x)$. Let us then combine Equations (X.4.2), (X.4.3), and (X.6.2) into a single system (of $2m+1$ equations). The latter system is readily seen to be microlocally integrable at $(0,u(0,0),u_x(0,0),u_x(0,0))$ $(\in \mathcal{M}\times\mathbb{C}^{2m+1})$. We have two sets of $\mathscr{C}^1$ solutions, (u,v,u_x) and (u,v,v), which, by Hypothesis (X.6.1), satisfy the same initial conditions. Theorem X.6.1 demands that they be equal in some open neighborhood $U' \subset U$ of the origin. But if $v = u_x$ Equation (X.4.3) reduces to (X.2.4). ■

X.7. Approximation by Smooth Solutions

We return to Equation (X.5.13). By the first inequality (X.5.16) we know that the solution $\mathbf{v}_\nu$ has the property that, if $(x,t) \in U''$, the point $(x,t,\mathbf{v}_\nu(x,t))$ lies in the region (X.5.12). It follows that (X.5.11) holds with $\theta = \mathbf{v}_\nu(x,t)$. Property (X.4.13) entails that, if U'' is sufficiently small, the $M\times M$-matrix

$$[\Xi_\theta - Z_\theta(\partial_z \mathscr{E}_\nu^{\mathbf{u}}(\chi\Xi^{\mathbf{u}})(Z,0)](x,t,\mathbf{v}_\nu(x,t))$$

is nonsingular. We derive from (X.5.13) that, in U″, $\mathbf{v}_\nu$ satisfies the system of nonlinear differential equations

$$\nabla\mathbf{v} = -\ [\Xi_\theta - Z_\theta(\partial_z \mathscr{E}_\nu^{\mathbf{u}}(\chi\Xi^{\mathbf{u}}))(z,0)]^{-1}[\nabla\Xi - (\nabla Z)(\partial_z \mathscr{E}_\nu^{\mathbf{u}}(\chi\Xi^{\mathbf{u}}))(Z,0)](x,t,\mathbf{v}). \tag{X.7.1}$$

We have denoted by ∇ the gradient with respect to (x,t). Since $\mathscr{E}_\nu^{\mathbf{u}} g(z,0)$ is an entire holomorphic function of $z \in \mathbb{C}^m$ we see that the right-hand side will be a $\mathscr{C}^\infty$ function of $(x,t,\mathbf{v})$ in some open neighborhood of $(0,\theta_0)$ (recall that $\mathbf{v}_\nu(0,0) = \theta_0$) in $\mathcal{M} \times \mathbb{C}^M$ independent of ν. It follows at once (cf. Corollary X.4.3) that every solution of (X.7.1) is a $\mathscr{C}^\infty$ function of (x,t) in some open neighborhood of 0 independent of ν. Combining this observation with Theorem X.5.1 allows us to state:

THEOREM X.7.1. *Suppose the system of quasilinear differential equations* (X.4.1) *is microlocally integrable at* $\mathfrak{p} = (0,\theta_0)$. *Then any* $\mathscr{C}^1$ *solution* $\mathbf{u}$ *of* (X.4.1) *in* U *is the uniform limit, in an open neighborhood* U″ ⊂ U *of the origin, of a sequence of* $\mathscr{C}^\infty$ *solutions of* (X.4.1).

In the $\mathscr{C}^\omega$ category the requirement that the system be microlocally integrable is superfluous. The right-hand side in (X.7.1) is an analytic function of $(x,t,\mathbf{v})$ in an open neighborhood of $(0,\theta_0)$ independent of ν; and therefore the analogue is true of every solution of (X.7.1).

THEOREM X.7.2. *Suppose the coefficients* a_{jk} *and the zero-order terms* g_{ji}, *in the system of quasilinear differential equations* (X.4.1), *are analytic functions in the open set* $\mathcal{O}$. *Suppose furthermore the involution condition* (X.4.10) *is satisfied.*

Then any $\mathscr{C}^1$ *solution* $\mathbf{u}$ *of* (X.4.1) *in* U *is the uniform limit, in an open neighborhood* U″ ⊂ U *of the origin, of a sequence of analytic solutions of* (X.4.1).

In the fully nonlinear case we shall content ourselves with proving the analogue of Theorem X.7.2:

THEOREM X.7.3. *Suppose the system of first-order DE,* (X.2.4), *is involutive and the functions* $f_j(x,t,\zeta_0,\zeta)$ *are analytic in* $\mathcal{O}$. *Then any* $\mathscr{C}^2$ *solution* u *of* (X.2.4) *in* U *is the* $\mathscr{C}^1$ *limit, in an open neighborhood* U″ ⊂ U *of the origin, of a sequence of analytic solutions of* (X.2.4).

PROOF. Let u be a $\mathscr{C}^2$ solution of (X.2.4) in U. We apply the considerations of the beginning of the present section to the quasilinear system (X.4.2)–(X.4.3) derived from the fully nonlinear system (X.2.4). Here $M = m+1$

(subscripts will run from 0 to m and not from 1 to $m+1$); the initial data are $\mathbf{u}(x,0) = (u,u_x)(x,0)$. The assumption that the functions f_j are analytic enables us to choose the "first integrals" Z_i and Ξ_j to satisfy (X.3.17). We note then that

$$\mathscr{E}_\nu^{\mathbf{u}}(\chi\Xi_0^{\mathbf{u}})(z,0) = (\nu/\pi)^{\frac{1}{2}m}\int e^{-\nu\langle z-y\rangle^2}\chi(y)u(y,0)dy,$$

$$\mathscr{E}_\nu^{\mathbf{u}}(\chi\Xi_i^{\mathbf{u}})(z,0) = (\nu/\pi)^{\frac{1}{2}m}\int e^{-\nu\langle z-y\rangle^2}\chi(y)(\partial u/\partial y_i)(y,0)dy,\ i = 1,\ldots,m.$$

There is an open neighborhood $\mathcal{O}$ of 0 in $\mathbb{C}^m$ and constants C, $c > 0$ such that, for all $\nu = 1,2,\ldots$,

$$|\mathscr{E}_\nu^{\mathbf{u}}(\chi\Xi_i^{\mathbf{u}})(z,0) - (\partial/\partial z_i)\mathscr{E}_\nu^{\mathbf{u}}(\chi\Xi_0^{\mathbf{u}})(z,0)| \leq Ce^{-c\nu}. \tag{X.7.2}$$

Instead of solving Equations (X.5.13) with respect to $\theta = (\theta_0,\theta_1,\ldots,\theta_m)$ we now solve the system of equations

$$\begin{aligned}\Xi_0(x,t,\theta) &= \mathscr{E}_\nu^{\mathbf{u}}(\chi\Xi_0^{\mathbf{u}})(Z(x,t,\theta),0),\\ \Xi_i(x,t,\theta) &= [(\partial/\partial z_i)\mathscr{E}_\nu^{\mathbf{u}}(\chi\Xi_0^{\mathbf{u}})](Z(x,t,\theta),0),\ i = 1,\ldots,m.\end{aligned} \tag{X.7.3}$$

Call $\mathbf{w}_\nu(x,t)$ the solution of (X.7.3) such that $\mathbf{w}_\nu(0,0) = (u(0,0),u_x(0,0))$. By the argument used in proving Theorem X.7.1 we know that $\mathbf{w}_\nu$ is a $\mathscr{C}^\omega$ function in a suitably small open neighborhood of the origin independent of ν. An important observation is that, if we put $t = 0$ in (X.7.3) we get

$$w_\nu(x,0) = \mathscr{E}_\nu^{\mathbf{u}}(\chi\Xi_0^{\mathbf{u}})(x,0),\ w_{\nu,i}(x,0) = [(\partial/\partial z_i)\mathscr{E}_\nu^{\mathbf{u}}(\chi\Xi_0^{\mathbf{u}})](x,0),$$

in other words,

$$w_{\nu,i}(x,0) = (\partial w_\nu/\partial x_i)(x,0)\ (1 \leq i \leq m). \tag{X.7.4}$$

Thanks to Proposition X.4.1 and to the relations (X.7.3) we see that $\mathbf{w}_\nu$ is a solution of the system (X.4.2)–(X.4.3); and (X.7.4) enables us to apply Proposition X.6.1: w_ν satisfies the fully nonlinear system (X.2.4).

Finally, from (X.7.2) we derive that, in some open neighborhood of 0 in $\mathcal{M}$ and for suitable positive constants C, c,

$$|\mathbf{w}_\nu - \mathbf{v}_\nu| \leq Ce^{-c\nu},\ \nu = 1,2,\ldots \tag{X.7.5}$$

By combining (X.5.16) and (X.7.5) we conclude that $w_\nu \to u$ in $\mathscr{C}^1(U'')$ for a suitable choice of the open neighborhood of 0, U''. ■

Notes

The approach to first-order nonlinear differential equations by Hamiltonian lift to the one-jet bundle, followed in chapter X, is an adaptation of the classical method of characteristics. As we deal with complex equations and allow complex-valued solutions we must deal with complex one-jet spaces; holomorphy with respect to the fibre variables must be preserved throughout the argument. Otherwise the Poisson bracket and the other symplectic features of the theory are directly borrowed from classical calculus of variation (see Caratheodory [1]).

The line of approach to uniqueness described in chapter X was initiated in Baouendi, Goulaouic, and Treves [1], where the semilinear case and the case $m = 1$, $n \geq 1$ (one "space" variable x, any number of "time" variables t_j) was treated. The general case, as well as the approximation of $\mathscr{C}^2$ solutions by analytic ones, was settled soon after, in Métivier [1]. In our treatment of those questions we have essentially followed the latter article.

The main shortcoming of the Hamiltonian lift method is that it requires one more degree of smoothness of the solutions than the problem seems to warrant. In dealing with fully nonlinear first-order differential equations it is natural to assume that the solutions are of class $\mathscr{C}^1$. Yet our method requires that they be of class $\mathscr{C}^2$, to ensure that the Hamiltonians be differential operators with $\mathscr{C}^1$ coefficients. Thus the problem of uniqueness (and of approximation) of $\mathscr{C}^1$ solutions remains open.

References

Akahori, T.

1. A new approach to the local embedding theorem of CR structures for n ≥ 4 (the local solvability for the operator $\bar{\partial}_b$ in the abstract sense), Memoirs of the A.M.S., **67** (May 1987).

Andreotti, A., Fredricks, G., and Nacinovich, M.

1. *On the absence of Poincaré lemma in tangential Cauchy-Riemann complexes*, Ann. Scuola Norm. Sup. Pisa, Sci. Fis. Mat. **8** (1981), 365–404.

Andreotti, A., and Hill, C. D.

1. *Complex characteristic coordinates and tangential Cauchy-Riemann equations*, Ann. Scuola Norm. Sup. Pisa, Sci. Fis. Mat. **26** (1972), 299–324.
2. *E. E. Levi convexity and the Hans Lewy problem, I and II*, Ann. Scuola Norm. Sup. Pisa, Sci. Fis. Mat. **26** (1972), 325–363, 747–806.

Baouendi, M. S., Chang, C. H., and Treves, F.

1. *Microlocal hypo-analyticity and extension of CR functions*, J. Diff. Geom. **18** (1983), 331–391.

Baouendi, M. S., Goulaouic, Ch., and Treves, F.

1. *Uniqueness in certain first-order nonlinear complex Cauchy problems*, Comm. Pure Applied Math. **38** (1985), 109–123.

Baouendi, M. S., and Rothschild, L. P.

1. *Normal forms for generic manifolds and holomorphic extension of CR functions*, J. Diff. Geom. **25** (1987), 431–467.
2. *Embeddability of abstract CR structures and integrability of related systems*, Ann. Inst. Fourier Grenoble **37** (1987), 131–141.
3. *Cauchy-Riemann functions on manifolds of higher codimension in complex space*, Invent. math. **101** (1990), 45–56.
4. *Transversal Lie group actions on abstract CR manifolds*, Math. Ann. **287** (1990), 19–33.
5. *Minimality and the extension of functions from generic manifolds*, Proceedings AMS Summer Institute in Several Complex Variables, Santa Cruz 1989.

Baouendi, M. S., Rothschild, L. P., and Treves, F.

1. *CR structures with group action and extendability of CR functions*, Invent. math. **82** (1985), 359–396.

Baouendi, M. S., and Treves, F.

1. *A local constancy principle for the solutions of certain overdetermined*

systems of first-order linear partial differential equations, in Math. Analysis and Applications, Advances in Math. Supplementary Studies **7A** (1981), 245–262.

2. *A property of the functions and distributions annihilated by a locally integrable system of complex vector fields*, Ann. of Math. **113** (1981), 387–421.
3. *A microlocal version of Bochner's tube theorem*, Indiana Univ. Math. J. **31** (1982), 885–895.
4. *Unique continuation in CR manifolds and in hypo-analytic structures*, Arkiv för mat. **26** (1988), 21–40.

Bloom, T., and Graham, I.

1. *On "type" conditions for generic submanifolds of* $\mathbb{C}^n$, Invent. Math. **40** (1977), 217–243.

Bochner, S.

1. *Analytic and meromorphic continuation by means of Green's formula*, Ann. of Math. **44** (1943), 652–673.

Bochner, S., and Martin, W. T.

1. Functions of several complex variables, Princeton University Press, 1948.

Brézis, H.

1. *On a characterization of flow-invariant sets*, Comm. Pure Applied. Math. **23** (1970), 261–263.

Brós, J., and Iagolnitzer, D.

1. *Causality and local analyticity; mathematical study*, Proceed. Conf. sur la théorie de la renormalisation, Ann. Inst. Poincaré **18** (1973), 147–184.
2. *Support essentiel et structure analytique des distributions*, Séminaire Goulaouic-Lions-Schwartz, 1974–1975, Exp. 18.

Caratheodory, C.

1. Variationsrechnung und partielle Differentialgleichungen Erster Ordnung, Teubner, 1935.

Cartan, E.

1. Les systèmes différentiels extérieurs et leur applications géométriques, Hermann, 1945.

Cartan, H.

1. *Séminaires E.N.S.*, 1951–1952.

Cohen, P.

1. *The non-uniqueness in the Cauchy problem*, O.N.R. Techn. Report **93**, Stanford, 1960.

Cordaro, P., and Hounie, J.

1. *On local solvability of underdetermined systems of vector fields*, Amer. J. Math. **112** (1990), 243–270.

Cordaro, P., and Treves, F.
1. *Homology and cohomology in hypo-analytic structures of the hypersurface type*, J. Geometric Analysis **1** (1991), 39–70.
Cordoba, A., and Fefferman, Ch.
1. *Wave packets and Fourier integral operators*, Comm. P.D.E. **3** (1978), 979–1005.
D'Angelo, J. P.
1. *Real hypersurfaces, orders of contact, and applications*, Ann. of Math. **115** (1982), 615–637.
De Rham, G.
1. Variétés différentiables, Hermann, 1955.
Dolbeault, P.
1. *Formes différentielles et cohomologie sur une variété analytique complexe, I, II*, Ann. of Math. **64** (1956), 83–130; **65** (1957), 282–330. (1957), 282–330.
Eckmann, B., and Frölicher, A.
1. *Sur l'intégrabilité des structures presque complexes*, C. R. Acad. Sci. Paris **232** (1951), 2284–2286.
Godement, R.
1. Topologie Algébrique et Théorie des Faisceaux, Hermann, 1958.
Greenfield, S. T.
1. *Cauchy-Riemann equations in several variables*, Ann. Scuola Norm. Sup. Pisa, Sci. Fis. Mat. **22** (1968), 275–314.
Greiner, P. C., Kohn, J. J., and Stein, E. M.
1. *Necessary and sufficient conditions for solvability of the Lewy equation*, Proceed. N.A.S. **72** (1975), 3287–3289.
Grushin, V. V.
1. *A certain example of a differential equation without solutions*, Mat. Zametki **10** (1971), 125–128. English translation in Math. Notes **10** (1971), 499–501.
Hanges, N., and Jacobowitz, H.
1. *A remark on almost complex structures with boundary*, Amer. J. of Math. **111** (1989), 53–64.
Hanges, N., and Treves, F.
1. *Propagation of holomorphic extendability of CR functions*, Math. Ann. **263** (1983), 157–177.
Harvey, R., and Polking, J.
1. *Fundamental solutions in complex analysis, I, II*, Duke Math J. **46** (1979), 253–340.
Helffer, B., and Nourrigat, J.
1. *Approximation d'un champ de vecteurs et application à l'hypoellipticité*, Ark. Mat. **2** (1979), 237–254.

Henkin, G. M., and Leiterer, J.

1. Theory of functions on complex manifolds, Birkhäuser, 1984.

Hermann, R.

1. *On the accessibility problem in control theory*, Internat. Sympos. Nonlinear Differential Equations and Nonlinear Mechanics, Academic Press, 1963, 325–332.

Hill, C. D.

1. *What is the notion of a complex manifold with a smooth boundary*? in Prospect of algebraic analysis, Academic Press, 1987.

Hörmander, L.

1. *Differential operators of principal type*, Math. Ann. **140** (1960), 124–146.
2. *Hypoelliptic second-order differential equations*, Acta Math. **119** (1967), 147–171.
3. *Propagation of singularities and semiglobal existence theorems for (pseudo-)differential operators of principal type*, Ann. of Math. **108** (1978), 569–609.
4. The analysis of linear partial differential operators, Springer Verlag, 1983.

Iagolnitzer, D., and Stapp, H. P.

1. *Macroscopic causality and physical region analyticity in S-matrix theory*, Comm. Math. Phys. **14** (1969), 15–55.

Jacobowitz, H.

1. *On the intersection of varieties with a totally real submanifold*, Proceed. A.M.S. **101** (1987), 127–130.
2. An introduction to CR structures, Math. Surveys and Monographs, Amer. Math. Soc., Providence, 1990.

Jacobowitz, H., and Treves, F.

1. *Nowhere solvable homogeneous partial differential equations*, Bull. Amer. Math. Soc. **8** (1983), 467–469.
2. *Aberrant CR structures*, Hokkaido Math. J. **22** (1983), 276–292.

Kohn, J. J.

1. *Harmonic integrals on strongly pseudo-convex manifolds, I, II*, Ann. of Math. **78** (1963), 112–148; **79** (1964), 450–472.
2. *Boundary behaviour of $\bar\partial$ on weakly pseudoconvex manifolds of dimension two*, J. Diff. Geom. **6** (1972), 523–542.

Kohn, J. J., and Rossi, H.

1. *On the extension of holomorphic functions from the boundary of a complex manifold*, Ann. of Math. **81** (1965), 451–472.

Koppelman, W.

1. *The Cauchy integral for differential forms*, Bull. Amer. Math. Soc. **73** (1967), 554–556.

Kuranishi, M.

1. *Strongly pseudoconvex CR structures over small balls, I*, Ann. of Math. **115** (1982), 451–500; *II*, **116** (1982), 1–64; *III*, **116** (1982), 249–330.

Lawson, H. B.

1. The qualitative theory of foliations, Reg. Conf. Series in Math., Amer. Math. Soc. **27** (1975).

Levi. E.

1. *Studii sui punti singolari essenziali delle funzioni analitiche di due o più variabili complesse*, Ann. mat. pura ed appl. **17** (1910), 61–87.

Lewy, H.

1. *On the local character of the solution of an atypical differential equation in three variables and a related problem for regular functions of two complex variables*, Ann. of Math. **64** (1956), 514–522.
2. *An example of a smooth linear partial differential equation without solution*, Ann. of Math. **66** (1957), 155–158.

Libermann, P.

1. *Problèmes d'équivalence relatifs à une structure presque complexe sur une variété à quatre dimensions*, Acad. Roy. Belgique Bull. Cl. Sci. **36** (1950), 742–755.

Lieb, I.

1. *Die Cauchy-Riemannschen Differentialgleichungen auf streng pseudokonvexen Gebieten*, Math. Ann. **190** (1970–1971), 6–44.

Malgrange, B.

1. *Sur l'intégrabilité des structures presque complexes*, Symposia Math. **2**, Academic Press, 1969, 289–296.

Martinelli, E.

1. *Alcuni teoremi integrali per le funzioni analitiche di più variabili complesse*, Mem. R. Acad. Ital. **9** (1938), 269–283.
2. *Sopra una dimostrazione di R. Fueter per un teorema di Hartog*, Comment. Math. Helv. **15** (1943), 340–349.

Melrose, R. B.

1. *Transformation of boundary problems*, Acta Math. **147** (1981), 149–236.

Mendoza, G., and Treves, F.

1. *On the local solvability in locally integrable structures of corank one*, Duke Math. J. **63** (1991), 355–377.

Métivier, G.

1. *Uniqueness and approximation of solutions of first-order non linear equations*, Invent. math. **82** (1985), 263–282.

Mizohata, S.

1. *Solutions nulles et solutions non analytiques*, J. Math. Kyoto Univ. **1** (1962), 271–302.

Moyer, R.

1. *Local solvability in two dimensions: Necessary conditions for the principal-type case*, mimeograph ms, University of Kansas, 1978.

Nagano, T.

1. *Linear differential systems with singularities and applications to transitive Lie algebras*, J. Math. Soc. Japan **18** (1966), 398–404.

Nelson, E.

1. *Analytic vectors*, Ann. of Math. **670** (1959), 572–615.

Newlander, A., and Nirenberg, L.

1. *Complex analytic coordinates in almost complex manifolds*, Ann. of Math. **65** (1957), 391–404.

Nijenhuis A., and Woolf, W. B.

1. *Some integration problems in almost-complex manifolds*, Ann. of Math. **77** (1963), 424–489.

Nirenberg, L.

1. *A complex Frobenius theorem*, Seminar on analytic functions I, Princeton, 1957, 172–189.
2. Lectures on linear partial differential equations, Reg. Conf. Series in Math., Amer. Math. Soc. **17** (1973).
3. *On a question of Hans Lewy*, Russian Math. Surveys **29** (1974), 251–262.

Nirenberg, L., and Treves, F.

1. *Solvability of a first-order linear partial differential equation*, Comm. Pure Appl. Math. **16** (1963), 331–351.

Øvrelid, N.

1. *Integral representation formulas and* L^p *estimates*, Math. Scand. **29** (1971), 137–160.

Rosay, J. P.

1. *Sur un problème d'unicité pour les fonctions C.R.*, C.R. Acad. Sci. Paris **302** (1986), 9–11.

Rothschild, L. P., and Stein, E. M.

1. *Hypoelliptic differential operators and nilpotent groups*, Acta Math. **137** (1977), 248–315.

Sato, M., Kawai, T., and Kashiwara, M.

1. *Microfunctions and pseudo-differential equations*, in Hyperfunctions and Pseudodifferential Equations, Springer Lecture Notes no. 187, 1971.

Schwartz, L.

1. Théorie des distributions, Hermann, 1966.

Sjöstrand, J.

1. *Note on a paper of F. Treves concerning Mizohata type operators*, Duke Math. J. **47** (1980), 601–608.
2. Singularités analytiques microlocales, Astérisque **95** (1982), Soc. Math. France.

3. *The FBI-transform for CR submanifolds of* $\mathbb{C}^n$, Prépublications Math. Depart. Univ. Paris-Sud (1982).

Spivak, M.

1. A Comprehensive Introduction to Differential Geometry, Publish or Perish, 1979.

Sussmann, H. J.

1. *Orbits of families of vector fields and integrability of distributions*, Trans. Amer. Math. Soc. **180** (1973), 171–188.

Treves, F.

1. Linear Partial Differential Equations with constant coefficients, Gordon & Breach, 1966.
2. Topological vector spaces, distributions and kernels, Academic Press, 1967.
3. *Local solvability in* L^2 *of first-order linear PDEs*, Amer. J. Math. **92** (1970), 369–380.
4. *Hypoelliptic partial differential equations of principal type. Sufficient conditions and necessary conditions*, Comm. Pure Applied Math. **24** (1971), 631–670.
5. Basic Linear Partial Differential Equations, Academic Press, 1975.
6. *Study of a model in the theory of overdetermined pseudodifferential equations*, Ann. of Math. **104** (1976) 269–324.
7. Approximation and representation of functions and distributions annihilated by a system of complex vector fields, Centre Math., Ecole Polytechn., Palaiseau, France, 1981.
8. *On the local solvability and the local integrability of systems of vector fields*, Acta Math. **151** (1983), 1–38.
9. *Approximation and representation of solutions in locally integrables structures with boundary*, in Aspects of Math. and Applications, Elsevier, 1986, 781–816.
10. *On the local solvability for top-degree forms in hypo-analytic structures*, Amer. J. Math. **112** (1990), 403–421.
11. Homotopy formulas in the tangential Cauchy-Riemann complex, Memoirs of the A.M.S. **87** (Sept. 1990).

Webster, S. M.

1. *A new proof of the Newlander-Nirenberg theorem*, Math. Zeit. **201** (1989), 303–316.
2. *On the proof of Kuranishi's embedding theorem*, Ann. Inst. Poincaré **6** (1989), 183–207.

Weierstrass, K.

1. *Über die analytische Darstellbarkeit sogenannter willkürlicher Functionen reeller Argumente*, in Mathematicshe Werke **3** Berlin, 1903, 1–37.

Index

Note: Boldface indicates where definition of term can be found.